Putnam's Geology

Professor William Clement Putnam, the original author of this book, died on March 16, 1963, at the age of fifty-four. He had taught geology for thirty-two years. This book is the result of his experience as a teacher, and of his love of teaching beginning students.

PUTNAM'S GEOLOGY

Fourth Edition

EDWIN E. LARSON
PETER W. BIRKELAND
University of Colorado

New York · Oxford
OXFORD UNIVERSITY PRESS
1982

Cover Photo: Zion Narrows of the Virgin River, Zion National Park, Utah. *Bill Ratcliffe*

Library of Congress Cataloging in Publication Data

Putnam, William Clement, 1908–1963.
Putnam's Geology.

Includes bibliographies and index.
1. Physical geology. I. Larson, Edwin E., 1931–
II. Birkeland, Peter W. III. Title.
IV. Title: Geology.
QE28.2.P87 1982 550 81-18762
ISBN 0-19-503002-8 AACR2

Printing (last digit): 9 8 7 6 5 4 3 2 1

Printed in the United States of America

PREFACE

In this revision of *Putnam's Geology,* we have again attempted to retain Bill Putnam's readable and understandable style while explaining rather complex geological matters. Much of the material has been revised and updated in view of new information and ideas that have been put forth in the last four years. We have increased our coverage with the addition of two new chapters—one on the solar system (Chap. 2) and one on energy and resources (Chap. 21). As in the last edition, summaries at the ends of the chapters and an extensive glossary at the end of the book are provided. Study questions have been added at the end of each chapter, and three appendices—a detailed chart on minerals and mineral identification by Frank Beck, a discussion of the use of topographic and geological maps by Vance T. Holliday, and a complete table of metric conversions—at the end of the text. A study guide, written by Richard C. Finch of Tennessee Technological University, is available for student use.

Introductory textbooks in physical geology show no one common pattern in sequence of topics. Nor is there general agreement on the order of presentation in the classroom. In the latest edition, we again keep the arrangement preferred by Bill Putnam, which he developed during his thirty-two years of teaching. Each chapter is self-contained, and thus the chapters can be taught in nearly any sequence. It does seem logical to us to proceed from a discussion of time and the solar system to the materials that make up the visible earth and on to the various processes responsible for wearing down and shaping the landscape, concluding with a panorama of earthquakes, mountain building, and plate tectonics—processes that generally continue to elevate portions of the earth's crust. Finally, we refer to material discussed in earlier chapters to focus on problems related to the development and use of energy and the discovery and use of mineral resources now and in the future. Plate tectonics still is a guiding theme. The theory, briefly introduced in the third chapter, is considered in other chapters relative to the processes that are described and explored in detail in Chapter 20. As in past editions, environmental issues form an integral part of the text.

We hope that this book will nurture an interest in the landscape that will become a permanent source of pleasure and satisfaction to the reader.

Many people have influenced out thinking during the course of this revision. Our fellow instructors, graduate teaching assistants, and students have been helpful in their comments over the years. We have benefited from a reading of the entire text by Richard Finch, and from comments and suggestions from more than 50 users of the Third Edition of *Putnam's Geology* from all over the country. A. A. Bartlett of the University of Colorado and Craig Bond Hatfield of the University of Toledo read and commented upon Chapter 21. Joyce Berry and Carol Miller of Oxford University Press did the editorial work, for which we are extremely grateful. Thanks are also due Oxford staff members Margaret Joyner, for her excellent work in the design of the book, and John Manger, for editorial advice. We alone, though, are responsible for errors or omissions.

Putnam's original text was known for its excellent photographs. We have tried to maintain those high standards by retaining the best photographs from the previous editions and by adding some of the best work, both past and present, from the U.S. Geological Survey collection as well as the work of many friends. We are especially grateful to William C. Bradley for access to his extensive collection and Janet Robertson, who photographed features in the local area. We also want to thank Edith Ellis and Paulina Franz for typing the manuscript drafts.

Boulder, Colorado E.E.L.
October 1981 P.W.B.

CONTENTS

I shall lift up mine eyes unto the hills
From whence cometh my help.

Psalms: cxxi, 1.

Fig. 1-1 The Grand Canyon, Arizona, was carved into the Colorado Plateau by the Colorado River over a period of millions of years. *USGS*

1

GEOLOGY

THE FOUR-DIMENSIONAL SCIENCE

Geology—the study of the earth—is not, strictly speaking, a fundamental science, but rather a multidisciplinary one. Thus, in studying the earth and the various bodies of our solar system, earth scientists apply the principles of such truly fundamental sciences as physics, chemistry, and biology, as well as mathematics and other multidisciplinary sciences, such as astronomy. Classically, geology has been more field oriented than laboratory oriented, and it maintains that tradition today.

To be sure, there is an overriding historical slant to geology. Most of the landforms and rocks (Fig. 1-1) we find today at the earth's surface are relics of past ages when the world and its living organisms were different. All we have left to tell us of those distant times is the fragmentary rock record or perhaps the landscape itself. Through careful scrutiny of the rocks and the land forms, and through a knowledge of the way geological events occur on earth today, geologists attempt to recreate such events from the past. Underlying their ability to draw conclusions and to advance hypotheses about bygone ages is the conclusion that the present is the key to the past, a concept called ***uniformitarianism.*** The term means only that the processes in evidence in the world today probably existed in the past. Perhaps the rates of those processes were different, but by and large, the processes themselves were the same. That is, rivers under the influence of gravity probably flowed in earlier days much as they do today; ancient winds probably blew in response to atmospheric disturbances, volcanoes erupted, and earthquakes shook the land, much as they do now. When one can observe that not far from a fossil ripple-marked sandstone (Fig. 1-2) the same type of ripple marking is being formed in the soft sand deposited along the banks of the Colorado River, it is difficult to escape that conclusion. The concept of uniformitarianism only reaffirms that the natural laws have prevailed on our planet continuously and unchangingly through the ages. The idea is basic not only to geology, but to all the other sciences as well: materials attract each other through gravitational, electrostatic, and magnetic forces the same way now as they always did; hydrogen bonds with oxygen to form water now as it did in the past.

Fig. 1-2 Ancient, ripple-marked sandstone, formed about 225 million years ago, northwestern Colorado, valley of the Colorado River. After deposition, these sediments were tilted to a steep angle. *Edwin E. Larson*

The alternative view is that all that surrounds us has no connection with the past. Such a view leads to utter confusion—a world in which there is no continuity, in which each age is characterized by processes and events never seen before and probably never to be seen again. All evidence seems to indicate, fortunately, that such is not the case.

In the past, the concept of uniformitarianism was not readily accepted. Before about 1700 many people, particularly those who adhered to Judeo-Christian principles, thought that the earth had been the scene of repeated, large-scale catastrophes, each of which had wrought extensive changes in the physical features of both the earth's surface and of plants and animals. By the mid-1700s, however, it became apparent to some that geological features, even those on a huge scale, could result from the continuing action of natural processes similar to those working today. The validity of that view, which was championed first by a Scotsman, James Hutton (1726–97), and subsequently by his friend and countryman, John

Playfair (1748–1819), was still being discussed almost a century later. But it was not until about 1830 that Charles Lyell (1797–1875), an Englishman sympathetic to the views of Hutton, documented the reality of uniformitarianism through many careful observations of rocks and landforms in western Europe. Because of his efforts in the field Lyell is generally considered the "Father of Geology" (Fig. 1-3).

In 1859, Charles Darwin (1809–82) did for biology what Hutton and Lyell did for geology—he established the idea of biological uniformitarianism. Apparently Darwin was greatly influenced by the work of Lyell. On the trip he made on the HMS *Beagle* during his scientifically formative years, he carried along and consulted a geology treatise written by Lyell.

Once the ideas of physical, chemical, and biological uniformitarianism were accepted, earth scientists could start to unravel the tangled web of the past.

Sciences such as physics and chemistry are considered to be basic or fundamental scientific sciences, ones in which the scientific method can be applied. First, experiments are performed, and the various physical and chemical parameters are then measured. Based on these measurements, a plausible explanation or hypothesis is advanced. Subsequent experiments test the hypothesis, and, if it is shown to

Fig. 1-3 Sir Charles Lyell in 1863. *George Eastman House*

be in error, it is discarded and another is advanced and tested. If an important hypothesis continues to meet the tests of experiments over a long period of time, it is declared to be a ***natural law***—the law of gravity is an example.

This is not generally true in geology. Rather than being black and white, the realm of geologists is usually in shades of gray. The "experiment"—commonly on an immense scale and of extreme complexity—has already been performed in nature, and all that can be seen is the result. Even the scope of observation is restricted because rocks at the earth's surface usually are not well exposed but are masked by soil and younger deposits.

Attempts to duplicate natural processes and geological formations in the laboratory are not always successful because of uncertainties about conditions occurring long ago in nature—particularly those relating to the composition of materials and the rates of physical and chemical change and the total length of time involved in geological processes. Earth scientists know, for example, that the formation of a mountain range hundreds of kilometers long is a complex process requiring tens of millions of years. Obviously, such a process cannot be duplicated in a laboratory. What can be gained, however, is a clearer understanding of the process—knowledge that may help other earth scientists to solve a particular complex problem. Even the monitoring of processes going on today is not simple. On Hawaii, for example, earth scientists have thoroughly studied the timing and nature of the intermittent volcanism, but we still know very little about what is actually going on deep beneath the island where the lava originates. In spite of such difficulties, earth scientists, from careful observation based on their knowledge of physical, chemical, and biological processes are able to draw conclusions and to advance hypotheses. Actually, they often advance several equally plausible hypotheses to fit their observations—an approach called the method of ***multiple working hypotheses.*** Then the search continues for evidence that will disprove one or another of the several hypotheses. As new evidence is turned up, all existing hypotheses may be discarded and a new, seemingly more plausible one advanced. Geological hypotheses, then, are much like present-day weather forecasts in that both are based on an estimate of probability.

GEOLOGICAL TIME—THE FOURTH DIMENSION

As is apparent by now, the study of the earth involves time. All of us are aware of time, and how our concept of it changes from childhood to adulthood. To a child, a day in the future is essentially the same as a year. As one gets older, however, an appreciation of past, present, and future time develops. To most of us, a few hundred years of history encompass a great deal of time. Fragmentary accounts from the days of the early Greek, Roman, and Egyptian civilizations we class as "ancient history." The gulf between our own culture and those civilizations seems so great that we find it difficult to relate to them. It has only been through the study of the earth that we have come to realize the true vastness of the fourth dimension, which extends back to the beginning of our world and beyond. In fact, as has been stated by A. Knopf, a well-known U.S. geologist, it is the concept of the immensity of geological time that has been perhaps the most significant scientific contribution made by geology. Whereas in the historical world we deal with hours, days, and years, in the geological one we deal with thousands, millions, and billions of years.

How can we stretch our imagination to conceive realistically of a length of time spanning many millions or billions of years? The truth is that we cannot. It is like trying to imagine the colossal distances that separate us from nearby stars or other galaxies.

But it is within the four dimensions that geological processes have been and still are working. Given enough time, it is possible for even the slowest process to bring about large-scale changes. Time, then, is most essential for the uniformitarian concept.

In order to estimate the passage of time, it is necessary to use some device that records the

Fig. 1-4 Coral, about 375 million years old, from the Sulphur Springs Range in Nevada; maximum width about 4 cm (1.5 in.). The growth lines on such corals have been used to determine the number of days in an ancient year. *C.W. Merrian, USGS*

occurrence of equally spaced events. We now use mechanical or electronic clocks as monitors; in older societies, people used sundials, sand clocks, or water clocks. The basis for all timepieces is found in the motions of our planet in the solar system. In essence, the sun and encircling planets are parts of a solar clock that is endlessly ticking in our corner of the Milky Way Galaxy. Yet even that natural clock, as we shall see, is not perfect and is prone to run down. The day is the obvious unit of time; its length is determined by the speed of rotation of the earth on its axis and is defined as the time between two consecutive passages of the zenith of the sun (***solar day***) or of a distant star (***stellar day***) past a particular spot on earth. The year, another basic unit, is designated as that time required for one passage of the earth around the sun. Inasmuch as the speed with which the earth travels around the sun varies throughout the year, the length of a solar or stellar day is variable, which has necessitated the establishment of the ***mean solar day***—one that is always of the same length, regardless of the season. As we well know, there are not an even number of days in a year; rather there are about 365¼. So that the calendar year and the planetary year continue to correspond, we need to add one day every four years, that is, each Leap Year. But even that adjustment is inadequate, and additional minor changes in the calendar are required about every 400 years.

An improved timepiece, based upon the electronic measurement of atomic vibrations in the element cesium, has been available for some time. This ***atomic clock*** is capable of recording variations in the length of a day of billionths and trillionths of a second. With it, scientists have become aware that the spinning earth is slowing down, a phenomenon recorded as a progressive, but slight—very slight—lengthening of each successive day. Most scientists agree that the slowing of rotation is primarily the result of gravitational interaction—a sort of drag—between the earth and moon.

Apparently, the slowing of the earth's rotation is not a recent phenomenon; it has been going on for as long as the earth has had its moon. The slowing can be documented by Babylonian records of solar eclipses and, beyond that, by studies of the number of growth lines deposited per lunar month on ancient sea shells, particularly corals (Fig. 1-4). The astonishing conclusion reached by the scientists is that 300 million years ago, a year would have been 440 days long.

Conventional clocks are fine for most purposes, but in recording geological time, unconventional methods are required. One unique attempt to determine the age of the earth was made in the mid-1600s by Archbishop Ussher (1581–1656) of the Irish Protestant Church. The Archbishop firmly believed that the Bible contained a complete record of the world's events since its inception. Therefore, by adding up all the genealogies recorded in the Old Testament, he was able to set the origin of the earth as the evening of October 22, 4004 B.C. Someone has jested that he failed to tell us if it were standard or daylight saving time. Ussher's estimate was a refinement of a figure similarly determined in 1642 by John Lightfoot, a distinguished Greek scholar and Vice Chancellor of Cambridge University. The date 4004 B.C. was referred to in the Great Edition of the English Bible published in 1701, and from then on it was incorporated into Christian thought. For a century thereafter, anyone who maintained that the earth's development took

Fig. 1-5 Sedimentary rocks exposed in Coal Canyon, Arizona. Tens of millions of years were required for the deposition of this thick section. Erosion rates commonly are measured in centimeters per thousand years, so you can estimate that millions of years were required to form a canyon of such grandiose proportions. *Tad Nichols*

more than 6000 years was considered a heretic. In effect, a theological straitjacket was placed on all thinking in the Western world concerning the antiquity of the earth. In order to explain the many geological features and documented events, it was necessary to invoke the occurrence of innumerable catastrophes. Only the patient efforts of Hutton and Lyell and the development of uniformitarianism as a concept and geology as a science dispelled this idea.

With the emergence of geology, scientists became aware that a tremendous amount of time must have passed to allow the changes that they knew had taken place to occur. But how much time—thousands or hundreds of thousands or millions of years? One might assume that if geologists knew the rate at which layers of sedimentary rocks were deposited, as well as their thickness, they could estimate the length of time erosion and deposition had operated on the surface of the earth (Fig. 1-5).

However, there are a number of things wrong with such an assumption. No one knew then—or knows now, for that matter—the rate at which sediments are deposited. Obviously, the rate must have been different for a conglomerate made of boulders a meter across than it was for a limestone deposit consisting of the remains of microscopic marine organisms. Furthermore, few if any sediments are deposited without interruption. There may have been times of accumulation separated by intervals in which no deposition occurred, or during which erosion of previously deposited sediments occurred. To add to the difficulties, no place was known, on the land surface of the earth, where strata representing the total age of the earth had been laid down continuously. Nonetheless, in 1883 the European geologist William Sollas (1849–1936) did make such an assumption, estimating that 26 million years had been required to accumulate the sequences of strata of which he was aware.

Another attempt to relate present-day processes to the age of the earth was made in 1899 by the Irish geophysicist John Joly (1857–1933). The amount of salt dissolved in the oceans had been determined fairly accurately by that time, and Joly reasoned that if a fair estimate could be made of the amount of salt that rivers contribute each year to the oceans, he could establish a value for the age of the sea, and accordingly, a rough estimate of the age of the earth. His best estimate was 100 million years.

Joly did not see the many pitfalls in his method. Immense quantities of salt now present in salt beds were removed from the oceans by evaporation of sea water in ages past. Furthermore, some of that salt is now being recycled—sent back to the sea—as the salt beds erode away. His estimate also does not take into account the probability that the amount of salt added to the sea each year through erosion has not remained constant throughout geological time. There is now evidence that times of relatively high mountains and rapid erosion have alternated with times of low relief and correspondingly slower rates of erosion.

Relative time

While some earth scientists were trying to establish absolute ages for the earth and certain geological features, most were involved in deciphering geological records in their local areas of study. They placed all the events recorded in the rocks into a ***relative time*** sequence—by determining whether an event occurred before or after another one. In our daily lives, we do much the same thing, stacking the events of the day into a succession based on the order of their occurrence.

Only a few methods are used to establish the relative time sequence. Some of them are so straightforward that they seem obvious; yet in the early days of geology, when little was known about the earth, such methods were extraordinary. Each represents a discovery made by an individual only after painstaking observation. For instance, Nicolaus Steno (1638–86), a Danish physician and theologian who became curious about the earth, arrived at the conclusion, in 1669, that in any sequence of layered rocks any one layer will be older than the layer above it and younger than the layer below it. He reasoned that before the stratum above could be laid down, the layer below must have already been deposited. That simple

Fig. 1-6 Rippled layers of sand, salt, and mud recently deposited along the bank of the Colorado River. The exposed vertical cut is about 1 m high. *Tad Nichols*

rule is called the ***law of superposition.*** It is used primarily to determine the relative ages of sedimentary rocks or lava flows.

Another rule established by Steno is that sediments were originally deposited horizontally or nearly horizontally (Fig. 1-6). Therefore, if one finds a sequence of strata tipped up at an angle, the event that produced the tilting must have occurred after the strata were laid down (Fig. 1-7). It has been found that exceptions to this rule, the ***law of original horizontality,*** do exist, but they are rare.

James Hutton established another method of relative dating when he determined that if one body of rock cuts across another body of rock, the latter must be older than the former. That is, one rock must already be in place for another, younger one to cut across it. For example, sedimentary rock layers not uncommonly contain younger, once-molten tabular bodies, called dikes, that cut across the pre-existing strata. Similarly, faults (fractures along which movement has occurred) and most cracks that cut across a rock body must be younger than the rock in which they are found.

Finally, weathering and removal of material by erosion can occur only near the surface of the earth and can only affect rocks already on the surface. In some areas, because of these processes, there may be a gap in the geological rock record—a place where the sequence of rock layers or units is interrupted. This break is called an ***unconformity*** because part of the rock record is missing and the sequence does not "conform." An unconformity results from a period of erosion in the geological past during which previously formed rocks were lost, followed by a period of deposition of additional

Fig. 1-7 Sedimentary rocks exposed near Bluff, Utah, looking north over the San Juan River. The originally horizontal rock layers were tilted to their present positions. *John S. Shelton*

layers of rocks. In the walls of the Grand Canyon, for example, several unconformities can be seen. One of the most obvious is the so-called Great Unconformity near the bottom of the canyon, in which the rocks below are older than the flat-lying sediments above by nearly three-quarters of one billion years (Fig. 1-8).

The features used in relative age determination are shown in Fig. 1-9, and by means of these criteria, it is possible to establish a temporal sequence of geological events for most areas on earth (see Fig. 1-8).

Although it is comparatively easy to establish a relative time sequence in local areas, it is usually much more difficult to correlate rocks in one area with rocks in another, even when the other area is close by. The best method devised thus far is to compare the physical features of the rocks themselves. If a sequence of strata made up of identifiable layers can be traced or mapped into another area, then correlation between the areas is easily done. Also, if each layer of a sequence of strata has diagnostic features that can be identified in the same sequence in a nearby area, then correlation can be made with some assurance.

Sometimes, however, strata change laterally in physical character, or the strata present in one area do not show up in another, nearby area. Correlation, then, becomes much more difficult. If one area is quite far removed from another, correlation is even more difficult.

Many sedimentary rocks contain ***fossils,*** the preserved remains or traces of prehistoric animals and plants (Figs. 1-10 through 1-12). In most cases, the hard parts of an organism (for example, bone or exterior shell) are fossilized. However, soft parts can also be preserved. Tracks, trails, burrow structures, and feces—essentially any trace of a living organism—may also be preserved and are also included under the broad definition of a fossil.

Fossils have proved to be of great value in correlating strata and in determining the relative age of the strata in which they are found. They are especially useful in correlations be-

Fig. 1-8 The south wall of the Grand Canyon is composed of almost 1.5 km (1 mi) of strata in which a number of geological events can be relatively dated. The oldest rocks–the Grand Canyon Series–were deposited from the bottom upward, apparently over a long period of time, tilted, and eroded to a nearly flat surface. Subsequently, the stack of flat-lying sediments above was deposited and eventually eroded into its present-day form. The Great Unconformity represents a time gap of nearly one billion years. *Edwin E. Larson*

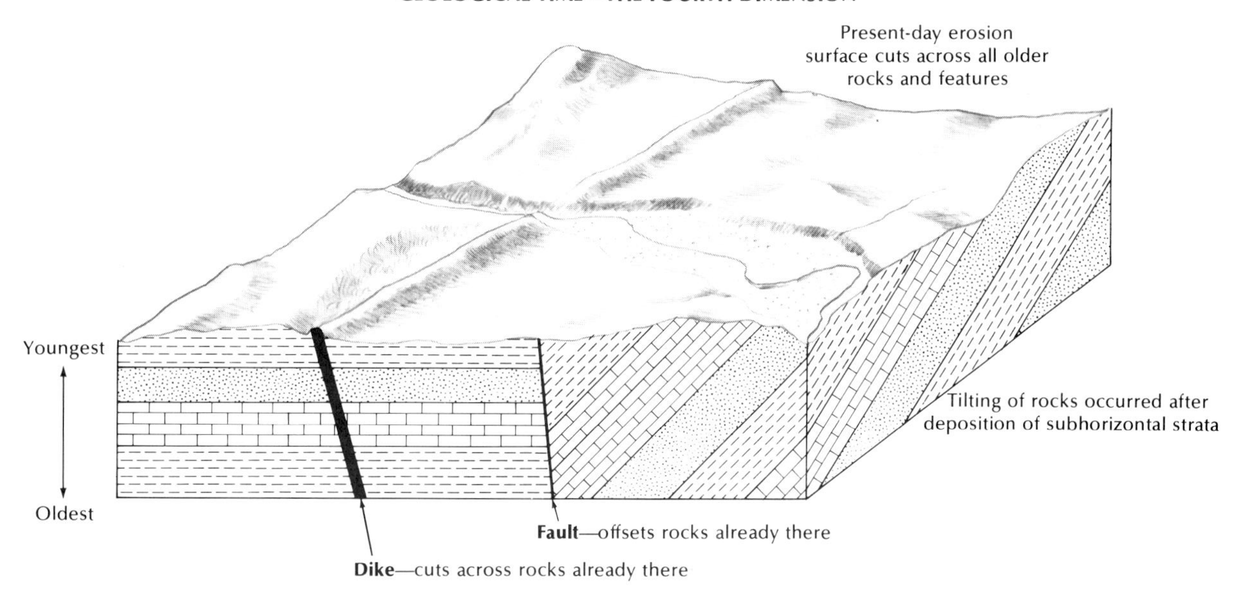

Fig. 1-9 Diagram of the features that enable relative age determinations.

tween widely separated areas, even on different continents. Fossils also form the basis for the relative time scale, a means of placing successions of fossil forms (and the strata in which they are found) in their correct order of occurrence. The scale clearly shows that life has changed continuously in form and kind throughout geological time.

Fossils have excited interest since prehistoric times. Fossils are shown in ancient drawings, some have been represented in heraldry, and for centuries they have been cherished because of the interestingly decorative effect they give to polished stone. The word itself, from the Latin *fossilus*, meaning something dug up, shows an awareness almost since the beginnings of language that a fossil was a thing of the earth.

Although such an awareness may have existed, it was accompanied by an abysmal ignorance of the true meaning of such testimonials to the existence of life in the remote past. To some, fossils were the creation of Satan, placed in the world to confuse all mortals. To others, such as Avicenna (A.D. 980–1037), the Arab philosopher responsible for the revival of interest in the works of Aristotle, they were inorganic petrifactions that grew within the rocks as a result of the workings of a creative force, the *vis plastica*. Only by chance did they come to resemble bones or shells of living creatures. Ignorance extended even to the discovery of the remains of prehistoric people, such as the Cro-Magnons and Neanderthalers, whose bones were sometimes carted to the village churchyard for a proper interment.

It remained for a singularly devoted, untutored, and eminently practical canal-builder, William Smith (1769–1839), to establish the relationship between stratified rocks and the fossils they contain. Fortunately for him, and for us, he lived and worked at the right time and place—during the period when many canals were being built in England. The advent of the Industrial Revolution, coupled with the post-Napoleonic prosperity of Britain, placed a pre-

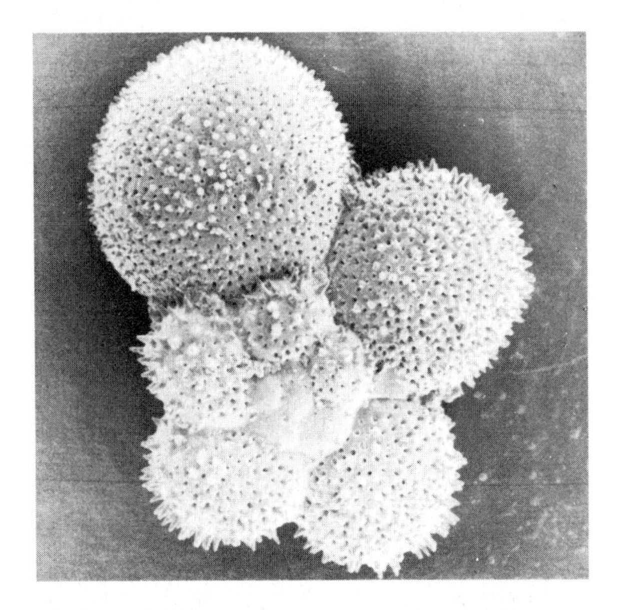

Fig. 1-10 Fossils of three species of one-celled animals (Foraminifera) from 80 million-year-old sedimentary rocks in Colorado. Each shell is about 0.5 mm in length. The photographs were taken with a scanning electron microscope. *Don Eicher*

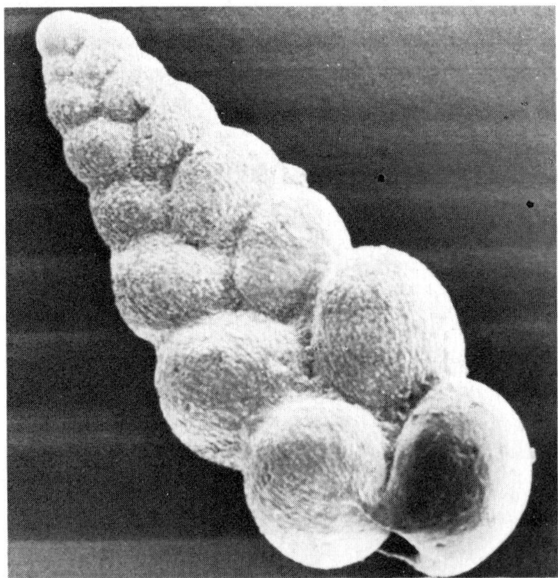

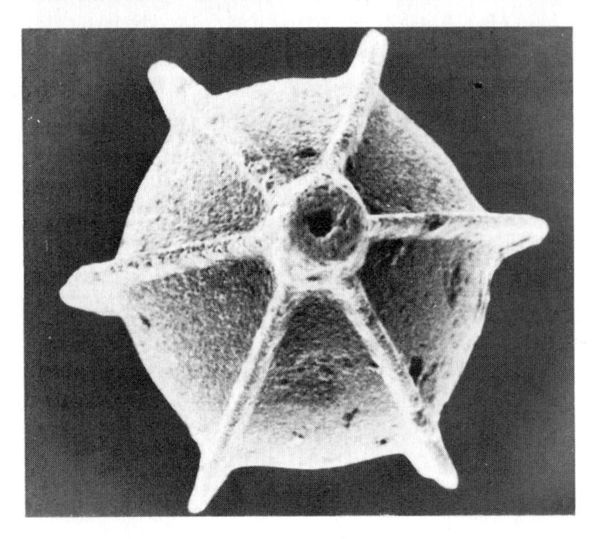

mium on the construction of such relatively cheap and modestly efficient transportation systems as canals.

In the construction of canals, the recognition of various rocks and an understanding of their physical properties are matters of prime importance. If the rocks are harder than expected, if they slump and cave readily, or if they allow water to drain away, then the canal may be a disaster. Fortunately, for Smith, not only is the rock structure of midland and eastern Britain relatively simple, but the rocks themselves are distinctive. Smith carefully noted the nature of all the kinds of rocks in which the canals were cut. His approach was empirical, and considering the rate at which canals were dug, by hand, he was in no hurry. All told, he tramped up and down the English countryside for 24 years making observations. When he was done, he had made the great fundamental discoveries that (1) the strata in southeastern England occurred in the same order—for example, chalk beds were always found above coal layers, unless disturbed structurally, and never vice versa, (2) different layers contained assemblages of distinctive fossils, and (3) the distribution of sedimentary rocks in southeastern England followed a logical pattern that could be represented on a map.

The geological map of England and Wales that Smith published in 1815 was the world's first. Since Smith's time, geological mapping has been found to be an effective means of depicting the geological features of any given area and today represents one of the most important tools of the working geologist.

Smith was not aware of the evolutionary significance of the fossils he collected so patiently over the years. To him they were little more than uniquely shaped objects. Yet, unknown to him, he had found the key by which strata could be correlated with one another, not only locally, but whole continents apart.

Such a method of correlation is no simple matter, however, because we are dealing with the remains of organisms that lived in environments as varied as those of today. Furthermore, during the long span of geological time, some groups of animals and plants gradually evolved into more complex creatures, while others, having achieved perhaps an uneasy balance with their environment and their enemies, remained essentially unchanged. Still others, such as the dinosaurs, became extinct.

This brings us to an important contribution that ***paleontology*** (the study of ancient life) makes to the whole realm of contemporary thought. It is the demonstration—from the fossils preserved in the thousands of feet of stratified rocks in many lands—that many forms of life have evolved from relatively simple to complex hierarchies of plants and animals. In fact, the fossil record was one of the stronger bod-

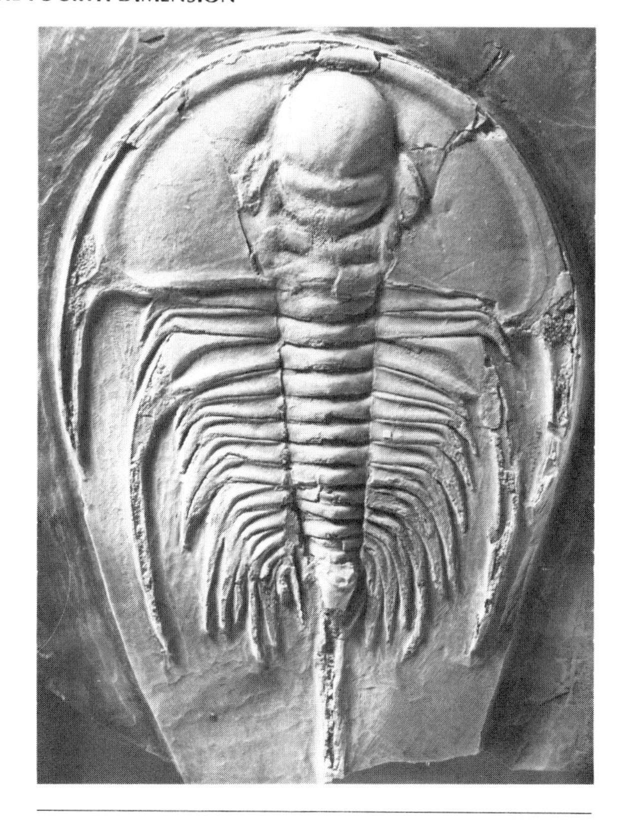

Fig. 1-11 *Olenellus fremonti,* a trilobite from the Marble Mountains, California. Its length is about 11.5 cm (4.5 in.). This animal, related to the hermit crab, lived in the oceans between 600 and 225 million years ago; then it disappeared from the face of the earth. *Takeo Susuki*

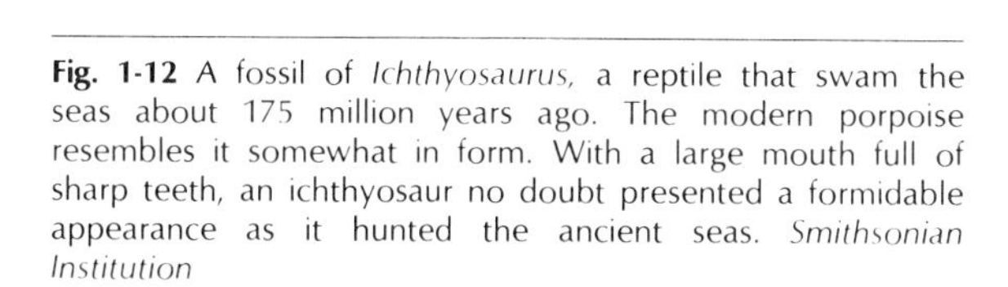

Fig. 1-12 A fossil of *Ichthyosaurus,* a reptile that swam the seas about 175 million years ago. The modern porpoise resembles it somewhat in form. With a large mouth full of sharp teeth, an ichthyosaur no doubt presented a formidable appearance as it hunted the ancient seas. *Smithsonian Institution*

ies of evidence advanced by Darwin and his followers in support of the theory of evolution.

Geological time scale

As earth scientists accumulated knowledge of strata in different parts of the world, as well as in a single region, they were able to compare rock layers on the basis of their fossil content. It was necessary first to determine the order of deposition in one region, and when that had been done, the succession of rocks there might be compared with a different succession of rocks—perhaps a whole continent away, and not necessarily the same kind of rocks at all, but ones containing fossils of the same geological age.

The most valuable fossils for correlating rocks are those with comparatively short histories, geologically speaking—yet which, in their limited time on earth, achieved a wide geographical distribution and underwent rapid evolutionary changes. Such ideal examples are called ***index fossils.***

Through the application of the law of superposition, it was possible to ascertain, at least locally, which fossils were older than others. And eventually, as more and more stratigraphic sections were correlated, it became possible to establish a chronological order on the basis of the fossils in the rock. That was the beginning of the relative geological time scale (Fig. 1-13).

Geological time is divided into units using much the same philosophy employed in dividing historical time. As a single example, in the Western world one of the longer time intervals is the Christian era. True, there is a unity within the nearly 2000-year chapter of history, but were we to find ourselves transported to ancient Palestine or Rome, it is very likely we should be more impressed by the differences than by the similarities.

It is a rare one-semester history course that, in attempting to cover such a broad panorama, succeeds in making it appear more lasting than the fleeting, multicolored patterns of a kaleidoscope. Most history courses, however, divide that enormous block of time, with all its crowded events, into smaller units; sometimes they become very short indeed, and a semester may be devoted to a five- or ten-year period.

If we carry the analogy with human history a bit further, the most intensely studied parts of the record are more likely to be the later rather than the earlier parts. For example, 2000 years of Egyptian history may be sketched in with broad strokes, whereas a large part of a semester may be devoted to the decades following the Treaty of Versailles.

Our view of geological time is similarly colored. Events closer to us leave more complete and decipherable records that distant events for which the records have grown increasingly fragmentary over a long period of time and through the repeated deformation rocks undergo in such processes as mountain building. The result is that more is known about events in the later parts of earth history than in the earlier, and that knowledge is reflected in the larger number of subdivisions of the last part of the geological time scale.

If we consider the names on the scale, the grand divisions, comparable in significance to such human episodes as the Stone Age, are the ***eras.*** In most cases they take their names from our concept of the dominant aspect of the life during each era. The most ancient, the ***Precambrian,*** was an interval of long duration, characterized by a paucity of life forms. The eras that followed are called the ***Paleozoic*** (ancient life), ***Mesozoic*** (middle life), and ***Cenozoic*** (modern life). In broad terms, the Precambrian was a time of no life or scant primitive life. The Paleozoic was a time in which invertebrates (snails, clams, and the like) and simple backboned animals, such as fish, amphibians, and primitive reptiles, were ascendant. The Mesozoic was high noon for the reptiles, and it is the chapter in earth history when the hegemony of the dinosaurs was complete. In the Cenozoic, which is the present era, mammals became the dominant land vertebrates.

The eras are divided into lesser units called ***periods,*** which are, in turn, divided into units called ***epochs.*** From the names of periods, it obviously is hard to find a common pattern. Two have a familiar ring to Americans—the Pennsylvanian and Mississippian—and we can

RELATIVE GEOLOGIC TIME			LIFE FORMS
ERA	**PERIOD**	**EPOCH**	
Cenozoic (recent life)	Quaternary	Holocene	Rise of mammals and appearance of modern marine animals
		Pleistocene	
	Tertiary	Pliocene	
		Miocene	
		Oligocene	
		Eocene	
		Paleocene	
Mesozoic (middle life)	Cretaceous	Late	Abundant reptiles (including dinosaurs); more advanced marine invertebrates
		Early	
	Jurassic	Late	
		Middle	
		Early	
	Triassic	Late	
		Middle	
		Early	
Paleozoic (ancient life)	Permian	Late	
		Early	
	Carbon-iferous: Pennsylvanian	Late	First reptiles
		Middle	
		Early	
	Carbon-iferous: Mississippian	Late	
		Early	
	Devonian	Late	First terrestrial vertebrates—amphibia
		Middle	
		Early	
	Silurian	Late	
		Middle	
		Early	
	Ordovician	Late	First vertebrates—fish
		Middle	
		Early	
	Cambrian	Late	Primitive invertebrate fossils
		Middle	
		Early	
Precambrian			Meager evidence of life

Fig. 1-13 Relative geological time scale, based primarily on superposition of beds and the character of the fossils in the strata.

use them to illustrate the philosophy behind the naming of such time units. Most periods are named for regions in which rocks containing fossils characteristic of their segment of geological time were found. This means that strata deposited in the later Paleozoic in an inland sea covering much of the eastern United States were named for their occurrence in Pennsylvania, just as strata approximately 20 million years older were named for their occurrence in the Mississippi Valley—not the state. Incidentally, neither of the latter terms is used in Europe; there the two periods are treated as a single unit, the Carboniferous, which acquired its name from the coal-bearing strata of England.

Other period names with origins that can be recognized readily are the Devonian, named for rocks that crop out along the southwestern tip of Great Britain in Devon and Cornwall, and the Jurassic, named for rock exposures in the Jura Mountains along the western border of Switzerland. A less familiar period is the Permian—named after the ancient Kingdom of Permia in Russia.

Only persons with a knowledge of Latin may recognize the ancient name of Wales, Cambria, as the basis for the name of the Cambrian Period. Two closely related periods, the Ordovician and Silurian, perpetuate the memory of Stone Age tribes whose homes had been in Wales—the Ordovices in the north and the Silures in the south.

Some of the other periods are named not for places, but for the physical characteristics of their rocks. The Cretaceous is a good example, since it is derived from the Latin word *creta,* chalk, and refers to the exposures of chalky rocks in the cliffed coast of southern England. In fact, most of the chalk of the world is of the Cretaceous Age. Many other kinds of rocks, however, were also formed during the Cretaceous. In the United States, Cretaceous rocks run from conglomerate through sandstone and shale to limestone and coal; or, as in the Sierra Nevada, they may include the enormous bodies of granite intruded during that period. The Triassic takes its name from the fact that, in Germany, the rocks of the period are divided into three distinctive layers—a limestone in the middle, with reddish sandstones and shales above and below.

The two periods of the Cenozoic Era are called the Tertiary and the Quaternary, and the rather strange names of their subdivisions, or epochs, are a special case. They were established, for the most part, by Charles Lyell according to the percentage of extinction of marine molluscan fossils. Thus, in Eocene (from the Greek words for dawn and recent) strata, about 1 to 5 per cent of the species found are still living. In the Miocene (Greek words for less and recent), 20 to 40 per cent of the fossil species are still alive. In the Pleistocene (Greek words for most and recent), 90 to 100 per cent of the fossil shells are those of species living in the seaways of the world today. Without perhaps fully realizing it, Lyell had divided the Cenozoic era of geological time statistically—one of earliest applications of statistics to a natural science.

The Pleistocene epoch is regarded by many geologists as being coincident with the Ice Age, and, although the length of time represented by the multiple advances and retreats of the ice sheets appears to vary from place to place, it probably is close to 2 million years.

The most accepted way of dividing the Cenozoic Era is into the Tertiary (meaning third)—which includes all but the Pleistocene and post-glacial time—and the Quarternary (meaning fourth). The division represents a sort of cultural lag, since we no longer speak of the Primary and Secondary rocks as our forebears did. With the first and second of a series gone, it seems strange to speak of a third and fourth, but you will see the terms employed consistently in most geological writing today; such is the force of habit.

Absolute time

The relative geological time scale provides us with a framework within which to view the events that have shaped the earth, but we cannot tell from it when a specific event actually took place. We know, for example, that the dinosaurs lived after the plants found in the coal beds in Pennsylvania; but how long afterward and for how long? Or, at what specific time or over what time span was the Sierra Nevada range formed? Earth scientists, as we have already learned, contrived in various ways to answer such questions, but up to the end of the nineteenth century no method proved reliable or easily workable. In 1895, however, the door was opened wide by the discovery of elements with unusual physical properties—a discovery that led to techniques that have since enabled us to obtain the goal of absolute age dating.

Few people alive in 1895 were aware of the eventual significance to the world that a simple experiment by French physicist Antoine Henri Becquerel (1852–1908) was to have. He had been intrigued by Roentgen's discovery, within

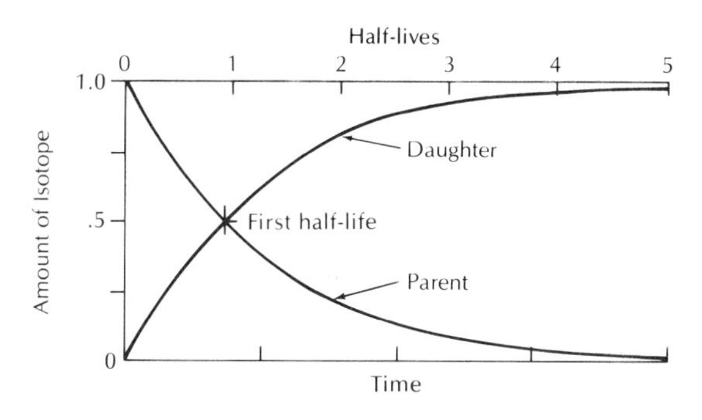

Fig. 1-14 Radioactive decay of a hypothetical isotope and growth of the daughter product. The time required for the isotope to decay to one-half the initial amount is called its half-life and is unique to each radioactive isotope. Note that the parent/daughter ratio changes continually with time.

the same year, of the X-ray, and wondered whether the phenomenon of phosphorescence bore any relationship to the strange rays. Various substances, he reasoned, might be able to pick up energy from the sun, and accordingly, he exposed a number of them to the sun's rays—with no discernible effect. By chance (serendipity has played a part in many scientific discoveries) he observed that when some salts of the heavy element uranium were placed on photographic paper, the paper darkened as though subjected to some sort of radiation, and that was true whether or not the uranium salts had been exposed to the sun's rays. Becquerel's discovery was the first demonstration of ***radioactivity.***

Among those first experimenting with radioactive substances was the distinguished physicist Lord Rutherford (1871–1937), born Ernest Rutherford, in New Zealand. In 1902, he discovered that when unstable radioactive atoms disintegrated into completely different atoms, radiation was given off. Radioactivity, then, has some of the essence of medieval science of alchemy, in which the attempt was made to change base metals to precious ones. Radioactive elements, however, undergo natural spontaneous disintegration with the formation of different, and generally not precious elements.

The initial work by several people, including the French chemists Marie and Pierre Curie, indicated that both uranium and thorium were radioactive elements. During the early 1900s it was also found that several forms with the same elemental properties could exist, each differing only in their atomic weights. To those variants of the same element the radiochemist F. Soddy (1877–1956) gave the name ***isotopes*** (from the Greek meaning same place). Eventually it was concluded that radioactive decay occurs within the nucleus of the atom and only in certain isotopes that are unstable. It was initially assumed, and later experimentally confirmed, that each radioactive element breaks down at its own rate and that for many elements the rate of decay is extremely slow, requiring many millions or even billions of years to decay completely. A radioactive element or isotope (called the ***parent***) will slowly and spontaneously disintegrate over a period of time to form an almost equivalent amount of decay product (called the ***daughter***). The ratio of daughter to parent changes continually with time, and can, therefore, provide a measure of the total time that has elapsed during the decay process. The time that it takes for a radioactive isotope to decay to one-half its initial quantity is called its ***half-life.*** Each isotope has a specific half-life. The process of radioactive breakdown is illustrated in Figure 1-14.

Rutherford recognized the importance of naturally occurring isotopes, and in 1905, in a series of lectures delivered at Yale University, he proposed the use of radioactive decay as a means of absolutely dating rocks. He pointed out that there was not one clock but as many as there were usable radioactive isotopes. In 1906, he announced the first ***radiometric date*** ever obtained, from a mineral contained in a rock found in Connecticut—500 million years. The American physical chemist Bertram Boltwood (1870–1927) followed in 1907 with dates for other rocks, ranging from 410 million to 2200 million years. A great many factors were not taken into account by the early age-daters, and many of the first dates were in error. The important thing, however, is that a method had been discovered that held high promise of providing absolute ages of minerals and rocks, and that even the first dates determined by the method indicated that the earth was hundreds of millions of years old.

The radioactive materials initially used in absolute dating were minerals containing three radioactive isotopes—uranium-238, uranium-235, and thorium-232. (The numbers indicate the atomic weight of the isotope.) The process of nuclear disintegration is long and involved, but each of the three reaction chains eventually produces one of the isotopes of lead as the stable end-product. The gas helium is also produced in many intermediate breakdown steps. Many of the first dates were based on the ratio of radioactive uranium or thorium to helium; but later dates, which proved to be much more reliable, were based on the ratio of uranium or thorium to lead.

All told, there are more than 1000 natural and artificial radioactive isotopes. Two naturally occurring isotopes, potassium-40 and rubidium-87, have also proved to be very useful in the absolute dating of rocks. Table 1-1 lists the radioactive isotopes commonly used in absolute dating and their rates of decay (half-life).

Obtaining reliable dates from rocks and minerals is actually very complicated. It requires elaborate laboratory procedures and sophisticated equipment to precisely measure the small amounts of parent and daughter isotopes. The work is carried out in relatively few laboratories, and a level of performance has now been reached such that little error is introduced into the dating from laboratory procedures. The principal source of error comes from the minerals and rocks themselves. If a mineral is to produce a reliable date, it must have remained a closed system since its origin. What that means is that no parent or daughter materials have entered or left the mineral during its entire existence. It is not always easy to tell how closely the restriction has been met. One simple precaution that eliminates some of the uncertainty is to collect only fresh, unweathered rocks and not to use rock samples that appear chemically altered. Rocks dated by the potassium-argon method are particularly liable to loss of the daughter product argon-40.

Table 1-1. Commonly used radioactive isotopes, their breakdown products and rates of decay

Radioactive isotope	*Daughter product*	*Half-life (years)*
Uranium-235	Lead-207	710 million
Uranium-238	Lead-206	4.5 billion
Thorium-232	Lead-208	14.1 billion
Potassium-40	Argon-40	1.3 billion
Rubidium-87	Strontium-87	47.0 billion
Carbon-14	Nitrogen-14	5570

Argon is a gas and as such is extremely difficult to keep trapped within the atomic latticework of the host mineral; it tends to diffuse slowly out of the mineral structure. And, as you might expect, if a rock is subjected to increased temperatures the likelihood of argon loss increases. Reduction in the amount of argon-40 leads to an underestimation of the age of the rock. If the temperature remains sufficiently high for a long enough period, all the previously trapped argon will escape. In effect, the radioactive clock is reset. Dating of such a rock will give the time of thermal resetting, which in some circumstances can be useful.

The best safeguard against hidden errors in radiometric dating comes through the use of consistency checks. One obvious check is dating a mineral or rock by more than one radiometric method. For example, if both rubidium-strontium and potassium-argon methods are applied to the same rock and if the results agree, the likelihood is high that the date is reliable. If the two dates differ appreciably, then additional dating methods must be used to determine which, if either, is the correct date. Another consistency test may be made through comparing the absolute and relative ages of rocks. For example, rocks that are known by physical correlation to be the same age should produce nearly identical dates.

Not all rocks can be dated by radiometric methods. First, the minerals of the rock must have formed within a short time, and must have remained as a closed system thereafter. Datable rocks, then, are principally those that have formed as a result of igneous or metamorphic processes (see Chap. 4). The time of the deposition of sedimentary rocks, which commonly are made up of fragments of igneous and metamorphic rocks and other sedimentary rocks, cannot generally be radiometrically dated, since any dating of the fragments would only reveal the age of the source rock—not the time of deposition. It is possible, however, to date igneous rocks, such as lavas or ash layers, that are interlayered with sediments or dikes that cut across sedimentary rocks. By using such methods, earth scientists can correlete sedimentary rocks, particularly those containing index fossils, with the absolute time scale.

Second, the rocks and minerals to be dated must initially have contained sufficient radioactive material to be measured. And third, enough time must have elapsed since the origin of the rock to allow measurable quantities of daughter products to accumulate. Generally, most rocks younger than about 250,000 to 500,000 years old cannot be dated.

Through the application of radiometric methods during the last fifty and particularly during the last twenty years, earth scientists have been able to refine the geological time scale, which was previously established on the basis of fossil data (Fig. 1-15). Not only that, but age dating has allowed geologists to unravel the Precambrian rock record, in which fossil remains are scarce or non-existent. It even shows that the Precambrian includes most of geological time. Through the use of absolute dating methods, geologists are now able to determine accurately such things as rates of evolution and the time spans over which mountains were formed.

The oldest dates obtained from the surface of the earth come from continental rocks and cluster around 3 to 3.2 billion years. Holding the distinction of being the very oldest—3.8 billion years—are rocks found in western Greenland. Some meteorites, however, have been dated at 4.5 billion years; they are the most ancient materials ever found. Geologists consider the latter figure to be the best estimate of when our earth and our solar system came into being—an estimate partially corroborated by dates up to 4.2 billion years obtained from moon rocks gathered during the Apollo missions. As old as our solar system appears to be, its age falls far short of that of the Milky Way Galaxy (which includes our solar system), which, on the basis of astronomical considerations, is estimated to be between 9 billion and 19 billion years old.

If we wish to understand better the immensity of geological time, we can represent the age of the earth (4.7 billion years) by a single year. Then the record of abundant fossils extends back a scant 40 days or so; humans have been on the earth a matter of hours; and all of recorded history amounts to about a minute. Most of the year is taken up by the Precambrian Era. Interestingly enough, because the rock record is so poor and fossils are essentially non-existent for that long span of time, we know the least about it. Without radiometric dating we would have known almost nothing.

Carbon-14 dating, dendrochronology, and lichenometry

One other radioactive isotope used in age dating requires discussion: carbon-14. It has been left to last because its role in establishing the overall framework of the geological time scale has been minimal. Carbon-14 decays to nitrogen 14 with a half-life of only 5570 years. With conventional dating methods, it can only be used effectively for obtaining dates back to 30,000 or 40,000 years.*

That was a time, however, when human activities, which began slowly, progressed rapidly. By means of ***carbon-14 dating*** (or ***radiocarbon dating***) we are now able to place archeological traces of earlier cultures in historical perspective. Its use has also aided in determining the details of geological history during the later phases of the Ice Age.

The discovery of carbon-14 goes back to cosmic ray research in the upper atmosphere by the geophysicist Serge Korff in 1939. He

*Through the use of improved laboratory methods, it now appears that the time range over which carbon-14 is usable can be extended to nearly 100,000 years.

GEOLOGIC TIME				ABSOLUTE TIME*
ERA	PERIOD		EPOCH	
Cenozoic	Quaternary		Holocene	
			Pleistocene	2
	Tertiary		Pliocene	12
			Miocene	26
			Oligocene	37–38
			Eocene	53–54
			Paleocene	65
Mesozoic	Cretaceous		Late	
			Early	136
	Jurassic		Late	
			Middle	
			Early	190–195
	Triassic		Late	
			Middle	
			Early	225
Paleozoic	Permian		Late	
			Early	280
	Carboniferous	Pennsylvanian	Late	
			Middle	
			Early	
		Mississippian	Late	
			Early	345
	Devonian		Late	
			Middle	
			Early	395
	Silurian		Late	
			Middle	
			Early	430–440
	Ordovician		Late	
			Middle	
			Early	500
	Cambrian		Late	
			Middle	
			Early	570
Precambrian			Late	
			Middle	
			Early	3800+

*Boundaries estimated radiometrically in millions of years before the present

Fig. 1-15 Geological time scale, with boundaries between major divisions absolutely dated by radiometric methods.

noted that cosmic rays produce secondary particles without charge (neutrons) in their initial collision with nitrogen gas molecules. Those neutrons, he predicted, when they collided with the abundant isotope nitrogen-14 would react to free a positively charged particle (proton), thus forming the radioactive isotope carbon-14. Carbon-14, then, would combine with oxygen to form carbon dioxide, which circulates in the atmosphere and dissolves in the ocean waters; it is absorbed by plants and animals at the earth's surface. Carbon-14 decays to nitrogen-14, but new carbon-14 is continually being formed in the atmosphere so that an equilibrium level is maintained. As long as an organism lives, it continually replenishes its

Fig. 1-16 Tree rings of a Douglas fir that began growing in A.D. 113 and was cut down in A.D. 1240. This specimen was used as a construction timber in a prehistoric structure south of Mesa Verde in southwestern Colorado. *Laboratory of Tree Ring Research, University of Arizona*

carbon-14 content; once it dies, however, the isotope that has accumulated in the body continues to decay, but it is no longer replenished. It will reach an unmeasurable level in about 50,000 years. By measuring its carbon-14 level, it is possible to determine how long an organism has been dead.

The basic assumption made by those who use carbon-14 dating, as noted, is that after an organism dies, it no longer takes up carbon-14 from its surroundings. Yet some circumstances render that assumption invalid. For example, if the organic matter to be dated is buried, circulating ground water may provide a ready source of carbon-14, which then may enter into partial equilibrium with the carbon-14 already

Fig. 1-17 Bristle-cone pine growing in the higher altitudes of California's White Mountains. Some bristle-cones in the area are more than 4000 years old; that is, they began their existence nearly 2000 years before the birth of Christ and had been living for about 3500 years when Columbus discovered the Americas.

Fig. 1-18 Radiocarbon–calendar age relationship as determined from carbon-14 dating of wood of a known calendar age.

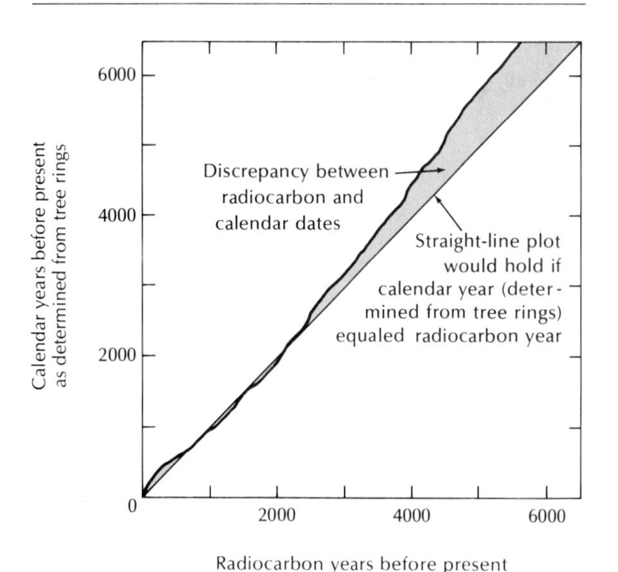

present. Or, grass roots, which can extend downward for two meters or more, may become so entangled with the organic matter that they cannot be completely separated from it. Because of such uncertainties and because of difficulties with the laboratory procedure, any carbon-14 dates, but particularly those over about 20,000 years, should be viewed with some reservation.

Carbon-14 dating is also based on the assumption that the amount of carbon-14 produced in the atmosphere has remained relatively constant during the last 40,000 years. If variations in the production of the isotope did occur, and lasted several hundreds or thousands of years, they would lead to systematic errors in the radiocarbon dates obtained from organisms that lived during those times.

In the early days of carbon-14 dating there was only a nagging suspicion that such might

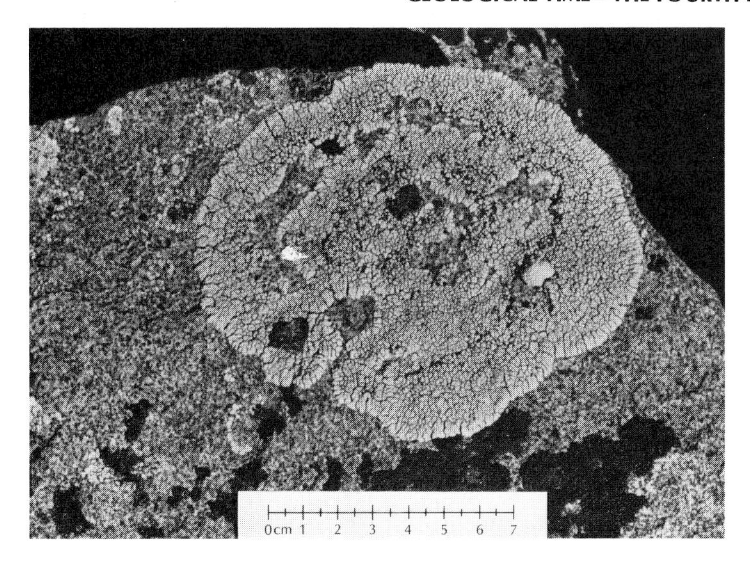

Fig. 1-19 Lichen growing on a rock above treeline, Colorado Front Range, near Boulder. The size suggests an age of several hundred years. *Janet Robertson*

be the case, but there seemed to be no way to obtain valid data to indicate the magnitude of possible errors. However, with the advent of ***tree-ring dating,*** or ***dendrochronology,*** a means for comparing carbon-14 dates with actual calendar dates became available. Each year that a tree lives, it normally adds one growth layer to its circumference. In a cross section from the tree, the growth layers appear as rings, and they can be counted (Fig. 1-16). When the rings of very old trees were counted and measured, it became apparent that in some successions the rings were close together, whereas in others they were wider apart. The spacing apparently reflects the rate of growth in response to the climatic conditions prevailing at the time. The spacing of the tree rings provided a "signature," and all trees that grew in the same locality during the same time span showed the same pattern of variation. Thus a key was provided for matching rings within a local geographical area, of similar climate. By counting and describing patterns in successively older trees, a tree-ring master pattern could be compiled for a particular geographical area. Today, through the use of tree rings from the long-lived bristle-cone pine (Fig. 1-17), one such tree-ring master template has been pieced together; it extends back in time, year by year, for about 7000 years. Dendrochronology has provided an alternative method of dating archeological sites whenever logs and tree trunks can be found in association with cultural relics. In addition, by carbon-14 dating the wood itself, it is now possible to determine systematic variations in the radiocarbon dates (Fig. 1-18). The results indicate that from the present to about 500 years ago, and from 2100 to 7100 years ago, radiocarbon ages are younger than the calendar ages, whereas from 500 to 2100 years ago the two chronologies are essentially the same. For dates near 7100 years ago, the error in the radiocarbon dates can amount to as much as 700 years. Most radiocarbon chronology laboratories now use that information to "correct" their carbon-14 dates.

Lichens are relatively primitive low-profile plants that grow on rock surfaces and look like splotches of paint (Fig. 1-19). The so-called moss-rock, often used in building fireplaces, is usually rock material coated with colorful lichens. These plants, particularly those that are circular in outline, can be used to date rock debris and rock surfaces less than several thousand years old.

The noted botanist Roland Beschel pioneered this unique age-dating technique when he observed that lichens on tombstones in Austria became progressively larger in diameter the older the tombstone. Other work

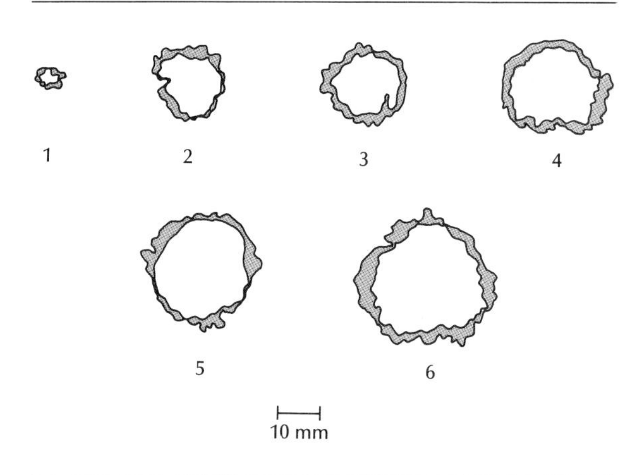

Fig. 1-20 Tracings of a fast-growing lichen species (*Alectoria minuscula*) made over a four-year period (1963–67) in the eastern Canadian Arctic. Note that the rounded shape is more or less maintained as size increases.

Fig. 1-21 Variation in growth rates with climate of the individuals of one lichen species.

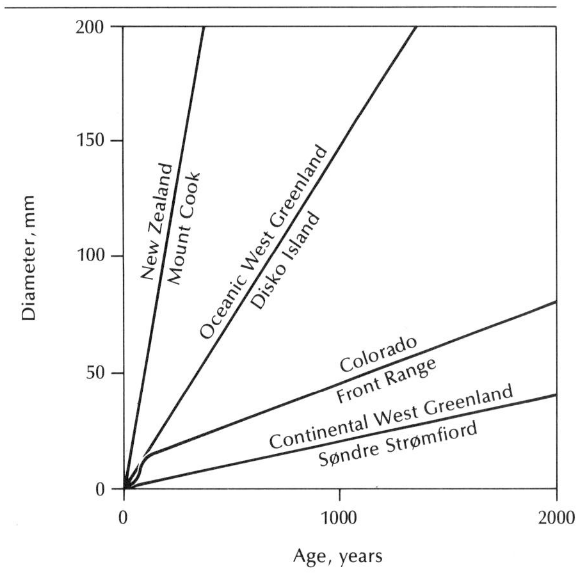

has verified his finding (Fig. 1-20). In many regions for which the lichen growth-rate is known (Fig. 1-21), the plants have been used to date young surface deposits, particularly glacial, rockfall, or mudflow deposits in alpine regions.

SUMMARY

1. *Geology,* the study of the earth, is such a broad field that it is undertaken by scientists with extremely diverse backgrounds. The information that they supply is combined to provide the total picture of the earth, as it is today and as it was in the past.
2. One of the basic concepts that permits the unraveling of geological history is *uniformitarianism,* which maintains that the processes we see in action today are probably the same as those of the past.
3. Within a local area of study it normally is possible to place the events recorded in the rocks into a *relative time* sequence on the basis of *original horizontality* and *superposition* of beds, *cross-cutting relationships,* and determination of erosion surfaces.
4. Through the observation of *fossils* (the remains and traces of prehistoric life forms), which form an evolutionary succession, a *relative geological time scale* was developed. From their fossil content it is possible, relatively, to date and correlate strata in widely separated areas, even on different continents.
5. The discovery of radioactivity in 1895 eventually led to methods for the *absolute dating* of minerals and rocks, and it was possible to base the geological time scale on absolute time. The oldest rocks yet found on earth are 3.8 billion years old, whereas meteorites are about 4.5 billion years old, which apparently marks the beginning of our solar system.
6. *Carbon-14 dating, tree-ring dating,* and *lichen dating* provide additional means of obtaining dates for very recent geological materials.

QUESTIONS

1. What are the main differences between a multidisciplinary science, such as geology, and a fundamental science, such as chemistry?
2. What is the difference between relative time and absolute time?
3. List all the possible methods that can be used to establish a relative time sequence.
4. How can fossils be used to establish the relative time scale?
5. Define the following: radioactivity, isotope, half-life, parent/daughter ratio.
6. Explain how radioactive isotopes are used to establish the absolute time scale.
7. What are the limitations to the use of carbon-14 in age dating?
8. Without using the book, draw a reasonable facsimile of the geological time scale, listing all the eras and periods and inserting absolute dates at each break between the eras and at the base.

SELECTED REFERENCES

Adams, F. D., 1954, The birth and development of the geological sciences, Williams and Wilkins, Baltimore (Reprinted by Dover Publications, New York, 1954).

Cohee, G. V., Glalssner, M. F., and Hedberg, A. D. (eds.), 1978, The Geologic time scale, American Association of Petroleum Geologists, Studies in Geology No. 6, Tulsa, Oklahoma.

Dalrymple, G. B., and Lamphere, M. A., 1969, Potassium-argon dating, W. H. Freeman, San Francisco.

Eicher, D. L., and McAlester, A. L., 1980, History of the earth, Prentice-Hall, Englewood Cliffs, New Jersey.

Hubbert, M. King, 1967, Critique of the principle of uniformity, *in* Uniformity and simplicity, Geological Society of America Special Paper No. 89.

Knopf, A., 1949, Time and its mysteries; series 3, New York University Press, New York.

Libby, W. F., 1961, Radiocarbon dating, Science, vol. 133, pp. 621–29.

McIntyre, D. B., 1963, James Hutton and the philosophy of geology, *in* The fabric of geology, Claude C. Albritton, Jr., ed., Addison-Wesley, Reading, Massachusetts.

McPhee, J., 1981, Basin and range, Farrar Straus & Giroux, New York.

Wendland, W. M., and Donley, D. C., 1971, Radiocarbon-calendar age relationships: earth and planetary science letters, vol. 11, pp. 135–39.

York, D., and Farquhar, R. M., 1972, The earth's age and geochronology, Pergamon Press, New York.

2

THE SOLAR SYSTEM

We live on earth, the third planet from our central star, the sun. With the aid of modern telescopes and other astronomical equipment, we can examine the eight other planets in our solar system. Beyond that, we can see that the sky is literally filled with stars, bright and faint, near and far. Most of the stars we see are part of a large, rotating pinwheel-shaped body of about ten billion stars called the ***Milky Way galaxy*** (Fig. 2-1) that takes about 250 million years to make one turn around its center.

The last time the Milky Way was in its present position, dinosaurs still existed on earth and mammalian-like animals (our ancestral line) had just come upon the scene.

As large and bright as we think the sun to be, it is but an average-sized star. Along with its planetary system, the sun is located not in the center of the galaxy, but relatively far out on one of the pinwheel arms.

It is truly humbling that our galaxy is but one small part of a visible universe that includes about one billion individual galaxies, each existing like islands of stars in the immense void of space. The size of the universe and the amount of material in it are beyond our comprehension. We do know that it is so large that light traveling at a speed of 300,000 km/sec (186,000 mi/sec) may take many millions and even billions of years to reach the earth from deep space.

So, as we look progressively farther out into the universe, we are looking progressively backward in time. The picture of the universe, then—except for our closest neighbors—is distorted because we see it as it was, not as it is. Modern astronomers, probing the vastness of the islands of stars, document the occurrence of extraordinary events—the births and deaths of stars and galaxies. Or, at least that is how they interpret the information coming to them. We still understand little about the universe—whether other intelligent life exists in other parts of the universe or even elsewhere in our own galaxy is still very much a mystery.

A solar system such as ours, astronomers tell us, probably is not unique in our galaxy. Of ten billion stars, there are likely to be many thousands with similar planetary systems. But we can never hope to learn much about the nature

Fig. 2-1 Spiral galaxy in the Ursa Major constellation, as seen through a telescope, closely resembles our Milky Way Galaxy. Individual stars in the galaxy are not discernible. The bright large and small spots are relatively nearby stars in our galaxy. *Hale Observatories*

of sister solar systems until we venture into the galactic realm.

The reason we have not been able to determine, telescopically, whether other solar systems exist, is that planets are dark bodies—they do not emit light directly but only reflect the light that falls upon them from bright central stars. So very distant planets remain invisible even to our most powerful telescopes. In the 1960s, the startling announcement was made that a nearby star, Barnards Star, has a planetary system consisting of at least two bodies, each about the size of our largest planet, Jupiter. Astronomers deduced the existence of the two large planets by observing small variations in the star's position. Yet, even if a planetary system does exist around Barnards Star, there is little chance we can learn more about it by observation from earth. More recently, there have been "sightings" of other planetary systems in the Milky Way.

Less than 20 years ago, knowledge of our solar system came primarily from direct observations of the earth, the study of meteorites that fell to earth, and telescopic observations of other planets, which began in the early 1600s. With the rise of space science, however, we have been able to augment that knowledge with direct observations of the moon and data telemetered back to earth from unmanned space probes. Data sent back from spacecraft are helping scientists to answer some of the age-old questions concerning the solar system and the earth, but they are also introducing new ones. We have arrived at a "golden age" in interplanetary travel.

An early astronomer, Aristarchus 220?–150 B.C., observed that the apparent position of the sun when viewed against the backdrop of the night sky did not change, no matter what the position of the observer. From that simple observation he concluded that the sun must be at a great distance from the earth and that it must be very large to appear as large as it does; so large, in fact, that it seemed more realistic for the earth to revolve around the sun than the other way around. Unfortunately, the views of Aristarchus were opposite to the earth-centered model of the solar system favored by Aristotle (384–322 B.C.), the dominant natural philsospher of his time whose influence lasted for a long time thereafter.

Fig. 2-2 Nicolaus Copernicus. Modern astronomy is based on his treatise, written around 1530, describing the sun as the center of a great planetary system. *Okregowe Museum, Toruń, Poland*

Subsequently, the intellectual darkness of the Middle Ages spread across Europe, and the Aristotelian views prevailed. The scientific legacy of Aristarchus and other scholars was all but forgotten.

In the period from the eleventh through the sixteenth century, a few people again pondered questions about the earth and the solar system. But now, a change in scientific philosophy occurred. In olden days, scientists were scholars who did "thought experiments," but never made observations of real objects to check on the validity of their conclusions. During this period of scientific resurgence, however, experimentation and observation became the primary basis for reaching conclusions.

Among the titans of the new wave of scientists was the Polish physician and astronomer Nicolaus Copernicus (1473–1543) (Fig. 2-2). He resurrected the heliocentric model of the solar system in which the planets move around the sun in circular orbits, and made observations to check his conclusions. Some of the details did not fit the model exactly, but they seemed close enough. His concept of the solar system was a minority view and contrary to the dogma steadfastly upheld by the church. Fearing the consequences, Copernicus held back his idea until shortly before his death, when he finally summoned up enough courage to publish it. Once published, his theory became the topic of conversation—much of it behind closed doors—throughout scientific circles in Europe. The model he constructed has proved to be basically correct. However, it did require refinement by those who came after, particularly the Danish astronomer Tycho Brahe (1546–1601) and the German astronomer Johannes Kepler (1571–1630).

Brahe was primarily an observationist. In 1576 he began building a "palace of astronomy" under the patronage of King Frederic of Denmark on the small island of Hven, near Copenhagen. For the next 20 years, he and his many followers and assistants were engrossed with making observations of the sky. Brahe's forte was his ability to make careful, exacting measurements of the angular positions of all the visible planets and the moon. When he died in 1601 he left the records of his observations to Johannes Kepler, who had been his assistant during his last years. Without those records, Kepler probably would not have been able to explain the real motions of the planets.

In about 1608, when Kepler was pondering the details of the Copernicus model, the first telescope became known in Europe, and soon thereafter observational astronomy made great strides through its use. The inventor of the first telescope has been lost in antiquity, but Galileo Galilee (1564–1641), a teacher of mathematics and astronomy in Padua, Italy, was quick to foresee the potential usefulness of the instrument, and constructed one in 1609 (Fig. 2-3). The early telescopes were very primitive, and magnification of distant objects by only about 30 times was the best that could be achieved. Nonetheless, Galileo turned his telescope skyward and examined the moon and some stars and planets. On the moon he observed craters and mountains; near Jupiter he discovered four "small stars," which seemed to move around Jupiter rather than the sun. He called these four bodies Medicean stars, but they have since been renamed, in his honor, the Galilean satellites. His observations, he soon realized, were at odds with the geocentric idea of Aristotle, in which all bodies of the solar system were perfectly spherical and revolved only about the earth. Galileo also observed the planet Saturn, describing it as a central globe flanked by two small globes. Later, when he looked again, the smaller globes had disappeared. He never was able to explain this perplexing turn of events and wrote a friend, "Had Saturn devoured his own children?" Galileo's planetary observations and his resulting ideas were not well received by Catholic theologians. He was summoned to Rome twice, the second time after writing a popular piece that disproved the doctrines of Aristotle on physics and motion. Called before the Inquisition, he was compelled in 1635 to completely rescind his heliocentric doctrine. Thereafter, Galileo concerned himself with other areas of science.

Meanwhile, Kepler enthusiastically supported the astronomical discoveries of Galileo. Kepler was an interesting mixture of scientist and astrologist. He had an intuitive feeling that the universe, the human mind, geometry, the theory of numbers, and even music, were interrelated if one could only find the true key. While striving to understand planetary motions, Kepler continued to assume, as Copernicus had done before him, that the planets track in perfect circles around the sun. The careful observations made by Brahe, however, did not support that idea. His calculations were tantalizingly close to the observed data, but something was lacking in his analysis. Eventually it came to him—he realized that he had been a victim of habit. Because Copernicus had believed that planets travel in circles, so did he. But actually they moved in oval, or ***elliptical*** paths, which only appeared to be circular—that had been the problem all along. In 1618, Kepler published his *Epitome Astronomiae,* the first complete manual of astronomy based on the new principles. Later, he wrote *Harmonice Mundi* (Harmony of the World) in which he finally realized his youthful dreams of relating planetary motions with geometry, number theory, and musical harmonies.

As long ago as 1611, telescopic observations of the sun revealed black, irregular spots on the surface that appeared to drift across the body. These observations challenged the old belief that the sun was a globe of pure fire and light. Once the sun spots had been sighted, it was thought that the sun was black within and covered with an ocean of fire. Some even took the sun spots to be cooler islands protruding above the fire ocean. As late as 1795 it was conjectured that the dark solar surfaces were inhabited by living beings.

The rise of astronomy and our increased understanding of our planetary system came from the later improvement of the telescope, which enabled us to extend the use of our limited visual power. Throughout the seventeenth and the eighteenth century, the moon was mapped, satellites were discovered orbiting many of the other planets, and the riddle of Saturn was solved. In 1659, using a handcrafted telescope, the brilliant Dutch mathematician and natural philosopher Christian Huygens (1629–95) became convinced that the smaller, periodically disappearing globes described by Galileo were actually reflecting portions of a flat ring-structure encircling saturn in the equatorial plane. Huygens also discovered a moon beyond the ring structure. Between 1671 and 1684 the structure was shown to be discontinuous, consisting of dark and light bands, and four more, smaller satellites were discovered revolving around Saturn.

Since Galileo's time, astronomers had observed the large moons of Jupiter whose movements constituted a kind of planetary clock. However, the clock seemed to keep variable time, running about eight minutes slower when Jupiter and earth were on opposite sides of the sun than when the two planets were on the same side of the sun. Most observers accepted this as one of the oddities of the solar system, but Ole Römer, a Danish astronomer, conjectured that the time discrepancy was evidence that light traveled with a rapid, but finite velocity. The Jupiter moon-clock seemed slow because light took an additional eight minutes to travel the increased distance to earth when Jupiter was on the far side of the sun. Until that time, everyone believed that the speed of light was infinite. But experiments since Römer's day have demonstrated that light travels in a vacuum at a velocity of some 300,000 km/sec (186,000 mi/sec).

Our moon has always been a favorite of stargazers. It has long been observed that the moon periodically goes through progressive changes in apparent size and shape, called phases. These phases are likely responsible for our seven-day week, and are certainly responsible for the length of our months. When the moon is full, its dark and light patches can be seen with the naked eye. With the telescope, the lunar surface was finally seen in relative detail and its features were depicted on maps. The best early map was published in 1647 by the German astronomer Johannes Hewelke in his treatise, *Selinographia* (Fig. 2-4). The names he applied to the described features have been largely replaced by those proposed by J. B. Riccioli, an Italian professor, in 1651. Mountains were named after famous astronomers and mathematicians—hence names like Tycho, Plato, and Aristarchus. To the dark areas, regarded as oceans or seas (in Latin, *mare* or *maria*), he applied character names with either geographical or meteorological significance—as in Mare Serenitatis (Sea of Serenity) or Oceanus Procellarum (Ocean of Tempests). Topographical maps of the moon were made in the early 1800s, and in 1850 the moon was first photographed through a telescope.

While many astronomers were preoccupied with observing and charting the heavens, a few were trying to understand the mechanics of the solar system—that is, what kept the planets in

Fig. 2-3 The Italian physicist Galileo, shown here demonstrating the use of the telescope, made fundamental contributions to our understanding of our solar system.

fixed orbits. Kepler had suggested gravity, but he provided no clear idea of what he meant. The person who finally provided the answer was the great English physicist and astronomer Isaac Newton (1642–1727), (Fig. 2-5). The ***law of gravitation*** he formulated is the essence of simplicity—as most natural laws have proved to be. The law states simply that for two bodies, of mass m_1 and m_2, the force of attraction between them, F_g, is proportional to the product of the masses divided by the square of the distance between them at their centers of mass. Written in mathematical shorthand,

$$F_g \propto \frac{m_1 m_2}{r^2}$$

Newton believed that this law applied to all bodies in the universe, large or small, close together or far apart.

Newton's law adequately described how bodies behave in response to their masses but did not explain what gravity was. After several

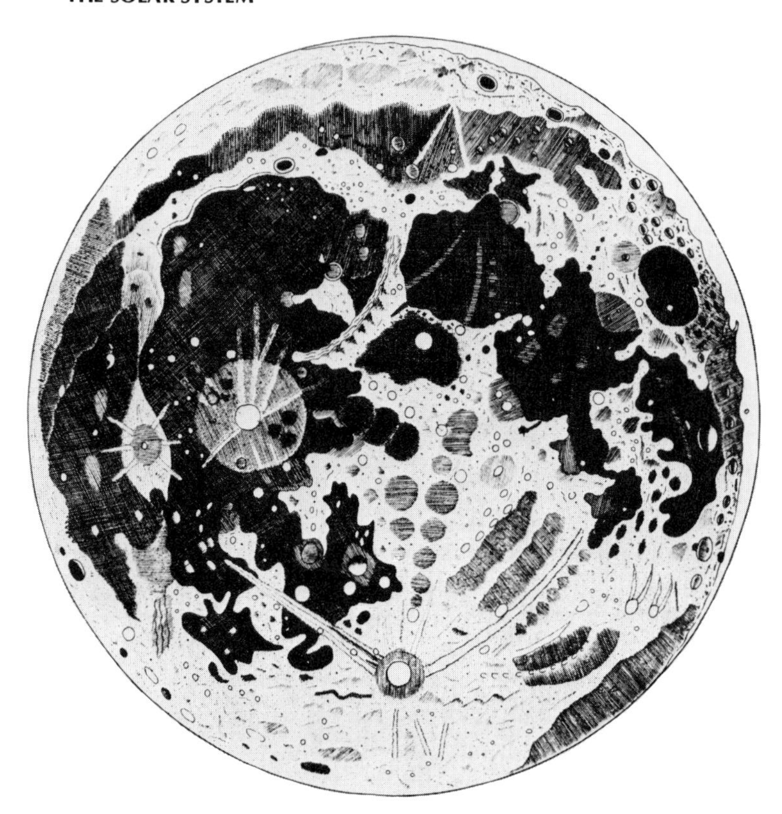

Fig. 2-4 An early map of the features of the nearside of the moon, by Antor. From *Selinographia,* published in 1647 by Johann Hevelius (the German astronomer Johannes Hewelke). The large, dark areas are maria; the smaller, circular regions, large impact craters. *The Bettmann Archive*

attempts to explain it metaphysically, Newton abandoned the challenge, and admitted that he could not explain it. Two hundred years later, Albert Einstein (1879–1955) wrestled with the same problem, but died before he was able to arrive at an answer.

With the formulation of the law of gravitation, the mechanical workings of the solar system were finally understood. For the next three hundred years, the telescope was still further improved, and astronomers were able to add more detail to their concept of the solar system. In the twentieth century, observational astronomy reached the point of diminishing returns. Earth-bound observations, even through the huge, highly sophisticated telescopes of the day, were limited by the atmospheric layer through which incoming light had to pass. To progress, astronomy had to move beyond the envelope of air around the earth. In the mid-1960s, balloonists carried instruments to more than 15 km (9.3 mi) above the earth, but the telescopes they used were necessarily small. The obvious solution was to establish relatively large artificial satellites orbiting the earth beyond the atmosphere and to send space probes to the reaches of the solar system. In the last 15 years, both objectives have been realized.

Today scores of satellites are in orbit around the earth, collecting data primarily about the earth below, such as the commonly seen images of the earth showing surface weather conditions.

Space travel truly began with the Apollo missions of the 1960s. In the initial phases, astronauts circled the earth testing the equipment and their physiological responses to weightlessness in preparation for the first manned mission to the moon in 1969.

But space exploration is not a simple matter. To place a human being on the moon was probably the most awesome engineering feat ever achieved. Much of the pioneer work had been accomplished by the American physicist Robert H. Goddard (1882–1945), and his associates, in the late 1800s and early 1900s. Goddard, from the time he was a young man, had been fascinated with the idea of traveling to Mars and the moon and he dedicated his life to it. He began his research before the invention of the airplane, concentrating on rocket propulsion. Although Goddard never lived to see his rockets carry a payload into space, his work allowed others, in the 1960s and 1970s, to initiate space travel.

Between 1968 and 1972, 24 men traveled to the moon, and of these, 12 men actually walked on its surface. Since 1972, the space effort has been aimed at exploration beyond the moon. Spacecraft have now visited Mercury, Venus, Mars, Jupiter, and Saturn. *Voyager II,* which passed Saturn early in 1981, is scheduled to approach Uranus in 1986.

The unmanned spacecraft of today are highly sophisticated vehicles, designed, for example, so they can take high-quality photographs and sense and measure the magnetic field and chemical composition of a planet and a wide spectrum of electromagnetic radiation. Special

Fig. 2-5 Sir Isaac Newton, perhaps the greatest scientist the world has yet known, demonstrated that the planets maintain fixed positions in the solar system because of gravitional interaction. *The Bettmann Archive*

Table 2-1. The planets and their statistics

Principal statistics	*Mercury*	*Venus*	*Earth*	*Mars*
Mean distance from sun (millions of kilometers)	57.9	108.2	149.6	227.9
Mean distance from sun (astronomical units)	.387	.723	1	1.524
Period of revolution	88 days	224.7 days	365.26 days	687 days
Rotation period	59 days	243 days retrograde	23 hours 56 minutes 4 seconds	24 hours 37 minutes 23 seconds
Inclination of axis	<28°	3°	23°27′	23°59′
Inclination of orbit to ecliptic	7°	3.4°	0°	1.9°
Eccentricity of orbit	.206	.007	.017	.093
Equatorial diameter (kilometers)	4,880	12,104	12,756	6,787
Mass (earth = 1)	.055	.815	1	.108
Volume (earth = 1)	.06	.88	1	.15
Density (water = 1)	5.4	5.2	5.5	3.9
Atmosphere (main components)	None	Carbon dioxide	Nitrogen oxygen	Carbon dioxide
Atmospheric pressure at surface, bars	10^{-6}	90	1	.006
Known satellites	0	0	1	2

radar units scan the surface features below clouds. Some spacecraft are designed to land gently on a planetary surface to implement experiments to determine the presence or absence of some form of extraterrestrial life.

FEATURES OF THE SOLAR SYSTEM

The principal components of our solar system (Table 2-1) are nine planets, differing in size and density, which orbit the sun in near-circular elliptical paths, all roughly in the same plane (Fig. 2-6). Most have their own satellite bodies, or moons, in orbit. The inner, so-called ***terrestrial,*** planets are composed chiefly of dense rocky material like the earth, whereas the outer ones, with the exception of Pluto, are large bodies composed chiefly of frozen gases. Between the two groups is a zone, the ***asteroid belt,*** in which quantities of "rocks," of various sizes, orbit the sun. Many of the meteorite showers observed on earth appear to originate in this zone. Finally, comets, composed of rock dust and ices, traverse the solar system in elliptical paths.

The sun

The sun, the center of our solar system, is a huge rotating sphere of white-hot gasses—330,000 times more massive than the earth and 109 times as large in diameter. More than 99.8 per cent of all the mass in the solar system is centered in the sun. Like most bodies in the solar system, it rotates on its axis, but because the body is fluid, not all of its areas rotate at

Jupiter	*Saturn*	*Uranus*	*Neptune*	*Pluto*
778.3	1,427	2,869	4,496	5,900
5.203	9.539	19.18	30.06	39.44
11.86 years	29.46 years	84.01 years	64.8 years	247.7 years
9 hours 50 minutes 30 seconds	10 hours 14 minutes	15 hours retrograde	22 hours or less	6 days 9 hours
3°5′	26°44′	82°5′	28°48′	?
1.3°	2.5°	.8°	1.8°	17.2°
.048	.056	.047	.009	.25
142,800	120,000	51,800	49,500	6,000(?)
317.9	95.2	14.6	17.2	.1(?)
1,316	755	67	57	.1(?)
1.3	.7	1.3	1.7	?
Hydrogen, helium	Hydrogen, helium	Helium, hydrogen, methane	Hydrogen, helium, methane	None detected
?	?	?	?	?
15 + rings	17 + rings	5 + rings	2	1

the same speed. Near the poles, one revolution takes 29 earth days; near the equator one revolution takes 25 earth days. The sun's mean density is 1.4 gm/cm³ (it is slightly more dense than liquid water), and its mass distribution is not uniform throughout—it is denser near the center.

What we normally see from the earth is the sun's ***photosphere,*** composed largely of non-charged gas atoms at temperatures near 5400°C. Much of the photosphere is in constant motion as heat is transported from below. Surrounding the photosphere is a changing zone of ionized gases (the ***corona),*** which is clearly visible only when the photosphere is blocked out, as, for instance, during an eclipse of the sun by the moon (Fig. 2-7). The gas particles in the corona, which was first witnessed in the mid-1800s, acquire their energy from deep within the sun and appear to have temperatures near 2.5 million degrees centigrade. The corona has no outer limit, no boundary. The ionized particles move out in space to form the ***solar wind,*** which sweeps outward at speeds of 300 to 600 km/sec (675,000 to 1,350,000 mi/hr). The solar wind causes comet's tails to be "blown away" from the sun, and it is this stream of particles that is trapped in the magnetic field of the earth, promoting auroral displays and causing radio interference.

If the outside of the sun is hot, the inside is even hotter. Physicists estimate that, near its center, temperatures climb to near 15 million degrees centigrade. Particles at that temperature move so fast that they tend to fly off from the sun. However, the huge mass of the sun

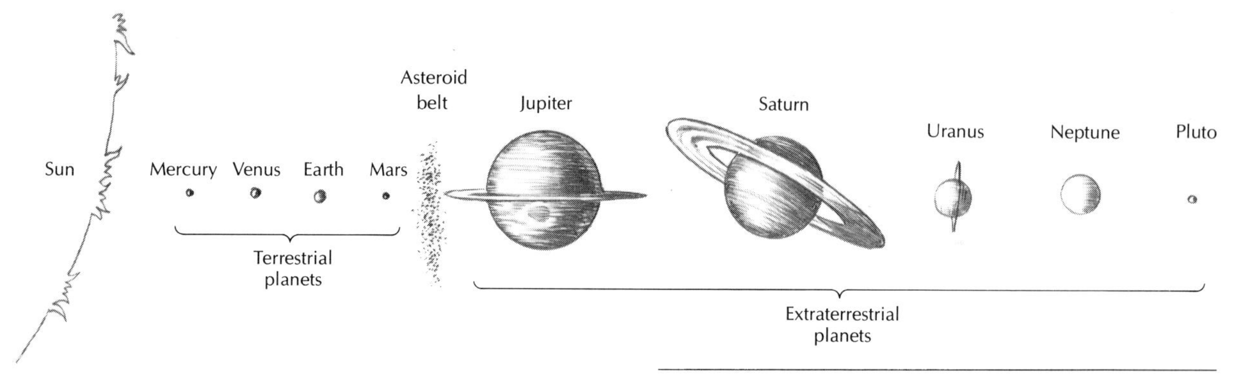

Fig. 2-6 Nine planets circle the sun, more or less in the same plane. The inner four planets are small and composed mostly of rocky material; four of the outer five are large and composed mostly of hydrogen and helium.

creates such a large gravitational pull that most particles are held toward the center. The size of the sun, then, is the result of an equilibrium between expansional and contractional forces. Pressures near the sun's center are 0.5 million million kg/cm² (7 trillion lb/in.²), which is equivalent to the weight of forty 100,000-ton aircraft carriers concentrated on an area the size of one of the perforations of a postage stamp. Of the ninety-two naturally occurring elements found on earth, more than 60 have been identified through spectroscopic analysis in the photosphere and corona of the sun. Hydrogen (60–80 per cent by weight) and helium make up about 95 per cent of the total.

The sun releases an immense amount of energy in all directions into space. The amount has been estimated to be roughly 5×10^{23} horsepower.* Today, most solar physicists believe that this incredible amount of energy can only be continuously supplied through thermonuclear reactions occurring deep inside the sun. A century ago, however, scientists were unsure of how the sun produced its energy. One of the popular ideas first advanced by the German physicist Ludwig von Helmholtz (1821–94), and later advocated by the English mathematician and physicist William Kelvin (1824–1907), involved the gravitational collapse or shrinking of a large body. This "Helmholtz contraction," could produce core temperatures near one million degrees centigrade in a body as large as the sun. It was calculated that a decrease in radius of only 42.5 m per year (140 ft/yr) could produce the amount of energy needed to keep the sun shining at its present rate, and that contraction of the sun to its present size from a large diffuse cloud would have provided a constant energy flow for about the last 50 million years. In the 1800s, when this theory was proposed, such a time-span appeared prodigious and quite adequate to fit most geological considerations. Later, however, through the use of radioactive clocks, it was demonstrated that the earth, and presumably the sun also, was several billion years old. Thus the Helmholtz hypothesis appeared to be inadequate.

In 1905, Einstein formulated a basic relationship between mass and energy: $E = mc^2$. In the equation, E stands for the amount of energy, m for mass, and c for the speed of light. Previously it had been held that mass and energy were fundamentally different. His formula, however, demonstrated simply that energy could be converted to mass and vice versa. The destruction of only a small quantity of mass produces an immense amount of energy. The development of the first fission-type atom bombs in the 1940s was based on this principle.

*This is 500,000,000,000,000,000,000,000 horsepower.

Fig. 2-7 The sun's corona, photographed through a telescope during an eclipse, is composed of ionized gases that radiate outward to form the solar wind. *Yerkes Observatory Photograph, University of Chicago, Williams Bay, Wis.*

When an atom of uranium-235 undergoes fission (splitting) into two atoms of lower atomic weight, mass is lost and energy is liberated. Although fission releases great amounts of energy, it is nowhere near that released when two lighter atoms are joined (fused) to form a heavier atom of slightly less total mass. We are all familiar with the destructive power unleashed by the detonation of a hydrogen bomb. The basis of this reaction is the fusion of two hydrogen atoms to produce a helium atom, with a reduction in mass and a release of energy.

Solar physicists now agree that fusion reactions are the basic source of the sun's energy. If only 0.45 kg (1 lb) of hydrogen protons and neutrons combined to form helium nuclei, 1000 kilowatts of power would be released continuously for 135 years. However, energy is also released when any of the lighter elements combine to form heavier ones—but only up to the element iron. Formation of elements heavier than iron uses rather than releases energy.

So the sun and apparently all stars are nuclear breeding pots in which different elements with lighter nuclei fuse to create more complex nuclei, and to give off heat and light.

Fusion reactions are difficult to initiate—temperatures in a newly developing stellar body (the ***protostar***) must be elevated to a threshold value before the reaction will occur. For example, fusion-type atom bombs are triggered by a small fission device, which reacts more easily, producing temperatures above the threshold value. It is conjectured that the sun's reaction was triggered by an initial collapse or shrinking (the Helmholtz contraction). If a protostar contains enough material, it will contract to raise internal temperatures to near one million degrees centigrade. Then, physicists believe, fusion reactions can occur, generating heat to continue and amplify the reactions. Thereafter, as long as enough nuclear fuel is present, the new star will continue to shine. It is estimated that there is still enough hydrogen in our sun so that it will continue to shine for another five billion years.

Observations of many other stars reveal that our sun is an averaged-sized star. The rate at which hydrogen is consumed in a star is thought to be directly related to its size: those much larger than the sun will burn brightly, at high temperatures, for perhaps a few million years. In contrast, stars much smaller than the sun will be relatively dim and cool and will last very much longer.

Many, perhaps most, stars in the heavens do not occur singly like the sun, but in groups of two or even three. Apparently, material sufficient to initiate and continue nuclear reactions was concentrated in two or three bodies. In our solar system, Jupiter, which is composed largely of hydrogen and helium, appears to have fallen just short of accumulating enough material initially to start the fires of its nuclear furnace.

Mercury

This small planet, which is not much bigger than our moon, is closest to the sun and receives about nine times as much sunlight as does the earth. As a result, its surface is extremely hot, so hot, in fact, that lead would melt there. Because of its small size the planet lacks an appreciable atmosphere and the huge fireball of the sun blazes down on it from a jet-black sky.

It was not until 1965 that Mercury's periodicity of rotation on its axis was accurately determined to be once every 59 earth days. That means that the planet rotates three times for every two complete orbits around the sun. The most detailed information about Mercury has come since the mid-1970s, when the planet was visited by *Mariner 10*. Probes revealed a noontime temperature of 427°C (800°F), a magnetic field 100 times weaker than that on earth, and a moon-like landscape pocked by impact craters of all sizes (Fig. 2-8). The strength of the magnetic field supported the idea that Mercury possesses a relatively large iron core (nearly as big as our moon) beneath a relatively thin (640 km, 400 mi) outer shell of rock.

Venus

Because Venus is relatively close to the earth and completely covered by thick, reflective yel-

Fig. 2-8 The cratered surface of Mercury, photographed from a spacecraft at a distance of 77,800 km (48,300 mi). The distance across the bottom of the photograph is about 600 km (400 mi). *NASA, JPL*

low-white clouds, it is, next to the sun and moon, the brightest body in our sky. The cloud cover has guarded the mystery of Venus, and only as recently as the 1960s was it determined, on the basis of radar data, that the planet rotates slowly on its axis—once every 243 earth days—in a direction opposite (retrograde) to the earth.

Since 1962, several U.S. and Russian spacecraft have visited Venus and sent back data concerning its features. The dense clouds begin at an elevation of about 65 km (40 mi) and extend, unbroken, down to about 35 km (22 mi) above the surface; below, the atmosphere is clear of clouds. Temperature increases steadily downward from −50°C (−60°F) at cloud-top elevations to about 475°C (890°F) at the surface. The atmosphere is almost entirely carbon dioxide (97 per cent) and nitrogen (2 per cent). The gas is so dense that the at-

Fig. 2-9 Meteorite impact craters on the nearside of the moon: illumination from right to left. There are many smaller craters; in comparison, the largest crater Goclenius is about 60 km (40 mi) across. The linear depressions in and near Goclenius appear to have resulted from fracturing. *NASA*

mospheric pressure on Venus is about 100 times that on earth at an equivalent elevation. It is thought that the extremely hot Venusian surface temperatures are the result of a super greenhouse effect, whereby long wavelength radiation given off by the surface rocks in response to solar heating is absorbed by the abundant carbon dioxide in the atmosphere and cannot escape.

Ground-surface radar imagery suggests the existence of large-scale, linear topographic highs and lows resembling earth-style mountains and lowlands.

The earth's moon

The earth is the closest planet to the sun that possesses a moon—a satellite body that orbits a planet. Its orbital plane is called the ***ecliptic plane.*** During its orbit around the earth, the moon revolves once on its axis. Therefore it always shows but one face to earthbound observers; the farside of the moon was never seen until the Apollo missions in the 1960s. Because of its relatively small mass (1/81 that of the earth), the moon retains but little atmosphere—in fact the density of the lunar atmosphere is several orders of magnitude less than the density that can be achieved in the best laboratory-created vacuum on earth. Perhaps the best-known feature of the moon is its cratered surface, which is easily observed through a telescope (Fig. 2-9). Early observers, however, were divided as to whether the craters were the result of meteorite impacts or volcanic eruptions in the airless surface environment. Finally, on July 20, 1969, U.S. astronauts first stepped onto the lunar surface. In the six missions, the last in 1972, our astronauts carried out a battery of scientific experiments, and brought 382 kg (840 pounds) of lunar rocks back with them. The data gathered from the Apollo missions have provided a great deal of information about the moon, and, additionally, the sun.

At each landing site, corner reflectors—reflecting devices shaped like egg cartons— were set out. They were designed to reflect any light falling on them, regardless of the angle of incidence, back to its source. Once in place, powerful light beams were directed at them from the earth and reflected back to the sending station. By determining the time required for transmission and reflection, the distance to the moon—400,000 km (240,000 mi)—could be determined within about 15 cm (6 in.). That would be equivalent to determining the 5000-km (3000-mi) distance between New York and San Francisco with an error of only 0.2 cm (0.1 in.). The corner reflectors will enable other scientists not only to chart variations in the moon's orbit, but to determine whether certain areas of the earth's surface are drifting laterally.

The astronauts on most missions also conducted a simple experiment to analyze the intensity and the chemical composition of the solar wind. These streams of atoms cannot easily penetrate the atmospheric shell around the earth, but they strike, with full force, the airless surface of the moon. To "capture" the wind, the astronauts unrolled a thin piece of aluminum foil, dubbed the "window shade." Solar-wind atoms burrowed into the soft aluminum on impact and were trapped. The aluminum foil with its trapped particles was carried to earth and examined. Only the non-reactive (***noble***) gases could be analyzed and the most abundant were found to be helium and neon. It was estimated that about 6 million helium and 15,000 neon atoms struck the foil per second.

Instruments were set out at each station to monitor tremors from moonquakes and the impact of meteorites. Moonquakes are both less common and smaller in magnitude than are quakes on the earth. From the data, scientists have conjectured that, like the earth, the moon consists of concentric layers. The outermost layer is about 1 km (0.6 mi) thick and composed of fragmented material resulting from meteorite bombardment. Beneath this, and down to about 60 km (37 mi), is a layer of solid rock probably composed of dark-colored lava (basalt) or the lighter-colored rock typically exposed in the lunar highlands. The next zone appears to extend to a depth of 1000 km (620 mi) and is a rocky layer rich in magnesium and

Fig. 2-10 Meteorite impact craters on the farside of the moon. Unlike the craters on the nearside of the moon (Fig. 2-9), only two (arrows) are filled with dark lunar lava. *NASA*

iron. There is evidence of a core, extending from 1000 km (620 mi) to the center at 1800 km (1116 mi), which differs from the zone above only in being partially molten; there is no clear-cut evidence to suggest the existence of a small iron core at the center.

Scientists are eager to know if a metallic core exists because of conflicting magnetic data obtained during the Apollo missions. Magnetometers aboard orbiting spacecraft had revealed an extremely weak lunar field. The permanent magnetism of samples returned to earth, however, was relatively intense—seemingly greater than that which could have been acquired in the present weak field. The question then arises, "Did the moon at one time possess a stronger field, generated (as was the earth's) in a molten iron core or was the moon elsewhere in the solar system, where the magnetic field was stronger, when the rocks were magnetized? The contradictory magnetic data constitute perhaps the most puzzling enigma of the lunar investigation. They have led to a number of possible explanations concerning the history of the moon, none of which is entirely satisfactory.

During each Apollo mission, while two astronauts roamed the lunar surface, one astronaut remained in the orbiting module conducting experiments and photographing the lunar surface below. In all, an area equal in size to the United States and Mexico was photographed in detail. Of particular interest were pictures of the moon's farside, which revealed that although large craters were common, the dark circular maria so common on the front side and visible to the earth were rare (Fig. 2-10).

Analysis of lunar rocks have added greatly to our knowledge of the moon's history. All the rocks and soil from the first mission were quarantined for 3 weeks, during which time they were examined for microscopic life. Cultures were taken, test animals and plants were injected with lunar dust, and the astronauts were kept under close medical surveillance. These precautions were taken to ensure that no organisms of extraterrestrial origin had been brought back from the moon. All the tests were negative, so apparently the moon is devoid of life of any kind. Small samples from the moon have been distributed to hundreds of scientists all over the world for extensive testing.

Rocks from the maria regions are predominantly a type of dark lava called basalt, which is common on earth. Lunar basalts, however, contain more titanium and iron than terrestrial basalts. Most scientists agree that the dark maria rocks were formed by large-scale eruptions of molten lava. Radiometric age dating of the rocks showed that the basalt floods occurred over large parts of the moon between 3.8 and 3.2 billion years ago.

Rocks collected in the lighter-colored highland regions contain more calcium and aluminum and less iron, magnesium, and titanium than do maria rocks. Instead of being surface lavas, they appear to have cooled and solidified from molten material beneath the lunar surface. Since formation, they have been forced to the surface, perhaps by the impact of large meteorites. All highland rocks have been dated at 4.2 to 4.0 billion years—ages that confirm the previously held idea that the highland rocks are older than the maria rocks. The oldest age, 4.2 billion years, is greater than that for earth rocks (~3.8 billion years), yet it falls short of the estimated time of the origin of the solar system of about 4.5 billion years ago. Perhaps later missions to the moon will uncover rocks of that age, or perhaps, for one reason or another, none exist.

The astronauts also brought back many samples of lunar breccia, a shock-welded jumble of angular fragments of pre-existing maria and highland rocks shattered during meteorite impacts. The largest meteorites appear to have bombarded the moon about 4 to 3 billion years ago, to produce huge cratered depressions. Later flooding by basalt flows filled these depressions and produced the flat-bottomed maria. Curiously, there are many large impact craters on the farside of the moon also, but only rarely are they filled with basalt.

Perhaps our latest phase of the investigation of the moon is best summed up by Bevan French, an American earth scientist who has been involved with lunar exploration since 1964, in *The Moon Book:*

What we have learned about the moon has also revamped our thinking about the earth. Although the earth and moon have different chemical compositions and different histories, the moon is still an important model of what the primitive earth may have been like. The moon clearly records a primordial melting and widespread chemical separation that produced a layered internal structure almost immediately after it had formed. It is likely that the present internal structure of the earth, including its iron core, also developed very early in its history, perhaps as a result of the accretion process that formed it.

The intense early bombardment recorded by the moon . . . may be a general characteristic of the solar system too. If a similar intense bombardment struck the earth at this time, it would help explain why no terrestrial rocks older than 4 billion years have been found.

The earth and moon provide two contrasting examples of how differently planets can develop, and in their contrast we can see some of the factors that control the evolution of planets. Size is important. A large planet can hold volatile materials like water, and it can also retain more internal heat to produce continuous geological changes. Chemical composition is also important; a planet without water, no matter how large, lacks the one substance that is essential for the only kind of life we know. The presence or absence of radioactive elements determines whether a planet will be hot or cold during its lifetime, and the amount of iron in a planet determines whether it can ever develop a strong magnetic field. The first two bodies we have explored, the earth and its moon, show two different lines of development. Although we think that the planets all formed in the same general way, it is almost certain that we will find further different planetary histories as we explore the solar system.

Mars

Mars, long known as the red planet, is not much larger than our moon. Because of its nearness to earth, Mars has long been the object of study. Features visible through the telescope convinced some of the early observers that Mars was inhabited by intelligent beings. Black lines first observed by the Italian astronomer Giovanni Schiaparelli in 1877 were interpreted as canals, dug to bring water from the polar regions to the arid equatorial region. In H. G. Wells's *War of the Worlds,* the Martians, realizing their ultimate fate, cast covetous eyes on earth and launched attacks to wrest it from the earthlings. Invasion of the earth by Martians was a popular theme all through the late 1800s and early 1900s. The fantasies came to an end in 1965 when *Mariner 4* flew within 10,000 km (6200 mi) of Mars and sent back pictures showing a bleak cratered landscape more reminiscent of the moon than the earth. Later views sent back by *Mariner 6* and *Mariner 7* showed no signs of the canals or any other evidence of intelligent life. The probes aboard the spacecraft also showed that the atmosphere, which is largely carbon dioxide, is so sparse that the air pressure is less than one-hundredth of that at sea level on the earth. In the frigid polar regions, the carbon dioxide freezes to form "dry ice," which falls to the ground like our snow. The martian polar caps, which shrink and swell seasonally, are essentially all dry ice.

When *Mariner 9* arrived at Mars in 1972, a planet-wide dust storm completely obscured the surface for weeks. As the dust finally began to settle, the views became spectacular. First to protrude through the dust shroud was the biggest volcano ever seen by humans (Fig. 2-11), since named Olympus Mons. It is nearly three times higher than Mount Everest (24 km; 78,000 ft) and is about 500 km (300 mi) across at the base, which is terminated by a steep, high cliff, which many scientists believe is the result of water-wave erosion. Were Olympus Mons in California, it would reach from Los Angeles to San Francisco. This volcano is one of a group of four huge cones. Other large volcanoes have been observed 5000 km away.

The cameras on *Mariner 9* also revealed a canyon unequaled on earth (Fig. 2-12). This chasm, called Valles Marineris, is 4000 km (2500 mi) long, up to 200 km (120 mi) wide, and up to 6 km (3.75 mi) deep. Remember that Mars is a relatively small planet with only one-seventh the volume of the earth. That makes the scale of the volcanoes and canyons on Mars truly awesome.

The presence of Valles Marineris and other valley-like features, and the steep cliff around the base of Olympus Mons, demonstrate a martian enigma. These and other features revealed in the photographs appear to be water cut; yet there is no free water anywhere on the planet's surface. Even if the little water vapor in the atmosphere were condensed out, it would barely fill one large lake basin. The answer seems to lie in the polar regions: it is thought that permafrost (frozen ground) existing beneath the polar surface can, on occasion, be melted by surface volanic activity, thereby producing flash floods that erode the landscape.

Perhaps the most dramatic aspect of the martian exploration was when the *Viking* spacecraft visited the planet to search for life forms. To accomplish this task, lander craft with television cameras were detached from orbiting vehicles and directed toward the surface.

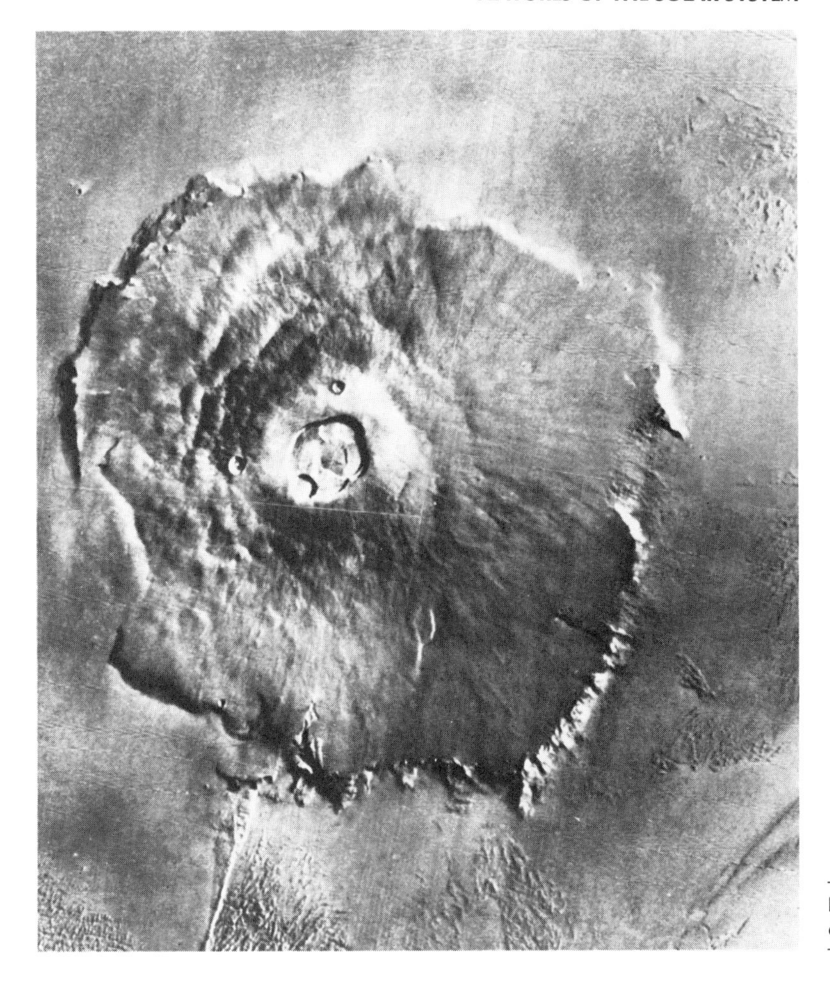

Fig. 2-11 Olympus Mons, the largest volcano on Mars. *NASA*

The first pictures relayed to earth showed a barren surface strewn with angular rocks of all sizes. There were no signs of life. Soon after touchdown, mechanical scoops retrieved some of the martian soil, and three separate experiments designed to detect the existence of microorganisms were performed on the material. The experiments were supposed to detect the presence of metabolic processes, including photosynthesis and respiration. The results of the three experiments are ambiguous, but most researchers have concluded that microorganic life does not exist on Mars. It is possible, of course, that the experiments were poorly designed—we were looking for life as we know it here on earth. Perhaps, martian life, if it exists, is quite different. Another visit to Mars is scheduled for 1984 and scientists are planning other experiments which they hope will clarify the situation.

Mars has two small moons, named Phobos (Greek for "fear") and Deimos (Greek for "hate"), which were photographed during the *Viking* missions. Both were shown to be heavily cratered angular, football-shaped bodies, only a few tens of kilometers in length. It is thought that both satellites might have been derived from the asteroid belt.

Jupiter

Jupiter, the largest planet in our solar system, is easily seen from earth with the naked eye. Its most conspicuous features are a large elongate

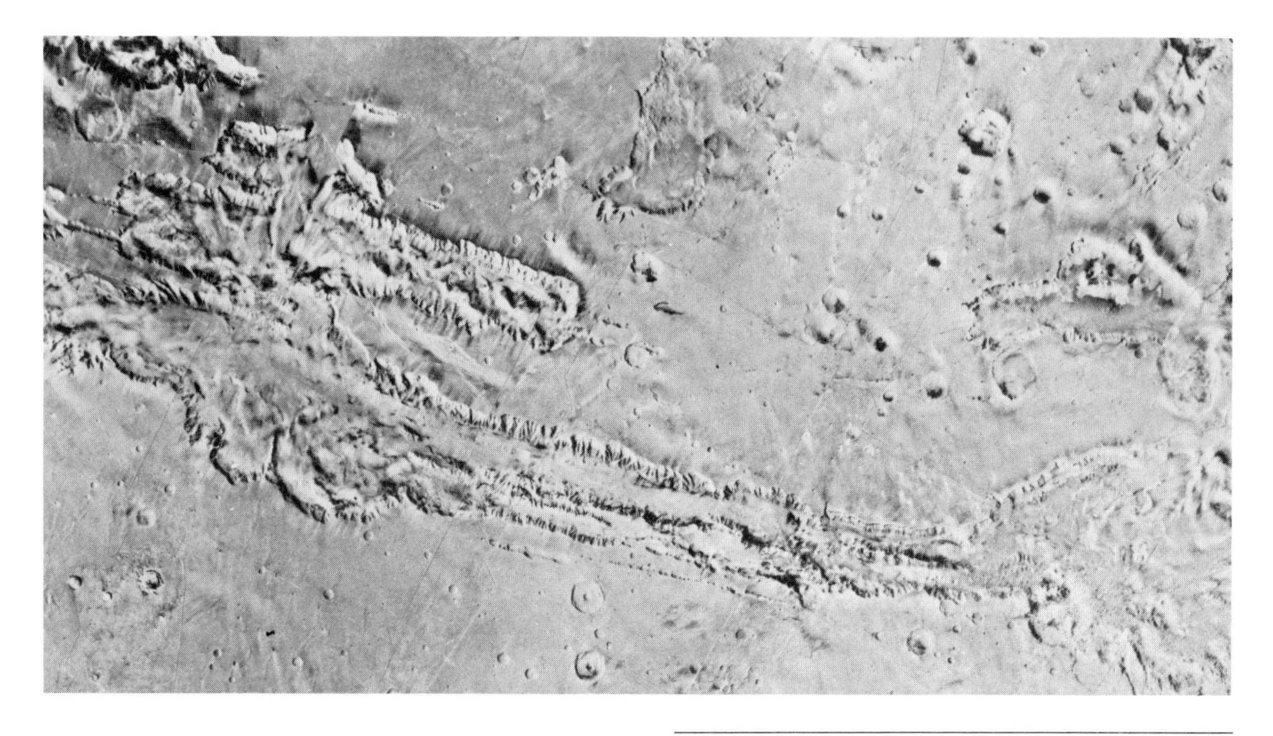

Fig. 2-12 View of the canyon Valles Marineris, on Mars. The part shown here is about 500 km (300 mi) long. *NASA*

red spot on its surface, first viewed in 1665, and a retinue of 15 moons, including the four large satellites first discovered by Galileo in the early 1600s. In 1979, *Voyager I* and *Voyager II,* after two years and 645 million km (400 million mi) of space travel, closely approached Jupiter and obtained measurements and photographs of the planet and five of its moons. As the two *Voyagers* approached the planet, dense cloud cover became particularly apparent in shades of red, orange, and white. The Great Red Spot was shown to be a vast atmospheric storm, twice as wide as the entire earth, rotating counterclockwise every six earth days. Temperature readings indicate that the center of the spot is cooler and presumably, therefore, protrudes above the margins into the cooler levels above. Photographs taken over a period of several months show variations in the pattern of the storm; they also indicate that the streaked cloud pattern is primarily the result of mid-latitude wind currents that move at cloud-top elevations at speeds reaching 120 m/sec (260 mph). Huge lightning flashes were seen crackling through the cloud tops.

One of the startling facts revealed by the *Voyager* missions was the existence of an equatorial ring-structure, which resembled that around Saturn. Jupiter's rings, however, are thin and diffuse and invisible from the earth. Probes verified that the planet's atmosphere is largely composed of hydrogen (~75 per cent and helium (~20 per cent). Small amounts of sulfur compounds appear to produce the reds, oranges, and yellows. Other sensors indicate that Jupiter possesses the strongest magnetic field in the solar system, tipped about 10° off the rotational axis, and that the planet gives off about twice as much heat as it receives from the sun. This latter fact supports the idea that heat trapped inside the body during its early days of formation is leaking away.

What seems truly remarkable about Jupiter is the long-term stability of its atmosphere. The red spot, for example, has been monitored for more than 50 years. Why this should be so is undoubtedly related to conditions in the deeper parts of the atmosphere. It is suspected that below the cloud-top region of the atmosphere, where the pressures are greater, a layer of highly compacted metallic hydrogen exists and, below that, a layer consisting largely of highly compressed hot liquid gases, such as methane, ammonia, and water. Probably at the center is a rocky or molten iron core about the size of the earth.

One of the main tasks of the *Voyager* mission was to photograph the Galilean satellites, which carry the names of friends and lovers of the mythical god Jupiter. Both Callisto, the outermost, and Io, the innermost Galilean moon, were named after maidens who enticed Jupiter and accordingly angered his wife, Juno. She, perhaps overreacting, turned Callisto into a bear and Io into a heifer. The two middle moons, Europa and Ganymede, received the names of two of Jupiter's friends—the former a beautiful maiden; the latter a handsome youth. Just like our moon in relation to the earth, each of the Galilean satellites keeps the same face toward Jupiter as it orbits the planet.

What the *Voyager* imaging team expected to see when the cameras focused on Io was the crater-pocked face of a dead moon. Instead, they saw perhaps the most sensational spectacle of the entire mission—eight different volcanic vents simultaneously erupting, some spewing plumes as high as 320 km (200 mi) above the surface (Fig. 2-13). Besides the earth, Io is the only body in the solar system that is actively volcanic. Later observations by *Voyager II* showed six of the volcanoes still erupting. As far as we can tell, these volcanoes are unlike their earthly counterparts in that the material being ejected is primarily elemental sulfur and sulfur dioxide—which accounts for the red-to-orange surface of the moon. Scientists suspect that the heat necessary to melt the material is supplied by continuous bending and stretching of the moon in response to tidal interactions between Jupiter and nearby Europa and Ganymede.

Europa, the brightest Galilean satellite, apparently mostly consists of silicates overlaid by a thin crust of frozen water. Its size and density are about the same as our moon. Unlike the other moons, Europa's surface is transected by a complex system of dark linear streaks indicative of fracturing (Fig. 2-14). Impact craters are relatively rare.

Ganymede, the largest moon in the solar system, is a little larger than our moon, but only one-half as dense, which indicates a composition of about 50 per cent water or ice, the rest being rock. Only part of Ganymede's surface shows extensive impact cratering. Scientists believe that glacier-like creep of the outer icy layers has smoothed out surface irregularities in the crater-free areas.

Callisto, the outermost of the four big moons, possesses the lowest density and is thought to consist largely of frozen water. Unlike the others, its entire surface is heavily cratered.

Saturn

This planet, next beyond Jupiter, is also one of the large, low-density planets. Despite its size, it is so far from earth that only rarely can it be seen by the naked eye. Its most characteristic feature is a well-developed set of equatorial rings. Along with the craters on the earth's moon and the red spot on Jupiter, the rings of Saturn are one of the favorites of backyard astronomers. In 1979, *Pioneer II* sped close to the cloud tops of Saturn, more than a billion kilometers distant from the earth where it had been launched six years before. Photographs show a more complex ring-structure extending farther out in space than previously thought. As *Pioneer II* passed through a part of the ring-structure, the ring pattern was only minimally disturbed, leading scientists to speculate that the rings probably consist of particles of ice, each particle being about several centimeters in diameter.

Detectors aboard the craft indicated that the atmosphere is, like that on Jupiter, mostly hydrogen and helium in about the same ratio as in the sun. The compositional layering on Sat-

Fig. 2-13 A spectacular eruption on the Jupiter moon of Io. The volcanic plume, composed largely of sulfur and sulfur dioxide, extends about 320 km (200 mi) into space. *NASA*

urn probably resembles that of Jupiter. Saturn's magnetic field is aligned exactly along its rotational axis, but it is much smaller than Jupiter's. At the cloud-top level, the field was found to be nearly equal to that at the earth's surface. Heat-sensor data indicate that Saturn also emits about three times as much thermal energy as it receives from the sun, and must, therefore, possess a source of internal heat.

Late in 1980, *Voyager I* made its closest approach to Saturn, and, with improved cameras, even more details about the ring-structure and number of moons (Fig. 2-15) were obtained. Whereas the photographs from *Pioneer II* showed only differences in color patterns in the cloud tops, those from *Voyager I* showed that the atmosphere is highly turbulent and possesses a reddish spot similar to that on Jupiter, but smaller. The ring-structure is larger and more complex than was first thought, consisting of many dark and light zones. Radial dark spikes across the ring-structure have not yet been explained. Additional moons were observed, bringing the total to 17 encircling the planet. Some of the inner moons circle the planet at distances that correspond to some of the abrupt dark/light transitions in the ring-structure. It appears, therefore, that in some instances, at least, the ring-structure is maintained and stabilized by the effect of the moons.

Titan, the largest moon on Saturn, was shown to be smaller than the Jupiter moon Ganymede. Titan's atmosphere appears to consist of hydrocarbon and nitrogen compounds. Temperatures on the moon are sufficiently low that lakes of liquid nitrogen may exist at the polar regions.

Uranus

Although Uranus is 67 times larger than the earth, it is so far from the earth that it escaped detection until 1781, when the German-born William Herschel (1738–1822), who had moved to England and became a distinguished conductor, composer, and astronomer, perceived an uncommonly bright "star" that seemed to be moving. He announced his discovery to the Royal Astronomer, who checked his observations and identified the body as a planet orbiting beyond Saturn.

It is possible to see Uranus with the naked

Fig. 2-14 These linear streaks on the surface of Europa appear to be fractures in a relatively brittle, thin outer layer of frozen water. *NASA*

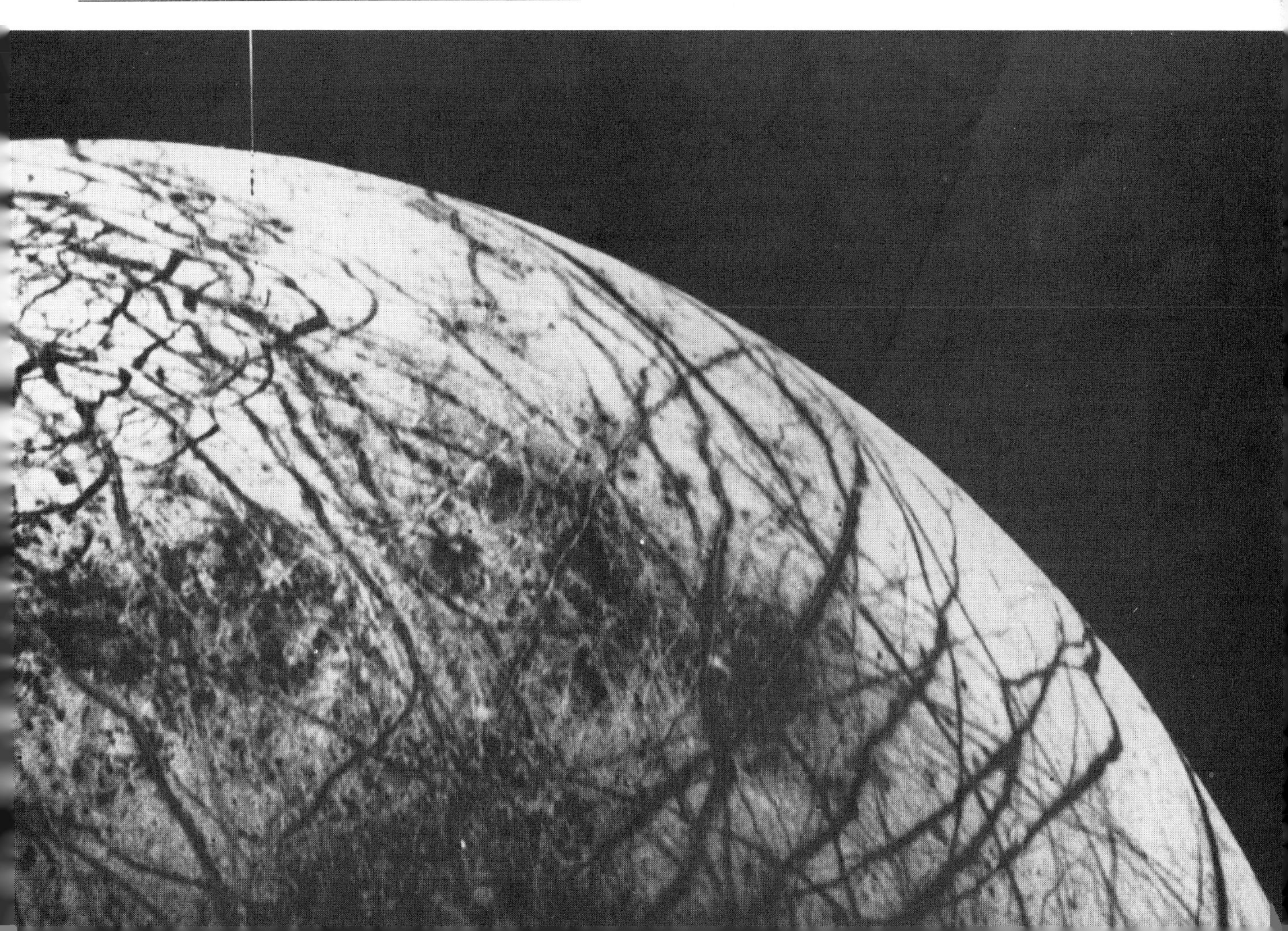

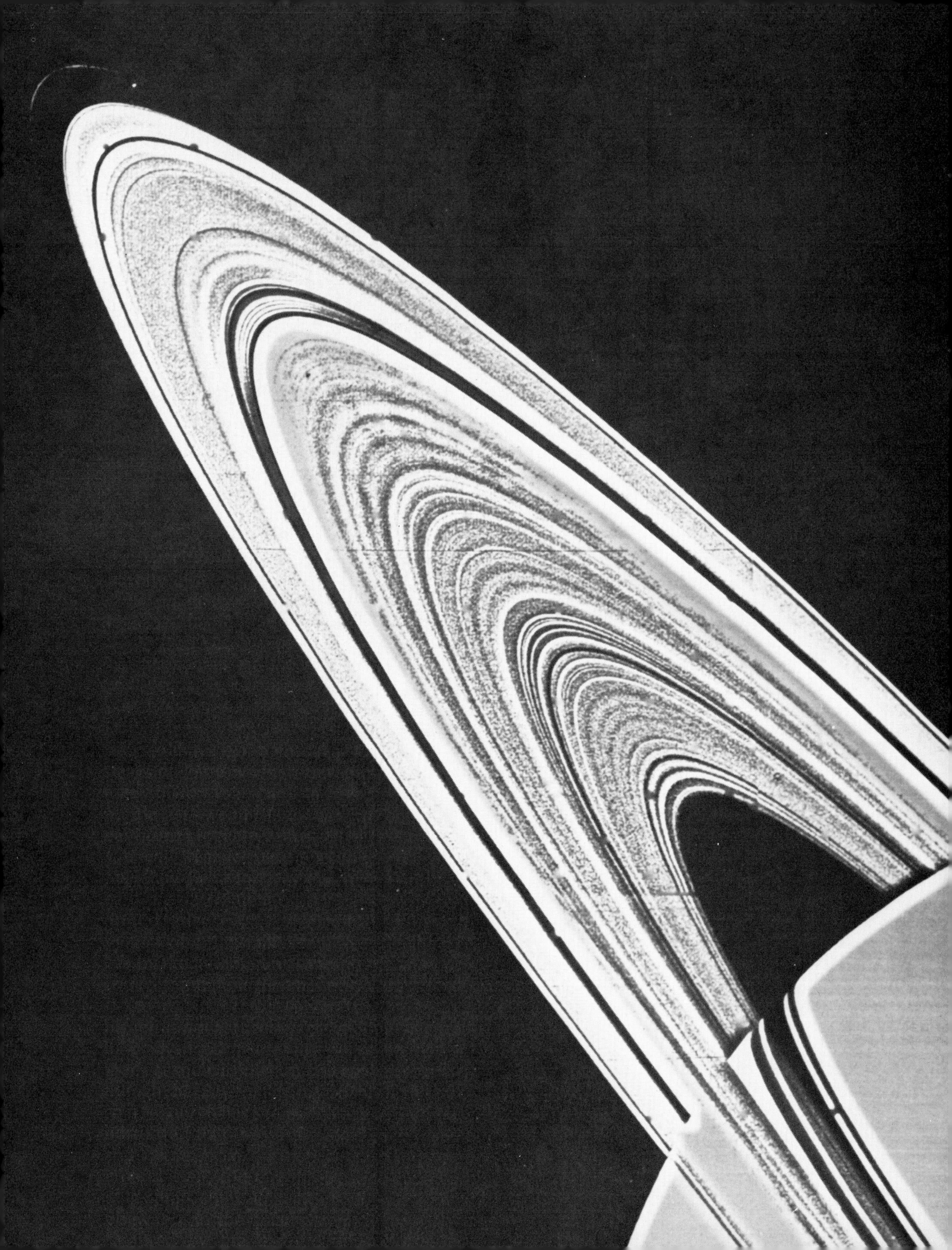

eye, but conditions must be optimal. The planet is so dim that it was plotted as a star on a score of different star charts before Herschel's discovery. The telescope reveals Uranus to be a greenish-blue orb surrounded by five moons (Fig. 2-16).

Unlike the other planets, the axis of its rotation is essentially in the plane of its orbit around the sun. As a result, the seasonal variation on Uranus will be greatly different from that which we are used to on earth. From an airborne study of the planet in 1977, during its passage in front of a faint star, scientists observed that the star blinked off and on five times prior to passage, and five times again after passage of the planet in front of the star. From these results scientists now are convinced that Uranus is surrounded by five equatorial rings, making it the third ringed planet.

In 1986, barring unexpected difficulties, *Voyager II* will approach Uranus, a day eagerly awaited by astronomers.

Neptune

This planet is nearly as large as Uranus, but it is so far out in the solar system that it was not discovered until 1846, nearly 100 years after the sighting of Uranus. Its discovery was an unusual event that represented, at the time, a scientific triumph for Newton's law of gravitation.

Since the discovery of Uranus, sky watchers had noticed that its orbit possessed irregularities, and some began to question whether the laws of gravitation applied so far out in space. Others became convinced that the orbital fluctuations were the result of a relatively close passage of another, yet unknown planet. The French astronomer Jean Leverrier (1811–77) calculated, on the basis of gravitational attraction, the likely position of the unknown planet and in 1846 asked the Berlin Observatory to examine that part of the night sky he had indicated. Comparison of star maps immediately confirmed a "foreign" star Neptune, only one degree from the predicted position.

Few details of Neptune have been established. Viewing Neptune from the earth through a telescope is like examining a dime, with the naked eye, at a distance of about 1.6 km (1 mi). Two moons, Triton and Nereid, are known to orbit the planet. Obviously we will learn more about Neptune only from spacecraft observation.

Fig. 2-15 The complexity of Saturn's ring-structure is shown in this awesome *Voyager I* photograph. *NASA*

Pluto

Pluto, the smallest planet, is the farthest planet from the sun, most of the time. Because of its small size and great distance from earth it was not discovered until 1930, by an observer at Lowell Observatory in Flagstaff, Arizona. Its orbit is relatively highly elliptical, so much so that at times it is closer to the sun than is Neptune.

In 1978, a moon of Pluto was discovered. Named Charon, after the mythical ferryman on the river Styx, it orbits Pluto once every 6 days, 9 hours, 17 minutes, which is exactly the time it takes Pluto to spin once on its axis.

Pluto is unlike most of the outer planets—it is small and dense. Because of this and because of its unusual orbit, it has been suggested that the planet was at one time a moon of Neptune, which was dislodged during a catastrophic solar event and subsequently went into orbit around the sun.

ORIGIN OF THE SOLAR SYSTEM

In this section, we will restrict our discussion to ideas about the formation of the solar system. Keep in mind that the event we will be discussing occurred around 5 billion years ago. Obviously our "reconstructions" are only conjectural.

As long ago as 1755, the well-known German philosopher Immanuel Kant (1724–1804) developed a theory that all matter in the solar system originally existed in a widely extended nebulous cloud, which by contraction and accentuation of angular motion, formed a rotating disc. In time, through further contraction, the matter formed into a central body (the

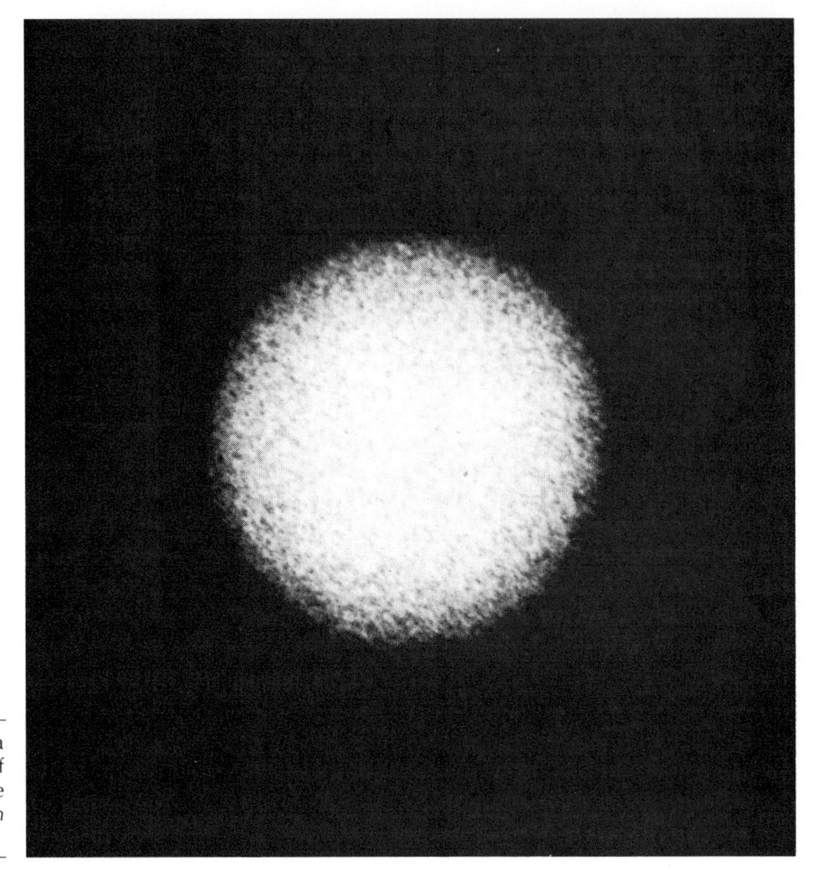

Fig. 2-16 Uranus, as observed through a balloon-borne telescope at an elevation of about 24 km (15 mi). At this distance, the planet is completely featureless. *Princeton University, NSF/NASA*

sun) containing most of the mass and several satellitic planetary bodies. Kant's ***nebular hypothesis*** was the first explanation based on scientific knowledge. The gravitational contraction necessary to produce the solar system seemed to fit with the ideas of contraction put forth by Helmholtz to produce the sun's radiant energy.

In the early 1800s Pierre Simon de Laplace (1749–1827), an exceptional mathematician and natural philosopher, provided an explanation essentially identical to the nebular hypothesis even though he was unaware at the time of Kant's earlier conjectures.

By the late 1800s, astronomers knew more about the solar system and many began to find difficulties with the Kant-Laplace hypothesis. In particular, it was pointed out that about 60 per cent of the angular momentum (which can be considered as the "quantity of rotation") in the solar system was possessed by Jupiter, even though it contained only one-thousandth of the mass of the system. By contrast the sun, containing more than 99 per cent of the mass, possessed only 2 per cent of the angular momentum. In 1906, the Americans Forest Ray Moulton (1872–1952), an astronomer, and Thomas C. Chamberlin (1843–1928), a geologist, advanced a hypothesis that the planetary system had originated from the passage of another star close to the sun. They imagined that droplets of the sun were gravitationally drawn out into space, acquired rotational motion, and went into orbit around the sun. The erupted star material condensed into small bodies, called ***planetisimals,*** which subsequently, through multiple collisions, coalesced to form the planets. Many scientists supported this view at the time. However, it was soon realized that since the sun is composed mostly of hot hydrogen and helium under pressure, liberation of this material into the vacuum of space should have led to an explosive dissipation of the gases rather than a condensation. Obviously, the starting material could not have been hot.

In 1944 a modified hypothesis that possessed fewer improbabilities was advanced. Termed the ***protoplanet hypothesis,*** it was first introduced by the German astronomer C. F. von Weizacker and later modified by the American astronomer Gerard Kuiper. Somewhat reminiscent of the earlier nebular hypothesis, there are some key differences. Initially, a vast nebulous cloud of gas, dust, and ice began to condense slightly. Temperatures would have been near absolute zero and no warming sun would have existed. Gravitational contraction continued, but was so slow that turbulent motions were smoothed out as the vast cloud began to rotate slowly. In time, the formless nebular cloud became a rotating disc-shaped body in which about 90 per cent of the mass was concentrated in the central bulge. Clumps of more densely packed material in the disc zone would occasionally collide with one another, eventually coalescing into larger clumps of material called ***protoplanets,*** most of which were restricted to the plane of the disc. The protoplanets, like the material in the central zone, would at this stage be composed largely of hydrogen and helium along with smaller amounts of the heavier elements. In the process of colliding and coalescing, some of the smaller protoplanets may have begun to circle the larger ones.

At about this time, contraction of the proto sun had created enough heat for nuclear reactions to take place. It began to glow and then to radiate intensely. The resultant rapid heating of the protoplanets would have essentially vaporized the lighter elements, especially hydrogen and helium, on the bodies near the sun and swept the material outward. Bodies far from the developing sun, however, would have remained sufficiently cool to retain most of the lighter matter. The inner planets, therefore, would tend to be largely rocky in nature and relatively dense and the outer planets to be composed largely of the low-density materials hydrogen, helium, and water.

Table 2-2. Compensated densities of the terrestrial planets and the moon

Planet	*Mean uncompressed density, gm/cm^3*
Mercury	5.4
Venus	~4.2
Earth	~4.2
Mars	3.3
Moon	3.35

Theories advanced during the last ten years still largely accord with the protoplanet hypothesis. New data from the recent spacecraft excursions and from modern astronomy have helped verify, refine, and modify the hypothesis. Astronomers have been studying the so-called T-Tauri class of stars, which appear to be in the early stages of stellar evolution, just after star ignition. Observations indicate that material is being radiantly ejected outward from such T-Tauri stars at rates of about a billion times greater than from our sun at present. Apparently, the early solar wind can sweep a great deal of inner material toward the margins of a newly developing planetary system.

It has been observed that the mean uncompressed densities of the inner four planets, if allowance is made for the greater degree of compaction of the interiors of the larger ones (Table 2-2), decrease with distance from the sun. The only abundant element that can affect such differences in density is iron, of which there appears to be a greater concentration in planets nearer the sun. Tiny Mercury, for example, is thought to have a relatively large iron core. The relationship of the radial density distribution of the planets to planet formation has not yet been resolved.

Because of differences in size and initial composition, each newly formed planet appears to have experienced a somewhat different "life cycle." Small planets (and moons), such as Mercury and Mars (and our moon), appear to have gone through their planetary stages of development relatively rapidly, within a few billion years, and to have been "dead" for the last several billion years. In contrast, the earth and probably Venus still have active subsurfaces. Distance from the sun can also affect planetary behavior. For instance, Venus and earth are nearly identical, it is believed, in composition, density, and size. Yet Venus, only 41 million km (25.75 million mi) closer to the sun, has markedly different atmospheric and near-surface conditions than has the earth.

The origin of the moons of the various planets is not particularly clear. Some scientists have advanced the idea that our moon is a fragment of the earth which has broken away, although, because of the improbably high rate of spin required, this seems unlikely. Others have postulated that the earth pulled rocky rubble and debris from the nebular disc into orbit around itself, and that this became the moon. Still others think that the moon was formed elsewhere in the solar system, and was later captured by the earth during a disruptive planetary event.

The asteroid belt is not particularly well explained in any of the hypotheses of the origin of the solar system. Most believe that it represents planetesimal material that failed either to merge into a single planet or to be swept into nearby developing planetary bodies.

SUMMARY

1. The earth is one of nine planets that orbit our central star, the sun, all in the same direction and roughly in the same plane. Our sun is but one of about ten billion stars that make up the large spiraling *Milky Way galaxy*. It is estimated that the visible universe consists of about one billion galaxies.
2. Many outstanding scientists, including Copernicus, Brahe, Kepler, Galileo, and Newton, provided the basis for our modern understanding of the solar system and beyond. Space probes of the nearby planets have greatly advanced our knowledge.
3. The sun, 109 times larger in diameter than the earth, possesses more than 99 per cent of all the mass in the solar system. The *photosphere,* composed of non-charged gas atoms at a temperature near 5400°C, is what we normally see from the earth. Beyond that is a tenuous zone, the *corona,* which consists of ionized particles that move rapidly away from the sun. Near the center of the sun, temperatures climb to about 15 million degrees centigrade. The sun's energy appears to be produced by the fusion of lighter atomic nucleii into heavier ones and the destruction, in the process, of a relatively small amount of mass.
4. Mercury, the closest planet to the sun, is small and hot and essentially lacks an atmosphere. Its surface is pocked by numerous meteorite impact craters.
5. Venus, similar in size to the earth, is completely surrounded by a dense hot atmosphere composed largely of carbon dioxide, in which surface pressures are about 90 times those on earth. Radar imagery seems to indicate the existence of earth-like mountains and lowlands.
6. The earth's moon is a layered body apparently composed entirely of a silicate material. An iron core seems to be lacking. The dark *maria* regions are composed of basalt that erupted between 3.8 and 3.2 billion years ago. The upland regions are underlaid by rocks richer in calcium and aluminum, but poorer in iron, magnesium, and titanium, than are lunar basalts. The highland rocks have been dated at 4.2 to 4.0 billion years.
7. Mars, a small planet, possesses only a thin atmosphere relatively rich in carbon dioxide. In the frigid polar regions, this gas freezes to form dry-ice polar caps. Spectacular surface features include huge volcanoes and immense canyons. Space probes thus far have shown no signs of life.
8. Jupiter, the largest planet, is largely composed of hydrogen and helium, with perhaps a small rocky or molten iron core at its center. We see only the tops of the atmospheric clouds, which are in constant mo-

tion. The Great Red Spot is a large circulating storm system. Jupiter has a diffuse ring-structure, the strongest magnetic field in the solar system, and 15 moons. On one of its moons, Io, active volcanism has been detected.

9. Saturn resembles Jupiter in many ways. Its ring-structure, however, is much more complex. Of its 15 moons, the largest one, Titan, possesses an atmosphere of hydrocarbon and nitrogen compounds.
10. *Uranus* and *Neptune* are large nonterrestrial planets about which we know very little. In overall aspects they resemble Jupiter and Saturn. Tiny *Pluto* appears to be a rocky planet and probably at one time was one of the moons of Neptune.
11. Our solar system probably started as a diffuse, cold cloud of gas, dust, and ice that slowly contracted. Most of the mass concentrated in a central bulge, whereas lumps and clumps of the remaining material (*protoplanets*) circulated in an equatorial plane. In time the central body became hot enough to allow nuclear fusion reactions to occur, turning eventually into our central star, while the protoplanetary bodies formed planets and satellite moons.

QUESTIONS

1. Describe the earth's position in the Milky Way galaxy.
2. Discuss the progression of ideas concerning the solar system as formulated through the fifteenth, sixteenth, and seventeenth centuries.
3. How does our moon differ from the earth?
4. What planet resembles our moon most closely and why?
5. What accounts for both the similarities and differences among the different planets?
6. In any hypothesis advanced to explain our solar system, what are the important points for consideration?

SELECTED REFERENCES

French, B. M., 1977, The moon book: Penguin Books, New York.

Gore, R., 1980, Voyager views Jupiter's dazzling realm: National Geographic, vol. 157.

Hartman, W. K., 1980, A climactic year in solar-system exploration clears up mysteries about the planets' satellites: Smithsonian, vol. 10, pp. 36–46.

Hodge, P. W. 1969, Concepts of the universe: McGraw-Hill Book Co., New York.

Kaufmann, W. J., III., 1979, Planets and moons: W. H. Freeman and Co., San Francisco.

Kopal, Z., 1979, The realm of the terrestrial planets: John Wiley and Sons, New York.

Mehlin, T. G., 1968, Astronomy and the origin of the earth: Wm. Brown Co.

Mission to Jupiter and its Satellites, 1979; Special Science Reprint, American Association for the Advancement of Science, Washington, D.C.

Pannekoek, A., 1961, A history of astronomy: Interscience Publication, Inc., New York.

Pioneer Saturn encounter, 1979: NASA, Ames Research Center, Moffett Field, Calif.

Rey, H. A., 1976, The stars, a new way to see them: Houghton Mifflin Co., Boston.

Wood, J. A., 1979, The Solar System: Prentice-Hall, Inc., Englewood Cliffs, New Jersey.

Voyager encounters Jupiter, 1979: NASA, Jet Propulsion Laboratory, Calif. Institute of Technology, Pasadena.

Fig. 3-1 The earth as seen from the *Apollo 8* spacecraft, with the surface of the moon in the foreground. From this distance, the earth appears blue. The streaky white swirls are large-scale cloud patterns. *NASA*

3

THE EARTH

The earth is a relatively dense and rocky planet roughly spherical in shape and three quarters covered by ocean waters. It revolves once per day on its axis, which is inclined at an angle of 23.5° to to the plane of its orbit around the sun. One complete orbit of the earth around the sun takes approximately 365¼ days at a mean distance of 148 million km (94 million mi). At this distance from the sun, there was sufficient radiant energy for life to evolve in abundance, and it now appears fairly certain that the planet earth is the only haven for life within our solar system.

The sun also provides the energy that evaporates fresh water from the oceans and carries it over the land, where it shapes surface features and makes life, as we know it, possible. Internally the earth is still restless. Energy from deep inside the earth, over geologically long intervals, can crush and melt rocks, build mountain ranges, and move continents. Some of these events, such as volcanic eruptions and earthquakes, can be extremely violent and locally destructive, but over long periods of time they have been responsible for shaping the surface of the earth.

What we know about the earth today has been general knowledge for only a relatively short period of time. When people first pondered the size and shape of the earth and its place in the solar system, the "world" with which they were familiar amounted to only a small portion of the earth's surface. Adequate clocks were not to be had, and even the simplest surveying instruments were unheard of. To most people, in fact, the earth appeared flat. Indeed it was believed that the sun and stars revolved around the earth, and this belief was maintained in most religions. Yet some observers noticed that the shadow the earth cast on the moon during an eclipse appeared round; that when ships approached port the sails always come first into view; and that when one traveled relatively long distances north or south the stars in the night skies changed in position. The Greek astronomer Eratosthenes (*c.* 275–195 B.C.) not only concluded that the earth was round, but on the basis of a very simple experiment determined its circumference to be 38,400 km (23,862 mi). Modern-day estimates do not differ greatly from that figure.

Exploration and travel for commercial gain

and for glory added to the knowledge of the earth in the reports of returning seafarers. They also led to the development of instruments necessary to accurate navigation and to the accumulation of accurate observations of the stars and their positions, which also is important in navigation.

The Phoenicians went to the edges of their known world before the birth of Christ, traveling as far as Cornwall, and down the west coast of Africa. The Vikings sailed as far as Greenland and North America in the eleventh century. However, it was not until the late fifteenth and the sixteenth century that Europeans developed a curiosity about the rest of the world and set sail across all the oceans. Perhaps the most heroic of the early adventures and the one that truly demonstrated the modern idea of the earth's size and shape was that of Ferdinand Magellan (1440?–1521) which began in 1519 (Fig. 3-2). After the reports of these seafaring men, it was again accepted that the earth was round, and, with the development of new and improved surveying instruments, its dimensions eventually were measured.

Fig. 3-2 Magellan, from a sixteenth-century allegorical painting.

Newton noted that since the earth rotated on its axis it should behave like any other spinning body. If particles are to turn with the body and not fly off into space, an inward pull, called a ***centripetal*** force (toward the center), must be exerted. Try swinging a rock at the end of a string—you will find that it is necessary to maintain an inward pull on the string or the rock will fly off in a straight-line trajectory. Because of the way forces behave, an outward, or ***centrifugal*** force is also generated as a reaction to the inward pull. Newton reasoned that the centrifugal force generated in the spinning earth would oppose the inward, attractive force of gravity and that it would therefore lessen the pull of gravity. This effect should be greater near the equator, since the distance to the spin axis is greater there than near the poles (Fig. 3-3). Because the earth's gravitational pull was most reduced near the equator the earth would tend to bulge outward in that region, resulting in a flattened shape at the poles.

By the mid-eighteenth century, field surveyors confirmed the idea that the earth was a flattened sphere. The shape corresponds to the mathematical figure that would be produced by revolving an ellipse around its short axis and, accordingly, is termed an ***ellipsoid of rotation.*** The distance from the center of the poles was determined to be 6356.9 km (3950.2 mi) and that from the center to the equator, 6378.4 km (3963.5 mi). The ellipsoid of rotation of those proportions, defined as the ***earth reference ellipsoid,*** corresponds essentially to the surface of mean sea level. Since the reference ellipsoid has a regular definable shape, the earth's volume could be calculated and was found to be 1,024,000,000,000 km^3 (246,000,000,000 mi^3).

The ellipsoidal shape of the earth can give us an idea of how strong the earth's rocks really are. Imagine for the moment that instead of solid rock material, the earth is composed of a liquid, such as water, that can easily flow and become deformed. What shape would a "liquid earth," rotating on its axis once per day, take on? Surprisingly enough, it would be virtually the same as that of the actual earth—which implies that over long periods of time the rocks in the earth's interior behave much as a liquid does, flowing and deforming in response to the forces acting upon them.

In reality, the earth does not have a perfectly regular shape like the reference ellipsoid of rotation: there are mountains, continents, and ocean basins that spoil the regularity (Fig. 3-4). It is difficult to take all such irregularities into account, and geophysicists have finally agreed on a generalized shape in which the topographic highs and lows are represented, but smoothed out and reduced. The form is called a ***geoid*** (meaning *earth shape*). (In referring to Mars or the moon, we might speak of a "marsoid" or a "moonoid.") The smoothed-out earth shape differs from the earth reference ellipsoid, but only slightly, and the differences usually concern only mapmakers and geophysicists. A geoid may be defined as a surface upon which the attraction of gravity is everywhere the same, a surface upon which water cannot run downhill, for there are no inequalities of gravitational attraction. In the ocean basins, the geoid corresponds to mean sea level. On land, the surface can be imagined as that which would connect the mean water-surface levels in a set of narrow canals, criss-crossing the continents and connecting the seas (Fig. 3-5).

Since the dawn of the Space Age, numerous satellites have been sent into orbit around the earth. Under the influence of the gravity field, which extends out into space, they circle the planet, and careful tracking of their paths has enabled scientists to determine the shape of the earth even more precisely. The results show that the earth is slightly pear shaped; there are three bulges that protrude a maximum of 66 m (216 ft) above the reference ellipsoid. In 1500 Christopher Columbus stated that the earth had *la forma de una pera*. We can only speculate on the source of his idea, but to some degree it has proved true.

MASS OF THE EARTH

The ***law of gravity,*** first stated concisely by Isaac Newton (see Chap. 2) can be used to describe the gravitational attraction between a body of mass (m_1) at the earth's surface and the mass of the entire earth (m_E) itself. In everyday

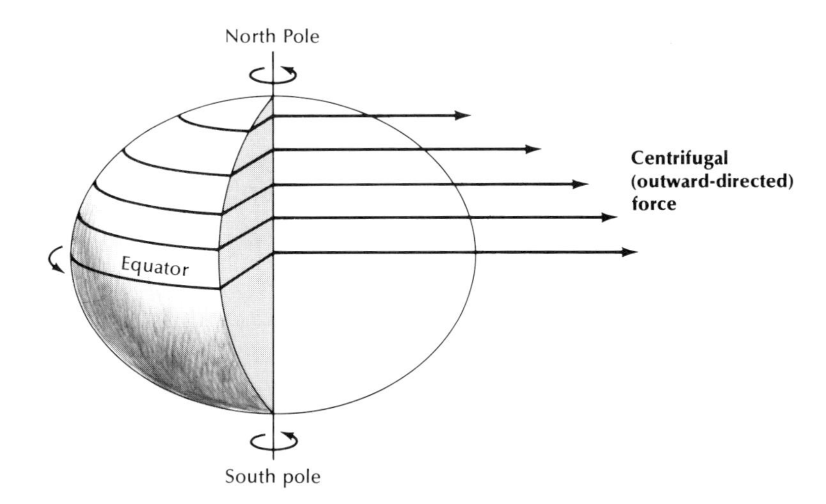

Fig. 3-3 Exaggerated ellipsoidal shape of the spinning earth. Centrifugal force is greatest at the equator, where distance from the spin axis is greatest, and decreases poleward.

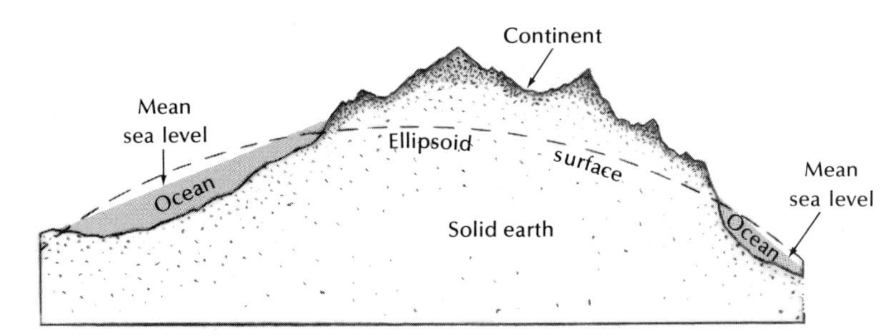

Fig. 3-4 Relationship of the earth's irregular shape to the surface of a reference ellipsoid, which provides a smoothed-out approximation of the earth's shape.

usage we call that attractive force the ***weight*** of the body. The mass of the earth acts as if it were concentrated at its center, at a distance of the earth's radius (R_E) away from the mass at the surface. The attractive force (F_g) between the earth and a surface body can be given in mathematical shorthand as

$$F_g = \frac{G m_1 m_E}{R_E^2}$$

where G in the equation is the universal gravitational constant; it is one of the fundamental properties of nature and is also related to the intrinsic properties of space, in a way that is not yet understood.

In Newton's time, neither the mass of the earth nor the value of the universal gravity constant ("big G") was known. It was not until 1797 that another English scientist, Henry Cavendish (1737–1810), finally determined the gravitational constant in a simple, elegant experiment (Fig. 3-6). Unfortunately Newton did not live long enough to witness it. Cavendish suspended from the ends of a balance arm two large lead spheres, each 30 cm (11.7 in.) in diameter. Next he fastened two smaller lead spheres, each 5 cm (1.95 in.) in diameter to the ends of a light, rigid rod, which was hung from its center on a very thin wire. Placing the horizontally suspended rod between the large spheres, he waited until the system had come to rest and then moved the large spheres closer to the smaller ones. Because the distance between the small and large spheres was shorter, the gravitational attraction between them increased, causing the rod assembly to rotate to a new rest position. Cavendish knew the elastic properties of the wire and was able to calculate the increase in gravitational force necessary to rotate the rod assembly. Since he

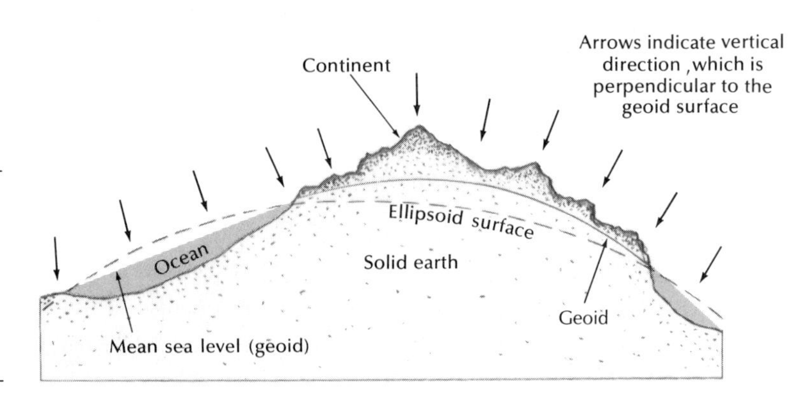

Fig. 3-5 Geoid surface, in relation to the reference ellipsoid and surface topography. In the oceans, the geoid corresponds to mean sea level; under the continents, it rises above the ellipsoid surface and approximates the topography. The geoid is the best representation of a level surface that is everywhere perpendicular to the vertical.

Fig. 3-6 Lord Henry Cavendish, an English physicist and chemist, used an extremely sensitive balance to determine the gravitational constant in 1798, fully 70 years after the death of Isaac Newton. *New York Public Library*

knew the force of attraction, the masses of the spheres, and the distance between them, he could use Newton's law of gravitation to calculate "big *G*." In spite of its simplicity the apparatus was incredibly sensitive; the force of gravity he measured was equivalent to the weight of about one ten-millionth of an ounce. The value of the gravitational constant he determined was not greatly different from that determined by modern methods. Once Cavendish knew "big *G*" he solved the Newtonian equation for the mass of the earth and found it to be 6×10^{24} kg or 1.3×10^{25} lb. Since he knew the approximate volume of the earth, he was able to calculate its density from the relation

$$\text{Density} = \frac{\text{Mass}}{\text{Volume}}$$

His figure was 5.448 gm/cm^3. As the methods for determining the mass and volume of the earth improved, the value has changed; today the accepted figure is 5.519 gm/cm^3. In comparison, the density of pure water is nearly 1 gm/cm^3, so that the density of the earth, as a whole, is about five and one-half times the density of water.

The average density of rocks exposed at the surface of the continents is about 2.7 gm/cm^3, or only about one-half that of the whole earth. Certainly some very dense material must exist in the earth's interior for the average density to be so high. Cavendish's experiment provided one of the first clues on the nature of the earth's interior. Since his time, quantitative information from other sources (see Chap. 18) indicates that some of the material near the earth's center is four to five times more dense than the average surface rock.

MEASUREMENTS OF GRAVITATIONAL ACCELERATION; ISOSTASY

Any object falling through the air toward the earth falls perpendicular to the geoid surface under the influence of gravity. As it falls, its velocity increases until frictional drag from the air causes it to reach ***terminal velocity***—the speed at which the pull of gravity equals the frictional force. In a vacuum, all objects, regardless of size or shape, fall at the same, ever-increasing velocity. The rate at which the velocity increases is called ***gravitational acceleration*** (commonly denoted as g), and it is directly related to the force of gravity—that is, the greater the force of gravity, the greater the gravitational acceleration.

In fact, the simplest way to measure the force of gravity at any spot on the earth's surface is to measure the gravitational acceleration. And the simplest device for that purpose is a pendulum, which was first used by Galileo. The story is told that the idea of the pendulum

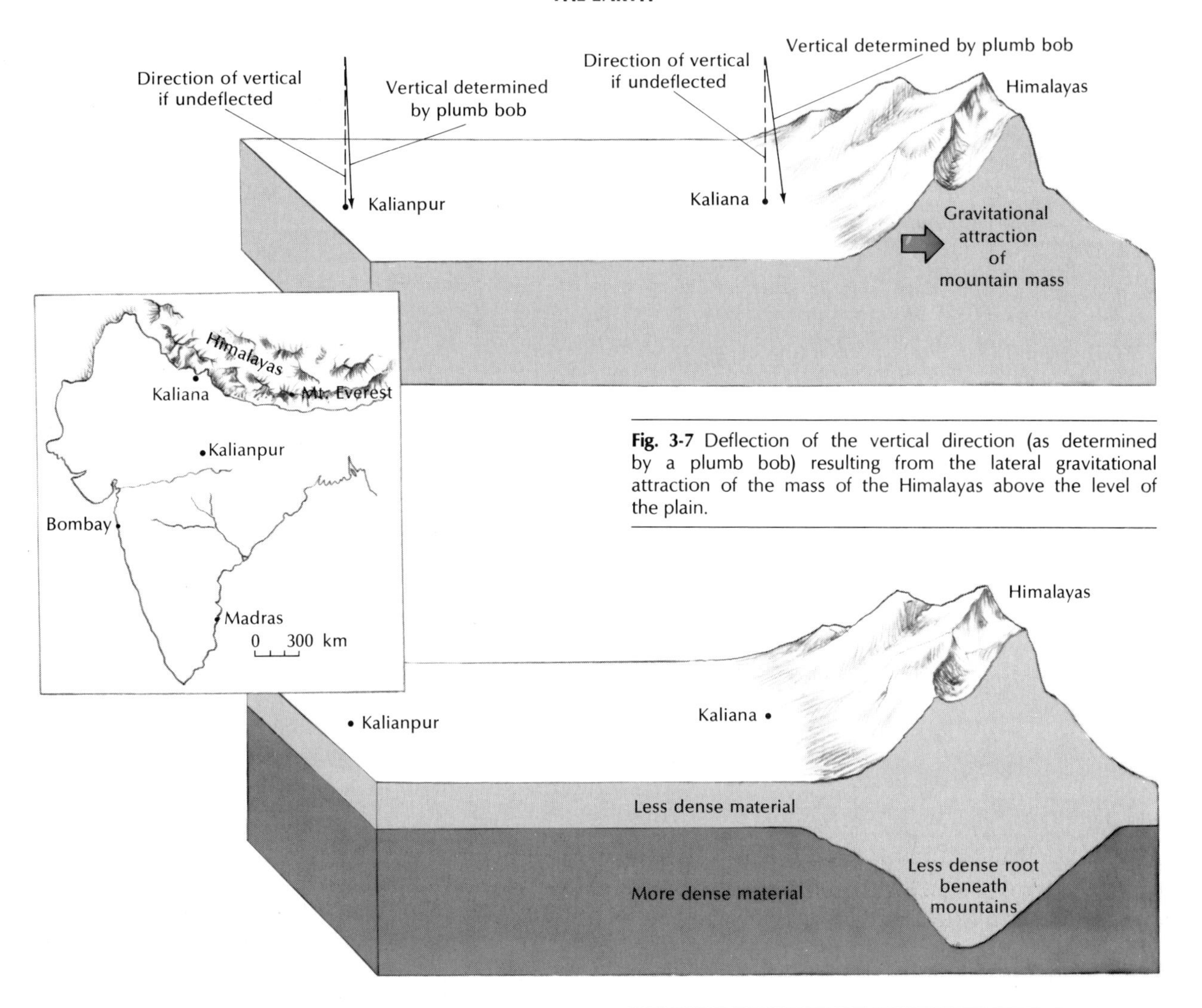

Fig. 3-7 Deflection of the vertical direction (as determined by a plumb bob) resulting from the lateral gravitational attraction of the mass of the Himalayas above the level of the plain.

Fig. 3-8 Airy's model, showing the root of the less-dense material that must exist beneath the Himalayas to account for the actual discrepancy of the vertical as measured at Kaliana and Kalianpur.

came to Galileo one day while he was sitting in church. Rather than listen to the lengthy liturgy, he watched the chandeliers sway to and fro. The period of oscillation (the length of time required for one complete swing of the pendulum) depends upon two factors: the acceleration of gravity at that location and the length of the pendulum. If one of the factors could be measured, the other could be calculated. In principle, then, by measuring the length of a pendulum and timing its period of oscillation, the force of gravity at that location could be determined. But timing presented a problem, for in the late seventeenth century the most accurate clocks were based upon the pendulum, and of course clock time varied with gravitational acceleration. Subsequently, as various precision chronometers came into use, the problem was solved, and the pendulum remained the principal means of determining acceleration of gravity until only recently. Today, even more accurate instruments, called ***gravity meters,*** or gravimeters, are used. They are so sensitive that they can detect differences in acceleration between two stations less than 1 m (3 ft) apart in elevation. Gravity meters have been carried all over the earth and even to the moon. Measurements taken with these instruments demonstrate that the acceleration of gravity is greater near the poles, where the distance to the center of the earth is much less than it is near the equator, where the earth's radius is greater and the centrifugal effect is more pronounced.

In addition to regular variations in gravitational acceleration due to the rotation of an ellipsoidal earth, there are variations related to irregularities in surface topography. A mountain peak like Mount Whitney in California is farther from the center of the earth than is the floor of a depression like Death Valley, the result being that gravitational readings on the peak are lower than those in the valley. It is possible to correct gravity-meter readings to take into account variations in altitude, and when such corrections are made, it becomes apparent that at any one locality gravitational acceleration is also affected by the density of the rocks close below the surface, which, of course, is determined in great part by the type of rock present.

Earth scientists have found that, in general, the higher, continental parts of the earth are associated with rocks of lower average density, whereas the lower oceanic portions are associated with rocks of higher average density. The rocks of our planet are not very strong over long periods of time. Rather, they flow and deform much as a liquid does. A continent of relatively low density, or a mountain range rising high above its surroundings, therefore, could not be supported for any length of time at the earth's surface unless it floated in a denser material, much as an iceberg floats in water. If an iceberg juts far above the surface of the water, it must also reach even further below the surface. It is reasonable to expect that mountains and continents, too, extend far downward below the surface, like the roots of many trees. And gravity data suggest that they do.

The first indication that mountains might have "roots" came from a mapping survey of India, in the mid-nineteenth century, directed by Sir George Everest (1790–1866), for whom the highest mountain on earth was named. During the survey the distance between two stations, one at Kaliana and the other at Kalianpur, about 600 km (373 mi) apart along a north-south line, was measured with great precision (Fig. 3-7). Kaliana was situated close to the base of the Himalayas, whereas Kalianpur was out on the plain. Two different methods were used to determine the separation. One was the standard surveying technique known as ***triangulation,*** in which intersecting lines are constructed and precision surveying instruments are used. The other was a method in which elevation angles measured from the horizontal to the polestar (North Star) were used to calculate earth distances. Although both methods were painstakingly carried out, the results differed by about 153 m (502 ft). The surveys were rerun, but the difference remained. Archdeacon John Henry Pratt, a British mathematician (d. 1871), pondered the problem and concluded that the difference in values was related to the measuring technique. In order to measure an elevation angle to a star, a surveyor must first determine the horizontal plane. But if the horizontal planes at the two different stations are not the same, an error in the measured distance could be introduced. Only a very slight angular difference (5 sec of arc) was needed to produce the differences in distance between Kaliana and Kalianpur. In his analysis of the problem, Pratt determined the vertical direction at each station using a plumb bob (a lead weight hung on a cord) rather than the horizontal (Fig 3-7). If the horizontal planes between the two sites were askew, so must be the vertical directions. Pratt reasoned that the Himalayas could behave as a separate mass (like one of the large spheres in the Cavendish experiment), exerting a lateral gravitational force on the plumb bobs at the two stations and pulling the weights toward the mountains (Fig. 3-7). The amount of deflection, of course, would be greater at the station closer to the mountain. Pratt was quite sure he had solved the problem, and to demonstrate the correctness of his approach he calculated the gravitational effect that the Himalayas would have at each of the two stations. In order to do so he first estimated the mass of the mountain block, based on its size and density. Much to his chagrin, when he had finished his computations he found that instead of 5 sec of arc, the difference amounted to 16 sec of arc. Apparently he was on the right track, but had failed to consider some other factor. Pratt concluded that the source of his error had to be in his estimate of the mass of the Himalayan block; it must have been less than he had originally thought.

Meanwhile, Sir George Airy (1801–92), a noted English astronomer, also became interested in the problem. He suggested that all blocks at the earth's surface are relatively low in density and that they "float" on the denser rocks below. Lower-density mountain blocks that stand high must extend far downward into the denser material (Fig. 3-8). The part of the Himalalyan block rising vertically above the surface of the plain is much denser than the surrounding air and would exert a lateral gravitational attraction on a plumb bob at any one station, deflecting it *toward* the mountains. But the root of the mountain block would also exert such a pull. Because the root is somewhat less dense than the surrounding subsurface rocks, the plumb bob would tend to be deflected *away* from the mountains. At Kaliana and Kalianpur, the gravitational effect of the mountain mass was greater than the effect from the root below, so that the plumb bob was deflected toward the mountain, but at a smaller angle than Pratt's simple model had allowed him to calculate. At about the time that Airy formulated his hypothesis, Pratt came to nearly the same conclusion.

Since the time of Pratt and Airy, the idea that mountains and continents had roots has been tested over and over again, and shown to be valid. Conclusive evidence has come from the beaming of sound waves downward through the outer layers of the earth. Upon encountering the base of the continent or mountain root, the waves are reflected back to the surface. Geophysicists know how fast the waves travel in the outer earth, and by observing how long it takes for their return, they can determine the depth of the root of a continent or a mountain mass. Under the Sierra Nevada range in California, for example, the root extends 60 km (37 mi) downward.

Pratt and Airy's theory—that the blocks at the earth's surface are in floating gravitational equilibrium— is of great significance. It has been given the name ***isostasy*** (from the Greek *isos,* equal and *stasis,* standing still). If the equilibrium is disturbed; the blocks will move up or down, and the denser material at depth will slowly deform and flow to allow the blocks above to readjust. Before considering isostasy in relation to the earth, let us again look at the closely analogous system of ice in water. If ice is piled on top of a floating ice cube, the entire mass will sink until it comes to a new rest position—the top will actually be higher than before and the root will extend deeper. If, on the other hand, ice is shaved from the top of the cube, the cube will rise to a new equilibrium position, but the top will not be quite so high as before, and the root will not project so far down. The same is generally true of the earth. If material is loaded on a portion of the earth's surface, that portion will sink to some degree; if material is removed, that portion will rise until a new isostatic equilibrium is reached.

During the ice ages of the recent past, portions of the continents were loaded with great masses of ice, up to about 3000 m (9840 ft) thick (see Chap. 13), and the earth's surface was pushed downward. Today most of that ice has melted away, and most of the former ice-laden areas have risen and will continue to rise until isostatic equilibrium is re-established. As shown in Figure 3-9, the upward rebound in northeastern North America has already amounted to as much as 275 m (902 ft). The rebound is still progressing, but its rate is decreasing. The last time the area was ice covered was about 10,000 years ago. Yet the large amount of rebound that has already taken place shows how quickly isostatic readjustment can occur.

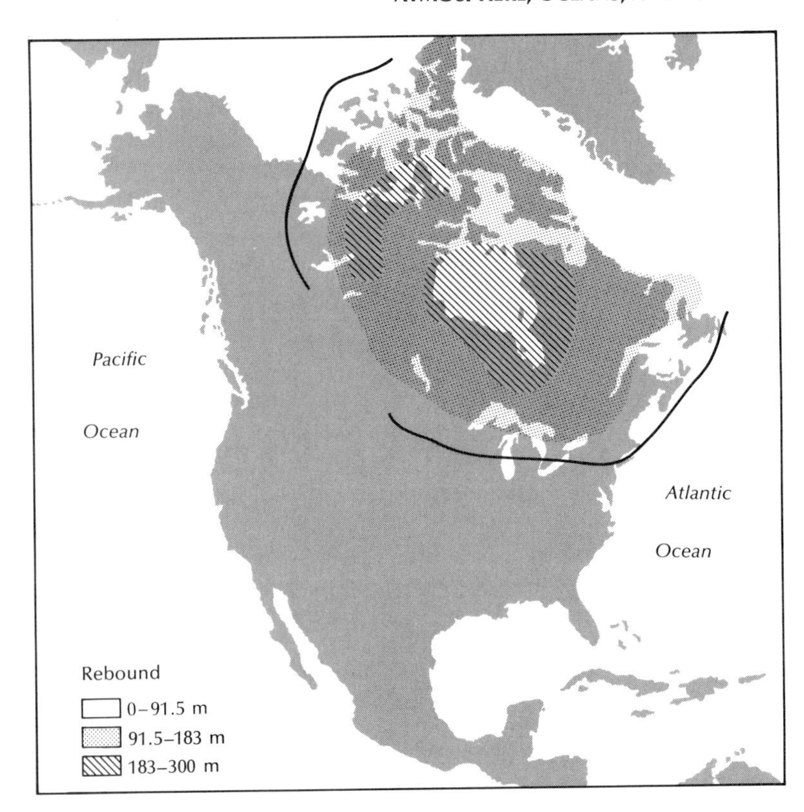

Fig. 3-9 North America, showing rebounding in the northeast after the rapid melting of the ice in the last glaciation. The regions of maximum uplift correspond to areas of maximum ice thickness.

High mountains are subject to high rates of erosion, and as material is carved from a mountain top, the block undergoes continual isostatic uplift. Because of isostatic recovery, it takes a considerable length of time, perhaps several tens of millions of years, to wear down a high mountain range.

ATMOSPHERE, OCEANS, AND THE SOLID EARTH

The features that stand out most noticeably when one looks at the earth from space are the oceans and the continents. Above the entire earth surface, water and land, circulates the tenuous invisible gaseous envelope we call the ***atmosphere.*** Yet the ocean waters and the atmosphere together comprise only a small fraction of the total mass of the earth. The bulk is made up of higher-density rocks, which lie beneath the continental land surface and the bottom of the ocean and, in some form, extend all the way to the center of the earth.

Atmosphere

Many of the beauties of nature result from the characteristics of the atmosphere—skies of deep blue, sunsets blazing in lurid shades of red and gold, multicolored rainbows after summer showers, the skies in tempest. But it is the scientific aspects of the gaseous envelope that concern us now.

The atmosphere is composed mostly of nitrogen (78 per cent by volume) and oxygen (21 per cent by volume). That leaves little room for the other gases, of which the most abundant are argon, carbon dioxide, and water vapor. The lower boundary of the atmosphere is the sea and land surface, but there is no sharp upper boundary. The air close to the earth's surface, compressed under the weight of the air above it, is much more dense than that higher up. Nearly all the air lies within 96 km (60 mi) of the earth; above that height there is less air than in the best vacuum we can produce in the laboratory. Above a height of 960 km (597 mi), the atmosphere consists mostly of hydrogen and helium, and above 2400 km (1491 mi), only small, fast-moving molecules of hydrogen are found. As the earth moves through space, some of the light, active gases (mostly helium and hydrogen) are lost into space, left behind to become part of the interplanetary gas mixture.

Although the atmosphere close to the earth is composed largely of nitrogen, it is the presence of oxygen, carbon dioxide, and water vapor that most interests geologists. Most organisms cannot survive without oxygen, and carbon dioxide is vital to plants. Water, which occurs as a vapor in relatively small amounts in the atmosphere, has several vital roles: it is essential to the existence of most plants and animals, it is the prime absorber of radiant heat in the atmosphere, and it is the principal substance involved in the processes of weathering and erosion.

The earth's atmosphere is unlike that of any other planet in our solar system. It appears to have evolved to its present state through geological time by the slow action of many processes, among them the loss into space of many of the less dense gases, the addition of gases from organic processes and of gaseous emanations associated with volcanic activity, and the accumulation of gases, such as argon and helium, which are radioactive-isotope decay products.

The atmosphere is in continual agitation and flow, as is quite evident from global weather patterns. The driving force for the circulation of the air is, in one way or another, related to the energy received from the sun.

Because of the spherical shape of the earth, heat from sunlight is more concentrated in the equatorial regions of the world than in the higher latitudes and polar regions (Fig. 3-10). Thus, above the Equator the atmosphere is warmer than it is over the poles. Figure 3-11 shows patterns of air flow on a model earth. In response to inequalities in the atmospheric heat budget, hot air moves toward the polar regions at high altitudes, while denser, colder air flows toward the equator at lower altitudes (Fig. 3-11A). Complications in the idealized pattern arise because the earth spins on its axis: air begins to follow a curved path (deflected, apparently, by the Coriolis force) toward the right in the Northern Hemisphere and toward

Fig. 3-10 The sun's rays on the face of the earth are equal in intensity (**A**), but because of the earth's curvature, they are less concentrated in the polar regions, so that the poles receive less heat than the Equator (**B**).

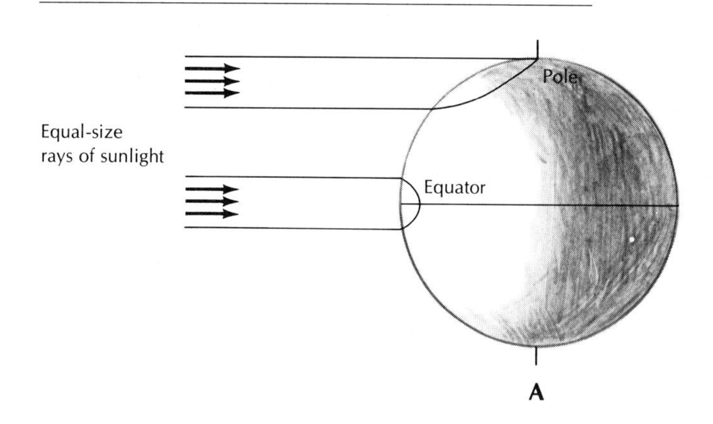

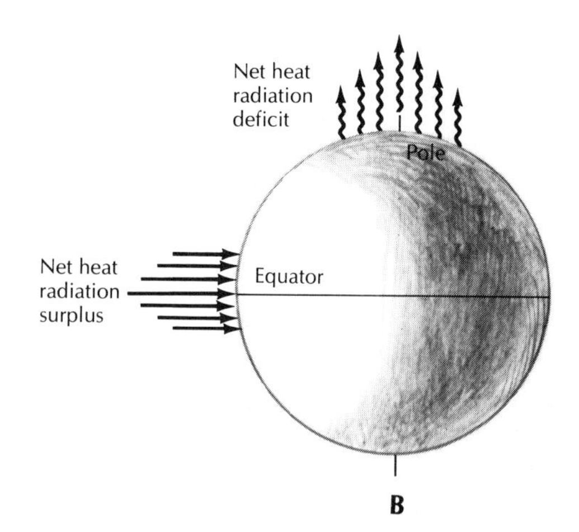

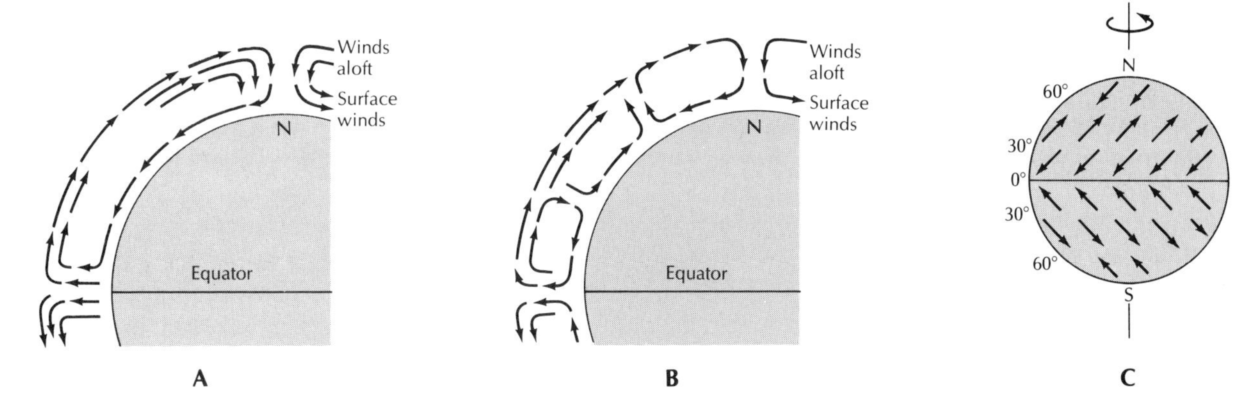

Fig. 3-11 Air flow patterns on a model earth. Cross section of flow on (**A**) a non-rotating earth and (**B**) a rotating earth. The surface wind pattern that would develop on a rotating earth is shown (**C**).

the left in the Southern Hemisphere. Circulation is short-circuited into the smaller air cells shown in cross section in Figure 3-11B. Figure 3-11C shows the surface wind pattern that would result.

The actual pattern of winds blowing across the surface of the earth conforms rather closely to the idealized pattern of Figure 3-11C. Differences between the actual and the idealized pattern are the result of the non-uniform distribution of the oceans and continents, topographic irregularities, and seasonal variations in temperature.

Oceans

The ocean, which covers about 71 per cent of the earth's surface, has many faces. It has long been a source of food and an avenue of transportation. It's ever-changing aesthetic qualities have captured the imaginations of humans since time immemorial. Life, as far as we know, began in a simple fashion in the dilute, but sustaining soup of the Precambrian seas. Today, after perhaps three and one-half billion years of evolution, life forms exist in all parts of the ocean basins in a bewildering array of life-styles.

The oceans supply water to the atmosphere, through evaporation by the sun's rays. Water vapor is carried by prevailing wind systems far inland, where it precipitates to recharge the surface and ground waters and to nourish plants and animals. Chemical weathering of rocks would be minimal except for the moisture that lingers at the earth's surface after the water has flowed away. And runoff of the excess water is without a doubt the prime agent involved in the erosion of the land.

Scientifically, the water realm is several worlds in one. We can study, for example, the chemistry of water, the dynamics of current and wave action, marine life forms, large and small, or the geology of the rocky basin that underlies the ocean. All are part of the field of ***oceanography,*** which began in earnest with the voyage of the British warship HMS *Challenger* in 1872. Before then, most of the lore of the oceans was collected from the tales of seafarers and consisted of myths and mysteries, with a dash of fact. During its three-and-one-half-year cruise, the *Challenger* sailed more than 110,000 km (68,354 mi) in most of the major oceans, accumulating data on their physical, chemical, biological, and geological aspects.

It was not until 1930 that centers for oceanography began to appear in the United States, and the rest of the world was just as slow to

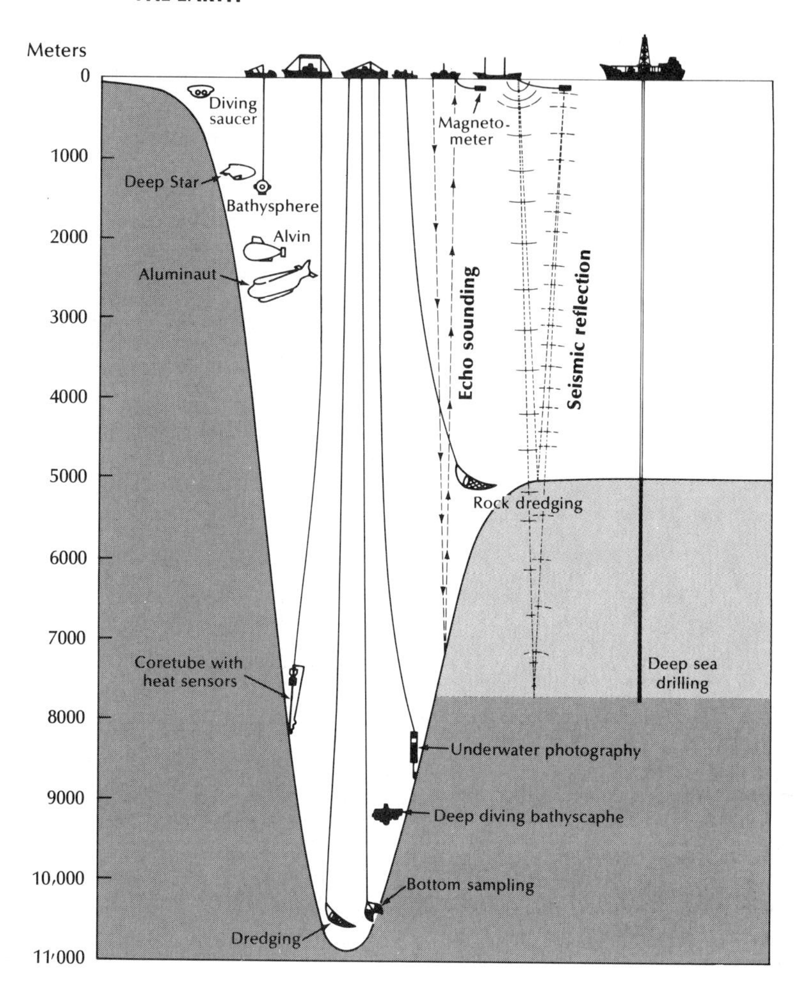

Fig. 3-12 Methods of sampling the ocean's waters and rock. Oceanographic exploration has become a highly technical field.

develop sea studies. Because of a lack of oceanographic vessels until 10 or 15 years ago, most geological information came from the continents. We knew almost nothing about the geology of over 70 per cent of the globe. Now a fair number of oceanographic institutes exist around the world, including, within the United States, such outstanding centers as Woods Hole Oceanographic Institution in Massachusetts, the Scripps Institute of Oceanography in California, and the Lamont-Doherty Geological Observatory in New York. Using the most modern equipment, today's oceanographers are adding a great deal to our knowledge of the ocean world; knowledge that has led us to reconsider some of the earlier ideas concerning the earth's large-scale geological features.

Because of their inaccessibility, the ocean depths remained virtually unexplored until modern technology provided the necessary sophisticated (and commonly expensive) equipment (Fig. 3-12). A single research vessel may take water samples from various depths, measure water temperatures and current velocities, and collect samples from the ocean bottom by dredging or coring. The sea floor can be examined visually by means of TV or photographic cameras or deep-diving submersibles (Fig. 3-13). One of the larger submersibles built in the United States has reached a depth of about 10,900 m (35,752 ft). A number of the smaller submersibles, each carrying two to four people can dive 600 to 2400 m (1968 to 7872 ft). Most have windows for direct observation and manipulator arms for sampling.

Virtually all oceanographic ships carry an ***echo sounder,*** a sonic instrument that records ocean depths. The device operates by creating a loud noise or "ping" under water; the sound travels to the sea floor and is reflected back to the ship. Since the speed of sound in water is almost constant, the time it takes a sound wave to journey to the sea floor and back is directly related to the depth. On most deep-sea expeditions, the echo sounder pings about every one-quarter second, resulting in a continuous visible recording of the bottom profile as the ship cruises along. In 1969 Henry W. Menard, a well-known American oceanographer, and a recent director of the U.S. Geological Survey, described it as follows:

> We cruised over hills and low mountains a mile and a half below and visible only on the echo-sounder. However, after a while the distance between the echo-sounder and the bottom is forgotten. As the marine geologist surveys a new range of undersea mountains, he senses them around him. I was once surveying with a captain new to the game. He re-

Fig. 3-13 The *Alvin,* one of several free-diving research submersibles now in operation, is capable of dives as deep as 2000 m (6560 ft). Equipped with a mechanical arm for sample retrieval, photographic and television cameras, and other devices, and powered by large batteries, the crew-operated miniature submarine has added greatly to our knowledge of the ocean basins. *Woods Hole Oceanographic Institution*

marked, as we headed again toward a peak on the map we were making, that he could not suppress a captain's feeling that we would hit it, even though he knew perfectly well it was a mile below the ship.

If the frequency of the sonic "ping" is increased, some of the sound waves penetrate the ocean floor and are reflected back from the rocks and sediment layers beneath it (Fig. 3-14). Marine geologists are able to "see" not only the shape of the bottom, but also the nature of the rocks below the surface. The process is called ***sub-bottom profiling.*** With every track of the ship, another continuous cross section of the rocks below the ocean bottom is obtained, from which the marine geologist must imagine the map of the sea floor. On land, geologists can see only the rocks on the surface and must imagine what they would look like in cross section.

Until only recently, ships obtained information on rocks many tens of kilometers below the ocean bottom by detonating explosive charges. Generally two ships, separated by several kilometers, were used in the operation, one to drop the depth charge and the other to "listen" to the sound waves as they bounced back to the surface. Today, because of environmental restrictions, the use of explosives in the sea has been curtailed severely. Most ships now produce alternate signals by means of a pressure gun or a sparker system. Waves from such sources can penetrate a maximum of about 10 km (6 mi) beneath the ocean floor. Gravity meters and magnetometers, to measure the earth's gravity and magnetic fields, respectively, are also carried on many geophysical research ships.

One of the great advances in oceanography began in the early 1960s when a shipboard rig actually drilled into the ocean bottom to a depth of 180 m (590 ft). Many technical problems had to be surmounted before that first probe of the deep-sea floor. Imagine a ship rolling and pitching in rough seas while the end of a drill bites into rocks that lie perhaps 3000 m (9843 ft) below. No ship could possibly be anchored in waters thousands of meters deep. Yet, the vessel must remain directly over the drill hole during the entire operation, a feat that is accomplished with a system of motors that push the ship this way and that in response to signals received from three sonar buoys attached, in triangular array, to the sea floor. The method works so well that it is possible for a drilling crew to pull an entire string—which might be compared to a 3000 m (9840 ft) piece of limp spaghetti—out of the hole, change the drilling bit, re-enter the same hole, and resume drilling.

The premier drilling ship in service today is

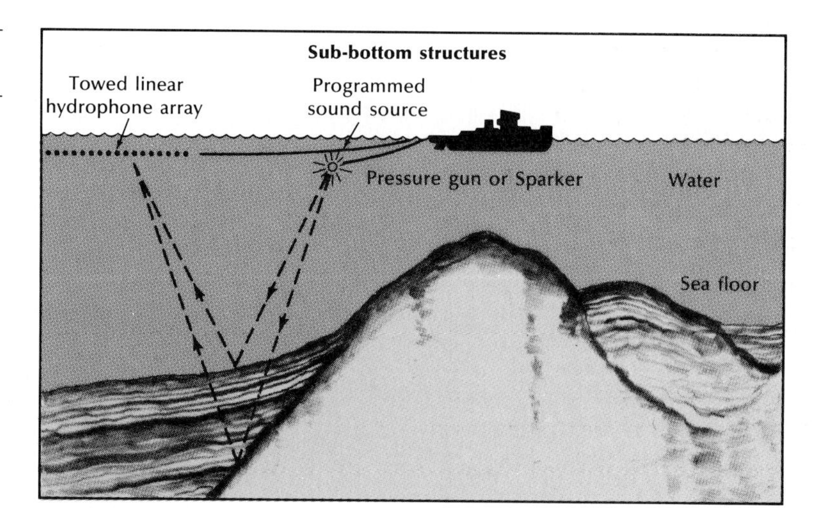

Fig. 3-14 The method of sub-bottom profiling, with the actual profile superimposed.

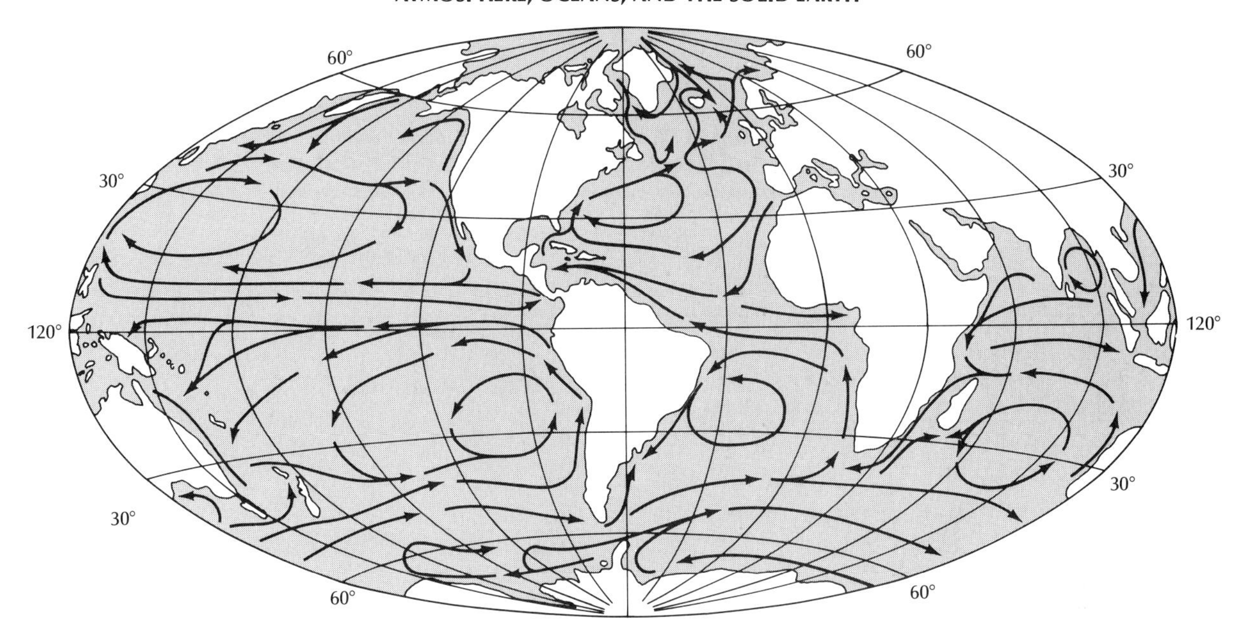

Fig. 3-15 Generalized world pattern of surface oceanic currents and gyres. The pattern in each part of an ocean basin is unique.

the *Glomar Challenger,* a U.S. research vessel that has voyaged to all parts of the world sampling rock material.* It was designed to drill in water depths up to 6000 m (19,680 ft) and to obtain samples from 750 m (2460 ft) below the ocean bottom. It has been in nearly constant service since it began operation in 1968. Core samples carrying a record of the ocean basins have been studied intensively, with geologically significant results. For example, the record of the rocks beneath the ocean floor goes back only to the Jurassic Period (about 150 million years), and in most parts of the oceans, it does not go beyond the Cretaceous Period (about 135 million years). It appears, then, that the ocean basins are relatively young, whereas the continents are at least 2 billion years old. We will consider that unusual result in more detail in Chapter 20.

*Another ship, the *Glomar Explorer,* is being constructed as a replacement for the *Challenger.* The new ship is larger, which would enable deeper drilling, particularly into the thick sediments of the deeper part of the continental margins, where oil might be found.

Sea water, as is well known, is salty. In fact, 100 lb (45 kg) of average sea water, when evaporated, leaves about 1.6 kg (3.5 lb) of salt. The word *salt* is a general chemical term and includes many compounds. The salt most familiar to us is table salt, ***sodium chloride,*** a compound formed of sodium (a metal) and chlorine (a toxic gas). The salt compound, although comprised of the two elements, does not resemble either of them. During formation of salt, the sodium atom loses a negatively charged particle (electron) and the chlorine atom gains one. The resultant atoms now carry a charge and are called ***ions.*** When table salt is dissolved in water the two ions split up and behave more or less independently.

The principal ions dissolved in sea water, making up more than 85 per cent of the dissolved material, are those of sodium and of chlorine. Other ions include sulfate, magnesium, calcium, potassium, bicarbonate, and bromine, in descending order of abundance. Traces of other ions, many of them required by marine organisms, are also present. Like the atmosphere, the chemistry of the oceans appears to have changed over long periods of geological time. Much, if not most, of the water itself probably came from deep inside the earth. It was released as steam, along with molten rock, during volcanic eruptions that occurred through the ages. For the most part, the salts in the ocean represent materials derived from the land through the process of chemical weathering. Some of the salts are used by organisms in constructing their shells, some are precipitated out of solution and fall to the ocean floor. A few times in the geological past, large-scale evaporation of sea water in restricted areas led to the precipitation of hundreds of meters of calcium sulfate, sodium chloride, and, to a lesser extent, potassium salts.

The waters of the ocean are constantly in motion. The currents, in some cases, are wide and feeble; in others, they are narrow and strong. The action of currents is not restricted to the surface layers. Flow patterns also exist at deeper levels and bring about vertical as well as lateral movement of the ocean waters. Deeper currents are generally completely unlike surface currents.

Surface flow is dominated by the prevailing winds. Winds push at the surface of the water, eventually creating a slow-moving ocean drift. Like atmospheric currents, water currents are also affected by the earth's rotation; as they flow they, too, continue to curve until large rotating current cells—called ***gyres***—develop (Fig. 3-15).

Most deeper currents are produced by differences in water density. Heavier water tends to sink, whereas lighter water tends to rise. Basically, two things control the density of sea water: temperature and salinity. The colder or saltier the water, the greater the density. Water temperature is determined by climate, whereas salinity is determined by evaporation and precipitation. Cold, dense polar waters tend to sink and move at depth toward the equator; warm, lighter waters in the equatorial regions move near the surface poleward.

It is usually assumed that deep currents are slow-moving ones. Recent photographic evidence, however, has indicated that in some places currents moving at depths up to 7318 m (24,003 ft) have been rapid enough to produce ripple marks in the bottom sediment (Fig. 3-16).

The solid earth

The earth is composed of higher-density, mostly solid material. Our best information about the nature of this material comes from the rocks at the surface. Drilling on the land and in the sea has allowed us to directly observe the earth to depths of about 9150 m (30,012 ft), and analysis of the products of volcanism gives us some idea of the temperature gradient and rock material down to 200 to 250 km (124 to 155 mi). For knowledge of the rocky substrate beyond those depths, we rely on geophysical studies—most of them from interpretations of the waves that emanate in all directions during an earthquake (see Chap. 18). Some useful data, however, have also been obtained from studies of the magnetic and gravitational fields at the earth's surface and of the regional loss of heat from the earth's interior.

The data suggest that the earth has a pronounced density gradient, with rocks near the center being four to five times more dense than surface rocks (Fig. 3-17). The greatest variety of rocks, which have an average density of only 2.7 gm/cm^3, are found on the continents and extend beneath them to maximum depths of around 70 km (44 mi). By contrast, the rocks just below the oceans have an average density of 3.0 gm/cm^3, and project downward only 5 to 6 km (3 to 4 mi). Because of the greater thickness of lower-density, continental rocks, they float higher isostatically than do the rocks of the ocean basins. Rocks down to about 2900 km (1802 mi) are rich in silicon and oxygen with lesser amounts of iron, magnesium, aluminum, calcium, sodium, and potassium. Below that depth, the earth's ***core*** is thought to be nearly pure iron (possibly with small amounts of silicon, nickel, and sulfur)—both in a liquid and in a solid state. The magnetic field of the earth appears to be generated in the liquid parts of the iron core (see Chap. 19).

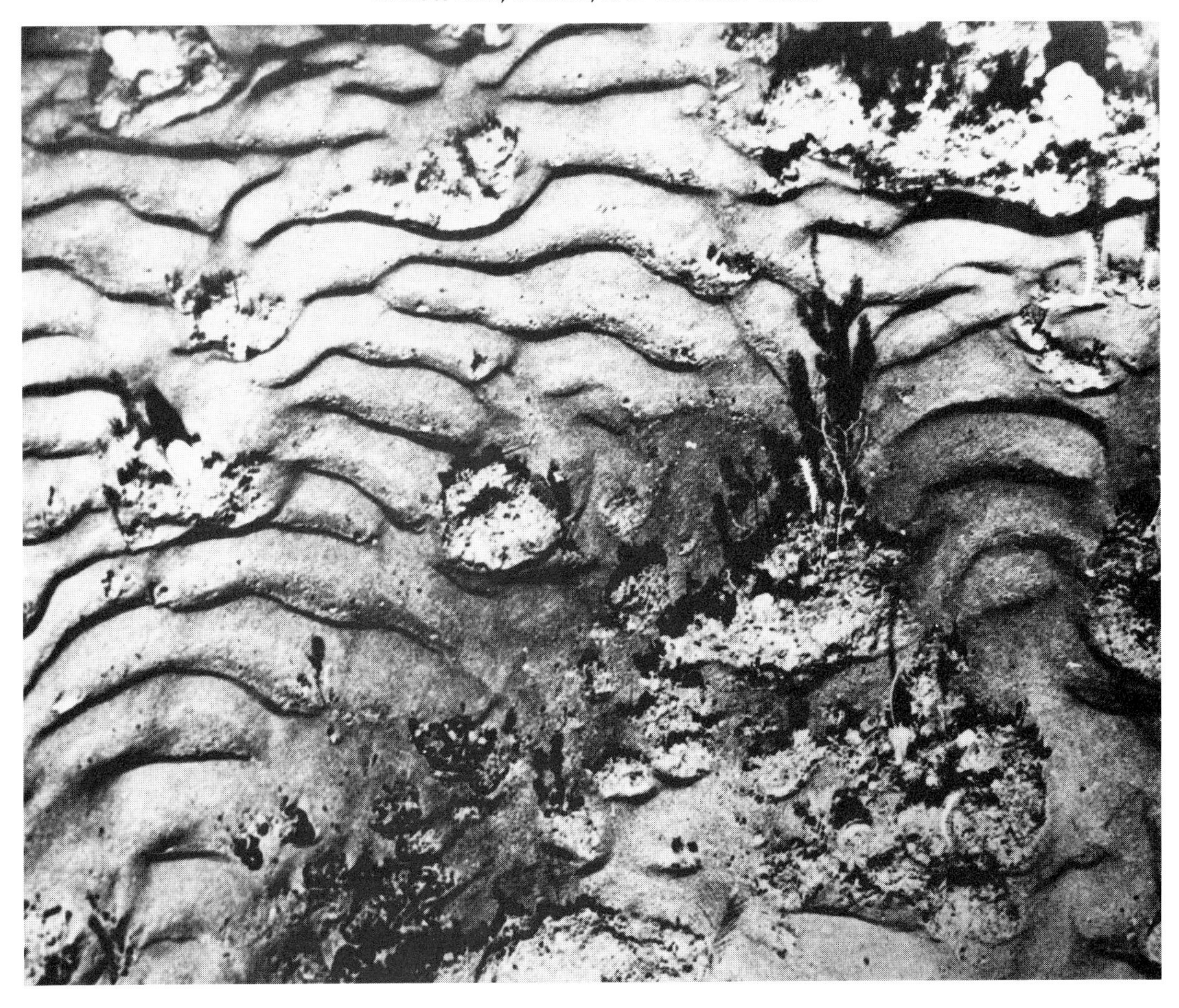

Fig. 3-16 Short-crested ripples on the ocean floor photographed at a depth of 867 m (2844 ft) on the Blake Plateau. *Bruce C. Heezen*

Exactly how the earth came to be layered will probably forever remain a mystery. Some geophysicists believe that ***density layering*** occurred quickly, as a result of the complete melting of the entire proto earth shortly after it came into being in the solar system. In the molten state, the heavier materials could, under the influence of gravity, easily make their way toward the center of the earth, and the lighter ones would tend to move outward. Some of the heat needed to produce melting could have come from decay of radioactive isotopes, which were more abundant initially, and some may have been released during gravitational compression of an initially larger, low-density proto earth into the smaller, more-dense earth we know. Yet many earth scientists feel that such heat sources would have been insufficient to completely melt the earth. Alternatively, if temperatures became high enough to melt only the iron, then metal droplets could slowly make their way to the center of the earth, and

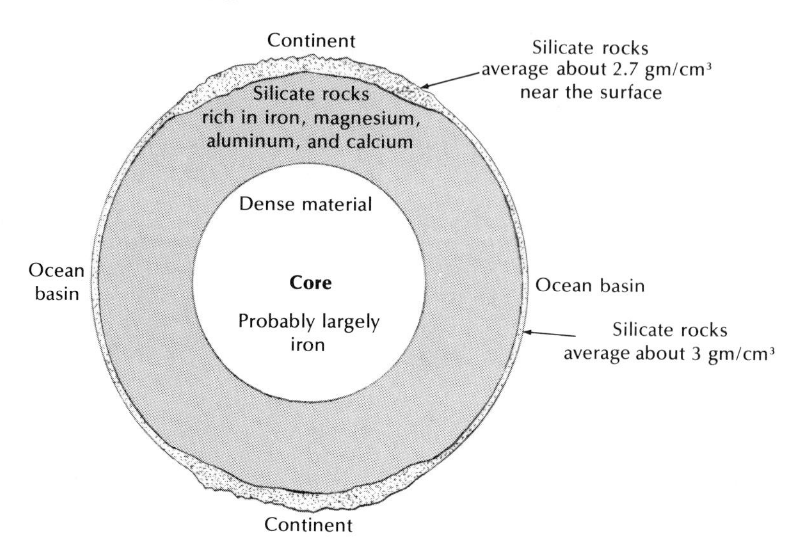

Fig. 3-17 Schematic cross section of a three-layered earth as determined by direct and indirect methods.

the density-stratification process would be a slow-acting one. Whatever the process, there is evidence that, by about 3 billion years ago, there existed a molten core large enough to produce a magnetic field.

Raymond Siever and Frank Press, two prominent American geoscientists, have called the process of density layering and the movement of iron to the center of the earth the "iron catastrophe." They believe that it may be the most important earth-shaping event that has ever happened. When it occurred, the earth not only became density stratified, but volatile substances, particularly water, were brought from the depths to form our oceans directly and, through a partial dissociation of the water, to produce indirectly one of the principal requisites of life—oxygen. Moreover, a natural subsurface "heat engine" was set into motion, a system that has continued to function, to keep the earth geologically active through its first five billion years. As far as we can tell, this geological activity shows few signs of abatement.

When we compare our planet with nearby solar bodies—Mercury, Mars, the moon—the present state of geological activity within the earth is unparalleled. It appears, as we discussed in Chapter 2, that the activity in large part depends upon the initial mass of the planetary body. In bodies considerably smaller than the earth, the activity may last only one or two billion years.

The part of the solid earth most important to humans is that near the surface of the continents. From it we derive soils, basic building materials, and fossil fuels. Mountains and deep canyons, in fact all the landforms on earth that we so cherish are carved or modeled from the solid earth.

Many of the earth phenomena we see or experience, such as volcanism, mountain building, lateral movement of continental blocks, and earthquakes, are the result of processes occurring far below the surface. Most of them appear to be particularly active at depths less than 250 km (155 mi), but some activity occurs at depths up to 700 km (435 mi). Many of the internal processes can be considered to be ***constructional;*** that is, the earth's surface is formed as a result of such processes. The internally generated constructional process is discussed in later chapters.

The earth's surface, however, is also subject to ***destructive*** activities, such as weathering, erosion, and deposition of sediments, that occur at or close to the surface and that tend to reduce the high spots and to fill in the low ones. In many of the subsequent chapters, we will discuss surface processes and how they level the earth.

The face of the earth has changed continuously down through the geological ages through the interaction of constructional and destructional processes. When land-leveling dominates for long periods, as for example during much of late Precambrian and early Paleozoic Eras, the continents are reduced to broad, low-relief features that barely rise above sea level. When, however, construction becomes dominant, as it has recently, the continental masses are built up until they stand, studded with lofty mountain peaks, high above the oceans.

As noted earlier, the principal energy source for the surface processes on earth, and life, is the sun. Our planet receives only a tiny fraction of the sun's radiant energy, and yet it has been enough to drive the destructive activities on the surface throughout geological time. There seems little likelihood that the solar power source will fail for several billion years to come.

Earth scientists have long searched for a unified theory or model that would neatly tie together all internally generated processes. Today, many geologists believe that the recently formulated theory of ***plate tectonics*** (from the Greek word *tektonikos,* meaning carpenter or builder) fulfills the requirement. According to the model, the outer shell of the earth down to about 100 km (62 mi) is rigid and is composed of a number of segments or plates—somewhat like the tiles on a bathroom floor—which can move with respect to each other in response to internal sources of power (Fig. 3-18). At some plate boundaries the plates push together, at some they pull apart, and at some they slide laterally past one another. It has been postulated that all the internal processes and the attendant surface features and phenomena result from plate motion. The plate boundaries include most of the world's earthquake belts, volcanically active regions, deep-sea trenches, island arcs, and some mountain chains.

Plate tectonics is undoubtedly the dominant theory for the internally generated processes of the earth—a theory that has revolutionized earth science. It does not provide an unequivocal answer to the question of what causes plate movement, however—in fact, some parts of the model have not yet been adequately explained—but virtually all geologists now accept the idea that continents drift and change in surface position, that the earth's outer shell is not fixed in space and time.

In the chapters in this book that deal with internal earth processes, we shall continually allude to plate tectonics, pointing out discrepancies as well as consistencies between the model and the geological facts. In Chapter 20, with all of the earlier chapters as background material, we shall examine the model in detail, comparing it with others that have been proposed to explain the "whys" and "wherefores" of activity beneath the surface of the earth.

Surface of the earth

The continents and the ocean basins constitute the two principal (or first-order) topographic features of the earth's surface. The contrast in elevation between these features reflects the fundamental differences in the kinds of rocks found beneath them. We are familiar with most of the major continental landforms, but the forms on the floor of the oceans are rarely seen. Both will be described in more detail, as will certain unique topographic features found at the junctions between continents and ocean basins.

Continents Secondary features found on most continents are plains, plateaus, and mountains. Smaller, more irregular landforms also abound, but will not be considered here.

Plains are broad areas of low relief that occur generally at relatively low elevations. In the United States one could point to the Eastern Seaboard from New York southward and most of the drainage basin of the Mississippi River as prime examples. ***Plateaus*** resemble plains in being broad, relatively flat or gently rolling regions, but they occur at relatively high elevations. Commonly, streams and rivers will have

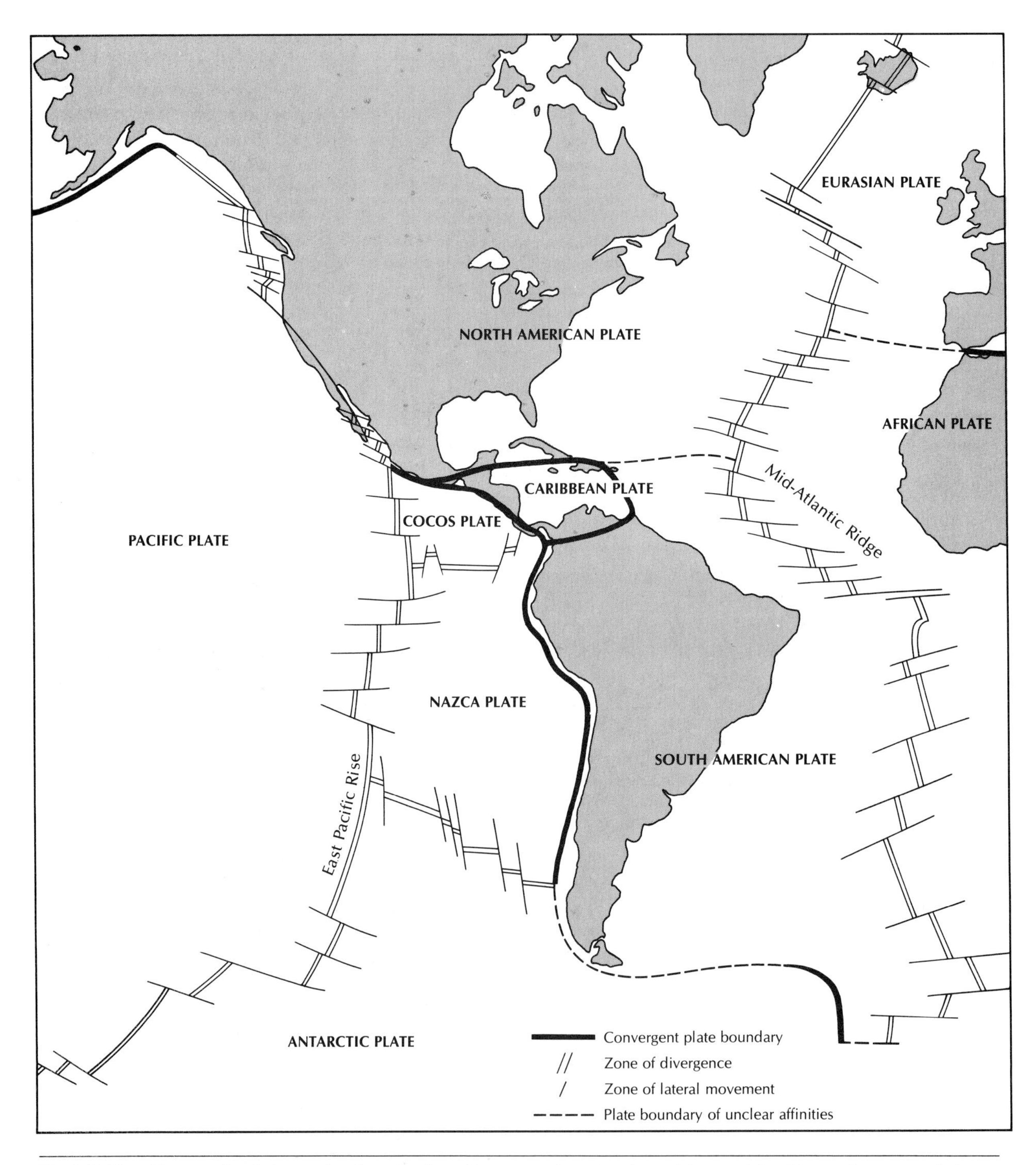

Fig. 3-18 The Western Hemisphere, showing the formally named rigid plates, bounded by zones of divergence, lateral movement and convergence. According to the plate-tectonic model, essentially all internally generated phenomena are brought about by plate-margin interactions.

dissected a plateau region to some degree (Fig. 3-19). The austere, but breathtakingly beautiful Colorado Plateau of the southwestern United States is a fine example of a plateau region. Incision by streams and rivers (principally the Colorado River and its tributaries) has produced color-banded, steep-walled canyons close to 2 km (1 mi) in depth. ***Mountains*** represent the highest parts of the continental masses. They can consist of single protuberances, but commonly they are found in linear groups of mountain blocks and peaks, called chains or ranges (Fig. 3-20). The highest peaks in the world are found in the Himalayas, where ranges rise in immense waves of sculptured stone. Swept by avalanches and riven by glaciers, they present supreme challenges to the best climbers in the world. In that mountain range, Mount Everest, at 8848 m (29,021 ft), stands as the highest topographic feature in the world today. The highest mountain in the United States, excluding Hawaii and Alaska, is Mount Whitney, in the Sierra Nevada Range in California—a peak only about one-half as high (4418 m; 14,491 ft) as Everest.

Fig. 3-19 Two stages in the dissection of a plateau. In the upper diagram, the canyons are narrow and the upland surface is dominant; in the lower diagram, the canyons have deepened and widened to the point where the upland surface is nearly obliterated. Adapted, by permission, from Fig. 32.13 in *Physical Geography* by Arthur N. Strahler. Copyright © 1969 by John Wiley & Sons, Inc.

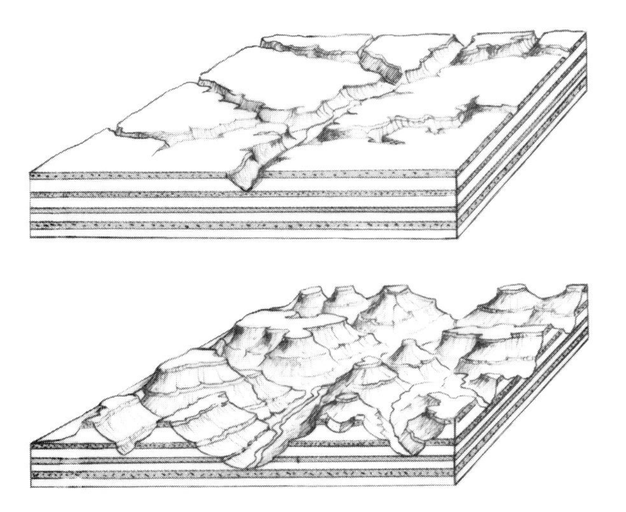

Ocean basins The common, second-order topographic features in the ocean basins include abyssal plains and hills, submarine plateaus, oceanic ridges and rises, fracture zones and cracks, and volcanic cones.

Abyssal plains are relatively low-relief surfaces at depths of 4000 to 5000 m (13,120 to 16,400 ft) (Figs. 3-21 and 3-22). They are the flattest extensive surfaces on earth—a result of a covering of sedimentary rocks that blanket a somewhat irregular surface below. The blanket of sediments is the result of a continued slow accumulation of sedimentary debris eroded from the continents and carried out to sea by oceanic currents.

In many of the deeper parts of the ocean basins there exist topographic protuberances, termed ***abyssal hills,*** which extend from 30 to 1000 m (98 to 3280 ft) above the sea floor. Although the origin of abyssal hills is not altogether certain, oceanographers generally consider them to be the result of volcanism.

Submarine plateaus, in contrast, are relatively low-relief features that stand far above the deeper parts of the ocean basin. One such topographic feature is the Blake Plateau, which lies in the Atlantic just east of Florida. The Melanesian Plateau, to the east of Australia, is one of the largest submarine plateaus on earth.

Ridges and ***rises*** are the elongate sub-oceanic mountain chains found throughout the ocean basins of the world (Fig. 3-23). In general, a rise is a more rounded, less sharply defined topographic feature than is a ridge. In the Pacific, the ridge and rise system (called the East Pacific Rise), runs nearly north-south, hugs the eastern margin of the ocean basin off South and Central America, and appears to run aground and disappear just off the tip of Baja California; in the Atlantic, the Mid-Atlantic Ridge runs down the center of the ocean basin and extends from the Arctic nearly to Antarctica; in the Indian Ocean, the Mid-Indian Ridge begins near the Red Sea on the north and extends southward for about 5000 km (3107 mi), where it splits to become the Southwest and Southeast Indian ridges. In a few places, such as Iceland, a ridge actually protrudes above the ocean's surface.

Fig. 3-20 Mount Tyree, Sentinel Range, Antarctica. Looking eastward across the range to the Edith Ronne Shelf. *USGS*

The cross section of the ridge is unlike that of continental mountains. In a traverse across a ridge, the ocean bottom does not rise continuously to the axis, but does so rather in a series of steps (Fig. 3-24). At the crest of some ridges and rises is a keystone-like depression called the ***median valley*** (Fig. 3-25). It appears that the ridge system is an elongate welt along which the sea floor has been warped upward and in the process has been stretched and cracked parallel to the elongation.

Fracture zones cut across the axes of the oceanic chains and extend laterally outward into the sea basins on either side (Fig. 3-25). In places, entire segments of the ocean floor appear to have been moved laterally along the fracture zones, slicing the ocean range into blocks sharply offset from one another.

Volcanic mounds and conical hills rise from the ocean floor either singly or in clusters; some are aligned along fracture zones (Fig. 3-26). From their shape and magnetic properties, and the nature of the samples dredged from some of them, they appear to be of volcanic origin. In a few cases, volcanic eruptions have been so voluminous that cones have been built above sea level. The most magnificent set of such volcanic islands is found in the Pacific and includes the Hawaiian Islands. Once the islands rise above the sea, the erosive work of the pounding surf can begin. If volcanism cannot keep ahead of erosion, the island will be

Fig. 3-21 Sub-bottom profile of the abyssal plain off the coast of Nova Scotia (see Fig. 3-22 for location). Note how flat the sedimentary layers are beneath the plain.

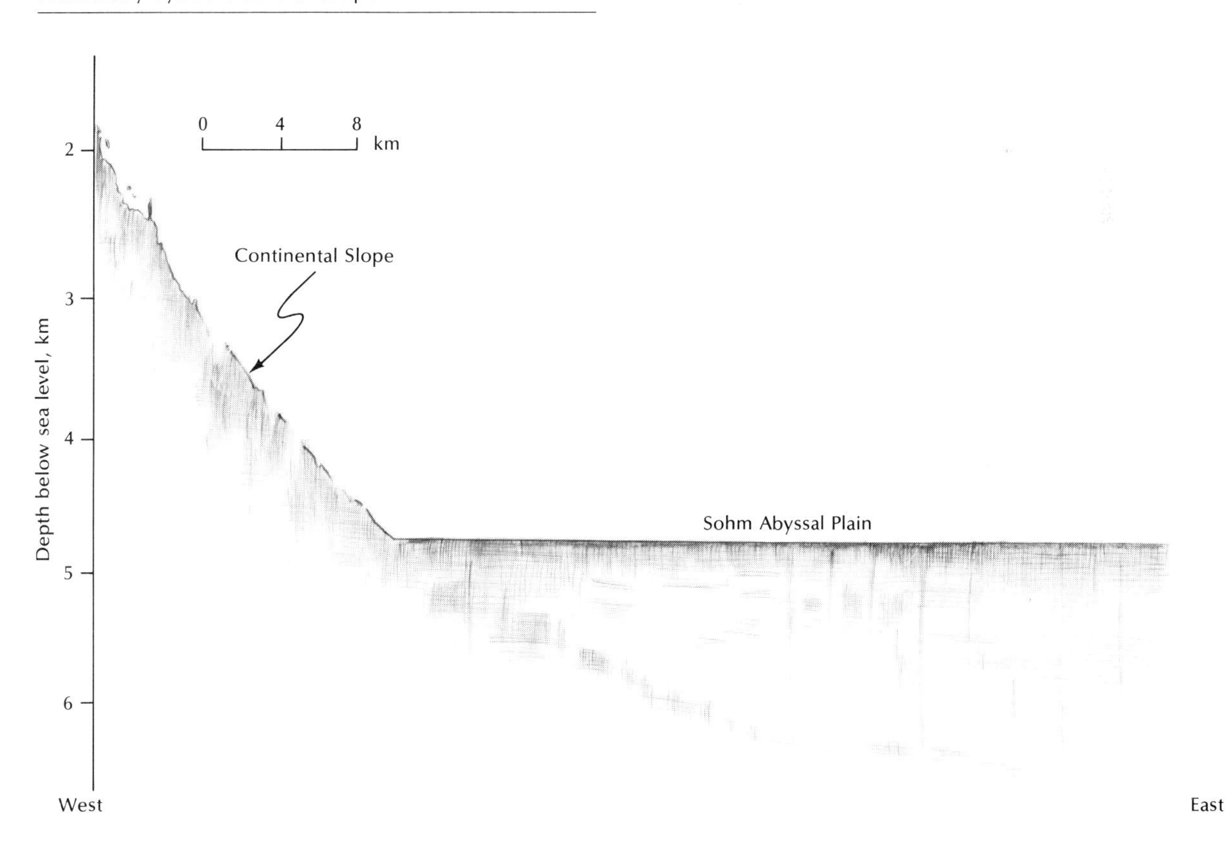

Fig. 3-22 Topography of the Atlantic Ocean floor, off the east coast of North America, including the Blake Submarine Plateau and an abyssal plain.

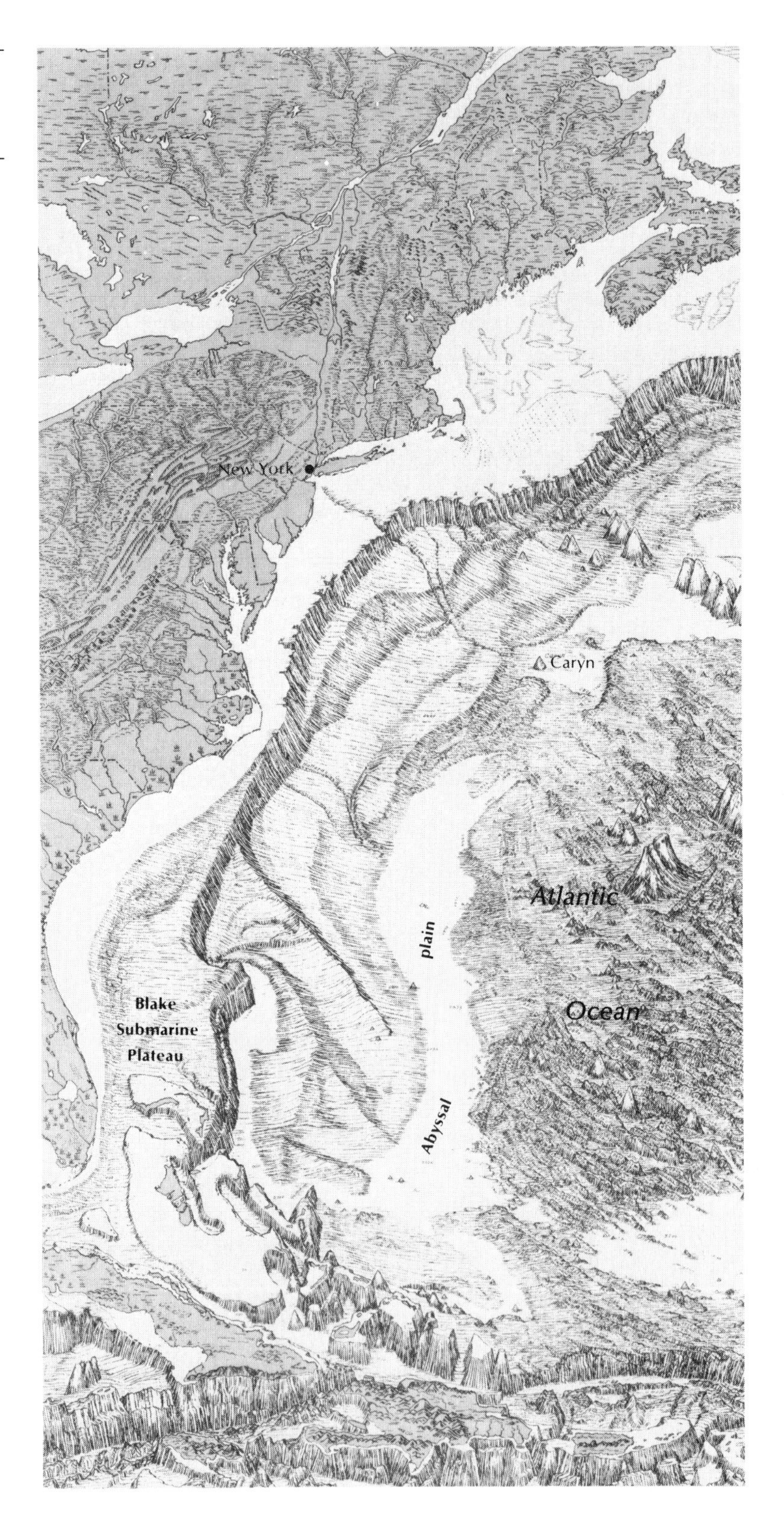

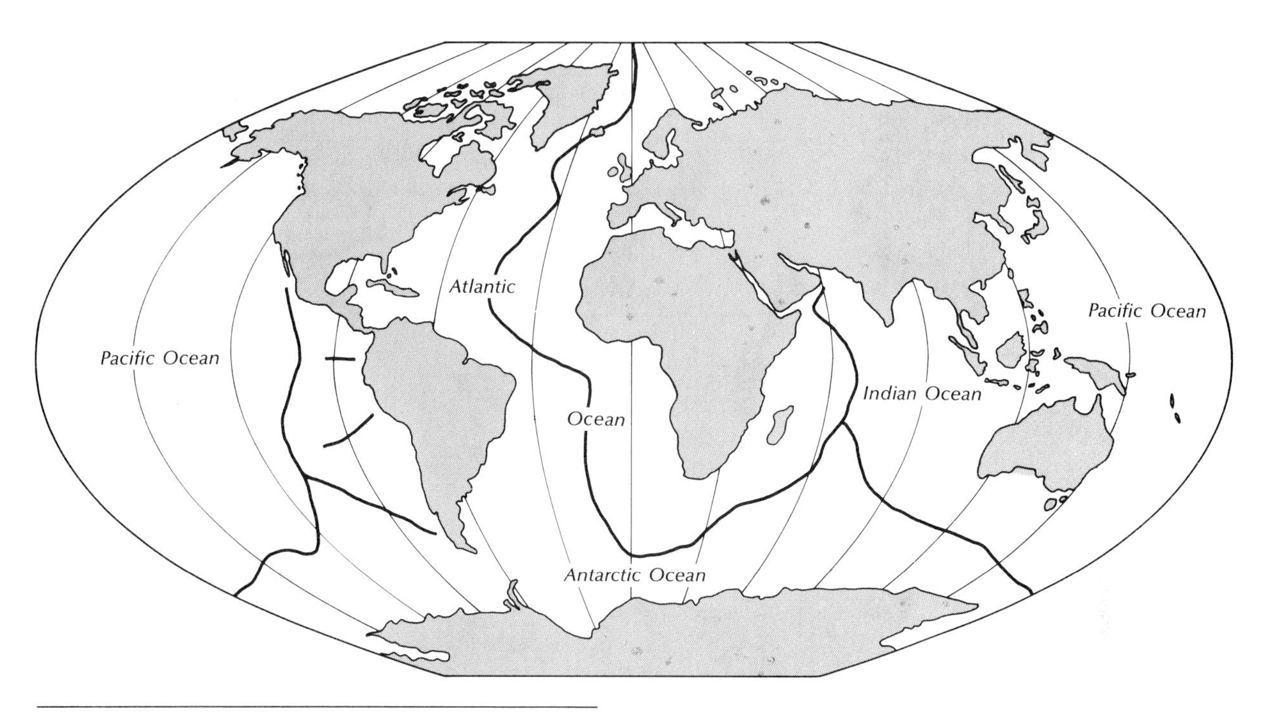

Fig. 3-23 General location (black line) of the oceanic ridge and rise system.

Fig. 3-24 Cross section of a typical oceanic ridge system, showing a general increase in elevation toward the crest and a relatively depressed crestal region (the median valley). Step-like changes in topography appear to be related to movement along faults that parallel the ridge axis.

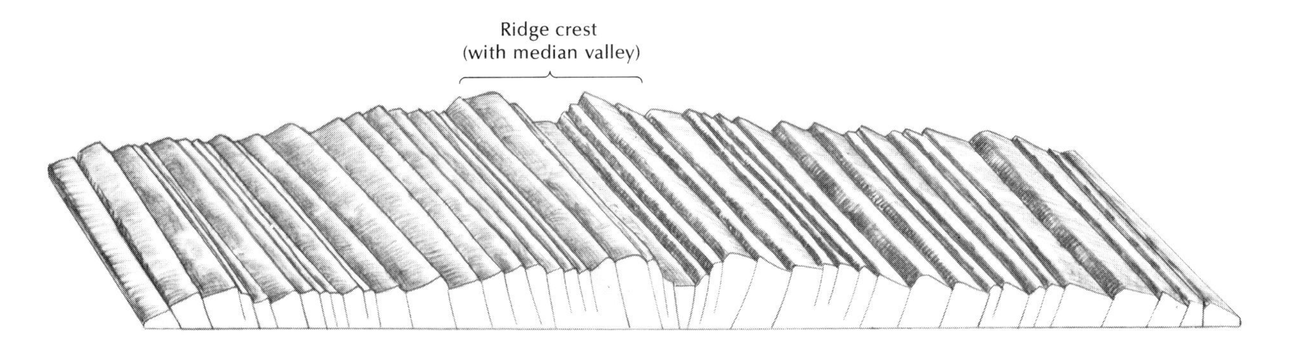

Fig. 3-25 The Mid-Atlantic Ridge in the North Atlantic, showing the fracture zones that transect and offset the ridge and the elongate median valley.

Fig. 3-26 The northwest Pacific Ocean basin, showing a multitude of volcanic cones, many of which project above sea level, and the narrow, sinuous deep-sea trenches along the ocean margin.

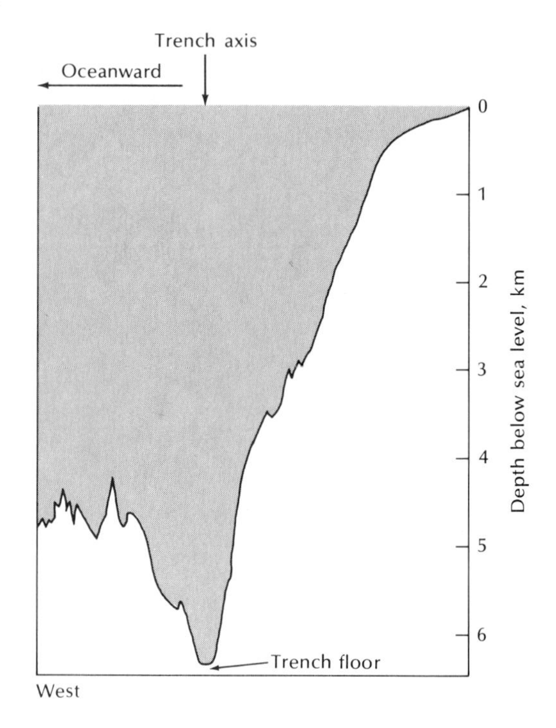

Fig. 3-27 Profile of the Peru-Chile Trench off the coast of Peru. The trench is nearly devoid of sediment.

planed down to sea level. Some of these planed-off cones are found far below sea level, attesting to relatively great changes in the level of the oceans.

Sea-margin features

Zones where the continents meet the ocean basins are generally characterized by island arcs and deep-sea trenches on the one hand or by continental shelves and slopes on the other.

Island arcs, which are linear archipelagos of volcanic islands (such as the Aleutians and the islands of Japan), and the elongate, narrow, and deep ocean-floor trenches that border them (Fig. 3-26) are common along some continental margins, particularly that of the Pacific Ocean. The deepest places in the oceans—at depths up to 10,700 m (35,096 ft)—are found in the trenches. Island arc-and-trench systems have a high level of earthquake activity. The deepest earthquakes—at depths up to 700 km (435 mi) beneath the surface—occur below the arcs.

Some continental margins, although volcanically active, deviate somewhat from this common pattern. For example, the west coast of South America, which is bordered by the Peru-Chile Trench, lacks an archipelago (Fig. 3-27). The active volcanic mountains (the Andes) are on the South American continent. The west coast of the United States is also unusual in that a volcanic chain—the Cascades—parallels the coast, but no ocean trench lies on the margin of that chain.

Continental shelves, relatively wide and shallowly submerged platforms on the margins of continents, are bordered seaward by a sharp increase in gradient, called the ***continental slope.*** For all practical purposes, continental shelves can be considered part of the continent, since they consist of continental-type rocks and are commonly a seaward extension of the coastal plain. Extensive continental shelves are typical of many coastlines along the Atlantic Ocean and are particularly well developed along the southern coast of South America (Fig. 3-28).

Fig. 3-28 Topography of southern South America and part of the South Atlantic basin. The nearly flat, gently sloping continental shelf extends along the entire eastern margin of South America and is separated from the deep basin by a relatively abrupt break in slope, the continental slope. At a slightly lower sea level, South America would be appreciably wider.

SUMMARY

1. The earth is an inner planet, held in its orbit around the sun by *gravitational attraction*. The law of gravitation pertains to all bodies in the universe.
2. The slightly flattened ellipsoidal shape is the result of the inward pull of the earth on all its particles of matter and of the rotation of the earth on its axis.
3. The average density of the earth is 5.52 gm/cm^3, about twice that of its surface rocks. The obvious inference is that some very dense material must be present below the earth's surface.
4. Several lines of evidence indicate that continents are of relatively low density and float *isostatically* on the higher-density mantle rocks. When loading or unloading of the earth's surface occurs, the crustal blocks will readjust until they are again in floating equilibrium.
5. The atmosphere is principally nitrogen (78 per cent) and oxygen (21 per cent). It also contains small amounts of other gases, many of which are important. The latter include carbon dioxide and water vapor. It appears that the composition of the atmosphere during the early stages of the earth's history was different from what it is today. Circulation of the atmosphere is driven by energy from the sun.
6. The earth's oceans cover 71 per cent of its surface area. The waters of the oceans probably came from inside the earth and have been accumulating over the entire span of the earth's history. Most salts in sea water are derived from the land and brought to the sea by rivers. Surface currents are principally wind driven, whereas subsurface flow is the result of differences in density between water masses.
7. The earth is layered. The outer layer, of relatively light silicate rock, extends down to about 6 km (4 mi) beneath the ocean basins and up to about 70 km (44 mi) under the continents. The next layer, which is largely silicon and oxygen with lesser amounts of iron, magnesium, aluminum, calcium, sodium, and potassium in a solid state, extends down to 2900 km (1802 mi). The core, the innermost zone, is mostly iron and is partly liquid and partly solid.
8. *Volcanism, mountain building, earthquake activity,* and *continental drift* rebuild the earth's surface. They are generated internally by processes that occur to a depth of 700 km (435 mi).

   Other processes, such as *weathering, erosion,* and *sedimentation,* act to level the surface and are powered by the sun's energy. The face of the earth has changed continuously through time in response to the interaction of *constructional* and *destructional* processes.
9. *Plate tectonics* forms the basis for understanding and interrelating the internally generated phenomena. The outer shell of the earth appears to be composed of a number of rigid plates that move: pushing together, diverging, or sliding laterally. Most of the constructional phenomena are developed at the plate margins; little within the plates.
10. The first-order topographic features—the *continents* and *ocean basins*—themselves consist of different kinds of secondary (second-order) topographic features. On the continents there are *plains, plateaus, and mountains;* in the ocean basins there are *abyssal plains, submarine plateaus, oceanic ridges* and *rises, fracture zones* and *cracks,* and *volcanic cones.* At the sea margins there are *island arcs* and *deep-sea trenches,* and *continental shelves* and *slopes.*

QUESTIONS

1. What are the principal physical features of the earth?
2. Venus, a planet similar to the earth in many ways, rotates very slowly on its axis. Would

you expect Venus to be ellipsoidal in shape?
3. Discuss each of the following terms and describe the relationships that might exist between them: reference ellipsoid, geoid, vertical, horizontal.
4. What is isostasy and what evidence is there that it exists?
5. What are the main currents in the oceans and atmosphere and what drives them?
6. How did the layering of the earth occur, and how does it differ from that of other planets and our moon?
7. What is the model or theory of plate tectonics?
8. Describe the principal topographic features of the continents and the oceans.

SELECTED REFERENCES

Bates, D. R., and others, 1957, The earth and its atmosphere, Basic Books, New York.

Beiser, A., and the Editors of Life, 1962, The earth, Time Inc., New York.

Buswell, A. M., and Rodebush, W. H., 1956, Water, Scientific American, April.

Cailleux, A., 1968, Anatomy of the earth, World University Library, McGraw-Hill Book Co., New York.

Donn, W. L., 1972, The earth: Our physical environment, John Wiley and Sons, New York.

Editors of the Scientific American, 1950–1957, The planet earth, Simon and Schuster, New York.

Gamow, G., 1958, Earth, matter, and sky, Prentice-Hall, Englewood Cliffs, New Jersey.

Gamow, G., 1960, Gravity, Anchor Books, Doubleday and Co., Garden City, New York.

Hamilton, H. C., 1854, The geography of Strabo, Henry G. Bohn, London.

Heezen, B. C., and Hollister, C. D., 1971, The face of the deep, Oxford University Press, New York.

Heiskanen, W. A., and Meinesz, V., 1958, The earth and its gravity field, McGraw-Hill Book Co., New York.

King-Hele, D., 1967, The shape of the earth, Scientific American, vol. 217, no. 4, pp. 67–76.

Krauskopf, K. B., and Beiser, A., 1973, The physical universe, 3rd ed., McGraw-Hill Book Co., New York.

Sverdrup, H. U., Johnson, M. W., and Fleming, R. H., 1942, The oceans, Prentice-Hall, New York.

Urey, H. C., 1952, The origin of the earth, Scientific American, vol. 187, no. 4, pp. 53–60.

Wilson, J. T., compiler, 1970, Continents adrift: readings from Scientific American, W. H. Freeman and Co., San Francisco.

4

MATTER, MINERALS, AND ROCKS

Philosophers, alchemists, and chemists have considered the composition of matter down through the ages. To many of the ancient Greeks, matter was known simply to be the substance of which any object is composed. At that time curious persons had no means, other than their own senses, for investigating the physical world, and it was only natural that knowledge about that domain was either nonexistent or rudimentary. By about 400 B.C., however, the philosopher Democritus concluded, on the basis of logic, that matter must be made up of small indivisible bits, which he called ***atoms,*** all of which are similar and eternal. About 50 years after that first pronouncement of the atomic theory, the highly regarded philosopher Aristotle (384–322 B.C.) proposed that the principal elements of terrestrial matter were fire, water, air, and earth, whereas the only element of the heavens was "quintessence." Earth was considered the essence that gave things the property of solidity.

As alchemists strived to attain the elusive goal of turning base metals into gold, it became apparent that the physical world was not as simple as Artistotle supposed (Fig. 4-2). Fire, water, and air were still thought to be elemental, but earth came to be recognized as a mixture of many things.

The work of other chemists experimenting with gases, as well as his own efforts, enabled French chemist Antoine Lavoisier (1743–94), during the late eighteenth century, to correctly identify 23 elements, which he defined as pure substances, and to indicate that other elements probably existed. He also stated that a chemical compound is a pure substance that consists of two or more elements in combination.

Today, there are 106 known elements, of which 92 are found naturally at the earth's surface. Spectral analysis of the sun shows that most of these naturally occurring elements exist there, although not in the same proportions.

In the early 1800s, the English chemist and physicist John Dalton (1766–1844) drew on the atomic theory of the Greeks when he hypothesized that all matter is composed of tiny invisible atoms of different weights and chemical properties. An element, he concluded, is made

Fig. 4-1 Rhombohedral crystal form of two intergrown dolomite crystals. Dolomite is a carbonate mineral closely related to calcite. *M. Halberstadt*

Fig. 4-2 A sixteenth-century rendering of the mysterious powers of an alchemist. *Prints Division, New York Public Library*

of only one kind of atom. He even went so far as to say that when two or more elements combine, their atoms form identical groups of atoms called ***molecules.***

The next milestone in the quest for the fundamental building blocks of matter came mostly from experiments carried out in the late 1800s and early 1900s, particularly with radioactive substances. Almost all atoms were found to be composed of three kinds of particles: negatively charged ***electrons,*** positively charged ***protons,*** and ***neutrons,*** which carry no charge whatever. In any atom there are as many negative particles as there are positive ones, and the electrical charge is balanced. In atoms of a particular element, the number of protons equals that of electrons and is always a fixed number (the ***atomic number***). But the number of neutrons can vary slightly from atom to atom of that element. The atoms of an element that differ only in the number of neutrons (and therefore in ***atomic weight***) are termed ***isotopes*** of that element—such as uranium-235 and uranium-238.

It was also found that electrons, protons, and neutrons are roughly the same size—about 10^{-12} cm (one-millionth of one-millionth of a centimeter) in diameter. An electron has much less mass than the other two particles; it is only 1/1836 and 1/1837 as massive as a proton and a neutron, respectively. From its mass and neutral charge it is assumed that a neutron is essentially a proton and electron, combined.

An ingenious experiment by the English physicist Lord Rutherford, in the early twentieth century, demonstrated that in any single atom all the protons and the neutrons exist in a very dense central kernel, or ***nucleus,*** that contains more than 99.9 per cent of the mass, but represents only about one-billionth of the volume of the atom. The nucleus is the center of an essentially spherical body with a diameter of about 10^{-8} cm (one-hundredth of one-millionth of a centimeter) in which one to many electrons—the number depending upon the particular element—move rapidly in continual agitation, now closer to the nucleus, now further away (Fig. 4-3). Atoms that contain many electrons resemble, to a degree, a swarm of gnats swirling around one's head. Moreover, when many electrons are present, certain ones are constrained to a path close to the nucleus, whereas others tend to circle near the outer margin of the electron "cloud." The outer electrons are the electrons that enter into chemical reactions, as when two elements combine to form a compound.

In their pursuit of the composition of matter, nuclear scientists have cracked the nucleus like a walnut to reveal many new particles, such as pions, mu mesons, and positrons, each particle with unique properties.

Because atoms are so tiny it requires large numbers of them to be visible to the eye. A one-half pound lead brick, for example, contains nearly 1×10^{24} atoms (1 followed by 24 zeroes). Another way of demonstrating the large numbers of atoms (and molecules) present in a small volume of matter is as follows: If you pour two quarts of water into the ocean, and "stir well," any cup of water subsequently taken from the ocean will contain *one* molecule from the two quarts you poured in.

Atoms normally can combine, or ***bond,*** with other elements. In some cases, the atoms that bond together are all the same element, as in the metal gold and the gas hydrogen. In other cases, however, atoms of two or more elements bond, in a definite proportion, to form a ***chemical compound.***

One atom of chlorine, a toxic gas, can bond with one atom of sodium, a highly reactive metal, to form sodium chloride (table salt or halite; Fig. 4-4), a non-toxic compound, except in large amounts. Alternatively, two atoms of chlorine will bond with one atom of calcium to form the compound calcium chloride, a salt used as a drying and dehumidifying agent.

BONDING

Two atoms bond through the interaction of one or more electrons from the outer parts of their electron clouds. Several types of bonding

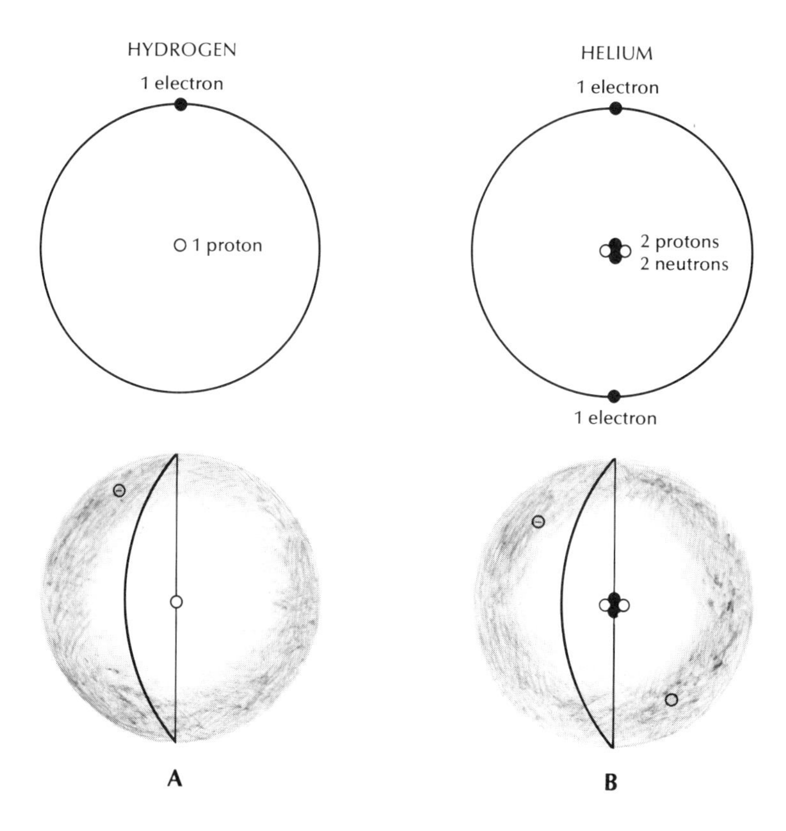

Fig. 4-3 A. The hydrogen atom contains only one electron and one proton. **B.** The helium atom contains two electrons, two protons, and two neutrons.

are possible, and the particular type that the atoms of any one element can undergo is an intrinsic property of the element, determined by the number of electrons in the atom.

In one type of bonding, there is an actual transfer of one or more electrons from the electron cloud of one atom to that of another—one atom loses electrons and the other gains them. The resulting positively and negatively charged particles are called ***ions*** (Fig. 4-5). The ion carrying a positive charge is called a ***cation,*** the ion carrying a negative charge an ***anion.*** Usually the positively charged cations are formed by the loss of one or more electrons from atoms of the metallic elements, such as calcium, sodium, or potassium. ***Ionic bonds*** result from the electrostatic attraction between positively and negatively charged particles and are normally quite strong.

In another type of bonding, the electron, rather than transferring completely from the donor atom to the acceptor, remains halfway between the atoms and is shared by both. This is called ***covalent bonding,*** and it too is quite strong. In such a way, two atoms of hydrogen will join to one atom of oxygen to form a water molecule (Fig. 4-6). In most cases of covalent

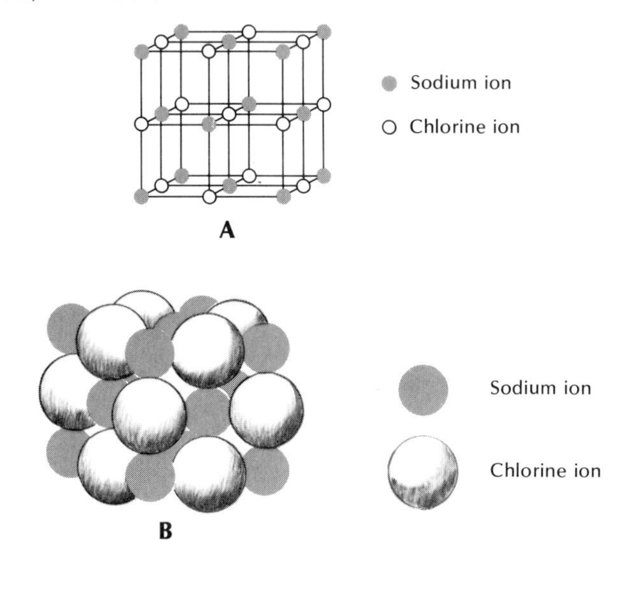

Fig. 4-4 The structural arrangements (**A** and **B**) of sodium and chlorine ions in the mineral halite. **A.** An exploded view showing the relative positions of the ions. **B.** This more closely represents the way in which the ions are actually stacked together. **C.** The external cube shape of the halite crystal consists of billions of sodium and chlorine ions and reflects its internal ionic arrangement.

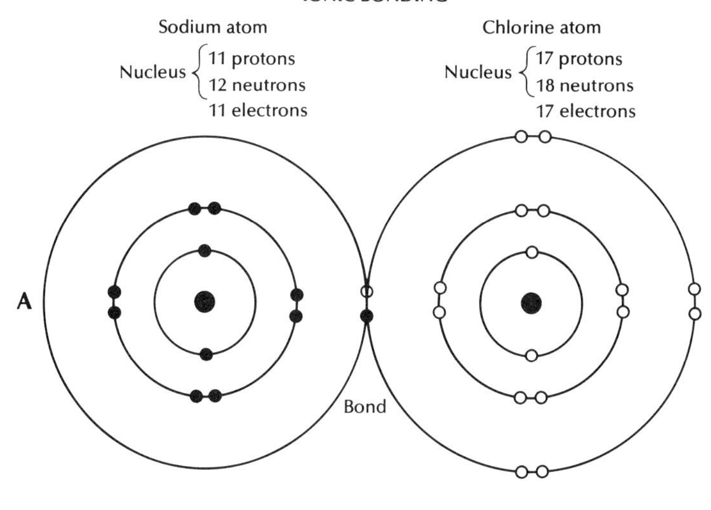

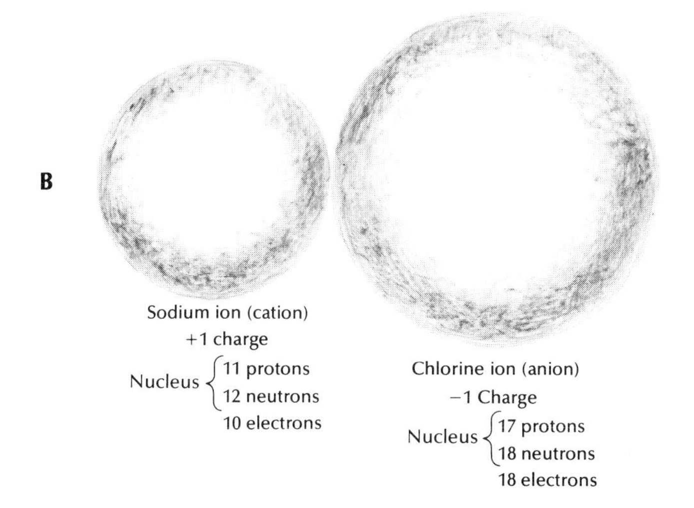

Fig. 4-5 A sodium atom has one loosely bound electron in its outer shell, whereas a chlorine atom has seven electrons, lacking only one from being completely full. When the two atoms come together, the outer sodium electron moves to the shell of the chlorine atom, producing a positively charged sodium ion (cation) and a negatively charged chlorine ion (anion). Electrostatic attraction holds the two ions together, and an ionic bond is formed. The relatively stronger attraction of the sodium nucleus pulls the remaining electrons in the cation toward the center, resulting in a decrease in size.

bonding, the electron does not remain equidistant from the two atoms. At times it will move into the electron cloud of one atom, and at times into the other; when it does so the bond will become, momentarily, ionic in type. The atoms in most solid geological substances are held together by such alternating ***ionic-covalent*** bonding.

A third type of atomic bond, less common than the previous two, is the ***metallic bond.*** As you might expect from the name, such bonding is found principally among the metals in their uncombined state. In a metal, all the atoms are exactly the same size, and they are packed around each other like marbles. The tight packing of the equally sized spheres results in each atom being surrounded by twelve other atoms (Fig. 4-7), so the electron cloud of

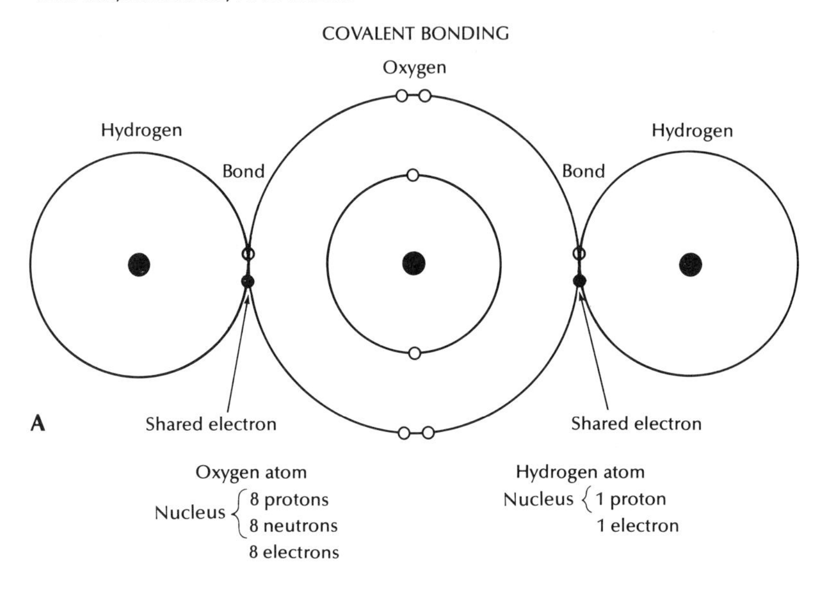

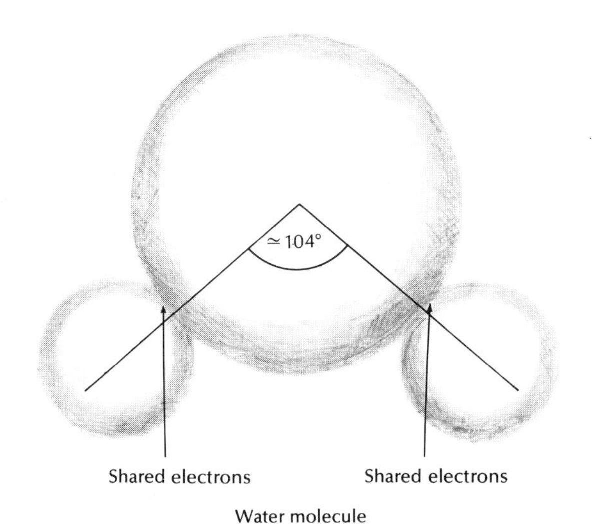

Fig. 4-6 Covalent bonding of two hydrogen atoms with one oxygen atom (**A**) to form a water molecule (**B**). Electrons from the three atoms are shared to form very strong bonds.

each atom must react simultaneously with those of its twelve close neighbors. The bonding tends to be covalent wherever the spheres touch each other. However, because more than enough electrons are available from all the atoms to complete bonding, those not in use at any moment tend to drift through the metal. An electron forming a covalent bond may be at any moment replaced by a drifting electron. The replaced electron is then free to move on to another atom. Because of the musical chairs played by the atoms, metallic bonding has also been called ***time-shared covalent bonding.*** It is the abundance of moving, loosely held electrons that gives the metals their characteristic abilities to conduct heat and electricity and to reflect light.

Molecules, that is, integral groups of atoms,

can also be bonded together. Such bonding does not rely on the sharing or exchange of electrons in this case; instead it relies on relatively weak electrostatic forces between molecules. Accordingly, such bonds are also generally relatively weak. In many compounds, particularly those involving atoms of more than two different elements, several types of bonds, or levels of bonding, coexist.

Fig. 4-7 Close packing arrangement of atoms of the same size; each sphere is surrounded by twelve others. The bond between any two atoms is covalent; since, however, any one atom can share electrons with any of its twelve neighbors, extra electrons are available for bonding. As a result, the electrons continually drift, and bonding electrons come from different neighbors at different times. Such time-shared covalent bonding is typical of the metals and is called metallic bonding.

STATES OF MATTER AND THE STRUCTURE OF MINERALS

Compounds exist in any of three different states—solid, liquid, or gas—depending on the strength of bond between the various atoms and on the temperature and pressure. Water can exist at the earth's surface in all three states. As you know, when water is cooled sufficiently it freezes; if heated sufficiently it becomes a gas. It is apparent, then, that the principal difference between the three states is one of heat content, a measure of which is temperature. From experiments on gases in the 1800s it became clear that temperature was also a measure of the average translational speed of the gas molecules: the higher the temperature, the greater the speed. When you blow up a balloon, the molecules move in all directions, in a random, mixing fashion, hither and yon—order is totally lacking. At any one time, many molecules are striking the balloon wall, creating an outward pressure. If the balloon is heated the molecules become more agitated, striking the wall harder and more frequently. The pressure thereby increases, resulting in an expansion of the balloon.

When a gas such as nitrogen is cooled, the atoms (or molecules) slow down until they no longer are independent of one another. Then, ***attractive,*** or bonding, forces become strong enough to cause the atoms or molecules to begin to stick to one another, and the liquid state is attained. Although relatively slow moving, each atom in a liquid still tends to move independently of every other one. There are moments, however, when two or more atoms jostle each other—and bonds are formed momentarily. When, in the next moment, those atoms resume their aimless wandering, the bonds are ripped apart. Around a particular molecule (or molecules) or within a small volume of liquid, therefore, there will be from time to time (when bonds exist) an ordered assemblage of molecules. ***Short-range ordering,*** as that state is called, is characteristic of all liquids. Because the atoms or molecules continually change partners, a liquid lacks rigidity, that is, it tends to flow. As it flows, some portions move more rapidly than others, and thus additional interatomic bonds are broken. That results in a resistance to flow called ***viscosity.***

Continued cooling of a liquid progressively slows down the atoms or molecules. Eventually, the bonds between them can remain firm and the change from liquid to solid will occur. In the formation of a solid, the atoms fit together in the closest way possible for their size and shape and the type of bonding they undergo. That is, a solid is not a jumble of atoms or molecules; instead each atom or molecule occupies a particular, set position with regard to those surrounding it. An ordered latticework of atoms is created—a little like the repetitive pattern on wallpaper, but in three dimensions. A ***long-range order*** exists and the material is said to be crystalline or to possess ***crystallinity.*** With few exceptions, only one particular atomic ordering is possible for each compound.

The word ***crystalline*** comes from crystal, the

Fig. 4-8 Cluster of naturally occurring quartz crystals from Crystal Springs, Arkansas. These transparent crystals possess a characteristic external form bounded by a number of relatively flat surfaces. *Smithsonian Institution*

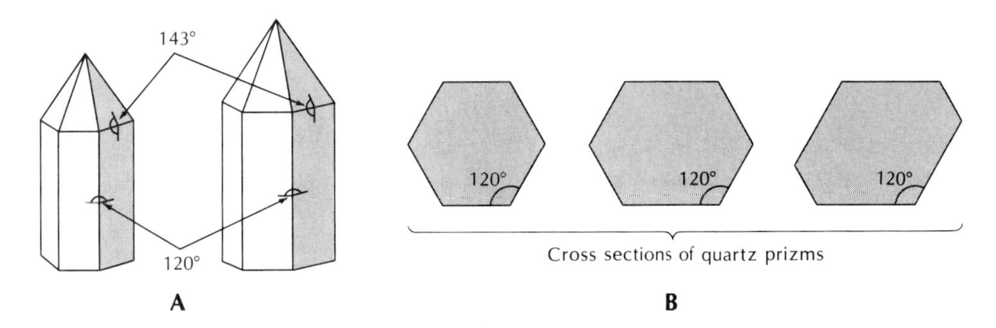

Fig.4-9 Commonly occurring external forms of quartz crystals: the six-sided prism (straight-sided form) and pyramid (pointed form). The angles between two adjacent prism faces is always 120°, regardless of the difference in size between crystals (**A**) or apparent distortions in shape (**B**). The fact that crystals of a particular mineral have similar external forms led early mineralogists to suspect an orderly atomic arrangement.

Anglicized Greek word for ice. Crystalline compounds and elements that occur naturally are called ***minerals.*** In fact, it was the study of minerals that eventually led to our knowledge of long-range ordering, or crystallinity, found in many compounds in the solid state. Today ***mineralogists*** (those who study minerals) have described and named about 2500 minerals. In the early days of history, however, only a few were known, generally those that were showy in appearance or that had been found useful to humans. For example, the common mineral quartz (Fig. 4-8), which often occurs in clusters of shiny, ice-like crystal spires, was well known in the Middle Ages, and certain ore minerals that yield iron, copper, lead, and silver have been recognized for thousands of years. Minerals were used by our early ancestors just as they were found. Clay was molded into bricks and made into poetry; flint and jade were fashioned into weapons and jewelry; the oxides of manganese and iron were made into pigments and paints; turquoise and amethyst were set into jewelry and other ornamentation; and gold, silver, and copper were used for ornaments or utensils. In time it was found that usable metals could also be extracted from certain minerals by heating or smelting.

By the eighteenth century the identification and understanding of minerals began to increase rapidly, largely because chemists, who were trying to understand the composition of matter, commonly used naturally occurring minerals in their experiments. Eventually they began to understand the chemical properties of the materials they worked with, but neither chemists nor mineralogists were able to explain why many minerals were bounded by planar crystal faces.

By 1669 Nicolaus Steno, the Danish physician discussed in Chapter 1, had demonstrated that the faces on a quartz crystal always meet at the same angle, regardless of the crystal size and shape (Fig. 4-9). Some time later, it was shown that the surface structure of crystals is always the same in any given mineral species. As the field of crystallography grew, the crystal forms of many kinds of minerals were studied. And, in spite of the large number studied, it became apparent that all the many crystal forms of all the minerals could be classified into just seven basic groups or systems. That seemed to speak of an underlying order, but no one knew what sort of order.

As early as 1611 Johannes Kepler (1571–1630), the famous German astronomer, aware that snowflakes were always hexagonally symmetric (Fig. 4-10), conjectured that their regularity in

Fig. 4-10 Naturally occurring ice crystals (snowflakes) possess a six-sided symmetry—as do all ice crystals—suggesting a regular internal ordering. *Moody Institute of Science*

form was probably due to the geometrical arrangement of their minute building blocks. Later, in 1784 Réné Hauy (1743–1822), a French crystallographer, concluded that any mineral exhibiting an observable crystal form must be made up of small polyhedral units, each of which must have the same symmetry as the whole crystal (Fig. 4-11).

By the early 1880s, following the announcement and acceptance of Dalton's atomic theory, mineralogists had concluded that minerals were chemical compounds with definite compositions, and that each compound must consist of a large number of atoms uniquely fitted together in a regular three-dimensional pattern. Crystallographers had finally come to realize that the planar faces so characteristic of each particular mineral must be the outward reflection of the internal ordering—but they had no way of proving it for some time to come.

The proof, which eluded scientists until 1912, came about (as is so often the case) as the result of an extraordinarily fortunate and essentially intuitive experiment performed by German physicist Max von Laue (1879–1960) and his associates. The X-ray had been discovered by William Roentgen in 1895, but little was known about it up to 1912. Von Laue and his companions conjectured that X-rays, like light rays, might be wave-like in character, but of exceedingly short wavelengths. But in order to demonstrate those wave properties they needed a better analytical device than the diffraction grating available at the time. A standard ***diffraction grating*** consists of a glass plate engraved with fine, closely spaced parallel lines. When light passes through the plate the grooves interfere with, or diffract, the light waves in such a way that they are dispersed into a spectrum of colors, each color corresponding to light of a different wavelength. Finally, von Laue hit upon the idea that the tiny atoms in a crystalline substance might be systematically arranged in layers spaced closely enough together to serve as a diffraction grating.

After the usual false starts and mishaps, he was successful in sending an X-ray beam through a crystal of copper sulfate and onto a

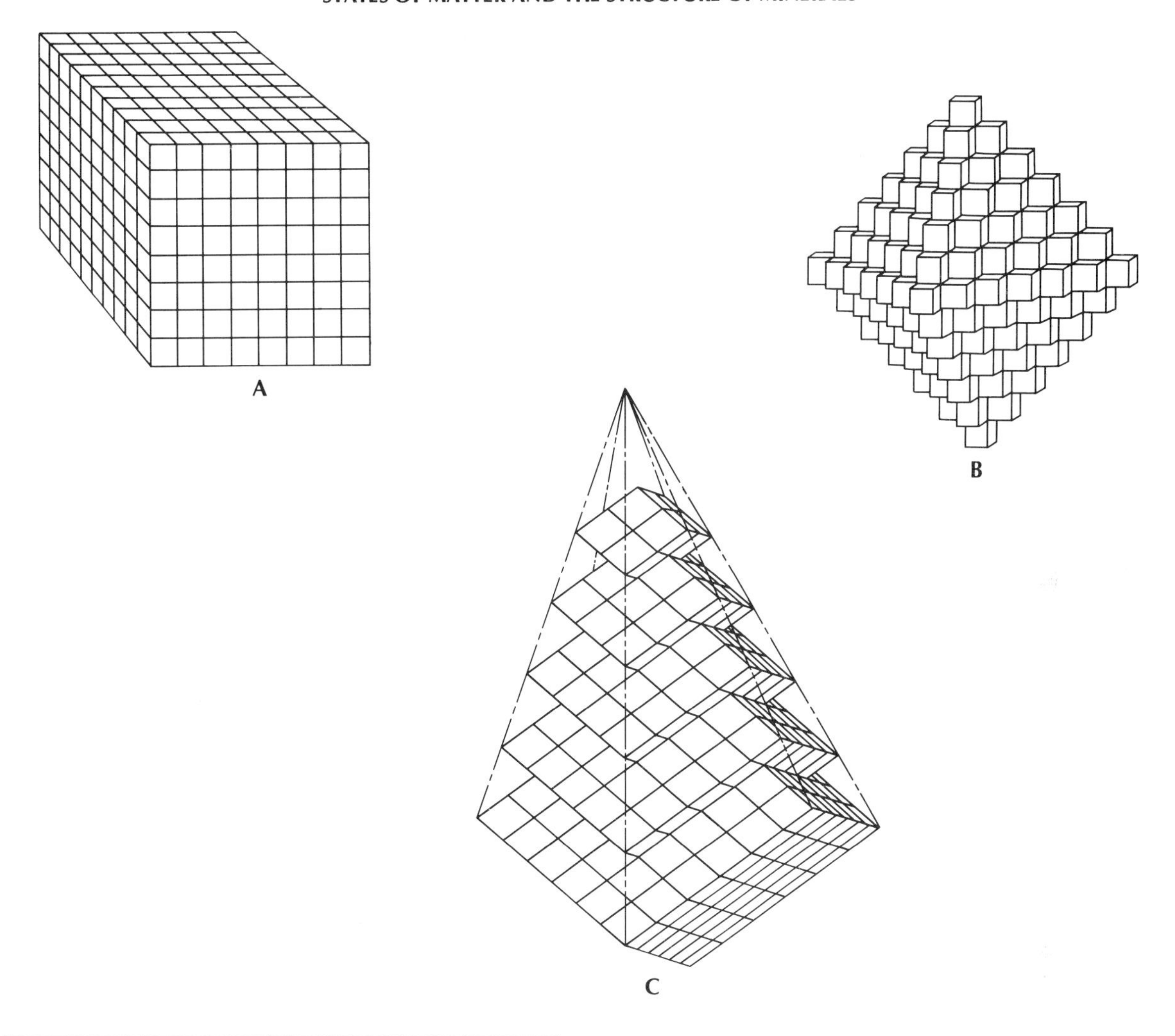

Fig. 4-11 Different crystal forms are made up of many small units having the same shape as the crystal itself. **A.** Cube; **B.** octahedron; **C.** scalenohedron.

photographic plate behind the crystal (Fig. 4-12A). When the plate was developed there appeared upon it a pattern of dots that von Laue interpreted as evidence of X-rays, which had been reflected from the electrons in the outer portion of the electron clouds of the crystal's regularly arranged atoms (Fig. 4-12B). Besides showing beyond a doubt that X-rays are wavelike in nature, the experiment revealed that the planes of atoms in a crystal are regularly spaced. X-rays had proved to be the key to the door of an unseen, ordered world in miniature, its very existence previously only inferred from surface measurements of interfacial angles and the regular geometry of crystal faces.

Later experiments with X-rays demonstrated that all minerals, whether they exhibit crystal faces or not, have a crystalline, long-range order. After von Laue's experimental breakthrough, many years and a great deal of effort were required before the "von Laue spots" for any particular mineral could be used to elucidate a crystalline structure. Even the simplest atomic lattices requires elaborate calculations and the application of complex theories of wave motion. Today, automatic, high-precision X-ray equipment and high-speed computers have made accurate crystal determination almost routine. In modern biochemistry and molecular biology, X-ray crystallography is having perhaps its finest hour. The complex structure of DNA (deoxyribonucleic acid—a molecule in the nucleus of a cell that carries the information necessary for exact cell replication) has finally yielded to X-ray analysis, and, one by one, the structures of the various protein molecules are coming to be understood. The solution of the atomic structure of minerals is like child's play when compared to the structural analysis of the simpler organic compounds, such as hemoglobin (a molecule that transports oxygen in the blood of many animals), which contains nearly 10,000 atoms in one molecule.

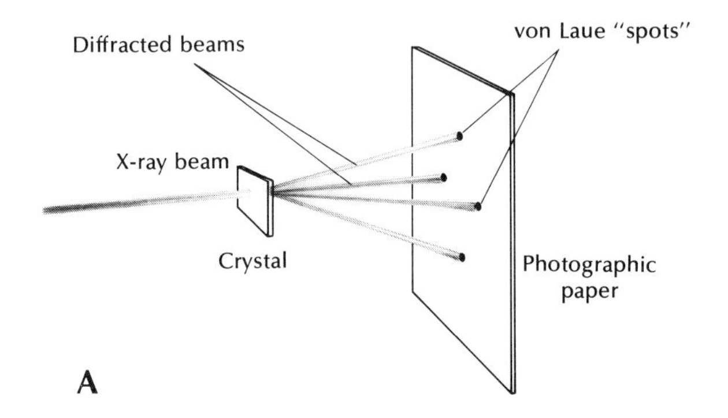

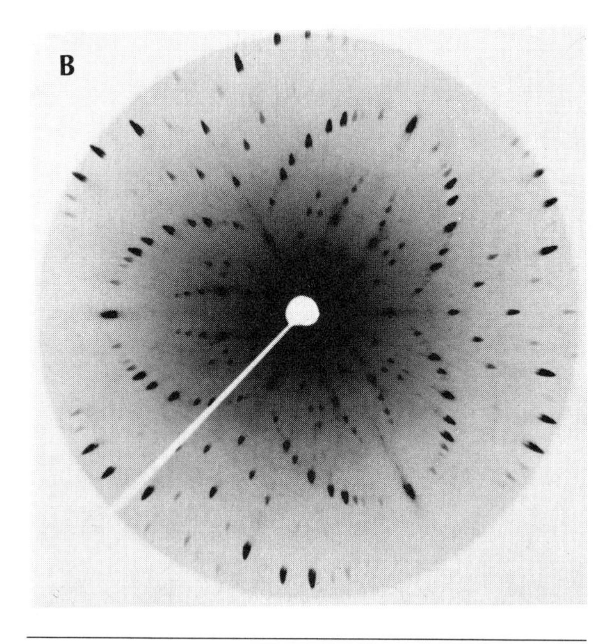

Fig. 4-12 Von Laue X-ray analysis. **A.** Technique in obtaining a Von Laue pattern. The X-ray beam is broken into a number of smaller pin-point beams by the internal crystal structure. **B.** Von Laue spots obtained by X-ray analysis of the mineral calcite. *Bernhardt J. Wuensch*

PHYSICAL PROPERTIES OF MINERALS

With few exceptions, the crystallographic structure of a particular mineral is always the same, no matter where the mineral comes from. It is, therefore, a definitive property of a mineral and can be used to identify it. The crystallographic structure of a particular mineral is also a natural consequence of the way in which the atoms of the elements making up the compound bond together. But it is only one of several properties that each mineral displays.

Take, for example, carbon, which can exist as either of two minerals—diamond or graphite. Diamond is very hard, so hard in fact that it is used as an abrasive; it shines brilliantly, can take and hold a polish, and breaks smoothly along certain planes. Graphite, on the other hand, is dull and soft and separates into small flakes that easily slide over one another.

The reason for such profound differences in minerals that are chemically the same substance lies in their completely unlike arrangement of atoms. In a diamond crystal each atom is in direct contact with four other atoms and covalently shares the four electrons in its outer shell with them. The bonds are exceptionally strong, resulting in the extreme hardness of the mineral. However, the geometrical arrangement of atoms is such that there exist four sets of parallel planes that mark zones of weaker bonding. It is the existence of these planes that allows the diamond cutter to extract several marketable smaller stones from one large one.

The carbon atoms in graphite, on the other hand, occur in regularly spaced thin, parallel sheets. The bonds uniting the atoms within each sheet are many times stronger than those between one atomic layer and its neighboring sheet above or below. Graphite, therefore, splits easily, parallel to the sheets, and the sheets slide easily over one another.

Color

Color is the most obvious property that minerals possess, but it is not always the most definitive. When white light falls on a mineral, some of the wavelengths will be absorbed and some reflected. The color of the mineral corresponds to the wavelengths of the reflected light. Black minerals, for example, absorb essentially all the light that falls on them. What causes a mineral to absorb only certain wavelengths is varied and complex. It can be a fundamental property, directly related to chemical composition, as in the blues and greens of some copper minerals. Or it may be unrelated to composition, depending instead on crystal structure and the type of bonding (as in diamonds and graphite). Commonly it is caused by foreign atoms contained in the crystal structure. Minerals, being natural compounds, are not as pure as the materials a chemist uses in research. Small amounts of foreign atoms usually occur in the lattice work as impurities. Pure quartz is colorless; the colored varieties of that mineral, so sought after by collectors, contain small quantities of impurities. Therefore, we cannot use colors with the same confidence in identifying minerals as in naming birds, for example. Considerable experience is needed to determine whether or not the color of a mineral is significant.

One way to avoid ambiguities is to grind the mineral into a fine powder and then observe its color. The intrinsic color will be seen, whereas colors due to impurities will generally be lost. A simple way to make the test is to rub the mineral on a piece of unglazed porcelain (a ***streak plate***), leaving a thin film of powder, or ***streak,*** of true color on the porcelain surface.

Cleavage

Cleavage is the ability of a mineral to split, or *cleave,* along closely spaced parallel planes (Fig. 4-13). Not all minerals have that ability; many fracture along widely spaced surfaces that can be relatively smooth, but are often curved or irregular and hackly. One notable (and familiar) type of fracture surface, commonly produced when glass is broken, roughly resembles the valve of a sea shell and, appropriately, was called ***conchoidal,*** from the Greek *konkolides,* meaning like a shell (Fig. 4-14).

Some minerals are characterized by only one set of cleavage planes; others by two, three, four, or (rarely) six. Because of crystal symmetry no minerals possess five, or more than six, sets of cleavage planes. Not only the number of cleavages, but the angles between them as well are characteristic for any mineral. Some cleavages meet at right angles, but many do not. If splitting occurs easily and produces a nearly unblemished flat surface, the cleavage is said to be *perfect*. In some minerals the cleavage is less perfect and can be called *good, distinct,* or *indistinct,* depending on the roughness of the resultant surface. The geometrically repetitive nature of cleavage, its planar character, and the distinctive orientation of the planes are strong evidence that cleavage, like the crystal form of minerals, is a property determined by the regular geometric ordering of the atoms in the mineral. Cleavage directions generally run parallel to those crystal planes in which the packing of ions is greatest.

Hardness

Mineral hardness, too, is related to atomic structure. Even a small scratch on the surface of a mineral requires the breaking of bonds and the separation of atoms, and the ease of separation will depend on the kinds of atoms and bonding inherent to the mineral. Hardness is an easy property to determine, and it can be definitive. A harder mineral will scratch a softer one.

The scale we use today to judge hardness was devised more than a century ago by an Austrian mineralogist, Frederich Mohs (1773–1839), and it bears his name. In the scale, ten minerals that essentially cover the range of hardness variability are arranged in order from the softest (1, talc) to the hardest (10, diamond). The hardness numbers, although in sequential order, are not equally spaced. For example, the actual interval between diamond, to which he assigned a value of 10, and corundum, 9, is greater than the rest of the scale combined. If absolute values were assigned to the various minerals used in the scale, diamond would be about 42. By means of the scale, the hardness of all minerals has been rated and listed, as in Table 4-1.

Fig. 4-13 Cleavage fragment of the common mineral calcite. Because of the characteristic regular arrangement of the atoms in the mineral, it always breaks, or cleaves, along three smooth plane surfaces, any two of which meet at an angle of 74°55′. The resulting shape is rhombohedral.

Fig. 4-14 Conchoidal fracture on a quartz fragment. Notice the nearly concentric arrangement of rounded grooves and ridges resembling the markings on a sea shell. *Janet Robertson*

Table 4-1. Moh's hardness scale

Mineral	*Hardness*	*Equivalent in hardness*
Diamond	10	
Corundum	9	
Topaz	8	
Quartz	7	
Potassium feldspar	6	glass, knife blade
Apatite	5	
Fluorite	4	
Calcite	3	penny
Gypsum	2	fingernail
Talc	1	

Luster

When a mineral is viewed in ordinary light, the amount of light reflected from its surface and the way it is reflected determines its luster. Essentially all lusters fall into two groups: ***metallic*** and ***non-metallic.*** The first term is applied to minerals that reflect light in about the same way that polished metals, such as iron or copper, do. In general a very high luster is characteristic of minerals possessing metallic bonding. Minerals with a metallic luster commonly are opaque (they will not allow light to pass through), even along thin edges held up against the light.

Non-metallic lusters are quite variable, but certain common ones have been singled out. If a mineral reflects light to about the same degree as glass, it has a glassy, or ***vitreous*** luster. Other terms that are essentially self-explanatory are earthy, greasy, waxy, dull, resinous, pearly, and silky.

Specific gravity

The specific gravity of a mineral may be defined as the weight of a specified volume of the mineral divided by the weight of an equal volume of water at 4°C (39°F)—the temperature at which water is most dense. It expresses how much more dense a mineral is than a common substance, water. For example, quartz has a specific gravity of 2.7, which means that its density is 2.7 times that of water. The specific gravity of any mineral is, within limits, characteristic of that mineral, and it is determined by the atomic weights of the elements that make up the compound and the degree of packing of the atoms.

A rough estimate of the specific gravity of a mineral can be made (after some experience has been gained) simply by hefting the mineral by hand. Some minerals feel light, whereas others, particularly those with metallic lusters, feel heavy.

Magnetic properties

A few minerals, particularly some made up of iron and oxygen or sulfur, are relatively strongly magnetic; most are not. A simple test of this property can be made by the use of a small hand magnet.

MINERAL DESCRIPTIONS

A mineral, to briefly review, is a naturally occurring inorganic compound (or element), crystalline in nature and possessing a definite chemical composition or a restricted range of compositions. Some materials are still considered minerals even though they do not meet all these requirements. Two notable exceptions are opal, a non-crystalline solid, and mercury, a liquid. More than 2500 minerals have been recognized on the face of the earth. Some of their names derive from Greek, Latin, Old English, or other tongues, and describe such properties as color, crystal form, density, or cleavage. Others were named for the geographical locality in which they were found. Most, however, have been named for people—mineralogists, crystallographers, scientists, explorers, mine owners, mining engineers, or public officials.

Minerals do not occur in equal abundance; some are relatively common, most are rare. The abundance of a mineral more or less reflects the quantities of the component elements that are available for its formation at or near the earth's surface (Table 4-2).

Surprisingly, only ten elements occur in sizable amounts, and they make up about 99 per cent of the total mass of the minerals at the

Table 4-2. Average chemical composition of the continental crust

Element	*Weight (mass) per cent*
Oxygen (O)	46.6
Silicon (Si)	27.2
Aluminium (Al)	8.1
Iron (Fe)	5.0
Magnesium (Mg)	2.1
Calcium (Ca)	3.6
Sodium (Na)	2.8
Potassium (K)	2.6
Titanium (Ti)	0.4
Hydrogen (H)	0.1

earth's surface. All the others, many of which are important to life and to the prosperity of humans, make up less than 1 per cent. It is strikingly evident that of the elements making up continental rocks, oxygen and silicon together make up nearly 75 per cent of the mass. The seven metals—aluminum, iron, calcium, sodium, potassium, magnesium, and titanium—make up most of the rest. The average composition of the crust beneath the oceans is only slightly different from that of the continents. In comparison to the other common elements, oxygen atoms (ions) are relatively large and make up about 94 per cent of the volume of the continental crust. Quite literally, when you walk on the earth, you are stepping on a thick carpet composed largely of oxygen ions.

Because of the overwhelming abundance of oxygen and silicon it is only natural that silicate minerals, which are composed of these two elements, are the most plentiful on earth. There are many ***silicates,*** most of which contain, in addition to silicon and oxygen, varying amounts of one or more of the seven relatively abundant metals and in some instances small amounts of hydrogen. Silicates are thought to predominate all the way to the earth's core to a depth of 2900 km (1802 mi). The core itself appears to be largely iron.

We are all familiar with the organic compounds known as hydrocarbons, which are mostly complex combinations of carbon and hydrogen atoms. The diversity of molecular combinations is primarily due to the chemical properties of carbon (and to a lesser extent hydrogen) and its ability to bond in many ways. In the inorganic world, the SiO_4 tetrahedron—the basic building block of the diverse silicate group—functions similarly. Through various combinations of it and other, like units (or metal ions), the entire complex silicate group can be formed.

The silicon atom is relatively very small, and it bonds covalently to the four oxygen atoms that can fit around it to produce a four-sided tent-shaped form called a ***tetrahedron*** (Fig. 4-15). The centers of the four oxygen atoms are at the points of the tetrahedron, each side of which is an equilateral triangle. The form can exist singly, or it can be joined to other tetrahedra by sharing a common oxygen atom to produce a bewildering array of single- and double-chain structures, rings, or three-dimensional networks (Fig. 4-16). Common metal ions fit into available spaces in the crystalline lattices, helping to bond the tetrahedron units together.

Although silicates make up the bulk of the common minerals, a few ***non-silicates,*** particu-

Table 4-3. Important rock-forming minerals

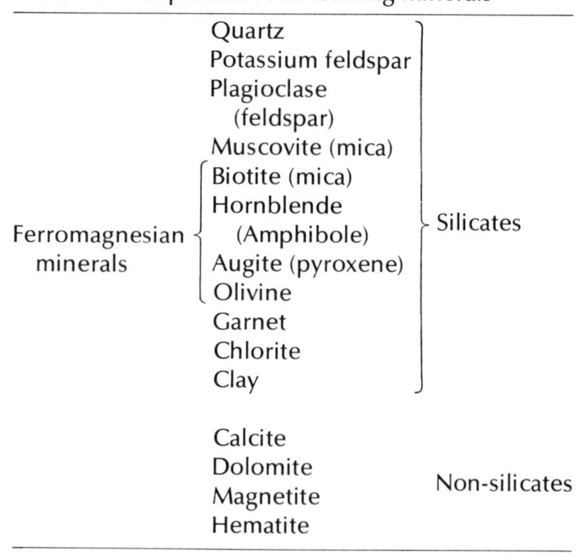

	Mineral	
	Quartz	Silicates
	Potassium feldspar	Silicates
	Plagioclase (feldspar)	Silicates
	Muscovite (mica)	Silicates
Ferromagnesian minerals	Biotite (mica)	Silicates
Ferromagnesian minerals	Hornblende (Amphibole)	Silicates
Ferromagnesian minerals	Augite (pyroxene)	Silicates
Ferromagnesian minerals	Olivine	Silicates
	Garnet	Silicates
	Chlorite	Silicates
	Clay	Silicates
	Calcite	Non-silicates
	Dolomite	Non-silicates
	Magnetite	Non-silicates
	Hematite	Non-silicates

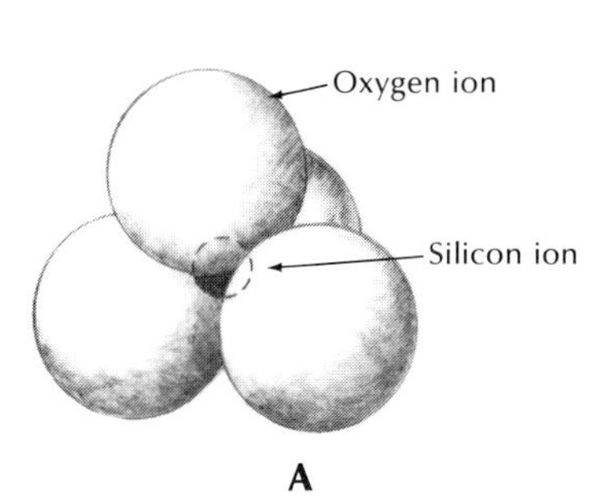

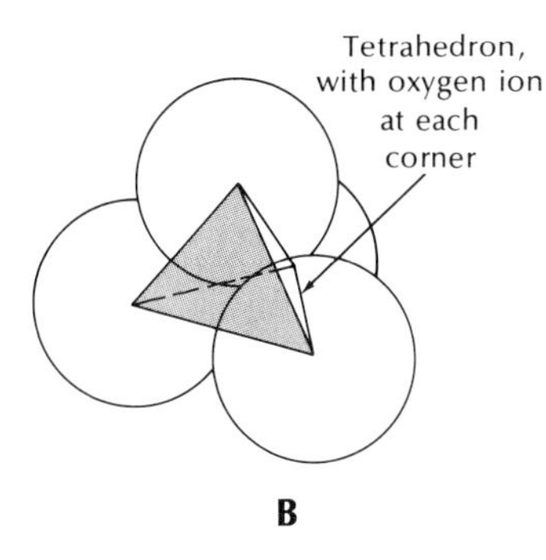

Fig. 4-15 A. The SiO_4 tetrahedron, consisting of one silicon ion surrounded by four oxygen ions. **B.** The same pattern with lines connecting the centers of the oxygen ions emphasizes their tetrahedral arrangement.

larly certain oxides and carbonates, are also abundant in some localities.

The minerals that make up most of the solid earth are only about fifteen in number. Their names are given in Table 4-3—and they are described in the following pages.

The minerals naturally fall into two groups: silicates and non-silicates. In addition, a subgroup of four silicates has been designated the ***ferromagnesian minerals.*** All four contain cations of iron and magnesium and tend to be dark in color.

Before moving ahead, we should take a moment to explain the shorthand notation used to indicate the chemical makeup of a mineral compound. Rather than spelling out each element composing a mineral, one- or two-letter symbols are used to represent each element. Silicon is Si, oxygen is O, and SiO_2 is the formula for silicon dioxide, the mineral known as quartz. The 2 subscript indicates that quartz is made up of two parts oxygen to one part silicon, by atomic weight. With the chemical formula, one can quickly see what the constituent elements of a mineral are and in what proportion they occur. The symbols for the other common elements are given in Table 4-2.

Silicates

Quartz (SiO_2) Quartz has a vitreous luster, a hardness of 7, and when pure, is clear and colorless. In fact, the Greeks thought it was a kind of frozen water. The name may be derived from a Saxon word meaning cross-veined. Quartz lacks cleavage, but it commonly fractures conchoidally (see Fig. 4-14). Should quartz grow free from interference, it crystallizes customarily in a six-sided crystal form, which is terminated by as harp-pointed pyramid at each end. If quartz grows into cavities, as it commonly does, it will possess only one pyramid at the end of the crystal that extends into the opening (see Fig. 4-8). Crystals that grow into openings may sometimes reach lengths of 0.3 m or more. Usually quartz occurs in association with other minerals as tiny grains, 2 to 3 mm across, that generally lack crystal faces. When they are fresh the disseminated grains may sparkle like tiny bits of glass.

Feldspar The feldspars are by far the most abundant of the common minerals, probably making up at least 50 per cent of the rocks at the earth's surface. The name comes from *feld,* the Swedish word for field, and *spar,* a mineral commonly found in fields overlying granite. The two most common ones are *potassium feldspar,* which is rich in potassium, and *plagioclase,* which is rich in sodium and calcium. The SiO_4 tetrahedra in these minerals are joined in a strong three-dimensional network that possesses planes of weakness in two directions at or nearly at right angles to each other (Fig. 4-17). Cleavage along the planes of weakness and a hardness of 6 (on the Mohs scale) provide two of the most characteristic properties of the feldspars. In many rocks they occur in well-formed crystals with a tabular form, so that on weathered or broken surfaces they often resemble small rectangular gravestones or fence laths.

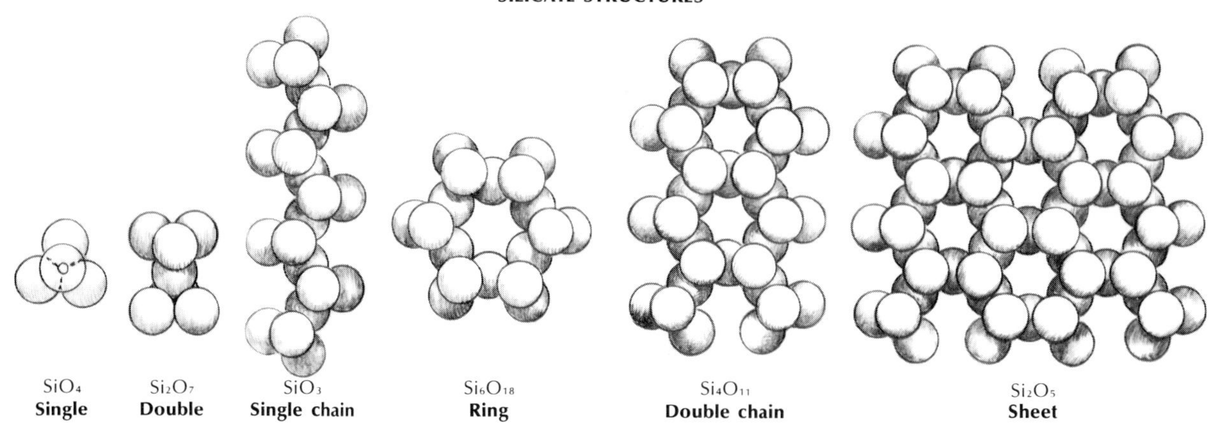

Fig. 4-16 The diversity of structures that can be formed by the basic SiO_4 tetrahedron.

Potassium feldspar ($KAlSi_3O_8$) There are several common varieties of potassium feldspar (among them orthoclase), which vary in the way their ions are arranged. For our purpose, little will be gained by differentiating between them. Potassium feldspar has a vitreous luster and may be colorless, but is usually milky white or flesh pink. It sometimes resembles unglazed porcelain, like the dull surface exposed on a chipped dinner plate.

Plagioclase ($NaAlSi_3O_8 \cdot CaAl_2Si_2O_8$) This mineral, like potassium feldspar, has a vitreous luster. Its color is most likely to be white or pale gray, although some varieties show a beautiful iridescence, or play of colors, much like those of a peacock's feathers. In some cases it can be almost glass-clear, like quartz. Plagioclase can probably be best distinguished from potassium feldspar or quartz by examining the crystal or cleavage surfaces of the mineral for ***striations***—a multitude of very closely spaced, parallel straight lines that look as if they had been engraved on the surfaces. Stria-

Fig. 4-17 Fragment of potassium feldspar typically displaying two cleavages at right angles. One cleavage plane parallels the front (illuminated) surface, the other parallels the upper surface. *Janet Robertson*

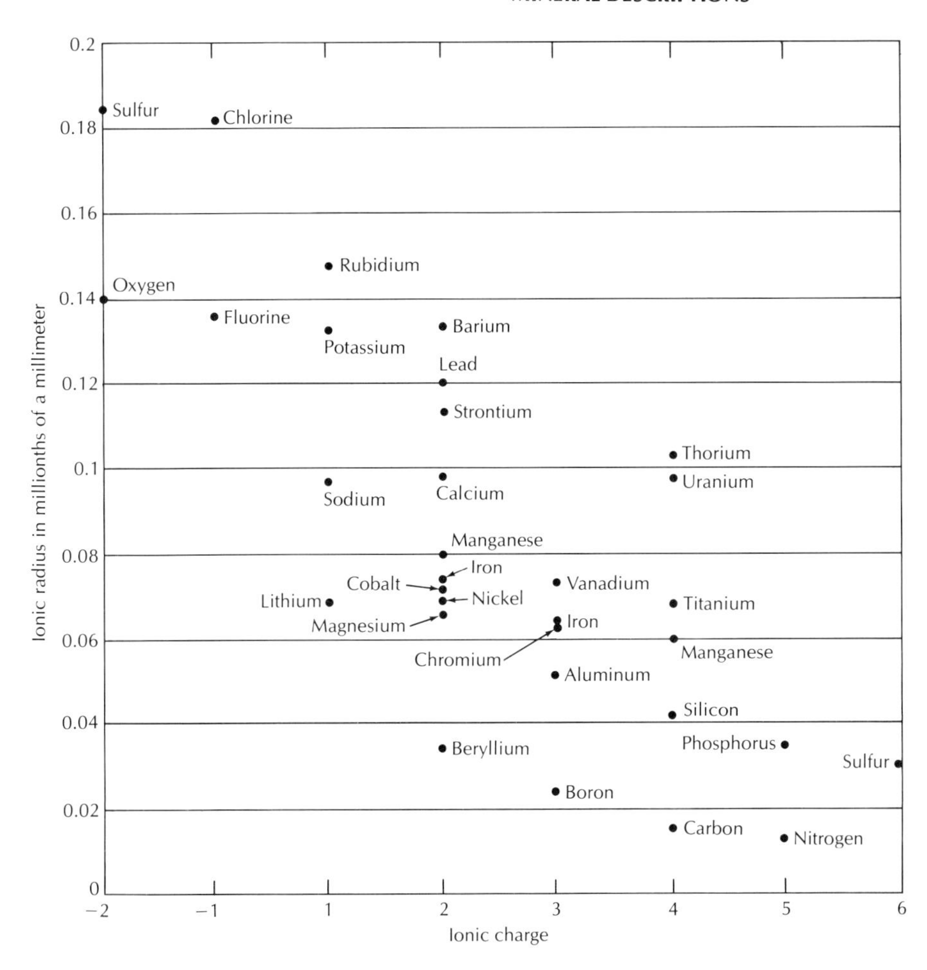

Fig. 4-18 Ionic charge versus ionic radius for some of the more common elements. For cations, the charge represents the excess of protons over electrons, and is positive. For anions, there is an excess of electrons over protons, and the charge is negative.

tions, which are well developed in plagioclase, are not found in potassium feldspar.

If we were to analyze the chemical composition of plagioclase taken from several different rocks, we would find that the mineral contained calcium, sodium, aluminum, and silicon as the principal cations, but that the proportions of each varied in each rock. The reason for the variability is easy to understand if we look first at the relative sizes of the four cations (Fig. 4-18). Those of sodium and calcium are relatively large, whereas those of silicon and aluminum are relatively small. In a way, we can think of the crystal lattice as "blind." During crystal growth in a mixed solution of ions, the lattice will accept any ion as long as it is approximately the same size as the others—for example, sodium and calcium cations or aluminum and silicon cations *could* simply interchange or substitute for one another. Cations, however, are positively charged particles, and during interchange, the neutrality of the electrical charge in the lattice must be maintained. The cations of sodium, calcium, aluminum,

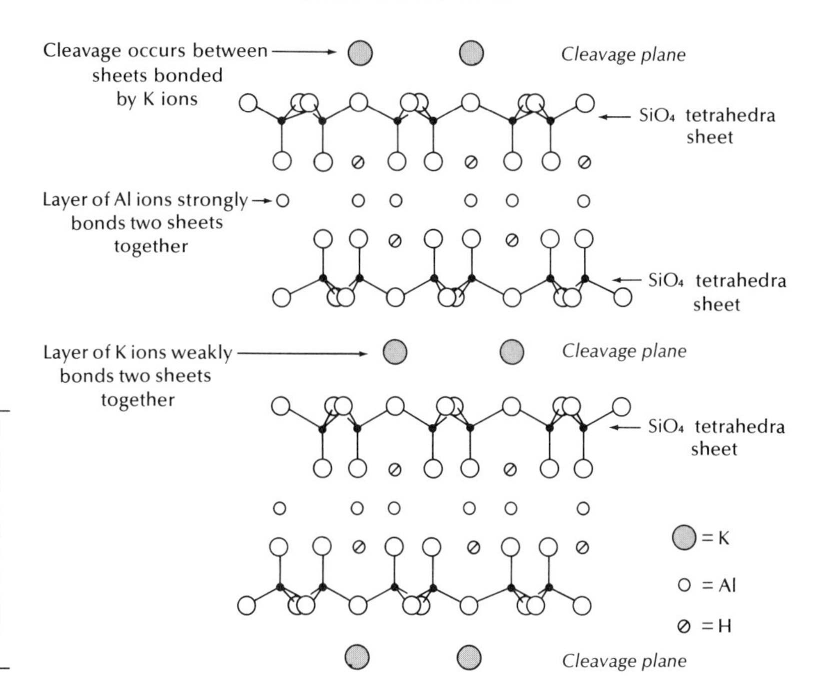

Fig. 4-19 The double-sandwich structure of muscovite. The points of the SiO_4 tetrahedra in the upper two sheets face each other and are strongly bonded by aluminum (Al) ions. The lower sheets are similarly bonded. The double layers, however, are bonded by potassium (K) ions, which form a relatively weak bond; the one perfect cleavage in mica occurs at and parallel to these planes of weakness.

and silicon carry electrical charges of +1, +2, +3, and +4, respectively (Fig. 4-18). Therefore, sodium (+1) cannot be substituted for a calcium ion (+2) unless at the same time a silicon ion (+4) is substituted for an aluminum ion (+3). The process of interchange of cations in minerals is called ***solid solution.*** In a case such as that just described, in which pairs of cations are involved, it is called ***coupled solid solution.***

Substitution of one metal for another is very common in the mineral world, and is one of the major reasons that minerals are not pure chemical compounds. A simple way of informing the reader that two components have the same charge, are nearly identical in size, and can substitute in any proportion for one another is to enclose the symbols in parentheses; for example, (Fe,Mg) indicates that iron and magnesium are freely interchangeable as in olivine $(Fe,Mg)_2SiO_4$.

Mica The micas include a number of closely related minerals, all of which possess one perfect or nearly perfect cleavage and have a hardness ranging from 2 to 3. X-ray analysis shows that the crystal of mica consists of parallel sheets, like the pages in a book, which are made up of SiO_4 tetrahedra strongly bonded together at their bases (see sheet structure in Fig. 4-16). The sheets themselves are bonded to each other less strongly, so that they can split apart easily—thereby the perfect cleavage so characteristic of that mineral group (Fig. 4-19). The two most common rock-forming micas are *muscovite* and *biotite.*

Muscovite ($KAl_3Si_3O_{10}(OH)_2$) The common name for muscovite, is *white mica;* and generally it is colorless, gray, or transparent, especially when split into thin sheets. Muscovite was used in the tiny windows of the houses of medieval Europe before the widespread use of glass brought more light to the gloomy interiors. The name is derived from Muskovy, the name given to Old Russia.

Muscovite has a silky or pearly luster, and in sunlight the cleavage plates of the tiny mineral grains, which are common in many rocks, shimmer and shine—the German name of *glimmer* for white mica conveys an impression of that property.

Biotite ($K(Mg,Fe)_3AlSi_3O_{10}(OH)_2$) Named in honor of the French physicist Jean Baptiste Biot (1777–1861), biotite is commonly called *black mica,* and its chemical formula indicates that it includes iron and magnesium (in solid solution). Thin cleavage sheets of it lack the degree of transparency of muscovite. Black mica ranges in color from dark brown to black, and in many rocks it occurs as jet-black flakes that shine like satin in the sun.

Hornblende ($Ca_2Na(Mg,Fe)_4(Al,Fe,Ti)_3Si_6O_{22}$-$(O,OH)_2$) This ferromagnesian mineral is the most common of a large and complex group, with similar physical properties. The ***amphiboles,*** hornblende, a dark mineral, is commonly dark green or jet black, and shines as brightly as a lacquered surface in the sun.

The crystals are, as a rule, long and narrow, and one of its most distinctive properties is its cleavage pattern (Fig. 4-20). The SiO_4 tetrahedra are joined in long, double chains that parallel the long axis of the crystal, and the two good cleavages, which intersect each other at angles of 56° and 124°, are parallel to planes of weakness (weaker bonds) that exist between groups of chains (Fig. 4-21).

Augite ($Ca(Mg,Fe,Al)(Si,Al)_2O_6$) The ferromagnesian mineral augite looks superficially like hornblende in that it is dark and possesses two good cleavage planes and a vitreous luster. Augite crystals, however, generally are stubbier, to the point of being nearly equidimensional, and their two cleavage planes meet nearly at right angles (87° and 93°). The name is derived from the Greek word for luster, in reference to its pearly sheen. Crystals seen in cross section are nearly square (Fig. 4-22). The cleavage planes in augite also parallel planes of weaker

Fig. 4-20 A. Drawing of a hornblende crystal, showing the characteristic flattened six-sided face, elongate shape, and two good cleavages at 56° and 124°. **B.** The crystal in cross section, as photographed under the microscope. *Edwin E. Larson*

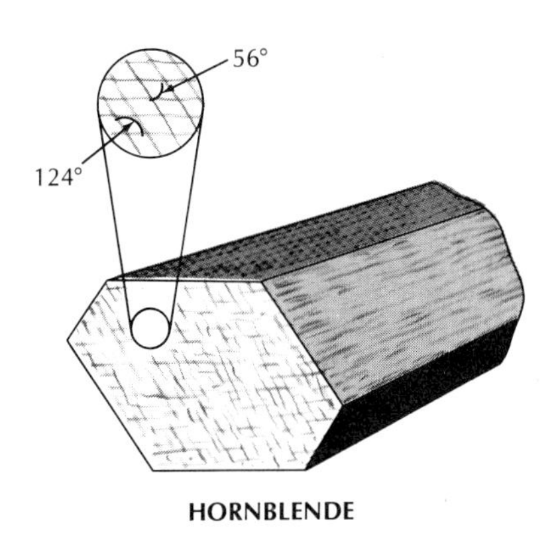

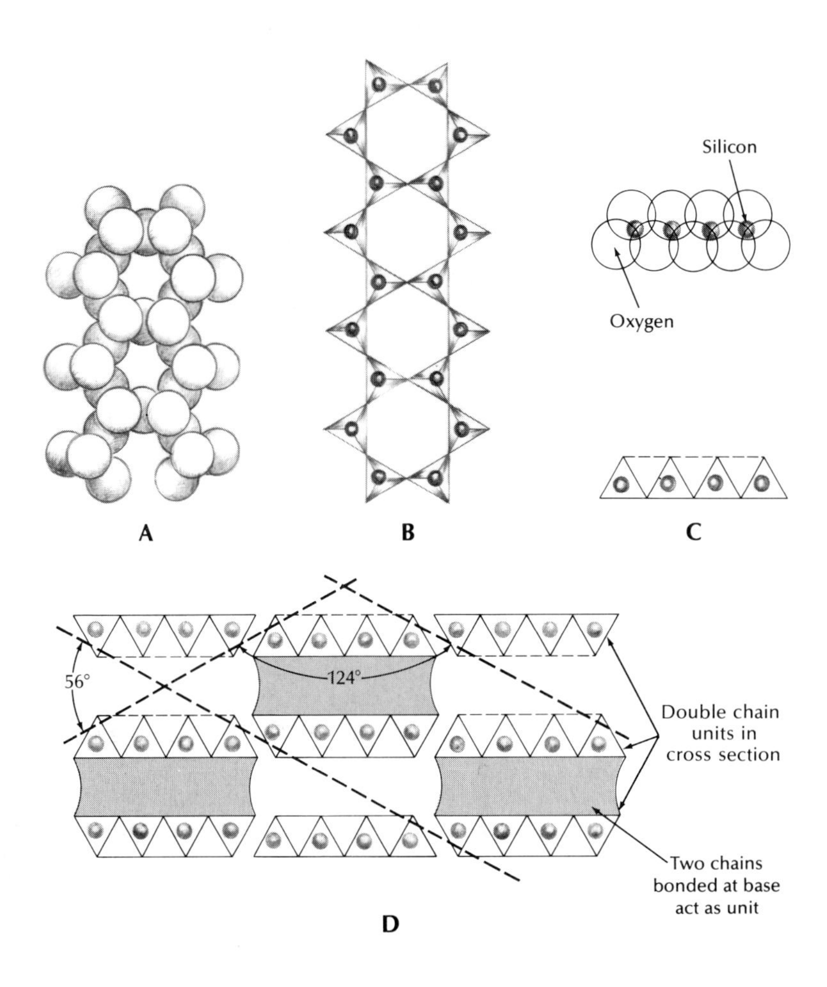

Fig. 4-21 In hornblende, the SiO_4 tetrahedra join together to form strongly bonded double chains. **A.** Chain network as seen from above (plan view). **B.** Schematic representation of **A** showing tetrahedral arrangement. **C.** Cross sections of **A** and **B. D.** The chains are parallel to the long axis of the crystal and are bonded to each other mostly by metal cations. The bonds are relatively weak, and the two characteristic cleavages occur along the planes of weakness, as shown in the cross section.

bonds between silicate chains, but in this case the chains are single rather than double (Fig. 4-23).

Hornblende and augite are the more common darker, rock-forming minerals. The principal distinctions between the two are (1) hornblende crystals tend to be long and narrow, whereas augite crystals are short and stubby; (2) hornblende has two cleavages parallel to the long axis of the crystal that meet at 56° and

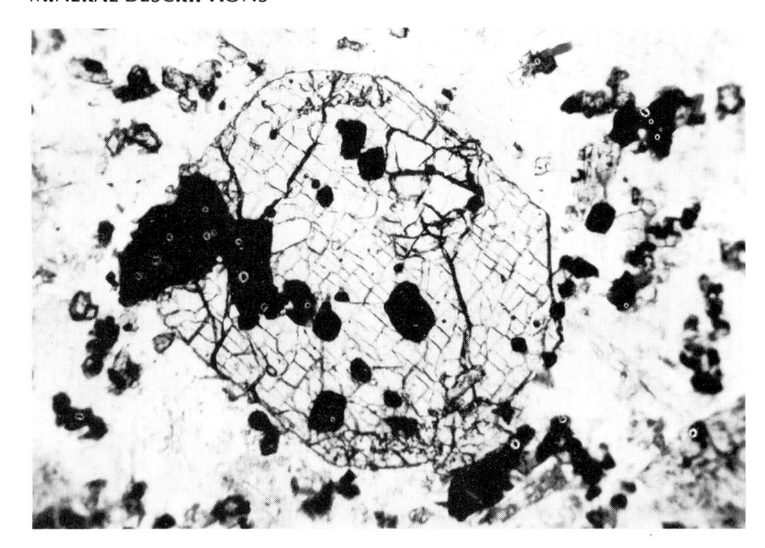

Fig. 4-22 A. Drawing of an augite crystal (pyroxene) showing the characteristic, nearly equant, eight-sided face, stubby form, and two good cleavages that meet nearly at right angles. **B.** The crystal in cross section, photographed under the microscope. *Edwin E. Larson*

AUGITE

124°, whereas augite has two that intersect each other at approximately right angles; (3) hornblende crystals seen in cross section approach a rhombic pattern, whereas augite crystals are more nearly square.

Olivine ($(Fe,Mg)_2SiO_4$) Another ferromagnesian mineral, olivine usually occurs as rounded, green, granular glassy crystals. When the crystals are large enough and free from blemishes they can be cut into attractive, though fragile gemstones. *Peridot* is the name given to the gem variety of the mineral.

Olivine crystals in fresh, dark lavas often look like tiny bits of dark- to light-green bottle glass. As indicated by the chemical formula, olivine is a solid-solution mineral, which can contain variable amounts of iron and magnesium. Most varieties have an intermediate composition and most commonly are richer in magnesium than in iron (Fig. 4-24).

Garnet Garnet is the name given to a group of minerals that possess basically the same complex silicate structure and mostly contain Ca, Fe, and Al. It has no cleavage and breaks with an uneven or conchoidal fracture. It possesses a resinous to a vitreous luster and a hardness of about 7. Colors of garnet are highly varied, but most commonly they are red, brown, or yellow. The name refers to the resemblance, in color, of red garnets to the seeds of pomegranates. Garnet almost always occurs in well-formed equidimensional crystals—perhaps its most distinctive property (Fig. 4-25).

Chlorite Chlorite, a complex group of hydrous silicates containing Mg and Al and, to a lesser degree Fe and other metals, usually occurs in scaly or thinly banded masses. Its luster is vitreous, it gives a greenish streak, and its grass-green to blackish-green color is one of its most characteristic properties. Its name derives from a Greek word for light green. Other properties are a hardness of only 1 to 2.5, and one perfect cleavage.

Clay Clay is the common name of a group of hydrous alumino-silicate minerals that result from the weathering of rocks. Basically white, it can be easily stained by impurities. Clay feels greasy to the touch, and often has a distinctive

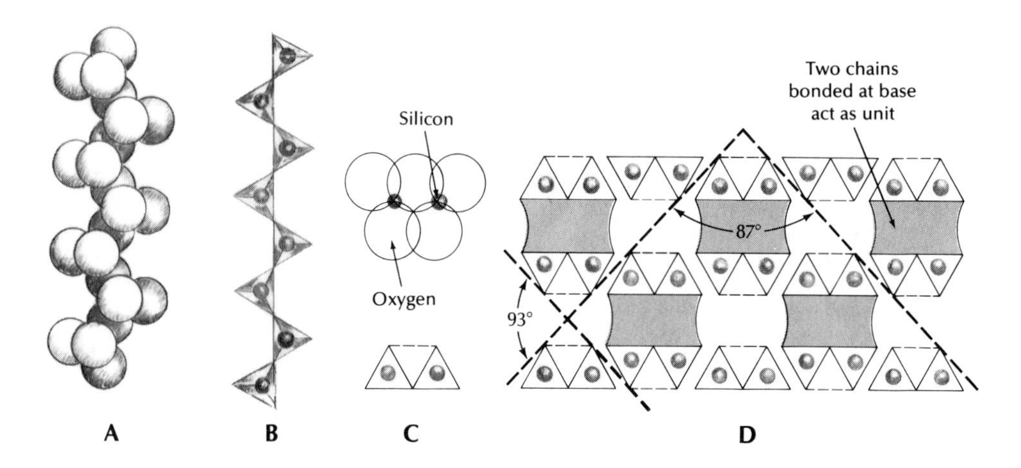

Fig. 4-23 A. In augite, the basic pattern is a single chain of SiO_4 tetrahedra (plan view). **B.** Schematic representation of A, showing tetrahedral arrangement. **C.** Cross sections of A and B. **D.** Metal cations bond chains together, but are relatively weak. Cleavage occurs at 87° and 93° along planes of weakness corresponding to planes of weaker bonds, as shown in cross section.

Fig. 4-24 Solid solution between magnesium and iron and its relation to the variation in density in olivine.

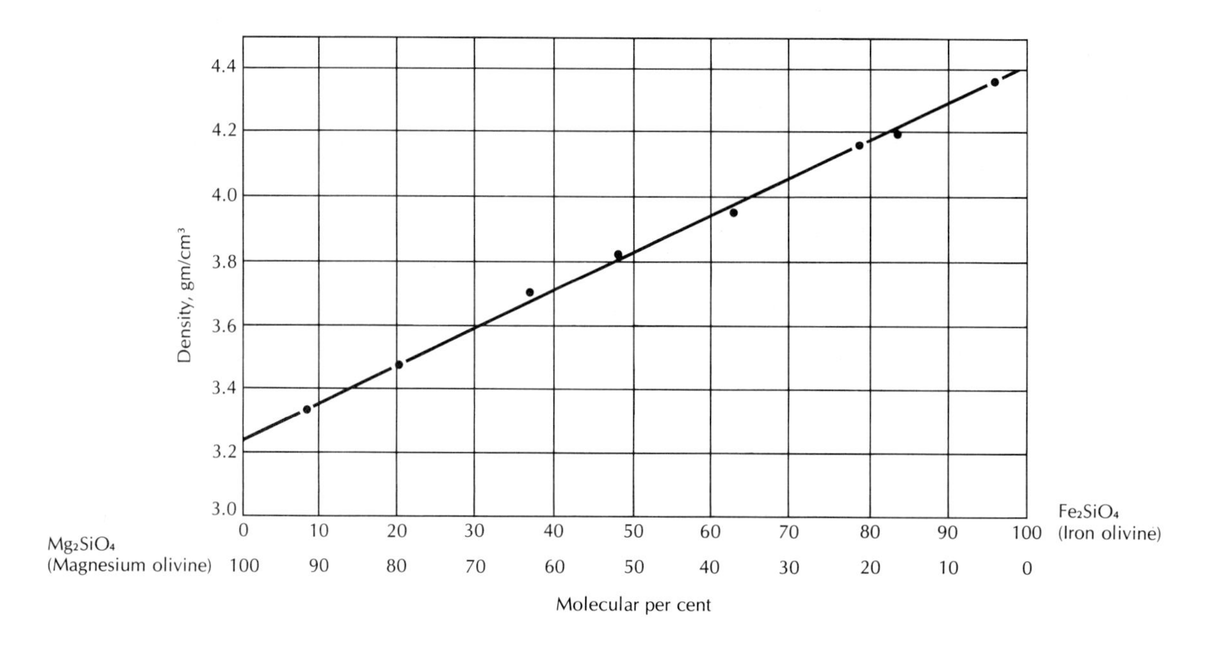

Fig. 4-25 Garnet crystals, about 5 cm across, showing the characteristic equidimensional form. From Québec. *M. Halberstadt*

odor—somewhat like the air just after the start of a summer rain. Some varieties adhere to the tongue or become plastic when moistened. Most commonly, clay occurs in soft, compact earthy masses.

Non-silicates

Calcite ($CaCO_3$) **and dolomite** ($CaMg(CO_3)_2$) Calcite, which normally is a light-colored (white or pale yellow) or colorless mineral, although—depending upon the amount and the nature of the impurities—the color may range across a color spectrum including yellow, orange, brown, and black. Its name is derived from the Greek word for limestone, *khálix*. Calcite has a vitreous luster, a hardness of 3, and occurs in a variety of crystal forms, which commonly are six-sided (Fig. 4-26). It possesses nearly perfect cleavage in three di-

Fig. 4-26 Typical scalenohedral crystal form of calcite; an intergrown crystal. The form takes its name from the fact that each face has the outline of a scalene triangle. *M. Halberstadt*

rections (see Fig. 4-13), and the intersections of the cleavage planes produce a rhombohedral pattern—that is, when the mineral breaks into fragments, each of the faces is rhombic or approximately diamond shaped.

Dolomite resembles calcite superficially. To distinguish one from the other, place the unknown mineral in cold, dilute hydrochloric acid. Calcite reacts readily, with a vigorous release of bubbles, whereas dolomite reacts much more quietly and slowly. Dolomite also has a slightly greater hardness (3.5), a higher specific gravity, a slight curve of its crystal faces, and a commonly pearly luster. The mineral was named after Sylvain Dolomieu, an eighteenth-century French geologist and mineralogist.

Magnetite (Fe_3O_4) **and hematite** (Fe_2O_3) Magnetite the black oxide of iron (and an important ore of iron) occurs most commonly as disseminated grains in rocks rich in ferromagnesian minerals. Opaque, it has a metallic to submetallic luster and lacks cleavage. Certainly its most distinctive property is its ability to be attracted to a strong magnet. The name derives from the locality in which the mineral was first found, the ancient Mediterranean kingdom of Magnesia.

Hematite, also a major ore of iron, which takes its name from the Greek word for "blood-like stone," looks like magnetite when it occurs in its granular form, but it is less magnetic and produces a red to red-brown powder on a streak plate. At times it occurs as a very fine pigment that colors an entire rock red, sometimes so intensely that it looks as if the rock had been painted. Some varieties possess a metallic luster; when cut and polished, they make fine semiprecious gemstones called black diamonds. Most hematite, however, has a dull, fibrous, or earthy luster. Some of the latter varieties are referred to as red ochre, a natural paint pigment.

ROCKS

Select a rock at the earth's surface and look at it closely (Fig. 4-27)—more than likely you will find that it is composed of a number of different mineral grains that differ in color, luster, grain size, grain shape, and so on. Because the mineral grains in most rocks are relatively tiny, many of the properties of the various minerals are hard to assess. For example, when the crystal in question is no larger than a grain of rice, it is difficult to determine how many cleavages there are and at what angles they meet. In time, however, if you persist, you will be able to identify nearly all the minerals in any common rock. And when that time comes, you will realize that all common rocks are almost entirely composed of the relatively few minerals just described. You will also become aware that rocks differ not only in the number and amount of their constituent mineral grains but also in texture, which pertains to grain size, both absolute and relative, and in how the grains fit together.

Geologists have found that most differences in mineralogy and texture are related to the way rocks were formed. By studying rocks and comparing their similarities and differences, we can obtain clues about the processes and events that formed them in the geological past. The study of rocks has undoubtedly become our principal basis for understanding the earth.

As an aid to comparing similarities and differences between rocks, geologists have devised a rock classification system—imposing an order where none existed before. Biologists have done the same thing for the animal and plant kingdoms. To be sure, the scheme is arbitrary, and there are the inevitable exceptions, borderline cases, and overlaps, but it works reasonably well for most rocks. Fundamentally the classification is a ***genetic*** one, based on the origin of the rocks. Each grouping represents rocks that have formed in much the same way, and each main group has been subdivided. The smallest divisions are based on a rock's texture, its mineralogy, and the proportion of its mineral constituents. Each division has been given a formal rock name, for example, *granite, shale, slate.* Once a rock has been classified, the name itself carries all the genetic, mineralogical, and textural information concerning that rock; which makes for ease of comparison of rocks of all ages from around the world, as well as ease of communication among earth scientists.

In the following section we will describe only the major subdivisions of rock classification. We shall leave for later chapters the detailed description of the rocks within each group and the processes that form them.

All rocks can be placed into three main groups: ***sedimentary, igneous,*** and ***metamorphic.***

Sedimentary rocks

Of the three genetic rock families, sedimentary rocks are perhaps the most readily comprehended because many of them closely resemble the materials from which they are composed; also many of the processes responsible for their formation occur before our eyes or in reasonably accessible environments. About 75 per cent of the earth's surface is covered by sediments. Even so, sedimentary rocks form only a thin, discontinuous veneer that is spread over the much more abundant igneous and metamorphic rocks, which are the true foundations of our continents and ocean basins.

Many sedimentary rocks can be thought of as secondary, or derived, rocks in that they are composed of bits and pieces of pre-existing rocks held firmly together by a cement. Such a texture is called ***clastic*** (Greek, broken), and the fragments can be called ***clasts.*** Examples of clastic sedimentary rocks are (1) ***sandstone,*** which consists of sand grains cemented together; (2) ***conglomerate,*** which consists of larger, rounded fragments cemented together; and (3) ***shale,*** which consists of very small particles. Some sedimentary rocks may result from chemical precipitation in lake or sea water (for example, rock salt); others may result from the accumulation of a variety of organic remains.

Sedimentary rocks accumulate on land or on the floors of lakes or seas. They are built up through the slow deposition of material, and so are typically formed layer upon layer. The layers are called ***strata;*** a single layer is a *stratum* (directly from the Latin—a blanket or pavement, derived from *stratus,* p.p. of *sternere,* to spread out). Individual layers may range from paper-thin sheets to massive beds tens of meters thick.

Igneous rocks

Igneous rocks are those that have solidified from a molten silicate material to which the name ***magma,*** or ***melt,*** is given. All magmas originate deep below the surface, where temperatures are relatively high. If the magma finds a path to the surface it erupts, cools, and solidifies to form ***volcanic*** rocks. Magma, which never reaches the surface, ultimately cools and crystallizes in the deep subsurface domain to form ***plutonic*** rocks. The word *plutonic* comes from the name for the Greek god of the lower world, Pluto.

In most igneous rocks, the mineral grains form an interlocking network of crystals, some perhaps with crystal faces, most without. The network, called the ***crystalline texture,*** results from the progressive and simultaneous growth of many mineral grains during solidifaction, or crystallization, of the magma. The overall mineralogy, average grain size, and differences in grain size give clues to the conditions under which crystallization occurred. For example, magma underground cools and crystallizes more slowly than magma at the surface, and for that reason the mineral grains in plutonic rock have a chance to grow much larger than those that crystallize quickly during the rapid cooling of a lava flow. In some cases, grains formed during the very rapid cooling of a lava flow are too small to be seen without a microscope.

Metamorphic rocks

Metamorphic rocks were a puzzle to the first geologists, and even today their origin is not clear. Such rocks do not form on the earth's surface, but appear instead to be products of the action of internal heat, pressure, and chemical activity of fluids through long periods of time—long at least when judged by our time. The above factors induce recrystallization, either partial or complete, of the preexisting minerals of the rock. The name *meta-*

Fig. 4-27 A rock of a number of different minerals, some black, some gray, some white. The individual grains of each mineral generally have similar properties, which differ from those of the other minerals. *Edwin E. Larson*

morphic means "change in form." New minerals appear, and they may develop a wholly new fabric or orientation. Instead of being randomly oriented and heading every which way, as is true of many igneous rocks, the minerals undergoing metamporphic recrystallization may align themselves parallel to one another as do cards in a deck of cards. Such layering is called ***foliation*** (from the Latin *folium* leaf), and it can be weakly to strongly developed and on a coarse to fine scale. Some metamorphic rocks have such a strongly developed foliation that it superficially resembles the stratification of sedimentary rocks. These metamorphic layers, however, consist of interlocking crystals segregated into layers of dark minerals (ferromagnesian) and light-colored minerals. In some cases, elongate minerals, like hornblende may become aligned with their long axes parallel, to produce a ***lineation.***

In most types of metamorphism, the rock undergoes little or no change in chemical composition as its minerals recrystallize. The elements present simply regroup themselves under conditions of higher temperatures and pressures to form new minerals, which are stable in the new subsurface environment. In some cases, however, new minerals are formed because heated gases and fluids circulating within the earth, and very often associated with plutonic igneous activity, have introduced new material.

Metamorphic rocks are almost certain to be complex because they have no single mode of origin; in some cases, temperature is the important factor, in others, directed pressure, and in still others, the nature of the fluid phases. They can be made from all manner of rocks: igneous, sedimentary, or even previously metamorphosed rocks. If they have any factor in common, it is crystallinity, and, like igneous rocks, their fabric is one of interlocking crystalline minerals.

SUMMARY

1. An *element* is composed of the same kind of atoms; a *compound* is composed of groups of different kinds of atoms.
2. Atoms consist of *electrons, protons,* and *neutrons*. In the atoms of a particular element, the number of electrons equals that of protons and is a fixed number (*atomic number*). But the number of neutrons can vary.
3. Atoms bond together through *ionic, covalent,* and *metallic bonding* processes that involve interaction of the outer electron shells. Molecules can also bond together through electrostatic forces.
4. Matter can exist in three different states: *gas, liquid,* and *solid.* The particles that make up a gas are *randomly ordered;* those in a liquid possess *short-range ordering;* those in a solid, *long-range ordering* or *crystallinity.*
5. *Minerals* are naturally occurring inorganic crystalline compounds and elements of a definite chemical composition or a restricted range of chemical compositions. The crystalline nature of minerals has been established through *X-ray diffraction* studies.
6. Every mineral has a definite set of properties that can be used to identify it. Those commonly used are *color, cleavage* (or lack of it), *hardness, luster, specific gravity,* and *magnetic properties.*
7. Oxygen and silicon make up almost 75 per cent of the minerals of the earth's surface. Most minerals are composed of those elements and are called *silicates.* The basic building block of silicates, the SiO_4 *tetrahedron,* can exist singly or in an array of chains, rings, or three-dimensional networks.
8. Important rock-forming minerals are relatively few in number and include the following silicates: *quartz, potassium* and *plagioclase feldspars, muscovite* and *biotite micas, hornblende, augite, olivine, garnet, chlorite,* and *clay.* Four of them are relatively rich in iron and magnesium: biotite, hornblende, augite, and olivine. Four common non-silicate minerals are *calcite, dolomite, magnetite,* and *hematite.*
9. Rocks are composed of aggregates of one or more minerals, the grains of which fit together in various ways (*texture*). Most mineralogical and textural differences are related to the way in which different rocks form.
10. The system of rock classification is genetic. The smallest units in the classification are based on the rock's texture and mineralogy and on the proportions of its mineral constituents. All rocks fall into one of three main categories: *sedimentary, igneous,* and *metamorphic.*
11. Most *sedimentary rocks* are made of fragments (*clasts*) of pre-existing rocks cemented together. Some are the result of chemical precipitation or accumulation of organic remains. The source of *igneous rocks* is a molten material called *magma.* If magma solidifies below ground the rocks are called *plutonic;* if it solidifies on the surface, they are called *volcanic. Metamorphic rocks* form from the solid-state recrystallization of pre-existing rocks under the influence of heat, pressure, and the chemical activity of fluids.

QUESTIONS

1. Describe the parts of an atom and outline the various ways atoms bond together.
2. What is a crystalline substance? How does it differ from a mineral?
3. Describe the various physical properties of minerals. Which are fundamental properties?
4. How are the crystal form and cleavage of any mineral related? Why is there a lack of any type of cleavage in quartz?
5. What are the most common elements of the earth's crust? How do these elements bond together to produce the common minerals?
6. List ten silicates and three non-silicates and give a thumbnail sketch of the physical properties of each.
7. What are the distinguishing characteristics of each of the following mineral pairs: potassium feldspar–plagioclase; muscovite–biotite; hornblende–augite; calcite–dolomite; magnetite–hematite; olivine–quartz; quartz–potassium feldspar.
8. Name the three classes of rocks and give the characteristic features of each.

SELECTED REFERENCES

Berry, L. G., and Mason, B., 1959, Mineralogy, W. H. Freeman and Co., San Francisco.

Bloss, F. D., 1971, Crystallography and crystal chemistry, Holt, Rinehart and Winston, New York.

Bragg, Sir Laurence, 1968, X-ray crystallography, Scientific American, vol. 219, no. 1, pp. 58–79.

Desautels, P. E., 1968, The mineral kingdom, Madison Square Press, Grosset & Dunlap, New York.

English, G. L., 1934, Getting acquainted with minerals, McGraw-Hill Book Co., New York.

Holden, A. and Singer, P. 1960, Crystals and crystal growing, Doubleday and Co., New York.

Hurlbut, Jr., C. S., and Klein, C., 1977, Manual of mineralogy, 19th ed., John Wiley and Sons, New York.

Mason, B., 1958, Principles of geochemistry, John Wiley and Sons, New York.

Pough, F. H. 1976, Field guide to rocks and minerals, 4th ed., Houghton Mifflin, Boston.

Sinkankas, J., 1966, Mineralogy: a first course, D. Van Nostrand Co., Princeton, New Jersey.

Turekian, K. K., 1972, Chemistry of the earth, Holt, Rinehart and Winston, New York.

Tutton, A. E. H., 1924, The natural history of crystals, E. P. Dutton and Co., New York.

Vanders, I., and Kerry, P. F., 1967, Mineral recognition, John Wiley and Sons, New York.

Fig. 5-1 Eruption of Cerro Negro, Nicaragua, which produced the lava flow in the middle background, and subsequently, the steep-sided cone. Large rocks, explosively hurled from the vent, fall near the crater and tumble down slope trailing white plumes of vapors (mostly water vapor). *USGS*

5

IGNEOUS PROCESSES AND IGNEOUS ROCKS

Almost all igneous rocks owe their existence to the crystallization of minerals from the progressive cooling of a hot molten liquid called ***magma,*** or, alternatively, ***melt.*** The existence of magma and the relatively rapid formation of one group of igneous rocks—volcanic rocks—can be readily documented during and following the eruption of a volcano at the earth's surface (Fig. 5-1). Geologists have come to realize that the magma that feeds the fiery fountains of a volcano originates deep below the surface, in pools called ***magma chambers***—pools surrounded mostly by solid rock. The subsequent cooling and crystallization of the magma that remains in the chambers occurs slowly and leads to the formation of generally coarser-grained plutonic rocks. Later, erosion may strip off enough of the overlying rocks to lay bare the deep-seated plutonic rocks.

About 200 years ago, however, when the science of geology was in its formative stages, earth scientists had other ideas concerning the origin of igneous rocks. One such person, the influential geologist Abraham G. Werner (1750–1817), gave inspiring lectures at the Freiberg Mining Academy in Saxony. In 1787 Werner published a general theory on how rocks of all kinds were formed. Volcanoes, he conjectured, resulted from the combustion of buried coal seams, which caused rocks near the earth's surface to melt. Since most coal is geologically of relatively recent origin, he considered volcanism to be a very young geological phenomenon. At the time, Werner's hypothesis was completely reasonable. He knew that plutonic rocks occurred below the coal seams and therefore could in no way be related to volcanic activity. From the evidence at hand, he concluded that plutonic and metamorphic rocks were part of the original crust of the earth, which had been precipitated out of the sea waters that once covered the globe. Werner's theory, called ***neptunism,*** survived for more than 50 years. The beginning of its demise was brought about, in 1795, by the efforts of the perspicacious Scot James Hutton (whom we earlier encountered in connection with the principle of uniformatarianism). Hutton had taken note of the existence of subterranean heat, which could be demonstrated in mines, and inferred from hot springs and volcanoes. In his tramps through the countryside, he found evidence suggesting that the coal seams had themselves been seared and charred by molten material—evidence in direct opposition with Werner's contention. Hutton also observed a body of rocks composed of the plutonic rock granite. After carefully inspecting its mineralogy, he decided that sea water could not possibly have contained simultaneously in solution all the different ions of which the granite was composed. Rather, it appeared to him that the granite resulted from crystallization of molten rock deep beneath the surface. That sagacious conclusion established the basis of modern thought concerning the origin of plutonic igneous rocks.

Unfortunately, when Hutton published his two-volume work, *Theory of the Earth with Proofs and Illustrations* (1795), he was ahead of his time. His theories directly opposed Werner's, whose influence was then at its zenith. Eventually, however, Hutton's views, which better fitted the observational data, came to be accepted.

Today we recognize many kinds of igneous rocks, each kind the result of the particular chemical and physical conditions that existed during its formation, and each with a different name. The name of an igneous rock depends primarily on its mineral composition and texture.

MINERAL COMPOSITION

The kinds and amounts of minerals found in any igneous rock depend on the chemical composition of the magma from which the rock crystallized and on processes taking place during crystallization. Essentially, as you might expect, all magmas are molten mixtures rich in oxygen and silica, with lesser amounts of the metal ions aluminum, iron, magnesium, calcium, sodium, potassium, and titanium. All contain a small percentage of such volatile constituents as water and carbon dioxide.

Table 5-1. Typical compositions of magmas

Oxide Component	*Felsic*	*Mafic*	*Ultramafic*
SiO_2	73.86	50.83	43.54
TiO_2	0.20	2.03	0.81
Al_2O_3	13.75	14.07	3.99
Fe_2O_3	0.78	2.88	2.51
FeO	1.13	9.06	9.84
MnO	0.05	0.18	0.21
MgO	0.26	6.34	34.02
CaO	0.72	10.42	3.46
Na_2O	3.51	2.23	0.028^+
K_2O	5.13	0.82	0.005^+

From Hyndman, 1972

Initial Composition of Magma

Most magmas, it is generally agreed, are generated in a poorly defined zone in the outer 100 to 300 km (63 to 188 mi) of the earth where, locally and intermittently, temperatures become high enough to bring about melting (Fig. 5-2). This zone is made up of several different kinds of rock material, as shown in Figure 5-2, and the initial compositions of different magmas are determined by the chemical composition of the parent rocks that melt to form them. Magmas relatively rich in magnesium, iron, and calcium are called ***mafic.*** The term derives from magnesium and the Latin word for iron, *ferrum* (hence 'ferrous' and 'ferric'), and refers to an abundance of these elements in the magma (Table 5-1). Those relatively rich in sodium, potassium, and silicon are called ***felsic*** (from *fe*ldspar and *si*licon) and contain larger amounts of feldspar-producing elements. Magmas that are transitional in composition between mafic and felsic are called ***intermediate,*** and those containing extremely large amounts of magnesium and iron are called ***ultramafic.*** Each of these terms can be applied not only to magmas, but to rocks as well.

As Figure 5-2 shows, felsic magmas are generated for the most part *within* the continental

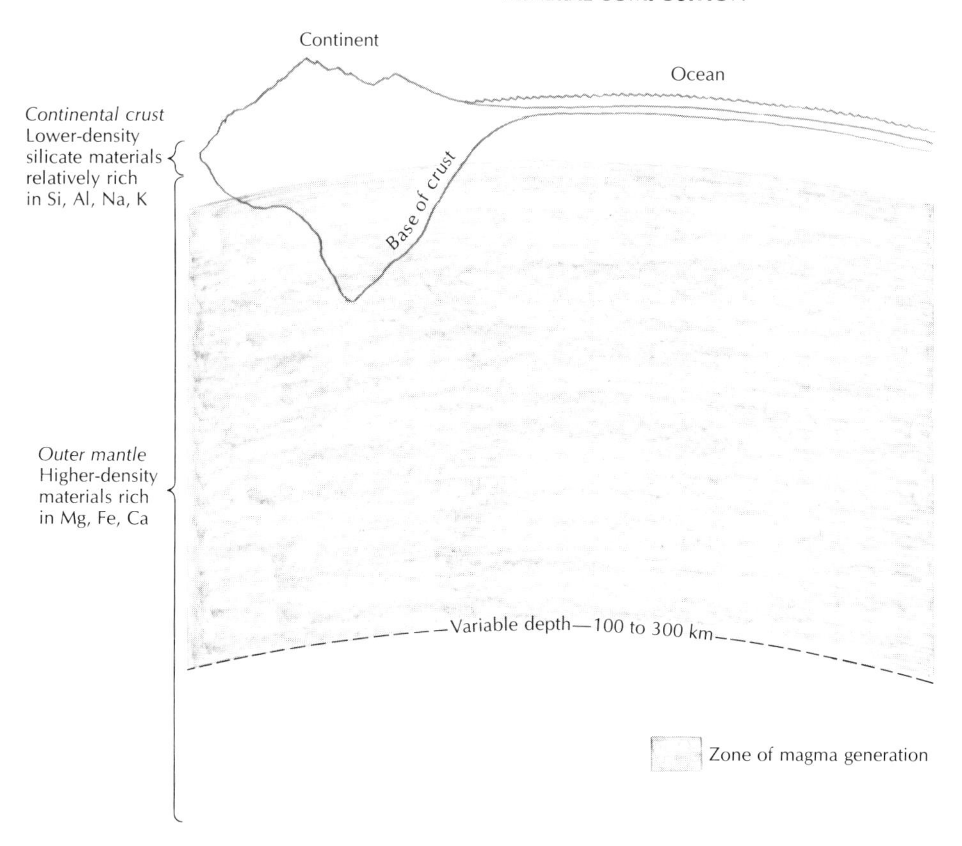

Fig. 5-2 Cross section of the outer portion of the earth, showing zone where melting and magma formation could occur.

crustal regions, where felsic parent rocks are abundant, whereas mafic and ultramafic magmas are likely to be derived from parent materials, rich in magnesium, iron, and calcium, that occur *beneath* the crust of both ocean basins and continents alike. Intermediate magmas commonly form, at depth, at the margins of continents.

Because of the thermal gradient within the earth (Fig. 5-3) the greater the depth from which a magma rises, the hotter it will be. Commonly the temperatures of mafic magmas are from 1200 to 1250°C (2192 to 2282°F), whereas those of felsic magmas are closer to 700°C (1292°F).

The initial composition of magmas also depends on the amount of parent rock that is melted. Most parent rock contains several different kinds of minerals, each with a different ***melting point.*** As temperatures rise locally, the minerals with lower melting points will begin to melt first. If there is a limited amount of heat available, melting will be only partial, and the resulting magma will be derived only from those minerals with lower melting points. For example, in the zone of potential melting beneath the continental and oceanic crusts, parent rocks appear to be made up of ultramafic and mafic materials in the ratio of 3 : 1. Because the melting points of the mafic mineral assemblage are generally lower, partial melting in this zone produces mafic magma, which moves toward the surface, leaving the unmelted ultramafic minerals behind.

The melting of parent rock is also affected by the presence or absence of volatile materials. Water, particularly, appears to act as a catalyst, reducing the thermal stability of the silicate bonds. In the presence of even very small amounts of water, the melting points of most silicate minerals are reduced by as much as 200 to 300°C. All igneous rocks contain some water, which suggests that it was available at the time of magma generation. If it were not for the presence of water, few magmas would probably form.

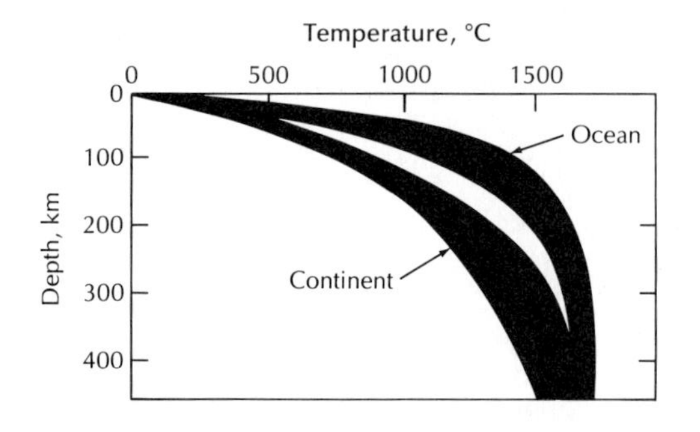

Fig. 5-3 Temperature versus depth for continents and oceans. The thickness of the lines represents uncertainty.

Crystallization of a Magma

After subsurface emplacement, or surface eruption, has taken place, a magma will subsequently cool and begin to crystallize.

In crystallization induced through cooling—the reverse of melting—temperatures drop until some ionic bonds are able to resist the disruptive thermal forces. At that point, crystals begin to form. The first-formed solids are those with the highest melting points (Fig. 5-4). As the magma continues to cool, minerals with successively lower melting points will crystallize until all of the magma has been consumed.

Experimental studies concerning the order of occurrence of the common silicate minerals throughout the process of initial crystallization and later reaction of the minerals with the melt were begun in the early 1900s by N. L. Bowen and his colleagues at the Geophysical Laboratories in Washington, D. C. Their work led to the formulation of an idealized succession, ***Bowen's reaction series,*** which has proved to be of fundamental value to the understanding of igneous rocks. However, the study of many rocks shows that this succession is not always realized in nature. Therefore, we will discuss the concept only in its broadest application to the formation of igneous rocks. In general, common silicate minerals with high melting points (in the range of 1050 to 950°C; 1922 to 1742°F) include olivine and calcic plagioclase; those with intermediate melting points (900 to 750°C; 1652 to 1382°F) augite, hornblende, and biotite, and intermediate plagioclase; and those with low melting points (750 to 600°C; 1382 to 1112°F) muscovite, sodic plagioclase, potassium feldspar, and quartz.

The particular succession of minerals that crystallize from any magma is determined primarily by the initial magmatic composition. For example, in a mafic lava, well-formed crystals of olivine and calcic plagioclase are commonly the first to form, followed at lower temperatures by augite, which forms in the spaces between the earlier-formed crystals (Fig. 5-5). By the time the augite has crystallized, no magma will remain. If, on the other hand, the starting magma is felsic, the cations magnesium, iron, and calcium are relatively scarce, and the first-formed crystals might be biotite and sodic plagioclase, followed, if any magma remains, by muscovite, potassium feldspar, and quartz (Fig. 5-6).

One complication of this rather simple picture of crystal succession is that some of the ferromagnesian minerals formed at higher temperatures, particularly olivine, augite, and hornblende, may react with the remaining melt, and change to one of the lower-temper-

Fig. 5-4 Crystals of calcic plagioclase (elongate, white) and olivine (equant, gray) are the first to form during crystallization of a mafic magma. The black material between the crystals is uncrystallized glass. Photographed under the microscope. *Edwin E. Larson*

Fig. 5-5 Drawing of basalt, as seen under the microscope. It shows well-formed grains of calcic plagioclase and slightly corroded olivine, with grains of augite that have crystallized out between the larger grains. A few magnetite grains are also present.

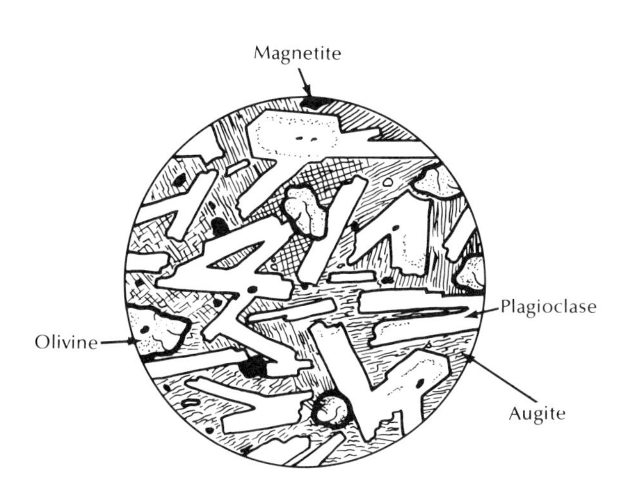

ature ferromagnesian minerals. Whether this happens seems to depend both on the rapidity of crystallization and the initial composition of the magma. In some cases, mineral grains may be caught in the act of transformation so that a core of a high-temperature mineral phase will be rimmed by a later-formed, low-temperature phase.

Plagioclase presents a special case of reaction behavior. Calcic plagioclase is stable at high temperatures, intermediate plagioclase at intermediate temperatures, and sodic plagioclase at low temperatures. As magma cools, therefore, the stability range of the precipitating plagioclase changes progressively, such that earlier formed, more calcic crystals are no longer in equilibrium with the melt at lower temperatures. To accommodate this, the earlier formed crystals, through reaction with the melt, undergo conversion to forms richer in sodium. If cooling is too rapid to permit completion of this process, however, the resulting plagioclase crystal will be zoned, with a calcic core formed at high temperature armored by layers that are progressively more sodic (Fig. 5-7).

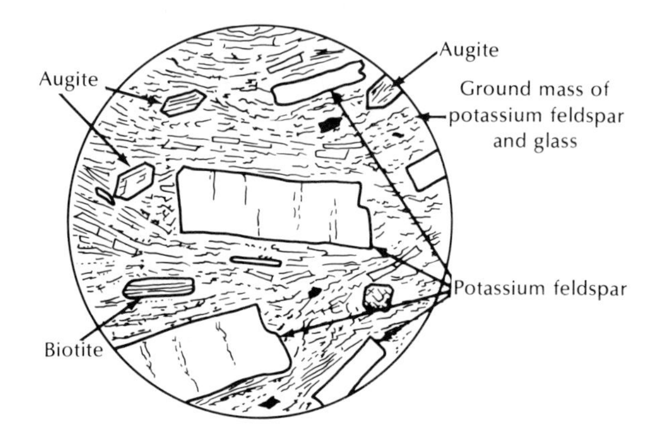

Fig. 5-6 Drawing of lava derived from a felsic magma, as seen under the microscope. The early, well-formed crystals of potassium feldspar, augite, and biotite are large. Fine-grained feldspar crystals and glass have formed between the larger crystals.

Magmatic differentiation

Processes that can occur during cooling and crystallization of magmas may also change the composition of the magma. ***Magmatic differentiation,*** the term used to describe these processes, is usually the result of either *compositional zonation* or *crystal fractionation*.

Compositional zonation Recently it has been found, in some large bodies of felsic magma,

Fig. 5-7 Zoned plagioclase crystal in a mafic lava flow, photographed under the microscope. Such a crystal has a core relatively richer in calcium and aluminum and a margin richer in sodium and silicon. *Edwin E. Larson*

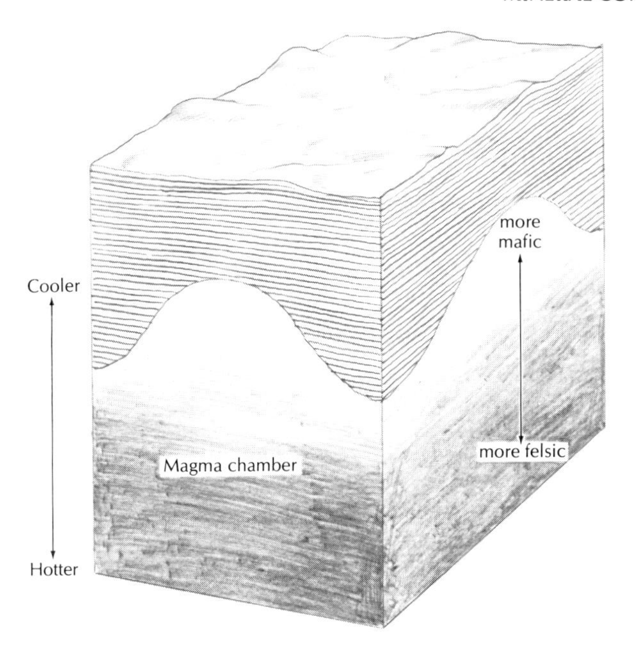

Fig. 5-8 Diagram of a compositionally zoned felsic magma chamber.

that the melt in the upper part of the chamber was relatively cooler and more felsic than that in the lower part (Fig. 5-8). The existence of magma partitioning such as this, termed ***compositional zonation,*** has been particularly well documented through the study of explosively erupted volcanic materials. Large-scale eruptions generally begin with material from the top of the chamber, tapping progressively deeper parts of the magma as they proceed. The study of materials from each phase of the eruption makes it possible to reconstruct the variation in chemical composition that existed in the magma chamber immediately before eruption.

Crystal fractionation As crystallization proceeds in any cooling magma, the earlier-formed crystals will generally differ in composition from the remaining magma. Certain ions will be selectively concentrated in the growing crystals and relatively depleted in the residual magma, and conversely, of course, elements not concentrated in the growing crystals will be relatively concentrated in the magma. Physical separation of the crystal portion from the residual magma, called ***crystal fractionation,*** will produce a secondary magma different in composition from the starting magma. Crystals can be separated from the residual magma in a number of ways. Some are heavy enough to sink through the melt and come to rest on the chamber floor, in which case armoring by successive layers effectively prevents further association of the crystals with the magma. In the Palisades, a large plutonic body exposed in cross section as a bold cliff along the Hudson River in New York and New Jersey, fractionation took place when crystals settled to the bottom of the chamber (Fig. 5-9). Fractionation can also take place if the residual magma drains away from the earlier-formed crystal mush.

Crystal fractionation leads to mineral variations in igneous rocks. For example, over a period of time, the lavas spewed from a volcano will commonly vary in composition: lavas rich in iron, magnesium, and calcium may be erupted first, followed by others less rich in

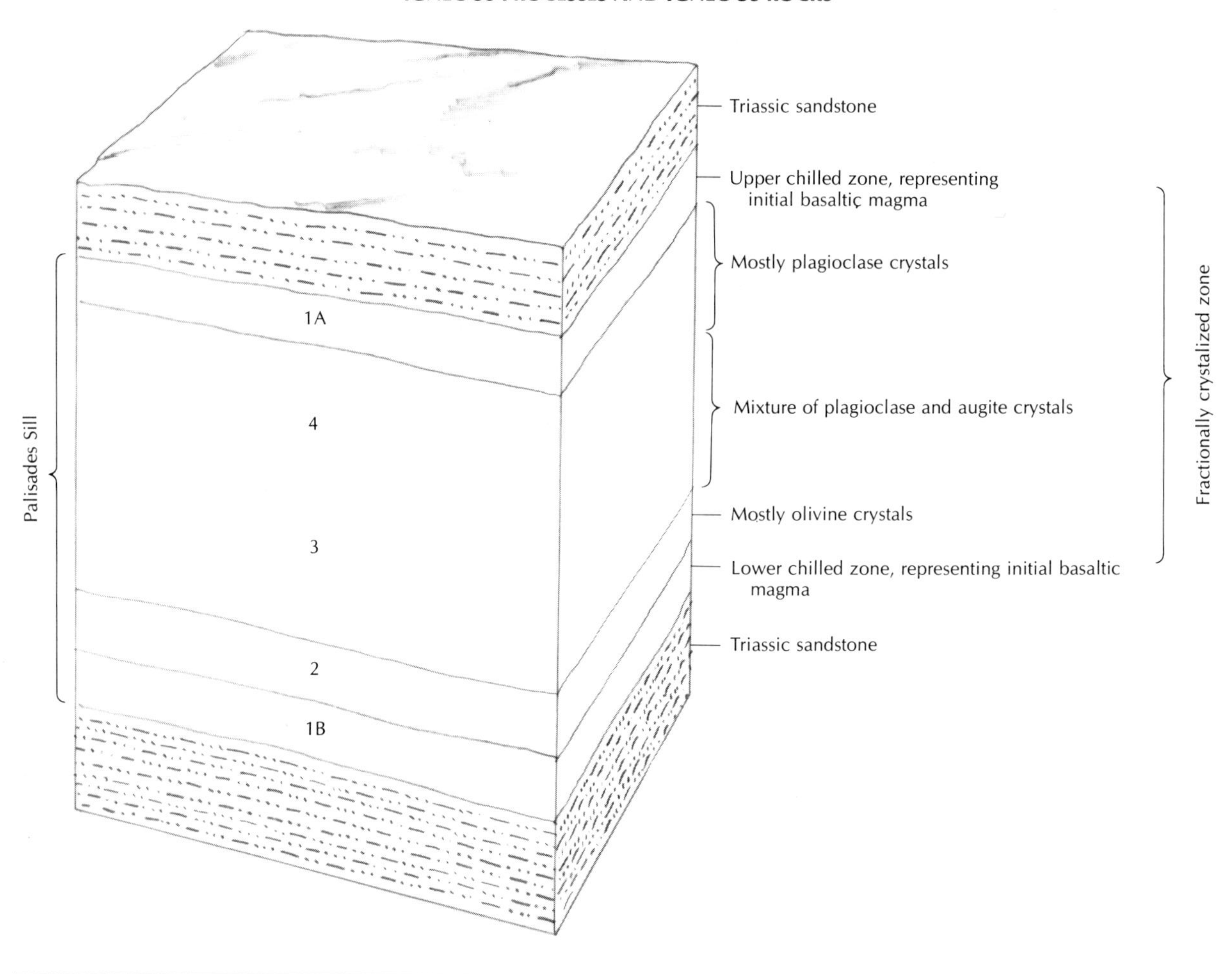

Fig. 5-9 Diagram of the Palisades Sill along the west bank of the Hudson River, New York. Upon initial intrusion of the basaltic magma into the Triassic sandstone, the upper and lower zones quickly chilled and solidified, producing a fine-grained basalt (zones 1A and 1B). Thereafter, during cooling, the remaining magma underwent fractional crystallization. Olivine, which formed first, settled to the bottom to form a layer (zone 2), followed, as cooling and crystallization proceeded, by a layer of augite (with some plagioclase) (zone 3), and finally by plagioclase (zone 4). The overall composition of zones 2, 3, and 4 is the same as that of the initial magma in zones 1A and 1B.

these minerals and more rich in sodium, potassium, and silica. During volcanic eruptions in Hawaii, for example, the most commonly erupted lava is mafic. It crystallizes quickly to form a dark rock composed mainly of olivine, calcic plagioclase, and augite. The remaining melt in the undisturbed cooling underground chamber may with time begin to form crystals of olivine and calcic plagioclase, which settle out and fall to the bottom; thus the magma is relatively depleted of calcium and magnesium. It is not uncommon, therefore, that during the next phase of volcanic activity, the first magma erupted is different from the average mafic composition—being relatively enriched in potassium, sodium, and silicon, and relatively depleted in magnesium and calcium. The longer the time between eruptions, the more changed is the initial lava from the original composition.

New magma of the original composition introduced into the magma chamber at any time during the middle or later stage of crystal fractionation will interrupt the process, producing a hybrid magma. The exact composition of the hybrid will depend on the amount of new magma introduced and the composition and amount of residual magma and first-formed crystals mixed with it.

Fig. 5-10 Coarse-grained plutonic igneous rock with a crystalline texture, photographed under the microscope. Note how the crystals have grown together to form a compact, interlocking system. *P.W. Lipman, USGS*

TEXTURE

The word ***texture*** is familiar to most people as having something to do with cloth; in fact it is derived from the Latin *textura,* a weaving. It can mean the arrangement and size of the threads in a woven cloth; for example, burlap has a much coarser texture than silk. But more basically, it means structure or system. Thus, when we apply the term to an igneous rock, we mean the size of the crystals as well as their mutual relationships.

Crystalline texture

An igneous rock in which the minerals form an interlocking system is said to be ***crystalline*** (Fig. 5-10). If the crystals are visible to the naked eye, the rock is said to have a ***fine-, medium-,*** or ***coarse-grained*** texture, depending on the size of the crystals. An igneous rock with a crystalline texture, as revealed by the microscope, but in which most of the crystals are too small to be seen by the eye alone, has an ***aphanitic*** texture. The size of the crystals has nothing to do with their mineralogy; the mineral makeup of rocks with completely different grain sizes could be all the same. Why, then, the differences in crystal size? The answer again lies in the way in which the magma solidifies. If it cools slowly and under relatively undisturbed conditions, large crystals have a chance to grow around the nuclei in the still-fluid magma. They may grow to a fair size, up to 1.3 cm (0.5 in.) or more, in what is essentially a sort of crystal mush, with the last-forming minerals filling the interstices between the earlier-forming materials when the whole mass finally solidifies. A crystal grows through the slow-acting diffusion, or movement, of the ions from the magma to the crystal surfaces, where they are added one by one. Should the magma cool rapidly, crystals grow around the floating crystal nuclei as they did in the slow-cooling magma, but the whole process speeds up. Diffusion of ions over long distances in the magma is impossible, so many centers form and each nucleus continues to grow from diffusion taking place over very short distances.

Fig. 5-11 Obsidian, the most common type of volcanic glass, showing the characteristic conchoidal fracture. *Ward's Natural Science Establishment, Inc.*

The result is a rock that although crystalline, is a tightly knit fabric of much smaller crystals.

The texture of igneous rocks is determined in large part by what is known as their mode of occurrence—whether they solidify above ground or below it. Or to phrase the statement another way—whether they are ***volcanic*** or ***plutonic*** rocks. Volcanic rocks cool relatively rapidly and therefore, for the most part, have fine-grained textures; intrusive, or plutonic, rocks cool more slowly, allowing the growth of larger crystals, and a coarse-grained texture is the result. In general, even the grain size of plutonic rocks is largely governed by the rapidity of cooling. If the magma chamber is small or close to the surface, it will cool more rapidly than a larger one or one at depth, and correspondingly, the rocks will have a finer grain size.

Two additional textures seen frequently in igneous rocks are ***glassy*** and ***porphyritic.*** They, too, provide information about the conditions that existed during cooling.

A glassy texture is typified by the volcanic rock ***obsidian*** (from the Latin *obsidianus,* after its describer Obsius), the most common type of volcanic glass. It is non-crystalline, because it passes quickly from the liquid to the solid state, leaving no time for the ions in the original magma to arrange themselves in ordered ranks as crystals. Obsidian commonly fractures conchoidally or irregularly into fragments with sharp edges (Fig. 5-11)—a property that made it an effective weapon when flaked or chipped into arrowheads, spear points, and knife blades by the people of ancient cultures.

Porphyritic texture takes its name from the Greek word *porphyra,* the word used for imperial purple—a highly prized dye extracted from an eastern Mediterranean shellfish. By extension, the name was applied to a very specific kind of rock—***porphyry,*** a dark, igneous rock from Egypt that contains small white feldspar crystals embedded in purplish, fine-grained crystals. Porphyry was greatly favored in the Roman world for busts of emperors, as well as their sycophants and the lesser dignitaries of the court. It made a strikingly regal contrast, especially when set off against a white marble toga.

Today, by a further extension of the original word, the term porphyritic is applied to any igneous rock with crystals of two markedly different sizes (Fig. 5-12). Such a texture is seen to indicate that the magma underwent two cycles of cooling, perhaps an earlier slow-cooling phase during which some crystals, with well-formed faces, grew to a relatively large size, followed by a later, more rapid phase when the smaller grains crystallized. In such a rock the larger crystals are called ***phenocrysts*** (from the Greek *phainein,* to show, combined with crystal), and the material in which they are embedded is called the ***groundmass.*** Commonly, the groundmass is finely crystalline and even glassy (when the last stages of cooling are very fast).

A porphyritic texture is common in igneous rocks that solidified in small bodies at shallow depths. Initially, well-shaped crystals began to

Fig. 5-12 Porphyritic lava, photographed under the microscope. Large light-gray tabular crystals of calcic plagioclase are set in a fine-grained groundmass composed of plagioclase (elongate, light gray), augite (dark gray), and magnetite (black). In most cases, a texture such as this is the result of two distinct phases of cooling, a slow one followed by a rapid one. *Edwin E. Larson*

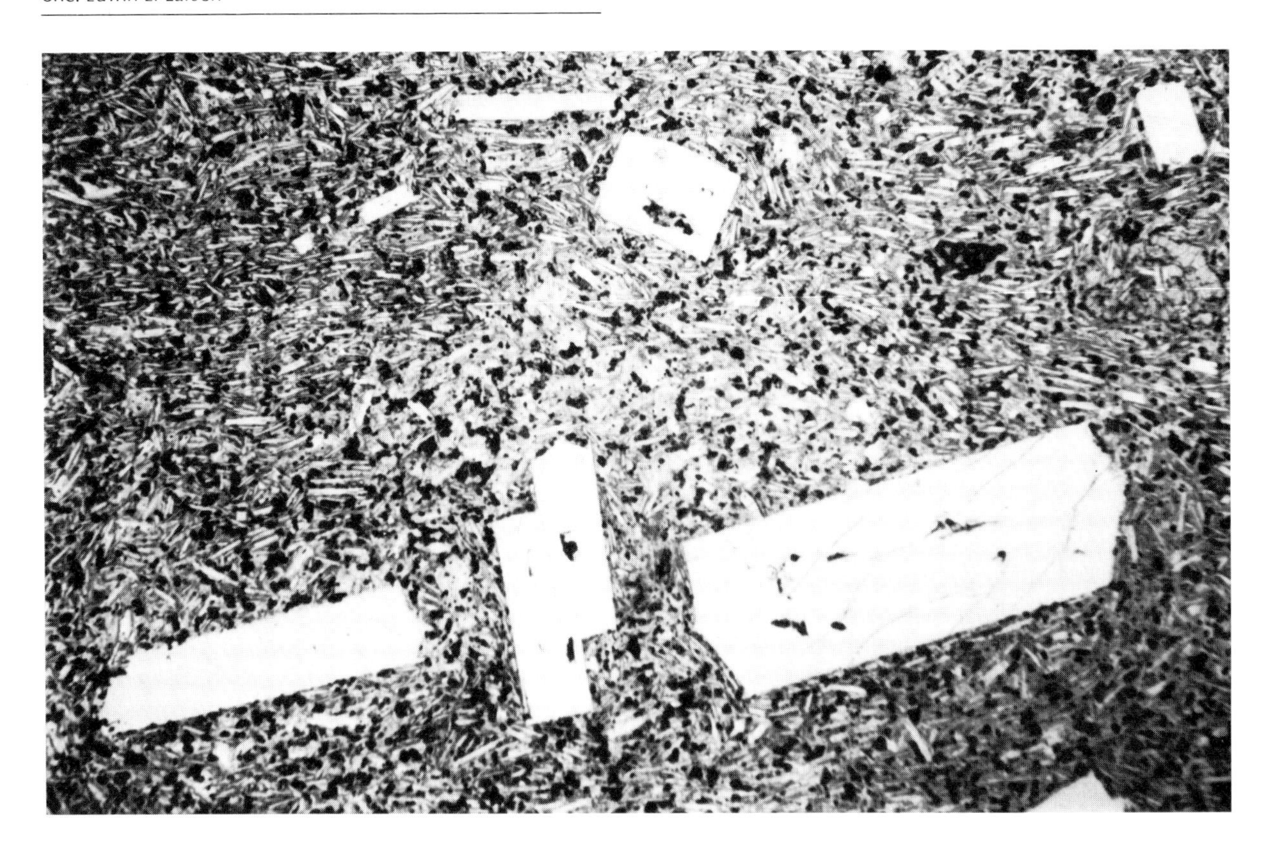

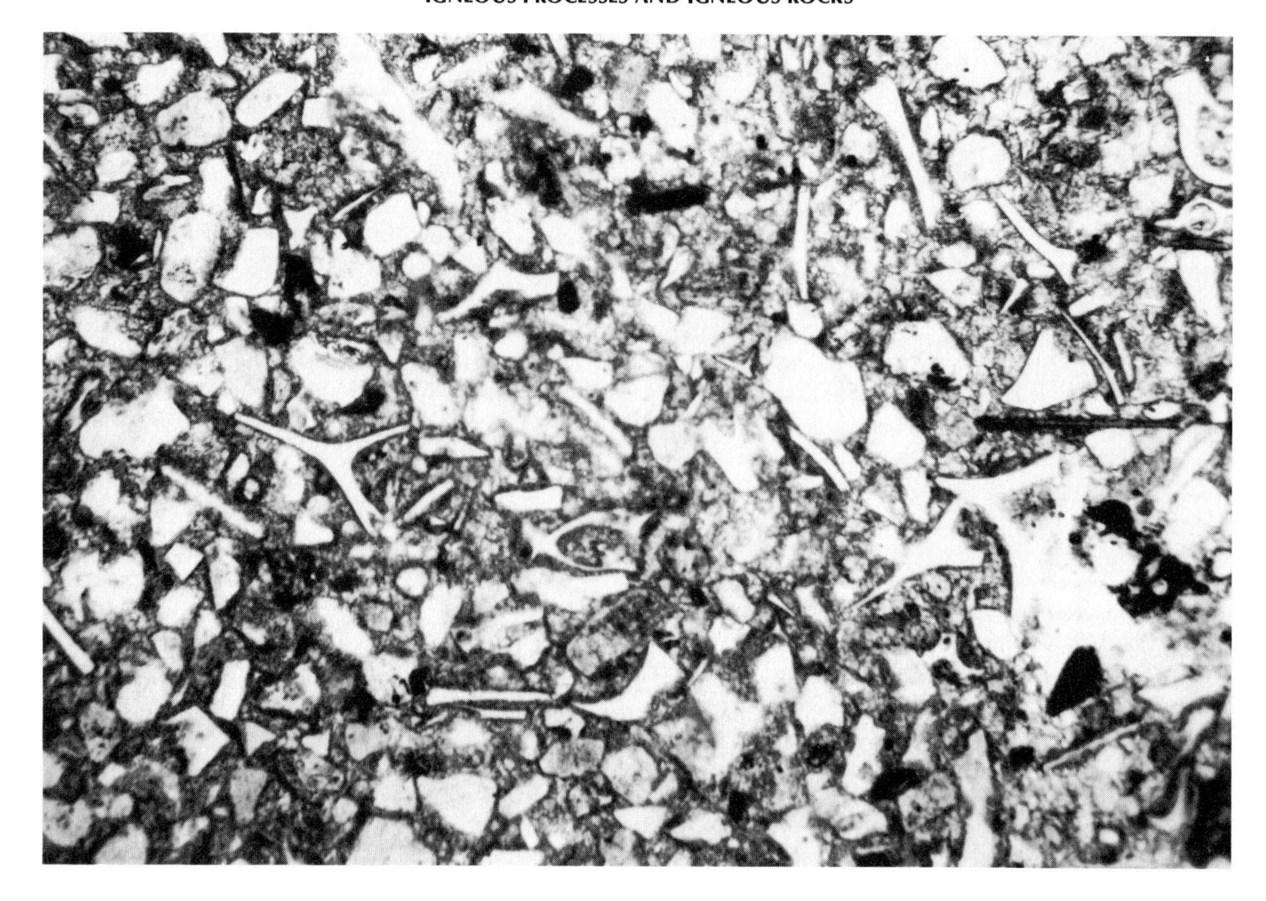

Fig. 5-13 Volcanic ash composed predominantly of larger glass shards (light gray) and biotite and magnetite grains (dark gray to black) set in a matrix of finer glass shards. *Edwin E. Larson*

form in magma in a deeper chamber. Movement of the magma to shallower depths then brought about a relatively rapid cooling, and the magma "froze" around the still-floating, larger and earlier-formed crystals.

Because the larger crystals in a porphyry form while the magma is still largely fluid, they grow without interference and often may achieve a nearly perfect crystal form. The minerals crystallizing later come out of solution more rapidly, and since the growth of each crystal is interfered with by its neighbors, very few develop well-formed crystal shapes.

Quickness of cooling and variations in cooling history probably are the main reasons for variability in igneous rock texture, but not the only ones. Volcanic rocks rich in potassium, aluminum, and silicon tend to be finer grained, even to the point of being glassy, than dark-colored volcanic rocks rich in iron, magnesium, and calcium. In fact, glass-rich varieties of ferromagnesiam and calcium-rich rocks are virtually non-existent. Quickness of cooling could not have been the principal reason for the difference in grain size, since both types of rocks are volcanic and must have cooled rapidly. The reason seems to be related to differences in the temperature of the lava when it was erupted and to the viscosity of the erupted molten material. As mentioned earlier, crystals grow through the process of diffusion. At higher temperatures, the process is more efficient than at lower temperatures; also, ions can move about more easily in a less-viscous magma. Lavas rich in iron, magnesium, and calcium originate at greater depths and are usually much hotter and therefore less viscous than those rich in potassium, aluminum, and silicon.

In plutonic rocks, differences in textures are also related to the amount of volatile constituents (such as water, carbon dioxide, and hydrogen chloride) in the magma. It has been found that, in any given magma, as the content of volatile materials is increased, initial crystallization takes place at a lower temperature and the efficiency of ionic diffusion is increased. The amount of dissolved gases in the magma is related to the amount contained in the rocks that melted to form the magmas, and to pressure. At greater depths, the confining pressure is greater, and magma can hold more volatile materials in solution than it can at lesser depths, and lower pressure. Larger crystals should form from magmas relatively richer in volatile materials, other things being equal.

Clastic textures A clastic (fragmental) texture is usually associated with sedimentary rocks; yet volcanic rocks can also be clastic.

Indeed, some clastic rocks have come about through the reworking of volcanic rocks by streams flowing on the flanks of a volcano. The deposits are composed mainly of rounded to subangular fragments, which can range from boulders to clay-size grains. Such rocks are difficult to classify. But, since they are associated with volcanism, we shall consider them as volcanic with a clastic texture.

Clastic rocks can also be formed as a direct result of volcanism. The term applied to most such rocks is ***pyroclastic*** (Greek, fire and broken), which differentiates them from normal sedimentary rocks and emphasizes their mode of origin. Pyroclastic rocks form from the volcanic material blown into the atmosphere during the explosive discharge of a vent and consist principally of fragments of volcanic crystalline rocks, volcanic glass, and phenocrysts (Fig. 5-13).

CLASSIFICATION OF IGNEOUS ROCKS

All igneous rocks can be classified on the basis of their mineralogical characteristics and their textures. More than several hundred names have been applied to igneous rocks, but we will discuss only a few major divisions here. The bulk of the named rock types are simply minor varieties of those listed below (Fig. 5-14).

Six of the major rock units, which comprise most of the igneous rocks, are shown. The names apply mostly to the crystalline igneous rocks. Clastic volcanic rocks are discussed separately.

Crystalline igneous rocks

Grossly, the three plutonic-volcanic divisions correspond to three different starting magmas. Gabbro and basalt form at high temperatures and from mafic melts; diorite and andesite at intermediate temperatures and from magmas of intermediate composition; and granite and rhyolite at relatively low temperatures and from felsic melts.

Felsic rocks

Granite (the origin of the name is lost in antiquity, but it is believed by some to be derived from the Italian adjective *granita*, grained) is a relatively light-colored, usually coarse- to fine-grained plutonic rock. Potassium feldspar and sodic plagioclase together make up most of the rock (Fig. 5-15). Quartz is normally present in amounts up to 25 per cent; it is the last mineral to crystallize and occurs as gray rounded to irregular masses filling the spaces between earlier-formed minerals. Muscovite is also a common constituent of some granites. Black minerals, chiefly amphibole and biotite, are present in small amounts only.

Since granite is a widely used rock for such structures as tombstones, monuments, and government and bank buildings (it has been used in the construction of stately edifices for thousands of years), almost all of us readily recognize the characteristically speckled appearance of this rock, with its white to gray background of feldspar and quartz flecked with dark spangles of mica plates and needles of hornblende as well as its coarse- to fine-grained texture. In the coarser varieties, some minerals (chiefly the feldspars) may be 1 cm (0.4 in.) or more across.

Rhyolite (from the Greek, *lava torrent* or *stream* + stone), a volcanic rock, has the same compositional range as granite, but a wholly different texture. Commonly, it is fine-grained, glassy, or porphyritic; if porphyritic, the groundmass may be so glassy or so fine grained that the minerals can be resolved only with the aid of a microscope, if at all. The usually small phenocrysts in the porphyritic varieties are sodic plagioclase, quartz, biotite, and, rarely, other ferromagnesian minerals (Fig. 5-16). Quartz can be an early-crystallizing mineral during the formation of a rhyolite, and when it is, the crystals commonly appear completely formed and often terminated by pyramidal crystal faces at both ends. Rhyolite, ordinarily light colored, may be white, light gray, or various shades of red. A characteristic textural feature is a streaked pattern known as ***flow banding*** (Fig. 5-17). As the name implies, the banding results from the concentration of colored material or glass in layers, during flowage of the highly viscous lava. In many cases it appears that the lava must have flowed very slowly, much as heavy molasses does.

Rhyolite magma can cool so rapidly that crystallization of minerals is impossible. The resulting ***volcanic glass*** (already mentioned in our discussion of texture and obsidian) lacks long-range internal crystal ordering. Thus, geologically, these glasses are considered to be supercooled or "frozen" liquids. Window glass, another supercooled liquid, is colorless because it has no metallic impurities. Obsidian, however, has such an abundance of impurities that the color is generally black, or less commonly red.

Volcanic glass tends to crystallize very slowly

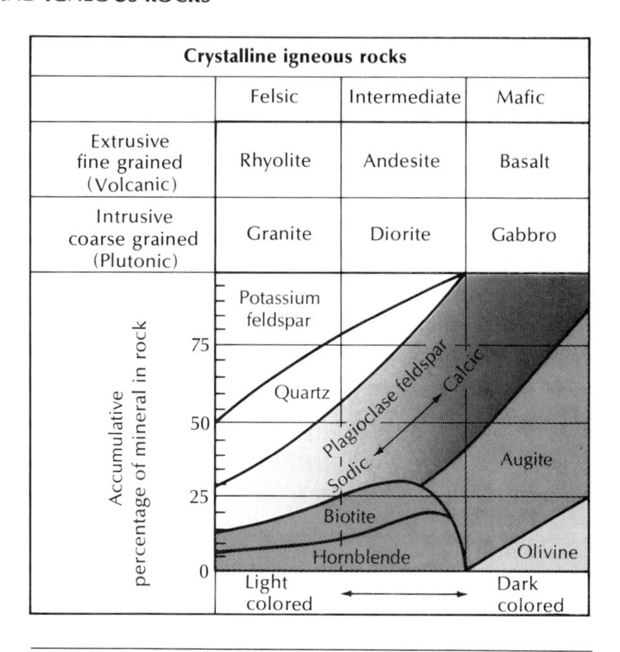

Fig. 5-14 The mineral composition and names of six common crystalline igneous rocks. The degree of shading corresponds to the color of the mineral. Felsic rocks contain only a small percentage of dark minerals, mafic rocks a large percentage.

at the earth's surface. Over an extremely long period of time, the ions in obsidian undergo solid-state diffusion (usually in the presence of water), and eventually crystallize. Since diffusion takes place slowly, only tiny crystals can grow. The process is called ***devitrification;*** the resultant rock is called devitrified obsidian or devitrified glass. As a result, true obsidians are virtually unknown in the rock record farther back than the Cretaceous Period (135 million years).

Pumice (an ancient name, from a Greek word meaning worm-eaten and mentioned as long ago as 325 B.C. by the philosopher and naturalist Theophrastus) is a special variety of rhyolitic volcanic glass, rather like petrified froth, resembling the foam on the top of beer (Fig. 5-18). Because of an abundance of gas cavities the rock is extremely porous and, accordingly, light in weight. Some pumice will even float on water; if pumice is blown out of coastal or oceanic volcanoes, it may drift for thousands of kilometers before becoming waterlogged and sinking to the bottom.

Fig. 5-15 Granite from San Bernardino County, California, photographed under the microscope. The large crystal (center) is potassium feldspar, the banded elongate crystals are mostly plagioclase, and the light- to dark-gray, equally spaced irregular grains are mostly quartz. *J.C. Olson, USGS*

Intermediate rocks

Diorite, a coarse- to fine-grained plutonic rock, has a mineral composition that places it about midway between granite and gabbro. Its name, which is derived from the Greek word meaning to divide, indicates its intermediate composition. It has an intermediate plagioclase feldspar as its chief constituent, with little or no quartz or potassium feldspar. Hornblende is its predominant dark mineral, although biotite is often relatively abundant. The dark minerals can together be nearly as abundant as the feldspar (Fig. 5-19).

Because it contains little quartz and potassium feldspar, and nearly equal amounts of plagioclase and ferromagnesian minerals, diorite tends to be a drab gray rock. It is not used as widely as granite for building stone because it is somewhat less abundant and possibly because its somber gray color is less pleasing.

Andesite, named for its occurrence in the Andean summit volcanoes of South America, is generally a gray to grayish-black, fine-grained volcanic rock. Commonly, its texture is porphyritic. In general, visible quartz is lacking, the chief feldspar is intermediate plagioclase, and the dark minerals are principally augite, hornblende, and biotite (Fig. 5-20). The same minerals also occur as phenocrysts in the por-

Fig. 5-16 Rhyolite from Rio Grande County, Colorado, photographed under the microscope. Phenocrysts are mostly quartz and potassium feldspar with a few plagioclase, biotite, and magnetite (black) crystals. The groundmass, originally glassy, has recrystallized to a network of extremely tiny grains of quartz and potassium feldspar. *P.W. Lipman, USGS*

Fig. 5-17 Flow banding caused by viscous flow in a felsic lava that largely congealed as obsidian. The blocks are at the base of a volcanic dome south of Mono Lake, California. *John Haddaway*

phyritic varieties. Andesite in general covers the same range of intermediate composition as its plutonic counterpart, diorite. Andesitic rocks are more abundant at the earth's surface than are rhyolites, but are less abundant than basalts.

Mafic rocks

Gabbro, an old name for many of the dark rocks used in Renaissance palaces and churches in Italy, is a plutonic rock that consists typically of a coarse-grained intergrowth of crystals of pyroxene and calcic plagioclase (Fig. 5-21). Many gabbros also contain olivine, and some contain small amounts of hornblende. Unlike diorite and granite, gabbro contains larger amounts of ferromagnesian minerals than of feldspars. There are exceptions, of course, and one gabbroic variety, ***anorthosite,*** consists almost entirely of calcic plagioclase interlocked in a coarse-grained texture (Fig. 5-22). Another variety, much sought after as a decorative building stone for storefronts and banks, contains large dark-purplish plagioclase crystals that give a wonderfully impressive play of colors, like those of peacock feathers.

Basalt, one of the most ancient names in geology, apparently dates back to Egyptian or Ethiopic usage, and one of the first references to it by name is by Pliny the Elder. It is by far the most abundant of all volcanic rocks. Many regions of the world, such as the plateau bordering the Columbia River in the northwestern United States, were once almost completely inundated by vast outpourings of basaltic lavas. The volume of basalt in the Columbia Plateau is estimated to be about 307,200 km^3 (73,701 mi^3). In addition, many oceanic islands, such as Samoa, Hawaii, and Tahiti, are volcanoes composed of basalt, and the igneous rocks directly underlying the oceanic sediments appear to be primarily basalt.

Basalt looks very commonplace for the dominant position it holds among volcanic rocks. When it is unweathered, it is ordinarily coal black to dark gray, and fine grained to aphanitic in texture. The two principal mineral constituents are pyroxene and calcic plagioclase, and many varieties contain olivine as well (Fig. 5-23). Any of the three minerals may be present as phenocrysts in porphyritic varieties. Basalt is commonly frothy and cellular and filled with innumerable small holes called ***vesicles***—gas bubbles trapped in the lava as it solidified. Such a structure, which is by no means confined to basalts, is denoted by the adjective ***scoriaceous*** (Fig. 5-24).

One common variety of basalt, ***diabase,*** has a

Fig. 5-18 The bubble structure of pumice, photographed under the microscope. The lensoidal light gray areas represent the centers of glass bubbles. The walls of the bubbles show as black. *Edwin E. Larson*

Fig. 5-19 Diorite from the Front Range, Colorado (left), photographed under a microscope. The large finely striped crystal in the center is intermediate plagioclase containing small crystals of quartz, hornblende, and biotite. The light-gray to white irregular grains are mostly potassium feldspar and quartz. *W.A. Braddock, USGS.* As seen in a hand sample (right), the light-gray patches are composed of intermediate plagioclase or, less commonly, quartz; the black areas are made up of grains of hornblende, biotite, and magnetite. *Janet Robertson*

texture in which the plates of calcic plagioclase occur in an interlocking system. Spaces between the crystals are occupied by pyroxene, which crystallizes later (Fig. 5-25). Diabase of a coarser-than-normal grain size commonly results from the relatively slow cooling of shallowly buried, relatively thick bodies of magma.

Basaltic melts, because of their high temperature and chemical composition, have relatively low viscosities and flow more rapidly than felsic melts. Thus, when basaltic lava forms pools or stops flowing, it is still largely molten except for a thin outer skin. As the flow cools, and begins to crystallize from the outside inward, it contracts, and stresses develop in a plane parallel to the top and bottom cooling surfaces. With further crystallization, a number of vertical joints, or fractures, begin to form. In a fairly uniform sheet of lava, such fractures tend to be relatively uniformly spaced and random in direction, resulting in polygonal (usually four to eight faces) vertical prisms, or columns, that look like giant fence posts, bunched together and stacked on end (Figs. 5-26 and 5-27). When the closely spaced columns are seen from above their pattern resembles that of hexagonal bathroom tiles (Fig. 5-28). Such fracturing is called ***columnar jointing*** and is most common in basaltic rocks, although by no means confined to them. If conditions are right, any sheet of magma, regardless of composition, can develop a columnar joint pattern upon crystallization. Some of the finest examples of columnar jointing in the world can be

Fig. 5-20 Porphyritic andesite from Conejos Peak, Colorado, photographed under the microscope. Most of the phenocrysts and small lath-shaped crystals are intermediate plagioclase. The groundmass is a mixture of glass and fine crystals of pyroxene. *P.W. Lipman, USGS*

Fig. 5-21 Gabbro from Victoria Land, Antarctica, photographed under the microscope. The large dark-gray crystals are pyroxene, the light-gray crystals are mostly calcic plagioclase, and the black crystals are iron oxides. *W.B. Hamilton, USGS*

Fig. 5-22 Anorthosite from Comanche County, Oklahoma, as photographed under the microscope. The rock is composed almost entirely of large tabular crystals of calcic plagioclase. *W.B. Hamilton, USGS*

seen at the Devil's Postpile in the Sierra Nevada Range in California and Fingal's Cave in Scotland.

Pyroclastic and volcanic-clastic rocks During the eruption of a volcano, in addition to liquid lava, which flows down the flanks to subsequently congeal into typically crystalline rocks, a great deal of lava is also hurled into the air above the vent (Fig. 5-29). Some falls back to earth near the erupting vent, some falls farther out on the volcano's flanks, and some is hurled high aloft to be carried long distances from the vent by prevailing winds. All such explosively ejected material is termed pyroclastic. Some consists of large and small incandescent clots of liquid lava, which cool and largely consolidate during their flight. Some are spindle shaped, but most occur in rounded, irregular forms. All such clots are called ***volcanic bombs*** (Fig. 5-30). Many have a crust a few centimeters thick that looks much like that of a loaf of French bread. Some have a flattened snout, which apparently results from impact with the earth's surface, indicating that the bomb did not solidify completely during its trajectory. Volcanic bombs cool rapidly and therefore are usually aphanitic to glassy in texture; they can be of any chemical composition, felsic to mafic.

Smaller volcanic rock particles are also commonly ejected from vents. Tiny fragments of chilled lava (about 2 cm, or 0.8 in., in diameter) are termed ***lapilli*** (an Italian word for little stones), and rocks composed dominantly of that material are called ***lapilli tuffs.*** Even smaller pieces, down to the size of dust, are called ***ash.***

Generally, bombs and cinders are so heavy that they cannot be blown far from the vent; they fall mostly on the volcano flanks. Deposits of angular bombs, cinders, lapilli, and ash, in an unsorted mixture, but possessing crude layering are called volcanic ***agglomerates*** or sometimes ***volcanic breccias.*** If the ashy material dominates they are called ***tuff breccias.*** Lapilli and ash can be blown high over an erupting cone; subsequently some will fall back on the cone flanks and some, particularly the ash, will be carried by prevailing winds and subsequently deposited in sizable amounts as far as 1000 km (621 mi) away. Ash that is carried into the air and that later falls back to earth to form layered deposits is called ***air-fall ash,*** and the consolidated rock composed of that material is called an ***ash-fall tuff.*** The latter, fine grained in texture, is composed of pieces of pumice and jagged fragments of volcanic glass and phenocrysts—usually potassium feldspar, quartz, amphibole, and biotite—and small angular clasts of volcanic rocks ripped from the throat of the volcano during the explosion. Highly explosive

Fig. 5-23 Basalt from the Columbia River Formation in Idaho, photographed under the microscope. The tabular phenocrysts and smaller lath-shaped, light-gray grains are calcic plagioclase, the dark-gray crystals are pyroxene, and the black grains are magnetite. *W.B. Hamilton, USGS*

volcanoes that pour forth great quantities of ash are usually those of felsic to intermediate composition.

One particular variety of pyroclastic rock composed largely of ash appears to have been extremely hot when it was deposited—so hot that the angular fragments of glass were welded together after they came to rest. Under their own weight the deposits, which may be up to 300 m (984 ft) thick and laterally extensive, settle and compact—even to the point that softened pumice fragments lose their porosity and completely collapse into dark-colored, flattened blebs of volcanic glass. The resulting rock, called a ***welded ash-flow tuff,*** is

Fig. 5-24 Cellular or scoriaceous basalt, a fine-grained to glassy basalt filled with tiny bubble holes (left). *Hal Roth.* (below) Cellular basalt, photographed under the microscope. *Edwin E. Larson*

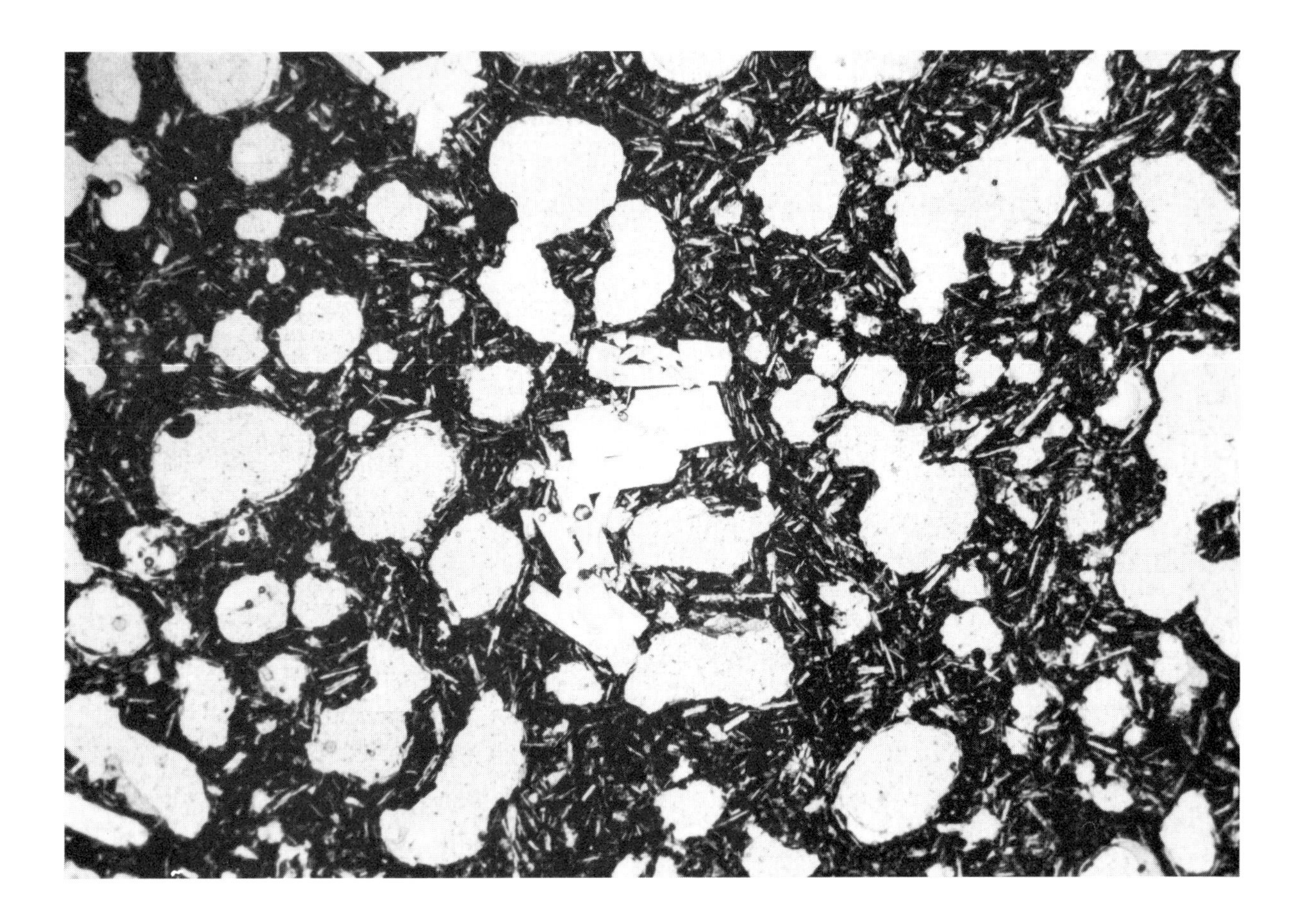

dense and strong, with an easily visible banding parallel to the flow top (Fig. 5-31). Some of the most impressive layers of welded tuff are found in the vicinity of Yellowstone National Park in Wyoming. Single beds, resulting from catastrophic emissions of immense volumes of hot ash, are more than 160 km (99 mi) long and up to 100 m (328 ft) thick.

Some volcanic deposits are bedded, composed of sub-angular to sub-rounded and rounded fragments, ranging in size from boulders down to silt and clay. Such deposits are the result of sedimentary processes acting on the slopes of the volcano. The upper slopes of a volcano commonly are piled high with relatively unstable pyroclastic debris and interlayered lava flows that are particularly susceptible to slumping and stream erosion. When saturated with rain (which may fall as a direct result of the volcanic activity), the slopes often become very unstable. Under the influence of gravity the rain-soaked material moves downslope, within the confines of stream channels, in torrents of muddy ash and coarser pyroclastic debris and out onto the gentler slopes at the perimeter of the volcano and beyond (Fig. 5-32). Such slides have been called volcanic mud flows—a name recently replaced by the Indonesian term ***lahar.*** On the islands of Indonesia, volcanic cones are numerous and lahars are common. Laharic deposits are found on the flanks of Mount Rainier in Washington, where they cover an area of 320 km^2 (124 mi^2) and have a volume of 2 km^3 (about 0.5 mi^3).

Also, through the continued activity of streams on the sides of the volcano, material can be eroded from the steeper slopes of the volcanic vent and deposited in aprons surrounding its base (Fig. 5-33). Such deposits, which are composed of stream-worn clasts, are truly sedimentary rocks and are so named. Often stream-deposited material will be interlayered with strata composed of lahars.

The term volcanic breccia, which we introduced earlier with respect to certain pyroclastic deposits, can also be applied to any volcanic rock layers in which the fragments are angular (Fig. 5-34).

Fig. 5-25 Moderately coarse-grained diabase from Victoria Land, Antarctica, photographed under the microscope. Lath-shaped calcic plagioclase crystals are present in an interlocking pyroxene crystal system (dark-gray). Magnetite (black) is also present. *W.B. Hamilton, USGS*

PLUTONIC BODIES AND IGNEOUS PROCESSES

It is now quite clear that molten material from volcanoes does not owe its existence to the melting of rocks through the combustion of coal seams, as Abraham Werner so strongly believed. Rather, volcanic activity appears to be

Fig. 5-26 Columnar jointing in the lower portion of a Tertiary basaltic lava flow at Fingal's Cave, Island of Mull, Scotland. *Popperfoto*

Fig. 5-27 Columnar structure in basaltic lava flow; The Devil's Postpole, California. *Sierra Club*

the surface manifestations of igneous processes and activity that go on deep below the surface, even to a depth of 250 km (about 155 mi). Although we can learn something about the deep-seated activity from the detailed study of volcanoes themselves—where they occur and what the eruptive products are like—we never know exactly what is going on at depth. The situation is analogous in a sense to a lawn sprinkler system. When it is working we see the fountaining sprays of water, but we have little idea of the complexity of the subsurface plumbing system or where the source (the pump or water main) is located. The only way to find out is to dig up the pipes. Such a venture is not very feasible when one considers

Fig. 5-28 Upper surface of the lava flow at The Devil's Postpile, California, showing the ends or cross sections of the columns seen in Fig. 5-27. Cooling of the lava causes it to fracture into columns of 4, 5, 6, or 7 sides. *Hal Roth*

Fig. 5-29 Cerro Negro erupting at night. Luminous tongues of fire, mostly of pyroclastic material, are explosively discharged from the vent. The white streaks represent trajectories of individual volcanic bombs. *Edwin E. Larson*

uncovering the "plumbing system" that carries magma to the volcanoes. Fortunately, however, natural processes have in part accomplished the task for us. Through erosion, acting over long periods of geological time, large quantities of surface rocks have been successively stripped away in many parts of the world. In some instances, erosion has cut down through several kilometers of surface rocks. Unfortunately, erosion has never cut deeply enough to expose the lower parts of the plutonic plumbing system, but at least we can examine the upper parts.

Remember, though, that when we observe crystallized plutonic bodies we are actually looking at a fossil, or petrified, system of conduits and pipes. The igneous body itself was molten long ago. An eruptive vent that might have been present at the surface has been mostly stripped away. The study of igneous rocks, then, presents an insoluble problem: when a volcano is active we cannot see the

Fig. 5-30 A large elongate volcanic bomb amidst other smaller-sized pyroclastic material found on the northeast flank of North Sister volcano in the Cascade Range. *Edwin E. Larson*

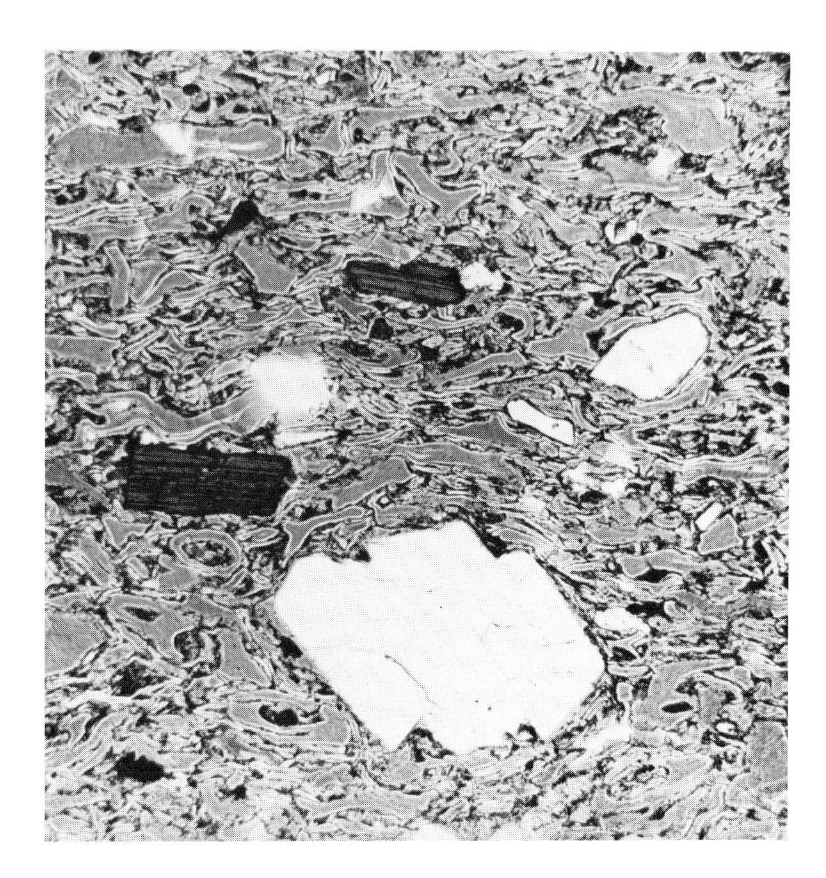

Fig. 5-31 Welded ash-flow tuff from near Gunnison, Colorado, photographed under the microscope. Partly broken phenocrysts of sodic plagioclase (white) and biotite (dark gray) are in a matrix of fused, angular fragments of glassy ash (light gray). Note that the crude layering in rock is parallel to the long axis of the biotite grains. *J.C. Olson, USGS*

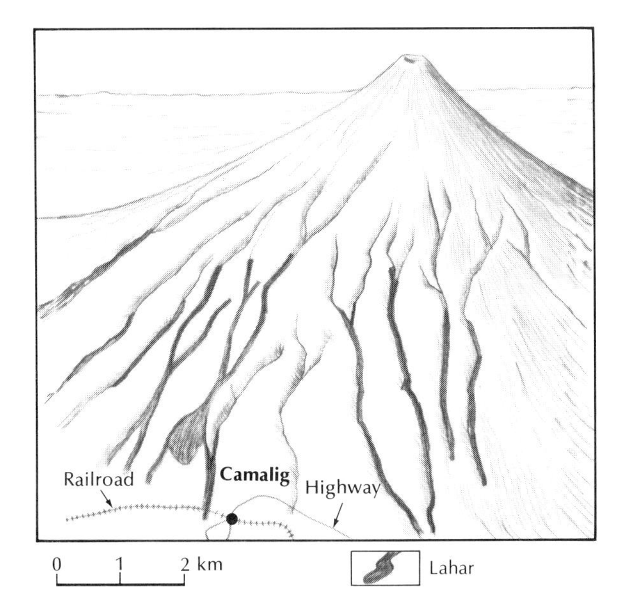

Fig. 5-32 Diagram of Mayon Volcano, Philippine Islands, showing tracks of lahars generated during the 1968 eruption.

Fig. 5-33 Laharic material that accumulated on the slopes of volcanoes in Yellowstone National Park about 40 million years ago, buried a stand of trees (only the petrified trunks remain). The character of the material can be seen in the outcrop above the geologist's head. Note the angularity of the clasts, the lack of good stratification, and the mixture of large and small particles.

Fig. 5-34 Volcanic breccia–angular fragments of andesite set in a matrix of fine-grained andesite fragments, ash, and mud. From Gunnison County, Colorado. *Janet Robertson*

conduit system below; when we can see the subsurface channels the volcano superstructure has long since gone.

Intrusive rock bodies

Bodies of plutonic rock are found in a great variety of shapes and sizes. Many of the largest are elongate bodies up to 1600 km (994 mi) in length. Some are characteristically tabular or cylindrical; some are irregularly ellipsoidal, circular, or even hot-dog shaped. All such bodies, regardless of size or shape, are called by the general name ***pluton.***

Most plutons did not result from the melting and later solidification of rocks in place. Usually their chemical composition is completely different from that of the host rocks (commonly referred to as ***country rocks***) that surround them. Rather, magma at depth, once formed, appears to be relatively mobile, and at times amazingly so. It can push aside pre-existing rocks and force its way into cracks, some of which it may actually produce—the process is called ***forceful injection.*** In some crack systems, magma has been able to move distances of at least 50 to 100 km (31 to 62 mi). It can also move ahead by (1) melting away the surrounding country rocks or (2) wedging or prying out solid chunks of rock, which it then replaces. When magma makes its way into a pre-existing rock, it is said to have ***intruded*** into the host rock. Plutonic rocks, then, can be called ***intrusive*** rocks, and the surface between the pluton and country rock, the ***intrusive contact.***

Magma can move and intrude both laterally and vertically. It is through vertical rise that it comes close enough to the surface to be uncovered by erosion. If magma can rise high enough or send out fingers, or ***apophyses*** (Greek, meanimg offshoot), that can do so, it will eventually break through at the surface to produce a volcanic eruption. Some outbreaks of mafic magma have originated at a depth of 250 km (155 mi).

The reasons that magma moves and intrudes where it does are varied and not always completely clear. One good reason for its tendency to rise, however, is that liquid magma is normally less dense than the solid country rocks

Fig. 5-35 A mafic dike, about 3 m (10 ft) thick, cuts across shaly sediments, Hance Rapids, Grand Canyon, Arizona. *P.E. Patterson*

around it. A buoyant effect is created such that given an opportunity (for example, the presence of a fracture) the magma will move upward.

Once magma has stopped moving, it becomes stagnate and begins, or continues, to cool. At great depths it cools slowly, to often produce a coarsely crystalline texture; material crystallizing near the surface, where cooling is more rapid, forms rocks with finer-grained textures. Magma that has risen very close to the surface may cool so quickly that the grain sizes of the resulting rocks are essentially the same as those of volcanic rocks.

Porphyritic textures are typical of intrusive rocks from shallow to intermediate depths where there may have been a considerable movement of magma from one environment to another in a complex conduit system—which would result in different rates of crystallization.

Small plutons Dikes and sills are tabular intrusions in which two dimensions are large compared to the third—they have about the same geometry as that of a thin pad of note paper. The two types of intrusions are set apart by the relation they have to the layering (commonly that of sedimentary rocks) of the surrounding rocks. Dikes are ***discordant***—they cut across the layers (Fig. 5-35). Sills are ***concordant***—they are more or less parallel to the layer (Fig. 5-36). Concordance does not mean that sills follow a single stratum only, since it is not at all uncommon for them to angle up or down from one stratum to another and to follow it, for perhaps hundreds or thousands of meters, before making another step-like change in level (Fig. 5-36).

Dikes and sills can be of more than geological importance. Two such intrusions played a decisive role in Pennsylvania during three hot summer days a century ago in July of 1863. One intrusion is the thick basaltic sill that underlies Cemetery Ridge, a resistant, low-lying escarpment, and the other is the narrow, but persistent dike that supports Seminary Ridge, both of which were the dominant elements of upland terrain at the Battle of Gettysburg. The Union Army held the former and the Confederate forces the latter. Both sides made the most of the so-called ironstone boulders that had weathered from basalt outcrops along the top of each ridge. Piled into fences, the rocks made excellent defensive positions against the round shot and Minié balls of that day. Cemetery Ridge, along which the Union brigades were deployed in full strength on the fateful

Fig. 5-36 Diabase sill on Banks Island, Northwest Territories, showing an abrupt change in level of intrusion along a fracture. All the rock layers now above the sill and those that have since been eroded away were uplifted during the emplacement of the sill. *Geological Survey of Canada*

third morning, made such a continuous rampart, with a nearly unbroken forward slope up which Pickett's command had to charge that the assault was virtually foredoomed to failure, and doubly so when his men were called upon to dislodge men sheltered behind a practically shot-proof basaltic barricade.

Dikes are seldom more than 30 m (98 ft) wide, but some are hundreds of kilometers long. An exceptionally long one, probably the longest in the world, is the so-called Great Dike of Rhodesia, in southeast Africa. More than 480 km (298 mi) long, it has an average width of only about 8 km (5 mi).

Dikes can occur singly, but commonly they occur in groups known as ***dike swarms*** (Fig. 5-37). Sometimes they are aligned on roughly parallel courses, or sometimes they radiate from centers, such as the host of basaltic dikes that lace the northern part of Great Britain, with some individual intrusions attaining lengths of 160 km (99 mi) or so. The focal point for a radial set of dikes usually is the throat, or conduit, of an extinct volcano or a cylindrical pipe-like intrusion.

Depending on their resistance to erosion relative to that of their host rocks, dikes may be distinctive features of the landscape. If they are more resistant, they stand up somewhat as continuous walls (Fig. 5-38); if they are weaker, they may be etched out.

Columnar jointing is also typical of dikes, but the columns generally lie horizontally if the dike itself is vertical. They look much like an immense stack of cordwood, since the columns grow inward horizontally from the dike's vertical side walls, which are the cooling surfaces. Dikes most commonly form when magma intrudes a fracture, the walls being pushed apart as it makes its way.

Sills may run for great distances across the country, with lengths comparable to those attained by dikes. An excellent example is the Great Whin Sill in Northumberland in northeastern England. It looms as a dark, north-facing ledge, about 30 m (98 ft) thick, dominating the country for much of its course. As result, the Romans, with their experienced eye for the military potential of the terrain, seized upon the natural barrier as a foundation for Hadrian's Wall, which was built to keep the Picts from ravaging northern Britain.

Certainly the most familiar of all sills in the United States is the abrupt cliff of the Palisades

Fig. 5-37 Swarm of subparallel dikes of felsic rock cutting across a dark-colored gneiss, in the face of Painted Wall, Black Canyon of the Gunnison National Monument, Colorado. The Gunnison River (left) flows at about 686 m (2250 ft) below the top of the cliff. *W.R. Hansen*

Fig. 5-38 Four wall-like dikes of different sizes cutting relatively less-resistant sedimentary rocks on the northwest side of Spanish Peaks, Colorado. *G.W. Stose, USGS*

that follows the New Jersey shore of the Hudson River. The Palisades sill is a large one; it is more than 300 m (984 ft) thick. It cooled slowly and without much internal disturbance, so that the crystals within it developed a noticeable layering, as previously discussed.

Because contraction also occurs in the solidifying magma of sills, columnar joints are characteristic of sills, too; since most sills tend toward a horizontal position at the time of their origin, the columns in such an intrusion are likely to be vertical, having developed at right angles to the cooling surfaces.

Distinguishing buried lava flows from sills that have intruded into sedimentary rocks is not one of the simpler field problems in geology. Igneous rocks in both categories may have about the same texture, and both may have well-developed columnar joints.

In many sills a "chilled" zone at the margins, where crystallization was the most rapid, is very common. Rock texture in that zone is fine grained and in some cases almost glassy. Additionally, there may be a "baked" zone in the invaded rocks immediately adjacent to their boundary with the sill, both above and below. Such rocks are fired or hardened in much the same way that clay is fired in a kiln to make bricks or pottery. In a sill, fragments of the overlying host rock may be present in the igneous material, generally in the chilled zone. If the igneous body is an interbedded lava flow, however, the upper baked zone will be lacking altogether. Also, more likely than not, fragments of rock eroded from the flow will be found incorporated in the covering strata,

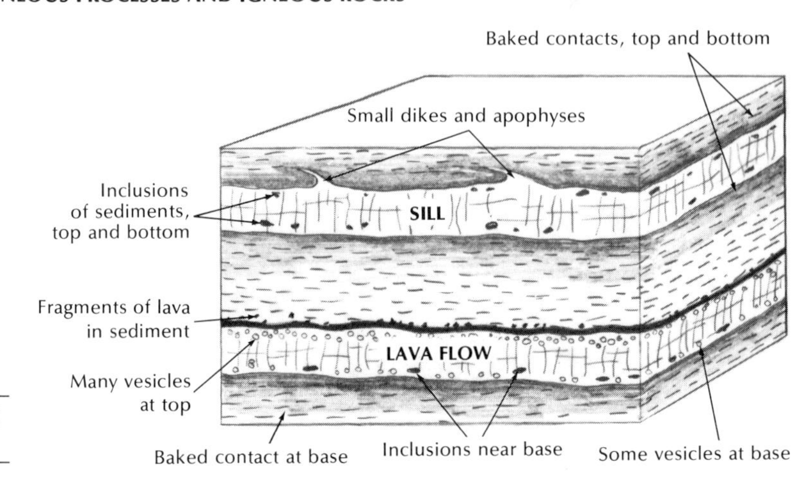

Fig. 5-39 Criteria used to distinguish a sill from a buried lava flow.

which, of course, was deposited after the lava solidified. Figure 5-39 graphically displays the criteria used to distinguish lava flows and sills.

Sills usually are found in the outermost part of the earth, particularly in the relatively thin sedimentary rock sequences that cover the crystalline rocks beneath. The American geologist M. R. Mudge has studied the occurrence of nearly 100 concordant structures (mostly sills) in the western United States and found that none has been emplaced below a depth of about 2300 m (7544 ft). Sills are confined to the outer layers because, when one of them intrudes, all the rocks above it must be lifted up by a distance equal to its thickness. The force necessary to lift such immense weights comes from fluid pressure in the magma body. Sill formation can occur only when the fluid pressure is equal to or greater than the overburden weight, a condition that can be met normally only when thin sections of overburden rock are involved, that is, in the outer skin of the earth.

Many if not most sills are offshoots from a dike. Magma intrudes upward until its fluid pressure exceeds the overburden weight; then sills branch off if the type of the layered rock section is conducive to sill formation.

Volcanic necks are the solidified magma of channels or conduits that once connected a volcanic vent on the earth's surface with a deep magma reservoir. Volcanoes are vulnerable to erosion; much of their interior consists of loosely consolidated ash and pyroclastic material, their steep slopes are easily gullied, and, if they are high enough, they intercept more snow and rain than does the surrounding countryside. The augmented runoff working on an exposed structure erodes many volcanoes relatively soon, geologically speaking, after they lapse into dormancy.

Their internal skeleton of radial dikes often stand up as partitions, and the solidified magma of the conduit forms a central tower known as a ***volcanic neck.*** A well-known example is Ship Rock in New Mexico (Fig. 5-40).

More often than not, volcanic necks are found in clusters, rather than singly. Prime examples in the United States are the buttes of the Navajo-Hopi country near the so-called Four Corners area, where more than 100 volcanic necks interrupt the surface of the plateau. Being dominantly dark rock, and projecting in jagged spires, they appear in sharp contrast to the red and white sedimentary rocks into which they are intruded.

Another group of volcanic conduits that have achieved a measure of fame are the so-called pipes of ***kimberlite,*** a rock of predominantly ferromagnesian minerals and one of the sources of diamonds. In South Africa the pipes are deeply weathered near the surface into what is called "blue ground," a sticky clay from which the diamonds were at one time separated by washing. The weathered pipe-like columns of dark rock were mined in the Kimberley pit to more than 1000 m (3280 ft) before those workings were abandoned. Since the pressures and temperatures necessary for diamonds to crystallize are not fulfilled short of a depth hundreds of kilometers below ground, the diamonds must be phenocrysts carried from deep below the surface. Many kimberlites contain fragments of foreign rocks that apparently were ripped from the walls of the pipe as the magma moved toward the surface (Fig. 5-41). From the presence of the diamonds and the nature of the foreign fragments, geologists have estimated that kimberlitic magma must come from depths of near 250 km (155 mi). In spite of this, most are only about 90 m (295 ft) across at the surface. Their general shape, then, is something like an extremely long drinking straw. Diamond-containing kimber-

Fig. 5-40 Ship Rock, New Mexico, is a volcanic neck with two radiating dikes. The volcanic cone that at one time probably existed above the site has been eroded away. The neck is intruded into sedimentary rocks, which are relatively more easily eroded than the neck. Throughout this part of the Colorado Plateau, such volcanic necks form starkly picturesque features. *John S. Shelton*

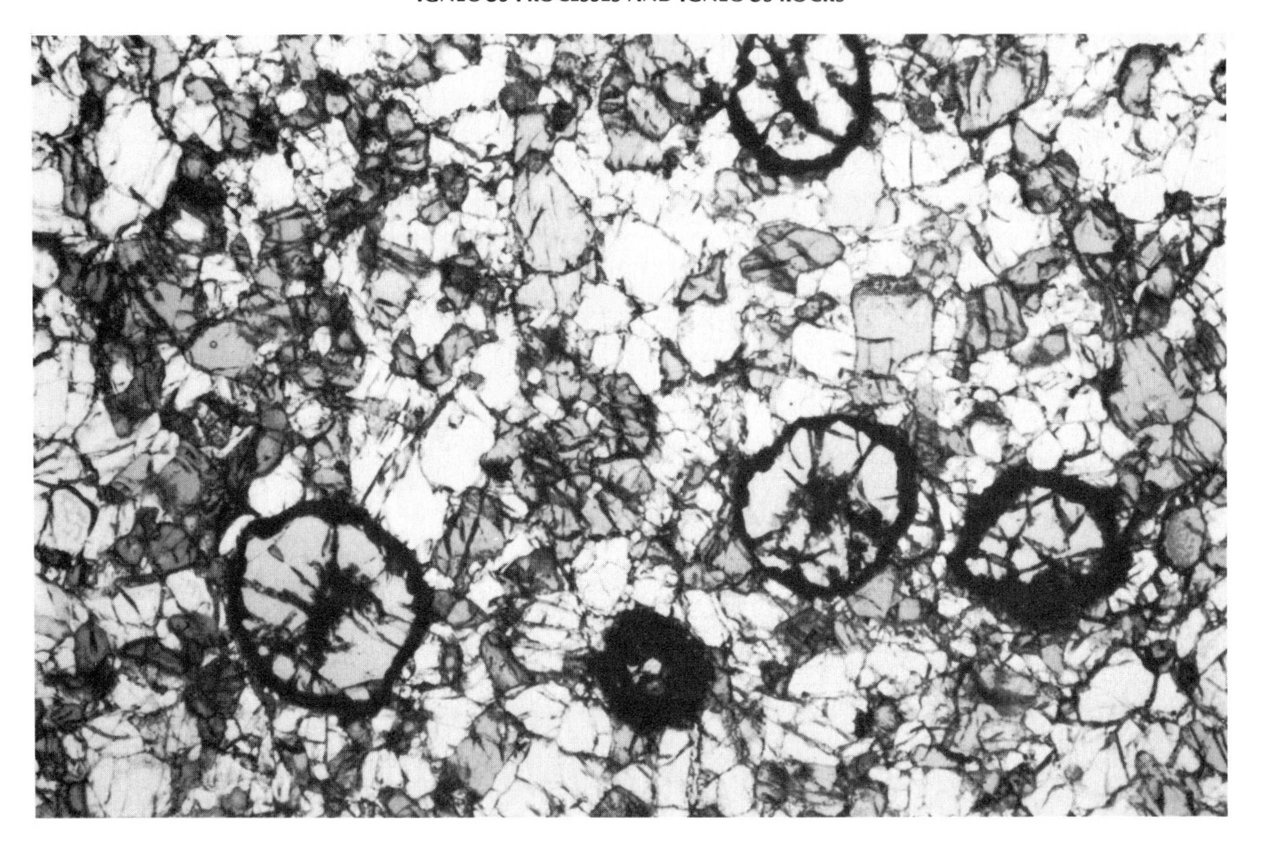

Fig. 5-41 Ultramafic rock, composed of garnet (black rimmed), olivine (light gray), and pyroxene (medium gray), derived from the mantle and carried to the surface during the intrusion of a kimberlite pipe, South America. *C. Stern*

lites are found in the United States in Arkansas and in an area near the Colorado-Wyoming border.

Laccoliths, concordant igneous bodies, are more or less circular in outline with a flat base and a dome-shaped upper surface (Fig. 5-42). They are generally up to 3 km (2 mi) across at the base, with maximum thicknesses near the middle of several thousand meters. Apparently laccoliths began development much like sills. However, the magma injected between the rock layers was so viscous that the lateral flow could not accommodate the amount of lava flow. More lava moved into the developing body than was transported maginally, and the body began to swell, or dome, upward, carrying the sediments above it into an arch. Because felsic magmas are generally more viscous than mafic ones, it would be expected that most laccoliths are felsic rocks, and observation bears this out. Commonly several laccoliths will occur in a group, and all appear to have been fed from a single magma body. In the United States, laccoliths are conspicuously

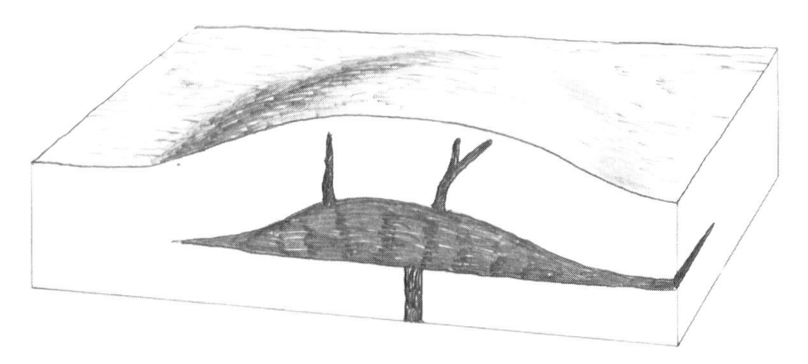

Fig. 5-42 Laccolith fed from a central pipe or dike. This type of occurrence is quite common.

developed in southwestern Colorado and eastern Utah. Some of them, for example, the LaSal and Henry mountains in Utah, stand in scenic beauty high above the surrounding countryside.

Figure 5-43 is a composite diagram depicting the characteristic mode of occurrence of dikes, pipe, sills, and laccoliths.

Large elongate or irregular plutons The largest bodies of plutonic rocks are granite and diorite. They are found on all continents and range in occurrence from Tertiary to Precambrian. In Labrador and northeastern Canada, where erosion has laid bare huge expanses of the earth's surface, plutonic rocks are found in bodies that may cover hundreds or even thousands of square kilometers. Large bodies composed mostly of granite with lesser amounts of diorite—we will lump them together and call them granitic rocks—are found also in the cores of many mountain ranges. The Coast Range pluton in British Columbia is an elongate body about 1600 km (994 mi) long (Fig. 5-44).

Such huge bodies of rock are called ***batholiths*** (a word introduced in 1895 and derived from the Greek words for depth and stone). In modern usage, a batholith is defined as a pluton with an exposure in excess of 100 km^2 (39 mi^2). If it is less than that, a pluton is called a ***stock.*** Many stocks, however, appear to be only the top of a batholith. If erosion were to proceed further, it is likely that a batholith would be unroofed and exposed.

Much has been learned about the physical configuration of the upper parts of batholiths because erosion has cut down to different levels at different places within the same body. Deep mining operations have also provided information. We lack definitive data, however, on the lower regions of batholiths. The walls of many of them are very steep sided, which has

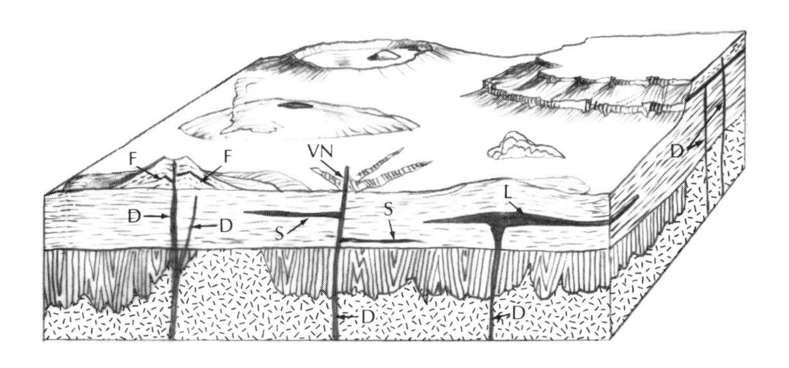

Fig. 5-43 The occurrence of dikes (D), sills (S), volcanic necks (VN), laccoliths (L), and flows (F).

led many geologists to hypothesize that such bodies extend downward to great depths. Gravity and earthquake-wave studies have given us some insight into their structure, but the results are often inconclusive. For instance, it is suspected that a batholith, perhaps in part still molten, lies just below the bubbling paint pots and geysers in Yellowstone National Park. In fact, it appears that this batholith has existed for at least 2 million years. Three voluminous eruptions from its chamber occurred 2, 1.2, and 0.6 million years ago. Geophysical investigations indicate the existence of an anomalous rock body, perhaps partially molten, that extends downward for at least 60 km (37 mi).

It is thought that a frozen batholith also lies under the pile of volcanic rocks that form most of the San Juan Mountains in southwestern Colorado (Fig. 5-45). From gravity measurements it is estimated to be 240 km (149 mi) long in an east to west direction by 160 km (99 mi) wide, north to south. The top of the pluton is thought to be 2 to 7 km (1 to 4 mi) below ground and to extend downward for about 19 km (12 mi).

In many cases batholiths are elongate (see Fig. 5-44). Many of the larger bodies are as long as 1600 km (994 mi), and 240 km (20 to 149 mi) wide. Commonly, the long dimensions parallel the trends of mountain ranges in the cores of which they are found.

Batholiths are not simple bodies; any single large mass consists of smaller bodies of different kinds of plutonic rocks. One part may be a white granite almost entirely lacking in ferromagnesian minerals; another may be a standard granite; and yet another may consist of diorite or rocks intermediate between granite and diorite. The nature of the boundaries between the rock bodies as well as radiometric dating generally indicate that the smaller bodies were formed at different times and that the formation of the batholithic complex, as a whole, took many millions of years.

An indication of the heat and volatile content of a batholithic magma is provided by the halo, or ***aureole,*** of metamorphosed rocks that surround most batholiths (see Chap. 8). The aureole represents the original country rock, recrystallized in place by the heat and chemical activity of the invading magma.

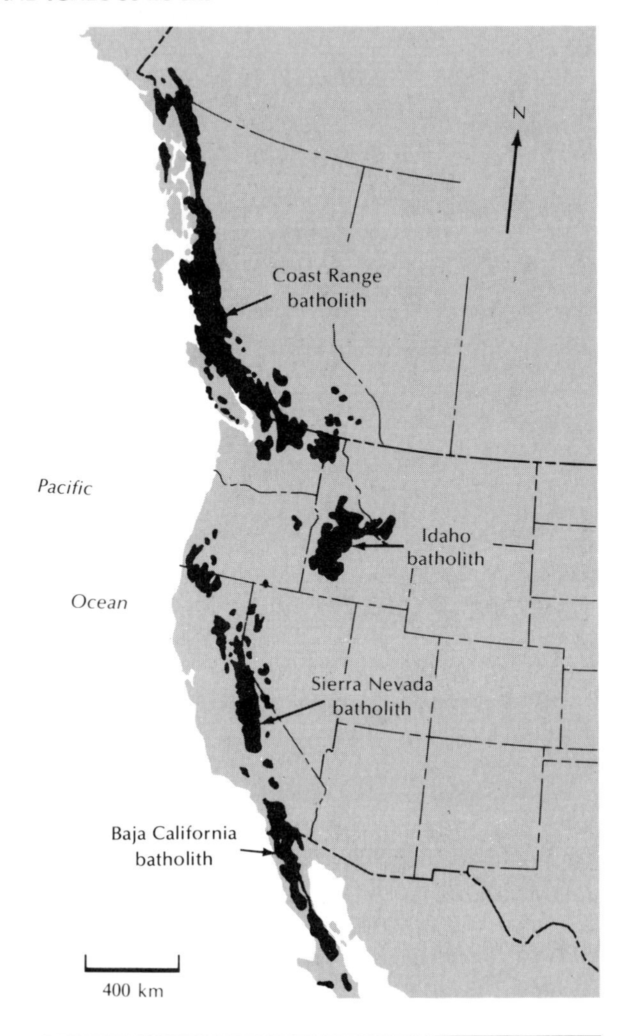

Fig. 5-44 Granitic batholiths (black) occur in large elongate bodies along the western margin of North America. Most of the plutonic rocks in these bodies was intruded during the Jurassic and Cretaceous periods.

As voluminous as batholithic magmas are, there is evidence to suggest that, once formed, they tended to move upward. In some instances, the magmas appear to have shouldered their way bodily into the country rock, displacing it. The boundary of the intrusion is knife sharp, and the granite is uncontaminated and homogeneous. The country rocks ob-

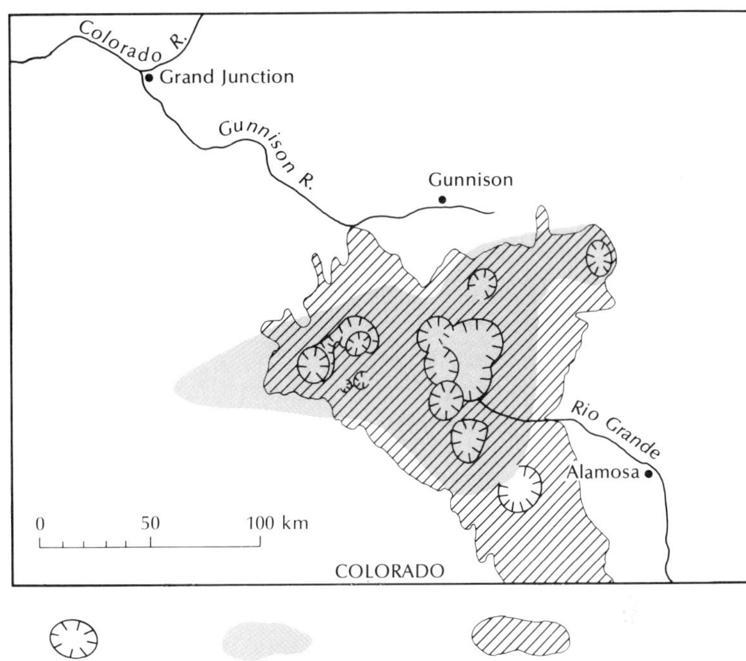

Fig. 5-45 Relation of a postulated batholith, calderas, and the San Juan Volcanic Field in southwestern Colorado. Many geologists think that the repeated explosive volcanism producing the calderas was the result of near-surface activity of the batholith.

Fig. 5-46 Sequential development of a salt dome, beginning with undeformed sediments overlying the salt layer (**A**) and progressing to a teardrop-shaped salt intrusion (**E**).

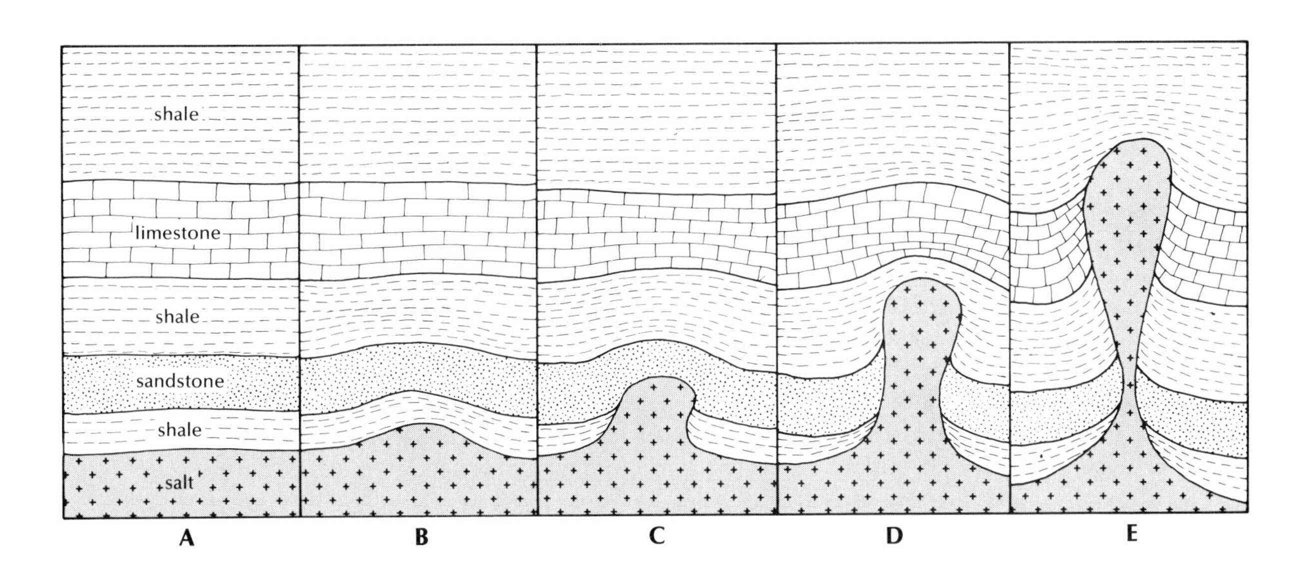

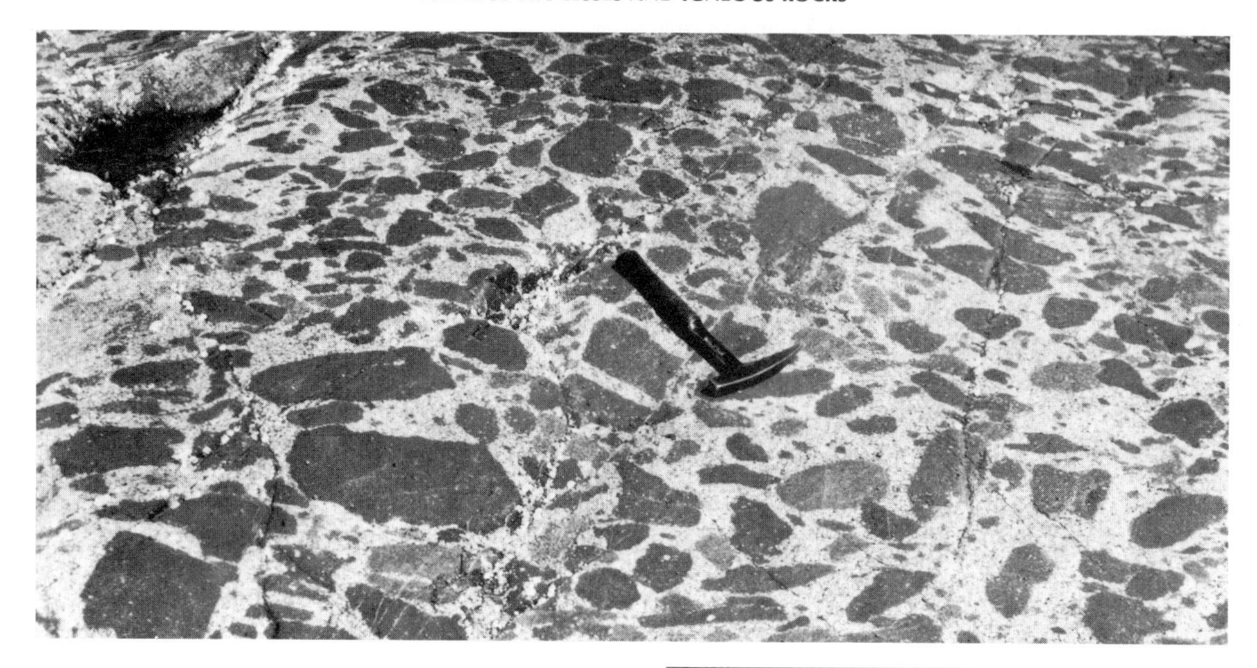

Fig. 5-47 Light-colored granite containing a host of angular equant-to-elongate dark blocks. Most likely, they were pried loose (stoped) from the walls of the country rock by the intruding magma. Subsequently, flowage caused subparallel alignment of the elongate or tabular fragments. *J.F. Seitz, USGS*

viously are gone, but gone where? The study of salt domes has given geologists an indication of how country rocks could disappear without a trace.

In some parts of the world, salt (mostly sodium chloride and calcium sulfate), which precipitated out from a restricted arm of the sea, lies in thick beds thousands of meters below clastic sedimentary strata. Salt can deform easily and has a relatively low density; therefore, it tends to rise (Fig. 5-46). After a while, bumps of salt begin to protrude into the sediments above. With more time, the bumps rise higher and higher, until an elongate, cylindrical mass of salt projects above the original beds. Sometimes the cylinder completely separates from the salt layer and continues to move upward, much like a drop of oil moving toward the surface in a vessel of water. Eventually, such low-density material will rise right to the surface. As the movement takes place, the surrounding sedimentary rock remains in the solid state, but slowly deforms to accommodate the rising plug of salt. Somehow, the salt intrusion is able to push aside the sediments. It is a slow process, but effective. Since granitic magma is less dense than the surrounding rocks and is easily deformable, many geologists feel that, given enough time and the right circumstances, the magma can displace the solid rock above, just as salt domes appear to do.

Is it also possible for batholiths to move by melting the rocks above? First, keep in mind that when a material changes from a solid to a liquid, heat is taken up, and that when it changes from a liquid to a solid, heat is given off. Many petrologists point out that apparently most melts contain little heat in excess of that required to keep the magma molten. Any melting that occurs in the country rock, therefore, must be accomplished by heat obtained from another source. Simultaneous crystalliza-

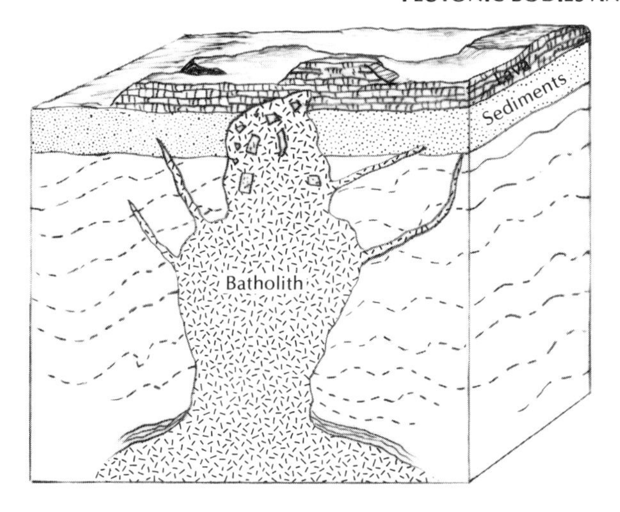

Fig. 5-48 Section through a hypothetical batholith depicting most of the ways a batholith intrudes into the country rock. At the bottom, on either side, the granite is shown without boundaries, indicating that either it reacts with the other rocks or it is produced when they melt. This would also be the zone of migmatite development. At the top, the batholith has penetrated sedimentary rocks and lavas, stoped blocks of which sink into the magma. The batholith can melt the country rock at any level.

tion (change of liquid into solid) in the magma could be that source. Normally, the first crystals to precipitate are more dense than the liquid magma and they would sink to the bottom of the chamber. Through continuous crystallization of the magma and melting of the wall rock, the magma chamber could move upward. Petrologists point out, however, that the process would only be effective under special conditions; most earth scientists, therefore, consider melting to be of only minor significance.

Some geologists have suggested, and there is some evidence, that magma could make a path for itself through the country rock by yet another process—***stoping*** of wall rock. Batholiths commonly contain, especially near their margins, angular fragments of rock that are totally different from the granitic rocks. Many of them are crudely aligned, as if by flow of the magma, and can be interpreted as small blocks that were pried loose, or stoped, from the roof or wall of the chamber and carried along during subsequent flow of the magma (Fig. 5-47). Apparently most of them contained minerals with higher melting points than the granitic magma and therefore did not melt. A diagrammatic summary of the various ways a batholith may make room for itself as it intrudes into the country rock is given in Figure 5-48.

The basal portions of some batholiths, especially those that were formed deep inside the earth, are surrounded with a wide marginal zone, and the transition between the invading and the invaded rocks is much less abrupt. The contact between the granite and its encasing shell may be blurred by a zone of ***migmatite,*** or mixed rocks—part igneous and part metamorphic. Migmatite has a typical ***lit-par-lit,*** or bed-by-bed, structure, which means that one layer of a rock may be granite, the next a metamorphic rock, the next granite, and so on (Fig. 5-49). Or knots and clusters of potassium feldspar, or other minerals typical of granite, may appear in the metamorphic envelope some distance out from the main body of granitic rock.

Such phenomena, taken together, have been interpreted by some geologists to mean that in some way the pregranitic rocks were digested, or replaced, by granite—in other words, they were converted in place to a brand-new rock that appears to be plutonic. The process is called ***granitization*** by its advocates—a vocal group—who strongly believe that most granitic rocks are formed in place through the alteration of large volumes of sedimentary rock by the chemical activity of solutions ascending from greater depths.

Some batholiths can, in fact, be seen to grade laterally from what appears to be plutonic rock in the center to plutonic metamorphic rock, then to metamorphic rock, and finally to essentially unaltered rock, or even sediments. Ghosts of the sedimentary layering can be made out in the "plutonic" rock. However, because such occurrences are so rare and because of experimental and theoretical considerations, most earth scientists do not believe granitization to be the major mechanism in the origin of granite.

Fig. 5-49 Migmatite from Summit County, Colorado, showing lit-par-lit structure. Light-colored bands, composed of quartz, potassium feldspar, and muscovite are plutonic; the dark layers, composed of intermediate plagioclase and biotite, are foliated metamorphic rocks. *M.H. Bergendahl, USGS*

Distribution of plutonic rocks

The occurrence of plutonic rocks around the world reveals a curious pattern. Mafic rocks are found almost everywhere, both on continents and oceans, but felsic rocks are restricted almost exclusively to the continents. There are no known granitic rocks in the ocean basins. Why? One might think that erosion on the ocean floor has been only minimal, or that batholiths close to the surface have not been unroofed. However, consider the fact that granitic rocks have lower densities (about 2.7 gm/cm^3) than do most of the rocks of the ocean basin (about 3.0 gm/cm^3). Large bodies, such as batholiths, would isostatically rise to protrude above the ocean floor as topographic highs—yet none do. Or changes in gravity would easily demonstrate the existence of such large, low-density masses under the sea—yet none have ever been so located. Moreover, volcanic activity in all the world's oceans has produced mafic rocks almost exclusively. Certainly the volcanoes are not being fed from a granitic magma. Studies of earthquake waves that pass through the outer layers of the earth have also failed to indicate the existence of large bodies of granitic material beneath the ocean basins.

Granitic rocks, then, must owe their existence to processes that originate marginal to, within, or below a continent. Most geologists would contend that nearly all granitic magmas are produced in the lower to upper-middle regions of the continental crust. Some would place their points of origin a few tens of kilometers beneath the continental masses.

Some light can be shed on the issue from a study of the distribution of coarse-grained felsic plutons in the continents. Many occur (in association with metamorphic rocks) in linear trends, or belts, and many of the geologically more recent ones are found in the cores of mountain ranges—leading to the conclusion that most of the large plutons are related to mountain-building processes. And when we find long chains of old batholiths that extend across a non-mountainous terrain, as in large parts of Canada, we probably are looking at the deeply eroded roots of fossil mountains. If we knew what produced mountain ranges (see Chap. 17) we could speculate on the origin of batholiths.

Plate tectonics (explained in Chap. 20 in more detail) is relevant to both problems. Briefly, the theory proposes that where the large, rigid plates of the earth converge, or come together, both mountains and igneous intrusions are produced. During the slow, convergent sliding of one plate against another, sufficient heat is produced to melt the rocks at depth, and magmas are formed all along the plate boundary. The visible effect of the activity at depth is the occurrence of linear chains of volcanoes—called ***island arc systems***—such as the islands of Japan and the Aleutian Islands. The materials that melt to become batholithic magmas are thought to be, in part, sedimentary debris eroded from the continents and trapped in deep, linear pockets along the boundary of collision and, in part, the rocky material that makes up the edges of the plates themselves. Because the sedimentary debris is relatively rich in potassium, aluminum, and silicon, the resulting magma is granitic. The theory seems to tie together many diverse phenomena, and most geologists have accepted it, at least in its general form.

Unfortunately, it is not possible to investigate the nature of the source magmas and their chambers until long after the volcanoes are dead and gone. It has been hypothesized that the Coast Range batholith in British Columbia and the elongate intrusions that run the length of the Sierra Nevada range and through Baja California probably represent plutons formed along a zone of convergence over a period of 90 million years, with the bulk of activity occurring between 100 and 80 million years ago.

Many zones of convergence today are marginal to continents. If that was true in the past, and if granitic rocks were formed periodically along the margin, it would seem that, with time, as successively younger chains of batholithic material are added on, the continents should be growing outward. In fact, some continents do tend to show a "younging" toward their margins, which has led many geologists to suggest that during the early stages of earth history the continents were small or even nonexistent, and have been undergoing continental accretion ever since.

Mafic plutons are a different matter. They almost never occur in bodies of batholithic proportions—at least as far down as we have been able to investigate. Since the rocks are too mafic to have been generated by the melting of most of the continental rocks themselves, they must come from below the continents. That view is supported by the observation that mafic rocks found on continents are not greatly different from those found in ocean basins.

Studies of earthquake waves and controlled explosions have provided much insight into the nature of subsurface rocks (see Chap. 18). The results strongly suggest that the earth below the continental masses and the ocean basins is made up of material that could, if melted, produce mafic magmas. Detailed study of earthquake waves related to magma movement deep beneath the island of Hawaii indicate that the magma feeding the fiery surface fountains must come from at least 50 km (31 mi) beneath the island. Geophysical investigations also have suggested that from about 100 to 250 km (62 to 155 mi) below the surface, some of the rock (perhaps up to 10 per cent) is molten (see Chap. 18). Above and below that depth range, the rocks are solid, raising the possibility that in some circumstances the molten material can pool together to form a magma, which can then begin its journey to the surface.

Plate tectonic theory has also been applied to the occurrence of mafic plutonic (and volcanic) rocks. If magma is to rise, the most likely place for it to do so is where the earth's brittle layers are being stretched, cracked, and pulled apart, thereby providing easier access to the surface. The most obvious place for intrusion of mafic magmas, then, would be along the diverging boundaries (***pull-apart zones***) of the world's plate system. Today, in fact, such zones are generally marked by higher-than-normal heat flow and mafic volcanism. Iceland, for example, which sits full astride the Mid-Atlantic Ridge, is composed largely of mafic dikes, sills, and lava flows. Moreover, most of the dike swarms that fed the flows run parallel to the trend of the Mid-Atlantic Ridge.

Many of the larger basalt fields of the earth, where erosion has cut sufficiently deep, show evidence of being fed by swarms of subparallel dikes. Many geologists have said, therefore, that the magma that fed the flows also could have moved into fractures or rifts that developed from cracking, brittle near-surface layers of the earth. But whether those fracture zones corresponded to extensive pull-apart zones, such as the Mid-Atlantic Ridge, or were isolated occurrences largely unrelated to plate boundaries is still open.

Those who do not subscribe to the theory of plate tectonics are faced with the problem of locating a source for the heat needed to melt large volumes of rock, particularly in the production of batholiths. Heat liberated from the decay of radioactive elements at depth is a possibility, but it seems improbable that there were enough radioactive elements to supply sufficient heat.

Another theory is that during the first days of the earth's formation, heat was trapped at great depths and is slowly moving toward the surface.

Yet, if we accept the plate tectonics theory we must ultimately ask: What is the source of energy that drives the plates? At the moment nearly everyone in earth science is searching for clues that will answer that question.

SUMMARY

1. The kinds of minerals and their proportions in an igneous rock depend on the initial composition of the magma and on differentiation that may occur as the magma cools and solidifies. The initial composition of magma depends on its location in the earth. *Felsic* magmas are generally restricted to continental regions and *intermediate* magmas to continental margins where plates collide; whereas *mafic* magmas are found both in continental and oceanic regions.
2. *Differentiation* of a magma can occur either by *crystal fractionation* or *compositional zonation*. The former is the result of the separation of earlier formed crystals that are generally more mafic and calcic, from the remaining more felsic magma. The latter is primarily restricted to large felsic magma bodies and is thought to be the result of density, thermal, or convectional stratification, with the lighter felsic constituents on top and the more-dense mafic constituents below.
3. Most igneous rocks are *crystalline* in texture; that is, the minerals have an interlocking structure. Grain size depends primarily on the cooling rate: plutonic rocks tend to be coarser grained than volcanic rocks. If cooling occurs so quickly that the minerals cannot crystallize, the resulting rock is a *volcanic glass*. A *porphyritic* texture implies two cycles of cooling, one in which cooling was relatively slow, followed by one in which cooling was rapid.
4. Some volcanic rocks possess clastic textures, either as a result of reworking of volcanic rocks by streams or accumulation of volcanic material (*pyroclastics*) blown into the air during volcanic eruptions.
5. The three most common plutonic rocks are *granite, diorite,* and *gabbro;* the three most common volcanic rocks are *rhyolite, andesite,* and *basalt.*
6. Magma originates in small-to-large pools inside the earth and, once formed, moves laterally and toward the surface.
7. Relatively small *plutons* can be subdivided into the following classes: discordant tabular bodies (*dikes*), concordant tabular bodies (*sills*), *volcanic necks,* and domed concordant bodies (*laccoliths*).
8. The largest plutons are called *batholiths* and *stocks* and consist almost exclusively of granitic rocks. They are found on all

continents, in bodies that may cover thousands of square kilometers. Commonly they are present in the cores of mountain ranges, in association with metamorphic rocks.

9. Batholiths intrude into the surrounding country rock in a variety of ways: *forceful injection, stoping,* and *melting*. Given enough time, they may rise through the surrounding country rock much as salt domes do through overlying sedimentary strata. Although there is some evidence to suggest that some batholiths originate through *granitization,* most appear to result from melting.
10. *Mafic* plutonic rocks are found both on continents and in the ocean basins, whereas *felsic* rocks are almost exclusively found on continents. Mafic rocks apparently originate in the mantle, perhaps in the layer between 100 and 250 km (62 and 155 mi). Large-scale granitic plutons owe their existence to processes originating marginally to, within, or closely beneath a continent.
11. Batholiths and associated metamorphic rocks in many cases appear to be the result of mountain-making processes marginal to a continent. Today, it is believed that that association occurs in convergence zones.

QUESTIONS

1. What is the basic system for naming igneous rocks? How well do you think the system works?
2. Which is more important in determining the type of igneous rock to be formed—initial magma composition or magmatic differentiation?
3. What evidence is there that fractional crystallization actually occurs?
4. Name as many rock textures as you can and give the important factors determining their occurrence.
5. What factors determine the viscosity of a volcanic rock?
6. How does rhyolite differ from basalt? From andesite? From granite?
7. Compare and contrast a sill, a dike, and a laccolith.
8. Describe the ways batholiths might move upward in the earth's crust?
9. How well do you think the plate tectonic theory explains the distribution of plutonic rocks?

SELECTED REFERENCES

Bowen, N. L., 1928, The evolution of igneous rocks, Princeton University Press, Princeton, New Jersey.

Billings, M. D., 1972, Structural geology, Prentice-Hall, Englewood Cliffs, New Jersey.

Chapin, C. E., and Elston, W. E., eds., 1979, Ash-flow tuffs, Geological Society of America Special Paper 180.

Daly, R. A., 1933, Igneous rocks and the depths of the earth, McGraw-Hill Book Co., New York.

Hamilton, W. B., and Myers, W. B., 1967, The nature of batholiths, U.S. Geological Survey Professional Paper 554-C.

Hyndman, D. W., 1972, Petrology of igneous and metamorphic rocks, McGraw-Hill Book Co., New York.

Mudge, M. R., 1968, Depth control of some concordant intrusions, Geological Society of American Bulletin, vol. 79, pp. 315–22.

Read, H. H., 1957, The granite controversy, Interscience Publishers, New York.

Simpson, B., 1966, Rocks and minerals, Pergamon Press, Oxford.

Turekian, K. K., 1972, Chemistry of the earth, Holt, Rinehart and Winston, New York.

Tuttle, O. F., 1955, The origin of granite, Scientific American, April.

Walton, M., 1960, Granite problems, Science, vol. 131, pp. 635–45.

Wyllie, P. J., 1971, The dynamic earth, John Wiley and Sons, New York.

Fig. 6-1 Mount St. Helens in eruption, May 18, 1980, looking north. The thick cloud extending north covers the region decimated by the volcanic blast, pyroclastic flows, and mud flows. *USGS*

6

VOLCANISM

Volcanic activity accounts for some of the most awesome displays of the forces of nature known to humans. Anyone who has ever witnessed a volcanic eruption or who has seen one on film, cannot help but be impressed by the immensity of the earthborn fireworks (Fig. 6-1). Of course, part of our fascination with volcanoes is related to their potential for destructiveness. For example, the eruption of Krakatoa off the coast of Java and Sumatra in 1883 killed nearly 37,000 people and that of Mont Pelée in the West Indies in 1902 destroyed nearly instantaneously the town of St. Pierre, with loss of all but two of 30,000 people within it.

Volcanism is worth studying because it provides us with our only direct source of information about the rocks and the conditions that prevail in the outer 250 km (155 mi) of the earth. Certainly the existence of liquid lava at the earth's surface is positive proof that rocks can and do melt in the natural environment.

From evidence contained in the rock record, it can be determined that the earth has undergone volcanism from its earliest days right up to the present—and there is no evidence to indicate that it is dying out. The activity has not been continuous through time; certain periods of earth history have been marked by more, and others by less, activity. Even a single, large volcano shows the same type of behavior; during its life history it lies dormant for long periods and then shows signs of growing activity.

More than 516 active or recently active volcanoes dot the earth's surface, yet until recently volcanism was a geological phenomenon alien to most North Americans—in sharp contrast to the phenomena related to ice, wind, and water.

In late March 1980, Mount St. Helens in the Cascade Range stirred from a century-long slumber and shot ash and steam clouds into the sky. Scientists and tourists alike flocked to the area. During its first six months of activity, the volcano erupted five times, destroying the volcanic superstructure and killing more than 60 persons.

The word *volcano* comes from Vulcan, the ancient Roman god of fire and metalworking, who manufactured the thunderbolts of Jupiter and the armor of the gods. One active volcano in the Mediterranean region carries the name Vulcano, and it was considered by the ancient

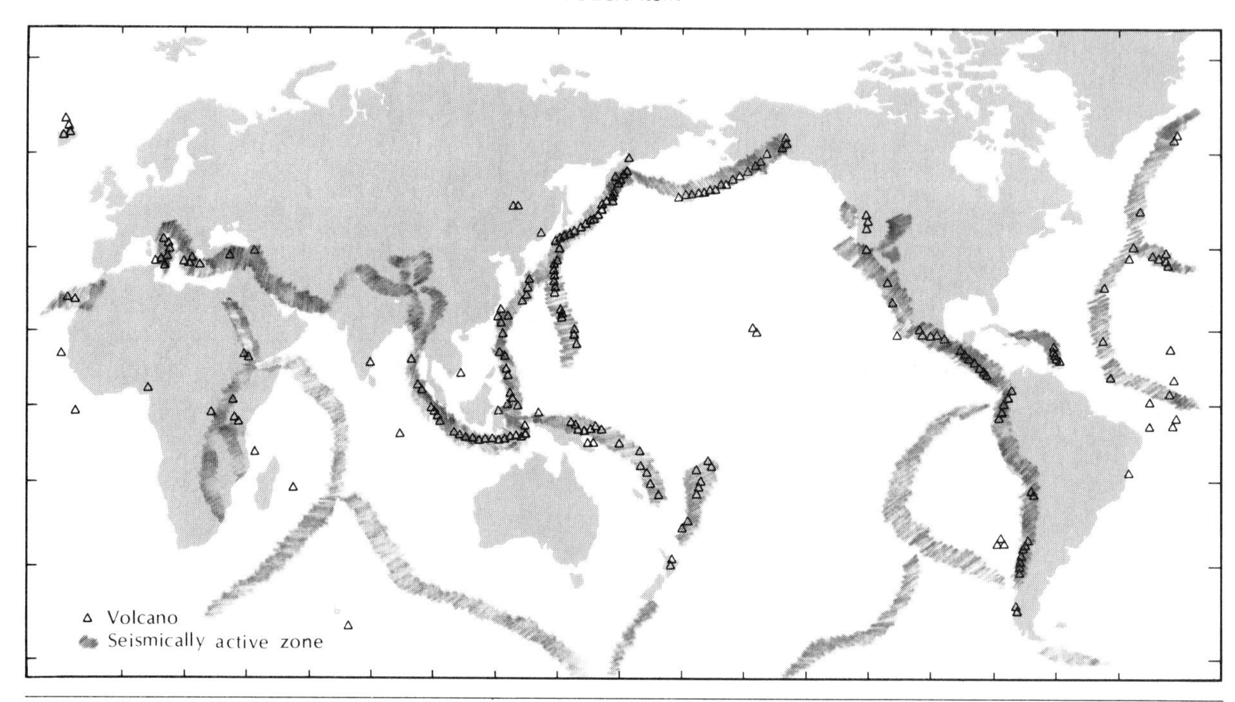

Fig. 6-2 Distribution of the active volcanoes of the world in relation to the active seismic zones. Each triangle represents one or more active or recently active volcanoes, of which there are about 516. Most volcanoes occur in belts marginal to the Pacific Ocean.

Romans to be the forge at which Vulcan practiced his blacksmithing. Several other cultures, steeped in ancient lore and superstition, have associated volcanism with one or more deities. To the Polynesians, volcanoes were under the control of the demigoddess Pele; in the ancient world of the Greeks, Hephaestus, the ugly son of Hera, became the principal blacksmith, who set up his forge beneath one or another volcano in the Mediterranean region. A popular idea during the Middle Ages was that volcanoes were the gateways to hell or the prison of the damned. The cacophonous and often eerie noises that sometimes emanated from the volcanoes were thought to be the screams and moans of tormented souls.

DISTRIBUTION

Volcanoes occur in many parts of the world, but their distribution is far from uniform (Fig. 6-2). Many occur within fairly well-defined zones or linear belts. Perhaps the most renowned is the ***Ring of Fire*** that girdles much of the Pacific Ocean. In that ring are found about four-fifths of the world's active volcanic centers. Most of the other active volcanoes are found in the Caribbean and Mediterranean regions, in Asia Minor and the vicinity of the Red Sea, and in central Africa.

On the map of the world's volcanoes (Fig. 6-2) is also shown the distribution of generalized zones of earthquake activity. It is remarkable that the two zones correspond as closely as they do, a fact that has led most geologists to believe that there is a definite relationship between the two phenomena. This is not to say that earthquakes cause volcanic eruptions, or vice versa; almost never are the two phenomena in evidence at the same locality at the same time. (An exception to that rule occurred in Chile in 1960, when two days after a disastrous

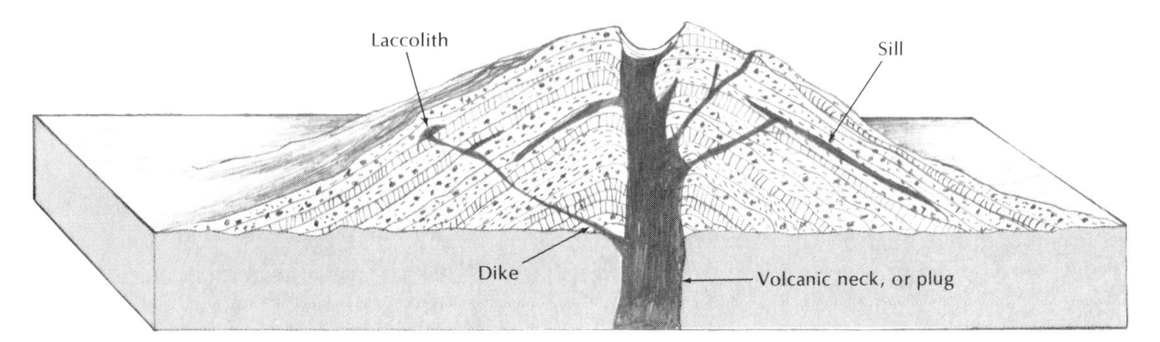

Fig. 6-3 Diagram of a composite cone (stratovolcano), showing volcanic neck, dikes, sills, and a laccolith.

Fig. 6-4 Plume of ash and vapor escaping from the summit of Shishaldin Volcano, Aleutian Islands, Alaska. Note the steep flanks of this stratovolcano. *USGS*

earthquake occurred, a small volcanic cone—Volcan Pujehue—went into mild eruption.) Rather, it seems that volcanism and earthquake activity are two different ways that the earth behaves in response to large-scale forces within it.

It can be pointed out that many of the linear belts of volcanoes more or less coincide (as do the earthquake belts) with the margins of the large and small plates that make up the outer rocky rind of the earth (see Chap. 20) and that fit together like the pieces of a gigantic three-dimensional jigsaw puzzle. Away from the edges of the plates, volcanism and earthquake activity are minimal. Where the plates converge, blocks push and slide past one another, and many volcanologists feel that the heat necessary for the melting of rocks and production of magma is created by the frictional energy released during such movement. Whatever causes the plate motion, it is thought, also accounts for the earthquakes in those zones. One dissimilarity between seismic and volcanic activity in zones of convergence is that earthquakes originate as far down as 700 km (435 mi), whereas magma sources extend no deeper than about 125 km (78 mi) at most, and many may be much shallower. The Pacific Ring of Fire appears to owe its existence to the convergence of plates. The volcanoes associated with that convergence zone usually erupt andesitic to rhyolitic lava and pyroclastic rocks, quite commonly, violently. The volcanoes of the Mediterranean region, so well known in history and mythology, also occur at a zone of convergence and also pour forth mixtures of lava and pyroclastic rocks, which are largely andesitic to rhyolitic.

Volcanic activity is also notable at the zones where plates are pulling apart (diverging). The most conspicuous zones of divergence, found in the ocean basins, stand out as elongate rises on the ocean floor. If the water were drained from the seas, we would see the oceanic ridge- and rise-system (see Chap. 3), a nearly continuous range that snakes its way through all the ocean basins and marks the zone of divergence. There is evidence that as the earth's rocky skin pulls apart, magma rises from beneath to help fill the void, and volcanic activity ensues. Since most of the volcanic activity is beneath the sea, we are not exactly sure of the way it occurs, or how often. But in a few places, such as the Azores and Iceland, eruptions have built volcanic edifices, upon oceanic rises, that project above the ocean's surface. At such places we have been able to study the nature of volcanic activity at a zone of divergence. Volcanoes in Africa's Rift Valley also appear to be associated with a diverging plate boundary. Unlike the Ring of Fire lavas, those associated with pull-apart zones are mostly basaltic. They seem to be derived from a magma source at depths of 50 to 70 km (31 to 44 mi).

Not all active volcanoes can be easily classified on the basis of convergence and divergence of plates. Some, like the volcanoes of the Hawaiian Islands, occur far out in the middle of the plates. No one yet has proposed a completely acceptable hypothesis for their origin. One thought, as discussed in Chapter 20, is that they are centered over "hot spots" in the outer part of the earth, like the top of a candle's flame, which have brought about the localized melting of rocks (see Fig. 20-24). Another view is that such volcanoes are at the leading edge of a large, active crack in the rocky skin of the earth that extends downward to an earth layer containing molten material. The crack would afford a channel for the molten material to move to the surface. In fact, seismic data provide strong evidence to support the idea of a partially molten layer at a depth between 100 and 250 km (62 and 155 mi). We cannot yet choose between the two explanations. Inasmuch as the roots and sources of all active volcanoes lie many tens and even hundreds of kilometers beneath the earth's surface, we probably will not determine fully the reasons for their character and distribution for some time to come.

TYPES OF VOLCANOES AND VOLCANIC ACTIVITY

Volcanoes come in all sizes. Most mark localized centers of eruption, virtually at one spot

Fig. 6-5 Mount Fujiyama, Honshu Island, Japan, displays the classic outline of a stratovolcano. This cone, which was last active in 1707, rises about 3000 m (1000 ft) above the surrounding countryside. *Fuji Photo Film U.S.A., Inc.*

at the earth's surface, which leads to the development of a ***volcanic cone.*** The cone is essentially a heaping or piling up of the products of eruption around a volcanic vent, or orifice. Some volcanoes have a variety of eruptive styles, pouring forth flows of lava at one time and volumes of pyroclastic material at another (Fig. 6-3). Such ***composite cones,*** or ***stratovolcanoes,*** as they are called, are rather steep sided (Fig. 6-4) and account for most of the scenically magnificent cone edifices in the world. Mount Fujiyama in Japan (Fig. 6-5), Mount Hood in Oregon, and Mounts St. Helens and Rainier in Washington are prime examples of stratovolcanoes. The rocks erupted from such volcanoes are mostly andesitic to rhyolitic.

In contrast, there are also cones built almost entirely of fluid lava flows, one flow upon the other. Cones built in such a manner generally are basaltic. With gently rounded profiles and a nearly circular outline, they resemble turtles

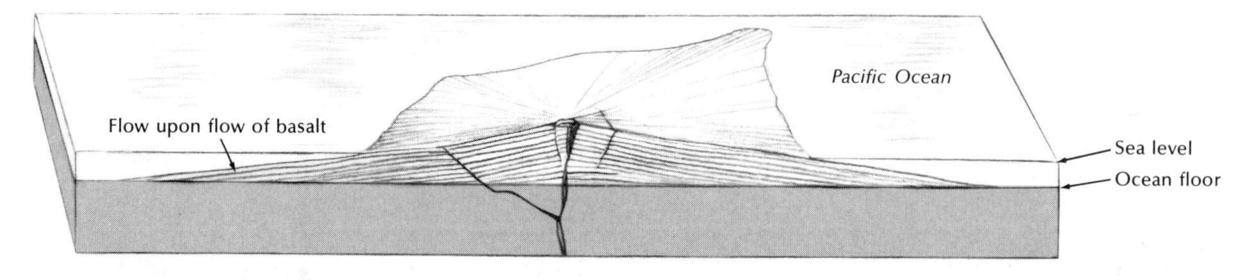

Fig. 6-6 Diagram of a Hawaiian shield volcano composed almost entirely of thin basalt flows. Because the magma is fluid, explosive activity occurs rarely, and pyroclastic material forms only a small part of the cone. As a result, the slopes are gentle and the volcano resembles a shield.

of colossal size or the shields once mounted along the gunwales of sea-roving Viking ships. As a consequence, they have been called ***shield volcanoes*** (Fig. 6-6). The volcanoes of the Hawaiian Islands and Kilimanjaro in Africa are some of the finest examples of shield volcanoes on earth.

In some cases, eruption is not restricted to a single, localized vent, but occurs simultaneously along a linear trend. The resultant constructional feature then forms an elongate welt rather than a symmetric cone.

Almost all composite and shield volcanoes are built up over a considerable period of time, although most require less than a million years for their construction. Some vents, commonly on the flanks of larger cones, go through a complete eruptive life cycle in only a few weeks, months, or years. Because there are few eruptive products, they generally attain a height of only 100 m (328 ft) or so. A particular type of small cone, steep sided and composed almost entirely of reddened scoriaceous blocks, lapilli, and bombs is termed a ***cinder cone*** (Fig. 6-7). It may be found as a solitary isolated feature, but quite commonly many cinder cones will occur in close proximity, each marking a single, short-lived eruptive event. In some cases the growth of a cinder cone is associated with the eruption of small lava flows. The birth and growth of a larger-than-normal cinder cone was recorded in Mexico beginning in 1943. From its inauspicious beginning as a small crevice from whence emanated hot gases and gurgling sounds, the cone that was to be called Parícutin began to grow. The eruption lasted about nine years and built a cinder cone almost 369 m (1210 ft) high. During its early stages, activity was confined to the explosive expulsion of blocks, bombs, and lapilli; later, torrents of lava broke through the flanks of the cone and flowed outward, eventually reaching the nearby town of San Juan Parangaricutiro and covering all but the church steeple (Fig. 6-8). Its last days were punctuated by strong detonations during which fiery bombs and blocks up to 90 metric tons (100 short tons) in weight were hurled past the base of the cone.

Nearly all volcanic cones are characterized by a relatively small, funnel-shaped depression—called a ***crater***—marking the top of the conduit through which the products of eruption were channeled (Fig. 6-9).

There is considerable variety in the way volcanoes erupt. Even a single volcano may go through several phases of eruption, each one somewhat different than the one before. Nonetheless, volcanoes in general, or any single eruption, can be categorized into broad types based on degree of explosivity. Some volcanoes erupt violently, with the explosive release of great clouds of pyroclastic material, whereas others do so relatively quietly with only the outpouring of fluid lava. Even the quietest eruptions are spectacular to behold. Of course there are eruptions that fall gradationally at all levels between the two extremes. In general, volcanoes that erupt andesitic to

Fig. 6-7 Classic form of a geologically recent cinder cone in the Cascade Range, northern California. The entire cone, which is an accumulation of cinders, lapilli, ash, and bombs, probably formed in a few weeks' time. In the background is Mount Lassen, which erupted last in 1914–17. *California Department of Water Resources*

Fig. 6-8 Pańcutin Volcano, Michoacán, Mexico. This photograph, taken in 1950, shows the still-active vent. *F.O. Jones, USGS*

rhyolitic material tend to be more explosive than those that erupt basalt. The most explosive volcanoes in the world are found in Indonesia and Central America; moreover, the degree of explosivity is much greater in the Pacific Ring of Fire than in other parts of the world.

Explosive eruptions are the result of the sudden release of large quantities of gases contained (dissolved) in the magma. All magmas contain gases. Our best estimates indicate a content of about 1 to 2 weight per cent in basaltic magmas and about 2 to 4 weight per cent in andesitic to rhyolitic magmas. By far the greatest proportion of the gases (commonly 50 to 80 per cent) is water vapor, accounting for at least some of the vapor that swirls in dense clouds above an erupting volcano (Fig. 6-10). Carbon dioxide, nitrogen, and sulfur dioxide also contribute measurably to the gas content, but in varying proportions. Small amounts of carbon monoxide, hydrogen, sulfur, chlorine, fluorine, and hydrogen chloride are also detectable.

Some of the dissolved gases are undoubtedly derived from the country rocks that the magma contacts or engulfs as it moves upward in the crust; some, however, are indigenous to the magma and represent ancient volatile material trapped inside the earth during the early stages of its formation. In 1955 the geochemist W. W. Rubey reached the rather startling conclusion that essentially all the water on the face of the earth and in the atmosphere was probably liberated from within the earth by volcanic activity throughout the history of the planet. Most present-day geologists concur in that conclusion.

More gas is soluble in magma at higher confining pressures than at lower ones. Therefore, magma deep within the earth can contain more gas than can the same magma at a shallower depth. As the magma moves upward and nears the surface of the earth, and as crystallization of silicates that do not contain volatile-rich minerals begins to occur, the gas pressure increases until the contained gases begin to escape. If there is an open channel to the surface, and if the magma is fluid, the gases easily escape into the atmosphere. If the magma is less fluid, and if a plug of solidified material blocks the vent, escape of the gases is impeded. With time, pressure builds up. If the vent is suddenly cleared, either by release of the gases under high pressure or by eruption of magma, the pent-up gases escape violently and explosively (Fig. 6-11). It is like popping the cork on a bottle of champagne. Expanding rapidly, the gases propel blocks of magma and solid material upward to great heights, as if shot from cannons. The nearly instantaneous release of the gases can cause the top of a magma chamber to froth, forming pumice or scoria, depending on the composition of the magma. In some cases, continued expansion of the gases within the vesicles in the chilled glassy froth can cause the fragmentation into fine ash and lapilli (Fig. 6-12). When large quantities of gas are catastrophically discharged, the magma chamber can be emptied nearly instantaneously.

Andesitic to rhyolitic magmas, which tend to be less fluid than basaltic magmas, are more apt to erupt explosively.

In the following pages we will document some of the possible variations in mode of

Fig. 6-9 Asama Volcano, Honshu, one of Japan's most active and dangerous volcanoes. Notice the sharp central crater marking the top of the active volcanic vent. Recent ash covers the flanks of the volcano. Andesite lava issued from the vent in 1783 to produce the molasses-like flow to the right. *USGS*

Fig. 6-10 Eruption of Hekla Volcano, Iceland. The billowing plume, carried windward, is composed of volcanic ash and volatile constituents, mostly water. *Thorsteinn Josepsson*

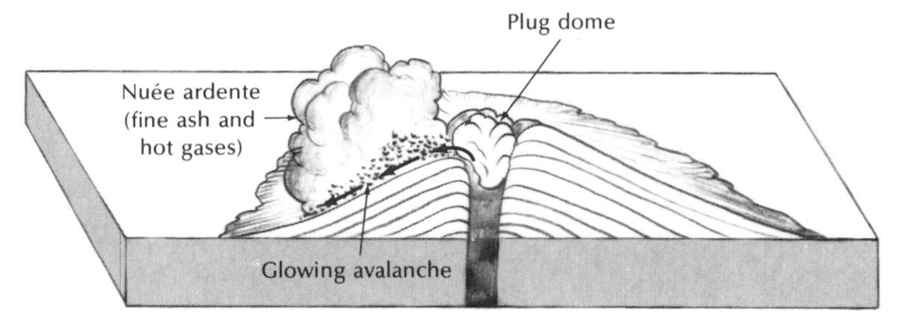

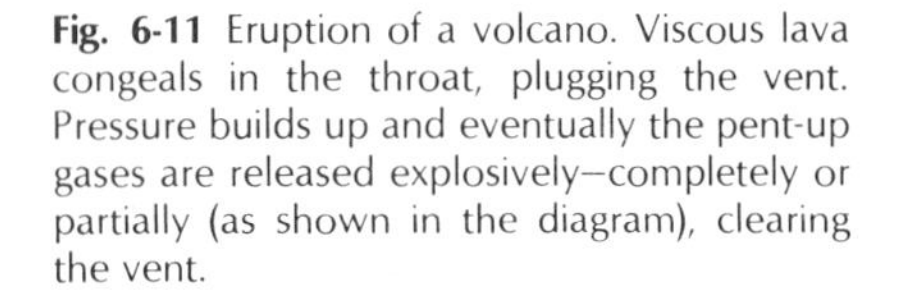

Fig. 6-11 Eruption of a volcano. Viscous lava congeals in the throat, plugging the vent. Pressure builds up and eventually the pent-up gases are released explosively—completely or partially (as shown in the diagram), clearing the vent.

eruption. Some of the information is taken from eyewitness accounts of historic eruptions, some from geological studies of prehistoric volcanic centers.

Explosive eruptions

Krakatoa Close to one hundred years ago, captains of sailing ships, beating their way through the Sunda Straits that separate the great islands of Java and Sumatra in the East Indies knew the island of Krakatoa well. Its conical, green-clad slopes rose uninterruptedly about 793 m (2601 ft) to the summit of the central peak. The straits were important since they were on the shortest sea route for the tea clippers en route from China to England. They were dangerous, restricted waters, haunted by sea-roving Dyaks who often attacked a becalmed vessel.

Although there had been volcanic activity on the island of Krakatoa since May 1883, it seemed innocuous enough to the crew of the British ship *Charles Bal,* tacking under all plain sail through the hot tropical Sunday afternoon of August 26, 1883, until they arrived on one heading at a point about ten miles south of the island. Minutes later the mountain exploded. The entire mountain disappeared in clouds of black "smoke," and the air was charged with electricity—lightning flashed continuously over the volcano, as it very often does during eruptions—and the yards and rigging of the ship glowed in St. Elmo's fire. Immense quantities of hot ash fell on the deck or hissed through the surrounding darkness into the increasingly violent sea. As the vessel labored through broken seas and squalls of mud-laden rain, the explosions continued, much like a never-ending artillery barrage, accompanied by a ceaseless crackling sound that resembled the tearing of gigantic sheets of paper. The last effect was interpreted as the rubbing together of large rocks hurled skyward by the explosions. With the dawn, the *Charles Bal* set all sail and, driven by a rapidly rising gale, was able to leave the smoking mountain far astern.

Paroxysms of volcanic fury continued to shake the mountain until the final culmination of four prodigious explosions came on Monday, August 27, at 5:30, 6:44, 10:02, and 10:52 A.M. The greatest of them, the third, was one of the most titanic explosions recorded in modern times—greater in intensity than some of our nuclear blasts. The sound was heard over great distances: at Alice Springs in the heart of Australia, in Manila, in Ceylon, and on the remote island of Rodriguez in the southwest Indian Ocean, where it arrived four hours after the explosion had occurred 4800 km (2983 mi) away.

The explosion seriously disturbed the atmosphere, and records of such a disturbance were picked up by barometers all over the world. They showed that a shock wave originating in the East Indies traveled at least seven times around the world before it became too faint to register on the instruments of that time.

A more impressive visual phenomenon was the huge cloud of pumice and volcanic debris that blew skyward. The steam-impelled cloud of volcanic ash is estimated to have risen to a height of 80 km (50 miles) on August 27, and to have blanketed an area of 768,000 km^2 (296,525 mi^2). The ash poured down as a pasty mud on the streets and buildings of Batavia—now Djakarta—133 km (83 mi) away. Pumice in far-reaching masses blanketed much of the Indian Ocean, and captains' comments recorded in logbooks of ships suddenly surrounded by great floating rafts of pumice far offshore make interesting reading.

Volcanic ash hurled into the upper levels of the atmosphere was picked up by the jet stream, and carried with it as a dust cloud that encircled the earth in the equatorial regions in 13 days. The ash spread across both hemispheres to produce a succession of spectacular and greatly admired sunsets over most of the world—even in areas as remote from Java as England and the northeastern United States—for the two years that it took the finer dust particles to settle. There is ample evidence to suggest that the presence of such large quantities of ash in the atmosphere effectively screens the sun's rays and produces a worldwide cooling of about 1°C. That does not seem like

Fig. 6-12 Well-bedded ash and lapilli exposed in a stream cut in southwestern Mexico. The ash layers are the product of innumerable eruptions. *K. Segerstrom, USGS*

much, but it has led some geologists to suggest that accelerated and explosive volcanism was the underlying cause of the ice ages in the recent geological past.

The violent explosion of the morning of August 27 also set in motion one of the more destructive sea waves ever to be recorded. It spread out in ever-widening circles from Krakatoa, much as though a gigantic rock had been hurled into the sea. About half an hour after the eruption, the wave reached the shores of Java and Sumatra, and on their low-lying coasts the water surged inland with a crest whose maximum height was about 37 m (121 ft). Since many of the people who inhabited the densely populated tropical coast lived in houses built on piers extending out over the water, nearly 37,000 people lost their lives.

The sea wave, after leaving the Sunda Strait with diminishing height, raced on across the open ocean. It was registered long after it was too faint to see as a train of pulses on recording tide gauges along the coasts of India and Africa and on the coasts of Europe and the western United States.

After the explosions, returning observers were startled to find that where the 793-m (2601-ft) mountain had stood there was now a hole whose bottom was 274 to 305 m (899 to 1000 ft) below sea level, and that the sea now filled the bowl-shaped depression. All that remained of the island were three tiny islets. Although the estimates vary, the loss of material associated with the eruption amounted to just under 21 km^3 (5 mi^3) in volume. In popular accounts of the eruption, the impression commonly given is that a volcanic mountain blew up and its fragments were strewn far and wide over the face of the earth. Were that the case, most of the debris covering the little islands that are the surviving remnants of Krakatoa would be pieces of the wrecked volcano, and the oversized crater, or ***caldera,*** now filled with sea water would be the product of a simple explosion.

However, few pieces of the original volcanic mountain are to be found, and instead the ground is covered with deposits of pumice up to 60 m (197 ft) thick. The pumice, which was also widely observed floating in great rafts over the nearly open ocean, is original magmatic material, frothed up by gases contained in the magma, and has nothing to do with the internal composition of the vanished mountain. Thus, the abundance of pumice and the absence of pieces of the mountain lead logically to the conclusion that the volcanic cone foundered, or collapsed on itself, rather than having been blown to bits.

That explanation was advanced by a Dutch volcanologist, Reinout van Bemmelen, in 1929, and refined by Howel Williams, of the University of California, in 1941. The accompanying diagram (Fig. 6-13), adapted from Williams, shows the sequence of eruptive events that very likely were responsible for the disappearance of a nearly 800-m mountain and the appearance of a deep caldera in its place.

Crater Lake Although Crater Lake, Oregon, stands at an altitude of 1846 m (6055 ft), its caldera is similar to that of Krakatoa (Fig. 6-14).

The diameter of either circular depression is disproportionately large compared to the dimension of the mountain of which it is a part, and in each case, most of the mountain has disappeared in the formation of the caldera. Crater Lake is situated where a volcanic mountain perhaps 3700 m (12,136 ft) high, and containing about 70 km^3 (17 mi^3) of material once stood; it has been called Mount Mazama. Crater Lake was formed about 6000 years ago at a time when the area had only a sparse Indian population. Later Indians carried a legend in which the evil god of fire living within Mount Mazama waged a terrible battle with the good god of snow who lived atop nearby Mount Shasta. Good triumphed, but during the conflict the top of Mount Mazama was destroyed. The legend was probably based on actual events. From careful study of the character of the rocks exposed in the roots of the old volcano and of the distribution and composition of the vast amounts of pumice and ash that surround the site, the events before and during the catastrophic eruption can be pieced together. According to Howel Williams:

> When the culminating eruptions were over, the summit of Mount Mazama had disappeared. In its place there was a caldera between 5 and 6 miles wide and 4000 feet deep. How was it formed? Certainly not by the explosive decapitation of the volcano. Of the 17 cubic miles of solid rock that vanished, only about a tenth can be found among the ejecta. The remainder of the ejecta came from the magma chamber. The volume of the pumice fall which preceded the pumice flows amounts to approximately 3.5 cubic miles. Only 4 per cent of this consists of old rock fragments. . . . Accordingly 11.75 cubic miles of ejecta were laid down during these short-lived eruptions, in part, it was the rapid evacuation of this material that withdrew support from beneath the summit of the volcano and thus led to profound engulfment. The collapse was probably as cataclysmic as that which produced the caldera of Krakatau in 1883.

Mont Pelée In 1902, as today, Martinique in the French West Indies was one of the picturesque links in the chain of islands reaching like green stepping stones across the Caribbean to join Cuba with the mainland of South America. Perhaps the most notable distinction of this small island is that it was the birthplace of Josephine, Empress of France and wife of Napoleon.

The island is mountainous; most of the interior is garlanded with a tropical forest; and the people then, as now, lived near the coast in villages, on plantations, and in the few large towns, of which St. Pierre (with a population of 28,000) was the most important and had a history of continuous settlement extending back to 1635. Ships customarily moored in a line stood off the beach and were kept headed seaward. St. Pierre was then the leading commercial town of the island and had a fair number of French and Americans in residence. Most of the natives were Carib Indians or descendants of Africans imported to work in the plantations, sugar centrals, and rum distilleries.

Mont Pelée, about 8 km (5 mi) north of town, was known to be a volcano, but it had smoldered quietly, for several centuries. On April 23, 1902, there began occasional rumblings, clouds of smoke, and spasmodic outbursts of ashes and cinders. The mild display rose to a more violent level of activity on May 4, when a flow of hot mud, steam, and some lava broke

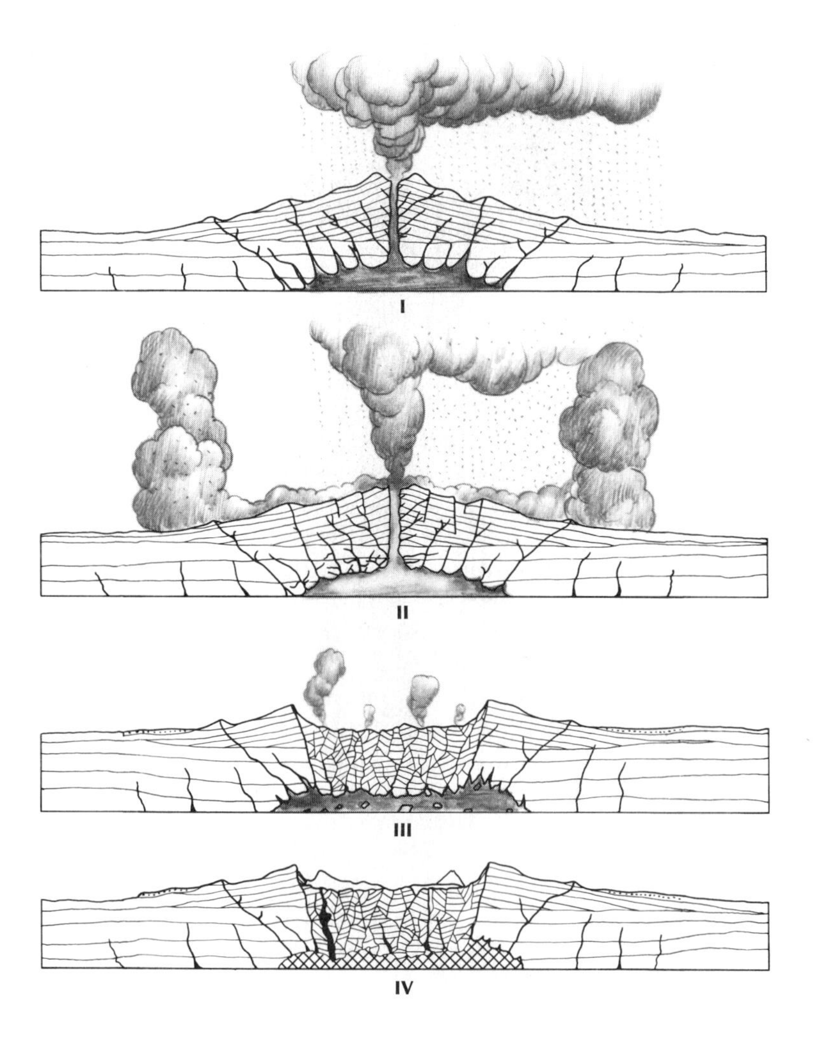

Fig. 6-13 Stages in the collapse of a volcano to form a caldera.

Stage I The cycle starts with fairly mild explosions of pumice. The magma chamber is filled and magma stands high in the conduits. As the violence of the explosions increases, magma is drawn off more and more rapidly.

Stage II The culminating explosions clear out the conduits and rapidly lower the magma level in the chamber. Pumice is blown high above the cone, or glowing pumice-laden clouds sweep down the flanks.

Stage III With removal of support, the volcanic cone collapses into the magma chamber below, leaving a wide, bowl-shaped caldera.

Stage IV After a period of quiescence, new minor cones appear on the caldera floor.

Fig. 6-14 Crater Lake, Oregon, looking southwest. The circular caldera is about 10 km (6 mi) across, the lake is nearly 610 m (2001 ft) deep, and the highest points on the rim are nearly 610 m above the lake. Wizard Island, a cinder cone that erupted on the floor of the caldera, is visible at the far right of the lake. Before the caldera collapsed, the area was the site of Mount Mazama, a steep-sided stratovolcano. *Ray Atkeson*

through the crater wall, coursed down one of the radial stream canyons, buried a sugar central, and killed 24 persons.

By that time St. Pierre was thoroughly aroused, and not even the presence of the governor and his retinue, together with the issuance of the usual proclamation, served to quiet the population. In fact, the city was kept in a continuous turmoil as country people and villagers, frightened into abandoning their homes, poured into town.

Early in the morning of May 8, 1902, at 7:45 A.M., according to the few eyewitnesses, the top of the mountain vanished in a blinding flash, and almost immediately thereafter a rapidly moving, fire-hot cloud engulfed the city, whose population, swollen with refugees, probably numbered more than 30,000 (Fig. 6-15). All but two—one of them in an underground dungeon—died in a blazing instant in a cloud at a temperature high enough to melt glass (650 to 700°C; 1202 to 1292°F).

About the only credible account of the eruption came from some of the survivors on ships in the roadstead. Eighteen vessels were in port at the time, and of them only the *Roddam,* with more than one-half her crew dead, was able to up anchor and escape. The cable ship *Grappler,* directly in the path of the incandescent cloud, capsized and blew up. The purser of the *Roraima,* then approaching the harbor from the sea, left the most complete narrative of any observer. The *Roraima,* enveloped in the wall of flame that incinerated the town, was hurled over on her beam ends. Her masts and stack were sheared off, her captain was blown overboard from the bridge and killed, and the ship herself burst into flames not only from the heat of the glowing cloud but from the thousands of gallons of blazing rum that poured through the streets of St. Pierre and spread out over the waters of the harbors. Through heroic efforts, 25 injured survivors out of a crew of 68 were taken off by the French cruiser *Suchet* in mid-afternoon.

Within the town itself only two human beings lived. All were dead except for Auguste Cyparis, the occupant of the dungeon, who languished there in a state of shock for four days until his rescuers, who had despaired of finding any living thing in St. Pierre, peered through the barred window of his dungeon when his cries attracted their attention; the other survivor, Léon Compère-Léandre, covered with burns made his way through the burning city and lived to tell his tale.

The escape of entrapped high-temperature gases was the chief reason for the catastrophically hot clouds that overwhelmed St. Pierre. Since there is no satisfactory English name for them, the French term **nuée ardente,** which might inadequately be translated as "glowing cloud," seems appropriate.

As was apparent to the trained observers who flocked to Mont Pelée after the disaster, the nuée ardente that swept through St. Pierre, searing everything in its path, was but a minor aspect of the total eruption. The major feature of the eruption was an avalanche of incandescent rock and pumice that shot out of the mountain top, cascaded down the canyon of the Rivière Blanche, and continued down to the ocean 3 km (2 mi) away. The nuée ardente was the hot dust and vapor cloud that rose above the ***glowing avalanche.*** (A glowing avalanche and the nuée ardente are together called a ***pyroclastic flow.***) Whereas the avalanche moved down the canyon, the fiery cloud jumped the ridge on the south side of the canyon and continued across the city. The destruction of St. Pierre was entirely the work of the dust and vapor cloud and the fires that followed it.

The 1902 eruption of Mont Pelée was important geologically because it provided an unparalleled example of a kind not too well understood before. No lava appeared in the early, violent phase of the eruption, as was possibly true also at Krakatoa.

When the violently explosive phase ended, a viscous, stiff lava was extruded into the summit crater, followed by the protrusion of a dome of blocky lava, encrusted with lesser spires and pinnacles. By September 1903, the spire had attained a height of perhaps 305 m (1000 ft) and a diameter about twice that. Estimates placed its volume at 100 million m^3 (3551 million ft^3).

Such ***volcanic domes*** are more common than many people think. Lassen Peak in northern California is an excellent example: a protrusion of blocky lava stands about 770 m (2526 ft) above the crater rim, with a volume of approximately 2.5 km^3 (0.6 mi^3). The mountain was last active in 1914–17, when steam explosions, after blasting a vent on the northern slope, melted the snow cap. The resulting mud and ash flows not only devastated the forest at the base, but swept 18-ton boulders for distances of 8 to 10 km (5 to 6 mi).

Among other examples of volcanic domes are the Mono Craters in east-central California (Fig. 6-16). Some of the volcanoes of the Valley of Ten Thousand Smokes in Alaska (Fig. 6-17), as well as the Puys of the Auvergne region of France, are domes. Of the latter, the Puy de Dome is perhaps the best known.

The nature of deposits dropped by turbulently flowing incandescent avalanches and nuées ardentes was uncertain before it was demonstrated at Mont Pelée. Commonly, their stratification may be chaotic. Large blocks are mixed with finer particles; the uniformly layered appearance characteristic of pyroclastic deposits when the ash settles more gradually through the atmosphere is lacking. Such deposits encircle Crater Lake (Fig. 6-18), for example; there, pyroclastic flows swept down the Rogue River canyon for 56 km (35 mi). They may have attained velocities of 160 km (99 mi) per hour and were capable of carrying pumice

Fig. 6-15 The charred remains of St. Pierre, Martinique, after the eruption of Mont Pelée in 1902. This capital city of 30,000 inhabitants was once a thriving port. *Library of Congress*

Fig. 6-16 Mono Craters, California, looking southwest into the Sierra Nevada. Mono Lake is in the foreground. At the lower right is Panum, a volcanic dome surrounded by a cone of pumice fragments. To the left is a larger dome of obsidian surrounded by a thick lava flow of obsidian bordered by a steep slope. Above and to the left of that is a lava flow of obsidian that has almost buried the dome that was its source. Above that are three more domes and a large flow of obsidian. The chain of eruption centers continues to the base of Sierra Nevada, forming a line of volcanoes nearly ten miles long. *Roland von Huene*

blocks 2 m (6.5 ft) in diameter a distance of at least 32 km (20 mi).

The principal difference between the eruptions of Mont Pelée and those of Krakatoa and of the ancient Mount Mazama (Crater Lake) is one of scale only. Krakatoa and Mount Mazama each erupted such large quantities of pumice and ash that their cone superstructures became weakened and collapsed along ring-shaped fractures to produce calderas several kilometers across. On the other hand, eruptions from Mont Pelée were restricted to the explosive discharge of small tongues of incandescent material, followed by intrusion of the sticky vent plug. The magma chamber was never emptied to any great extent and collapse of the cone could not occur.

Yellowstone and the San Juan Mountains As spectacular as the catastrophic eruption must have been that produced the Crater Lake caldera, it pales by comparison with those that

Fig. 6-17 Novarupta Rhyolite Dome in the Valley of Ten Thousand Smokes, Katmai National Park, Alaska. *USGS*

must have occurred not once, but three separate times in the Yellowstone National Park region at the eastern edge of the Snake River Plain, Idaho, during the last 2 million years. Careful mapping, in the land of grizzlies and geysers, by members of the U.S. Geological Survey has uncovered three large calderas, the smallest of which measures 29 km (18 mi) wide by 37 km (23 mi) long and the largest 64 km (40 mi) wide by 80 km (50 mi) long. The latter is apparently the largest caldera yet recognized on the face of the earth. Calderas of such dimensions are commonly called ***volcano-tectonic depressions.*** Yellowstone Lake itself occupies a portion of the youngest of these calderas. The formation of each caldera was associated with the cataclysmic eruption of hundreds of cubic kilometers of pumice-rich ash from the magma chamber. The temperature of the ash as it left the vent must have been in the range of 750 to 800°C (1382 to 1472°F). The ash traveled outward from the eruptive centers as far as 80 km (50 mi) so quickly that there was still sufficient heat left in the pyroclastic blanket after it came to rest—even at its far ends—to bring about the softening and sticking together of the glassy ash fragments and the collapse of the porous pumice fragments. The result was the formation of a dense, resistant rock called a ***welded ash-flow tuff*** (Fig. 6-19). Fine-grained ash blown high into the atmosphere during each of the eruptions was carried eastward by the prevailing winds and has been recognized in deposits as far east as Kansas. Rivers, such as the Yellowstone, incised canyons into the three different ash-flow tuffs, revealing that, in most places each deposit is more than 60 m (197 ft) thick. A somewhat sobering aspect of the Yellowstone eruptions is brought out by noting the sequence of the eruptions, as established by potassium-argon dating methods (see Chap. 1). The first took place 2 million years ago, the second 1.2 million years ago, and the third 0.6 million years ago. There was a lull of about 800,000 years between the first and second eruptions, but only 600,000 years between the second and third. Since the last eruption, 600,000 years have passed. Could the bubbling hot springs and fountaining geysers in Yellowstone National Park area be the harbingers of the next eruption? Any event of the immensity of even the smallest Yellowstone eruption in the past would bring about untold loss of life and property and a decline in economic prosperity. Because of that threat and because of the potential for geothermal power, the region around Yellowstone has been the subject of extensive geophysical investigations during the last several years. The results seem to indicate the presence of a rock body with anomalous physical properties just below the surface, 64 km (40 mi) in diameter and extending downward for 50 km (31 mi) or more. Most earth scientists have interpreted the body to be a magma chamber of batholithic proportions. The volcanic fireworks and caldera formations of the last 2 million years appear, in light of that finding, to be only the surface manifestations related to the development of the large, igneous body just below the surface. The question that remains unanswered is: How much of the body is still molten?

In the San Juan Mountains of southwestern Colorado, about 15 large calderas, each with one or more associated sheets of welded ash-flow tuff formed about 30 million years ago (Fig. 6-20). Like those at Yellowstone, they appear to have been closely associated with the development of a large, granitic batholith that lay close to the surface beneath them.

Fig. 6-18 Ash-flow tuff of consolidated volcanic ash, with lapilli and pumice blocks. These are the deposits of the pyroclastic flows that preceded the collapse of Mount Mazama to form Crater Lake. Since their eruption, the relatively nonresistant deposits have been eroded into pointed spires. The view is of The Pinnacles, along Sand Creek in Crater Lake National Park. *Oregon State Highway Dept.*

Mount St. Helens In the Cascade Range in the Pacific Northwest, many major composite cones are strung like beads on a north-south

Fig. 6-19 (above) Welded tuff from an area east of the Sierra Nevada, California, showing characteristic streaky layering resulting from vertical compaction of the ash flow after it comes to rest. The elongate black blebs, now obsidian, were initially porous fragments of light-colored pumice. *Karl Birkeland*

chain that extends more than 1000 kilometers (600 miles) from Mount Lassen in northern California to Mount Garibaldi in southern British Columbia. Within the range are such well-known cones as Mount Shasta in California; Crater Lake and Mount Hood in Oregon; and Mounts Adams, Rainier, Baker, and St. Helens in Washington. Four of the larger cones were named by the British naval commander George Vancouver, who surveyed the Pacific Coast of North America from 1792 to 1794. One peak was named after his third lieutenant, Joseph Baker; two others, Mounts Hood and Rainier, for admirals in the Royal Navy; and the fourth, Mount St. Helens, for a British diplomat, Alleyne Fitzherbert, Baron St. Helens. The chain is locally and intermittently active. Mount St. Helens was sporadically active for several decades during the last century, the last eruptive period ending in 1857. Minor eruptions had occurred at Mount Lassen from 1914 to 1917, and both Mounts Baker and Rainier in Washington are known to have produced steam and ash during the last 60 years. In 1975, vapor clouds issuing from vents near the summit of Mount Baker became so voluminous that scientists feared an eruption was close at hand, but none occurred. However, the mountain is being monitored. Studies of the volcanoes within the Cascade Range, particularly by scientists of the Volcanic Hazards Group of the U.S. Geological Survey, indicate that eruptions of moderate volume are likely to occur somewhere in the range as often as once every 1000 to 2000 years, and large eruptions may occur once every 10,000 years.

It was no real surprise, therefore, when scientists began to record numerous tremors in late March 1980, which emanated from directly beneath the 2950-m (9677-ft) Mount St. Helens. As the number and the intensity of tremors increased, it became increasingly clear that an eruption was imminent, and on March 27, 1980, the first cloud of steam and ash was explosively ejected from the relatively small crater at the apex of the cone. The exact moment

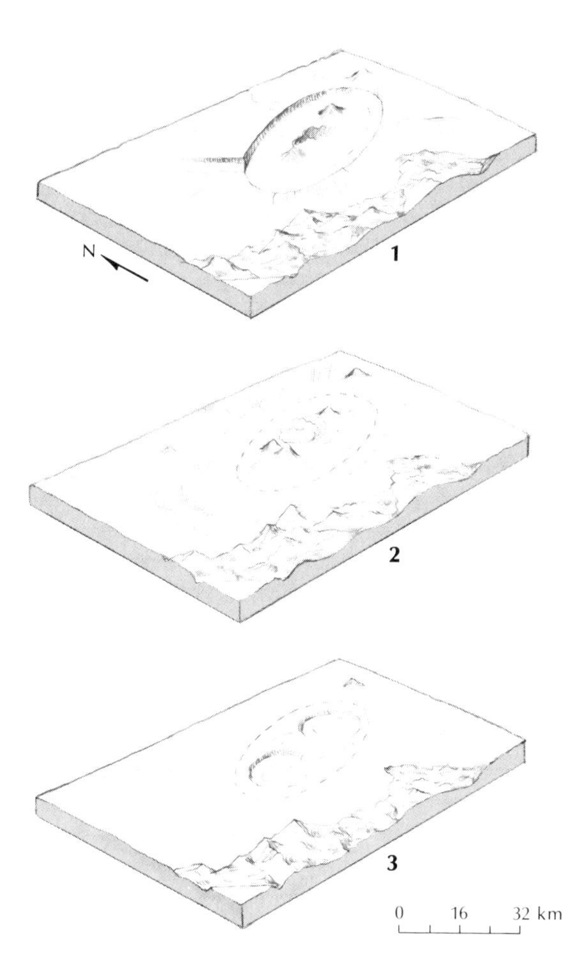

Fig. 6-20 A. (left) The development of three calderas in the San Juan Mountains, Colorado. First formed was the San Juan caldera (1), which was subsequently nearly buried by volcanic products (2). Collapse within the central part of the San Juan caldera formed the Silverton and Lake City calderas (3).

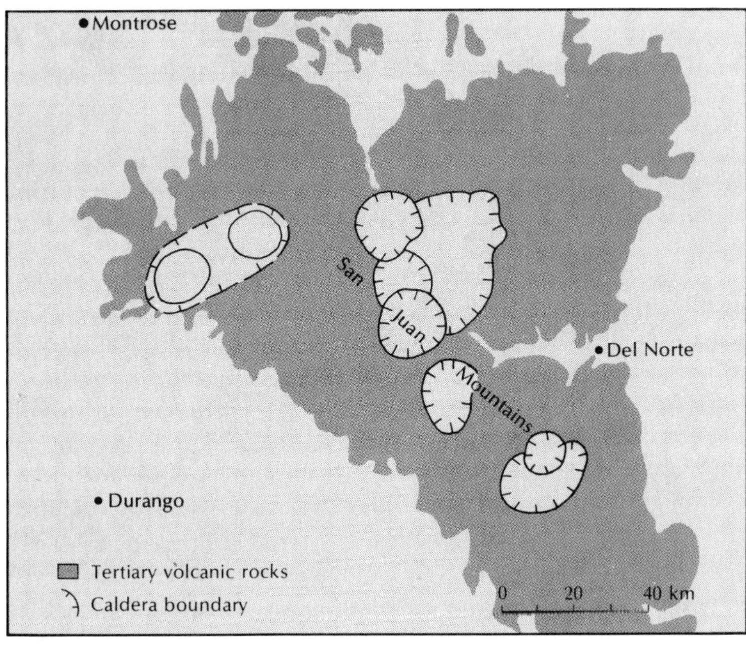

Fig. 6-20 B. Distribution of Tertiary volcanic rocks and calderas in the San Juan region, Colorado.

Fig. 6-21 The cracked, swollen North Bulge, Mount St. Helens, from the northeast, photographed on May 17, 1980, one day before the catastrophic eruption. *USGS*

eruption began is unknown because the volcano lay shrouded in clouds. Members of the U.S. Forest Service in the area, however, reported a loud boom at 12:36 P.M. A second explosion occurred during the early morning hours the next day, associated with the eruption of an ash and steam cloud that soared a mile above the cone.

During the first days of the eruption, vents close to the summit spewed volcanic ash, steam, and other gases; ash and ash-laden snow avalanches slid down the east flank; and tremors, apparently centered a mile beneath the northwestern part of the volcano, continued. Ash samples contained no newly formed magmatic material—only fragments of "old" cone material pulverized and blown out by the steam explosions. Light dustings of the ash, blown by the prevailing winds, fell at least as far away as Bend, Oregon, 240 km (150 m) to the south.

On April 1, the first "harmonic tremors" were

recorded on nearby seismographs, signaling underground movement of magma. Whereas normal earthquake tremors produce a burst of wave energy, flow of magma in cracks and conduits below the surface is associated with a near-continuous release of wave energy, termed ***harmonic tremor***. A stronger, more lengthy episode of this type of tremor on the following day, alerted scientists to the likelihood that full-fledged magmatic eruptions might soon take place.

Almost from the first days of activity, the volcano's shape had been carefully watched. By the end of April, it was confirmed that the upper part of the north flank had been pushed up or out about 82 m (270 ft) since the first eruption. This region, called the North Bulge, continued to grow laterally at an average displacement of about 1.5 m (5 ft) per day. Scientists interpreted the swelling to be the result of magma pushing its way upward into the cone and foresaw potential danger throughout the area of the north slope. An updated hazard warning was issued on April 30, 1980. As a result of that warning, access to the area around the volcano and particularly the north slope was restricted. Residents were evacuated and only a few, such as the tough old man of the mountain, Harry Truman, at Spirit Lake, refused to leave. During the next two and one-half weeks, the North Bulge continued to grow, and by May 17 it projected outward nearly 122 m (400 ft). The completely changed outline of the volcanic cone was evident even to the most casual observer.

Sunday morning, May 18, dawned bright and clear; at about 8:31 A.M., a strong quake occurred beneath the cone and, from eyewitness accounts, almost coincidentally, the unstable North Bulge began to quiver and then to slide, en masse initially, down the north slope. Almost immediately the volcano exploded violently. David Johnston, a geologist for the U.S. Geological Survey, watched from a position about 8 km (5 mi) from the peak as the sequence unfolded and radioed his last words to listeners, "Vancouver, this is it! . . . She's going!" Dr. Johnston along with scores of others died in the holocaust on the north face. The following passage provides a vivid account of the event:

> On May 18 a powerful explosion occurred from Mount St. Helens at 8:32 A.M. that was heard 200 miles away. Remarkable photographs, taken as the explosion began, show the north flank uplift peeling away from the volcano as a large vertical cloud began to rise from the summit. The [pyroclastic] cloud rose very rapidly to more than 10 miles above sea level, passing through the tropopause at 7 miles. Winds blew the cloud to the east. Ashfall at Yakima, 90 miles away, totaled as much as 4 to 5 inches and caused respiratory problems for some residents. By midafternoon, the ash had reached Spokane, reducing visibility to only 10 feet, although only half an inch was deposited there. Almost 2 inches of ash were reported from areas of Montana west of the Continental Divide, but only a dusting fell on the eastern slopes. Slight ashfall occurred in Denver on May 19. The ash blew generally eastward for the next several days, causing some problems for aircraft over the Midwest.
>
> The survey identified three components of the initial eruptive event in addition to the vertical cloud. A directed blast leveled the forest on the north and northwest flanks for a distance of up to 15 miles from the former summit. The blast swept over ridges and flowed down valleys, depositing significant quantities of ash. Although the blast was hot, it did not char fallen or buried trees. Many persons are known to have been killed by the blast, and others in the devastated zone are missing.
>
> The second component was a combined pyroclastic flow and landslide that carried the remnants of the north flank uplift across the lower slopes and about 17 miles down the Toutle River valley, burying it to depths as great as 180 feet. Large quantities of mud, logs, and other debris clogged several valleys around Mount St. Helens and rendered some shipping lanes impassible in the Columbia River.
>
> The third component was a pumiceous pyroclastic flow, funneled northward through the breach formed by the destruction of the north flank bulge. This flow dammed the outlet of Spirit Lake, trapping a large quantity of water.
>
> The volcano maintained an eruption column 10 miles high until a relatively sudden diminuition of activity occurred in the early morning of May 19. The altitude of the top of the column declined to about 2.5 miles. Activity continued to weaken through May 22.
>
> A new elliptical crater about a quarter of a mile deep had been formed by the explosion.

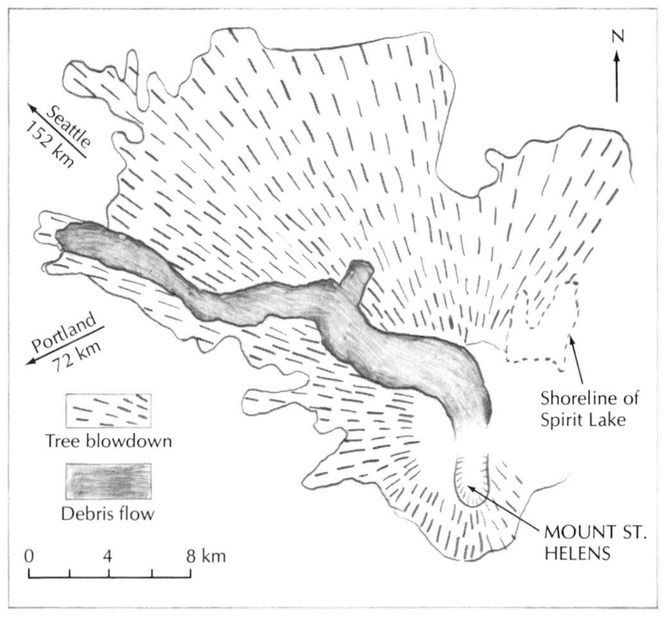

N
Seattle
152 km
Portland
72 km
Tree blowdown
Debris flow
Shoreline of
Spirit Lake
MOUNT ST.
HELENS
0
4
8 km

Preliminary analysis of seismic and deformation data indicates that there was no immediate warning of the imminence of a large explosion. A magnitude 5.0 earthquake occurred essentially simultaneously with the explosion at 8:32 A.M. Records of the only surviving tiltmeter, on the south flank, show that rapid inflation began at the same time as the explosion at 8:32 A.M.*

The force of the explosion shattered the cone and blew away about 2 km³ (0.48 mi³) of material, producing a snaggle-toothed cone remnant, 400 m (1300 ft) lower in elevation with a horseshoe-shaped crater 3.2 km (2 mi) long, 2.4 km (1.5 mi) wide, and 1.6 km (1 mi) deep (Fig. 6-22). The event was unprecedented in what is known of the history of the volcano. As pieced together by the observing scientists, the unusual event began with the earthquake on the morning of May 18, which triggered two landslides, one after the other, on the unstable oversteepened north flank of the volcano. A near-instantaneous release of pressure enabled pent-up gases to escape explosively, much like the bursting of a tire once the wall has been breached. As the eruption continued, what remained of the volcano's summit collapsed piecemeal into the vent, was blown into bits and pieces, and was carried outward and upward with the rest of the pyroclastic material.

Loss of life from that eruption stands at 60, perhaps more. It has been conservatively estimated, however, that if the hazard warning had not been issued and enforced, fully 10,000 lives would have been lost.

Since the eruption, activity has slowed, but by no means abated. Several times, sticky magma has pushed into the cone to produce a lava dome, resembling crusty rising bread, which subsequently has been fragmented and destroyed during explosive events. For the fourth time within a 48-hour period in mid-October 1980, the volcano erupted, each time sending a plume of steam and ash at least 8 km (5 mi) into the air. At this time, the volcano shows little sign of ceasing activity. If the last eruptive period during the last century is any indication, the volcano could remain active for several decades.

**Earthquake Information Bulletin,* Vol. 12, 1980. U.S. Geological Survey.

Fig. 6-22 A. (opposite above) Mount St. Helens, on May 18, 1980, from the northwest, looking across the jagged rim of the collapsed crater. Mt. Hood is in the background. *USGS* **B.** (opposite below) Damage in the vicinity of the volcano resulting from the eruption of May 18.

Eruptions of intermediate explosivity

Vesuvius One of the world's most famous volcanoes, Vesuvius is the only one active on the European mainland today (Fig. 6-23). Its renown probably results from its well-publicized eruption of A.D. 79 with the accompanying destruction of the cities of Pompeii, Herculaneum, and Stabiae. Although the mountain had been active in prehistoric times, it had been so long dormant that the Romans were unaware of its real character. Strabo, who visited the volcano about the beginning of the Christian era, surmised its volcanic origin from what he interpreted as burned and fused rocks near the summit, but he determined that the volcano was extinct. In fact the 1220-m (4002-ft) mountain we see today is superimposed in large part on the wreckage of the older and lower pre-A.D.-79 crater, to which the name Monte Somma is given. Its inactive, vine-covered bowl, encircled by steep cliffs, briefly made a stronghold for the gladiator Spartacus and his fellow slaves when they defied the power of Rome until they were slain by the legionaries of Marcus Licinius Crassus in 71 B.C. In A.D. 63, the volcano showed some signs of life when a succession of earthquakes commenced and caused some of the damage still to be seen around Pompeii. That, however, was but a prelude to the historic eruption of August 24, A.D. 79.

Fortunately, a relatively complete description of the destructive event has come down to us through two letters from the 17-year-old Younger Pliny to his friend Tacitus, the Roman historian. The letters were written primarily to describe the death of his uncle, Pliny the Elder, a leading philosopher of his day and also, rather surprisingly, an admiral of the Roman navy.

While Pliny the Younger turned to study of his books, the Elder Pliny, soon to achieve the distinction of being the world's first volcanologist, and a Roman of the old school, marched forth to his death on the mountain. Parts of Pliny the Younger's letters are cited here because they are such good examples of straightforward reporting, quite unlike the exaggerated, impossible version in Bulwer-Lytton's novel, *The Last Days of Pompeii*, in which most of the populace dies while watching a gladiatorial combat in the arena.

Parts of Pliny the Younger's letters follow:

> Gaius Plinius sends to his friend Tacitus greeting.
>
> You ask me to write you an account of my uncle's death, that posterity may possess an accurate version of the event in your history . . . .
>
> He was at Misenum, and was in command of the fleet there. It was at one o'clock in the afternoon of the 24th of August that my mother called attention to a cloud of unusual proportion and size . . . A cloud was rising from one of the hills which took the likeness of a stone-pine very nearly. It imitated the lofty trunk and the spreading branches. . . . It changed color, sometimes looking white, and sometimes when it carried up earth or ashes, dirty or streaked. The thing seemed of importance, and worthy of nearer investigation to the philosopher. He ordered a light boat to be got ready, and asked me to accompany him if I wished; but I answered that I would rather work over my books. . . .
>
> Ashes began to fall around his ships, thicker and hotter as they approached land. Cinders and pumice, and also black fragments of rock cracked by heat, fell around them. The sea suddenly shoaled, and the shores were obstructed by masses from the mountain. . . .
>
> My uncle, for whom the wind was most favorable, arrived, and did his best to remove their terrors. . . . To keep up their spirits by a show of unconcern, he had a bath; and afterwards dined with real, or what was perhaps heroic, assumed cheerfulness. But meanwhile there began to break out from Vesuvius, in many spots, high and wide-shooting flames, whose brilliancy was heightened by the darkness of approaching night. My uncle reassured them by asserting that these were burning farmhouses which had caught fire after being deserted by the peasants. Then he turned in to sleep. . . .
>
> It was dawn elsewhere; but with them it was a blacker and denser night than they had ever seen, although torches and various lights made it less dreadful. They decided to take to the shore and see if the sea would allow them to embark; but it appeared as wild and appalling as ever. My uncle lay down on a rug. He asked twice for water and drank it. Then as a flame with a forerunning sulphurous vapor drove off the others, the servants roused him up. Leaning on two slaves, he rose to his feet, but immediately fell back, as I understand choked by the thick vapors. . . . When day came (I mean the third after the last he ever saw), they found his body perfect and uninjured, and covered just as he had been overtaken. . . .

Pliny's letters clearly indicate that the destruction of Pompeii and Herculaneum resulted from the fall of hot ash, and in that shroud were buried 2000 of the 20,000 inhabitants who perished. Most of the dead were slaves, soldiers of the guard, or people who were too avaricious to leave their worldly goods. Most were suffocated by falling ash, by hot volcanic mud, or by volcanic gases; the temperature of the ash was high enough that their bodies charred away. Centuries later, when plaster of paris was poured into the cavities once occupied by their bodies, allowed to harden, and then excavated from the ash, their shapes, as well as those of dogs and cats, loaves of bread, and all sorts of objects in similar cavities, stood revealed. Hundreds of papyri were preserved in the library, along with murals on the walls of houses, and they give a most revealing insight into the interests and pursuits of those long-vanished Romans. The two cities of Pompeii and Herculaneum slept undisturbed for nearly 1700 years until the discovery of one of the outer walls in 1748 ushered in the period of modern archeology.

Vesuvius has continued its activity from A.D. 79 to the present; in A.D. 472, ashes drifted from its crater as far east as Constantinople. An especially violent eruption in 1631 is estimated to have killed 18,000 people and came after a period of quiescence that lasted long enough for the volcano to be overgrown by vegetation.

Fig. 6-23 Mount Vesuvius during the eruption of 1944. What appears to be a white-capped peak to the left of Vesuvius is a curved ridge known as Monte Somma, the margin of a caldera formed during the development of a prehistoric volcano. *Brown Brothers*

A large number of minor eruptions have been recorded, but major ones occurred in 1794, 1872, 1906, and 1944, in the midst of the Italian campaign of World War II. Lava then overwhelmed the village of San Sebastiano, but the most destructive effect, as far as the allied military effort was concerned, came from the introduction of glass-sharp volcanic ash into airplane engines.

The first lava is said to have appeared at Vesuvius in A.D. 1036, and its appearance has been a standard accompaniment of most of its eruptions ever since. Because eruptions in the current phase of the volcano's life history commonly include both the upwelling of large quantities of lava and violent explosions that blast great quantities of ash, cinders, bombs, and blocks skyward, the volcanic edifice that has built up since A.D. 79 is composed in part of solidified lava from flows and internally from dikes and conduits and in part of pyroclastic material blown out explosively. Most stratovolcanoes are similar to Vesuvius.

Quiet eruptions

Shield volcanoes, Mauna Loa and Kilauea The Hawaiian Islands, surely one of the most idyllic archipelagos in the world, owe their entire existence to volcanism. They are a chain of extinct, dormant, and active volcanoes built up from the depth of the sea and trending southeastward across the Pacific for 2560 km (1591 mi) from Midway on the north to the largest island Hawaii, on the south in an arc bowed slightly to the northeast. The name Hawaii is believed to be from *Hawaiiki,* the ancestral home of the Polynesians. The eight largest islands are at the southeastern end, and the relative erosional age of their landscapes generally decreases southeastward. That means that Hawaii, the only island with active volcanoes, appeared above the sea more recently than Oahu, the island on which Honolulu stands, as seen in the more advanced state of stream erosion, valleys, cliffs, and canyons on Oahu as well as the deeper soil cover that has developed there. Even more substantial evidence of the progressive "younging" of the islands toward the southeast is provided by potassium-argon dating of the flows on the eight largest islands of the chain. Kauai, the northwesternmost of the eight, is about 5.3 million years old, whereas Hawaii, at the southeast end, is 750,000, at most.

There are five major volcanic centers on Hawaii. Two of them include immense volcanoes: Mauna Kea, 4205 m (13,792 ft) above sea level, and Mauna Loa, 4170 m (13,678 ft). The bases of those mountains rest on the ocean floor, about 4600 m (15,088 ft) below the surface, and they are as tall as Everest, but of enormously greater bulk; the circumference of Mauna Loa is about 320 km (199 mi) at the base. They are by far the largest volcanoes in the world.

Although some pyroclastic material is included in the mass of the huge volcanic piles, for the most part they consist of thousands of superimposed, relatively thin flows of basalt. Many of the flows were extremely fluid at the time they were erupted. The result is that the slopes of classic shield volcanoes are gentle because they are made up of thousands of overlapping, tongue-like sheets of once-fluid material, rather than loose piles of heaped-up volcanic fragments.

The fires of Mauna Kea are banked for the time being, but Mauna Loa still maintains a high level of activity, although most of the historic lava flows have broken out on the flanks of Mauna Loa rather than simply overflowing from the summit caldera, known as Mokuaweoweo. In fact, activity in the caldera today is at a minimum. The caldera itself originated as the result of foundering through removal of support from below.

Mokuaweoweo has no circular crater; its very steep walls, which are about 180 m (590 ft) high, enclose a sink approximately 5.6 km (3.5 mi) long by 3.2 km (2 mi) wide. The long dimension trends northeast-southwest on the same line as the so-called Great Rift Zone, out of which so many of the historic flows have issued. The caldera itself has grown through the coalescence of several once-independent pit craters on the summit of the mountain, and it is Howel Williams's belief that the coalescence results from the collapse brought about by the

Fig. 6-24 Fire fountain at Kilauea, November 18, 1959. The lighter-colored parts of the fountain are glowing fluid masses of basaltic lava. The lava cools in the air and turns dark at the top of the fountain. *G.A. MacDonald*

draining away of lava from the magma reservoir within the mountain through fissures on its flanks.

The theory is supported by the pattern followed by typical eruptions. They may begin with volcanic dust being blown from the summit crater and a column of steam standing over it; the glow of light illuminates the clouds at night. Somewhat later, lava may break out on the flanks, almost always on the Great Rift Zone, either to the northeast or to the southwest of the summit.

Lava flows on Mauna Loa seldom issue from a single vent, but almost always break out from great cracks, or ***fissures.*** The first phase of such an outbreak may be the appearance of a line of ***fire fountains,*** or geyser-like columns of lava, that may spurt as high as several hundred meters into the air, forming a nearly continuous curtain of fire along the fissure (Fig. 6-24). The basalt that streams from the fissures is at a high temperature, close to 1200°C (2192°F) and consequently is usually extremely fluid. It may flow down pre-existing stream courses with velocities approaching those of the rivers themselves; where there are sharp irregularities, the lava plunges over them like a waterfall. When the lava stream reaches the sea, immense clouds of steam boil upward, the sea seethes like a gigantic cauldron, and part of the lava is quenched so abruptly that it froths up as a tawny, cellular sort of volcanic glass (Fig. 6-25).

Not all the Hawaiian basalts flow in torrential streams; blocky, ponderously advancing flows are common, too. They move forward much like a tank, or a caterpillar tractor, when the surface crusts over and is carried ahead by the still-molten interior. The advancing crust breaks up at the leading edge of the flow, and the blocks cascade over the front to make a carpet, or track, over which the flow can advance. The top and bottom of such a flow will become volcanic breccia when the whole flow has solidified, and for the most part the interior will be uniformly textured, homogeneous basalt.

Hawaii has given us two Polynesian terms to describe the surface character of lava flows, and they have now won such acceptance that they are commonly used in the literature of geology. Basalt with a rough, blocky appearance, much like furnace slag, is called by the remarkably brief name of ***aa*** (Fig. 6-26), whereas more fluid varieties, with smooth, satiny, or even glassy surfaces that are commonly contorted and wrinkled, are given the more euphonious name of ***pahoehoe*** (Fig. 6-27). It is not uncommon for a pahoehoe flow, as it loses gases and cools, to turn into an aa flow as it moves downslope away from the vent.

Kilauea is like Mauna Loa in some respects and very different in others. For one thing it is at a much lower altitude, about 1220 m (4002 ft); it is also no longer an independent mountain, but a partially buried satellite on the flank of the higher volcano. Perhaps within the next few millennia it will be inundated by flows from a flank eruption of Mauna Loa. Kilauea is the better known of the two volcanoes; a paved road leads directly to its rim, and it has had a steady stream of visitors for more than a century. Kilauea has had more than 50 eruptions in historic time.

The caldera of Kilauea is always a surprise to the first-time visitor, for much the same reason the Grand Canyon is. Both are in extreme contrast with their nearly level surroundings. The elliptical caldera of Kilauea, approximately 5 km by 3 km (3 mi by 2 mi), is countersunk with almost vertical walls into the very gently sloping surface of the old volcano (Fig. 6-28). The nearly level bottom of the caldera is made up of only very recently solidified lava, which spreads like a tarry stream over the entire floor. Activity today is confined to only part of the caldera—the volcanic throat, or fire pit, of Halemaumau, which bears a relationship to the larger caldera much like that of a drainpipe in the bottom of a washbasin. Basaltic lava rises and falls inside the fire pit. At times it spills over Halemaumau's rim onto the caldera floor; at other times it sinks more than several hundred meters below the surface. Then the floor of the pit is filled with long talus aprons of basalt blocks that have broken away from the vertical walls.

Commonly, lava swirls and seethes within Halemaumau without violent explosive activity, but occasionally there are departures from that pattern. Such a departure was the 1924 eruption, in which the sequence of events was as follows: (1) in January the lava lake was especially active and the level rose to within about 30 m (98 ft) of the rim; (2) in February it started to subside, and by May it had dropped to around 180 m (590 ft); (3) meanwhile the epicenters of a whole succession of minor earthquakes migrated steadily eastward along the line of the Puna Rift, accompanied by ground subsidence until almost certainly there was an eruption on the sea floor southeast of Hawaii;

Fig. 6-25 (opposite above) Basaltic lava, pouring over a cliff and into the sea during an eruption on the island of Hawaii in 1955. As the extremely hot lava meets the water, copious amounts of steam are produced. *G.A. MacDonald*

Fig. 6-26 (opposite below) Blocky, or aa, lava flow, slowly advancing over a field during the 1955 eruption on Hawaii. The flames and smoke are from burning vegetation. Inside the blocky interior the lava is still liquid. Such flows move ahead much like a tank; the cooled angular blocks cascade down the steep flow front and are overrun by the advancing flow. *G.A. MacDonald*

Fig. 6-27 Contorted surface of pahoehoe lava on Footprint Trail, Kilauea Volcano, Hawaii. *USGS*

(4) a great number of lava blocks avalanched from the walls into the fire pit; (5) finally, those blocks and much of the debris that had accumulated on the floor of Halemaumau were hurled out in a series of violent explosions between May 11 and 27.

An interpretation of that sequence of events is that the lava column dropped because lava was being drained away through fissures from beneath Kilauea—the entire level of which dropped by about 4 m (13 ft)—southeastward along the Puna Rift. That permitted ground water to move into the area left by the sinking column of lava. When a sufficiently high pressure was built up, and the lava column subsided below sea level, the ground water was converted into steam under cover of the blocks of rock that had fallen into the fire pit. Then the pressure rose to a point high enough to shatter the rocks, hurling them out of the pit in what was a succession of steam explosions rather than explosions produced by primary magmatic gases. To such a secondary eruption the name of ***phreatic explosion*** (derived from the Greek word for "water well") is given, and such explosions are characteristic of minor eruptions the world over: in Iceland, New Zealand, Japan, and possibly in California (the 1914–17 eruptions of Mount Lassen).

When lava enters the sea over long periods of time or when eruptions occur, as they sometimes do, beneath the surface of the sea, ***pillow lavas*** may be formed. As the name suggests, pillow lavas are lobes of lava material stacked one upon the other and resembling a pile of bed pillows (Fig. 6-29). They almost always form from pahoehoe-type basalt flows. The exact mechanism for their formation is not clear, but it appears that small extrusion toes, pushing out from cracks in the advancing pahoehoe flow, chill, when they come in contact with water, to form an elastic, glassy rind. The rind continues to grow into a pillow form, which either remains attached to the flow front or detaches from it. In the latter case pillows can bounce and roll down to the base of the slope. This result of the meeting of water and lava is relatively common in the geological record throughout the world.

The island of Hawaii has been the most intensely studied volcano in the world. The Hawaiian Volcano Observatory of the U.S. Geological Survey sits at the edge of the Halemaumau fire pit and is run by a staff who not only study the nature of the flows and gases that pour forth during each eruption, but also monitor almost undetectable changes in the surface shape of the volcano and earthquakes from deep beneath the island. Principally from earthquake data, scientists have concluded that the magma starts about 50 km (31 mi) or more beneath the surface and moves upward periodically to collect in shallowly buried magma chambers within the volcanic cones themselves (Fig. 6-30). As magma fills the upper chamber, the cone just above them inflates—that is, balloons upward and outward—which causes a distinct and measurable tilting and stretching of the volcano's surface. The stretching continues to a point at which eruption occurs and the tumescence declines, in response to the draining of the upper magma chambers. Scientists are now able to predict when an eruption is likely to begin, how long it will last, and if the eruptive phase has definitely ended or has merely become quiescent.

Fissure eruptions

Several regions on earth have been inundated by vast floods of lava that obviously did not come from a shield volcano or even from a chain of such volcanoes. Rather, the vast flows appear to be the result of eruption from innumerable cracks or fissures, commonly sub-parallel in their orientation, that extend over a considerable area. Prominent examples of ***flood lavas,*** which are universally of basaltic composition, are found in western India inland from Bombay, in South America near the Paraná River, in Antarctica, in South Africa, and in the area around Lake Michigan and in the Pacific Northwest in the United States (Fig. 6-31).

In the Pacific Northwest, one of the most extensive lava floods known is designated the Columbia lava plateau. It covers an area of 200,000 km^2 (78,125 mi^2). In places it is about 2 km (1 mi) thick, but individual flows are much thinner, only a few being as much as 120 m (394 ft) thick (Fig. 6-32). Their composition is remarkably uniform, especially in view of the fact that the enormous volume of basalt was erupted over a time span of about 3 to 4 million years. The total volume of lava erupted, amounts to more than 307,200 km^3 (73,701 mi^3), and the surface of the plateau covers a very broad area—extending from the Rocky Mountains on the east to the Cascade Range and Pacific border to the west. The hundreds, if not thousands of fissures that the lava welled up through are seen today as subparallel basaltic dikes trending north-south to northwest-southeast. The land buried by the lava floods was of moderate relief. As shown in the walls of canyons that cut across the basaltic plateau, the individual flows filled valleys and depressions, overtopped ridges, and ultimately coalesced to form a nearly uniform plain. They buried a wholly different sort of world. Such

Fig. 6-28 Kilauea Caldera, looking westward to the summit of Mauna Kea, Hawaii. The depression is about 5 km (3 mi) long by about 3 km (2 mi) wide. The smaller, sharply defined countersunk oval within the main caldera is the fire pit of Halemaumau. Mauna Kea has the classic rounded form of a shield volcano. *USGS*

Fig. 6-29 Pillow lavas of middle Cambrian age, Trinity Bay, Newfoundland. Some of the bulbous masses merge one into the other; others are separate. *Geological Survey of Canada*

tremendous outpourings of relatively uniform basaltic composition require an extensive source, of uniform composition. Most geologists feel that the lava must have come from the outer mantle of the earth (see Chap. 18), most likely from depths between 70 and 250 km (44 and 155 mi).

The nearest counterpart ever reported of a fissure eruption was a minor episode by comparison, impressive as it undoubtedly must have been to those who witnessed it. It was the eruption of the Icelandic volcano Skaptar Jokull that began on June 8, 1783. A stream of basalt amounting to about 3 km^3 (0.72 mi^3) in volume poured out from a fissure about 24 km (15 mi) long over a two-year time span.

Iceland has one of the most dramatic landscapes on earth, with more than 100 volcanic centers, of which at least 20 are active, a score of glaciers, and the all-encircling sea. In few other regions is the elemental contrast of fire, ice, and water more stark. In the long and remarkable cultural history of the island, extending back to A.D. 874, there have been many such encounters between outpourings of red-hot lava and streams of ice. The usual outcome is the melting of much of the ice, with a sudden, devastating flood of water and mud.

That is exactly what happened in the disastrous eruption of 1783. With basalt issuing along the length of the huge fissure, a broad flow of lava poured down the slope, filled the deep canyon of the Skapta to overflowing, and completely displaced a lake that lay in its path. The eruption continued for two years, and the two major lava flows it produced were 64 and 80 km (40 and 50 mi) long, respectively. Their average depth was 30 m (98 ft), but where canyons were filled to overflowing, they were as much as 185 m (607 ft) thick. Where the lava overtopped a stream valley and spread out across the plain it advanced along a front 19 to 24 km (12 to 15 mi) wide. The flow is estimated to have covered 90 km^2 (35 mi^2).

The eruption was one of the greatest disasters in the history of the island. The lava, blocking and diverting rivers and melting snow and ice, caused huge floods, which destroyed much of the island's limited agricultural land. Twenty villages were overrun by the lava, and many others were swept away in the floods. About 10,000 people, or 20 per cent of the population, died; 80 per cent of the sheep (190,000), 75 per cent of the horses (28,000), and more than 50 per cent of the cattle (11,500) perished.

Almost all the North Atlantic was obscured in the dust cloud, a phenomenon that greatly in-

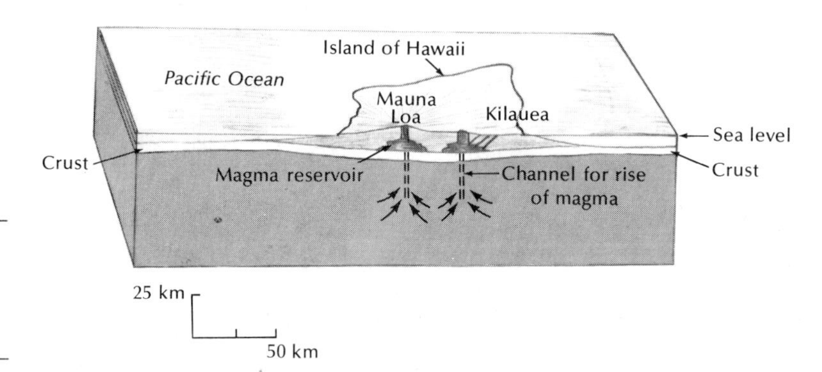

Fig. 6-30 Probable substructure of two volcanoes on the island of Hawaii. Magma, which is derived from depths greater than 50 km (37 mi), moves up into the crust and pools in larger chambers within the cones.

terested Benjamin Franklin. He wrote a brief description on the effect of the so-called dry fog in America (Griggs, 1922):

During several of the summer months of the year 1783, when the effects of the sun's rays to heat the earth in these northern regions should have been the greatest, there existed a constant fog over all of Europe, and a great part of North America. This fog was of a permanent nature; it was dry, and the rays of the sun seemed to have little effect toward dissipating it, as they easily do a moist fog rising from the water. They were indeed rendered so faint in passing through it that, when collected in the focus of a burning-glass, they would scarcely kindle brown paper. Of course, their summer effect in heating the earth was exceedingly diminished.

Hence the surface was early frozen.

Hence the first snows remained on it unmelted, and received continual additions.

Hence perhaps the winter of 1783–4 was more severe than any that happened for many years.

In Europe, the ash cloud had a most deleterious effect on the weather, too, so much so that it was called the "year without a summer." Crops failed in Scotland, 965 km (600 mi) away; fumes and ashes damaged crops in the Netherlands, and the ash cloud was reported from such widely scattered points as North Africa, Syria, eastern Russia, and Sweden. It rose to an altitude higher than the Alps, and the monks at the pass of St. Bernard correctly interpreted it as smoke, not haze. As might be expected, the event had a profound effect on people of the time, and among other things, inspired the following passage of Cowper (Krakatoa Committee, 1888):

Fires from beneath, and meteors from above—
Portentous, unexampled, unexplained. (*The Task,* BOOK II)

CLOSING THOUGHTS

In this chapter we have explored many of the more evident aspects of volcanism—the general distribution of active volcanoes, the char-

Fig. 6-31 The extent of **(A)** the Deccan basalts in India, and **(B)** the Columbia River basalt and its equivalents, northwestern United States.

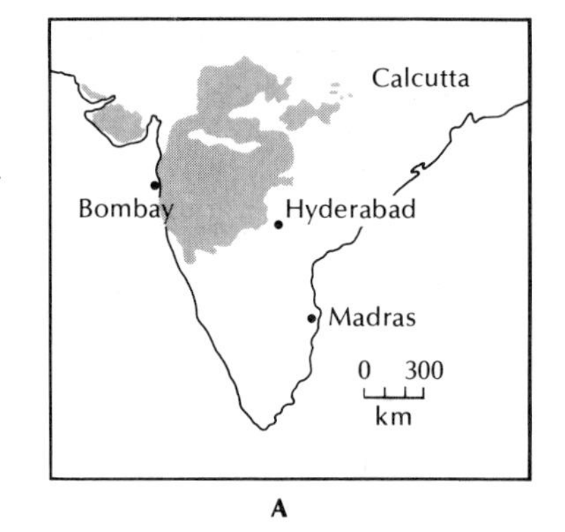

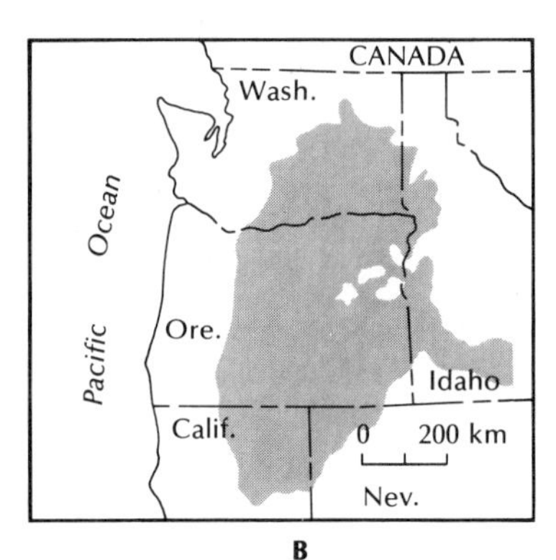

Fig. 6-32 Flow upon flow of the Columbia River basalt, eastern Washington. Note how the individual flows pinch and swell. Some are marked by columnar jointing. *Ray Atkeson*

acter of their cones and eruptive products, and the nature of volcanic eruptions. Whenever possible, we have tried to point out reasons why volcanoes are located where they are and why they behave as they do. As is evident from the text, there remain many unanswered or only partially answered questions. The basic reason for our uncertainty is clear; volcanism is only the surface manifestation of large- and small-scale processes that occur, unseen, relatively deep within the earth.

The major unanswered question concerns the ultimate cause of volcanic activity. The plate tectonics model has provided us with a framework in which to view the phenomenon of volcanism, and there is little doubt that the distribution and composition of many volcanoes are associated with activity at plate margins. But the nature of the subsurface activity and the depth at which it occurs, is imprecisely known. And of course the occurrence of many volcanic centers within the large plates themselves has required the formulation of axioms, not always well documented, that have had to be tacked onto the basic model. The process that leads to the generation of andesitic volcanism at converging plate boundaries is still a point of considerable debate and controversy. Many geologists believe it is the result of partial melting of rocks of basaltic composition at depths around 125 km (78 mi); others believe that, additionally, it requires melting of large amounts of rocks in the continental crust itself at depths of only a few tens of kilometers.

Another problem involves accounting for the immense volumes of magma erupted as basaltic flood lavas of extremely uniform composition, such as those found in the Columbia lava plateau. Some geologists have suggested that the layer between about 100 and 250 km (62 and 155 mi), which appears to be partially molten, provides a source for such volumes. If so, is it possible that basaltic magmas derived from different parts of the layer represent ***parent magmas*** from which all others are derived—by mixing of magmas, contamination by rocks above the source layer as the magma makes its way to the surface, and by differentiation? Those three processes might explain the great diversity in composition of lavas that erupt at the surface, even from the same vent or from closely related conduits.

And further, we do not know the exact nature of a magma chamber or how molten material initially moves and coalesces at depth to form a body of magma.

These are but some of the problems related to volcanism. We can only hope that as we continue to learn more about the earth's interior and the processes that operate therein, we will be able to better resolve them.

SUMMARY

1. Today there are more than 516 active or recently active volcanoes on earth. Most occur in linear bands, principally around the Pacific Ocean, but also in the Caribbean and Mediterranean regions, in Asia Minor and central Africa, and near the Red Sea. Zones of volcanism generally parallel belts of earthquake activity. Volcanic activity that occurs at convergent plate margins produces material that is andesitic in composition, whereas material produced at divergent zones is basaltic. Some volcanoes (such as those of the Hawaiian Islands) occur away from the edges of a plate and cannot be explained readily according to the plate-tectonic model.
2. Most large volcanoes can be classified as *composite cones* (*stratovolcanoes*) or *shield volcanoes*. Composite cones are steep sided as a result of the eruption of lava and pyroclastic material, and most are felsic to intermediate in composition. Shield volcanoes form from the accumulation of fluid basaltic lava and have gently rounded profiles.
3. In relatively viscous magma, gas discharge is inhibited. Commonly the pressure builds until it is released explosively. In some cases, pressure release causes the entire upper part of a magma chamber to froth, forming pumice or scoria.
4. In 1883, the island of Krakatoa erupted explosively. Sea waves, generated by the eruption, inundated the nearby coastal regions

killing about 37,000 people. The eruption was related to *caldera* formation. Crater Lake also owes its origin to the foundering of ancient Mount Mazama during a catastrophic eruption about 6000 years ago. The eruption of Monte Pelée (1902) was the result of a relatively small explosion of gas-charged particles (*glowing avalanche* and *nuée ardente*). Gigantic explosive eruptions and extensive caldera formation have occurred in the Yellowstone National Park region and in the San Juan Mountains. In both areas, the volcanic activity was related to a large batholith that existed at shallow subsurface depths. The recent explosion at Mount St. Helens was unusual in that it was laterally directed. It was triggered by a landslide initiated by an earthquake.

5. The Hawaiian Islands are shield volcanoes, with their bases on the ocean floor 4600 m (15,088 ft) below sea level. When they erupt, almost no explosivity accompanies the release of dissolved gases from the fluid basaltic magma.
6. In some cases, basaltic flows are fed not by a single vent, but by a *fissure* that might extend for several kilometers. One such fissure eruption was witnessed in Iceland in 1783. In the geological past, fissure eruptions produced so many lava flows in such a short time that in several extensive areas (such as the Columbia Plateau) the basalt is nearly 1.5 km (0.9 mi) thick in places.

QUESTIONS

1. What is the spatial distribution of volcanoes of various kinds throughout the world? Is their distribution related to other geological features and phenomena?
2. Compare and contrast the shape, internal structure, and chemical composition of a composite cone and a shield volcano.
3. What is the difference between a crater and a caldera?
4. What role do volatile materials play during volcanic eruptions?
5. In what ways was the eruption at Krakatoa similar to the one that produced Crater Lake?
6. What is a volcanic dome?
7. Outline the sequence of events leading up to the catastrophic eruption of Mount St. Helens on May 18, 1980.
8. Compare a quiet eruption with an explosive one.
9. What sorts of at-depth processes produce voluminous flood basalts?
10. Is the production of large plateaus, consisting of flow-upon-flow of basalt, geologically more significant than the generation of an island-arc chain of composite cones?

SELECTED REFERENCES

Bullard, F. M., 1962, Volcanoes in history, in theory, in eruption, University of Texas Press, Austin.

Francis, Peter, 1976, Volcanoes, Penguin Books, New York.

Jackson, Kern C., 1970, Textbook of lithology, McGraw-Hill Book Co., New York.

MacDonald, G. A., 1972, Volcanoes, Prentice-Hall, Englewood Cliffs, New Jersey.

Rittmann, A., and Vincent, E. A., 1962, Volcanoes and their activity, Interscience Publishers, John Wiley and Sons, New York.

Rubey, W. W., 1955, Development of the hydrosphere and atmosphere with special reference to the probable composition of the early atmosphere, *in* Crust of the earth, by A. Poldervaart, ed., Geological Society of America Special Paper 62, pp. 631–50.

Atlas of volcanic phenomena, 1972 U.S. Geological Survey, U.S. Government Printing Office, Washington, D.C.

U.S. Geological Survey, 1980, Earthquake Information Bulletin, vol. 12 (July–August), U.S. Government Printing Office, Washington, D.C.

Williams, H., 1941, Crater Lake: the story of its origin, University of California Publications, Bulletin, Dept. of Geological Sciences, vol. 25, pp. 239–346.

Williams, H., 1951, Volcanoes, Scientific American, vol. 185, no. 5, pp. 45–53.

7

SEDIMENTARY ROCKS

Widely spread over the surface of the earth, a relatively thin blanket of sediment has been consolidated into rock through slow-acting processes that are relatively simple to understand. Sedimentary rocks form in land or sea environments much more familiar to us than the deep crustal realm, where metamorphic and igneous rocks originate. Sedimentary rocks constitute 66 per cent of the area of the continents and, considering both continents and oceans, their average thickness is about 2 km (1.2 mi).

For the most part, sedimentary rocks are secondary, or derived, rocks. The major category of sediments consists of layers made up of clay, sand, or gravel particles derived from the disintegration or decomposition of pre-existing rocks. Layered rocks made of such fragmental material are called ***clastic*** sediments (Fig. 7-1).

Another large and economically important category of sedimentary rocks is chemically precipitated in such bodies of water as evaporating lakes or shallow embayments of the sea. A well-known example of that category is rock salt. Closely akin to it in origin are such well-known substances as gypsum and borax—both also chemically derived.

Organic sediments are a third category, and an important one to human beings. Coal, an important fossil fuel, is in this category, as are the oil shales now being studied as a source of oil. Another familiar kind of organic sedimentary rock is limestone, and, of its many forms, several represent the slow accumulation over many centuries of deposits made by plants and animals.

In this chapter, we will describe the main properties of sedimentary rocks and comment on how those properties can be related to the environment in which sedimentary rocks were formed. Although such a relation may not seem particularly important, it can be of the utmost importance to a geologist hunting for oil, minerals, or even groundwater. For example, a typical property, or set of properties, possessed by a sedimentary rock formation may indicate the presence of an important deposit in that formation or in similar formations nearby.

Fig. 7-1 Sedimentary rocks commonly are characterized by horizontal layering, which is usually most visible in arid regions. The layering in this aerial photograph closely simulates the contour lines of a topographic map. *William A. Garnett*

ORIGIN OF SEDIMENTARY MATERIALS

Many of the materials that make up sedimentary rocks are derived from weathering reactions (see Chap. 9). A much simplified weathering reaction is:

Rocks and minerals + water from rain and acids from carbon dioxide and organic matter $\xrightarrow{\text{weathering}}$

smaller solid particles + clay minerals + ions

The smaller solid particles—gravel, sand, and silt—are derived directly from the source area and thus reflect the type of rock outcrop there. For example, a source area in which basalts predominate will produce sands rich in olivine, calcic-plagioclase, and augite, whereas a granitic area will produce sands rich in quartz, potassium-feldspar, and biotite. Weathering on hillslopes in the source area before material is transported to its deposition site, however, can alter the mineralogy of the deposited sands. In climates in which intense chemical weathering is going on, more quartz is seen in granite-derived sediments than in the granite itself because quartz is more resistant to weathering than other minerals in granite.

Clay minerals form in soils and in other near-surface, low-temperature environments. They are fine-grained minerals, with sheet-like atomic structures like the micas (see Chap. 9). Each of the many clay minerals recognized reflects the environment in which it forms; however, if it is transferred to an appreciably different environment, it can change to another clay mineral. The interpretation of clay minerals in sedimentary rocks thus, is complex and depends, in part, upon the rocks and climate of the source area and the chemical conditions underground where sediment is slowly being transformed into rock.

Any weathering reaction produces ions or salts that dissolve in water. Some common ions are calcium (Ca^{2+}), sodium (Na^{+}), carbonate (CO_3^{2-}), and chloride (Cl^{-}); they and others are responsible for the mineral taste of certain waters. Under appropriate conditions, commonly in areas of fast evaporation, such ions precipitate out of solution. The minerals that they form are complex and are determined in large part by the chemical composition of the parent waters. Thus, precipitates divulge little information about the rocks or climate of the source area.

Still other kinds of sediments are made up of the largely insoluble shells of organisms. Such shells are formed, in part, by ions, which the organisms extract from their watery environment.

ENVIRONMENTS OF DEPOSITION

Sedimentary rocks accumulate in a wide variety of environments—about as many as there are different kinds of landscapes or climates. Most such environments, however, commonly occupy the lower parts of the landscape—the parts to which material moves from higher areas. The two major realms of sedimentation are (1) the sea (marine) and (2) the land (continental). Several kinds of sediments fit one category as well as the other; for example, the silts and clays that accumulate in the deltas of large rivers could be either marine or continental. Our purpose now is to describe briefly the main depositional environments; Chapters 10 through 15 will describe them in greater detail. It will be noted that each environment produces a more-or-less characteristic sediment, which later becomes a sedimentary rock.

A major concern of geologists is to examine rock outcrops, from which they attempt to identify the depositional environment. The task is not an easy one—often a field geologist must examine miles of valley walls or mountainsides before an identity can be established. Further, when we consider how widespread the processes of erosion are, it is not unreasonable to suppose that some of the rocks that might have provided the best evidence of a depositional environment have been eroded away. It is an excellent exercise in the method of the multiple working hypotheses to keep one's mind open, and to keep seeking more clues.

Before continuing our discussion, we should describe a property of sedimentary rocks that helps to distinguish one environment from another. That property is called ***sorting,*** and it refers to the degree of similarity in particle size in a sediment. In well-sorted sediment, for example, most particles are about the same size. In contrast, a poorly sorted sediment contains a wide assortment of particle sizes. Moderate sorting refers to an intermediate stage between the two.

Continental deposition

The ways in which sediments may be trapped on land are familiar to many of us. In the passages that follow, the main kinds of environments and deposits will be discussed, starting with mountainous terrain and progressing toward the seacoast:

Glacial deposits Glaciers, most of which today are confined to higher mountains or to the far distant Arctic and Antarctic, were once more widespread; their deposits—usually more disordered than those laid down by streams or in the sea—blanket much of North America and northern Europe. A good example of a typical glacial deposit is ***boulder-clay,*** which is literally that—rocks the size of boulders set in a clayey matrix.

Poor sorting is a major characteristic of glacial deposits, but such deposits still can be confused with other poorly sorted deposits.

Floodplains Although one of the major characteristics of rivers is the transport of sediment to the ocean, large amounts of material are commonly deposited along their margins, the floodplains (Fig. 7-2). Floodplains are those flat surfaces adjacent to rivers, especially in lowlands, but in mountainous regions as well, over which rivers spread in times of flood. During each flood a new layer of sediment is deposited. These depositional sites range from the extensive plains bordering the Nile, the Yangtze, and the Mississippi, to narrow strips of land bordering small streams. Sediment size will vary with local conditions, but it varies from small clay particles to the largest of boulders. Sorting varies from good to moderate.

Alluvial fans These landforms are typical of arid and semiarid regions. When a stream comes rushing from mountains or hills carrying a great deal of rock debris and suddenly reaches a basin, its sediments are dumped in a spreading fan-like form (Fig. 7-3). Excellent examples are the fans bordering the mountains of our southwestern deserts. Some are so poorly sorted that they can be easily mistaken for glacial deposits. The two deposits, however, form under quite different temperature regimes.

Lakebeds Lakes are sure traps for sediment. Glaciers, rivers, and wind action all move sediment into lakes. However, lakes are a relatively temporary feature by the geological time-scale, being formed mainly by glaciers, landslides, volcanic activity, and faulting. Eventually they fill in with sediment. Lake sediments generally are well sorted.

Deltas Deltas form where sediment-laden streams enter bodies of relatively still water, such as lakes or gulfs (Fig. 7-4). Although they form mostly below water level, their top surfaces eventually rise above the water and so become floodplains. Deltas are complex environments because they have their origin in a combination of river-lake or river-ocean processes. In that much of the sediment is laid down in relatively quiet water, the sorting is good.

Wind Locally abundant sand dunes and virtual seas of sand testify to the effectiveness of wind in those parts of the world where such factors as an abundance of sand, little vegetation with which to stabilize it, and strong winds to move it about occur together (Fig. 7-5). Such combinations are likely to be encountered in deserts, along many of the world's coasts, and along the floodplains of large rivers, of which the Volga is an excellent example. Wind also sweeps along a fine dust—called loess or silt—

Fig. 7-2 The present floodplain of the Rakaia River, New Zealand, is the lighter-colored gravel strip adjacent to the river. Overall the floodplain makes up one-third to one-half of the valley floor, the rest being a slightly higher terrace—the former floodplain. *Peter W. Birkeland*

that may pile up in vast windrows. A thick blanket of feebly consolidated dust dominates the tawny landscape of northern China near Peking. Unlike many other deposits, those moved and laid down by the wind are very well sorted, be they sand or silt.

Marine deposition

Some of the factors affecting the distribution of sediment in the sea are (1) distance of the deposit from land, (2) depth of the water, and (3) the physical and chemical properties of the water, and the types of plant and animal life. In order to simplify the story, we may say that sediments accumulate in four major zones and that the sediments themselves are well sorted.

Just seaward from the land is the ***shore zone***—the place where the surf breaks against the shore. On many coasts with a large tidal range, deposits are visible along the broad expanse of sea floor adjacent to the land that is laid bare at low water. Of the sediments carried

Fig. 7-3 Alluvial fans flank the ranges in the Mojave Desert, California. The fan shape is best seen at the lower right; elsewhere fans have coalesced. White saline deposits of temporary lakes lie between the ranges. *USAF*

Fig. 7-4 Delta built by streams issuing from glaciers on the south side of Inugsuin Fiord, Baffin Island, Northwest Territories, Canada. *J.D. Ives*

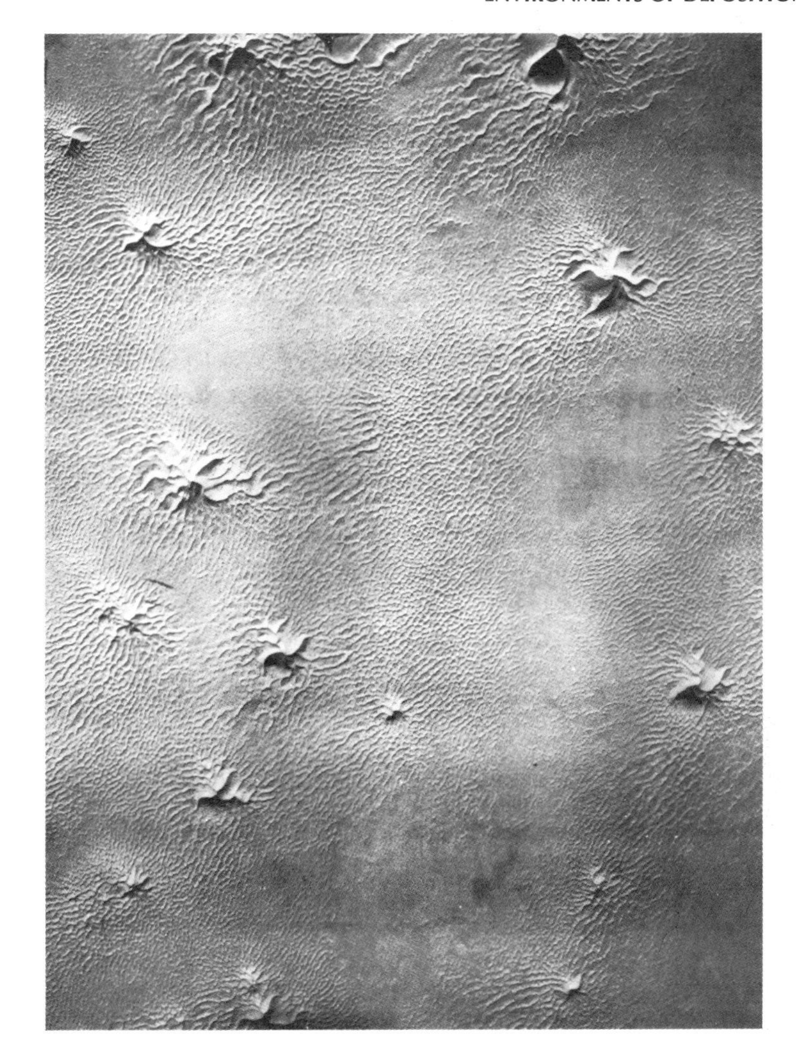

Fig. 7-5 Sand dunes in the Sahara. The forms range from relatively small ridges to star-shaped hills that may be as high as 100 m (328 ft). *H.T.U. Smith*

to the marine environments, the coarsest—sands and gravels—usually are trapped in this high energy zone, whereas the fine-grained sediments are carried offshore.

The ***continental shelf*** is much broader than the shore zone, extending seaward to an average depth of 133 m (436 ft). As a rule, the shelf is the region where most land-derived sediment comes to rest, although it makes up only 7.5 per cent of the area of the oceans. Most of the marine sediments we see exposed today were deposited on the continental shelf; hence the importance of studying that environment. The sediments are usually silts and clays, with some sands.

In some shelf areas, carbonate-secreting organisms—examples are algae and coral—contribute greatly to the formation of sedimentary rocks. Inorganic precipitation of calcium carbonate ($CaCO_3$) also is an important process in such environments. As a rule, these areas are characterized by shallow water at fairly warm temperatures and few continent-derived sediments. The remains of organisms, and other carbonate materials, build up on the ocean floor and harden to form limestone.

Fig. 7-6 This monastery at Meteora, Greece, surmounts bluffs of layered, cemented gravel deposits (conglomerate) of Tertiary age. The area has undergone tilting since deposition, as shown by the fact that the layers are no longer horizontal. *Jean B. Thorpe*

Seaward from the edge of the continental shelf, the topography steepens to form the ***continental slope*** which continues on down to the deep floor of the sea, or the ***abyss***—a region accessible only to deep-diving submersibles; it is now possible, however, to photograph this dark realm. There on the floor of the open sea accumulate the finest sediments, calcareous and siliceous oozes, composed of the remains of free-floating and swimming microscopic plants and animals, and the extremely fine-grained clays that carpet the abyssal plains. Only under special circumstances can coarser-grained, land-derived sediments be transported so far from land and to such great depths.

FEATURES OF SEDIMENTARY ROCKS

Sedimentary rocks have several key features that distinguish them from the other major rock types. Some of those features have even survived metamorphism and thus provide clues to the past history of metamorphic rocks. Sorting has already been mentioned; the other features will be discussed now.

Stratification

Most sedimentary rocks are made up of particles, ranging from very large to submicroscopic, that settled out through a medium such as air or water. In addition, most of them are layered (Fig. 7-6), and the layers, too, have a great range—from those measurable in millimeters to those measurable in meters.

Such depositional layers are called ***strata;*** an individual layer is a ***stratum.*** In everyday language, the layers commonly are called beds if their thickness is greater than 1 cm (0.4 in.). If the layers are less thick than this, they are better called **laminae** (from the Latin, *lamina,* for thin plate, leaf, or layer). The term is used here much the same way as we use it in speaking of the laminations in plywood.

In "quiet water" depositional environments,

Fig. 7-7 Cross section of an ancient river channel in Utah filled with conglomerate. *W.R. Hansen, USGS*

Fig. 7-8 Varved sediments from an ancient glacial-fed lake, northwestern Ontario. Each varve consists of a lower, light-colored silty layer, overlain by an upper, dark-colored clayey layer. *L.D. Ayres*

such as the bottoms of lakes or the ocean floor, strata are laid down almost horizontally, in layers of varying or uniform thickness. In contrast, the thickness of river deposits can be more variable as the deposits are traced laterally, and in places, ancient river channels can be seen (Fig. 7-7).

Layering can occur through variations in the energy of the transporting medium or through variations in the size, amount, mineral composition, or color of the sediment being delivered to a particular depositional site. A high-energy flood, for example, could wash large particles out onto a lake bottom. Following the flood, silt and clay—the usual sediment carried to the lake—might be deposited on top of the flood debris. The result is two layers of contrasting particle size. Discontinuous deposition also can cause layering, because when deposition begins again, chances are slight that the newly deposited material will have exactly the same properties as the sediment that came before it.

Some sediments and sedimentary rocks are characterized by the repetition of distinctive layers. Well-known examples are ***varves,*** which seem to form best in glacier-fed lakes. Each varve consists of two layers—a coarser silty layer overlain by a finer-grained layer of silt and clay (Fig. 7-8). The varve is thought to represent a seasonal event. The coarser fraction in the lower layer is thought to be laid down during the spring and summer, the seasons of active snow and ice melt. The finer-grained upper layer, however, represents quiet sedimentation during the fall and winter, when the ice stops melting and the surface of the lake is frozen over. Because each varve represents a year of record, varves can be used with some degree of success to determine the age of a particular lake. Thus, as a means of dating deposits or events, varves are not too different from tree rings.

Some sediments, ranging from coarse to fine grained, show a very different sort of stratification. The particles of the individual layers, instead of being distributed uniformly throughout, are graded—the larger particles concentrated at the bottom, the smaller at the top (Fig. 7-9). Such layers are said to have ***graded bedding***—a feature that occurs when a mass of sediment is discharged suddenly into a relatively quiet body of water. The larger particles drop out quickly, followed by the medium-sized ones, and finally the finest particles. The change from one dominant particle size to the other is gradational, and the whole makes up the graded bed.

An excellent example of how graded beds are formed may be seen where the excessively muddy Colorado River flows into Lake Mead, which is backed up behind Hoover Dam on the boundary between Nevada and Arizona. The muddy river water seems to disappear as if by magic, and anyone who has seen the dark, blue-green water of Lake Mead cannot fail to be impressed by the contrast. An explanation for the disappearance of the murky water is that, with its higher density, the mud sinks to the bottom of the lake where it moves as an underflow known as a ***density*** or ***turbidity current.*** If the mud remains undisturbed, it can eventually lithify to form a sedimentary rock called a ***turbidite.*** Every child has, at one time, formed a density current by poking at the edge of a mud puddle with a stick, causing the disturbed liquefied mud to flow out toward the middle of the puddle along its bottom.

As discussed in Chapter 1 (the law of superposition), in layered sequences of rocks, the younger layers rest on top of the older layers, rather like a layered birthday cake. This is a very important point to keep in mind when trying to determine the relative ages of rocks in a sedimentary rock sequence.

Roundness of grains

An important property of clastic sediments is the roundness of the grains. How are they rounded off? One explanation is that the distance of transport affects the degree of rounding. Assume that a rock tumbles from a cliff to an adjacent river bottom. Initially the rock has very sharp corners, but as it slowly moves downstream it collides with and scrapes against neighboring rocks and the corners become more rounded. Some of the more rounded gravels are found in beach and wind environments (Fig. 7-10), and good-to-moderate rounding characterizes river environments. In contrast, glaciers and mud flows do not allow much grain-to-grain contact; consequently, rounding of the coarse materials is generally poor.

Color

Igneous rocks, unaltered by exposure to the atmosphere, typically are shades of gray or black, because those shades prevail in the most abundant constituents of the rocks, feldspar and ferromagnesian minerals. Sedimentary rocks are often more colorful, however, with the materials that color them either filling the voids between the grains or coating them.

An important pigment in sediments is iron released by the weathering of iron-bearing minerals. If the pigment contains hematite (iron oxide, Fe_2O_3), the resulting rock is likely to be red. Hematite is the source of most of the rich red coloring in the walls of the canyons of Utah and Arizona. Other forms of iron may

Fig. 7-9 Sequence of graded beds in a playa in northern Sonora, Mexico. Each bed consists of sand or silt grading upward to fine clay and each probably records a flood. The bed at the center is about 2 cm (1 in.) thick. *W.B. Bull*

stain a rock brown, or even shades of pink and yellow. Iron may even be partly responsible for the purple, green, or black shades of some sedimentary rocks, but often the origin of coloring matter simply is not known. Pigments may have been carried to the site of deposition along with the sediments or produced later by chemical weathering of the original sedimentary grains throughout eons of deep burial.

Many of the dark sedimentary rocks owe their color to the organic material they contain. Coal, an excellent example, is of organic derivation, and its very name is a synonym for black. Depending upon the amount of organic material, sedimentary rocks may range from light gray to black. Some black muds, however, owe their color to their content of finely divided iron sulfide rather than to carbonaceous matter.

The range of colors that sedimentary rocks may display is one of their more intriguing properties; in dry areas, where vegetation is lacking and the soil cover is sparse, the true colors of the rocks stand revealed in striking array, as in Grand, Zion, and Bryce canyons and in Monument Valley, Canyon de Chelly, and the Painted Desert. It is the brilliant coloring of their sedimentary rocks as much as any other attribute that makes those places so renowned.

Fig. 7-10 A. (above) Well-rounded gravel clasts are characteristic of many beach environments. *William Estavillo* **B.** (left) Rounded river boulders of basalt stacked up to form a wall on a farm in southern Idaho. This amount of rounding occurred in 13 km (8 mi) of stream transport. *H.E. Malde, USGS*

Fig. 7-11 Mud cracks, Colorado River, head of Lake Mead. Depressions are due to gas vents. *Tad Nichols*

Mud cracks

When wet, clayey mud is exposed to the air it dries, shrinks, and then cracks—generally in a nearly uniform pattern of polygons (often four sided) that in some ways resemble the tops of lava columns (Fig. 7-11). In lava, cracking is caused by contraction upon cooling; in wet muds, by dehydration. As drying continues, the mud layers on the tops of the polygons may curl up at the edges, so much so that at times they form complete rolls, much like cardboard tubes or fancy pastries.

Mud cracks indicate that the sediment of which the mud was once a part was alternately wet and dry; thus such cracks are very typical of mud-bottomed, shallow lakes that on occasion dry up. They are not so characteristic of muddy tidal flats because the time of exposure at low tide is too brief for much drying out to occur.

Ripple marks

Nearly everyone has noticed the characteristic corrugated surface made by currents flowing across the sandy bottom of a lake or stream (Fig. 7-12) or has seen photographs of virtually the same pattern produced by the wind blowing across a desert sand dune. Such ripples, called ***current ripples,*** develop at right angles to the current and are likely to have steep slopes on the down current side but gentle ones upcurrent. The asymmetrical form results when an air or water current rolls sand gains upslope and gravity pulls them down the opposite side, or ***slip face.*** That slope stands at an inclination known as the ***angle of repose*** (the maximum slope at which sand grains will remain stationary without sliding down the slope). Current ripples in solid rock can be used to establish the direction once taken by ancient currents in the atmosphere or under water. In the past, it was thought that such ripples in water-laid sediments indicated shallow depths, but in recent years underwater photography has shown ripple patterns on the sea floor at depths of several thousand meters.

Another type of ripple has symmetrical sides, sharper crests, and more gently rounded troughs than current ripples have. Such corrugations are called ***oscillation ripples*** and presumably are the result of surface waves (called ***waves of oscillation***) stirring up the sandy bottom of a shallow body of water.

Cross-bedding

Earlier in the chapter, the point was made that sedimentary rocks usually are deposited in essentially parallel, horizontal layers. However, in several varieties of stratification, the layers are inclined at steep angles to the horizontal plane. Such layering is known as ***cross-bedding.***

One kind of cross-bedding can be seen in sand dunes (Fig. 7-13). Each layer of the dune at some time past was part of the surface, and, since the dune's configuration was established largely through a balance of wind transport upslope and gravity-sliding downslope, most of the layers are sweeping curves, which more often than not are concave upward. Because sand dunes are ephemeral landforms, which change in position and orientation with the inconstant wind, it is not surprising that the sweeping, shingled layers may intersect one another in complex patterns such as those in the walls of Canyon de Chelly, Arizona, or in Zion National Park.

Another kind of large-scale cross-bedding occurs in deltas. Streams carrying a fairly large load of moderately coarse debris deposit their sediment rapidly when they reach a body of water such as a lake or gulf. There, the sediment deposited by the stream constructs a leading edge out into the water, much as a highway-bridge fill is built out into a canyon by end-dumping from gravel trucks. The outer slope of the delta, like the slip face of a sand dune or a current ripple, also stands at the angle of repose. When such sediments are con-

Fig. 7-12 Ripple marks on a bedding plane in the Dakota Sandstone, Colorado. *J.R. Stacy, USGS*

Fig. 7-13 (opposite) Giant cross-bedding in ancient sand-dune deposits at Checkerboard Mesa, Zion National Park, Utah. *Ray Atkeson*

solidated into rock, three distinctive layers may result. At the top and bottom of a deltaic deposit are horizontal strata, known as ***topset*** and ***bottomset*** beds, respectively, and the steeply inclined layers in the middle that were once the delta front as it advanced out into the water are called ***foreset*** beds (Fig. 7-14).

River deposits also can be cross-bedded, but usually on a much smaller scale than those discussed above (Fig. 7-15).

Fossils

One of the more characteristic properties of sedimentary rocks is the presence of fossils (Fig. 7-16). They are the remains of once-living things that, on their death, were buried in sand, silt, lime, or mud. Much of their organic matter was gradually replaced over the centuries by inorganic matter, until, to use petrified wood as an example, many of the woody fibers were replaced by silica. Many species of plants and animals have been preserved as fossils. Even such improbable creatures as jellyfish, whose composition must be more than 95 per cent water, or such fragile structures as the compound eyes of flies or the delicate tracery of dragonfly wings have been preserved. Those creatures are the exception, however, because the organisms most commonly preserved as fossils are those that have such durable elements as shells, bones, and teeth. In fact, most

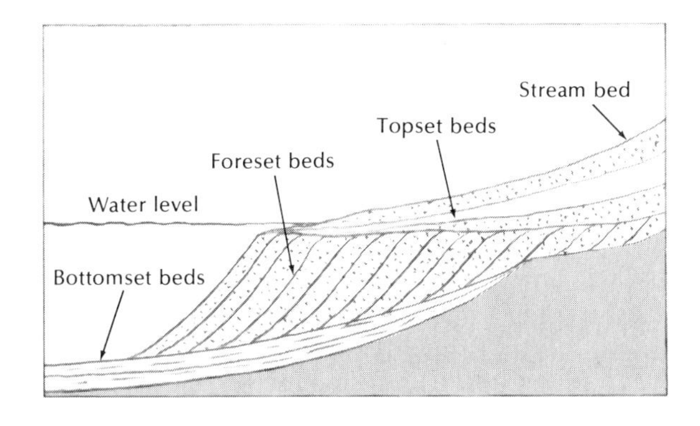

Fig. 7-14 Cross section through a delta illustrating the major sets of beds.

Fig. 7-15 Cross-beds in sands and gravels, Washington. *F.O. Jones*

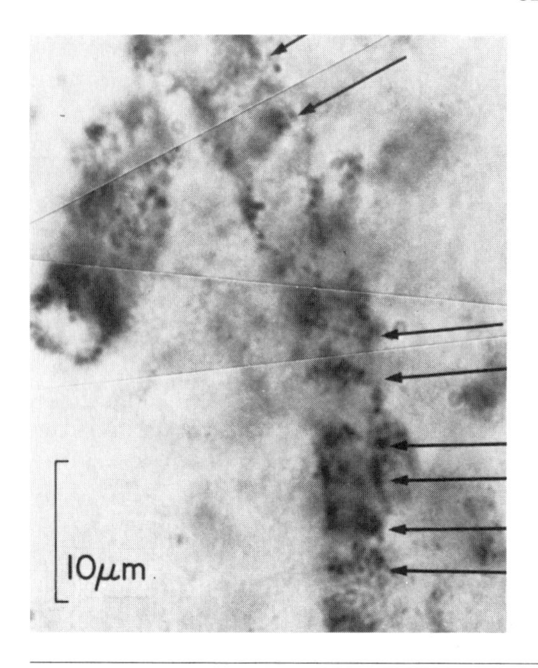

Fig. 7-16 Some examples of fossils preserved in rocks. **A.** (above, left) Microfossil cells, said to be about 3.5 billion years old, found in sedimentary rocks in Western Australia. *J.W. Schopf.* **B.** (above, right) This object, identified as the egg of a vertebrate, is the oldest known such egg. It was found in early Permian sandstone beds (red beds) in Texas. *Museum of Comparative Zoology, Harvard University.* **C.** (opposite, above) Brachiopod fossils in the Trenton Limestone of Ordovician age, Watertown, New York. *Smithsonian Institution.* **D.** (opposite, below) Leaf impressions preserved in Tertiary volcanic tuffs near Corvallis, Oregon. *Janet Robertson*

fossils are the remains of shells or skeletons. In some instances, an entire rock may consist of organic matter. Coal is made up of plant fragments, and some limestones are the remains of coral or of calcareous algae, or may be a mass of seashells. In addition to the remains of organisms, footprints, tracks, trails, and burrows also are considered fossils.

Fossils have many significant uses in geology. For example, they can be used to indicate how long life has existed on earth. In 1980 it was announced that the oldest fossil cells, said to be about 3.5 billion years old, had been unearthed in sedimentary rocks deposited in an ancient saline lake in Western Australia. These organisms lived a billion years or so after the earth formed and are 1.2 billion years older than the fossils previously confirmed as the oldest form of life. Fossils are also extremely important to geologists in dating rocks and in correlating rocks from areas to area. Finally, they provide vital clues to the depositional environments of the rocks in which they are found. Shell collectors are well aware of this—if they are to have a varied collection, they must search many environments for their specimens.

Concretions

Round, or almost round, solid bodies are sometimes found within sedimentary rocks. ***Concretions,*** as most such bodies are called (Fig. 7-17), are composed of material that solidifies around a small hard nucleus after the sediment was laid down. Any small particle—a grain of sand, a piece of shell, even a small insect—can act as a nucleus. A cement, which can be limy, collects around the nucleus, eventually binds all the particles together and gradually enlarges the concretion. Some may reach 1 m (3 ft) or more in diameter. Most are much smaller, however.

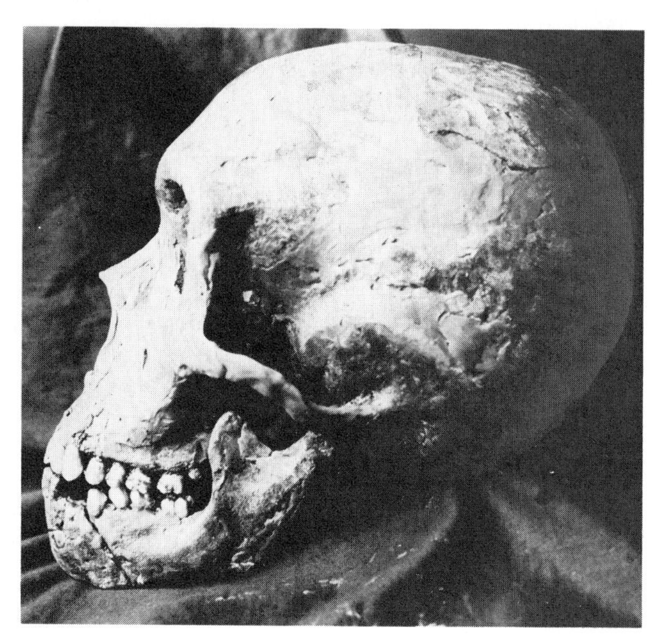

R32
R20
R28

Fig. 7-16 (continued) **E.** (opposite, above, left) Fossil stump of the tree *Sigillaria* from Carboniferous beds on Cape Breton Island, Canada. *Geological Survey of Canada.* **F.** (opposite, above, right) Reconstructed skull of Neanderthal man, a race of *Homo sapiens* that lived in Europe and Asia about 100,000 years ago. *American Museum of Natural History.* **G.** (opposite, below) Dinosaur bones excavated at Como Bluff, Wyoming, in the late 1800s. *Western Research Center, University of Wyoming.* **H.** (above) This fossil jaw of a shark was found in Idaho. Although present-day sharks shed their teeth as new ones grow in, the older teeth of this shark were buried in a complex spiral in its jaw. *Janet Robertson, Geological Society of America*

CONVERSION OF SEDIMENTS TO SEDIMENTARY ROCKS

Thus far, most of our discussion has had to do with sediments and the process of sedimentation, and very little has been said about the way in which sediments are converted into solid rock. What process, for example, converts loose sand, which at the beach can be idly sifted through the fingers, into a rock such as sandstone, which may be almost as unyielding as granite?

Is it pressure? The answer is emphatically no. If enough pressure were applied to cause sand grains to adhere to one another it would crush them into smaller and smaller particles. Pressure does play a role, however, in the process of ***compaction,*** which is the squeezing together of the particles in a sediment. If, for instance, enough pressure is applied to fine-grained muds, such as clay or silt, most of the interstitial water is squeezed out, the sediment shrinks markedly, and if clay is a dominant constituent, the particles tend to adhere to one another.

The partial closing up of the spaces between particles through compaction is an important precursor for ***cementation***—the most significant process involved in the conversion of sediments into sedimentary rock (Fig. 7-18). Fundamentally the process involves the deposition from solution of a soluble substance such as calcite ($CaCO_3$) and its building up as a layer of film on the surface of sand grains, silt particles, or clay flakes, as the case may be, until much of the pore space separating them is filled. Such a limy cement is precipitated in much the same way, although at a lower temperature, as the scale that forms inside a tea kettle.

Calcite is one of the most abundant natural

Fig. 7-17 Concretions weathered out of limestone are left scattered on the desert floor in the Kharga Oasis, Egypt. *Tad Nichols*

cements because it is among the more soluble of the common substances that may be dissolved in ground water. Another important natural cement is silica (SiO_2), which is also soluble, although less so than calcite. Iron oxide (FE_2O_3), too, is a cementing agent.

WHICH WAY IS UP?

The deformation of many sedimentary rock layers is so slight that it is no problem determining if the sequence is right-side up or up-side down (see Fig. 7-7). But pity the poor geologist who tries to unravel the complex structural geology of, for example, some of our western mountain ranges or the Alps. In places, the rocks have been so intensely folded that they are now upside down (Fig. 7-19). Some sedimentary rock features can help determine which way is up (Fig. 7-20). Baked zones associated with volcanic rocks also are an indicator of the up direction. Interpretation of up versus down is considerably more difficult in metamorphic rocks, especially if intense metamorphism has taken place, and impossible in plutonic rocks.

TYPES OF SEDIMENTARY ROCKS

The point was made earlier in this chapter that there are three major categories of sedimentary rocks: ***clastic,*** or fragmented rocks; ***chemical*** precipitates; and ***organic*** deposits. Like many classifications of natural phenomena, those categories are more distinct than the actual rocks. Not only are there gradational types from one category to the other, but there are also varieties that might just as logically be placed in one as another, as well as a few that fit into none.

Clastic sedimentary rocks

The clastic rocks are truly secondary rocks because they are fragments of pre-existing rocks that range in size from blocks the size of boxcars down to particles so fine that they remain in suspension almost indefinitely. Because clastic rocks consist of fragments of other rocks, they are very likely to show a wide range of composition. So much so, in fact, that in classifying them, the first property to be considered is the *size* of the particles that are cemented together rather than the material of which they are made.

Take sand, for example. To most people the word has two connotations: (1) it is a size term—all of us are conscious of the fine grittiness of sand in a bathing suit or between our toes; and (2) it has a compositional meaning—the beach sand that usually comes to mind ranges from white to a tawny yellow, and is likely to be thought of as consisting of quartz grains. Actually many beach sands contain mostly feldspar grains as well as a liberal sprinkling of other sand-size rock particles and mineral grains. Sand can consist of almost any material of sufficient durability. Along some of the rivers of the Atlantic states, there are sandbars of coal fragments. On some of the beaches of Hawaii the sands are coal black, too, but they are ground-up basalt. In the islands of the South Seas, the straw-colored sands of their fabled shores consist of fragmented coral heads, pieces of shells, and other organic debris. It would be confusing, therefore, if all sands of slightly different composition were given different names. There are many names for sandstones because the geologists seldom pass up the opportunity to name something. They even have their own dictionary of geological terms, 4.5 cm (1.8 in.) thick. However, that multitude of names will be left to professional geologists; in this book we will generally limit the vocabulary to the most important terms.

A widely accepted classification is a modified form of one originally proposed in 1922 by C. K. Wentworth (Table 7-1). It has the advantages that almost all the terms used are everyday words and that the size ranges are close to the ones in common usage. The actual dimensions, however, are arranged so that they fol-

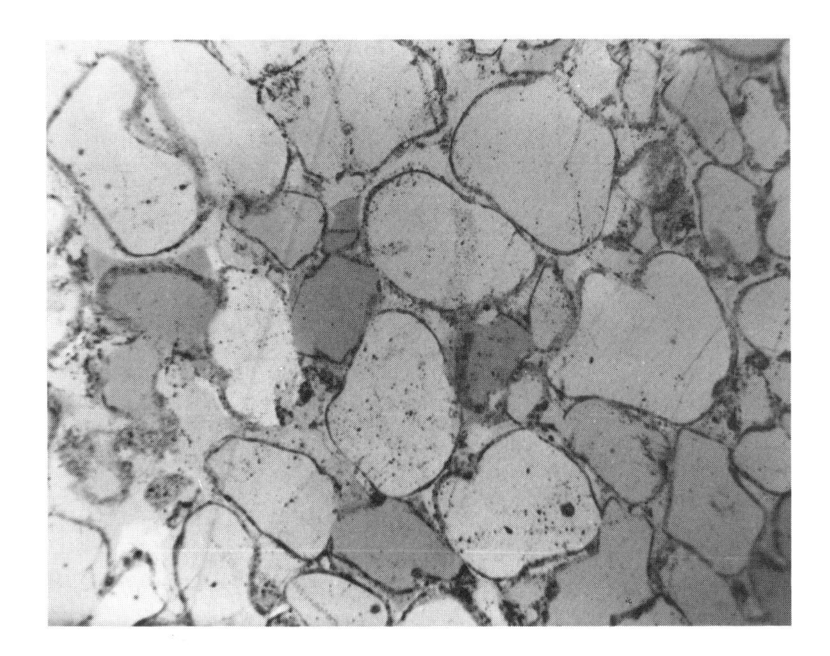

Fig. 7-18 A thin section of the Potsdam Sandstone from New York, photographed under a microscope. The rounded quartz grains are cemented together by fine-grained silica that fills the void between the grains. The dark material is hematite and other impurities. *Edwin E. Larson*

low in geometric progression. The term *clay* as used in the classification refers only to particle size. It should not be confused with the term clay mineral, which refers to very small minerals with mica-like structures, some of which are described in Chapter 9, and most of which are of clay size. To create some order out of the confusion, most clay minerals are of clay size, but not all clay consists of clay minerals.

Environmentally, particle size can be related to the energy of the transporting medium. Large gravels, for example, are moved by glaciers, swift rivers, and debris-laden mudflows. Less energy is required to move sand, and so sand is common to sand dunes and to some beaches and rivers. Because silts and clays settle so slowly, they commonly settle out only in quiet water, thus their association with lake and marine environments.

Table 7-1. Classification of clastic sediments and sedimentary rocks

Sediment	*Particle term*	*Limiting grain size (in mm)*	*Rock*
Gravel	Boulder	256	Conglomerate
	Cobble	64	
	Pebble	4	
	Granule	2	
Sand	Very coarse sand	1	Sandstone
	Coarse sand	0.5	
	Medium sand	0.25	
	Fine sand	0.125	
	Very fine sand	0.625	
Mud	Silt		Shale or mudstone
	Clay	0.0039	

Conglomerates These rocks are cemented gravels, and the larger fragments may range in size from boulders, several meters in diameter, to particles the size of small beads (2 mm). More often than not, the interstices, or pore spaces, between the larger boulders, cobbles, or gravel are filled with sand or mud; the whole mass of sediment is then cemented together to form a single rock (Fig. 7-21).

Breccia, a variety of conglomerate, has angular rather than rounded fragments. The same word was used for pyroclastic volcanic rocks in (Chap. 6). And the principle applied to volcanic rocks also applies to conglomerates: if most of the large fragments in the rock are angular rather than rounded, the rock is a breccia—the adjective sedimentary or volcanic is usually added to indicate its origin.

Fig. 7-19 Tightly folded and overturned sedimentary rocks in the Alps.

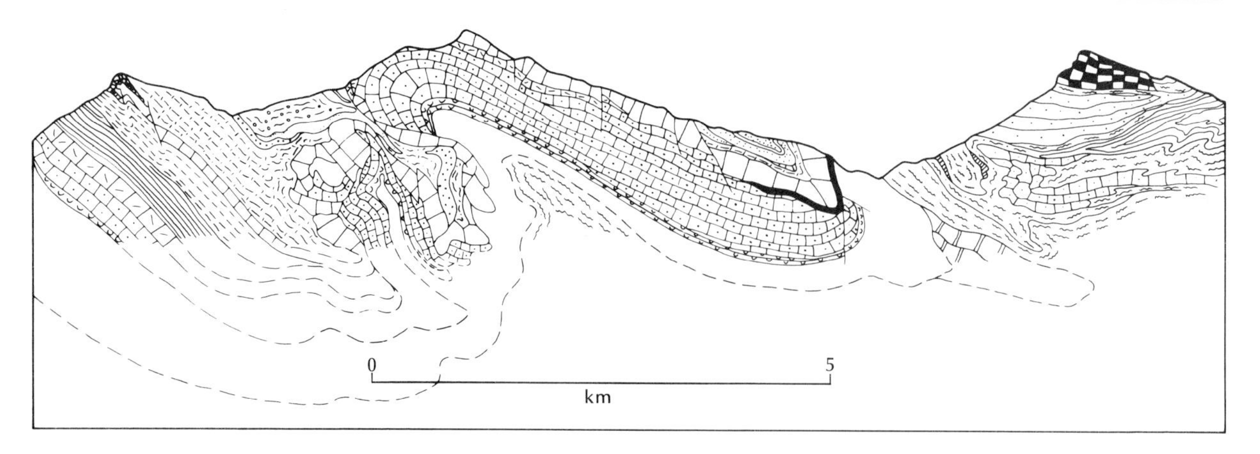

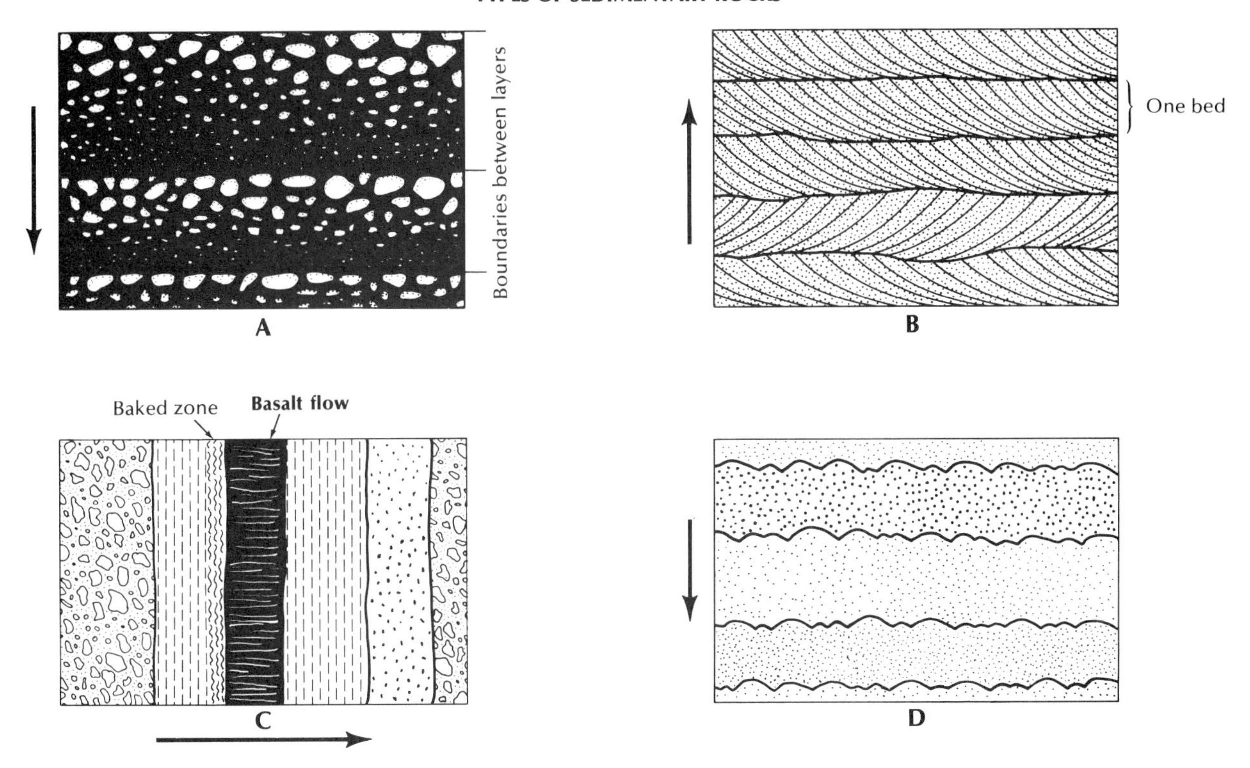

Fig. 7-20 Various ways to tell which way is up. For a true sense of direction tilt the book one way or the other. In each case the arrow points in the original "up" direction. **A.** Graded beds, with pebbles and sand shown in white; silt and clay in black. The sequence is upside down because each graded-bed layer is coarse at the bottom, gradually becoming finer at the top. **B.** These cross-bedded sediments are right-side up, because the base of each younger bed truncates the cross beds, which are concave upward. **C.** Basalt flow interbedded with sediments. Because the flow bakes only the sediments beneath it, the up direction is to the right. If the basalt were a sill rather than a flow, could you determine up from down? **D.** Beds of sand with oscillation ripple marks. Because such marks are concave upward when formed, the sequence is upside down.

Sandstones The sedimentary rocks called ***sandstones*** consist of cemented particles with diameter between 2 mm and 0.0625 mm. Very commonly they include shale layers, or beds of sandstone may alternate regularly with beds of shale or lenses of conglomerate.

Pure, well-sorted sandstone was often used as a building material before the advent of prestressed concrete or of lightweight aggregate. Quite a number of college campuses are adorned with examples of academic gothic—frequently, it seems, inhabited by the geology department—which were hewn from sandstone blocks. The White House and the Capitol are both built of sandstone quarried a short distance down the Potomac from Washington, D.C. In the Victorian Era—especially the General Grant period—one of the favorite construction materials was the so-called brownstone—a brownish red sandstone. Many of Europe's celebrated landmarks are made of clastic sedimentary rocks—the castles at Heidelberg and Salzburg and most of the great ducal palaces of Florence are but a few. Sedimentary rocks were greatly preferred over granite by builders in the past because such stratified rocks split more readily along their bedding planes, and thus could be worked far more easily with the hand tools of the time.

The cement determines the degree of induration, or hardness, of sandstones and, therefore, how well they perform as building stone. In some sandstones the cement is weak, and individual grains separate readily from their neighbors; in others the cement may actually be tougher than the material it holds together, and when the rock breaks, it breaks across the grains. Some cements dissolve readily, and the rock seems to crumble away, leaving a residue of sand grains.

Compositional differences affect the appearance of sandstone. Among the many kinds of sandstones, two leading varieties are ***arkose*** and ***graywacke.*** Arkoses are made up largely of quartz and feldspar grains and are commonly reddish to pale gray or buff. They result, typically, from the erosion of granitic rocks and require relatively rapid transportation and deposition for their formation.

Graywackes (from the German *grauwacke,* gray stone) were originally named for distinctive sandstones in the Harz Mountains of Germany. They are darker than arkoses, and, althought they commonly contain quartz and feldspar minerals, they have a much higher content of rock fragments—chiefly of the darker varieties of igneous and metamorphic rocks. The sand-size particles, in many instances, are set in a clayey or silty matrix, which, at the time of deposition, was essentially a muddy or clayey paste. Characteristically, graywackes are dense, tough, well-indurated rocks, dark green, gray, or black. Some of them appear to have been deposited in the sea, close to a steep mountain range. Muddy water carrying a large volume of sediment, including sand, was moved but a short distance from its source and deposited so rapidly that sorting is not always well expressed.

Shale The original constituents of this fine-grained rock were clay-size grains and silt particles, which matured into a typically laminated rock that splits readily into thin layers (Fig. 7-22). Shale is an ancient term in our language; it comes from the Old English word *scealu,* meaning scale or shell. When we use such an ancient word in geology it usually means we are dealing with a property so distinctive that it was recognized early enough to make its way into the rootstock of our native tongue.

Since shales are composed of clay grains and of individual mineral grains or rock particles less than 0.0625 mm in diameter, few of the constituents can be distinguished by the unaided eye. Under the microscope they can be resolved; there we can see that most shales are made of minute grains of quartz, feldspar, and mica and of larger rock fragments along with the ubiquitous clay. Nearly one-half of all sedimentary rocks are shale.

Many shales are shades of dark gray or even black, especially if they contain organic matter. Others are dark red or green or parti-colored, depending upon how much iron or other kinds of pigment they contain.

Although ***fissility,*** or the ability to split along well-developed and closely spaced planes, is an important property of shales, it is by no means characteristic of all shales. Some varieties, with comparable composition and grain size, are not fissile at all, but break in massive chunks or small compact blocks. They are best given the descriptive name of ***mudstone.***

Precipitated sedimentary rocks

In addition to the clastic rocks consisting of fragments and mineral grains derived from preexisting rocks, there is a second large group of sedimentary rocks, of chemically precipitated materials. In the following pages chemically formed rocks are discussed according to their composition and their mode of origin. Such an approach, although unavoidable, can be confusing, since some varieties of rocks—specifically, the carbonates—may have similar compositions, but unlike origins, and thus, of necessity, the same term appears more than once in the classification.

Evaporites Rocks or minerals that result primarily from the evaporation of water containing dissolved solids are known as ***evaporites.*** As the water becomes saturated, the ions come out of solution to form a crystalline residue.

Halite, or common table salt (sodium chloride, NaCl), is the most familiar evaporite min-

Fig. 7-21 Conglomerate forms this hanging rock at the foot of Echo Canyon in Utah, photographed by Andrew Joseph Russell about 1868. *Beinecke Rare Book and Manuscript Library, Yale University*

eral. Commonly it is formed when evaporation in an arm of the sea occurs faster than the inflow of water. Judging from the great thicknesses of the renowned salt deposits of the world, the process of influx of water and evaporation must have been repeated many times over. The evaporation of an inland water body, such as Great Salt Lake, can also produce a very flat surface as anyone knows who has seen the nearby Bonneville Salt Flats—widely known for the ideal surface the salts provide for speed trials.

Layers of salt deposited in the geological past are sometimes interbedded with other sedimentary rocks, and where such layers are near the surface, salt licks may be found. From earliest times salt has been a highly prized commodity. Today we take it for granted, but in ancient times people gave their lives in battle to win control over salt deposits or to seize the trade routes over which salt was transported. Famous among historic deposits were those of northern India—the center of a flourishing trade before the time of Alexander—as well as those of Palmyra, in Syria, from whence salt moved by caravan to the Persian Gulf. The salt mines of Austria are deservedly famous, and in the Salzkammergut region around Salzburg they were in operation at least as early as 2000 B.C.

Gypsum ($CaSO_4 \cdot 2H_2O$) is closely related to halite in origin, for it, too, is a product of the evaporation of sea water. Along with it is found an anhydrous (water-lacking) calcium sulfate ($CaSO_4$), called ***anhydrite.***

As a body of water evaporates, the minerals that precipitate out do so in a specific order. The order is determined by the solubility of the minerals in water—the less soluble first, the more soluble, which stay in solution the longest, last.

Gypsum is less soluble than halite and thus precipitates first when sea water is evaporated. However, great quantities of water must evaporate before either mineral forms. Both gypsum and anhydrite come out of solution after about 80 per cent of the sea water has evaporated, and halite appears after 90 per cent has evaporated. Following the precipitation of halite, the very soluble salts appear in such forms as sodium bromide (NaBr) and potash (potassium chloride, KCl).

Some salt deposits have astonishing proportions—in places several hundred meters thick. Yet there is a vexing problem concerning such deposits: the evaporation of 300 m (984 ft) of sea water will produce a bed of salt about 4.5 m (15 ft) thick, of which 0.1 m (0.3 ft) should be gypsum and anhydrite, 3.5 m (11.5 ft) halite, and the remaining 1 m (3 ft) potassium- and magnesium-bearing salts. Do several hundred meters of salt deposits represent the evaporation of thousands of meters of water without a fresh supply being added? Hardly, for that would require the evaporation of our deeper oceans, an event that certainly has not taken place.

How then, do we explain the great thicknesses of evaporite deposits? Consider the fact that gypsum and anhydrite make up strata many hundreds of meters thick in west Texas and New Mexico. A great amount of sea water must have been evaporated there in the geological past. But what took place was not simple evaporation, since the extensive bodies of sodium chloride that should be associated with the gypsum beds are absent. An explanation advanced by the American geologist Philip B. King is that water in a shallow, sun-warmed lagoon reached the temperature and concentration at which calcium sulfate precipitated out and settled on the floor of the lagoon. As sea water was continually replenished, the precipitation process continued. Were such a basin a subsiding one, a thick layer of evaporites could accumulate without the water necessarily being deep.

There are many other kinds of evaporites, of minor significance in amount, but highly significant economically. Among them are borax ($Na_2B_4O_7 \cdot 10H_2O$) and potash (KCl), both of which are found in lakes, or ancient lake deposits, of such desert regions as the Mojave Desert in California. Years ago, picturesque 20-mule teams hauled those salts across the incredibly rough desert floor (Fig. 7-23) in large wooden boxcars.

Carbonate rocks Rocks known as ***carbonates*** are made up of calcium or magnesium compounded with carbonate, generally in the form of calcite ($CaCO_3$) or dolomite ($CaMg(CO_3)_2$). If calcite predominates, the rock is called ***limestone;*** if dolomite predominates the rock is called ***dolomite.*** The latter term is virtually the only one used to designate both a mineral and a rock. In order to avoid confusion, some label the rock ***dolostone.*** Limestones are for the most part organic in origin, and very few are true chemical deposits. Limestone and dolomite look very much alike; the most practical field distinction between the two is the test described in Chapter 4.

Sea water is very nearly saturated with calcium carbonate ($CaCO_3$). That means that very slight changes in the temperature of the water or in its chemical composition can precipitate calcite. Usually, the initial mineral precipitated

Fig. 7-22 Thin, even bedding is characteristic of shale, which makes up much of this cliff in Utah. *Tad Nichols*

is ***aragonite,*** an unstable form of $CaCO_3$. There is a continuing argument as to how important inorganic processes are in forming the main limestone deposits of the world.

Travertine is a good example of a limy deposit that appears to have been deposited by spring waters saturated with calcium carbonate. It is of no great geological significance, but is economically significant, since it is greatly favored as an architectural material. Soft and readily worked, it has an interesting array of colors—generally pale yellow or cream, if pure; brown and darker yellow if it contains impurities—and often shows pronounced banding in wonderfully complex curving patterns. ***Tufa,*** or ***calcareous tufa,*** as it is sometimes called to distinguish it from volcanic tuff, also forms in springs and lime-saturated lakes, although its deposition seems to be fostered, to some degree, by lime-secreting algae. Tufa

Fig. 7-23 Twenty-mule team on its way to the Lila-C mine in 1892. *U.S. Borax*

and travertine, when cut and polished, are much favored for the lobbies of banks, building and loan associations, and the large railway terminals of a past era. Great quantities of tufa are imported from Italy and, as might be anticipated, much of monumental Rome is built of tufa, including Bernini's columns that nearly encircle the piazza in front of St. Peter's.

There seems to be little doubt about the inorganic origin of one curious type of limestone known as ***oolite.*** It is made of minute spherical grains of calcium carbonate the size of fish roe, called ***ooliths,*** from the Greek *oo* for egg and *lithos* for stone (Fig. 7-24). The grains form as layers of calcium carbonate around a nucleus—perhaps in much the same way that layers of pearl shell are built up. A well-formed oolith is the result of grains rotating again and again in a bottom current, and a logical place for such tumbling would be in a tidal area. Common environments of present-day formation are the coasts of Florida and the Bahamas.

The origin of dolomite has been and continues to be debated. The argument centers around whether dolomite forms as a primary precipitate from sea water or whether it forms from calcite or aragonite crystals. In some places, dolomite masses cut across limestone layers or follow fracture patterns, cutting the limestone in very much the same way that some igneous dikes do. It is widely held that such masses are the result of partial replacement of calcite in the limestone by magnesia-bearing solutions. In other places, however, dolomite occurs as widely spread layers, or beds, interlayered with limestone strata. The origin of such deposits is less clear cut. Some geologists believe that the dolomite was precipitated directly on the sea floor. Others take the view that the dolomite layers represent selectively replaced layers of limestone—a view that elicits further questions: (1) Was the original limy material replaced by magnesia very shortly after deposition? (2) Did chemical alteration occur long afterward, when the limestone was completely formed?

The favored theory at present states that dolomite does not form directly from water. In support of this theory, it should be noted that dolomite is very rare in modern sediments. Also, studies of a desert lake have shown that calcite forms first, probably being converted to dolomite over the decades or centuries it rests on the lake bottom. The "replacement" origin certainly looks more likely.

Organic sedimentary rocks

Rocks made of the remains of organisms, both animals and plants, are called ***organic.*** Coal is an excellent example because it consists of decomposed remains of land plants. Much coal contains the remains of small plants, despite the popular view that it is the residue of a jumble of fallen trunks and tangled roots, once set in a marsh alive with monsters winging through a canopy of bizarre trees or slithering over the murky floor of the swamp. Although both coal and oil shale are organic, or partly organic, sedimentary rocks, we will defer discussion of them to Chapter 21.

Limestone The most abundant organic sedimentary rock is limestone, and probably most

Fig. 7-24 A. (left) Ooliths from the Bahama Banks, as photographed under the microscope. **B.** (right) A thin section of oolite from Cambrian sediments in Pennsylvania, as photographed under the microscope. Although silica has replaced the original carbonate, the original layering of the ooliths is well preserved. *N.C. Nielsen*

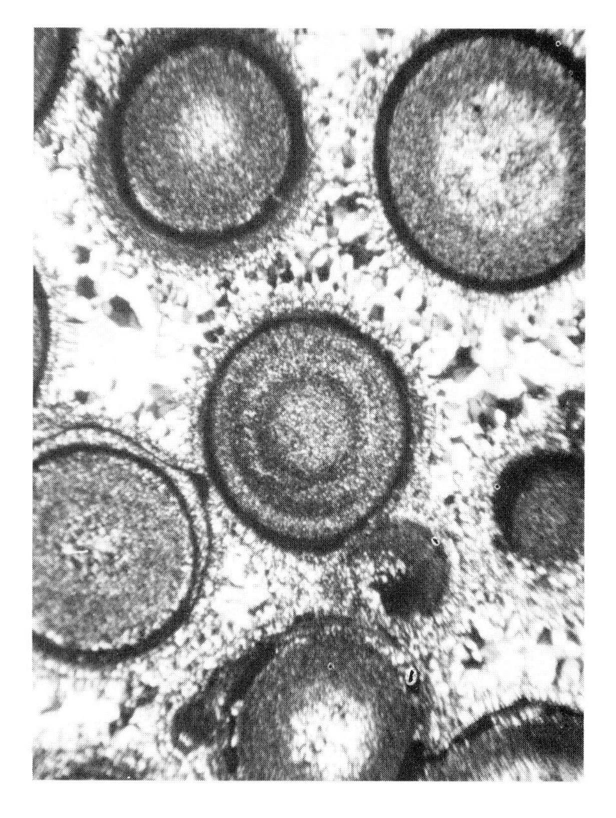

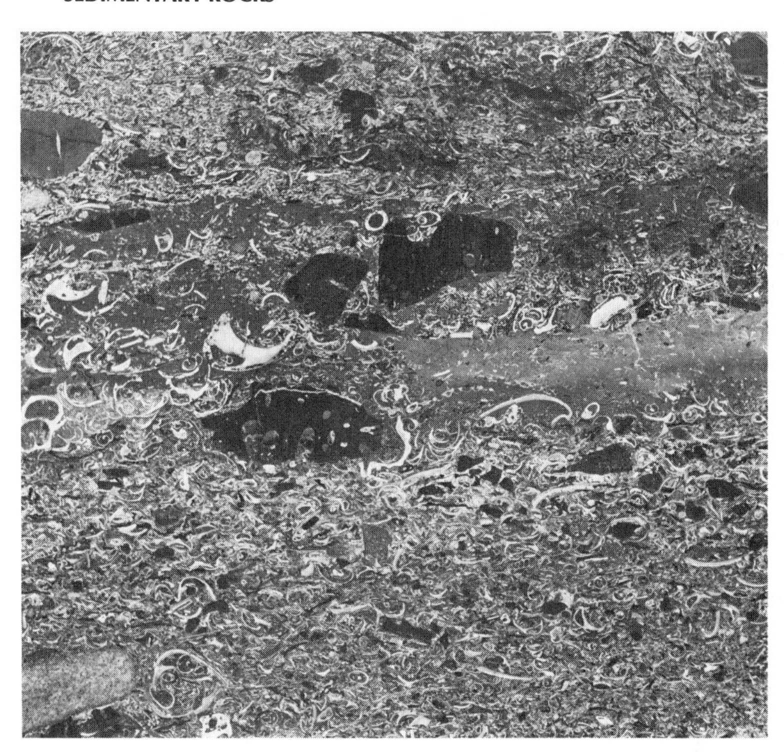

Fig. 7-25 Polished slab of shelly limestone probably deposited as a submarine mud flow generated by an earthquake several million years ago in the Los Angeles Basin, California. *Janet Robertson, Geological Society of America Collection*

limestone is truly organic and not precipitated chemically. By "truly organic," we mean that the deposits are either fossils of organisms in growth position or calcareous shell material that has been reworked and moved about by currents or both. In a strict sense, limestones made up of shell debris that has been reworked by currents should be classified as clastic sedimentary rocks (Fig. 7-25). Of limestones from large fossils and fossil fragments, perhaps the most impressive are the reefs—and certainly the most impressive modern reef is Australia's Great Barrier Reef (Fig. 7-26). That great carbonate barrier lies just off the coast, stretching a distance of 2000 km (1242 mi), and was the scourge of early sea captains; witness the experiences of Captain James Cook and his crew:

June, 1770, found him cautiously sailing northwards in his tiny ship, the 70-foot *Endeavour Bark*. The hazardous journey ultimately led to parts close to those discovered by the Spanish explorer, Luis Vaes de Torres (1605). Going finally ashore on a sea-girt spot of land near the tip of Cape York, Cook named this Possession Island, and there took formal possession of the east coast of Australia for Britain.

Never before had a ship sailed so dangerous and unknown a sea as that skirting the Queensland mainland. Cook found the waters dotted with islands, shoals and coral banks. His way was through twisting passages and shallows into a strange world of mystery and beauty. He was for a long time unaware of a great barrier to seawards that was closing in upon his track. At a spot near the present site of Cooktown, the coral banks crowded in on his ship in baffling array. She finally ran aground and was all but lost on a treacherous patch. The stirring story of that mishap and the masterful saving of ship and crew is one of the highlights of Australia's early history.*

Modern reefs are made up largely of corals and carbonate-secreting algae (Fig. 7-27). In order to grow so profusely, they require an environment free of most land-derived sediment and shallow, warm water that is both agitated

*From K. Gillett and F. McNeill, 1959, The Great Barrier Reef and adjacent isles: Coral Press Pty. Ltd., Sydney, Australia.

by wave action and high in nutrients. Those conditions are met today off many coasts within a 30° latitude of the equator. The reefs most familiar to North Americans form the Florida Keys and the Bahama Islands, and part of the Hawaiian Islands.

Reefs are extremely complex in their structure, but three main zones are commonly recognized in cross section (Fig. 7-28). One is the reef itself, which is formed of organisms building upward from the shallow sea floor to sea level. Such creatures can grow in the face of the prevailing winds, for these wind-driven currents bring the nutrients necessary for growth. Waves constantly crash against the reef, breaking off pieces of the delicate organisms, and forming an apron of steeply dipping organic debris—a sort of submarine talus of

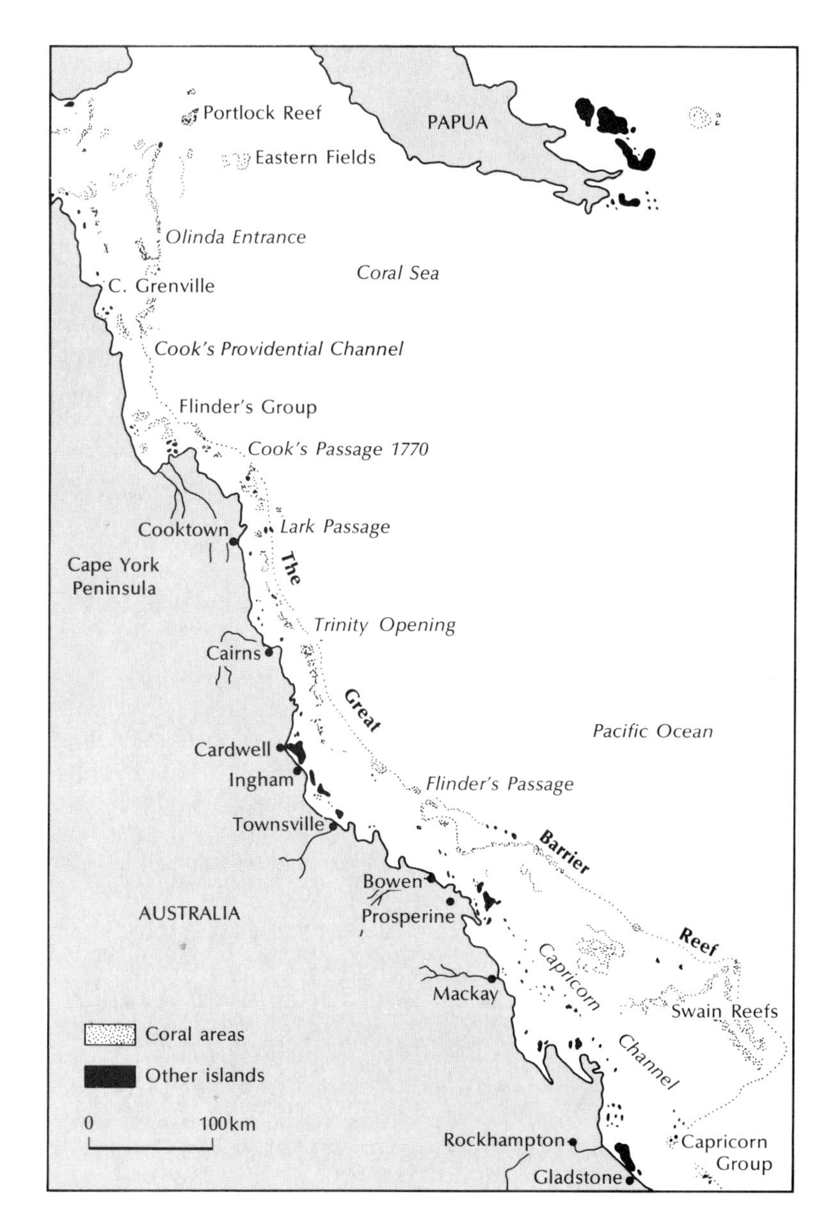

Fig. 7-26 Map of Australia's Great Barrier Reef.

Fig. 7-27 This patch reef community at Mosquito Bank, Florida, is characterized by a wide variety of coral and harbors a diverse and beautiful assemblage of reef fish. *H.G. Multer*

reef material, on the windward side. Landward of the reef is a quiet-water lagoon in which carbonate and clastic muds accumulate.

Other limestone deposits are quite different and consist of nothing more than the heaped remains of microscopic limestone fossils. They commonly are of minute, free-floating, single-celled animals, the ***foraminifera.*** Some chalk deposits, such as those near Dover, England, are 100 million years old, and the truly remarkable thing about them is how little alteration or recrystallization has occurred in all that time.

Siliceous rocks Such rocks are largely silica (SiO_2). In discussing their formation we will repeat the by-now familiar arguments of organic versus inorganic origin.

The most widely occurring siliceous rock is ***chert,*** a name used to cover a host of varieties of very dense, hard, nonclastic rocks made of microcrystalline silica. One familiar form is ***flint,*** which is dark-colored because it contains organic matter. Since flint is uniformly textured, has a conchoidal fracture much like obsidian, and is easy to chip while at the same time retaining a sharp edge, it proved to be the ideal strategic material for arrowheads and spear points in the Stone Ages of Europe and in the United States. In what was perhaps a braver time than ours, flints were essential for survival on the frontier, not only to strike sparks from steel for fire, but also to fire the flintlock gun of the eighteenth and the nineteenth century. Red varieties of the same rock commonly are called ***jasper,*** the color being derived from the iron oxide (Fe_2O_3) they contain.

By volume, the most important bodies of chert are thick and well bedded (Fig. 7-29). Although some of them may have precipitated directly from the sea, there is generally not enough silica in sea water for direct precipitation. If somehow, part of the sea is enriched with silica, as may happen in the vicinity of an underwater volcanic eruption, then chert might form as a direct precipitate.

Thick cherts are most likely to be organic. Microscopic marine animals, such as the ***radiolaria*** and the ***sponges,*** and plants, such as the ***diatoms,*** build their shells from silica extracted from sea water (Fig. 7-30). In time, in an

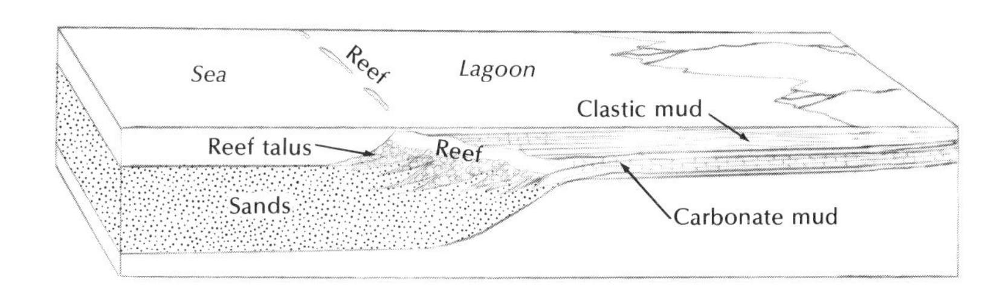

Fig. 7-28 A. (above) Diagram of the zones commonly associated with an offshore reef. **B.** (below) Reef deposits about 350 million years old on northern Banks Island, Northwest Territories, Canada. The light-colored reef deposits, which are embedded in clastic deposits, are relatively resistant to erosion and so form prominent outcrops. *Geological Survey of Canada*

Fig. 7-29 Well-bedded chert in the Cook Inlet region, Alaska. Originally horizontal, the chert has been subsequently folded, and a fault trends diagonally across the outcrop, from the man's right foot. *G.K. Gilbert, USGS*

environment in which both carbonate shells and clastic material are excluded, a thick deposit of chert can build up.

Two other sources of siliceous deposits, although minor, should be mentioned. One is the direct precipitation of silica from hot springs, for example, the pedestals at the bases of active geysers in Yellowstone National Park. Such deposits, called ***sinter,*** are spongy and porous. Another source has to be sought for the irregularly shaped, modular masses of chert contained in limestones. Because such deposits cut across the bedding planes, they clearly replaced part of the limestone at a later date. The process is similar to that in which silica replaces woody fibers to form petrified wood.

SEDIMENTARY FACIES

Now that the major sedimentary rocks have been described, we should issue a warning—the basic properties of any rock unit probably will change as the unit is traced laterally. For example, suppose we are examining a thick layer of conglomerate in a canyon wall. Over a distance of several kilometers, the layer might change gradually to a sandstone and then to shale. The reasons are fairly simple: in any depositional basin, such conditions as sediment supply, current velocity, and biological by-products can vary laterally. The lateral differences in rock types within a single unit are called ***facies*** changes; hence, as depositional conditions vary, so do the resulting sediments (Fig. 7-31).

In order to further understand such changes, let us look at some present-day examples. A good example is a carbonate reef, which builds upward and seaward, but the deposit changes further seaward to the reef talus and landward to lagoonal deposits (Fig. 7-28A). Consider also going seaward from a gravelly beach. The gravels occur in the "high-energy" area, where the waves are hitting the shore, but farther offshore, where the deeper and quieter waters can move fewer particles, there are sands and, eventually, silts and clays. Another example of facies changes is glaciated terrain, where glacial deposits grade downvalley to stream deposits, which in turn might grade laterally to marine deposits (Fig. 7-32).

There are also lateral changes in desert basins and thus in the resulting sediments. Streams deposit gravels close to the mountain front, but with increasing distance from the front they deposit sand. In the center of the

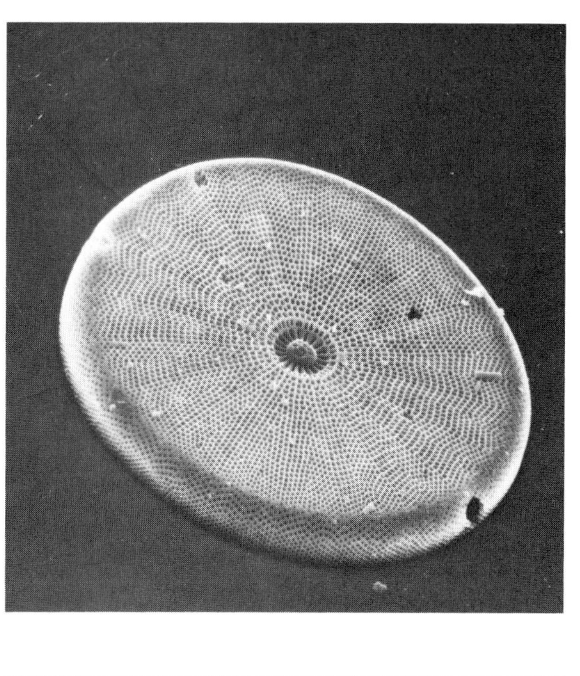

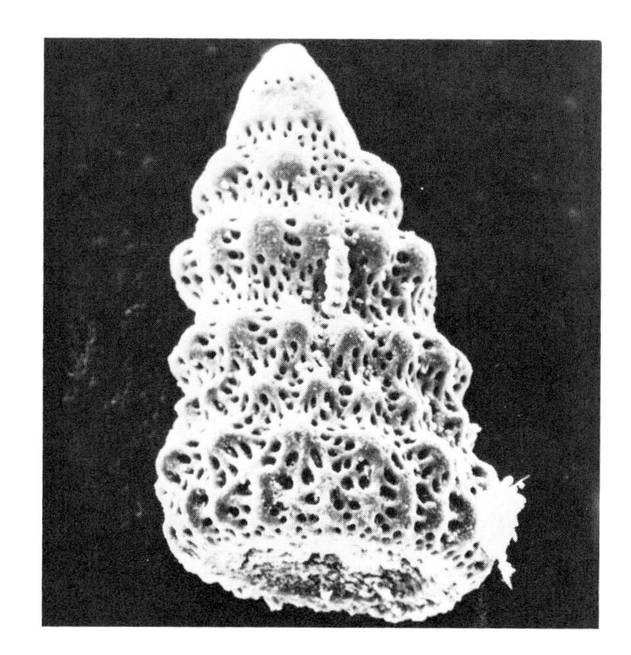

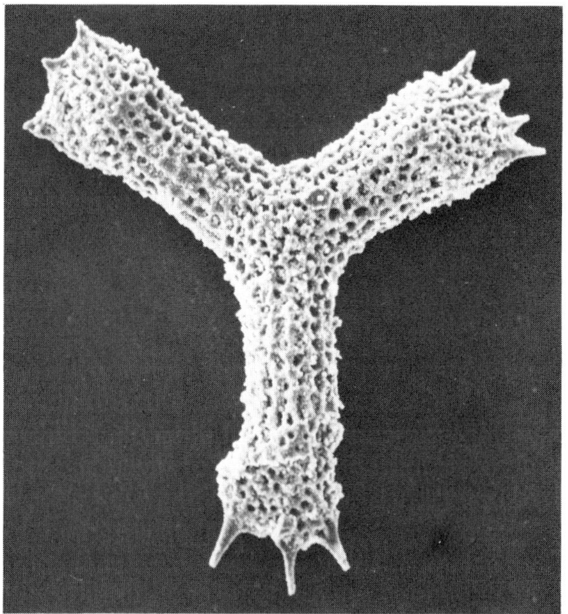

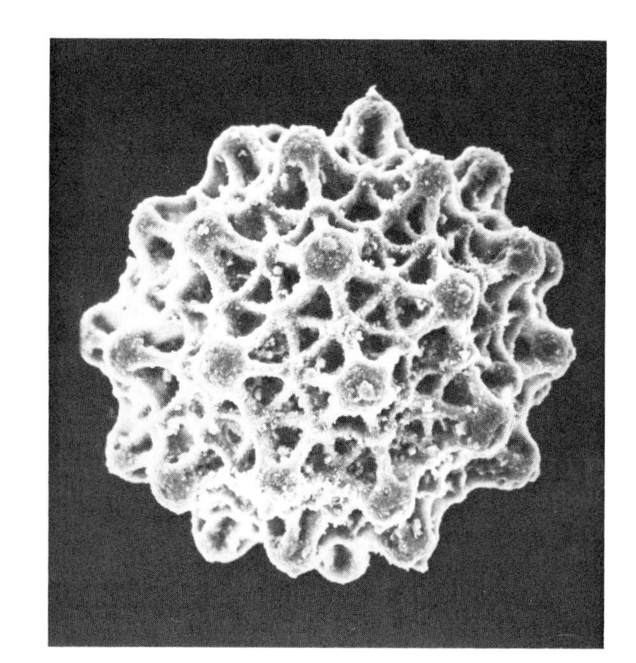

Fig. 7-30 Scanning electron micrographs (SEM) of fossil marine organisms from sediments from the California coast. The fossil on the upper left is a diatom; the rest are Mesozoic radiolaria. *AIME* and *Emile Pessagno, Jr.*

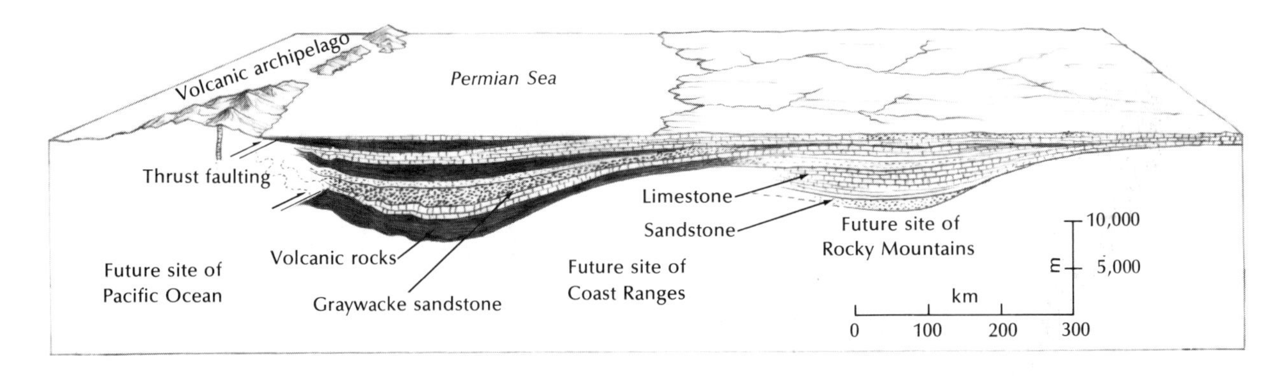

Fig. 7-31 Reconstruction of facies relationship of Paleozoic rocks in southeastern Alaska and British Columbia at the close of the Permian Period. Note that the thickness of the sediments varies. Also, the dominant volcanic rocks and graywacke sandstones to the west grade eastward to limestones and sandstones with distance from the volcanic archipelago.

Fig. 7-32 Examples of facies changes in recent sediments, Ekalugad Valley, Baffin Island, Northwest Territories, Canada. Glaciers occupy the uplands, and in places tongues of ice penetrate to the valley bottom. A ridge of glacial deposits (a moraine) has been left behind with the recent retreat of the ice. The ridge grades downvalley to sandy and gravelly stream deposits and these, in turn, grade into silts and clays deposited in the quiet waters at the head of the fiord. Although all the sediments are the same age, their character is determined by the environment of their deposition. *National Air Photo Library, Canada*

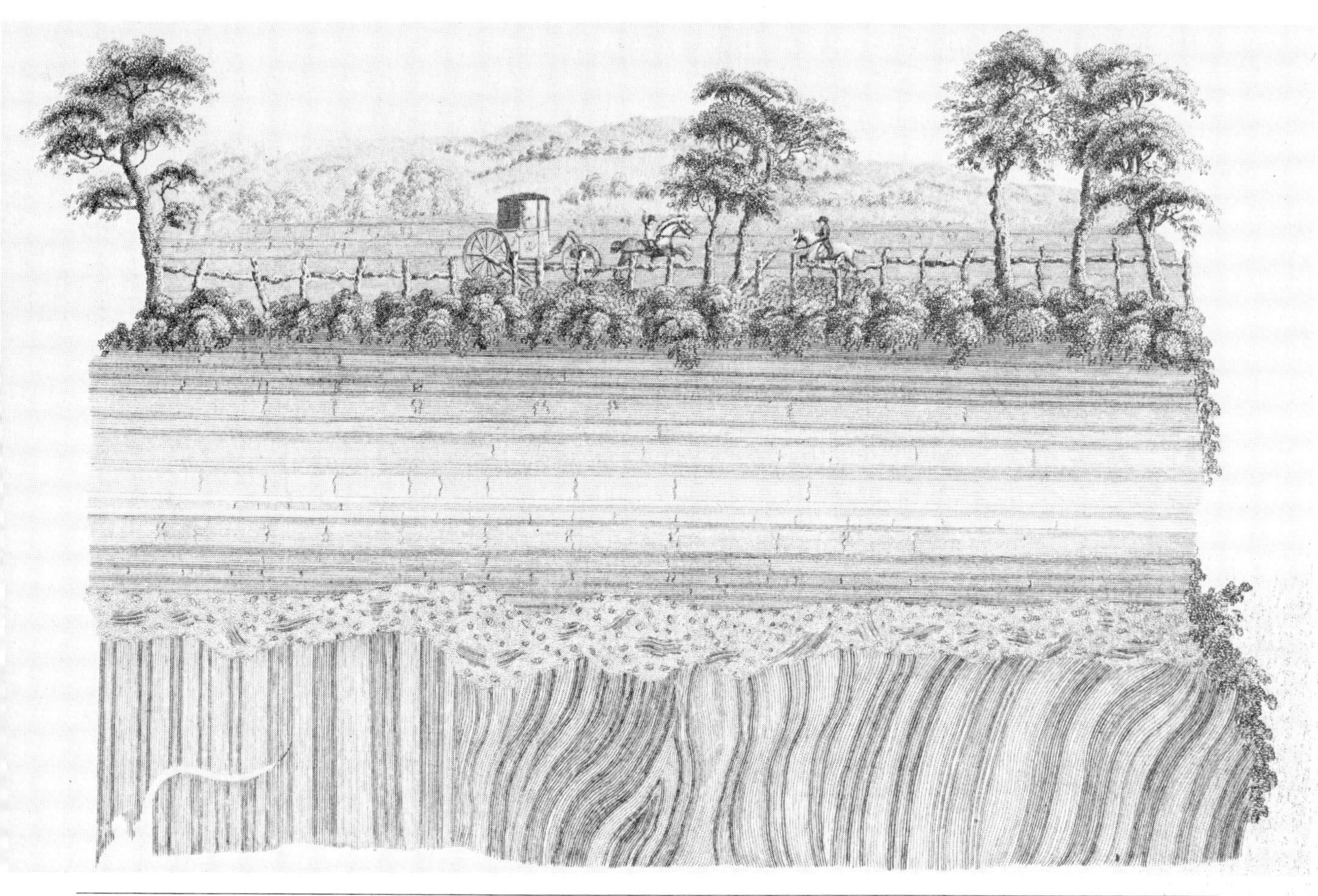

Fig. 7-33 An eighteenth-century engraving showing a distinct angular unconformity in which horizontal beds rest on vertical and near-vertical beds. The sequence of events is as follows: (1) horizontal deposition of lower beds, (2) tilting of beds during an episode of mountain building, (3) erosion of the beds to form a relatively flat surface, and finally (4) deposition of younger, flat-lying beds. The unconformity shown here could represent millions of years.

basin, silts and clays may be laid down, forming the floors of temporary desert lakes. The lowest parts of the basins sometimes contain salt deposits, formed when the temporary lakes evaporate. The examples are endless—just look around at the various sediments being formed in the wide variety of surface environments near you.

The concept of facies was exceedingly important to the early geologists; without it they would not have been aware that a limestone sequence in one place was the same age as a sandstone sequence 100 km (62 mi) or so away. It is hard enough to follow facies changes where outcrops are good and rocks are not deformed. Consider the plight of the geologist working with strongly deformed sediments in forested terrain with few rock outcrops. Only with hard work and a fertile imagination can such a rock record be unscrambled.

One can well appreciate the problems involved in correlating disparate sequences of sedimentary rocks using fossils found in those rocks; in some strata the fossils may reflect evolutionary change, whereas in others they may reflect environmental change.

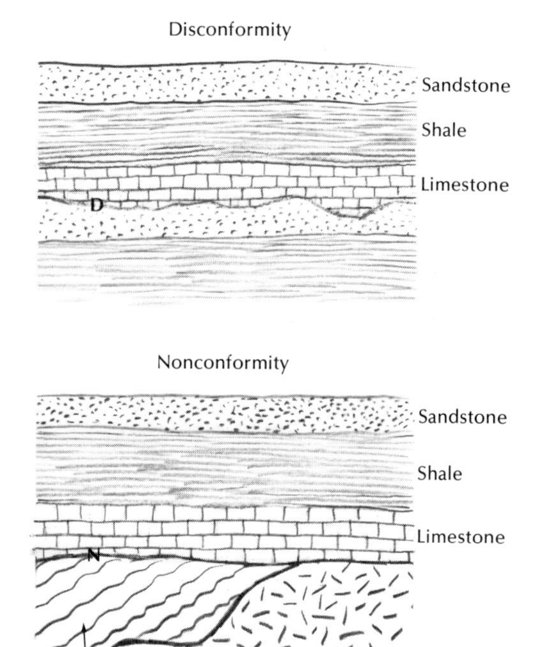

Fig. 7-34 Examples of a disconformity (D) and a nonconformity (N).

UNCONFORMITIES

Sedimentation is not a continuous process—commonly it is interrupted by other environmental changes. We would expect the oceans, for example, to be area of continous sedimentation, but areas of the seas shift and change—what was once an area of marine deposition can be raised above sealevel for a while, and eroded, to be followed, at a later date, by another cycle of marine deposition. Fairly long-term gaps in the sedimentation record of rocks are called ***unconformities*** (see Chap. 1); erosion commonly accompanies times of non-deposition, especially if associated with uplift in a particular region.

Several kinds of unconformities are recognized. By far the easiest to identify is an ***angular unconformity,*** in which the beds above and below the unconformity are not parallel, but meet at an angle (Fig. 7-33). Obviously, to arrive at this geometric pattern, the older beds must themselves have been deformed during a previous upheaval (see Chap. 17), subsequently eroded, and finally buried by younger sediments. In a ***disconformity,*** the rock layers are parallel above and below the unconformity, but the lower beds can have considerable relief (Fig. 7-34). Finally, a ***nonconformity*** is characterized by sedimentary rocks resting on igneous or metamorphic rocks (Fig. 7-34); obviously much time, erosion, and geological history are represented where such contrasting rocks are seen together.

Unconformities are usually localized. The three types can occur alone or blend laterally with one another, and even laterally to areas where sedimentation occurred without any telltale signs of interruption. Being of regional extent, they can aid in the correlation of widely separated rock units, provided we have some idea of when they were formed.

SUMMARY

1. Sedimentary rocks result from the weathering, erosion, transporting, and deposition of any pre-existing rocks. Rock and mineral particles that are the result of weathering make up the *clastic* sedimentary rocks, and the ions from weathering recombine to form the various sedimentary rocks that are known as *chemical precipitates. Organic* deposits form the third major subdivision of sedimentary rocks.
2. The source areas for sediments are usually mountains or hills, whereas the desert basins, river floodplains, lakes, and oceans are the common depositional sites.
3. The sediments have a set of field characteristics that can be used to identify the environment in which they formed. Such characteristics include stratification, sorting and roundness of grains, cross-bedding, fossils, mud cracks, and ripple marks. In most environments, the major stratification layers are horizontal.

4. Once sediments are deposited, *calcite,* or *silica,* may precipitate from the circulating ground water and cement the grains together to form sedimentary rock.
5. In classifying rocks, the primary criterion for clastic sediments is grain size; for chemically precipitated rocks, it is composition. *Organic sediments* are the products of the accumulation of plants or animals, and the common ones are limestone, coal, and oil shale.
6. The concept of sedimentary facies is essential to interpreting the origin of sedimentary rocks. In short, it means that the characteristics of sediments laid down in a specific time period will change laterally because the environments of deposition change laterally. Hence, stream deposits may grade laterally to clastic marine deposits and then to carbonate reef deposits.

QUESTIONS

1. What are the arguments against using the mineral content of sandstones or the rock type of conglomerates as the major criteria for classifying the clastic sedimentary rocks?
2. List the features of sedimentary, plutonic, and igneous rocks that can be used to determine the up direction in a sequence of rocks.
3. Do enormously thick sequences of rocks—for example, shales or evaporites—mean that the sea in which they were deposited was deep?
4. Examine all the ways in which a limestone rock can form from a reef complex offshore to the deep ocean.
5. What sedimentary facies can you recognize in Figures 11-28 (the Nile), 12-12 (Death Valley), and 17-24 (Appalachians)?

SELECTED REFERENCES

Blatt, H., Middleton, G., and Murray, R., 1980, Origin of sedimentary rocks 2nd ed.: Prentice-Hall, Englewood Cliffs, New Jersey.

Eicher, D. L., and McAlester, A. L., 1980, History of the earth: Prentice-Hall, Englewood Cliffs, New Jersey.

Gillett, K., and McNeill, F., 1959, The Great Barrier Reef and adjacent isles: Coral Press Pty. Ltd., Sydney, Australia.

LaPorte, L. F., 1968, Ancient environments: Prentice-Hall, Englewood Cliffs, New Jersey.

Pettijohn, F. J., 1975, Sedimentary rocks, 3rd ed.: Harper & Row, New York.

Fig. 8-1 Banded metamorphosed sedimentary rocks, near Convict Lake on the east side of the Sierra Nevada, California. *John Haddaway*

8

METAMORPHIC ROCKS

A geologist studying igneous and sedimentary rocks has an advantage over one investigating metamorphic rocks (Fig. 8-1), in that some of the former originate on the earth's surface in environments where they can be observed.

One difficulty in understanding the origin of metamorphic rocks is that no one has ever seen a metamorphic rock being formed, and for that reason many of our ideas about them are pure conjecture. This is not to say, however, that it is as fanciful as the speculations of science fiction; there are physical and chemical limits within which any theory of metamorphism must operate.

Between a depth of about 20 km (12 mi), where under certain circumstances the temperature is high enough to melt rocks, and the earth's surface, where weathering takes place, there are wide ranges of temperature, pressure, and chemistry. As a result of geological processes, such as mountain building, plate convergence, and sedimentation, some of the rocks in that zone are prone to change. Rocks that are mineralogically stable in one ***physicochemical environment*** generally tend to become unstable when they are transported to another such environment (Fig. 8-2). How the rock actually changes depends on the new conditions and the chemistry of the starting rock. We call the altered assemblage of minerals a ***metamorphic*** rock and the process by which it is transformed ***metamorphism*** (taken from the Greek, to change in form). It should be emphasized that during metamorphism the minerals do not melt, but remain largely in the solid state.

As in the case of plutonic rocks we can investigate the results of the metamorphic process only after the overlying rocks have been removed through erosion, and long after the actual metamorphic process has run its course. Geologists have to infer the nature of premetamorphic materials and the physicochemical environment during metamorphism: field and laboratory studies of the altered rocks themselves, augmented by a knowledge of chemical reactions, help them in this.

It has become apparent from studies of the rocks themselves that metamorphism is basically an ***isochemical process.*** That is, in most instances, the overall chemical composition of the rock is nearly the same before and after re-

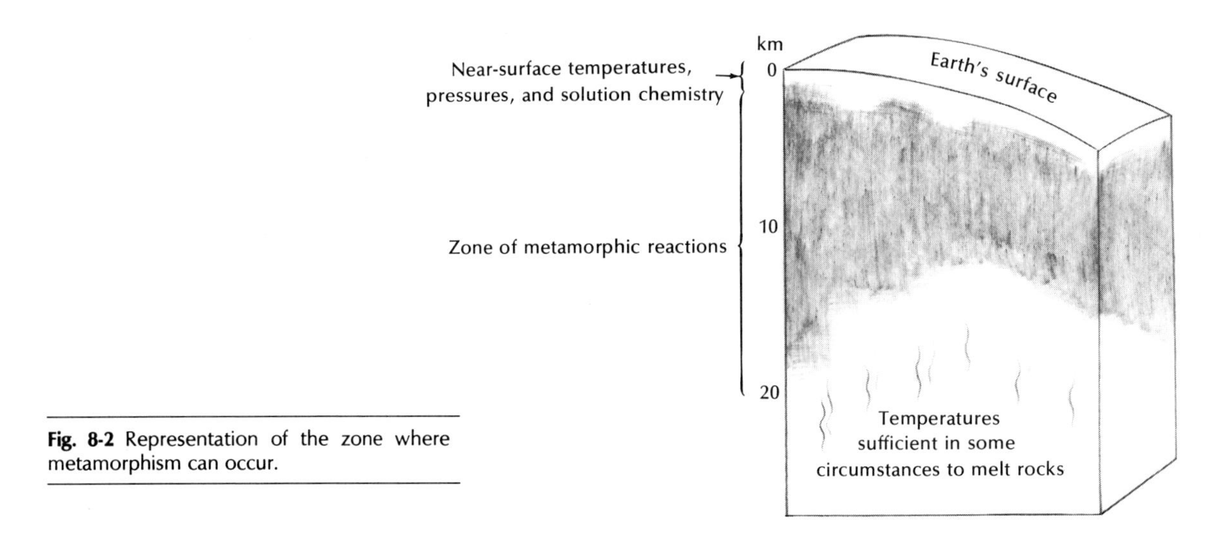

Fig. 8-2 Representation of the zone where metamorphism can occur.

crystallization. Little new material is added or lost during the change. (Exceptions to the rule involve circulating fluids, commonly hot, which add other chemicals to the rock being transformed.)

Because of the isochemical nature of the recrystallization process, the chemistry of the original rocks largely determines what can be formed under any set of physio-chemical conditions. For example, if limestone, composed solely of grains of calcite ($CaCO_3$), is metamorphosed, the resultant rock will be composed only of $CaCO_3$; whereas if a shale, composed predominantly of clay (hydrous aluminum silicate) and quartz (SiO_2) particles, is metamorphosed, a number of silicate minerals will be formed, but no calcite.

FACTORS IN METAMORPHISM

As one might expect, a rich variety of metamorphic rocks exists. Any kind of rock—igneous, sedimentary, or metamorphic—can serve as the starting material, and the combinations of temperature and pressure and the chemicals present that can bring about its alteration are limitless.

Heat

Thermal energy is probably the principal factor involved in the metamorphism of rocks. By increasing a rock's temperature, there may be enough energy to activate recrystallization of minerals. As the temperature rises higher and higher, ions diffuse more quickly, increasing the speed and efficiency of the metamorphic transformation. Moreover, as a rock becomes hotter, minerals containing such volatile components as water and carbon dioxide become less stable and eventually break down, releasing some of those components. The recrystallized minerals are, of course, proportionately less rich in volatile components.

It is also true that as the temperature of a rock is increased, the rock tends to decrease in strength, that is, it is more likely to deform or even to flow plastically in response to directed stresses. Many metamorphic rocks are highly contorted and look like they had been squeezed through a toothpaste tube.

If temperatures are high enough, rocks can actually melt. At first only certain minerals in which the ***melting point*** is exceeded will melt, but eventually, if enough heat is available, a nearly total fusion of the rock body and the formation of a magma reservoir will take place.

Pressure

Under increasing pressure, the mineral grains in a rock are subjected to a squeezing action. Such pressure can cause recrystallization and the formation of minerals with more tightly packed atomic structures and greater density than the initial mineral grains.

Also, if a rock consisting of many separate mineral grains is compressed, much of the stress is concentrated upon irregularities along the boundaries of the individual grains. Such uneven pressure leads to local spot-melting at the points of high stress and, simultaneously, reprecipitation of material in the areas of low stress.

Much of the pressure to which a rock is subjected comes from the load of the rocks above it. The thicker the rock body above a metamorphosing rock, the greater the pressure. The situation in a swimming pool is similar: the deeper you go, the more the pressure increases, in response to the increasing load of the water above. At any point in the pool, the pressure is equal in all directions—otherwise the water would tend to flow. The all-sided pressure in water is called hydrostatic pressure, and the analogous pressure created at depth in the earth from the rock load above is called ***lithostatic pressure,*** or ***confining pressure.***

Some of the pressure exerted on rocks results from the release of the volatile components (mostly water and carbon dioxide) in response to thermal activation, as discussed earlier. Such pressure commonly is called ***pore-fluid pressure,*** and it too is an all-sided pressure.

In many instances, ***directed pressure*** also plays a part in forming metamorphic rocks. It operates in a particular direction, and apparently is generated in response to mountain-building activity. No one seems to know quite how intense directed pressure may be, but all agree that it is important to the development of the foliated and lineated rock fabrics that are so characteristic of many metamorphic rocks. It appears that in a rock recrystallizing under directed pressure, many of the newly forming mineral grains—especially of platy species like mica and chlorite—are constrained to grow with their cleavage planes more-or-less parallel to each other and perpendicular to the direction of the pressure. Elongate minerals, such as hornblende, also commonly crystallize in bands (and, in some cases, with their long axes in parallel alignment) under directed pressure.

Chemical activity

The chemical environment is also of great importance to the recrystallization of rocks, but its exact role in the process is hard to evaluate. In most cases, chemical activity is related to the presence of small amounts of volatile components (mostly water and carbon dioxide) that permeate and fill pore spaces within and between grains in the metamorphosing rock. Some of these are derived, as we have discussed, from the breakdown of minerals as metamorphism progresses. Some of them, however, undoubtedly come from fluids already present in the cracks and pore spaces in the unmetamorphosed rock. Sediments, which are formed mostly in aqueous environments, particularly contain an abundance of initial pore fluids. In some way the permeating fluids increase the efficiency of the recrystallization process, that is, they ***catalyze*** the metamorphic reactions. In a progressively hotter environment, the fluids become correspondingly more active in their catalytic role. All earth scientists agree that without the presence of the pervasive, hot pore-space fluids, metamorphism would proceed at very slow, perhaps insignificant, rates. Laboratory studies have supported that view.

METAMORPHIC ROCKS

Metamorphic rocks are found throughout the continents and, to a much lesser degree, in the ocean basins of the world. A few occur in bodies of limited extent and include the ***contact, hydrothermal,*** and ***cataclastic*** metamorphic rocks. Most of them, however, occur in large, often elongate bodies that are exposed over thousands of square kilometers and include the ***dynamothermal*** metamorphic and mixed metamorphic-plutonic rocks. In the formation of each of the various kinds of metamorphic rock, different factors are important (Fig. 8-3). Heat plays the dominant role in contact metamorphism; chemically active solutions and heat are necessary for the formation of hydrothermally metamorphosed rocks; and localized pressure, for the formation of cataclastic metamorphosed rocks. In the formation of dynamothermal and mixed metamorphic-plutonic rocks, heat, confining and directed pressure, and chemically active fluids are necessary.

Metamorphic rocks of local extent

Contact metamorphic rocks At the margins of smaller plutonic bodies, the enclosing country rocks are commonly heated to such high temperatures that local metamorphism can occur in a relatively thin sheath, or ***aureole,*** as it is called (see Chap. 5). The recrystallized rocks are called ***contact metamorphic rocks,*** and their formation is primarily the result of the thermal energy derived from the hot magma within the plutonic body. Directed pressure plays an insignificant role in the recrystallization process and therefore such rocks generally lack foliation or banding.

The thinnest contact-metamorphic zones generally are found at the edges of dikes and sills. Because of their relative smallness and the rapidity with which these igneous bodies cool, the contact zones are usually only a few centimenters wide at most. Were the magmas injected into a shale, however, the sheath can be baked or hardened over distances of from several centimeters up to a meter or more. The clay minerals in the shale bake in much the same way that clay does when it is fired in a kiln to make bricks or pottery.

Larger intrusions, such as batholiths and stocks, which take many thousands to a million years to cool, make their influence felt over a wider area. In these cases, the country rocks may be converted into a dense, hard, nonfoliated rock called ***hornfels*** (Fig. 8-4).

Most hornfels rocks are generally aphanitic to fine-grained and are composed of nearly equidimensional recrystallized minerals. Some of the newly formed mineral grains are merely the result of recrystallization of previous grains on a one-to-one basis; others are the result of chemical reactions between two or more mineral species during recrystallization. If, for example, a rock containing dolomite ($CaMg(CO_3)_2$) and quartz (SiO_2) is heated, the

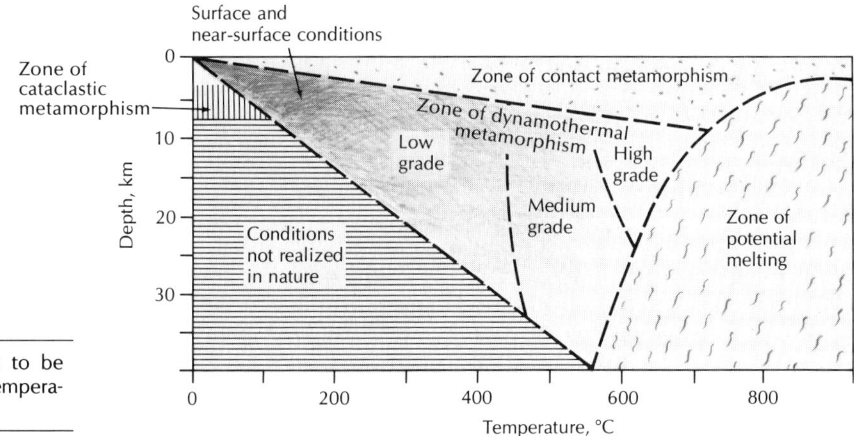

Fig. 8-3 The types of metamorphism to be expected under given depth and temperature conditions.

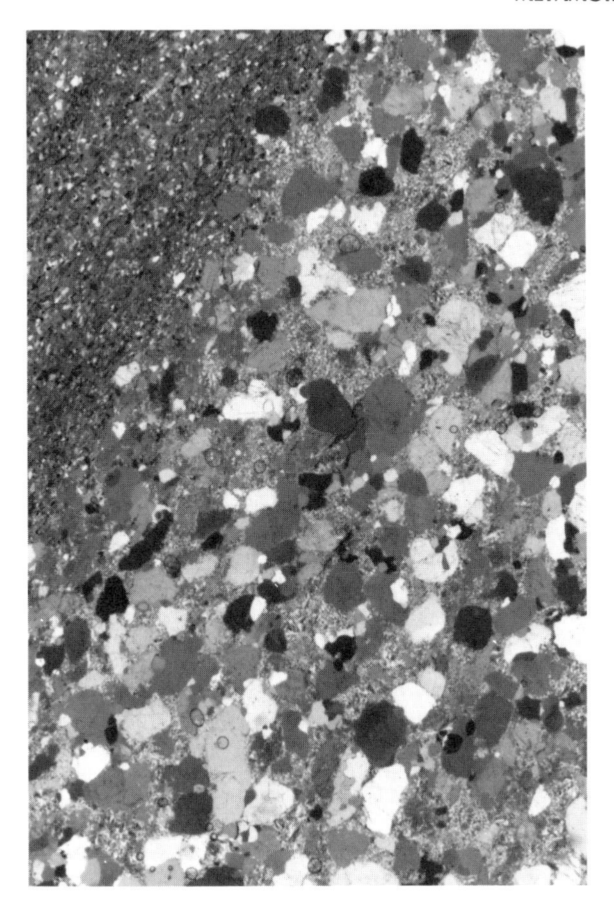

Fig. 8-4 Fine-grained hornfels, as photographed under the microscope. Quartz and feldspar (the larger grains) have been recrystallized on their margins only, whereas the original clayey matrix between grains has been totally recrystallized to muscovite (small grains in the background and upper left). This contact metamorphism occurred adjacent to a thick sill, Victoria Land, Antarctica. *W.B. Hamilton, USGS*

reaction between the two grains will result in the production of olivine (Mg_2SiO_4) and calcite ($CaCO_3$), with the liberation of carbon dioxide (CO_2). The reaction can be represented in the shorthand of chemistry:

$$2CaMg(CO_3)_2 + SiO_2 \rightarrow 2CaCO_3 + Mg_2SiO_4 + 2CO_2$$

Hydrothermal metamorphic rocks As some batholiths are intruded, there is a release of a great deal of hot, water-rich fluids, which travel long distances through enclosing host rocks carrying in solution ions derived from the magma body. Such liquids and gases (called ***hydrothermal solutions***) commonly are chemically very active and react readily with many of the minerals in the country rock with which they come in contact. The hot fluids not only act as catalysts in the recrystallization of the country rock, they also bring in new chemicals that are incorporated in the recrystallizing minerals. In some instances, the hydrothermal solutions react with the host rock to remove chemical constituents, which are then carried away as the fluids pass through and beyond the rock. Not all solutions that enter into the recrystallization process come from the magma itself. Some fluid is ground water that is heated and driven in convective circulation (like water boiling in a coffee pot) by the hotplate-like action of the nearby hot magma body.

Many minerals are particularly susceptible to alteration by hydrothermal fluids. For example rocks rich in olivine ($(Mg,Fe)_2SiO_4$) commonly are altered to serpentine ($Mg_3Si_2O_5(OH)_4$). The reaction, called ***serpentinization,*** is principally one of water enrichment by the solutions. Many geologists now feel that the formation of serpentine is particularly prevalent in the vicinity of the oceanic ridge systems, where hot, chemically active solutions rise along the faults and fractures associated with the ridge.

During serpentinization of olivine-rich rocks, there is an increase in volume and thus of internal stresses. Serpentine is a relatively low-strength rock, and when excessive stress is applied, it starts to slide, flow, and fracture. Many of the multitudes of cracks that permeate a body of serpentine may be filled in later with white veins of dolomite, which are in bold and striking contrast to the prevailing dark-green color of the rock. Because of its complex interweavings of white, green, and black, in a bygone age serpentine was greatly favored for the walls of florist shops, funeral parlors, and the lobbies of small-town hotels—it bears the picturesque name *verde antique*. The more common varieties of serpentine lack the white dolomite veinings and generally run to somber dark-green, black, or red colors. They nonetheless take high polishes and have been used as decorative facings in building lobbies and storefronts. In the United States, they may be seen in the imposing columns of the rotunda of the National Gallery of Art in Washington, D.C., as well as in the United Nations building in New York.

Cataclastic metamorphic rocks In localized zones near the surface of the earth, stresses may build up to the point where rocks shear and are crushed and granulated along planes known as faults (see Chap. 17). The resultant rock completely lacks cohesion. Most shearing and crushing occurs relatively close to the earth's surface, where, as a result of low temperatures and confining pressures, the rocks are brittle and can be easily pulverized.

The same stresses also lead to localized granulation at somewhat greater depths, where rocks are hotter and under higher pressure. There, however, temperatures are sufficiently high to allow some of the more finely ground materials to recrystallize (Fig. 8-5). In all likelihood, these materials never lose cohesion during shearing. The process that produces partially recrystallized granulated rocks is termed ***cataclastic metamorphism*** (Greek, broken down). Is is also called ***dynamic metamorphism.*** Cataclastic rocks show markedly parallel lens-shaped and banded patterns (Fig. 8-6).

Fig. 8-5 Early stages of cataclastic deformation of a granite from the Front Range, Colorado, as photographed under the microscope. The rock is cracked and broken, and in some places recrystallization, chiefly of quartz and potassium feldspar, has occurred. *W.A. Braddock, USGS*

Not all the minerals in rocks are equally resistant to shearing stress—some are unyielding, others are readily molded. In a rock reshaped by dynamic metamorphism, nonresistant, platy minerals, such as mica, may be drawn out in parallel streaky bands or layers, whereas the more resistant ones, such as feldspar and quartz, are pulverized. Some particularly resistant minerals may stand out as rounded or even lenticular, eye-like clots—in fact they are called by the German word for eyes *augen* (Fig. 8-6).

A common example of cataclastic metamorphism is the rock ***mylonite.*** The name comes from the Greek word for mill, and it is rock that, in a figurative sense, was caught in the geological mill and ground to powder. After the minerals in the original rock were crushed and pulverized, the tiny fragments were partially to completely recrystallized into an interlocking pattern, which left the rock as hard and durable as flint. In some cases, mylonite is transformed into a fine-grained glass-like material; with its dark, streaky appearance, it superficially resembles obsidian.

A microscope usually is needed to demonstrate that mylonite consists of angular, minutely brecciated mineral fragments, which have been recrystallized to form a metamorphic rock (Fig. 8-7).

Metamorphic rocks of regional extent: dynamothermal metamorphic rocks

This term refers to metamorphic rocks that characteristically are exposed on the earth's

Fig. 8-6 Sheared, streaked-out appearance of a cataclastic rock from Mount Redoubt, North Cascades National Park, Washington. Some of the larger, whitish, equant mineral grains and rock fragments are broken, rounded, and rotated. *M.H. Staatz, USGS*

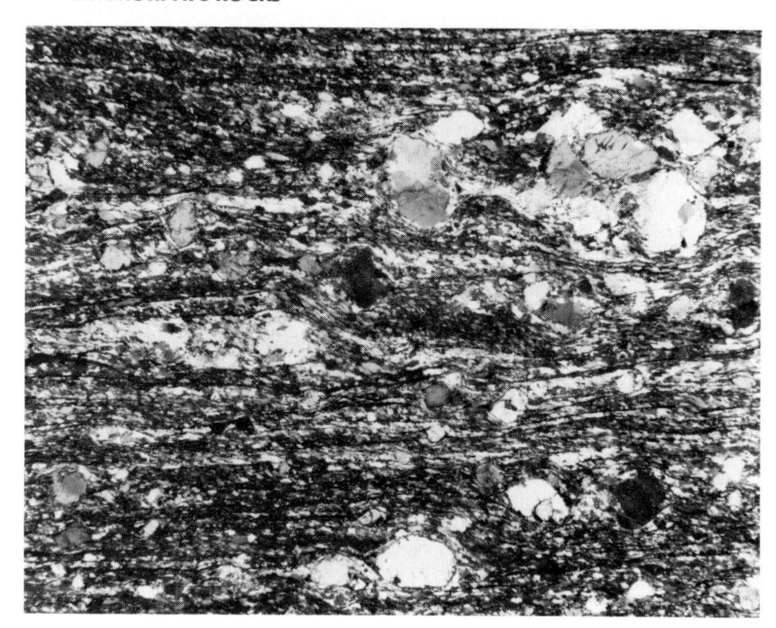

Fig. 8-7 Mylonite, as photographed under the microscope, from the Front Range, Colorado. The large broken grains (augen) are quartz and feldspar; most of the streaked-out matrix is composed of very small crystals of biotite and broken grains of quartz and feldspar. *W.A. Braddock, USGS*

surface over broad areas, sometimes many thousands of square kilometers. Commonly they are found in close association with large plutons (batholiths and stocks), and, like those igneous bodies, are often found in the axes of large mountain ranges.

On essentially all the continents on earth, leveling by erosion has produced large, flattened expanses of land floored almost exclusively by plutonic and dynamothermal metamorphic rocks. Such wide expanses of crystalline rock—both igneous and metamorphic—were given the name ***shields*** many years ago, and they were regarded as the foundation of the continents. Two well-known examples are the Canadian Shield, that broad expanse of igneous and metamorphic rocks marginal to Hudson Bay and extending southward into Minnesota and Wisconsin and eastward across Labrador, and the Fenno-Scandian Shield, which includes most of Finland, Sweden, and Norway. Such rocks, with so wide a distribution, must have a more general cause than the heat generated by a single, igneous intrusion, the reactions produced by chemically activated water, or the grinding effect of movement along a fault plane. Recrystallization is, in fact, brought about by heat and pressure (lithostatic and directed) working along with, to a lesser degree, chemically active solutions. In most large bodies of dynamothermal metamorphic rocks, it is apparent from the minerals they contain that temperatures were not uniform during recrystallization. Where temperatures were higher, high-temperature mineral associations were formed. At a distance away from such "hot spots," progressively lower-temperature mineral associations were formed, creating a zonation from ***high-grade*** metamorphic rocks to ***low-grade*** metamorphic rocks (Fig. 8-8).

If, as is common, the starting materials were essentially the same throughout the region in which metamorphism occurred, it is possible to document progressive changes in texture and mineral association with a progressive increase in regional metamorphism. Overall, an increase in metamorphic level brings about a progressive coarsening in grain size of the individual minerals—from submicroscopic to coarsely crystalline. Individual grains may be 1 cm or more across. There also tends to be an

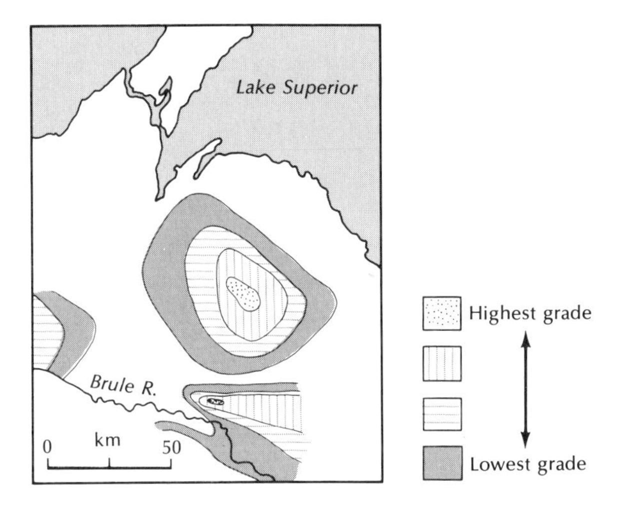

Fig. 8-8 Centers of dynamothermal metamorphism in northern Michigan. Relatively low-grade metamorphic rocks are found toward the outside and high-grade metamorphic rocks toward the inside of each center. Apparently the centers represent "hot spots" of thermal activity.

increase in segregation of certain mineral grains into distinctive bands. Notably, the nonferromagnesian silicates tend to separate from the ferromagnesian silicates.

While these changes are occurring, there is a progression in mineral associations, reflecting differences in the stability of minerals to the changes in temperature and pressure. Low-grade regional metamorphic rocks generally are similar to the initial starting materials and may contain a quantity (the amount depending on the initial rock chemistry) of fine-grained recrystallized muscovite and chlorite. At a higher level of metamorphism, this mineral assemblage becomes unstable and recrystallizes isochemically to produce a different set of minerals appropriate to the temperature, pressure, and pore-fluid conditions and the chemical composition. At even higher levels of metamorphism, recrystallization to another, more appropriate assemblage will occur, and so on.

Fig. 8-9 Biotite gneiss from Uxbridge, Massachusetts, showing light and dark banding, or foliation. Such dynamothermally metamorphosed rocks characteristically show rock cleavage or foliation. *Ward's Natural Science Establishment, Inc.*

Fig. 8-10 Relation of rock cleavage to the original bedding. Rock cleavage that parallels the hammer handle cuts across bedding in folded Cretaceous slate. Hidalgo, Mexico. *F.S. Simons, USGS*

Particular mineralogical changes have been found to be generally indicative of temperature, pressure, and pore-fluid conditions. In the field, such changes can be identified and depicted on a map, providing detailed information about physical conditions in different parts of the metamorphosing rock body.

Because of the role played by directed pressure most dynamothermal rocks are layered, or foliated (from the Latin *foliatus,* leaved). Some, however, are relatively massive, or nonlayered, whereas others, which are relatively unusual, are a closely layered mixture of high-grade metamorphic and plutonic rocks.

Foliated rocks All foliated rocks are characterized by the parallel orientation of their tabular minerals and varying degree of banding, or color layering (Fig. 8-9); some also can be split, or cleaved.

Low-grade metamorphic rocks One of the most familiar foliated metamorphic rocks is ***slate,*** which takes its name from the Old French word *esclate,* or *slat.* It has been used for centuries for roofing material and blackboards. Two properties are responsible for its desirable attributes: (1) it is dense and of uniformly fine texture, and (2) it can be cleaved along smooth, closely spaced parallel surfaces. The latter property is called ***rock cleavage*** to distinguish it from mineral cleavage (Fig. 8-10).

Rock cleavage in slate is the result of an extremely fine foliation that develops during low-grade dynamothermal metamorphism. Viewed under the microscope, the cleavage planes of tiny recrystallized flakes of muscovite mica, which are everywhere present in the slate, are seen to be all parallel (Fig. 8-11). Splitting of a slate along one planar surface is the result of sequential cleaving of numerous tiny mica flakes in that plane.

Slate is usually derived from finely laminated

sedimentary rocks, such as shale, but it can be formed from other finely textured rocks, such as volcanic tuffs.

The original stratification planes in shale or tuff are usually still easily recognizable, and, in most cases, do not coincide with the rock cleavage planes (Figs. 8-10 and 8-11). Somehow, during metamorphism, the small planar clay grains in the shale (which were aligned parallel to the bedding planes) are recrystallized to mica platelets, all with their cleavage planes parallel, but at an angle to the bedding. How the change comes about is not clear. Some believe that as the mica plates form they are constrained to grow with their cleavages at right angles to the directed stress. Others feel that, once formed, the plates are rotated mechanically (that is, deformed) in response to directed stress. And still others believe that the clay particles themselves are rotated during deformation and then recrystallize.

Fig. 8-11 The fine structure of rock cleavage, as photographed under a microscope. The bedding, demarcated by dark, nearly horizontal bands, is cut at a high angle by the closely spaced cleavage planes. Note that the alignment of recrystallized flakes of mica is parallel to the cleavage planes. *W.B. Hamilton, USGS*

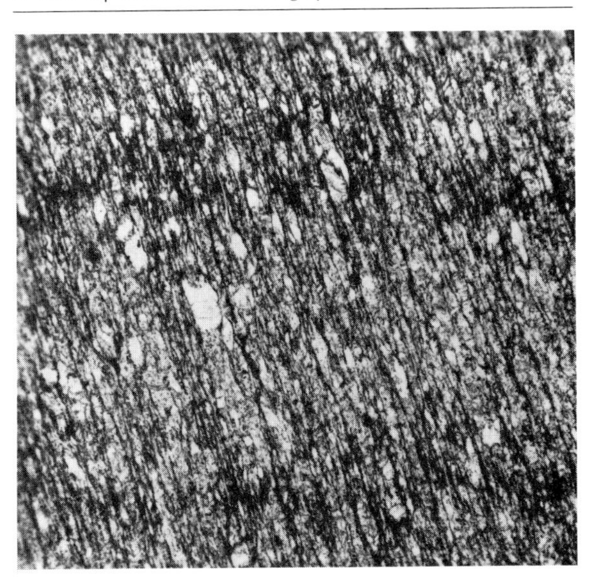

Recognition of the original bedding, the fineness of the foliation, and the smallness of all the individual mica grains are three bits of evidence indicating that slate is formed by a very gentle dynamothermal metamorphism at relatively low temperatures. The metamorphic rock looks superficially very much like the original sediment from which it was derived.

Phyllite (from the Greek, *phyllon,* a leaf), also a common low-grade metamorphic rock, is generally more crystalline than slate and intermediate in grain size between slate and schist (described below). The amount of aligned fine-grained platy minerals, mostly muscovite, may reach 50 per cent or more. As a result, cleavage is well developed, and the rock characteristically possesses a silky sheen (Fig. 8-12).

In rocks that contain fewer clay minerals than shales, and are accordingly less susceptible to alteration, low-grade metamorphism produces little outward evidence of having occurred except for the development of poorly defined rock cleavage and the growth of some grains of mica in subparallel alignment.

Intermediate-grade metamorphic rocks Probably the most widely occurring metamorphic rock is ***schist*** (from the Greek word *schistos,* cleft, or *schizein,* to split). Slates grade into schists with increasing grain size. Whereas the platy minerals in slate are unrecognizable without a microscope, those in schists generally are visible to the unaided eye. All schists include tabular, flaky, or even fibrous minerals, and the extent to which those minerals developed in parallel orientation determines to a considerable degree whether ***schistosity,*** the characteristically wavy or undulatory rock cleavage, will develop. Many schists split readily into tabular blocks. They are the familiar flagstones so widely used throughout Europe in courtyards and for castle walls, and in North America for fireplaces, patios, and barbecues.

The individual folia laminations of schist are, for the most part, regularly spaced at distances up to about 0.5 cm (0.2 in.). The spacing is the principal attribute that separates schists from the more coarsely layered gneisses (to be described next). Sometimes adjacent folia bands

Fig. 8-12 Well-developed rock cleavage in a phyllite, glistening in the sun, eastern Nevada. *Karl Birkeland*

possess different mineralogies: one band may contain mostly such platy minerals as biotite, muscovite, or chlorite and the adjacent one mostly quartz and feldspar. Generally the flaky minerals make up more than 50 per cent of the rock (Fig. 8-13).

Original bedding or other sedimentary structures are almost never identifiable in schists—indicating that their metamorphic grade is much higher than that of slaty rocks. From the mineral associations common to schists, and from the relatively large size and alignment of the grains, *metamorphic petrologists* (those who study the origin of metamorphic rocks) generally agree that most schists were formed over a range of intermediate temperatures under the influence of directed pressure (and to a lesser degree, chemically active fluids). Almost any rock can be transformed into a schist, but likely candidates are shale, siltstone, and muddy sandstone.

High-grade metamorphic rocks ***Gneiss*** (pro-

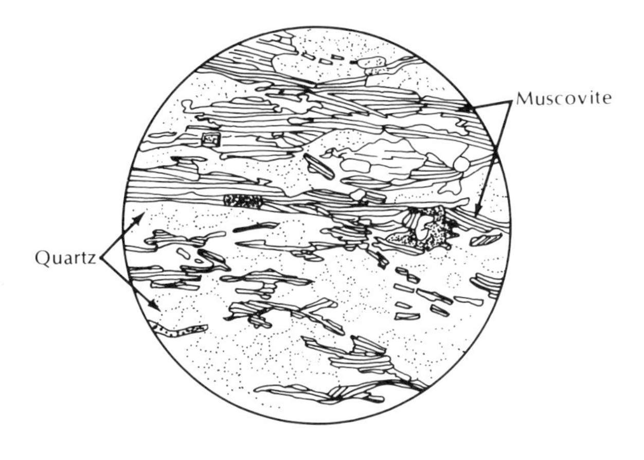

Fig. 8-13 Drawing of schist from Edge Hill, Pennsylvania, as seen under the microscope. It is composed almost entirely of grains of quartz and muscovite in sub-parallel alignment.

Fig. 8-14 Gneiss from the north rim of the Black Canyon of the Gunnison River, Colorado. The light-colored bands are composed mostly of quartz and feldspar, and the dark-colored layers mostly of quartz and biotite. Note the crenulations and folds, indicative of plastic folding and flowage. *W.R. Hansen, USGS*

nounced as though it were actually spelled nice) is an old Saxon miners' term for a rock that is rotted or decomposed. It is a banded rock, usually with layers of light-colored minerals alternating with dark-colored layers (Fig. 8-14) and is much coarser grained than schist or slate. In fact, the size of the quartz and feldspar crystals it contains is about the same as that in granite. During recrystallization of the rock under directed stress at relatively high temperatures, the minerals have been rearranged so that most of the light-colored ones are in one layer, whereas the dark ferromagnesian ones are in another; the light- and dark-colored bands alternate through the rock. The bands, unlike the cleavage planes of slate, do not make uniformly parallel planes that continue for long distances; more commonly, they are strongly distorted. They probably were deformed plastically; that is, although the rock was still in the solid phase it was able to flow, in about the same way that butter can without becoming liquid (Fig. 8-15). The common occurrence of plastic deformation in gneiss argues strongly for high temperatures during its formation—temperatures high enough to soften the rock and to allow plastic flow to occur.

Gneisses do not have the highly developed rock cleavage of slate or schist. In spite of the roughly uniform spacing of their bands, they break in about the same unpredictable fashion as a piece of granite does when struck with a hammer.

Non-foliated rocks Non-foliated rocks are shaped by the same processes that shape the foliated varieties, but as a consequence of their initial mineralogical composition, they are not banded. Two leading examples are *marble* and *quartzite*.

Marble is the finely to coarsely crystalline equivalent of the sedimentary rock limestone and accordingly consists largely of the mineral calcite ($CaCO_3$). In the transformation of limestone to marble at relatively high temperatures and pressures, the bedding and the visible organic shell material are largely obliterated and the result is a rather sugary textured rock com-

Fig. 8-15 Intricately folded quartz veinlet in fine-grained gneiss, Washington, D.C. area. *J.C. Reed, USGS*

posed of a fine- to coarse-grained aggregate of nearly equidimensional calcite grains (Fig. 8-16). None of the grains are bounded by crystal faces. Because marble is one mineral only, and the grains are essentially all the same size and equidimensional, there is little possibility for any foliation or banding.

We should point out that not all marble is the result of dynamothermal metamorphism. Some results from recrystallization during contact metamorphism when, for example, a limestone bed is intruded by a plutonic body of at least moderate size. One of the finest white marbles comes from a quarry at Marble, in west-central Colorado. It owes its existence to the contact metamorphism that took place when a Tertiary laccolith intruded a Mississippian-age limestone.

Pure marble is snow white, and one of the most highly prized varieties from ancient times down to our day comes from the quarries of Carrara, on the west coast of Italy. Carrara marble is a remarkably uniformly textured rock that is ideal for sculpture because it is pure and has a hardness of no more than 3 on the Mohs scale.

Not all marble is snow white, which is immediately clear to anyone who has observed it in banks, building facades, lobbies, public lavatories, and on old-fashioned tables and dressers. In general, the black and gray areas in marble are colored by carbonaceous matter, brown and red areas by iron oxide, and green areas by various iron- or magnesium-bearing silicate minerals.

Metamorphosed limestones are not always just calcite. Commonly the sediments contain clayey or sandy material dispersed throughout the limey material. When such rocks are metamorphosed, reactions between the calcite and the other material can produce new minerals. For example, in the metamorphism of a limestone containing quartz sand, a new mineral, ***wollastonite*** ($CaSiO_3$), is formed, and carbon dioxide is liberated. The reaction, in shorthand notation, is

$$CaCO_3 + SiO_2 \rightarrow CaSiO_3 + CO_2$$

Wollastonite is a colorless mineral, commonly arranged in fan-shaped radiating needles that penetrate the granular marble host. It is representative of a wide variety of related minerals that develop with increasing clay and sand content in impure limestone to yield, after metamorphism, the so-called ***calc-silicate*** rocks (Fig. 8-17).

Quartzite is metamorphosed quartzose sandstone, and in it the pore spaces once separating the individual grains are filled with newly crystallized quartz. The ghost-like boundary separating the original quartz grain from the silica added to it may be barely discernible. The silica that fills the pore spaces of the original sandstone commonly proves stronger than the

Fig. 8-16 Marble, consisting of medium-sized equant grains of calcite, photographed under the microscope. *Edwin E. Larson*

sand grains themselves, and when struck a hammer blow, the rock may break through the grains, rather than around them—as quartzose sandstones usually break. Most quartzites are the result of recrystallization at intermediate to high temperatures. Because they are predominantly one mineral, which occurs in equidimensional grains, they tend to show no foliation—regardless of the presence of directed pressure during their development.

Quartzites are nearly always light colored; light pinks or reds are very characteristic. Many are white or light gray, but with increasing amounts of impurities their colors darken. Very often they are interbedded with marble, calc-silicate rocks, and other rocks derived from sedimentary sources. Relic sedimentary structures, such as cross-bedding, are sometimes preserved; they are emphasized by slight color differences that superficially may resemble the banding in a gneiss.

Mixed high-grade metamorphic and plutonic rocks At great depths, temperatures can be so high—between 600 and 800°C (1112 and 1472°F) that minerals can melt.

If a gneiss consisting of light-colored bands of quartz and potassium- and sodium-rich feldspars that alternate with dark-colored bands of ferromagnesian minerals is heated until melting begins, the non-ferromagnesian minerals possessing relatively lower melting points (as

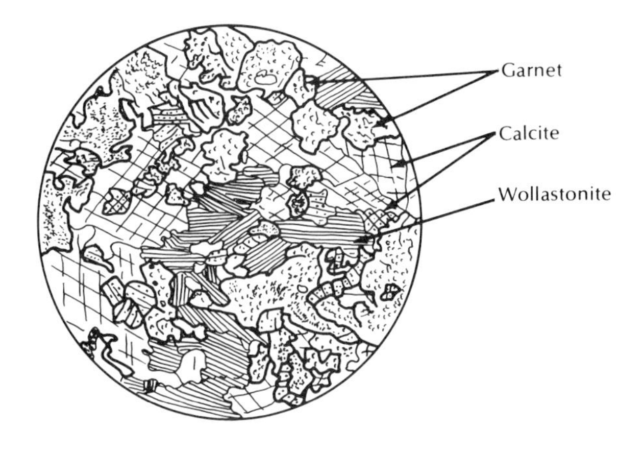

Fig. 8-17 Drawing of calc-silicate rock, Chihuahua, Mexico, as seen under the microscope. Wollastonite and garnet grains are embedded in grains of coarsely recrystallized calcite. Calc-silicate rocks generally are the result of metamorphism of a sandy or silty limestone.

discussed in Chap. 5), will be the first to melt. Layers rich in ferromagnesian minerals, with relatively higher melting points, may become plastic and perhaps flow and fold, but they will remain solid. If the temperature were to stabilize at that point and then fall, the resulting rock would be made up of bands of dark metamorphic rock (composed largely of ferromagnesian minerals) alternating with light-colored granitic plutonic rock (Figs. 8-18 and 8-19).

Such mixed plutonic/high-grade metamorphic rocks are termed ***migmatites*** (Greek, to mix) and they are common throughout the world wherever high-grade metamorphic rocks exist.

Some rocks that superficially resemble migmatites have been formed by the intrusion of sill-like granitic bodies into pre-existing high-grade metamorphic rocks.

Fig. 8-18 Close view of migmatite, a mixed plutonic-high-grade metamorphic rock, Sawatch Range, Colorado. The light layers are granitic; the dark layers, biotite gneiss. Rock flowage causes the crenulation of the layers. Note the penknife for scale. *Edwin E. Larson*

Not all geologists agree with the explanation of migmatites given above. Some contend that the granitic layers themselves are metamorphic rocks, formed by a process in which solid-state recrystallization is accompanied by large-scale diffusion of ionic material in and out of the layers, which finally become "granite-like." Accordingly, this process has been called ***granitization,*** or alternatively, ***metasomatism*** (Greek, to change the body).

If, in our example above, the temperature had not stabilized, but had continued to rise, the entire rock unit would have eventually melted to form a body of magma with a composition representative of the entire rock body—perhaps equivalent to a diorite. The common association of granitic rocks with high-grade metamorphic rocks and migmatites lends a great deal of support to this model of their origin. Again, some geologists believe that most granitic plutons are the result of large-scale granitization, and they can point to examples to support their belief. However, such examples are relatively rare and most field, laboratory, and experimental studies favor the magmatic origin of granite.

Realm of dynamothermal metamorphism Contact, hydrothermal, and cataclastic metamorphic rocks are of local significance only, and the reasons for their formation are fairly well understood, at least generally, if not specifically.

On the other hand, earth scientists have yet to understand clearly why large belts of dynamothermal metamorphic rocks form where they do. We can note that they are commonly found in elongate bodies, are often closely associated with deep-seated plutonic rocks, like those in batholiths, and are found commonly at the axes of mountain chains. Most of them show evidence of exposure to moderate to high temperatures and confining pressures, and most also show indications of directed pressure during their recrystallization.

It has also been found, from radiometric-dating studies, that large metamorphic belts remain sporadically active for anywhere between 200 and 1000 million years, with localized high

Fig. 8-19 Highly contoured layers of mixed plutonic-high-grade rocks in the Coast Mountains, British Columbia. This type of rock is commonly associated with the lower parts of batholiths and appears to form deep inside the earth where the temperature falls just short of that necessary for melting. *Geological Survey of Canada*

spots of activity that run from 10 to 40 million years. The times of increased activity are separated by quiescent intervals from 20 to 250 million years. The active periods may or may not coincide with peak periods of pluton formation. In the Caledonide Mountains of Scotland, for example, peaks in metamorphic activity are recorded at 730 million years, between 500 and 475 million years, 440 million years, and between 420 and 390 million years. Plutonic activity, however, was restricted to about 530 million years, between 410 and 395 million years, and at 365 million years.

Most geologists believe that dynamothermal metamorphic rocks were formed at moderate to fairly great depths and that their formation was closely associated with the formation of plutonic rocks. Both processes are thought to be directly related to mountain building.

Many geologists now feel that all three processes go on in zones of convergence where two plates are pushing together. The heat and directed pressure needed for the formation of dynamothermal metamorphic rocks are the result of the crushing action that occurs along the plate hinge line, and the rocks that represent the original pre-metamorphic materials are igneous and metamorphic rocks from the plates themselves, as well as sediments that accumulate in elongate basins marginal to the continents, along the zone of convergence.

Just as in the case of the formation of batholiths, the plate-tectonic model looks promising. Unfortunately, we can only study metamorphic rocks at the surface long after the converging plates have become inactive and after erosion has removed thousands and tens of thousands of meters of overlying rocks. And in the zones of convergence of today we have no idea whatsoever of what metamorphic processes are at work deep beneath the surface.

SUMMARY

1. As a result of such processes as mountain building, plate convergence, and sedimentation, some rocks in the earth's outer 15 km (9 mi) are subjected to different *physicochemical environments*. *Recrystallization* of the minerals may therefore occur. The process, which occurs while the rocks are in the solid state, is called *metamorphism*.
2. Metamorphism is brought about by *heat*, *lithostatic* and *directed pressures*, and the *chemical activity* of pore-fluids.
3. Metamorphic rocks of limited area include *contact metamorphic* rocks, *hydrothermally metamorphosed* rocks, and *cataclastically metamorphosed* rocks. Contact metamorphism is primarily a thermally induced process occurring at margins of plutonic bodies. The resultant rocks are called *hornfels*.

   Hydrothermal metamorphism is the result of interaction of hot, chemically active, water-rich fluids with the rocks surrounding large plutons.

   Cataclastic metamorphism represents recrystallization of materials that have been sheared and crushed along fault zones. A common cataclastic rock is *mylonite*.
4. Metamorphic rocks of wide extent are called *dynamothermal* (*regional*) *metamorphic* rocks. Recrystallization in such rocks results from heat, pressure, and the action of chemical solutions. Because of directed pressure most dynamothermal rocks are *foliated*.
5. *Slate* is perhaps one of the best-known *low-grade metamorphic* foliated rocks. Its *rock cleavage* results from the alignment of tiny recrystallized mica grains.

   Foliated rocks of *intermediate grade* are called *schists*. Their *schistosity*, or foliation, is determined by the parallel orientation of tabular, flaky, or fibrous minerals. Shale, siltstone, and muddy sandstone are the rocks most often changed to schists.

   High-grade metamorphism produces the foliated rock *gneiss*. It is coarser grained than schist, with a dark-and-light banding in which ferromagnesian grains are concentrated in the dark layers, and quartz and feldspar grains in the light layers.
6. *Marble* (recrystallized limestone) and *quartzite* (recrystallized quartz sandstone) are two common non-foliated rocks.
7. At great depths, temperatures are so high

that melting of lower-melting-point minerals takes place, forming migmatite, with bands of light-colored plutonic rock that alternate with dark-colored bands of metamorphic rock. Such mixed rock represents the highest grade of metamorphism.

8. Dynamothermal (regional) metamorphic rocks are commonly found near batholithic rocks and occur in the cores of mountain ranges. Many geologists think that such rocks are formed in zones of plate convergence.

QUESTIONS

1. What is meant by isochemical metamorphism? Are there any cases in which recrystallization is not isochemical?
2. Discuss the main factors involved in the metamorphism of a pre-existing rock. Which do you consider to be the most important factor?
3. Name and briefly describe the principal types of metamorphism. Where do each of these types occur?
4. Describe the differences in rock mineralogy and texture corresponding to the different types of metamorphism.
5. Which general kinds of metamorphic rocks are the most abundant worldwide?
6. How does a slate differ from a schist? From a gneiss?
7. Does rock cleavage have anything to do with mineral cleavage?
8. What are migmatites, and how are they related to regional metamorphism?

SELECTED REFERENCES

Barth, T. F. W., 1962, Theoretical petrology, John Wiley and Sons, New York.

Ernst, W. G., 1969, Earth materials, Prentice-Hall, Englewood Cliffs, New Jersey.

Harker, Alfred, 1930, Metamorphism, Methuen and Co., London. (Repr. 1976, Halsed Press, New York.)

Hyndman, D. W., 1972, Petrology of igneous and metamorphic rocks, McGraw-Hill Book Co., New York.

Mason, B., 1966, "Metamorphism and metamorphic rocks," Chap. 10 *in* Principles of geochemistry, 3rd ed., John Wiley and Sons, New York.

Miyashiro, A., 1973, Metamorphism and metamorphic belts, John Wiley and Sons, New York.

Ramberg, Hans, 1952, The origin of metamorphic and metasomatic rocks, University of Chicago Press, Chicago.

Ramberg, Hans, 1960, "Metamorphism," *in* Encyclopaedia Britannica, vol. 15, pp. 321–26.

Simpson, B., 1966, "Metamorphism," Chap. 22 *in* Rocks and minerals, Pergamon Press, Oxford.

Spry, A., 1969, Metamorphic textures, Pergamon Press, Oxford.

Turner, F. J., and Verhoogen, J., 1960, Igneous and metamorphic petrology, 2nd ed., McGraw-Hill Book Co., New York.

Tyrrell, G. W., 1929, The principles of petrology, E. P. Dutton and Co., New York.

Williams, H., Turner, F. J., and Gilbert, C. M., 1954, Petrography, W. H. Freeman and Co., San Francisco.

Winkler, H. G. F., 1979, Petrogenesis of metamorphic rocks, 5th ed., Springer-Verlag, New York.

Fig. 9-1 More than 600 years of weathering have taken their toll on the Great Wall of China. U.S. Army trooper at the wall in 1929. *Vance T. Holliday*

9

WEATHERING AND SOILS

In the days of the Pharaohs a cherished status symbol was the obelisk. Those hieroglyph-bedecked stone columns early became collector's items for a procession of conquerors of the Nile, beginning with the Caesars and ending with Napoleon. Even the United States collected an obelisk in 1879. After great effort, involving among other things cutting a loading port in the bow of one of the steamships of the time (whereupon it nearly foundered), the obelisk was finally set up in New York's Central Park (Fig. 9-2) to take its place among similar, far-wandering artifacts in such cities as Paris, London, and Rome—in which alone there are twelve of them.

New York's climate is considerably more humid than that of Egypt. No wonder that, in about 100 years, many of the hieroglyphs spalled off and the whole surface of the obelisk started to disintegrate, while its counterparts still standing in Egypt have survived nearly unscathed beneath the desert sun for almost 4000 years.

That brief story makes the point that climate is one of the leading factors in determining the rate and manner in which rocks disintegrate or decompose, or, as we say, ***weather***—using the word in about the same way we do when we speak of a weather-beaten face. Another crucial factor in determining the effectiveness of weathering is the kind of rock. Evidence for that can be found in New England graveyards, where slate headstones carrying the salty epitaphs beloved by some of our forebears survive from the 1700s, whereas the words carved on limestone or marble markers of much more recent vintage may be partly or wholly obliterated (Fig. 9-3). We can use data derived from tombstone studies to help us select the most durable rock for building construction in various areas. For the most part, carefully selected rocks will last the life of the building. Yet the effect of air pollution, especially in industrial areas, is greatly accelerating the natural rate of rock decay. In parts of Europe, for example, centuries of weathering only slightly modified some fine stone sculptures, whereas in the past century, with its intensive industrialization, these works of art have been almost destroyed (Fig. 9-4). Yet weathering can transform rocks into works of art (Fig. 9-5) or add stark beauty to the landscape (Fig. 9-6).

Fig. 9-2 A. (left) The surface of this obelisk, still standing in the desert at Karnak, Egypt, is scarcely marred by weathering over four millennia. *Jean B. Thorpe* **B.** (right) The obelisk of Thothmes III, from the temple of Heliopolis, Egypt, now in Central Park, New York. The lower part of the granitic column shows a loss of detail as a result of weathering. The monument was brought to New York in 1879, where most of the weathering took place in a few years. *Metropolitan Museum of Art*

Observant travelers have probably noticed how different the soils are in various parts of the world. In humid areas, where all traces of original rock structure have been obliterated, the soil may be dark colored at the surface, grading to reddish hues at greater depth. In contrast, soils in dry regions are quite light colored and thin, and the rock structure is readily visible. Climate and vegetation have long been known to be responsible for regional variations in soils, but rock type and length of time of formation are major factors as well.

Soil is perhaps the most valuable mineral resource on earth. Without it life as we know it would be impossible, and needless to say, it is a resource rapidly gaining in importance as the world population expands. Yet in many areas of rapid growth, housing tracts and industrial communities are built in low-lying flat areas that are our most productive agricultural lands. Common sense would dictate that such lands should remain agricultural and that any construction should take place on adjacent land with less productive soils. Although some land-use planning is directed toward that goal, it is too slow in coming, and all too often it is simply ignored. Although we could survive without a number of other substances, such as gold or diamonds—we would never survive without plain dirt.

In this chapter we shall examine the various processes that cause rock to weather and to form soils. We shall also look at the overall affects of the environment on the kind of soil that forms and then review some of the practical reasons why we should study soils.

Fig. 9-3 Nineteenth-century grave markers in a cemetery in Damariscotta, Maine. **A.** (left) Weathering has nearly obliterated the inscriptions on this marble stone. *John Manger* **B.** (right) Inscriptions are still visible on this granite marker of an earlier date. Weathering has slightly rounded the corners of the stone, and lichens growing on the surface have no doubt speeded up the weathering process. *John Manger*

Fig. 9-4 Decay of a seventeenth-century sandstone sculpture from the Rhein-Ruhr region, Germany. **A.** (left) The sculpture as it appeared in 1908. **B.** (right) The same sculpture in 1969. **C.** (below) A plot of the amount of decay with time. *Landesdenkmalamt Westfalen-Lippe, Münich*

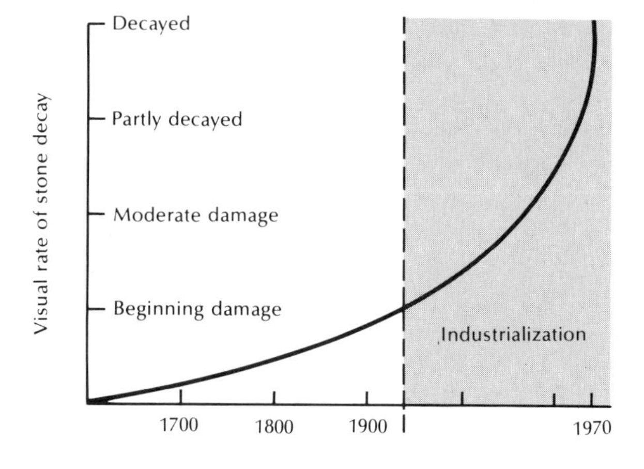

WEATHERING

Most rocks found in the top several meters of the earth's crust are exposed to physical, chemical, and biological processes much different from those prevailing at the time the rocks were formed. Because of the interaction of those processes, the rock gradually changes into soil-like material. Collectively the changes are called ***weathering,*** and two main types are recognized: chemical and mechanical. In humid, warm regions where vegetation flourishes and organic acids are abundant, ***chemical weathering*** is dominant, and rocks are prone to decompose or to decay. In harsher climates, where frost action may dominate, rocks break up mechanically, or disintegrate, without undergoing much chemical alteration, and the process is known as ***mechanical weathering.*** When rocks decompose, they are changed into substances with quite different chemical compositions and physical properties than those of the original rock. In contrast, if rocks disintegrate mechanically, they break up into smaller

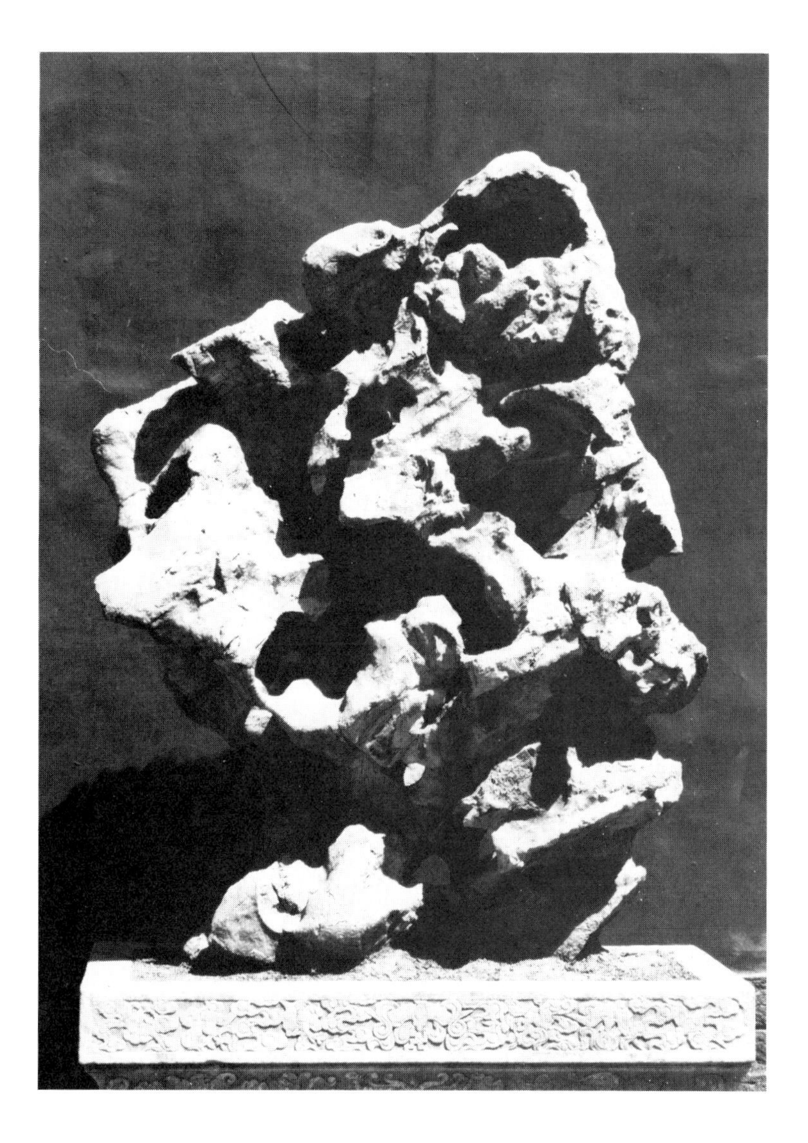

Fig. 9-5 Chemical weathering of a limestone breccia produced this rock displayed in the Imperial Palace, Peking, China. The sculpture, with its carved base, dates from the Ming Dynasty (1368–1644). *H.E. Malde, USGS*

Fig. 9-6 Different rates of weathering were responsible for carving Balanced Rock in the Garden of the Gods, Colorado. Photographed about 1880. *W.H. Jackson*

fragments, much as if they had been struck a hammer blow. There are few areas where only chemical weathering or only mechanical weathering operates alone but there are many areas where one or the other predominates.

Mechanical weathering

Some aspects of mechanical weathering are irritatingly familiar to all of us, such as the wedging apart of sidewalks, foundations, and walls by the roots of grass, trees, and shrubs. The same process goes on in the mountains, where a common sight high on the slopes is an isolated pine clinging to a sheer granite ledge. With no soil in which to take hold, the tree's roots force their way into crevices, and root-volume expansion with growth tends to push the rocks still farther apart. The process is much like the one known millennia ago to Egyptian slaves, who pried out granite blocks for obelisks, using water-soaked wooden wedges.

Almost all rocks are cut by cracks, large and small—sometimes as closely spaced as a fraction of a centimeter, at other times many meters apart, as they are in the stupendous cliff of El Capitan in California's Yosemite Valley. Such cracks are called ***joints,*** and they provide entry for roots and organic acid-bearing waters to penetrate far into the rocks and cause weathering (Fig. 9-7).

Freezing and thawing Water has highly unusual properties. Of greatest significance to mechanical weathering is the expansion of water when it crystallizes to ice. The expansion amounts to about 9 per cent, by volume. Should water freeze in a confined space, it can deliver an enormous outward pressure against its containing walls—as anyone who has contemplated a cracked engine block or a rup-

Fig. 9-7 Vertical and horizontal joints in granite in the Sierra Nevada, California. The joints provide entry for moisture (which may aid in mechanical disintegration when it turns to ice) and also enable plants to send down their roots, which wedge blocks apart. *Cedric Wright Collection, The Sierra Club*

Fig. 9-8 Angular blocks produced by frost wedging of well-jointed granitic rock in the Sierra Nevada. *Cedric Wright Collection, The Sierra Club*

tured radiator knows. Few people realize, though, how great the force actually is. Under confined conditions at −22°C (−7.6°F), pressures can reach 2100 kg/cm^2 (4.3 million lb/ft^2).

For such great pressure to build up in nature, there must be a completely enclosed system with no air—a condition seldom realized in an environment such as a crack in a rock. Consequently, the pressure created by the freezing of water in rocks is less than the above figure, but high enough to crack most rocks exposed in high mountains. Water in an open crevice freezes from the top down and thus is sealed in with ice. Then, if confined in the lower part of the crevice, it can act as a wedge to pry rocks apart along planes of weakness such as joints or bedding planes. The process is called ***frost wedging*** and is thought to be most effective when temperatures rise above the freezing point by day and drop below it by night. It is thought to be a rather prominent weathering process near and above timber-line on high mountains. As a result, summit uplands may be carpeted with frost-shattered angular joint blocks (Fig. 9-8). However, if temperatures remain far below freezing for much of the year, as in some high latitude polar areas, frost wedging proceeds at a much slower rate.

Not all block fields at high altitudes or in the high latitudes have been formed by present-day frost wedging. Indeed, many fields seem to be inactive, as shown by the weathered character of the blocks, the lichens that grow on them, and the well-developed cover of vegetation between them. Such fields were formed some time in the past and probably testify to more rigorous mountain or polar climates during former ice ages.

Salt crystal growth A mechanical weathering process somewhat similar to frost wedging is the result of the growth of salt crystals in rock. In arid regions, ground and soil water, as well as water within rock pores and cracks, commonly contains dissolved salts in ionic form. As the water evaporates, the salt ions are left in the remaining water and their concentration increases. In time, the concentration reaches the point at which salt minerals crystallize from the solution. Pressures accompanying the crystallization can be quite great—certainly great enough to dislodge individual minerals or flakes of rock or to shatter objects (Fig. 9-9).

Another salt-related mechanical weathering process is ***hydration,*** or the expansion of salt minerals when water is added. If the salts are part of the rock, the pressures generated can break up the rock. An excellent example is the weathering of the obelisk in New York (see Fig. 9-2). Before it was brought to the United States the monolith rested on its side in Egypt for about 500 years. During that time, salt-laden ground water penetrated the column, and salt minerals crystallized out as the ground water evaporated into the hot desert air. But in the humid climate of New York, the same minerals absorbed the water from the atmosphere and expanded. The expansion from within caused considerable mechanical weathering in the form of flaking of surface fragments over a short period of time. Freezing and thawing also may have contributed to the weathering of the obelisk.

Surface unloading In many parts of the world, the rock close to the surface is cut by joints that more or less parallel the surface, giving it the appearance of an onion skin (Fig. 9-10). Several processes, collectively called ***exfoliation*** from the Latin *exfoliatus* meaning stripped of leaves, or sheeting, explain the onion-layered appearance of these rocks. A common process is the upward expansion of rock as an overlying or confining rock burden is removed. At depth, the rock is under high confining pressures equivalent to the weight of the overlying mass. As erosion removes the overlying rock, the remaining rock can expand—usually upward or toward the valley walls. This expansion produces joints oriented at right angles to the direction of the release, hence they are usually parallel to the land surface (Fig. 9-10B). Many of the rock domes of the world are thought to have been formed in this way (Fig. 9-11). Workers in mines and quarries know the expansive properties of rock all too well. Not uncommonly, as new rock faces are exposed, the pressure release is instantaneous, and the resulting ***rock bursts*** send dangerous missiles flying through the air.

Temperature changes A generation ago, much was made of the theory of rock disintegration as a result of alternate expansion and contraction induced by severe temperature changes, especially in deserts. There, it was said, rocks expanded drastically under the noonday sun and contracted sharply with the falling temperature at night. Presumably the changes were greatest on the surface of a rock and least in its interior, because rocks are such poor conductors of heat. The result could be exfoliation on a small scale.

There is no doubt that exfoliated rocks exist; their number truly is legion, but there is uncertainty about the way in which they are formed. The peeling off of concentric rings of heated surface layers from a cooler interior by differential expansion is an appealing theory, but how can it be explained then, that, for exam-

Fig. 9-9 A. (right) Telephone pole damaged by salt crystallization, Bonneville Salt Flats, Utah. Saline groundwater is drawn upward in the pole, and, as the water evaporates, the salts crystallize, shattering the wood. *W.C. Bradley* **B.** (below) Salt-shattered bedrock south of the Dead Sea, Israel. *W.B. Bull*

ple, in the Sahara and the Arabian desert many stone monuments have survived for 4000 years with scarcely any blurring of their inscriptions?

Perhaps the most conclusive evidence that temperature changes alone cannot disrupt rocks comes from an experiment performed by the American geologist D. T. Griggs. He alternately heated and cooled the surface of a highly polished block of granite—for five minutes the heat was on, for ten minutes a fan cooled the block. The process subjected the surface to a temperature range of 100°F and was repeated 89,400 times, or the equivalent of 244 years of weathering, if each 15-minute cycle were considered a day. But even in the fiercest desert, the diurnal range is less than 100° F; if we were to consider the experiment in more realistic terms, 1000 years of actual weathering may have been more closely approximated.

What happened as a result of the punishment the rock received? Nothing. The surface of the rock remained unblemished, retaining its original bright polish throughout the entire ordeal.

Griggs then produced a little rain, as it were, by introducing a fine spray of water during the cooling cycle. Water was used for only ten days (the equivalent of two and one-half years of weathering), and in that brief time a number of notable changes occurred. The granite lost its polish, the surface of feldspar crystals became clouded, and exfoliation cracks started to appear.

That isolated experiment supports observations made in Egypt by an American geologist, Barton. He noted that almost no discernible change was visible on granite inscriptions that faced the sun, whereas those that were in the shade, and thus relatively damp, showed spalling of rock surfaces and hieroglyphs. An alternative hypothesis, however, states that in the hot deserts the amount of time required for weathering due to temperature changes is more than the amount represented by the age of the monuments Barton observed in the field or more than that represented by Griggs's laboratory experiment.

The majority conclusion is that temperature changes, in themselves, cannot cause rocks to exfoliate and that water plays an important role in exfoliation. A plausible explanation is that, in many deserts, no matter how arid they may appear to be, a little water is available from sporadic showers or from nocturnal dew. When the water combines chemically with the more susceptible minerals in a rock, they swell. It is the increase in volume that may initiate the process, lifting off the outer layers of a rock in concentric shells. To summarize, some exfoliation appears to be essentially a mechanical or disintegrative process, accomplished, however, by chemical alteration.

Very high temperatures associated with fires also seem capable of bringing about exfoliation, as exfoliated rocks around campfire sites indicate. Surface rocks in forested areas commonly show signs of exfoliation, probably the result of occasional forest fires, whereas rocks in non-forested areas do not (Fig. 9-12).

Chemical weathering

Chemical weathering occurs in all environments, but is dominant in hot and humid lands where temperatures are high, large amounts of water are available, and vegetation flourishes. Organic acids, which are potent agents of rock decay, are readily generated and rocks that can resist their chemical action are rare. Carbonic acid, a common organic acid, results from a union of water and carbon dioxide:

$$H_2O + CO_2 \rightarrow H_2CO_3$$

Generally water is readily available from atmospheric sources, and the carbon dioxide is derived partly from atmospheric sources and partly from root respiration and the decay of organic matter. Although relatively weak as acids go, carbonic acid is common enough in most natural environments to be an effective weathering agent.

Of the manifold processes involved in chemical weathering, three of the most important are ***solution, oxidation,*** and ***hydrolysis.***

Solution Solution is perhaps the easiest process of chemical weathering to visualize be-

Fig. 9-10 A. (above) Exfoliation sheets in Independence Rock, Wyoming, a noted landmark for early travelers in the area. *W.H. Jackson, USGS* **B.** (below) Exfoliation sheets can form relatively quickly, as shown by these that conform to the lower walls of a small canyon tributary to the Colorado River. *Edwin E. Larson*

Fig. 9-11 Granite domes are well developed in Yosemite National Park, California. Perhaps the most famous of them is Half Dome, with a vertical cliff on one side (lower center). *USGS*

Fig. 9-12 Exfoliation of granite that took place during a 1976 forest fire near Boulder, Colorado. High temperatures associated with the fire caused the thin white shell of rock to expand and peel off. *Scott Burns*

cause a rock may literally dissolve away, much like a sugar cube in coffee, but at a slower rate (Fig. 9-13). Limestone is especially susceptible to solution, as shown by the following reaction:

$$CaCO_3 + H_2CO_3 \rightarrow Ca^{2+} + 2HCO_3^-$$

Here, a solution containing carbonic acid reacts with calcite, the chief mineral in limestone, to form one calcium and two bicarbonate ions. Those ions are removed from the site of weathering by percolating water. Thus, where a layer of limestone once was, there may well be nothing, because the calcite has been completely dissolved. The process accounts for the profusion of caverns, underground channels, and disappearing rivers in limestone regions. Indeed, the growth of such underground voids sometimes leads to the sudden collapse of the overlying rock into a cavern (see Chap. 16).

Oxidation Rusting is a process familiar to most of us. In anything but the most severe climates, such as the central Antarctic ice sheet, all unprotected objects made of iron will rust away within a lifetime. In a rainy tropical climate, the struggle to maintain steel bridges, ships, rails, and automobiles is a relentless one.

Most rocks contain some iron-bearing minerals. When they are exposed to atmospheric attack, like an old Volkswagen frame in an auto graveyard, they rust. The rocks, originally gray, are stained a wide variety of colors, such as red, yellow, orange, or red-brown if weathering occurs in an environment with ample oxygen. A simplified reaction describing the process is

$$\underset{\text{ferrous iron oxide (gray-green)}}{4FeO} + \underset{\text{oxygen}}{3O_2} \rightarrow \underset{\text{ferric iron oxide (rust-colored)}}{2Fe_2O_3}$$

Indeed, discoloration of rock to yellowish-brown and red is one of the first visible signs of chemical weathering. It is thought that the exact color is determined by the kind of iron oxide that forms.

Hydrolysis The common rock-forming minerals weather by a process called ***hydrolysis.*** The product of hydrolysis is much different from the minerals being weathered. The weathering of feldspar (an aluminum-silicate mineral) is a common example of hydrolysis and is shown by the following reaction:

$$\underset{\text{feldspar}}{2KAlSi_3O_8} + \underset{\text{hydrogen ion from carbonic acid}}{2H^+} + \underset{\text{water}}{9H_2O} \rightarrow$$

$$\underset{\text{clay mineral (kaolinite)}}{H_4Al_2Si_2O_9} + \underset{\text{silicic acid}}{4H_4SiO_4} + \underset{\text{potassium ion}}{2K^+}$$

Feldspar eventually breaks down completely in the presence of carbonic acid, and the aluminum and some of the silicon combine with the hydrogen ions to form a clay mineral. The silicic acid and potassium ions remain in solution, and may be carried away, leaving only the clay mineral behind. Some of the potassium, however, might be used by plants or become part of other clay minerals.

Clay minerals—of which there are a great variety—are so small that they can be identified only by X-ray. The kind of clay that forms is determined by the environment. Kaolinite, the clay mineral formed by the reaction shown above, commonly forms in areas with humid climates, where large amounts of water move through the soil, leaching from it the ions released by weathering (Fig. 9-14). In dry regions, where little leaching occurs, the most common clay mineral is montmorillonite. Clays vary in their physical and chemical properties. Some are stabilized by heat and are used to make pottery and ceramics, some are especially useful as muds used in drilling wells, and others enhance the fertility of soils. Indeed, some are sold for facial mudpacks.

Granite weathering—an example of the hazy boundary between mechanical and chemical weathering

As mentioned earlier, the boundary between mechanical and chemical weathering is not always clear. The weathering of granite illustrates the point. In many deep road-cuts, small fragments of loose and broken granite (mainly the size of the individual mineral grains) can be seen. Because evidence of chemical weathering—notably the presence of clay minerals and iron oxide colors—may be missing, such weathering has been commonly ascribed to a mechanical process. Recently, however, care-

Fig. 9-13 In an experiment to demonstrate chemical weathering, acid water was dripped on these rocks for six months. Gypsum (right) weathers the most rapidly, and limestone (center) shows substantial weathering. In contrast, granite (left) is so resistant to weathering that its effects are barely discernible. *Janet Robertson*

Fig. 9-14 A. Kaolinite crystals, photographed under the scanning electron microscope, were formed from the alteration of feldspars at depth in ancient alluvial fan deposits, Front Range, Colorado. *T.R. Walker*

ful X-ray studies of the minerals in granite indicate that the most common alteration is a subtle change of the biotites to a very similar mineral of greater volume. Apparently the little water that does penetrate granite first reacts with biotite. The accompanying expansion of all the biotites produces enough internal pressure to break up the rock. Thus, volume expansion as a result of a subtle form of chemical weathering mechanically shatters the rock.

Research in the western United States suggests that about 100,000 years of weathering are needed to bring about such a change—an example of the snail's pace at which weathering takes place. Reactions that proceed so slowly are extremely difficult to duplicate in the laboratory. Some processes can be speeded up (as in Griggs's experiments), but it is difficult to know how closely they approximate what actually goes on in nature.

Relative weathering rates of minerals and rocks

Not all minerals and rocks weather at the same rate, as shown by the study of tombstones. Examples of ***differential weathering*** are numerous (Fig. 9-15). Rocks that weather more rapidly erode more rapidly, so it is not uncommon to find that the less resistant rocks form slopes and low areas, whereas the more resistant rocks form prominent ledges or cliffs (Fig. 9-16). Many of the common rock-forming minerals are relatively susceptible to chemical weathering. Calcite, for example, weathers the most rapidly because it dissolves so readily in water. The atoms of silicate and aluminosilicate minerals, are more tightly bonded, however, and thus are better able to resist weathering; yet they weather at different rates (Fig. 9-17). Quartz and feldspar appear to be the most resistant to weathering, and olivine and calcium-plagioclase the least resistant. The resistance of quartz helps to explain the predominance of quartz grains in many sandstones. With repeated cycles of weathering and transportation, the durable quartz grains persist, whereas the less durable grains weather away, often forming clay minerals and ions.

Rocks weather chemically according to the rate at which their constituent minerals weather. Hence granite, because it contains an

Fig. 9-14 B. Extreme alteration of a pyroxene grain, originally rectangular in shape, has produced these delicate needle-like structures. The alteration took place long after sediments were deposited in basins in southern New Mexico. *T.R. Walker*

assortment of resistant minerals, weathers chemically much more slowly than gabbro (Fig. 9-17). Grain size also is important, since fine-grained rocks weather more slowly than coarse-grained rocks of the same mineral composition.

SOIL

The products of weathering can be rearranged into layered materials that are quite different physically, chemically, and biologically from the parent rock. That material is soil. Because soil is so fundamental to life, it has been studied intensively for more than a century. Leaders in the study have been the Russians V. V. Dokuchaev (1846–1903) and K. D. Glinka (1867–1927), whereas in the United States E. W. Hilgard (1833–1916), C. F. Marbut (1863–1935), and Hans Jenny have carried out the basic research. In the United States, the emphasis of study is on soil genesis, classification, and the relationship between soils and environment, as well as crop production. Most recently, there has been a trend to use soil data to help solve environmental problems.

Many soils consist of three basic layers, or ***horizons,*** the sum total of which comprises the ***soil profile*** (Fig. 9-18). At the surface is the ***A horizon,*** a dark-colored layer rich in decomposed organic matter, or humus, derived from the decomposition of surface vegetative debris and roots. Beneath the A horizon is the ***B horizon,*** which characteristically displays the greatest amount of weathering in the profile. The B horizon is enriched in both clay and iron oxides derived from the weathering of minerals contained within the layer itself and from the A horizon above. Because of the iron-oxide enrichment, B horizons form the brownest or reddest layer within each profile. Beneath the B horizon is the ***C horizon;*** beneath that the parent material, which is thought to have been present in the position of the soil profile before soil formation. The C horizon is only slightly altered parent material.

Factors of soil formation

From the early work of Dokuchaev and Hilgard, it became apparent that soils differ from place to place because environmental condi-

Fig. 9-15 A. Arches National Park, Utah. Different rates of weathering have sculptured these narrow bedrock ridges into forms that rival works of art. *Jesse Kumin © 1976*

tions vary from place to place. But it was the work of Hans Jenny in the 1930s and 1940s that made possible a clear understanding of the effect of the environment on soils. Five factors are generally considered in describing the soil environment: (1) climate, (2) vegetation, (3) time, (4) parent material, and (5) topographic position. Because the influence of topography is mainly local, we will not discuss its effect on soil formation.

Climate and vegetation Climate has the greatest impact on soil properties from place to place, and, because the effects of vegetation are not easily separated from those of climate, both factors will be covered here.

Pedalfers are the common soils in humid temperate regions, under either grass or forest vegetation. The word is derived from the Greek *pedon* for "ground" and the symbols Al and Fe for the aluminum and iron such soils contain. Pedalfers consist of relatively thick A, B, and C horizons with a fairly high content of organic matter in the A horizon (Figs. 9-18 and 9-19A). Mainly in the cooler climates toward the northern limit of trees, but also in some warm humid climates, a special kind of pedalfer, known as a ***podzol,*** forms (Figs. 9-18 and 9-19B). Podzols are characterized by a whitish

Fig. 9-15 B. Indians took advantage of differential weathering in sandstone cliffs by building their dwellings in places that offered protection from the elements and from hostile neighbors. These structures in Mesa Verde National Park, Colorado, were inhabited from about A.D. 1200 to 1300 and then abandoned for unknown reasons. *Jesse Kumin © 1976*

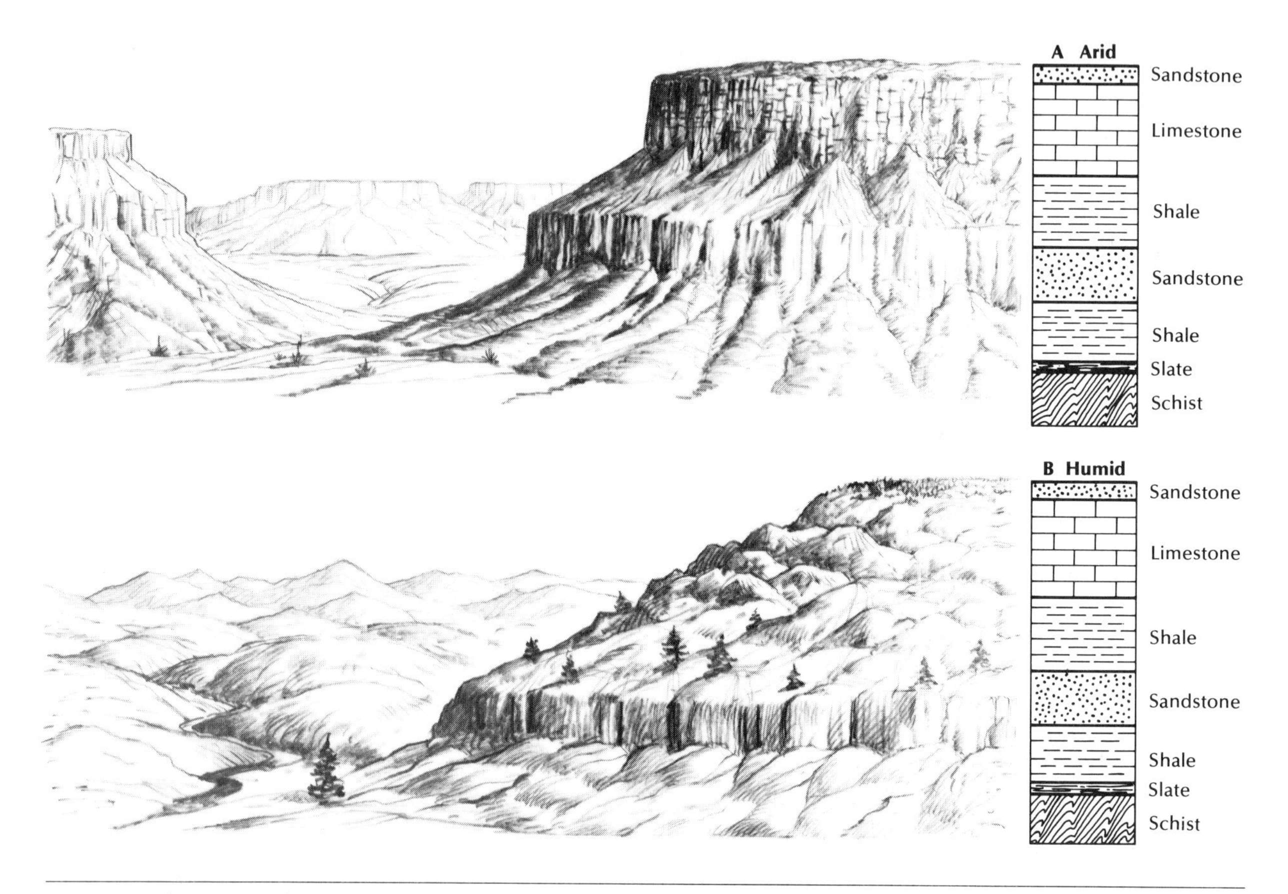

Fig. 9-16 Rocks vary in their resistance to weathering and subsequent erosion. Whether a rock forms a steep cliff or a gentle slope depends partly upon climate, however. In an arid climate (**A**), limestone and sandstone are cliff-formers and shale is a slope-former, often covered by talus. In a humid climate (**B**), sandstone also is a cliff-former, but limestone weathers by solution to form irregular slopes. Again, shale is a slope-former, often covered by a thick soil.

Fig. 9-17 Relative rates of weathering of various common minerals and rocks.

	Relative rate of mineral weathering		Relative rate of rock weathering	
			Coarse Grained	Fine Grained
	Quartz			
Most resistant	Feldspar	Na-plagioclase	Granite	Rhyolite
	Biotite			
			Diorite	Andesite
	Hornblende			
	Pyroxene	Ca-plagioclase	Gabbro	Basalt
Least resistant	Olivine			

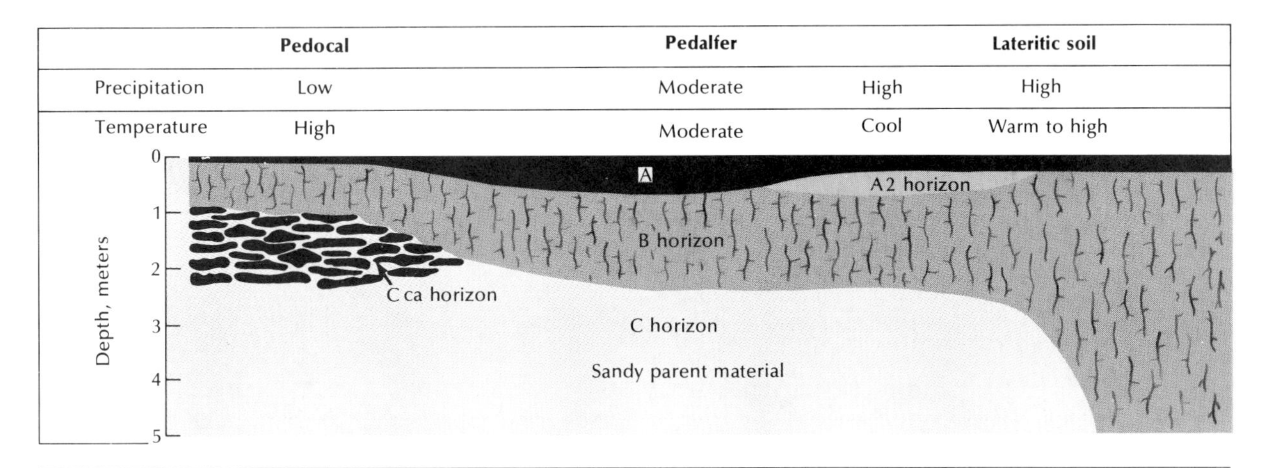

Fig. 9-18 Transect from a dry to a humid environment, showing soil profiles that are characteristic of each environment on fairly old landscapes. The desert profile could be found in the valleys of the western United States, the humid region profiles in the Great Lakes region and the Northeast, and the warm, humid region profile in the Southeast.

layer, called the ***A2 horizon,*** that lies between the A and B horizons, and from which most of the iron oxides have been removed by downward-percolating waters. The B horizon of podzols is characterized by iron enrichment rather than clay enrichment.

On old landscapes in humid warm climates, especially tropical climates, we see the end product of extreme weathering—the ***lateritic soils*** (Figs. 9-18 and 9-20). For the term laterite we are indebted to Buchanan Hamilton, an observant Scotsman who, while traveling in India in 1807, was greatly impressed by the ease with which the red-brown tropical clay could be transformed into a building material. Hindu laborers simply excavated the clay and shaped it into bricks that needed only case-hardening in the sun before they could be used. The bricks that Hamilton described were much like the sun-dried blocks, or ***adobe,*** of the arid Southwest, even though very different in origin.

Because the tropical clay could be used readily as a construction material, Hamilton called it ***laterite*** (from the Latin word *latere,* or brick). Its origin was a source of wonderment to him. As a construction material, laterite has served to build enduring monuments; among others, much of the long-forgotten city of Angkor Wat in Kampuchea (Cambodia) is built of laterite (Fig. 9-21). Despite a humid tropical climate, the buildings of that city not damaged by warfare in recent years are well preserved; this alone is an indication that laterite is virtually insoluble.

There are few soils that harden into actual laterites. But lateritic soils, deep, highly weathered, and reddish, are widespread in tropical climates. A lateritic soil can be regarded as a "soil skeleton," since such soluble elements as calcium, sodium, and potassium have been leached out, and even such a relatively insoluble substance as silica (SiO_2) has been removed. What remain are mainly iron oxide (Fe_2O_3), derived from the weathering of iron-bearing minerals, and quartz, if the parent rock contained any. The crystallized iron oxides form the bricks, or true laterite. Should the rocks from which lateritic soils are derived have a high content of aluminum, then bauxite ($Al_2O_3 \cdot 2H_2O$), a chief ore of aluminum, can form.

In some parts of the tropics, soils may be high enough in iron content to be mined as ore. Cuba's iron mines and those on the Suri-

A horizon Dark brown to black due to high amount of organic matter, and well aggregated so that rainfall readily enters soil with little surface runoff.

B horizon Brown to reddish brown due to iron oxides, and high clay content as shown by vertical shrinkage cracks that develop on drying. Grades downward to slightly altered C horizon material.

Fig. 9-19 Two different kinds of pedalfers. **A.** Grassland soil formed from loess in Iowa. The section shown is about 1 m deep. *Roy W. Simonson.* **B.** (opposite) Forested soil (podzol) formed from sandstone in New Zealand. *P.T. Tonkin*

gao Peninsula on Mindanao are examples. Bauxite has a very similar origin, except that it is formed from the weathering of rocks richer in aluminum than in iron. The bauxite ores of Little Rock, Arkansas were formed under climatic conditions that probably were very much like those of the tropical savanna today. However, most of the aluminum ore now processed in North America comes from the high-alumina clays of Surinam, Jamaica, and other South American and Caribbean lands.

A much different soil forms in semi-arid to arid regions. It is named a ***pedocal*** because an accumulation of calcium carbonate occurs at some depth (Figs. 9-18 and 9-22). Since rainfall is slight and vegetation scanty, the A horizon is thin and not too rich in organic matter, and the common clay mineral in the B horizon is montmorillonite. Beneath the B horizon, at about the depth to which water from the annual rainfall penetrates, is a accumulation of white calcium carbonate ($CaCO_3$) known as a ***Cca horizon.*** On some old landscapes, there is so much calcium carbonate that it forms a highly indurated layer known as the ***K horizon,*** or ***caliche.*** The carbonate (CO_3) in calcium carbonate is derived from the reaction of carbon dioxide (CO_2) and water, whereas the calcium ions are derived from the weathering of calcium-bearing minerals or from atmospheric or groundwater sources.

The three main types of soil—pedalfers, la-

teritic soils, and pedocals—are all found in the United States. The 100th Meridian is the approximate boundary between pedalfers to the east and pedocals to the west. In the West, however, pedocals occur only in dry basins; the adjacent mountains, in contrast, receive enough rainfall for pedalfers to occur. In the East, podzols occur in the Northeast and in the Great Lakes area, and lateritic soils are widespread in the Southeast, south of the glacial boundary.

A new soil classification scheme, called **Soil Taxonomy,** has been introduced by the U.S. Soil Conservation Service and is being used increasingly by professionals in the field. Soil Taxonomy, a highly quantitative and rather complex scheme, recognizes ten major subdivisions, called orders, which are described in Table 9-1. A simplified version of the older classification is used in this book, however.

Time All the key properties of soil take considerable time to form (Fig. 9-23). The A horizon forms most rapidly—several centuries in humid environments to several thousand years in arctic and alpine environments. B horizons take much longer to form because minerals weather so slowly. Discoloration due to the weathering of iron-bearing minerals shows up in about 1000 years, and the strongest red colors may require 100,000 or more years to de-

Fig. 9-20 Road-cut about 7 m (23 ft) deep near Rio de Janeiro, Brazil, exposes highly weathered lateritic soil formed from gneiss. The rock is so thoroughly decomposed that it can be readily cut with a knife or a shovel, yet quartz veins (diagonal white lines) are essentially unaltered. *Roy W. Simonson*

velop. Clay-rich B horizons, if formed from sandy parent materials, require at least 10,000 years for initial detection and more than 100,000 years for maximum development. In arid regions, the strongly cemented K horizons may take 100,000 to 500,000 years to form. Deep lateritic soils take the longest to form, and our best guess of how long is in the order of one million years or more. No wonder, then, that we should protect our soils from erosion! Once they are lost they are probably lost forever.

Parent material Parent material influences both the rate of weathering and the rate at which clay is produced. Granite, for example, produces a fairly deep sandy soil because, although biotite weathering breaks the rock down to sand-size particles, the minerals undergo chemical alteration very slowly to clay

Fig. 9-21 The building blocks of laterite that compose this temple at Angkor Wat, Kampuchea, are highly resistant to weathering because they are composed chiefly of residual iron oxide, all the other original materials having been removed by chemical weathering. In contrast, the sandstone columns and statues exhibit considerable weathering. *Leonard Palmer*

minerals. In contrast, soil on basalt in the same area would be fairly thin because fine-grained rocks weather fairly slowly. What soil there is, however, could be richer in clays because the minerals in basalt are less resistant to chemical weathering.

Benefits of soil research

We can benefit greatly from the study and mapping of soils. Some of the benefits may seem academic to us, but others are of immediate concern.

The study of soil development on various de-

Fig. 9-22 Desert soil (pedocal) formed from alluvium in New Mexico. *L.H. Gile*

posits has aided earth scientists in estimating the age of those deposits. In areas of glacial deposition, for example, soil studies not only helped to develop the idea of multiple glaciations in the United States, but also helped to determine how often ice ages occur. Armed with such knowledge, we might attempt to predict future climatic changes.

We can also estimate the frequency of geological hazards, such as landslides, rockfalls, and major flooding, from the study of soils. For example, areas in which such events took place in the recent past would have poorly developed soils, whereas better developed soils would be associated with areas in which hazards occurred long ago. Soil dating can also provide key data for building sites; and sites for nuclear power plants are probably the most crucial. The installations must be located in areas free of the ground breakage (faulting) associated with earthquakes, and soil studies can aid in detecting recently active faults.

Soils are also valuable in determining the nature of past climates (Fig. 9-24A). Buried pedocals in a humid region tell us that at one time there must have been a drier climate there. In the more arid parts of Australia, pedocals cover young deposits, but on ancient landscapes lateritic soils are found. The change seems remarkable—from tropical landscape to parched desert in a few million years, a relatively short time in the earth's history. Why did it occur? We are still developing theories for that and other climatic reversals in various parts of the world. The solution to the problem may help us to predict the march of climatic change.

Soils vary in productivity and manageability for crop growth. Maps produced by the U.S. Department of Agriculture help us to locate the best—and worst—soils for agricultural use. People involved in land-use planning should consult such maps when laying out subdivisions, to keep the most productive soils for agriculture. Less productive soils should be utilized to the maximum benefit for all. Once a good soil is paved over or covered by a house or factory, it is a resource lost. And when crops are planted on less productive land, it only serves to drive the cost of foodstuffs higher and higher. Citizens and politicians should take heed of those arguments.

Although lateritic soils are no longer considered a major problem in tropical agriculture (they make up 7 per cent of the tropical land area), we still need a better understanding of how they behave under cropping. Such soils are extensively weathered and thus are low in the nutrients necessary to plant growth. Cropping only reduces the nutrients further, and fertilizer is required to restore them. Some lateritic soils harden if dried out (Fig. 9-24B); once hardened, they can no longer be tilled. Intensive research, and fitting crops to soils rather than soils to crops, is needed to help solve the problems of food production in tropical areas.

Some soils (Vertisols in Table 9-1) swell and shrink as moisture varies seasonally. The reason for such behavior is the high content of clay in the soil; the clay minerals are mostly montmorillonite, which is prone to swelling on wetting and shrinking on drying. Recognition of soils that shrink and swell is of vital importance to the construction industry, for it accounts for more than $2 billion of damage in the United States alone each year.

Fig. 9-23 Time necessary in the western United States to develop diagnostic soil horizons from a sandy parent material.

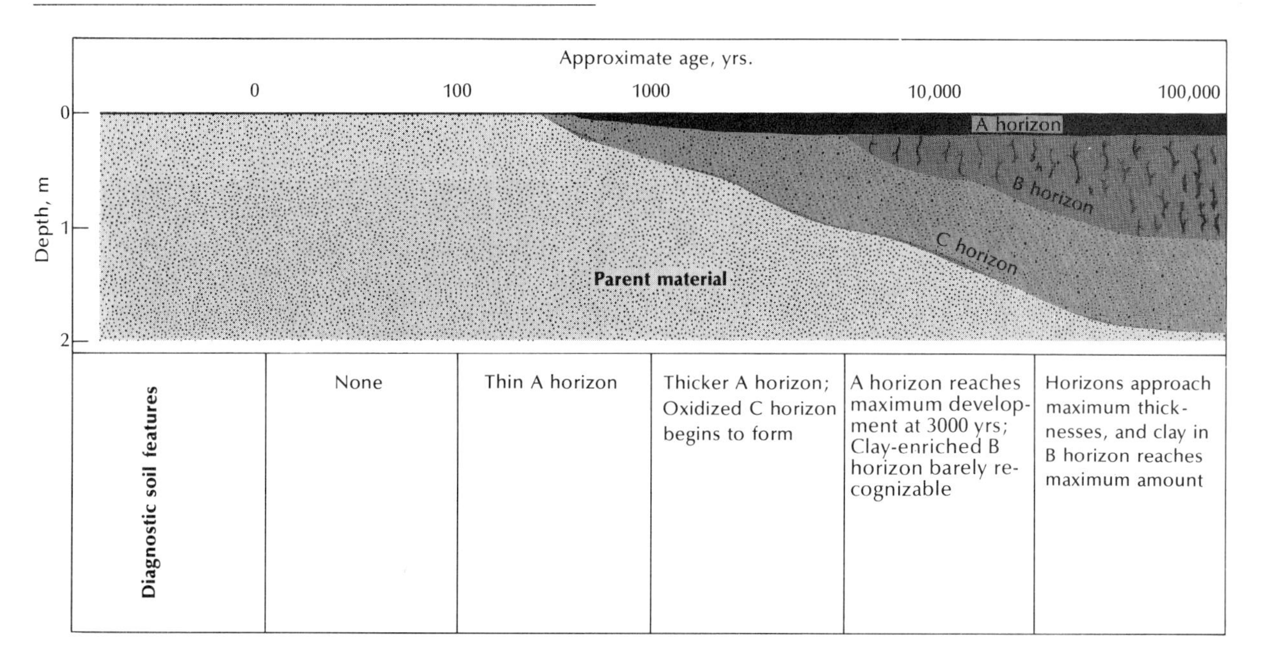

Fig. 9-24 A. The study of buried soils (S) can help to determine the nature of past landscapes and environments. *Lubbock Lake Project, Museum of Texas Tech University*

An adequate soil is necessary to protect land from accelerated erosion, as we learned so dramatically in the Dust Bowl in the 1930s (Fig. 9-25). The organic matter, clay particles, and ions in soil have the ability to bind, or to cause the soil particles to aggregate, to form small clumps, or blocks, of various sizes (see Fig. 9-19A). Water will penetrate such a sponge-like structure more readily than it will run off the surface. And the more water that sinks into the ground, the less the runoff and hence the less the surface erosion. The properties of the A horizon are especially critical to the ability of the soil to take up water. If that horizon is removed or compacted, runoff and erosion may be accelerated (Fig. 9-26). There has been considerable erosion of soil throughout the United States, yet most people do not seem to be aware that an important natural resource is being lost (Fig. 9-27).

Table 9-1. Orders of the new soil classification scheme used in the United States, compared to the older, less detailed scheme

Order	*Generalized properties*	*Old classification*
Entisol	Minimal development, an A horizon may be present	
Inceptisol	Weak development, with A horizon, a B horizon that lacks clay enrichment, with or without a Cca horizon	Pedalfer and Pedocal
Mollisol	Thick, dark A horizon, high in organic matter, a B horizon that may or may not be clay enriched, with or without a Cca horizon	Pedalfer and Pedocal
Alfisol	Relatively thin A horizon overlying a clay-enriched B horizon, with an A2 horizon separating the A and B layers in places	Pedalfer
Spodosol	Highly organic surface horizon above an A2 horizon, which in turn rests on an iron-enriched B horizon	Podzol
Ultisol	A horizon over highly weathered B horizon	Pedalfer and Lateritic
Oxisol	A horizon over an extremely weathered B horizon	Lateritic and Laterite
Aridisol	Thin A horizon above a relatively thin B horizon, some of which are clay enriched, with a Cca or K horizon at depth	Pedocal
Histosol	Peaty soil	
Vertisol	Very high content of clays; shrinks and swells with seasonal moisture variation	

Fig. 9-24 B. Lateritic soil hardened to a rock-like substance, eastern Australia. The soil formed several million years ago in a more humid climate, and probably hardened in later, drier climates. *Peter W. Birkeland*

Fig. 9-25 A. (right) Dust storm in Prowers County, Colorado, photographed in the 1930s. **B.** (below) Abandoned farmland near Stillwater, Oklahoma; note the accelerated erosion. *Soil Conservation Service*

Fig. 9-26 Gully developed on a slope in the Coast Range north of San Francisco. Excessive grazing caused increased runoff and gullying. The sharp topographic angle between the gully wall and the original land surface means that gullying was still active when the photograph was taken in 1906. If erosional deepening subsides, the gully walls will slope more gently. *G.K. Gilbert, USGS*

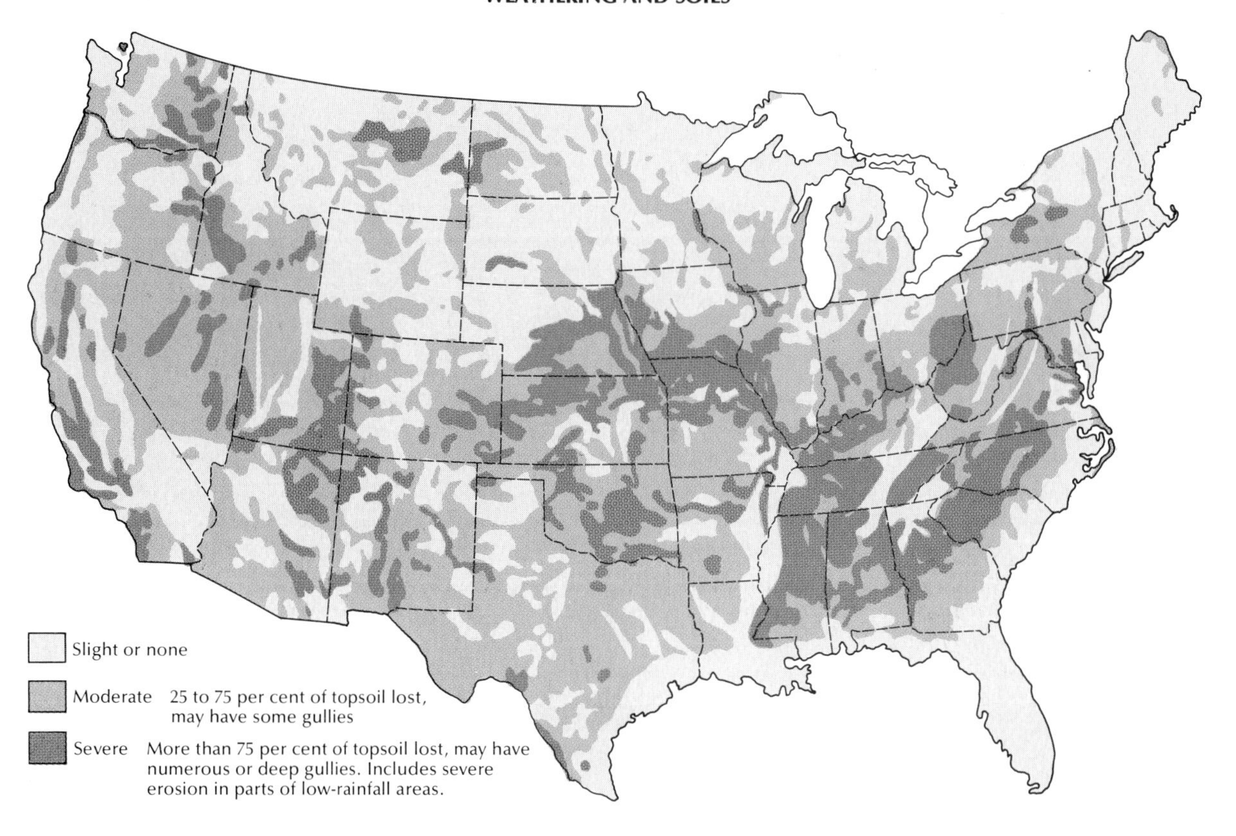

Fig. 9-27 Various classes of soil erosion in the United States.

SUMMARY

1. No rocks or sediments are stable at the earth's surface. With time they will alter or weather to products more stable in the surface environment.
2. Two main kinds of weathering are recognized. *Mechanical weathering* is the disintegration of particles to smaller size without an appreciable change in either the chemistry or mineralogy of the original material. In contrast, through *chemical weathering,* the chemistry and mineralogy of the original material are changed. Clay minerals found in soils and sediments are the result of chemical weathering.
3. Soils are layered bodies consisting of an *A horizon,* which is rich in organic matter, overlying a clay-enriched *B horizon* that rests on the only slightly altered *C horizon.*
4. Soils vary from site to site. The main factors controlling the variation are parent material, climate and vegetation, topographic position, and time. Most well-developed soil profiles have taken thousands or tens of thousands of years to form, and the very deep ones of the warm humid areas may have required a million years.
5. Soils can provide much information on the geological history of the earth, and their properties and distribution should be taken into account in all land-use decisions.

QUESTIONS

1. What properties of a rock are important in determining its resistance (1) to mechanical weathering and (2) to chemical weathering?
2. How would you go about investigating the relative rate of physical weathering between frost action and a combination of salt crystallizations and hydration?
3. Parts of the arid portions of Australia have old lateritic soils. What are several working hypotheses that might account for this?
4. How would you use soil information to date a surficial deposit, such as a beach, glacial, or river deposit?
5. How would you determine which minerals in the soil weathered to form the clay minerals?

SELECTED REFERENCES

Bartelli, L. J., and others, eds., 1966, Soil surveys and land use planning, Soil Science Society of America and American Society of Agronomy, Madison, Wisconsin.

Birkeland, P. W., 1974, Pedology, weathering, and geomorphological research, Oxford University Press, New York.

Bridges, E. M., 1970, World soils, Cambridge University Press, London.

Buckman, H. O., and Brady, N. C., 1969, The nature and properties of soils, The Macmillan Co., London.

Buol, S. W., Hole, F. D., and McCracken, R. J., 1973, Soil genesis and classification, The Iowa State University Press, Ames.

Carter, V. G., and Dale T., 1974, Topsoil and civilization, University of Oklahoma Press, Norman.

Hunt, C. B., 1972, Geology of soils, W. H. Freeman and Co., San Francisco.

Jenny, Hans, 1941, Factors of soil formation, McGraw-Hill Book Co., New York.

Jenny, Hans, 1981, The soil resource, Springer-Verlag, New York.

Sanchez, P. A., and Buol, S. W., 1975, Soils of the tropics and the world food crisis, Science, vol. 188, pp. 598–603.

Winkler, E. M., 1973, Stone: Properties, durability in man's environment, Springer-Verlag, New York.

10

MASS MOVEMENTS AND RELATED GEOLOGICAL HAZARDS

The movement of large masses of rock and debris downslope is an event known to many of us, especially if life and property are destroyed.

One such event occurred on the night of October 9, 1963, in Italy, when a torrent of water, mud, and rocks plunged down a narrow gorge, shot out across the wide bed of the Piave River and up the mountain slope on the opposite side, completely demolishing the town of Longarone and killing 2600 inhabitants in it and in adjoining towns (Fig. 10-2). It has been called history's greatest dam disaster, but when it was over, the Vaiont Dam in the narrow gorge was still intact. What could have caused the water in that reservoir, which was not even one-half full, to rise up over the dam and proceed on its destructive course? One clue is that one shoulder of the dam was supported by Monte Toc, nicknamed *la montagna che cammina*—"the mountain that walks"—by the local inhabitants. Despite assurances by engineers regarding the safety of the dam and the expensive work done to stabilize its slopes, Monte Toc did not walk that night in October, it galloped. About 250 million m^3 (330 million yd^3) of mountainside slid into the lake behind the dam. One wave rode 260 m (850 ft) up the valley wall opposite the slide; another rose 100 m (330 ft) above the dam and dropped into the gorge below. There, constricted by the narrowness of the gorge, the water, carrying tons of mud and rocks, raced on its destructive path. It was all over in seven minutes. The dam was strong enough, for it withstood the onslaught of tremendous forces, estimated at 4 million metric tons, and is still standing today.

California has long been known as earthquake country, but it also is renowned for its landslides (Fig. 10-3). A major one occurred in 1956 and for the following three years in the Palos Verdes Hills near Los Angeles, where a development had been built directly over an

Fig. 10-1 Rockfall on the headwall of a cirque in western Colorado. The rock is well jointed and the frequent rockfalls in the summer may be due to loosening of the blocks by a freeze-thaw process. *Peter W. Birkeland*

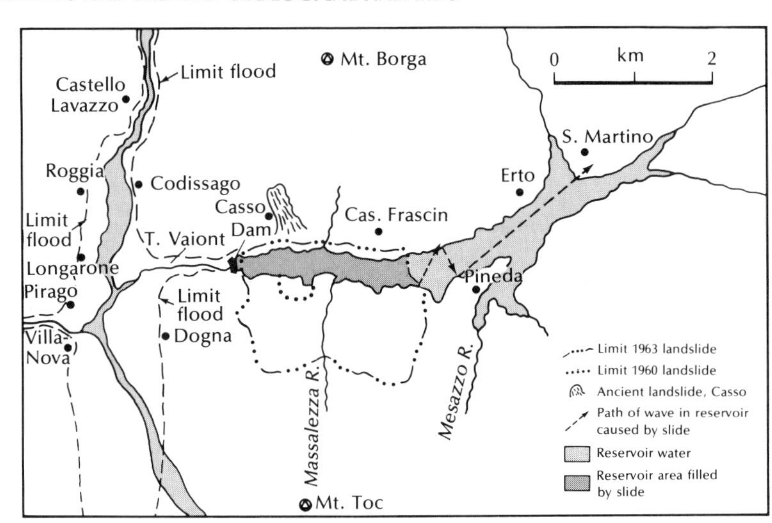

Fig. 10-2 Map of Vaiont Reservoir area, Italy, showing limits of the landslide, the area of the reservoir filled by the slide, and the extent of downstream flooding.

old landslide that started to move again. Many homes were totally demolished and the value of property destroyed ran over $10 million. Eventually damage suits of several million dollars were collected against the county. The result of that and similar events was that planners and developers began to consider more seriously local geological settings before attempting to build houses or roads on certain kinds of terrain. The cost of such tragedies should convince the skeptics; for example, in one two-year period, landslides accounted for $21 million worth of damage in eight San Francisco Bay counties. Human activities have been contributory to 80 per cent of the landslides in Contra Costa County, California, and to 90 per cent of the landslides in Allegheny County, Pennsylvania. Much of the work of geologists, especially those working for governmental agencies, involves the recognition of areas of potential landslides as well as those of other geological hazards.

Material at the earth's surface, be it weathered or fresh rock, can fail under special circumstances and move downslope under the pull of gravity. Such movements can take place very slowly and with little apparent evidence or they can happen so quickly and involve so much material that they stagger the imagination. In this chapter we will discuss the major kinds of gravity movements, assess their causes, and point out some remedies.

Mass movement is responsible for the downslope transfer of material to rivers, which then act as conveyor belts to carry it away. The way in which the walls of the Grand Canyon flare outward from the Colorado River is the result largely of the gravity transfer of rock fragments and mineral grains downslope to the river and its intricate network of tributaries, which then carry the material out of the Colorado Plateau.

How much or how little material will be shifted downhill by gravity and how rapidly or how slowly it will move are a consequence of many factors. These include (1) climate, especially rainfall per storm and per season, (2) amount of water in the material, from rainfall and groundwater, (3) kind of bedrock and the presence or absence of bedding planes, joints, or faults, (4) vegetation, for roots can strengthen slopes, (5) amount and kind of weathering, which can weaken rock and produce clay, (6) the steepness of the slope, (7) local relief, and (8) the presence or absence of earthquakes, for quakes can trigger devastating mass movements.

CLASSIFICATION OF MASS MOVEMENTS

Mass movements are difficult to classify, and

probably the term ***landslide*** is sufficiently general for many such movements that occur along a specific surface or combination of surfaces. It is common practice today to subdivide the kinds of mass movements according to (1) type of movement, (2) type of material involved, and (3) rate of movement (Fig. 10-4). However, the transition from one type to another is not always clear cut.

Rockfalls

When rock material drops at nearly the velocity of free fall, the event is called a ***rockfall*** (Figs. 10-1 and 10-5). It may range from the plummeting of an individual block to an avalanche of hundreds of thousands of tons. After such a fall, individual blocks commonly come to rest in a loose pile of angular rocks, or a ***talus*** (a term borrowed from medieval times for the slope at the base of a fortification wall), at the base of a cliff (Fig. 10-6A). Not all talus has such a catastrophic origin, however. In many places, talus cones are still forming. In other places the cones are being eroded (Fig. 10-6B) and perhaps cyclic episodes of talus deposition and erosion could be a response to climatic change. Should large blocks of rock drop into a standing body of water, such as a lake or fiord, highly destructive waves may be set in motion. Such waves are particularly feared in Norway, where small deltas often provide the only flat land at sea level. Should a rockfall-induced wave burst through a village, destruction is likely to be as complete as it is sudden, since the waves often range from 6 to 9 m (20 to 30 ft) in height.

A spectacular wave was set off by a rockfall

Fig. 10-3 In Laguna Beach, California, the Bluebird Canyon landslide of October 1978 destroyed 23 houses. Its potential for destruction had been predicted nearly 10 years earlier. Movement occurred over an area of 1.5 hectares (3.7 acres), along seams of clay-rich siltstone and sandstone, as a result of abnormally high rainfall. The slide, which may have been active in prehistoric times, has now been stabilized. *James Krohn, Slosso & Associates*

TYPE AND RATE OF MOVEMENT | BEDROCK | UNCONSOLIDATED SEDIMENTS AND SOIL

Falls

Falls: Mass travels most of the distance through the air and on impact leaps, bounds, and rolls. Rate of movement is extremely rapid

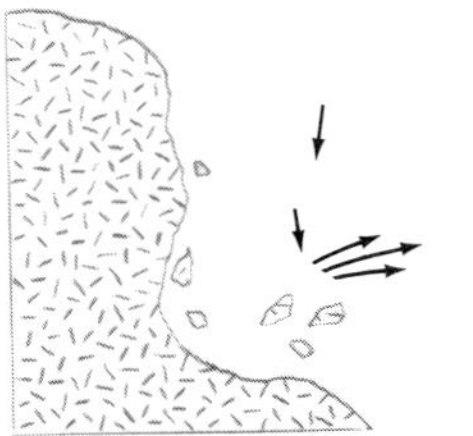

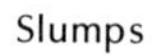

Similar to rockfall, but involves unconsolidated sediments and soil

Slumps

Slides: May include movement on one or several surfaces or narrow zones. Slump movements are along curvilinear surfaces, whereas glide movements are along more or less planar or undulatory surfaces. Rate of movement is extremely slow to moderate

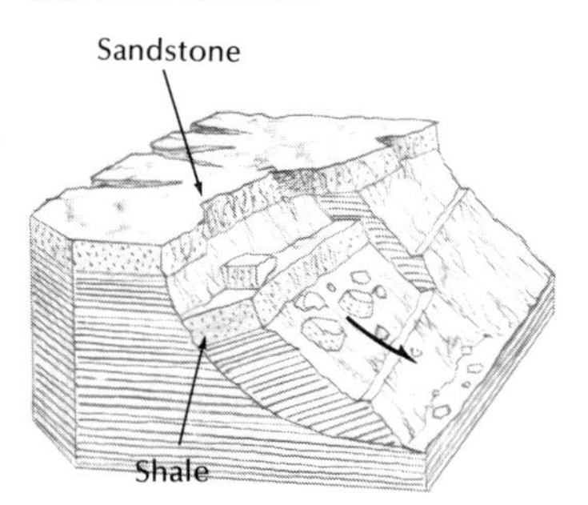

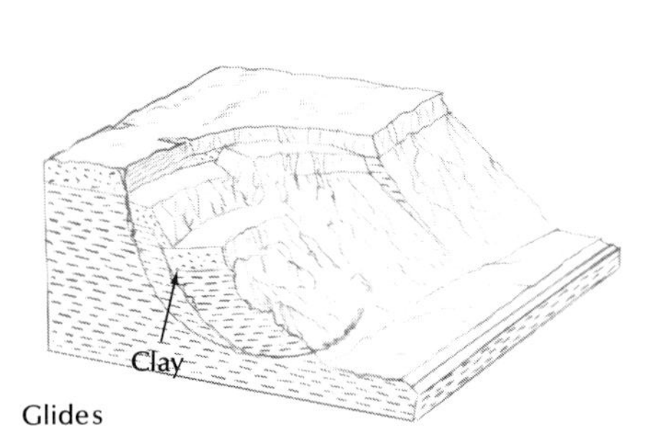

Glides

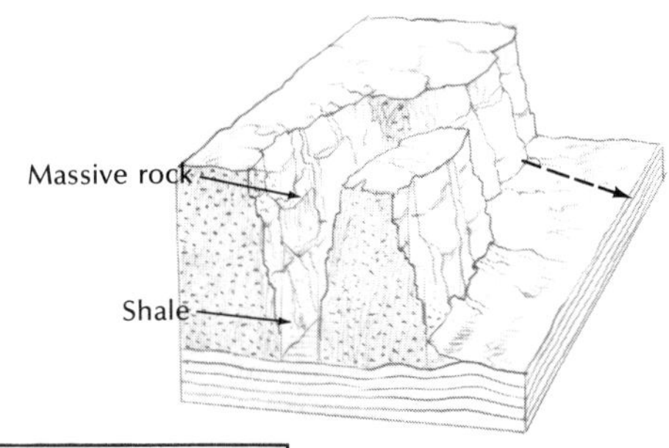

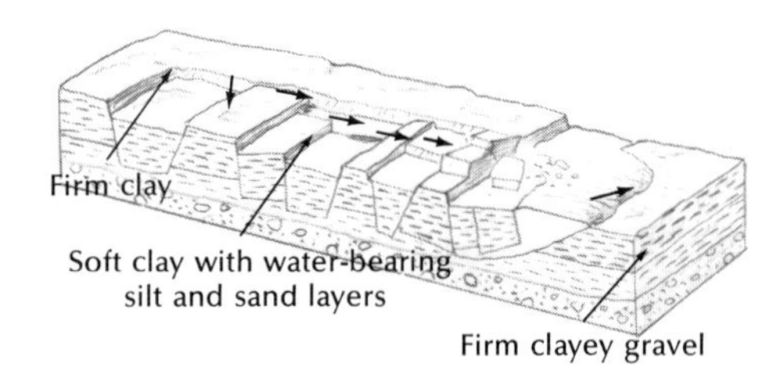

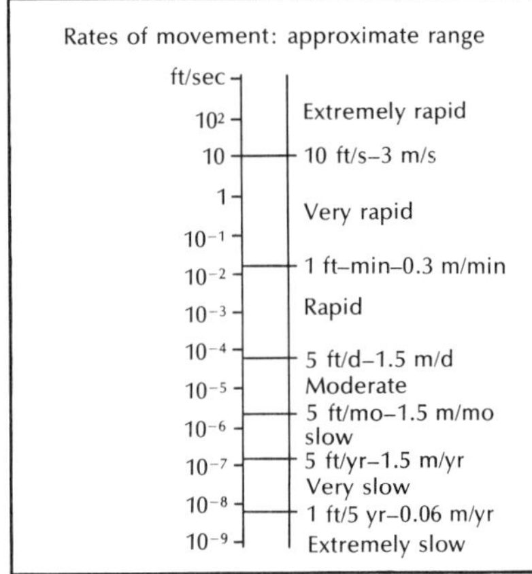

Fig. 10-4 Classification of mass movements.

TYPE AND RATE OF MOVEMENT

	BEDROCK	UNCONSOLIDATED SEDIMENTS AND SOIL

Flows: In bedrock, rate of movement is extremely slow and discontinuous. In soil and unconsolidated sediments, movement resembles that of viscous fluids and can be very rapid (except for creep)

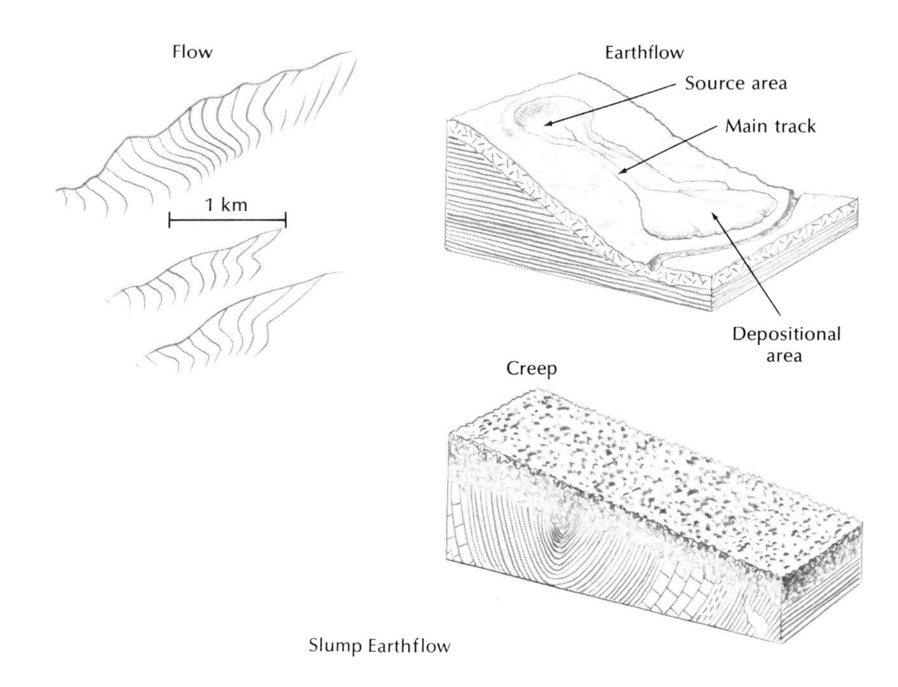

Complex: A combination of two or more of the above types of movements

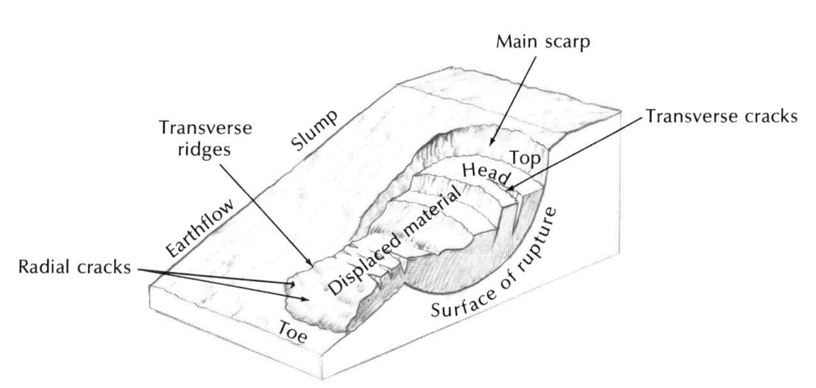

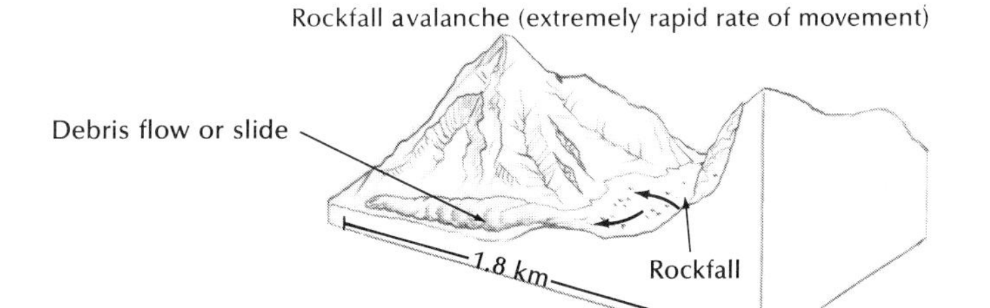

Fig. 10-5 This rockfall from the Flimerstein, Switzerland, occurred on April 10, 1939. It buried not only forests and arable land, but also a building and 11 persons, all in a moment of time. *Swissair-Photo*

at the head of Lituya Bay, Alaska, on July 9, 1958. The rockfall, set in motion by an earthquake, began at an altitude of about 900 m (2952 ft) and involved some 31 million m^3 (41 million yd^3) of rock. Water surged to a maximum height of 530 m (1738 ft) above the level of the bay at its head, and moved to the mouth of the bay at about 160 km/hr (99 mph). Don Miller of the U.S. Geological Survey has described the experience of a couple who were anchored at locality B (Fig. 10-7):

Mr. and Mrs. Swanson on the *Badger* entered Lituya Bay about 9:00 P.M., first going in as far as Cenotaph Island and then returning to Anchorage Cove on the north shore near the entrance, to anchor in about 4 fathoms of water near the *Sunmore*. Mr. Swanson was wakened by violent vibration of the boat, and noted the time on the clock in the pilot house. A little more than a minute after the shaking was first felt, but probably before the end of the earthquake, Swanson looked toward the head of the bay, past the north end of Cenotaph Island and saw what he thought to be the Lituya Glacier, which had "risen in the air and moved forward so it was in sight. . . . It seemed to be solid, but was jumping and shaking. . . . Big cakes of ice were falling off the face of it and down into the water." After a little while "the glacier dropped back out of sight and there was a big wall of water going over the point" (the spur southwest of Gilbert Inlet). Swanson next noticed the wave climb up on the south shore near Mudslide Creek. As the wave passed Cenotaph Island it seemed to be about 50 feet high near the center of the bay and to slope up toward the sides. It passed the island about 2½ minutes after it was first sighted, and reached the *Badger* about 1½ minutes later. No lowering or other disturbance of the water around the boat was noticed before the wave arrived.

The *Badger,* still at anchor, was lifted up by the wave and carried across La Chaussee Spit, riding stern first just below the crest of the wave, like a surfboard. Swanson looked down on the trees growing on the spit, and believes that he was about 2 boat lengths (more than 80 feet) above their tops. The wave crest broke just outside the spit and the boat hit bottom and foundered some distance from the shore. Looking back 3 to 4 minutes after the boat hit bottom Swanson saw water pouring over the spit, carrying logs and other debris. He does not know whether this was a continuation of the wave that carried the boat over the spit or a second wave. Mr. and Mrs. Swanson abandoned their boat in a small skiff, and were picked up by another fishing boat about 2 hours later.*

Slides

A multiplicity of downslope movements is included in the broad term "landslide," and no useful purpose is served here by reviewing the many schemes for classification proposed by geologists, engineers, and other specialists. That so many people are concerned indicates how dangerous landslides are. Many landslides, and the problems they create—and they are considerable as well as expensive—are largely our fault. Without our disturbance of natural slopes, landslides would occur chiefly in remote mountainous terrain or on hill slopes underlain by notably unstable rocks. Today, landslides are a problem of increasing magnitude as urban sprawl continues and the demand for high-capacity expressways mounts. Both trends require larger excavations for building foundations and deeper cuts and higher fills for highways. Oversteepening of slopes is a likely cause of ground movement.To give an idea of the costs involved, almost $50 million each year is spent to repair federal highways damaged by landslides.

A landslide may involve the bedrock alone, or it may be limited to the overlying soil mantle, especially if the latter is deep and water saturated. Usually, however, it involves both soil and rock, and it can remain more or less intact during movement or it can break apart (Fig. 10-4).

Two main categories of slides are recognized: (1) ***glides*** (***translational*** slides) and (2) ***slumps*** (***rotational*** slides). In a glide, slippage is dominantly planar; that is, a large mass of rock may become separated from other rocks and glide outward and downward along the surface of an inclined bedding plane (Fig. 10-8). In contrast, the motion of a slump is rotational—usually along a concave-upward slip-plane—so that the upper part of the landslide is dropped down below the normal ground level and the lower part bulges above it (Fig. 10-9).

*From pp. 58–59, U.S. Geological Survey Professional Paper 354-C.

Slumps have a very characteristic form. Most of them start abruptly, with a crescent-shaped scarp, or cliff, at their head (Fig. 10-10). Lower down there may be a number of lesser scarps, which on a plan view almost always appear concave downslope. Between the indidivual scarps the surface of the slide customarily is tilted or rotated backward against the original slope of the ground. The backward-rotated wedges cause more instability, since they create collecting basins in which small lakes or ponds can form. Another factor that greatly increases instability is the seepage of water along the margins of the slide. The concave-upward slip-plane (surface of rupture) down which the jumbled mass of soil and rocks moves may approximate a cross section of a cylinder with an axis that parallels the contour lines on the ground surface, if the slide is sufficiently broad. Otherwise, the surface of rupture is likely to be spoon shaped (Fig. 10-11). The slide may advance downslope from the point where

Fig. 10-6 A. (opposite) Cones of rocky debris, or talus, at the foot of steep slopes built up by successive rockfalls in the Sierra Nevada, California. *Tom Ross* **B.** (above) These eroded talus cones are all that remain from a once-continuous talus apron south of the Dead Sea, Israel, an area with less than 5 cm (2 in.) annual rainfall. *Peter W. Birkeland*

the surface of rupture intersects the ground as a glacier-like lobe of jumbled debris with a surface that is typically a chaotic pattern of hummocks and undrained depressions. If the slide moves down a forested slope, it often creates a desolate scene of broken trees. The inexorable thrust of the foot of a slump against a building or other structure almost invariably leads to its collapse, and the foot commonly is responsible for blocking canals, highways, and railroads and for engulfing other types of excavations.

Among such slumps, the immense ones that closed the Panama Canal at Culebra Cut shortly after it was opened in 1914 and that kept it closed more or less continuously until 1920 are impressive examples. Of the 128 million m^3 (167 million yd^3) excavated in the Gaillard Cut, landslides made necessary the removal of at least 55.5 million m^3 (73 million yd^3). Great masses of loose, unstable volcanic ash, shale, and sandstone slid on a gently inclined rupture surface toward the canal excavation. One unexpected result was that the bottom of the canal was heaved upward—once as much as 9 m (30 ft)—until what had been the canal bottom appeared as an island in mid-channel.

FLOWS

Flows, which are a very common mass movement phenomenon, differ from slides in that their movement resembles that of a viscous fluid, such as hot tar. It is common for flows in

Fig. 10-7 Lituya Bay, after the giant wave of July 1958 washed over it. The path of forest destruction around the bay is readily seen. (R) marks the rockslide, (D) the maximum altitude (524 m or 1720 ft) of forest destruction, and (B) the location of the *Badger* before it was carried over the spit. *D.J. Miller, USGS*

unconsolidated materials to be broken up internally, whereas many slide blocks retain their internal character, such as bedding, intact. However, the boundary between the two is fuzzy indeed. The rates of movement in flows vary from imperceptible to very rapid.

Creep

This descriptive word is used for the slow, glacier-like movement of shallow soil material downslope. We are likely to be oblivious of such movement, although we may observe building foundations thrown out of line, power and telephone poles tilted, bowed tree trunks (Fig. 10-12), and sidewalks and retaining walls cracked. Cuts made in hillsides generally will reveal active creep, commonly shown by bent-rock layering—look carefully for such layers should you ever think of building a hillside house (Fig. 10-13).

Several mechanisms account for creep. One is subtle expansion and contraction of the ground surface, which results in an imperceptible movement of material downslope (Fig. 10-14). Expansion and contraction can result from a wet-dry cycle in clayey material or from freezing and thawing. Another important mechanism is the saturation of the ground with water, such as occurs during rainstorms. Water adds weight to the slope and causes it to lose some of its cohesiveness. Gravity then takes over.

Fig. 10-8 Glide, or translational, landslide at Point Fermin, California. Although there are minor slump features at the rear of the slide, most of the movement was along sedimentary strata that dip gently seaward. The glide surface is located just below sea level. The maximum average movement was 3 cm (1.2 in.) per week. *John S. Shelton*

Fig. 10-9 Cross section of an ancient slump exposed by a recent landslide on the shore of Franklin D. Roosevelt Lake, Washington. Although little surface form is visible, the surface of rupture (A) and rotation of the beds (B) are clearly visible. *F.O. Jones, USGS*

Fault scarps modified mainly by creep show promise in dating fault movements. In parts of the western United States, recent fault scarps are fairly common, appearing as abrupt rock walls, several meters high, of rock or unconsolidated debris that zig-zag across the landscape (Fig. 10-15). Over time, however, creep smooths out the sharp top edges of the scarp, and the slope angle of the scarp is reduced. The method must be carefully applied, but geologists are using scarp modification to measure the age of the fault, and to determine the frequency of faulting. These data can then be used to help to assign a hazard risk to various construction sites. This is just another example of applying basic research, here the laying back of a slope with time, to solve a practical problem.

Earth flows

Earth flows are more rapid than creep, yet slower than mud flows or some landslides. They usually have a spoon-shaped sliding surface with a crescent-shaped cliff at the upper end and a tongue-shaped bulge at the lower end (Fig. 10-16). Thus they differ from creep deposits, which have little or no form and no sharp boundary with material at depth that is not moving.

Some earthflows move quite rapidly, but only under particular circumstances. They involve special clays, aptly called ***quick clays.*** Quick clay is composed primarily of flakes of clay minerals, and it has a water content that can often exeed 50 per cent by weight. It is commonly part of the debris deposited on the

Fig. 10-10 Crescent-shaped scarp at the head of a slump, Madison County, Montana. At the front of the scarp the original surface rotated backward during movement. Tilted trees also are a sure sign of landsliding. *J.R. Stacy, USGS*

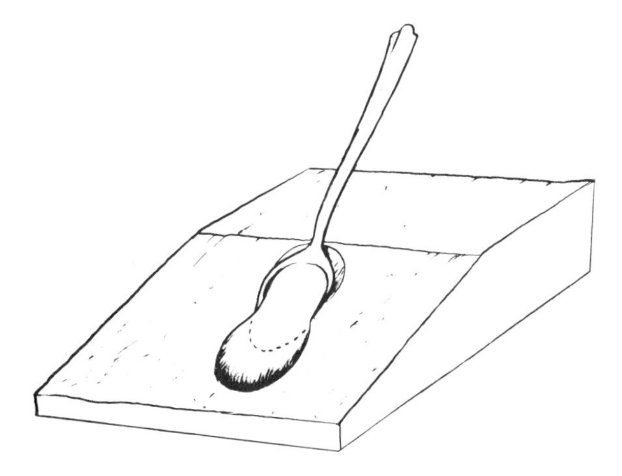

Fig. 10-11 The surface along which slump movement takes place is often spoon shaped.

Fig. 10-12 Bent or bowed tree trunks are often cited as evidence of active creep. The trees are rooted at some depth in material that is not moving or is only slowly moving downslope. Closer to the surface, however, the soil moves more rapidly, pushing against the trunks and bowing them downslope. The trees respond by maintaining vertical growth in their upper parts—hence the curved trunks. *Peter W. Birkeland*

sea floor adjacent to glaciers, so it is not surprising that in Norway, Sweden, and parts of eastern Canada several earthflows that involve quick clays take place every year. Quick clay has a most amazing and treacherous quality: ordinarily it is a solid capable of supporting 1 kg/cm^2 (14 lb/in.^{2}) of surface, but the slightest jarring motion immediately turns it into a flowing liquid. When quick-clay layers were originally deposited, generally in salt water, they contained sodium ions that kept the fine particles together. But when the clays are exposed to weathering, the sodium ions are leached out by rainwater and their cohesive effect is lost. Any sudden shock can produce liquefaction. In one slide in Sweden, the trigger is believed to have been the hammering of a pile driver. The result was that 32.3 million m^3 (42 million yd^3) of soil and gravel slid down into the nearby river, carrying with it 31 houses as well as a paved highway and a railroad it picked up along the way. One person was killed, 50 people were injured, and 300 homes were destroyed in less than three minutes.

The most damaging quick-clay earthflow on record occurred in 1893 in Norway where a 9-km^2 (3.5-mi^2) area was wrecked, killing 120 persons. And extensive damage resulted from the 1964 Alaska earthquake, in which the city of Anchorage and other coastal towns were es-

Fig. 10-13 A common example of creep as shown by bent strata in sedimentary rocks. The effects are more pronounced toward the surface because creep is mainly a near-surface phenomenon. *W.C. Bradley*

Fig. 10-14 The mechanism of creep due to expansion and contraction. Before an expansion-contraction cycle, particles on the original surface rest at (A). On expansion, the surface is elevated, and all particles move upward at right angles to the original surface (A → B). On contraction, the surface more or less assumes its original position as all particles move vertically downward under the influence of gravity (B → C).

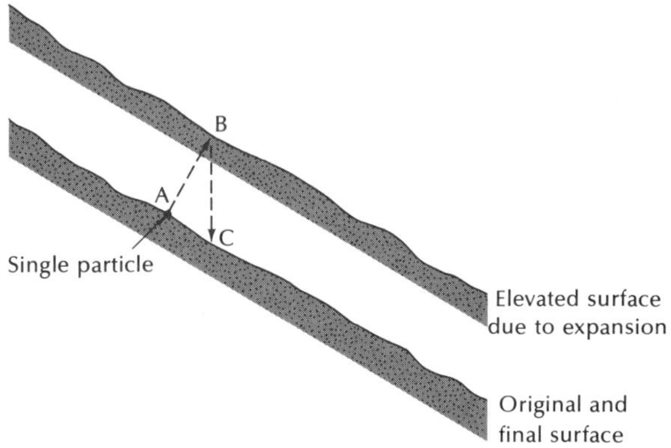

B
A
C
Single particle
Elevated surface due to expansion
Original and final surface

Fig. 10-15 Movement along a fault at the base of the Stillwater Range in Nevada produced this scarp between the tilted cabin and the toppled outhouse. At the time of formation, most fault scarps are steep, and the junction between the upper part of the scarp and the land surface is sharp. Erosional or depositional forces mute the angle and the scarp with time. *Perry Byerly,* from A.D. Howard and I. Remson, 1978, Geology in environmental planning, McGraw-Hill Book Co.

pecially hard-hit. Much of the damage was due to liquefaction of a quick clay (Fig. 10-17). In that case the liquefying shock was the earthquake.

DEBRIS AND MUD FLOWS

Debris flows and ***mud flows*** are an intermediate type of mass movement—with increasing water and velocity and less load, the movement becomes an ordinary stream flow, whereas with decreasing velocity it grades into an earthflow. The distinction between debris flows and mud flows is subtle and based on the size of material moved, with the latter usually restricted to rapid flows with at least 80 per cent sand, silt, and clay. A typical mud flow is a streaming

Fig. 10-16 Slumgullion earthflow in southwestern Colorado. Derived from highly altered volcanic rocks at 3500 m (11,480 ft), the flow has descended to 2500 m (8200 ft), where it has dammed the valley and formed a lake. A younger, active earthflow may be seen advancing over the older, stable one. The rate of movement varies from 6 m (20 ft) per year at midpoint on the flow to less than 1 m (3 ft) per year at its lower end. *C.W. Cross, USGS*

Fig. 10-17 Turnagin Heights, Anchorage, Alaska, after the 1964 earthquake. During the quake, a bed of quick clay close to the surface liquefied and flowed, causing the massive breakup. *U.S. Army*

Fig. 10-18 Mud flow from the distant Owlshead Mountains, southern Death Valley area, California (1970s). *W.C. Bradley*

mass of mud and water moving down the floor of a stream channel, such as a desert arroyo. Such a viscous mass, with a specific gravity much higher than clear water, very often carries along a tumbling mass of boulders and rocks, some of which may be as large as automobiles. The huge rocks are often found on the floor of desert basins far beyond the base of a bordering mountain range. There they remain, long after the enclosing mud that once rafted them out beyond the mountains has been eroded away.

Mud flows are an impressive feature of many of the world's deserts (Fig. 10-18). In arid lands, normally empty stream courses may fill almost at once with a racing torrent of chocolate-colored mud, following a cloudburst. Where arroyas are shallow, the flow may exceed the channel's capacity and spill out over the desert.

Debris and mud flows not only are capable of transporting large natural objects, such as house-size boulders, but may trap and sweep along trucks, buses, or even locomotives. Houses inundated by mud streams have been buried all the way up to the eaves (Fig. 10-19).

Arid or semi-arids lands are by no means the only regions where debris flows and mud flows

Fig. 10-19 Cabin buried to the eaves by a debris flow in 1941 at Wrightwood, California. The flow resulted from rapid snowmelt in the headwaters of the drainage. Cement-like materials surged down the valley some 24 km (15 mi), at velocities that averaged close to 3 m/sec (10 ft/sec) in the more fluid parts of the flow. *R.P. Sharp*

may be seen. They are characteristic of alpine regions, too, and are likely to be exceptionally destructive in areas where a combination of steep slopes, a large volume of water freed by melting snow, and a great mass of loose debris prevails. Geologist R. R. Curry describes such a flow in the Colorado mountains in 1961:

Direct observations of the mudflows were hampered by very intense rain and by the fact that the author was about 900 m from the cirque headwall, at the rain gauge, at the time the flows began. At about 4 P.M. on August 18 a loud roar became clearly audible above the thunder. A series of what appeared to be rockfall avalanches were noted in four different localities around the cirque headwall. These appeared confined to areas previously covered with talus cones and, even though the talus had been soaked by 48 hours of intense rain, large rock-dust or water-vapor clouds accompanied the disturbances.

. . . Individual flows occurred as a series of lobate pulsations which, in the case of the largest unit, lasted for 1 hour. This unit was made up of ten more or less distinct flow pulses, each traveling with a maximum surface velocity of 915–980 m/minute near the center of the flow at an elevation of 3750 m. The velocity of the flow pulses dropped to 1 m/minute or less where the flow went out onto the valley floor beyond the base of the talus slope or where a relatively small pulse breached the side of the 0.6–0.8 m high natural levees and slowly flowed out over porous talus. Intervals of 4–15 minutes of relative quiescence elapsed between the flow pulses. Velocity measurements were made by timing the travel of a given flow front between two reference points slightly less than 300 m apart and by analysis of 8 mm motion pictures.

Curry was interested in obtaining samples of the moving material and he did so

by forcing wide-necked glass liter-sized jars into the slowly moving side of the flow. The jars were inserted 30–45 cm into the flowing debris about 1 m above the base of the flow. Some bottles were broken and it was generally difficult to force the bottles through the armoring surficial boulders moving up to 150 m/minute, but once beneath the outermost boulders the flow was liquid enough to fill most of the jars. A total of almost 3 liters of matrix of a median diameter of 5 cm or less (a limit imposed by the size of the jars) was collected.

Curry's observations were important to the understanding of these processes. He was also fortunate in having been at the right place at the right time. How many of us, though, would have rushed *to* the mud flow rather than *away* from it on that rainswept August afternoon?

Mud flows are a common occurrence on the steep slopes of andesitic volcanoes because loose material is abundant, steam can turn the volcanic rocks into a clayey goo, and rainfall or rapid snowmelt can set huge masses in motion. Volcanoes in the Pacific Northwest are no exception. One very real hazard on any of those volcanoes is the rapid melting of snow or ice by the eruption of steam or lava. Vast amounts of water would be unleashed, which could pick up surface debris and turn into a mud flow downvalley. More than 55 mud flows have originated at Mount Rainier in the last 10,000 years (Fig. 10-20), and future occurrences are a major threat. Because mud flows can travel so far and with such high speed, it has been recommended that valley floors within 40 km (25 mi) of Rainier should be evacuated in the event of an eruption. Residents of those areas would be hard pressed to get out if a mud flow were already in motion. For example, one such flow in Japan traveled at a velocity of about 90 km/hr (56 mph). In Indonesia, volcanic mud flows are called ***lahars,*** a term now accepted and used in many parts of the world.

Another associated hazard is that mud flow debris might quickly fill in reservoirs in valleys flanking the mountain, displacing the water and producing downstream floods. An ingenious way of controlling smaller mud flows is to empty the reservoir, trapping the mud flow in it. There is no such hope of containing the large flows, however.

COMPLEX MOVEMENTS

It has been stressed that there is no clear-cut boundary between types of movements. Where one or the other clearly dominates, the classifications described thus far can be made. Often, however, several types are involved and

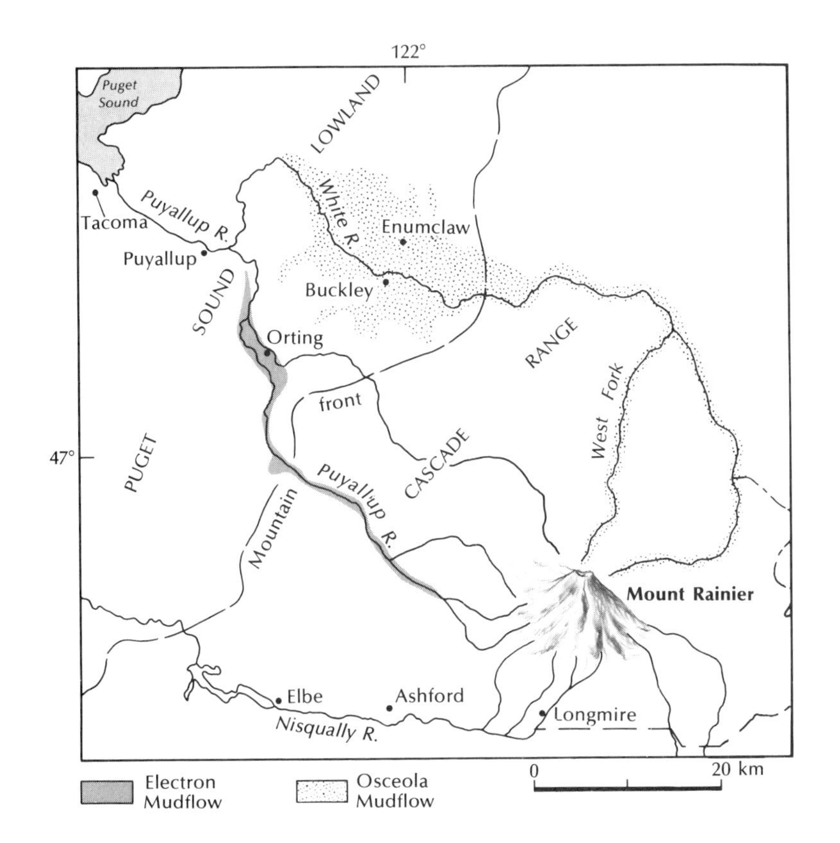

Fig. 10-20 Extent of two mud flows that originated on the flanks of Mount Rainier, Washington. The Osceola Mud Flow, which is 5800 years old, extended far from the mountain and buried the area now occupied by Enumclaw under 21 m (69 ft) of debris. Five hundred years ago, the Electron Mud Flow moved downvalley and deposited 5 m (16 ft) of debris near the present site of Orting.

Fig. 10-21 Complex slump/earth flow blocking Highway 24 near Oakland, California. *USGS*

these are classified as *complex* movements.

Perhaps the most common complex movement is the combination of slump and earthflow. Here the upper part has all the attributes of a classical slump block, but beyond the toe of surface rupture, flowage takes place (Fig. 10-21).

Some large rockfalls grade downvalley into very rapidly moving debris flows, the result being stupendous mass movements called ***rockfall avalanches.***

Such events start as rockfalls high on a mountain face, plunging down slopes and sweeping across valleys with a velocity of 160 km/hr (99 mph) or more. An excellent example is the prehistoric Blackhawk Slide in California (Fig. 10-22), which started as a rockfall in the source area. Its initial velocity probably was between 200 and 270 km/hr (124 to 168 mph), and, after leaving the canyon walls, it spread out across the valley floor at velocities of no less than 120 km/hr (75 mph), overtopping a 60-m (197-ft) hill. The front of the slide now rests 9 km (6 mi) beyond and 1100 m (3608 ft) below the source.

A celebrated example of a rockfall is one that occurred at Frank, Alberta, in 1903. At the crest of Turtle Mountain, a mass of strongly jointed limestone blocks, possibly undermined by coal mining carried on below the thrust fault at the

Fig. 10-22 The Blackhawk Slide, on the north slope of the San Bernardino Mountains, California. The surface form of such deposits and those of glacial moraines are somewhat similar and both deposits are poorly sorted. How would you go about telling the two landforms apart? *John S. Shelton*

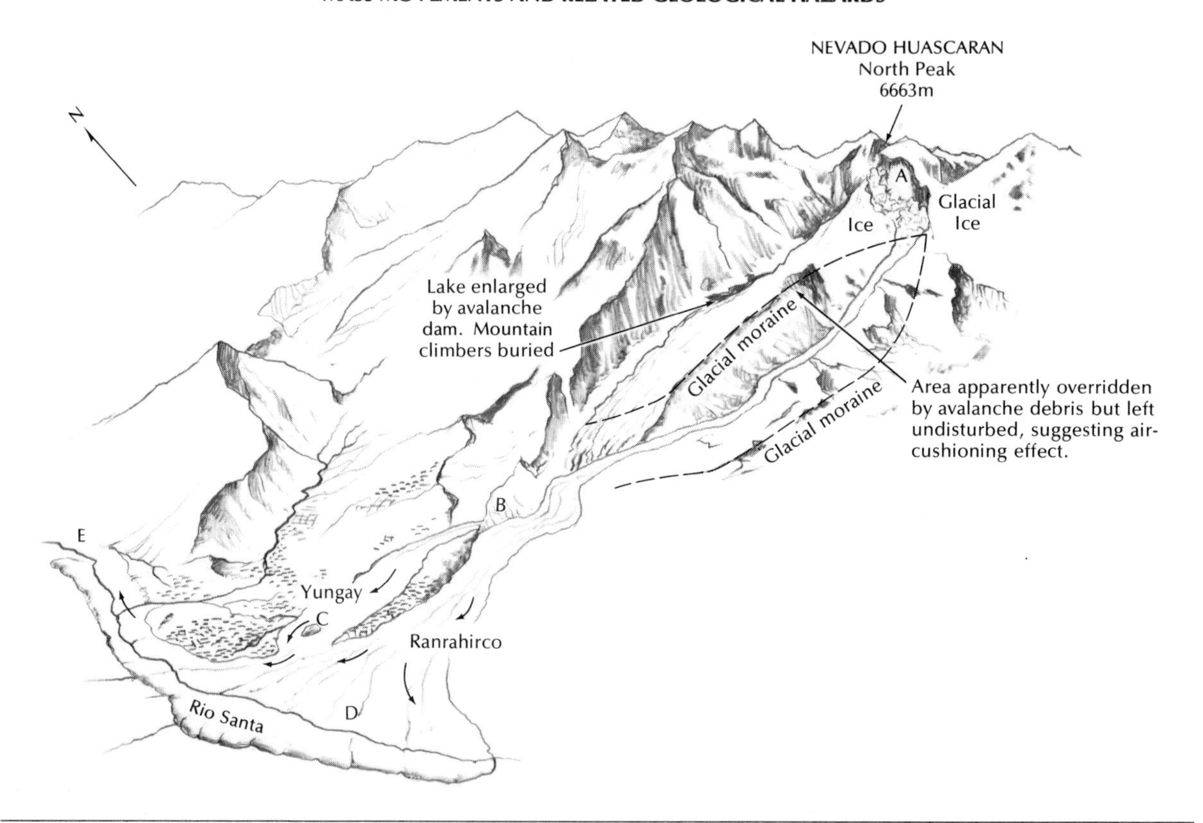

Fig. 10-23 Area affected by the rock and debris avalanche of May 1970 that originated on the north peak of Nevado Huascarán (A). Before descending on Yungay, the avalanche rode over a ridge 180 to 240 mi (591 to 787 ft) high (B). The only place spared in Yungay was the cemetery (C). Debris reached the banks of the Rio Santa (D) 14.5 km (9 mi) from the source in about 4 minutes. Part of it flowed more than 2 km (1.2 mi) upstream (E), but most of it flowed downstream toward the Pacific, causing more destruction in the floodplain of the river.

foot of the mountain, broke loose and plunged down the steep escarpment. About 37 million m^3 (48 million yd^3) thundered through the little coal-mining town of Frank—killing 70 people on the way—and swept to a high point 122 m (400 ft) above the valley floor on the facing slope.

The great rockfall-rockslide at Gohna, India, in 1893, remains as one of the most impressive falls of modern times. There an enormous mass of rock, loosened by the driving monsoon rains, dropped 1200 m (3936 ft) into a narrow Himalayan gorge. A huge natural dam was formed by the mass of detritus—perhaps 275 m (902 ft) high, 915 m (3001 ft) across the gorge at the crest, and extending for 3350 m (10,988 ft) upstream and downstream. That pile of broken rock, about 3.6 billion m^3 (4.7 billion yd^3), impounded a lake 240 m (787 ft) deep by damming the waters of the river.

The British engineers, then in India, proved to be a remarkably foresighted lot. They predicted the dam's failure within ten days of the time it actually occurred, over the two years of its existence. While the dam still existed, all bridges were removed downstream, the river channel was cleared of obstacles, a telegraphic warning network was set up, and everything was prepared for the imminent flood. When it came it set a record. Around 280 million m^3 (366 million yd^3) of water were discharged in four hours, causing a flood with a 75-m (246-ft)-high crest. Interestingly enough, after the flood was over, the river channel close to the dam, instead of being deepened, was raised 70 m (230 ft) by the sand and gravel deposited after the flood crest had passed and the river flow had returned to normal.

The western United States has been the site of several recent rock avalanches. In Montana one occurred in the canyon of the Madison River where, in 1959, an earthquake jarred loose enough rock from the canyon walls to dam the valley and form a lake. As at Gohna, there was flood danger to downstream areas from rapid spillover across the rock dam. Bulldozers were pressed into service to cut a spillway and to control the discharge from the lake.

In 1963, rocks cascaded down the slopes of Little Tahoma, a subsidiary peak on the flank of Mount Rainier, and quickly moved some 6 km (4 mi) downvalley on a lower slope. The debris descended some 2000 m (6560 ft), and the velocity was estimated at 130 to 145 km/hr (81 to 90 mph). Those falls may have been triggered by volcanic steam explosions.

The most tragic rock avalanche recorded, however, was related to the 1970 earthquake in Peru. About 15 km (9 mi) east of Yungay, a town of 20,000 inhabitants, rise the lofty Peruvian Andes and the snow-covered peak of Nevados Huascarán, which soars to about 6663 m (21,855 ft). Ground motion from the quake shook the mountain and broke loose a huge block of ice, snow, and rock from the top of the precipitous north face (Fig. 10-23). The mass fell in free flight and impacted 1000 m (3280 ft) below, where it knocked loose a large volume of rock and burst into thousands of smaller fragments. All the debris continued to cascade down the north face. Frictional heating of the ice during impact caused some melting. The entire mass was transformed into a rapidly moving debris flow. It raced along previously established avalanche and stream channels, and by the time it had moved away from the mountains, it contained an estimated 50 to 100 million m^3 (65 to 130 million yd^3) of water, mud, and rocks. Its speed has been estimated at about 210 to 280 km/hr (175 to 210 mph). Just a little upstream from Yungay the channel in which the flow was racing made a sharp bend. Most of the surging mass of mud and boulders made the turn. However, some rode over the ridge, and within seconds the entire city was buried. About 18,000 residents were interred almost instantly (Fig. 10-24). The mud flow continued downslope where it buried another town—Ranrahirca (1800 people)—and extensively damaged road and rail routes, power and communication lines, and the Hualanca hydroelectric plant (Fig. 10-25). Eventually it reached the bottom of the valley and the Río Santa, where its momentum carried it across the river as far as 60 m (197 ft) up the opposite bank. It finally came to rest far downstream.

How do we account for such high velocities and such long distances of transport over relatively flat surfaces, as evidenced by the events just described? Some believe that the great velocities are due to air entrapped and compressed beneath the falling mass of debris. The material is temporarily buoyed up in much the same manner that air temporarily buoys up a sheet of plywood dropped onto a flat surface. When the falling debris is pitched into the air, a compressed air cushion forms beneath it, and some geologists have noted the occurrence of such geological "ski-jumps," in the field. The eventual loss of the air cushion ends this remarkable transport mechanism. Another theory is that countless semi-elastic collisions of blocks with each other may keep the debris acting as a fluid, causing it to move great distances beyond the cliffs from which it was derived.

CONDITIONS FAVORING LANDSLIDES

Most slides and flows are the result of the forces of gravity acting upon earth materials in an unstable condition or position. Although the movement itself may be extremely rapid (or imperceptibly slow), a landslide does not suddenly spring into being, but rather develops gradually, step by step, with one element triggering another.

Instability does occur naturally, but human activities often act as the trigger in an unstable area when they disrupt precarious balances. The natural process of erosion can create an unstable condition by oversteepening a slope, as, for example, when a cliff is undercut along a river or by the action of the sea. The material making up the upper part of the slope is held in place by the weight of the material at the base of the slope, and when the anchoring material is removed, the upper part is rendered unstable. Or, if erosion wears away the toe of an ancient landslide, it could start to move again.

Highway engineers are particularly sensitive to the problems of oversteepened slopes. In making highway cuts, they are called upon to create slopes steeper than the original ones—a rather perilous business, for the slope must be not only steep, but stable as well. Shallow artificial cuts are commonly avoided because more material must be removed, and that is considerably more costly. Improperly designed highway fills can also fail (Fig. 10-26), and cuts in hillslopes for house foundations can create stability problems on slopes.

Fig. 10-24 The statue of Christ at Cemetery Hill was all that remained of Yungay. *USGS*

Inclination of rock stratification relative to the ground surface is another common cause of landslides. If layers of rock slant down toward a road-cut or railroad-cut, rocks may slide along the bedding planes between the layers. Or, in the absence of an artificial cut, a normally stable area may become unstable when a heavy load, such as a building or excavated material, is placed upon it. If layers of rock slant into the hillside, of course, the chance of sliding is usually minimal.

Water is the hidden devil in the ground, for it is the cause of many landslides (Fig. 10-27). When natural drainage conditions are changed, either by natural process or by human activity, the groundwater situation can become potentially dangerous. An unusually high content of water adds considerable weight to the soil or rocks, making it difficult for those containing clay to stick together. Water also acts rather like a lubricant along potential sliding planes.

Excess water can be added to slopes in a variety of ways—some obvious and some not so obvious. We can start by saying that many natural slopes that are not patently sliding are relatively stable. Human activity can easily tip the balance, however, and usually unbeknown to those involved. In some cases the mere watering of a sloping lawn can trigger a slide by increasing the water content to the critical point, at which the slope becomes unstable. Clearcutting forests on slopes can have a similar effect. Trees might be thought of as water pumps, returning the water to the atmosphere by transpiration. Remove the pumps, however, and the slopes probably will retain more moisture—an invitation to instability. An artificial reservoir can also be the cause of a landslide;

if enough rising water soaks into the banks, they lose much of their strength. Of course, a natural cause, such as climatic change involving greater precipitation, can also set landslides in motion. We should be aware of both the natural and artificial causes of such events, because if personal property is damaged the case might easily find its way to court. Finding the real culprit, however, is no easy task.

What caused the Vaiont Reservoir disaster (see Fig. 10-2), for example? In the first place, the valley is steep sided, due to a relatively recent river downcutting (Fig. 10-28). Engineers commonly seek such narrow gorges for dam sites, however, because less concrete is needed to plug the valley. But at Vaiont, the bedrock is made up of limestone and layers of slippery clay, and the bedding planes are in the worst possible configuration—toward the valley axis. Further, the limestone itself is rife with underground solution caverns that serve as collection basins for water that helps to saturate the ground. The valley has been the site of other landslides also; witness the ancient one near Casso and the 1960 slide into the reservoir along its south side. In 1963, the trigger was excess water introduced into an already unstable system. Some seeped in laterally from the reservoir and some was provided by the downpours. Creep in the slide area had increased from about 1 cm/wk (0.4 in./wk), the average rate after the reservoir was built, to 80 cm/day (31 in./day) on the day of failure. Even animals grazing on the slopes sensed the potential danger and moved away in time. What could have been done to prevent the disaster? The only thing that comes to mind is that the geological setting of the area could have been studied more carefully. Certainly, sites that in a geological sense have so many things wrong with them should be avoided. No amount of engineering could have saved the reservoir, at least within acceptable costs.

A side effect of the Vaiont slide was that towns near the head of the reservoir were also damaged. The destruction here, however, was from slide-generated water waves that ricocheted from one valley wall to the other, wreaking havoc in their path (see wave path in Fig. 10-2).

STABILIZING SLIDES

Civil engineers have devised many ingenious methods of controlling unstable slopes, although not enough people are aware that it is

Fig. 10-25 Block weighing about 630 metric tons that was transported to the former town of Ranrahirca. The largest block carried by the avalanche is estimated to weigh 12,800 metric tons. *USGS*

Fig. 10-26 Vertical aerial photograph of a landslide that removed part of a highway in Jefferson County, Ohio. Although the highway was built on artificial fill, the slide originated in the underlying natural materials, which had been saturated by springs outcropping on the hillsides. *Ohio Department of Transportation*

easier, cheaper, and safer to apply the methods of stabilization before, and not after, mass movements take place. Consolidation of unstable materials is one method of control, and an unusual example of its use was the freezing of surficial materials in a dormant landslide that threatened to start moving again during the construction of Grand Coulee Dam in Washington.

Another common way of stabilizing landslides is to regrade the slide area to a low angle, one that is stable under the newly imposed conditions, be they natural or artificial. At the same time, the amount of water in the slope should be reduced drastically, either by surface or subsurface drainage or by covering the surface of the slide with impermeable material. An example of slope treatment and drainage is provided by the history of the large slides—some covering as much as 64 hectares (158 acres)—that interrupt the hilly terrain of the Ventura Avenue oil field in southern California. In the rainy winter of 1940–41, one 24-hectare (59-acre) block slid as a single unit for a distance of approximately 30 m (98 ft). Because the slides sheared oil wells as they moved [in the 1940–41 episode, 23 wells were cut off as deep as 30 m (98 ft) below the ground surface],

Fig. 10-27 Fresh earth flow in the Berkeley hills, California. Although some small ground cracks are thought to have opened during the 1906 earthquake, most of the movement took place during the next rainy season, when the ground was soaked. *G.K. Gilbert, USGS*

unusually extensive and expensive efforts to curb the movement were made. Partial success was achieved by covering the surface with tar and by drilling horizontal drainage holes into the slides—64 km (40 mi) of them. Vertical wells were also drilled through the slides to a porous layer of sandstone, which served as a conduit to carry water away from the slides and into adjacent solid ground. There the excess flow could be pumped out.

Large corporations can afford such expensive remedial measures, but most homeowners cannot. Often the owner's only recourse is to cover the landslide with sheets of plastic—a tactic that works, but one that hardly enhances the beauty of a homesite.

Another common stabilizing method is retention of the toe of small slides. Steel or concrete walls are commonly put up to hold back the force of the slide, or, in places where bedrock is easily reached, the toe is weighted with large blocks of rock in an attempt to prevent future movement.

EDUCATING THE PUBLIC

Because conditions favorable to gravity mass movements build up slowly, it should be possible to determine ahead of time those areas that could become troublesome. As the population continues to increase and as construction follows growth, predictions become ever more important for the prevention of human anguish. Obviously, the more we know about the geology of a region, the easier it will be to point out the regions where earth movements are likely to occur. Such information is gathered by geologists and recorded on geological maps showing the distribution of different types of rocks on the surface and in cross sec-

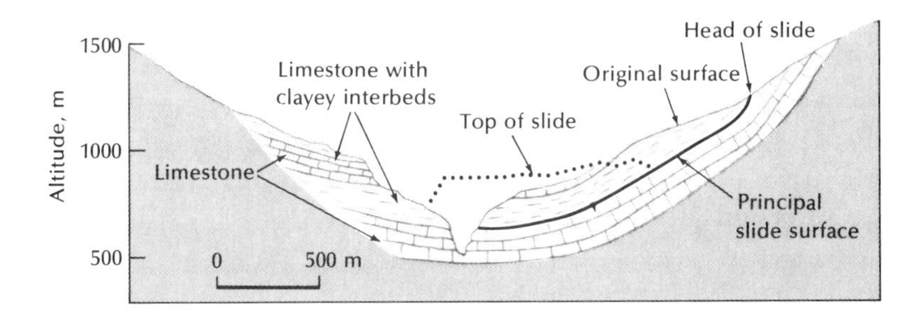

Fig. 10-28 Cross section of the valley in which the Vaiont Reservoir is situated, showing the geological setting and the pre- and post-landslide topography.

tions, which are informed guesses of the distribution of rocks below the surface. When such work is related to geological hazards—landslides, earthquakes, earth subsidence, and the like—it is called "urban geology" or "environmental geology," and it is becoming increasingly important as a direct application of science and technology to the humanizing of our surroundings.

Not only is geological knowledge of hazardous areas vital, but the knowledge must be available to other scientists and to those persons in authority who can and will use it properly. Cooperation among geologists, engineers, city planners and managers, and the public can prevent many mass movements and predict the possibility of others, thus ensuring proper land usage as land itself becomes more and more a rare and valuable commodity. M. R. Hill says "We must focus our efforts on converting the unforeseen to the predictable, transforming the predictable into the preventable, preventing the preventable, and restraining the foolish."

PERMAFROST

Mass wasting seems to be particularly active in high northern and southern latitudes, and at high altitudes. The reason is that close beneath the surface the ground is frozen solid. In places, the ground has been cold enough to preserve animals in near-perfect condition for many thousands of years (Fig. 10-29). In the following section, we examine frozen ground, some of the landscape features associated with it, and a few of the engineering problems that result from building on such a precarious surface.

Ground that remains frozen from one year to the next was called ***permafrost*** during World War II, and all efforts to substitute a term that sounds a little less like a trade name for a refrigerator have been successfully resisted. Permafrost is defined only on the basis of temperature, it being ground that remains below 0°C (32°F). Hence, the content of ice in permafrost can vary from very small to very large proportions. A knowledge of how much ice is present and where it is located is extremely important, as we shall see, to engineers and builders working on such terrain.

Permafrost, which is more widely distributed than many people realize, underlies almost 20 per cent of the land surface on the earth, including about 85 per cent of the land area of Canada and the USSR (Fig. 10-30A). It reaches its maximum thickness around the margins of the Arctic Ocean in Alaska, Canada, and the Soviet Arctic. Siberia holds the record for maximum thickness: 1400 to 1450 m (4592 to 4756 ft); whereas sites of maximum thickness in Alaska and Canada are about one-half that. In a general way the thickness decreases southward until finally it thins to nearly zero at about the southern boundary shown on the map. Permafrost also formed in ice-free areas during the Pleistocene, and that which now occupies former glaciated areas could have formed after the ice melted.

Two major permafrost distributions are recognized (Fig. 10-30B). To the north the distribution is continuous, whereas to the south it is discontinuous, occurring only in patches. It is hard to predict where those patches are located, and as a result land-use planning in the south can be more difficult than in the north. Permafrost also occurs in high mountains south of the border shown on the map. It has been reported, for example, on the summit of Mount Washington in New Hampshire, in Colorado's Rocky Mountains, and even on Mauna Kea, Hawaii.

Overlying the permafrost is a thin layer of soil in which ***ground ice*** (subsurface ice) thaws in the spring and freezes in the fall, remaining frozen throughout the winter. That soil is the active layer, and it varies from less than 1 to about 3 m (3 to 9 ft) in thickness. The upper surface of permafrost is called the ***permafrost table,*** and below that the pore spaces are filled with ice. Hence water in the active layer cannot sink underground—one reason why much of the Arctic tundra is so boggy and water soaked. Precipitation over large segments of the Arctic is very slight, although the many lakes and muskegs seem to belie that fact. Water remains at the surface because it cannot sink under-

Fig. 10-29 Carcass of a juvenile mammoth, nicknamed Dima, recovered in 1977 from permafrost near the headwaters of the Kolyma River, Siberia, in a gold-mining area. Dating indicates that the carcass was buried, frozen, for about 44,000 years. *N.A. Shilo* and *C.B. Tomirdiaro, USSR Academy of Sciences*

ground. Further, the evaporation rate is relatively low.

Patterned ground and solifluction

A bizarre, but common manifestation of ice-churned ground in permafrost areas is a curiously regular patterned surface (Fig. 10-31). Sometimes from the air the ground looks like a gigantic tiled floor. Some geometrically shaped polygonal areas are thought to be the result of ***frost heave,*** which is much more prominent in fine-grained soils than in coarse. When frost heaving takes place year after year in a soil of mixed composition, coarse materials, such as boulders and gravel, are gradually shoved radially outward from the central area, and finer materials remain behind and become concentrated. Other patterned ground results from a network of vertical ice wedges (Figs. 10-32 and 10-33). In some places, patterned features were formed in the past and are not ac-

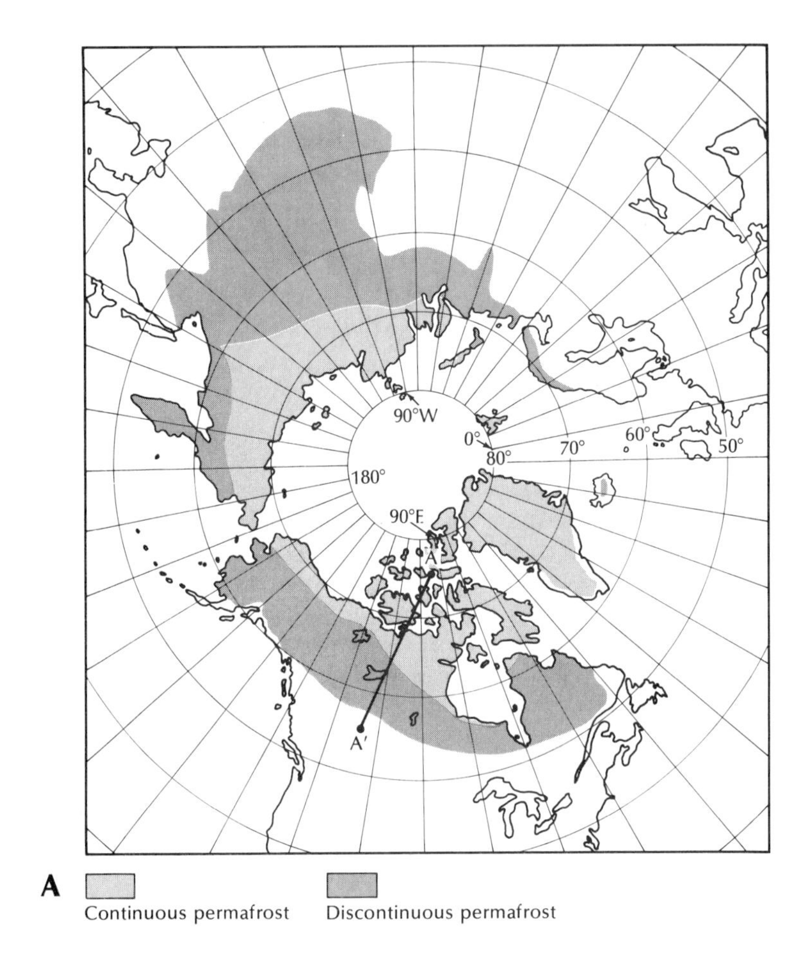

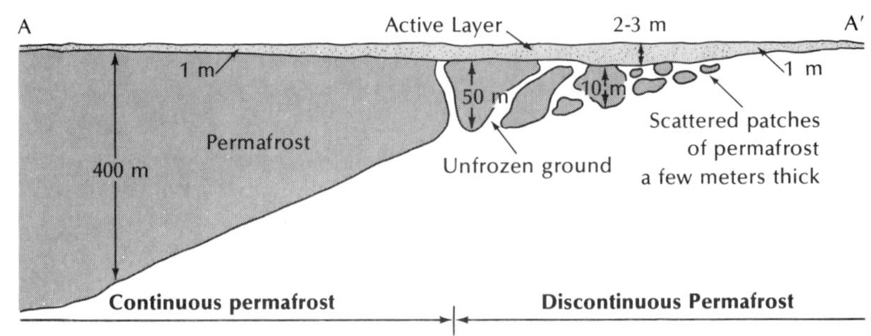

Fig. 10-30 A. Extent of permafrost in the Northern Hemisphere. **B.** Cross section of permafrost along the line A-A′.

tively forming today. By studying them, we can estimate Pleistocene climates (Fig. 10-34).

Solifluction is an extreme sort of creep that reaches a maximum development in cold climates (Fig. 10-35). Hilly terrain underlain by permafrost exemplifies it best, for although the surface layers freeze and thaw, the permafrost table remains constant. As we have noted, surface water cannot sink into the permafrost, so that water that would normally percolate far down into the ground is concentrated in the active layer. The active layer, then, is far more

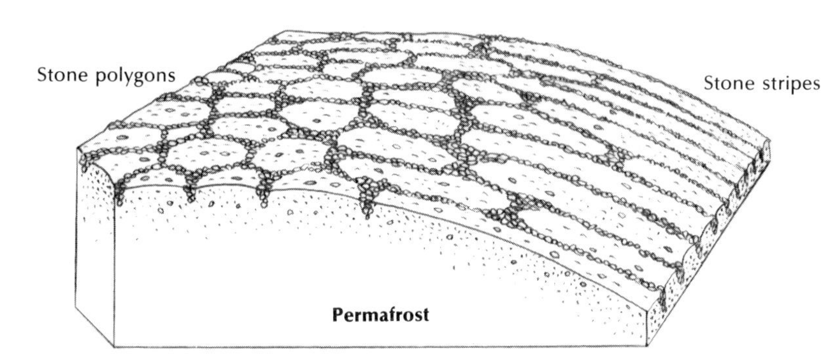

Fig. 10-31 Patterned ground comes in many shapes and sizes. On level terrain, stone polygons form, whereas on sloping terrain, where creep occurs, the polygons are stretched out, forming stone stripes.

susceptible to creep than similar terrain would be in a more temperate climate because (1) the opposing forces of ice crystallization and melting of ground ice are more active, (2) it holds more water than it would under a similar precipitation regime without permafrost at depth, and (3) the water-saturated, unstable ground rests on a frozen base over which surface it can readily slide.

Active solifluction produces a landscape that bears some resemblance to the wrinkled hide of an aged elephant. Different parts of the water-saturated surface layer creep downslope at different rates, so that hillsides where solifluction is active are festooned with soil lobes, or tongues, some of which advance rapidly and some slowly (Fig. 10-36). Maximum rates of lobe movement approach 40m/1000 yr (131 ft/1000 yr). A curious aspect of solifluction is that it tends to produce a rounded, smooth terrain that stands in strong contrast to the very rugged terrain in glaciated valleys.

Engineering problems in permafrost terrain

Before 1942, in the United States, little attention was given to permafrost and its manifold problems, although the Russians had studied the phenomenon intensively for more than one-half a century. Visitors to the interior of Alaska, especially to the area in and around Fairbanks, are invariably surprised by the tilted houses and the "drunken forests," with their tipsy-looking trees thrown out of line by melting of the permafrost or by its upward movement into the active zone.

From the scores of problems that permafrost can produce, only a few need be mentioned. Heated buildings in the Arctic are likely to thaw out the ground beneath them and to melt their way down into the soggy, unstable mush under the foundation (Fig. 10-37). Usually the structures sink unevenly, so that floors sag, the walls tip, and the doors stick. It appears that the most practical solution to the problem is to build houses and barracks on stilts, so that cold air could circulate under the structures. Disturbance of the permafrost would then be minimal.

On the other hand, unheated structures, especially such large ones as hangars or warehouses, *insulate* the ground surface below them so that the active layer does not thaw in the summer. The result is that the permafrost table rises, possibly blocking the flow of ground water through the active layer. If the ground water is forced to the surface, where it freezes, a phenomenon known as ***icing*** occurs. Icing may be spectacular, converting the interior of an unheated building into a huge block of ice, with ice cascading from the doors and windows.

Vegetation is so critical to the thermal properties of permafrost that the slightest alteration of delicate plant life can have disastrous effects. Even driving a vehicle across the tundra can result in a scar that will not heal for generations (Fig. 10-38). Indeed, road building without proper care for the permafrost terrain could result in the world's longest and narrowest lake.

In order to diminish the problems inherent

Fig. 10-32 Foliated ice wedge in late Quaternary silt along the Aldan River, central Yakutia, USSR. *T.L. Péwé*

to permafrost, special construction procedures are necessary in the building of roads, railroads, and airfields. The simplest method is to insulate the paving surface with a layer of gravel so that melting does not take place. Quite commonly, however, such insulation is so effective that the permafrost rises beneath the gravel layer and may even penetrate its base. Depending upon conditions in the active layer and the flow of ground water on the permafrost table, large-scale icing could result on the paved area. However, with proper foresight and engineering practice such embarrassment can be avoided.

The list of problems permafrost can create seems endless, but the degree to which Arctic

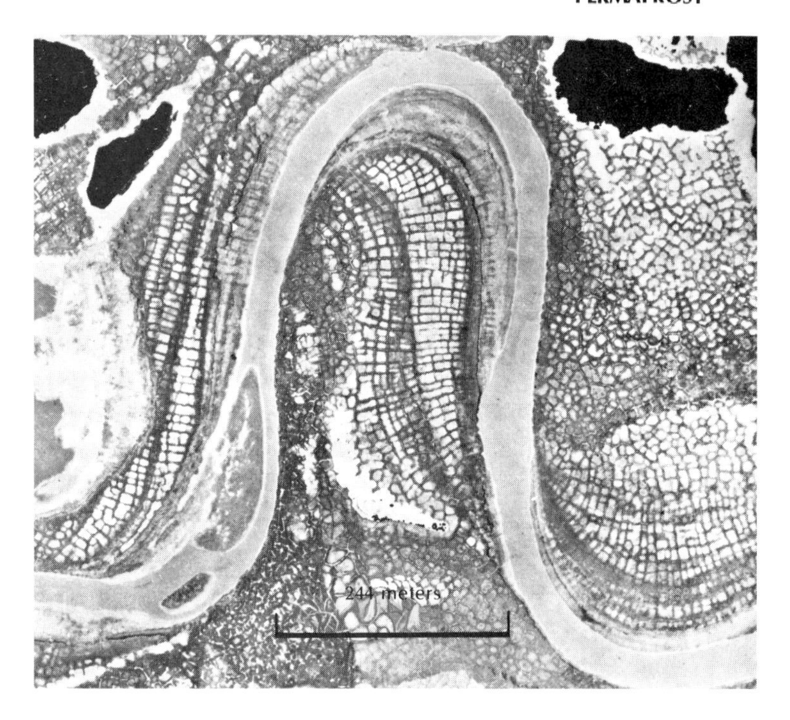

Fig. 10-33 Patterned ground along a river southeast of Barrow, Alaska. The pattern of interconnecting ice wedges is formed in extremely cold climates when ice-cemented permafrost cracks. The wedges are widest at the top and narrow downward. Evidence for former ice wedges may be seen in many countries just south of the Pleistocene glacial boundary. *O.J. Ferrians, Jr., USGS*

pioneers have overcome them is a testimonial to their ingenuity and perseverance. Even such a simple thing as developing a water supply in a permafrost area can become a major frustration. Ground water in the active layer is available only during the summer, and is usually at so shallow a depth that it is readily contaminated by surface wastes. Although there may be ground water below the permafrost, it is deep, and well sections drilled through the frozen ground are almost certain to freeze. Delivery of water also poses a problem. If water pipes are buried underground, they freeze; if placed above ground, they freeze, too, and are likely to be thrown out of line as the ground under them heaves when it freezes and sinks when it thaws. Expensive insulation is the only solution.

One solution to water supply in permafrost areas is to build dams and collect the summer snowmelt for year-round use. Yet climatic conditions can create difficulties for dams and reservoirs. The Russians, for example, ran into trouble when a far northern dam started to leak shortly after the reservoir behind it was filled. The dam was built on volcanic rock with tiny cracks that were permanently filled with veinlets of ice. Ordinarily, ice-filled rock below the permafrost level can be treated as solid rock. In the Russian case, however, the filled reservoir with its insulating layer of ice on the surface acted as a heat trap. The ice veinlets melted, and the bottom of the dam became virtually a sieve. Newer dams built in similar cold areas are refrigerated by pumping cold air into them to prevent such melting.

Sewage disposal is perhaps the ultimate problem. Septic tanks and leach fields freeze, and in the absence of bacteria, waste does not decay, and hence, disappear, as it does in warmer climates. At Point Barrow in Alaska the unsightly (but practical) solution is to heap everything atop an ice floe during the winter. In the summer the ice cake floats out into the Arctic Ocean, melts, and the waste sinks. Waste disposal is even more of a problem now that oil has been discovered in northern Alaska. The rapid development of the area will make that method of waste disposal and its attendant pollution intolerable.

Permafrost and the discovery of oil in the north slope of Alaska also led to a heated con-

Fig. 10-34 A. Mounds near Fairbanks, Alaska, that formed when a mosaic of ice wedges melted, causing the ground beneath them to subside. The intervening areas were left as mounds. The melting began when the trees were cleared for agriculture. *R.F. Black and T.L. Péwé, USGS* **B.** Low mounds in the southern part of the Great Valley of California. Similar forms occur in places far removed from former glaciers, leading to speculation of widespread cold climates and permafrost in the past. Many theories of origin have been put forth to explain the mounds, with construction by gophers gaining in popularity. *F.E. Matthes, USGS*

Fig. 10-35 Solifluction distorts railroad tracks along a tributary of the Yukon River, Alaska. *W.W. Atwood, USGS*

Fig. 10-36 Solifluction lobes on a 14° slope in the Ruby Mountains, Yukon Territory. *Larry W. Price*

troversy over the Alaska pipeline, which extends from the petroleum fields south to the ice-free port of Valdez, a distance of about 1300 km (808 mi). The oil flowing through the pipeline, which is more than 1 m (3 ft) in diameter, has a temperature of 70 to 80°C (158 to 176°F). It was argued that if the pipe were buried, difficult problems would arise from melting of the permafrost. It has been calculated, for example, that within the first decade after burial a cylindrical area 6 to 9 m (20 to 30 ft) in diameter would be thawed around the pipe. In successive decades the thawing would continue, but at a diminishing rate.

The major construction problem was the condition of the permafrost before thawing. If it were dry, thawing should have only a slight effect. However, permafrost is in large part

Fig. 10-37 A geologically famous roadhouse built in 1951 on the Richardson Highway, Alaska. By 1962 the right side of the house, which was heated, had settled considerably. An unheated porch (left), however, subsided much less. Damage was so extensive by 1965 that the building was razed. *T.L. Péwé, USGS*

composed of fine-grained sediments with a high ice content, and thawing of the material forms a water-saturated slurry into which the pipe could sink. If the pipeline were on a slope, the slurry could flow out onto the landscape, allowing the pipe to settle even deeper—into still-frozen ground—and melting would continue. Stress caused by such processes could cause the pipe to rupture, leading to oil spills that would rival in seriousness those that occur at sea. It was vital, therefore, to identify all potential problems before the pipeline was constructed.

After detailed investigations were made, it was found that the pipeline could be placed below ground for about one-half of its length without serious consequences. Conventional burial procedures were used. Because of the problems associated with the melting of the permafrost, and to insulate it against the arctic cold, the other half of the pipeline has been built above ground, on platforms about 15 to 21 m (50 to 70 ft) apart (Fig. 10-39). Areas disturbed during the construction were to be revegetated, returning the environment to a stable condition in which the permafrost, protected by the vegetative cover, is prevented from melting.

Fig. 10-38 Tractor trail bulldozed on the north slope of Alaska near the Canning River during the winter of 1967–68. The string of ponds formed during the summer of 1968, when the photograph was taken, and as thawing continued they grew larger. Notice the patterned ground and vehicle tracks. *Averill Thayer, Bureau of Sport Fisheries and Wildlife*

Fig. 10-39 Bears amble, undisturbed, alongside and under an insulated, above-ground section of the trans-Alaska pipeline. The permafrost is protected from melting and damage by a layer of gravel. *Alyeska Pipeline Company, Standard Oil Company of California*

SUMMARY

1. Gravity operates on all sloping ground, so that material moves to lower positions on the slope.
2. The transfer rate of material on a slope varies from millimeters per year to meters per second. Similarly, the amount of material involved in the transfer varies from small amounts of material to entire mountainsides.
3. *Slides* and *earthflows* are the mass movements that most often destroy property and dwellings. They can occur naturally or be produced by human activities, and once set in motion, they are difficult to stop. Slopes prone to slides or earthflows should not be used as building sites.
4. A rock avalanche is the most spectacular kind of mass movement, because so much material is transferred downslope so quickly, perhaps on a cushion of air.
5. Mass movement takes place at a fairly rapid rate in areas of permafrost. Such terrain is extremely sensitive to human manipulation; any engineering projects on it must be undertaken with great care.

QUESTIONS

1. Discuss the role of water, rainstorms, and mean annual rainfall in initiating slides and flows.
2. What kinds of mass movements can be triggered by earthquakes? Discuss the processes involved in such movements.
3. What are the geological factors involved in slides and flows in an area of tightly folded, interlayered sandstones and shales?
4. Are mud flows always an arid region phenomenon?
5. Discuss the problems of building houses and roads in permafrost terrain underlain by (1) granite and (2) interbedded limestone and shale.

SELECTED REFERENCES

Crandell, D. R., and Mullineaux, D. R., 1967, Volcanic hazards at Mount Rainier, Washington, U.S. Geological Survey Bulletin 1238.

Curry, R. R., 1966, Observation of alpine mudflows in the Tenmile Range, Central Colorado, Geological Society of America Bulletin, vol. 77, p. 771–76.

Ericksen, G. E., and Plafker, G., 1970, Preliminary report on the geologic events associated with the May 31, 1970, Peru earthquake, U.S. Geological Survey Circular 639.

Ferrians, O. J., Jr., Kachadoorian, R., and Greene, G. W., 1969, Permafrost and related engineering problems in Alaska, U.S. Geological Survey Professional Paper 678.

Kerr, P. F., 1963, Quick clay, Scientific American, vol. 209, no. 5, pp. 132–42.

Kiersch, G. A., 1965, The Vaiont Reservoir disaster, Mineral Information Service, vol. 18, no. 7, California Division of Mines and Geology, Sacramento.

Lachenbruch, A. H., 1970, Some estimates of the thermal effects of a heated pipeline in permafrost, U.S. Geological Survey Circular 632.

Péwé, T. L., 1966, Permafrost and its effect on life in the North, Oregon State University Press, Corvallis.

Price, L. W., 1972, The periglacial environment, permafrost, and man, Amer. Assoc. Geogr., Commission on College Geography, Resource Paper 14.

Robinson, G. D., and Spieker, A. M., 1978, "Nature to be commanded . . .", Earth science maps applied to land and water management, U.S. Geological Survey Professional Paper 950.

Schuster, R. L., and Krizek, R. J., eds., 1978, Landslides, National Academy of Sciences, Washington, D.C.

Sharpe, C. F. S., 1938, Landslides and related phenomena, Columbia University Press, New York.

Shreve, R. L., 1968, The Blackhawk landslide, Geological Society of America Special Paper 108.

Washburn, A. L., 1973, Periglacial processes and environments, St. Martin's Press, New York.

Fig. 11-1 A river meanders across the Iranian desert in the land of Elam and flows past the ancient city of Susa, where Esther was chosen queen. *Aerofilms, Ltd.*

11

STREAM EROSION, TRANSPORTATION, AND DEPOSITION

Few natural phenomena are more intimately involved with human affairs than rivers. In centuries past, such streams as the Nile, the Tigris, and the Euphrates literally were the givers of life as they threaded their way across a desert land (Fig.11–1). Ancient civilization depended on such waters for irrigation; through that communal enterprise many of the attributes of modern urbanized society arose. Mathematics, surveying, and hydraulics began in the designing of dams and canals. One of the earliest projects was a long dike built about 3200 B.C. on the west bank of the Nile, with cross dikes and canals, to carry flood waters into basins adjacent to the river.

Boundary disputes and ownership problems logically led to a system of codes and usages that evolved into a pattern of laws and courts much like ours today. The Code of Hammurabi (c. 1900 B.C.) included a provision that if a landowner damaged his neighbor's land through neglect of a portion of a canal that was his responsibility, he was liable for all the damage.

Rivers have long played a role as natural barriers—and two that were of decisive importance in Roman times were the Rhine and the Danube. The crossing of the Danube by the barbarians commonly is cited as one of the events heralding the fall of the Roman Empire.

In contrast with their role as barriers is the function rivers serve as communication routes. The Mississippi packet boat, with its flashing wheels and double columns of smoke, is gone forever. Its place is taken by the vastly more powerful diesel-propelled towboat (which actually pushes its load) and its broad acreage of heavily laden barges driving against the current. The endless parade of diesel-powered barges up and down the Rhine is an impressive sight to European travelers.

Rivers from time immemorial have been routes from the sea to the interior. Explorers have followed them; most of the world's leading cities are built on their banks. They are identified indissolubly with the history and national aspirations of almost all the lands that border them. It would be difficult to conceive of Germany without the Rhine, Austria and Vienna without the Danube, or Russia without the Volga or the Don.

In this chapter we will discuss first stream flow—how water flows down a valley—and

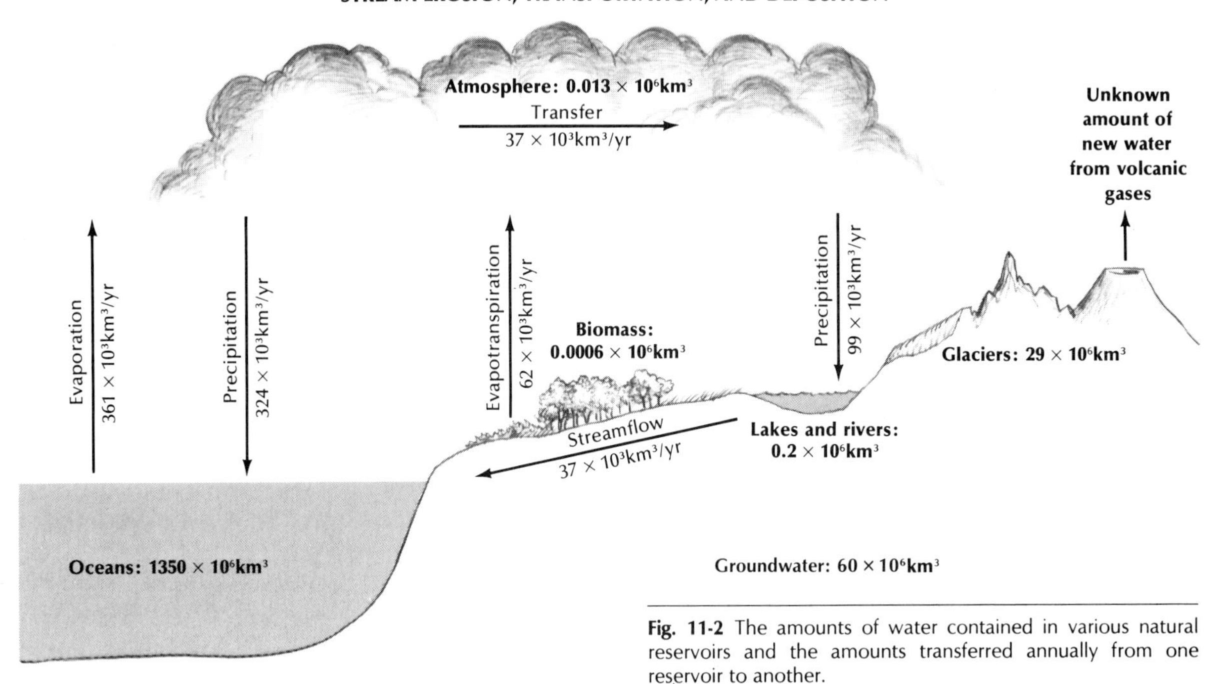

Fig. 11-2 The amounts of water contained in various natural reservoirs and the amounts transferred annually from one reservoir to another.

then relate it to the geological work a river is capable of performing, most of which takes place during times of major flooding. We will then discuss the ways of deciphering river history and the role of rivers in landscape evolution. Throughout, the environmental problems involved in manipulating rivers will be pointed out. We have become aware of such problems through mistakes, and only by knowing how and why we erred can we avoid future problems.

THE HYDROLOGIC CYCLE

Before discussing stream flow, we should consider first the amount of water on earth and how it moves from place to place (Fig. 11-2). There are several water reservoirs—the oceans, the rivers and lakes, the glaciers, voids in underground rocks, the atmosphere, and the biomass—and of them the oceans contain by far the most water. Although the volume content of these natural reservoirs has changed, as during major glaciations when considerable amounts of ocean water were transferred to glaciers, most geologists believe that the total amount of water on earth has been more or less constant for the past billion years. New water may have been added to the earth's surface by steam condensation during volcanic eruptions, but the amount is insignificant, overall.

Water is transferred from reservoir to reservoir each year, but, the entire budget is in balance. The cycling process is known as the ***hydrologic cycle.*** The water that is annually evaporated from the ocean (approximately equivalent to 1 m, or 3 ft, in thickness) exceeds precipitation on the oceans, and the excess is transferred to the lands via the atmosphere. On land the opposite occurs—that is, more water reaches the ground as precipitation than leaves through ***evapotranspiration.*** The latter term encompasses water losses through evaporation as well as through transpiration from plants and animals. In fact, in well-vegetated areas, the main loss of water to the atmosphere is through transpiration from plants. The cycle is completed and balanced as stream flow removes the excess from the land.

Stream flow performs work that continually changes the shape of the land. The geologist A. L. Bloom expresses the amount of available stream energy in a down-to-earth way:

The average continental height is 823 m above sea level. If we assume that the 37,000 cubic kilometers of annual runoff flow downhill an average of 823 m, the potential mechanical power of the system can be calculated. Potentially, the runoff from all lands would continuously generate over 12 billion horsepower. If all this power were used to erode the land, it would be comparable to having one horse-drawn scraper or scoop at work on each three-acre piece of land, day and night, year around. Imagine the work that would be accomplished! Of course, a large part of the potential energy of the runoff is wasted as frictional heat by the turbulent flow and splashing of water, but we will see that the "geomorphology machine" is really quite efficient, and in fact does erode and transport rock debris down to the sea almost as fast as if horse-drawn scrapers were hard at work on every small plot of land, over all the Earth.

Fig. 11-3 This car, suspended from a cable, allowed geologists to measure the velocity of the stream or to take water samples. Photo taken in 1890 on the Arkansas River near Canon City, Colorado. *USGS*

STREAM FLOW

Although streams play such a vital role in our lives, and their control has engaged the efforts of people for centuries, many aspects of their behavior are still unknown. Great impetus has been given in recent years to the study of stream flow because of its importance in the design of high-head hydroelectric plants, dams and spillways, and increasingly complex irrigation systems. Every industrialized nation is actively engaged in research into the nature of stream flow (Fig. 11-3), and the majority of such nations maintain large and well-equipped hydraulics laboratories. The largest laboratory in the United States is the U.S. Waterways Experiment Station, operated by the U.S. Army Corps of Engineers at Vicksburg, Mississippi. There, elaborate models of the Mississippi have been constructed and an immense amount of data collected and analyzed in order to find ways to bring that unruly river and its tributaries under control.

From laboratory studies and field investigations around the world, and, as anybody who has rafted a wild river will testify, water flow is mainly turbulent (Fig. 11-4). In turbulent flow, the water particles go every which way. Familiar examples of turbulent flow are the tumultuous rush of water through Niagara Gorge downstream from the plunge pool at the base of the Falls or the maelstrom of white water at the bottom of the spillway at Grand Coulee Dam. Despite the random paths of individual water particles, the main thrust of the water is forward, downslope in the direction the stream is flowing. Sometimes the particles swirl upward like autumn leaves, or like dust devils in

Fig. 11-4 Turbulent flow patterns in the Colorado River. *Martin Litton*

the desert—at other times they descend just as violently in the vortices of whirlpools and eddies. In part, it is the erratic flow pattern that makes the actual velocity of a stream so difficult to measure.

In very general terms, the velocity of a stream can be defined as the direction and magnitude of displacement of a portion of the stream per unit of time. Customarily, we measure it in meters per second (m/sec), feet per second (ft/sec), miles per hour (mph), or kilometers per hour (km/hr). Few streams, however, attain velocities in excess of 30 km/hr (19 mph), and velocities of less than 6 km/hr (4 mph) are more likely to be the rule. The velocity as much as any single factor is responsible for determining the size of the particles a stream can transport, as well as the way in which it carries the particles, or load.

At what point is an "average velocity" likely to be located within a stream? Different parts of the water in a stream move at different rates and, like glacier flow, the center moves faster than the sides and the top faster than the bottom because of friction along the channel sides and bottom, or perimeter (Fig. 11-5). Data show that the average velocity of a river is approximated by that velocity at 0.6 of the distance from the surface of the river to its bed, about in midstream. How, then, do we obtain that value, short of swimming? Actually, a reasonable figure can be obtained by throwing a stick into the middle of the stream, timing its travel over a known distance, and multiplying that velocity value by 0.8.

The average velocity of a stream depends on such factors as the gradient (or downvalley slope), the cross-sectional shape of the channel, the roughness of the sides and bottom of the channel, the discharge of the stream, and the amount of sediment the stream is carrying. We will look at the effect of each factor separately, and then at how each can vary along the length of a river.

Increased gradient obviously speeds up a stream's flow. Where the gradient reaches zero (when the stream empties into another river or a lake, for example), its velocity drops. When slopes are vertical, as in a waterfall, the velocity approaches that of free fall. So it is, in princi-

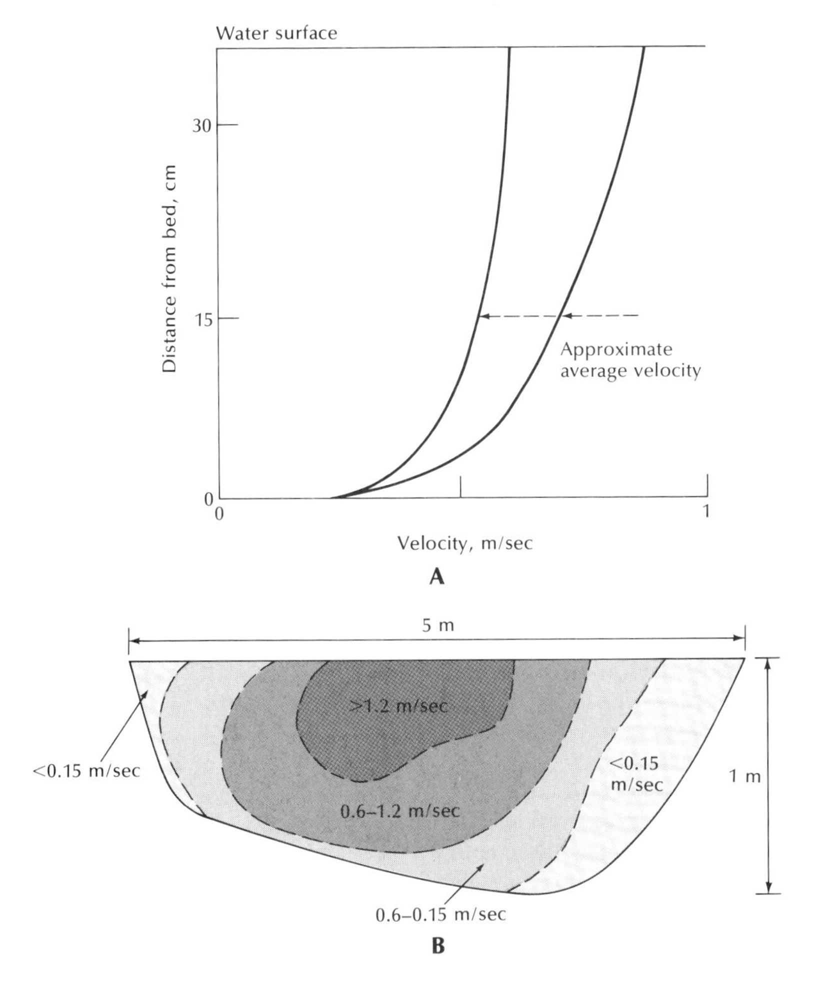

Fig. 11-5 A. (above) Vertical velocity profile for two rivers in Wyoming. Note that the velocity is zero at the bed of each stream. **B.** (left) Velocity distribution in a cross section of a river channel in Wyoming.

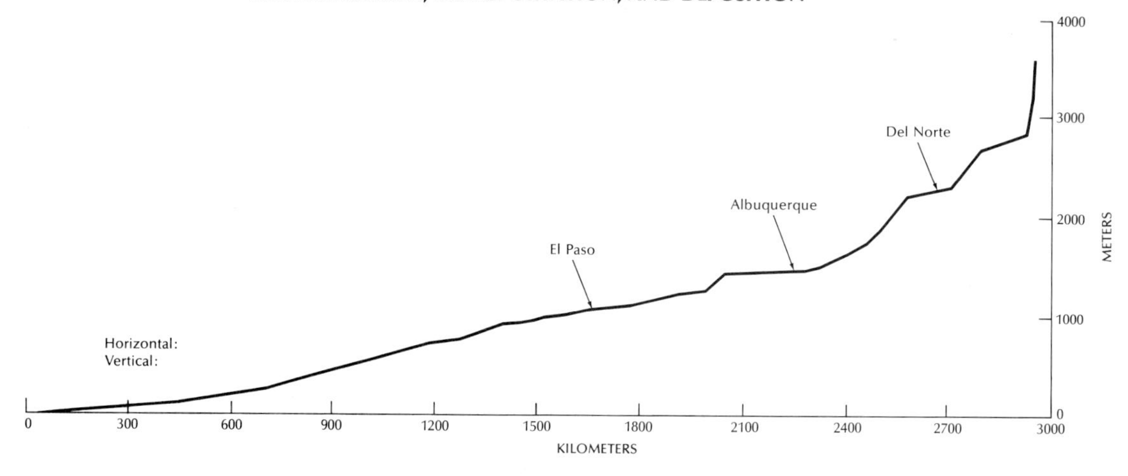

Fig. 11-6 Gradient of the Rio Grande from its headwaters in Colorado to its mouth at the Gulf of Mexico.

ple, for almost all streams. Where the gradient is low, a stream lazes along; where the gradient is high, the water leaps and quickens in its flow to the sea.

The cross-sectional shape of the channel has an effect on the velocity, too. In wide, shallow rivers, much of the water is in contact with the channel perimeter; hence frictional retardation is great and velocity low. The channel shape that allows for the greatest velocity resembles a semicircle—a shape that gives the maximum cross-sectional area and the lowest channel perimeter. The lower the perimeter, the less the frictional retardation and the greater the velocity.

The effect of roughness on velocity is obvious. A smooth, clay-lined channel will promote an even, uniform flow, whereas an irregular, bumpy, boulder-filled one induces enough turbulence to decrease significantly a stream's velocity.

Discharge is the quantity of water that passes a point in a given interval of time in cubic meters or feet of water per second. The discharge of most rivers is far from constant. In northern rivers, it fluctuates with the melting of snow and ice. Tropical rivers, too, especially those in monsoonal regions, show large seasonal variations, determined by the cycle of wet and dry seasons throughout the year. Streams of the arid southwestern United States show as great a range as any. Throughout most of the year they may have no surface flow at all, but during a sudden cloudburst they can become raging torrents, filling the channels from bank to bank.

The Amazon, by far the world's largest river, discharges an average of 156,000 m^3/sec (5.5 million ft^3/sec) to the sea. Just one day's volume would satisfy the water needs of New York City for about five years. For comparison, the discharge of the Amazon is four times that of the Congo, the world's second largest river, and ten times that of the Mississippi River.

Now, back to the effect of discharge on velocity. During a low discharge stage, a river is shallow, much of the water is close to the channel perimeter, and the velocity is low. Increase the discharge, though, and less water per unit volume comes in contact with the perimeter. In such a situation, frictional retardation is less and the river flows downvalley at a higher velocity.

An increase in sediment load and a corresponding decrease in the percentage of water has a strong breaking effect on velocity. That is readily understandable because the more sediment a stream receives, the muddier it becomes, until finally it may become a mud flow. The ratio of solids to water in a muddy, viscous stream may be so high that the viscosity will increase to the point that the stream can no longer flow. Such an effect is demonstrated sometimes by the ephemeral streams produced by short-lived thunder showers in the desert.

All the factors affecting velocity must vary in a downvalley direction in a fairly systematic manner. For example, if the velocity of the upper part of a river were much faster than the lower part, it would overtake the lower part. Or, if the lower part ran much faster than the upper part, it might run away from the latter. Of course, this does not happen, for rivers continually adjust themselves in order to remain intact ribbons of water flowing toward the sea.

Consider, for example, a river heading in high mountains and flowing across vast lowlands. In the mountains, the gradient is steep, but the velocity is slowed by the boulder-strewn rough channel as well as by a relatively low discharge. Progressing downstream, the gradient commonly lessens, but not necessarily the velocity. Indeed it may remain constant or even increase. Commonly, discharge increases as more tributaries join the main stream and as the channel becomes smoother—both contribute to a greater velocity. The tendency for greater velocity, however, could be offset by a lesser gradient, and the result could be constant velocity. The point to be made here is that determining the velocity of a stream is a complex matter, depending on many variables, none of which is easily measured. The interplay of these factors commonly results in a downvalley decrease in stream gradient (Fig. 11-6).

STREAM TRANSPORTATION

The roiled sediment brought down by the Mississippi River clouds the waters of the Gulf of Mexico far seaward of the river's mouth. In the Southwest, within a generation, some fairly large reservoirs have silted up completely, and the lakes once backed up behind the dams have been converted into dreary expanses of muddy or dusty silt, depending on the season. After a heavy rain in these regions, many streets and sidewalks are slippery with a coating of mud or are scattered with stones. Other examples could be cited to show that the land is inevitably wasting away, and that much of it is being swept to the sea.

That annual wastage can be an imposing amount is demonstrated by a single river, the Mississippi, which every day carries about one million metric tons of sediment to the Gulf of Mexico, and even more when in flood. That colossal drain of the central lowlands of the United States removes 5 cm (2 in.) of soil every 1000 years and has resulted in the construction, within the past million years or so, of a broad platform of sand, silt, and clay that covers an area of around 31,000 km^2 (11,966 mi^2) at the river's mouth, with a central thickness of at least 1.5 km (1 mi).

Most people know that streams carry a heavy burden, but they may not know exactly how it is moved. Even experts in hydraulics are unsure of some of the dynamics involved. One problem is that most rivers transport most of their load during flood stages, and sediment sampling at such times can be rather perilous. However, geologists generally agree that a river moves its load in three major ways—in part by solution, in part by suspension, and in part by bodily movement (rolling and sliding) along the bottom of the channel.

Dissolved load The dissolved material rivers carry is supplied largely through the leaching out of ions derived from the weathering of minerals and rocks in soils. It is this invisible dissolved load that gives some river water—especially the water in western U.S. rivers that cross arid or semi-arid regions—its distinctive taste. When such water evaporates, it leaves behind alkali salts, a white residue that covers the ground; from a distance, extensive patches of it look like fields of snow. Such accumulations, which are fatal to nearly all crop plants, are the bane of many irrigation districts. Keeping soluble residues from accumulating in the soil is an unceasing struggle and one that has been lost in a number of reclaimed areas.

Dissolved loads go wherever river water goes, and only precipitate out if conditions permit. They may be determined through an analysis of the water. Error can creep into the interpretation of data, however, because pollution can account for as much as 50 per cent of the dissolved load of some rivers.

Suspended load The contrast between a clear trout stream in the high mountains and the muddy, roiled water sluicing through an arroyo as the aftermath of a desert cloudburst is largely a function of the suspended load: the cloud of sediment in the water. If sediment is very finely subdivided, it may remain buoyed up in the moving water almost indefinitely. That is likely to be true for such particles as clay-size grains; but not so true for silt or sand or for larger particles. How long such material keeps afloat depends on several factors: (1) the size, shape, and specific gravity of the sediment grains, (2) the velocity of the current, and (3) the degree of turbulence.

The effect of these factors is obvious. Flat mineral grains, such as mica flakes, will sift down through the water much like confetti, when compared with the more direct way in which nearly spherical grains settle out. Specific gravity is important, too, because denser substances, such as gold nuggets, with a specific gravity of 16 to 19, are deposited far more rapidly than feldspar grains of the same dimensions, but with a specific gravity of about 2.7.

One of the most important processes that keep particles in suspension is turbulence. If a particle is settling and then is caught in an upward swirl of water, it may be whisked up suddenly in much the same way that tumbleweeds spiral upward in a swirling wind eddy in the desert. So it is unlikely that individual grains of sediment will settle out at a uniform velocity along the entire course of a river. Rather, each particle follows a complex path, drifting down the river with the moving current, here and there swirling erratically—much like a sheet of paper caught in a vagrant wind.

Typically, not all a stream's suspended load is distributed uniformly. Uniform distribution is likely to be true of the finer grain sizes, such as silt or clay, but the greatest concentration of larger grains, such as sand, remains closer to the bottom. Sand-size grains settle out more rapidly than do clay-size particles, and more turbulence and a stronger current are needed to lift sand from the bottom and keep it in suspension. In fact, the sand grains may be in suspension only briefly after which they sink back to the bottom of the channel to become part of the bed load.

Measurements of suspended sediments in a river are fairly easily taken. All that is necessary is to collect a bottle of river water and then to separate out and weigh the solid sediment. If the average discharge is known, the annual suspended load can be calculated.

Bed load Part of a river's burden—the ***bed load***—rolls or slides along the bottom either as individual particles or as aggregates of particles (Figs. 11-7 and 11-8). In terms of work accomplished by a stream in cutting down or widening its channel, the bed load does the lion's share. The bombardment of sedimentary particles against the sides and bottom of the channel wears it away as effectively as though it had been worked over by a harsh abrasive, such as carborundum. In rocky streambeds smooth hollows, called ***potholes***, may be formed (Fig. 11-9).

Individual particles may move by sliding or rolling or by ***saltation*** (from the Latin word, *saltare,* to jump). Saltation could be compared to the game of leapfrog. A sand grain may be rolling along the bottom, or may even be stationary, when it is caught up by a swirling eddy. Then it bounds or leaps through the water in an arching path. Should the velocity be great enough, it may be swept upward to become part of the stream's suspended load temporarily; if not, the particle sinks again, either to remain stationary or perhaps to continue downstream by leaps and bounds.

It is virtually impossible to measure the bed load of a natural stream, as compared to the dissolved load or suspended load. The latter two loads are diffused throughout the main body of the river, but the bed load moves along the most inaccessible part of the stream—the bottom—mainly during times of major floods, a time when the river's bed is least accessible for study. Bed loads can be trapped in the deltas of downstream reser-

Fig. 11-7 Boulders near Manzanar, California, which were once swept along as bed load by flash torrents from the distant canyon. *Ansel Adams*

voirs, and the simplest way to determine annual bed-load transport is to measure the volume of the delta.

Relative importance of dissolved versus solid load Rivers vary in the ratio of dissolved load to ***solid load*** (suspended load plus bed load) for a variety of reasons. It might come as a surprise, however, to hear that about one-half of the total world-wide load transferred to the oceans is in the dissolved form. In short, much of the continents is dissolving away through the effects of rainfall and weathering.

Rivers in the United States show a complete

range in the ratio of the dissolved to solid load (Fig. 11-10). The Colorado River is basically a solid-load river, whereas rivers of the southeastern states carry approximately equal proportions of solid loads and dissolved loads. Climate and relief can explain many of the differences. The Colorado cuts across an arid to semi-arid landscape, a terrain prone to rapid surface runoff that quickly picks up solid-load material. Also, because of the high relief of the region much solid material is being constantly transported to the valley bottoms. In the Southeast, rivers carry less solid material because erosion is less pronounced in an area of low relief and dense vegetation. Larger amounts of rainfall also ensure high rates of chemical weathering, which results in a high dissolved load.

Another factor that affects the amount of solid load delivered to a stream is the interaction between vegetation and precipitation. For example, in an arid region with little vegetation, the relative few occurrences of rain result in enough runoff to transfer a small amount of sediment out of the drainage basin (***sediment yield***). Up to about 38 cm (15 in.) of precipita-

Fig. 11-8 This huge granite boulder, weighing 8 metric tons, stands in the atrium of the national headquarters of the Geological Society of America in Boulder, Colorado. The rock, once carried as bed load by the Big Thompson River, was rounded and smoothed during transport. *Janet Robertson*

Fig. 11-9 Bed load moving across a river bed not only polishes the bedrock, but in places scours out potholes. *G.K. Gilbert, USGS*

tion, the sediment yield progressively increases. Beyond 38 cm, however, grasses become common and protect surfaces from erosion; so the sediment yield starts to decrease. At still higher rates of precipitation, forest vegetation becomes dominant and may retard erosion and sediment yield even further. The curve obtained by plotting these relationships (Fig. 11-11) is a valuable tool for predicting the responses of a stream to climatic change or to changes effected by humans, such as forest clear-cutting. Although the curve is derived from parts of the United States, it may be applicable in other countries.

Putting it all together, we can estimate how long the soil in the United States and the rest of the world will last. The average rate of erosion of the United States, before people began to disrupt the landscape, has been estimated at 3 cm/1000 yr (1.2 in./1000 yr). Humans, through construction and agricultural practices, have caused that to increase

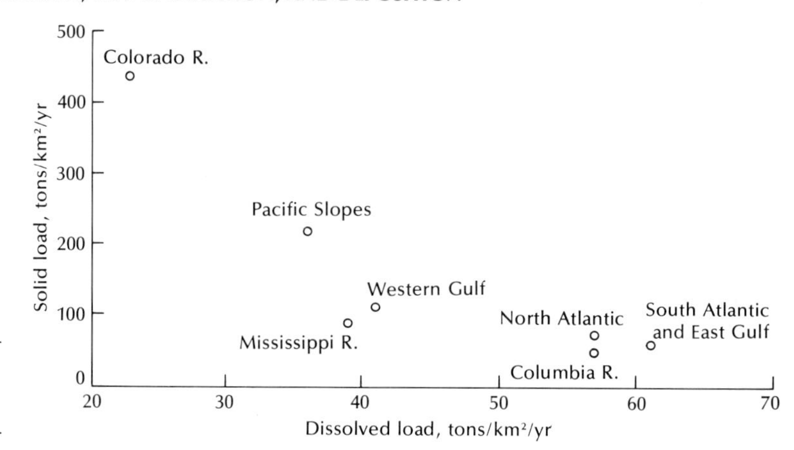

Fig. 11-10 Amounts of dissolved and solid load of various rivers and regions in the United States.

dramatically in places—increases of one order of magnitude or greater (Fig. 11-12). Taking the United States as a whole, however, the present rate is about double the prehuman rate given above.

The prehuman worldwide rate of erosion has been calculated at slightly over 9 billion metric tons/yr, equivalent to a lowering of the total land surface at a rate of about 2.4 cm/1000 yr (1 in./1000 yr). Human intervention has increased the rate to some 24 billion metric tons/yr. At the natural rate, the continents would be swept to the sea in a little over 300 million years, and humans are merely hastening the process along. It is doubtful that all the lands would be reduced to sea level, however, because the geological record clearly indicates that such an event has never happened in the past. Highland areas are surely eroded down, but other lands, formerly low, are pushed up to form highlands, and become good sources of sediment, as is shown by the fairly young marine sediments high on Mount Everest.

Competence and capacity The size of sedimentary particles a stream can transport, or its ***competence,*** depends primarily upon velocity. At low velocities many streams may run clear, and the sediment grains on their beds rest relatively undisturbed. With increasing velocity, the water becomes more and more roiled, and larger and larger particles are picked up. Finally, even such impressive loads as railroad locomotives may be swept along—several were rafted away, out of the roundhouse and into oblivion, during the Johnstown Flood (Pennsylvania) of 1889. In a cloudburst in the Tehachapi Valley of California in 1933, a behemoth of the rails, a Santa Fe steam freight-locomotive and its fully loaded tender, was swept several hundred meters downstream from the tracks and buried in the stream gravels.

The most stupendous example in the United States of the transporting power of running water probably was provided by the failure of the San Francisquito dam in California in 1928. When the 62.5-m (205-ft)-high concrete structure collapsed in the darkness, a wall of water 38 m (125 ft) high swept down the canyon with a velocity of perhaps 80 km/hr (50 mph), enough to raft 18-m (59-ft) blocks of concrete, weighing as much as 10,000 metric tons, almost 1 km (0.6 mi) downstream and to partially bury them in gravel and boulders.

The relationship of load size to stream velocity is not a simple one (Fig. 11-13). Research has shown that in order to initiate particle motion in sand or larger bed-load sizes, a greater velocity is required than that needed to keep the particles in motion. It is the added force of momentum that helps to keep the particles moving. A somewhat surprising result of the research is that the velocity required to initiate motion of clay is similar to that required for sand. The reason for the apparent anomaly is twofold: (1) clays rest at the bottom of the stream where velocities are nil, and (2) clay particles are held tightly together by cohesive forces. High velocities are required to produce the conditions needed to disrupt those forces, setting the fine-grained material in motion. Once in motion, the fine particles stay in suspension over a wide velocity range, quite unlike the narrow velocity range for bed-load material.

Capacity is a term for the potential load a stream can carry, and like competence, it depends in part upon the velocity of the current. A sluggish stream meandering across a swamp is capable of moving very little material, compared to a boulder-rolling mountain torrent. Capacity also depends upon discharge. Obviously, a rivulet of water moving with the same velocity as the mile-wide Mississippi can move but a fraction of that river's immense burden.

If we remember that capacity measures what a stream theoretically can do and that load measures what a stream actually is doing, then we can keep the two terms straight.

Fig. 11-11 Generalized relationship between annual sediment yield, mean annual precipitation, and vegetation.

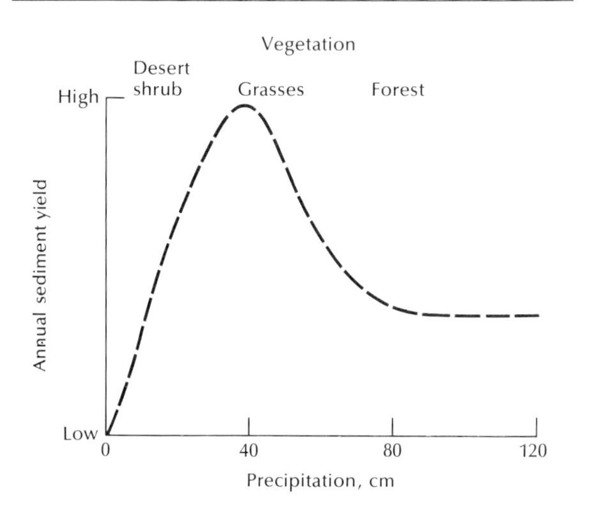

Graded streams and their disruption

A stream with a load that exceeds its capacity will drop its overload as abruptly as a Peruvian llama deposits its burden if it is convinced that it is beyond its carrying capacity. When a stream deposits its excess load on the channel bottom, we say that it is ***aggrading*** its channel, and that may happen when too much sediment is supplied or when the particle size exceeds the stream's competence. Conversely, an underloaded stream—one with a capacity greater than its load—is likely to pick up an additional quantity of material by ***degrading,*** or eroding, its channel.

When a stream is balanced between the two extremes and has achieved equilibrium, so that its slope and discharge give it sufficient current to handle the load, it is ***at grade.*** A ***graded stream*** was aptly defined by the geologist J. H. Mackin:

> A graded stream is one in which, over a period of years, slope is delicately adjusted to provide, with available discharge and with prevailing channel characteristics, just the velocity required for transportation of the load supplied from the drainage basin. The graded stream is a system in equilibrium; its diagnostic characteristic is that any change in any of the controlling factors will cause a displacement of the equilibrium in a direction that will tend to absorb the effect of the change.

A short discussion of Mackin's statement is in order. First, it is important to point out that equilibrium is reached only over a period of years. A river might carry a capacity load only a few days or weeks of the year, during times of flood. The rest of the time it might just be loafing along with little tendency to work.

Another important aspect is that if the equilibrium of a graded stream is disrupted, the stream will react quickly to counteract the disturbance. For example, if more load or larger particles are imposed on a stream, the stream might not be able to handle it. It will deposit the load in the stream bed where it will remain

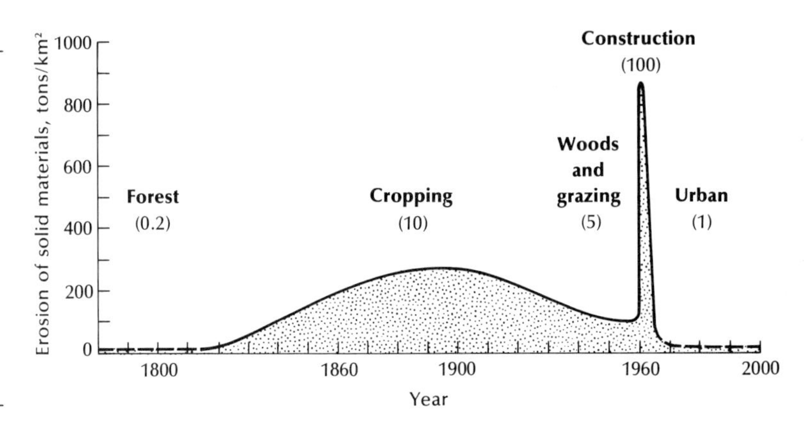

Fig. 11-12 Erosion rate related to land use in the Middle Atlantic region of the United States. The numbers in parentheses are the approximate erosion rate in cm/100 yr for the various time periods. The first increase in the erosion rate coincided with the clearing of forests for agriculture in the early 1800s. The rate decreased in the early to mid-1900s, when the land was used for grazing and partly returned to forests. Construction in the 1960s bared much of the land to runoff and erosion, but that was soon halted as paving, lawns, and other covers that inhibit erosion were added to the landscape.

until a steeper slope that can handle the load is built and a new grade attained. A classic example occurred in the gold rush days in California, when hydraulic mining in the mountains greatly increased the loads streams had to carry (Fig. 11-14). The effects of the mining were felt far downstream in the Great Valley, where the rivers filled in or aggraded part of their channels. Because the channels then were narrower and shallower than before, floods became more frequent.

In the past, when major glaciers formed in the headwaters of mountain streams, they had a similar effect on stream equilibrium. Because glaciers are such effective erosive agents, they greatly increased the amount and size of the load in the rivers draining from their lower ends. The streams reacted by dumping part of their load and steepening their slopes, to attain a new equilibrium.

The removal of part of a load from a river can bring about downcutting, or degradation. For example, when glaciers melted, the loads of many rivers decreased, and the rivers were able to downcut enough to handle the postglacial load.

Altering discharge also can greatly disrupt the equilibrium of a stream because discharge and velocity are closely related. A decrease in discharge lowers both stream competence and capacity, and aggradation results, whereas the opposite conditions and degradation accompany an increase in discharge. How can discharge be altered? Nature can do it in the form of climatic change, or humans can do it in the form of water-diversion projects.

The side effects of damming a river vividly point out what happens when stream equilibrium is disrupted. Above the reservoir, the stream carries its usual load, but when it meets the quiet reservoir waters the load is dumped to form a delta (Fig. 11-15). Because the delta has a low slope, some material is deposited on its surface, and some is swept out and deposited on the delta front. As the delta continues to build out into the reservoir, the valley upstream must aggrade to maintain the equilibrium slope.

Things are not much better downstream from the dam. There the stream has been deprived of its load—part of which now forms the delta—and the river degrades. In any economic analysis of a proposed dam, such factors must be considered. So also must the projected life-span of the reservoir, since it will eventually be filled with sediment.

Another common cause of stream disequilibrium is a change in ***base level.*** Base level is the low point to which most streams flow—the ocean. If the ocean level changes up or down, the streams will aggrade or degrade, respectively. Many coastal streams responded in such a way during the rapid fluctuations in sea level that accompanied the major Pleistocene glaciations.

FLOODS

When rains are heavy, and fall day after day, people begin to consider the possibility of floods, especially if they live along a river (Fig. 11-16). The ground will soak up a certain amount of water, but if rainfall continues, the ground exceeds its capacity to absorb more, just as does a sponge held under an open faucet. At such a point, runoff increases, and the water in the channel might swell to flood proportions (Fig. 11-17A). Increased discharge means an increase in velocity, so that the river bed is scoured and the channel is enlarged. Part of the floodwaters is thus accommodated. But if discharge continues to increase, the ability of the channel to contain the rampaging river is exceeded, and the excess water spills out over the floodplain—that low-lying ground that periodically is inundated with floodwaters.

Earth scientists have estimated how often floods of certain discharges will recur, and terms such as "5-year flood" or "25-year flood" are used. That means that it is estimated that floods of those discharges are thought to occur once in 5 years, or once in 25 years. Flood maps, which show to what extent floods of various sizes might inundate a floodplain, are then made up. Such maps are an integral part of wise land-use planning, and, if they are available, they should be consulted if construction close to a river is being considered.

Flood maps must be updated fairly often. One reason is that each new flood provides new data, which refines the statistics on flood-recurrence intervals. Another is that urbanization tends to increase the incidence of floods of a particular size. The reason is simple: the construction of roads and houses covers ground that once soaked up rainwater. The consequence is, of course, that the runoff for a given-size storm is increased, which, in turn, increases the incidence of floods of a given size (Fig. 11-17B).

Few people driving through the eastern Washington desert realize that part of their

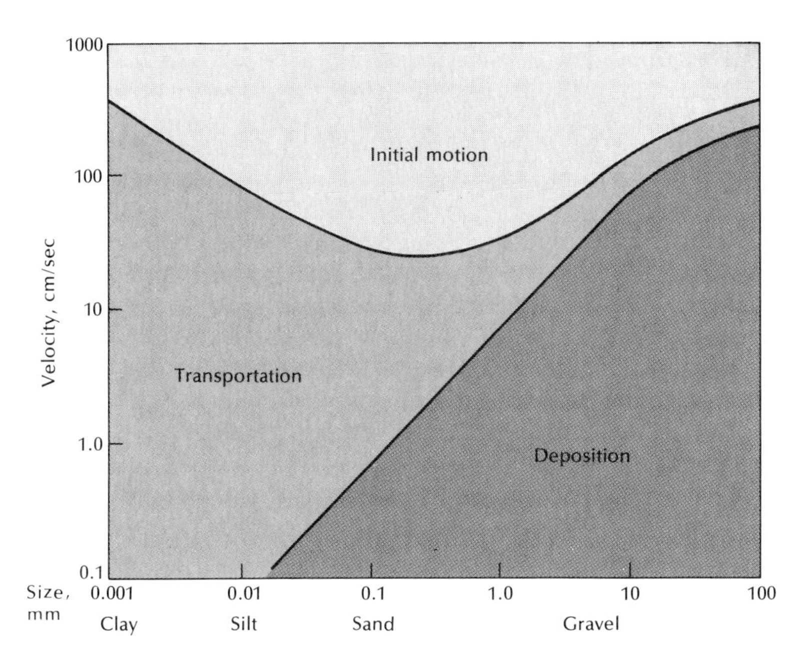

Fig. 11-13 Relationship of stream velocity to initial motion, continuing transportation, and deposition of particles of various sizes.

Fig. 11-14 Hydraulic mining for gold in Nevada County, California, in the late nineteenth century. The excess sand and gravel were washed to the rivers, thus greatly increasing their loads. *Bancroft Library, University of California*

route is along the course of an ancient flood, quite likely the largest on this planet. Called the Spokane Flood, it left such a clear mark on the Columbia Plateau that it appears in photographs taken from orbiting satellites, even though it took place thousands of years ago (Fig. 11-18). Geologist J. Harlan Bretz, more than anyone else, put together the geological pieces on which the fascinating story is based.

The story begins less than 20,000 years ago, when an enormous glacier pushed south along a wide front from Canada into eastern Washington, Idaho, and Montana. Ice blocked major drainages and formed a large lake known as Lake Missoula. Ice can hardly be said to make a stable dam, for it floats on water. Conditions changed, the dam burst, and an amount of water equivalent to at least ten Amazons or one hundred Mississippis was unleashed on the basaltic plateau. Peak discharge lasted a day, and in a week or two the lake had drained. Maximum water depths exceeded 200 m (656 ft), and velocities up to 75 km/hr (47 mph) seem possible.

The area was devastated. Many square kilometers of basalt were reamed out into a landscape appropriately called the "channeled scablands." One of the most spectacular erosional features is Dry Falls, a cataract large enough to encompass several Niagara Falls (Figs. 11-19 and 11-20). Boulders up to 30 m (98 ft) long were moved as far as 3 km (2 mi) by the floodwaters. Gravel bars formed, and some of the larger ones were more than 3 km (2 mi) long and 30 m (98 ft) high. Streams commonly form small sand ripples on their beds, but the Spokane Flood formed giant gravel ripples up to 7 m (23 ft) high, spaced more than 100 m (328 ft) apart.

The scabland features are so different from the expected norm that it is no wonder that Bretz had trouble convincing his colleagues, in the 1920s, of a great prehistoric flood out West. He persevered, gathered more data, and eventually convinced them. Today few geologists doubt his interpretation. In fact, when in 1965 a group of geologists from many nations were taken through the flood area, they wired Bretz a message that concluded with the following words: "We are now all catastrophists."

Similar floods, but on a much smaller scale, occur when a dam fails. On 5 June 1976, the Teton Dam in southern Idaho failed and released 3.4 billion m^3 (4.4 billion yd^3) of water in eight hours (Fig. 11-21). A total area of 480 km^2 (185 mi^2) was inundated by the flood, in which

Fig. 11-15 Classic delta formed by a small tributary to the Snake River, near Huntington, Oregon. *Omar Raup*

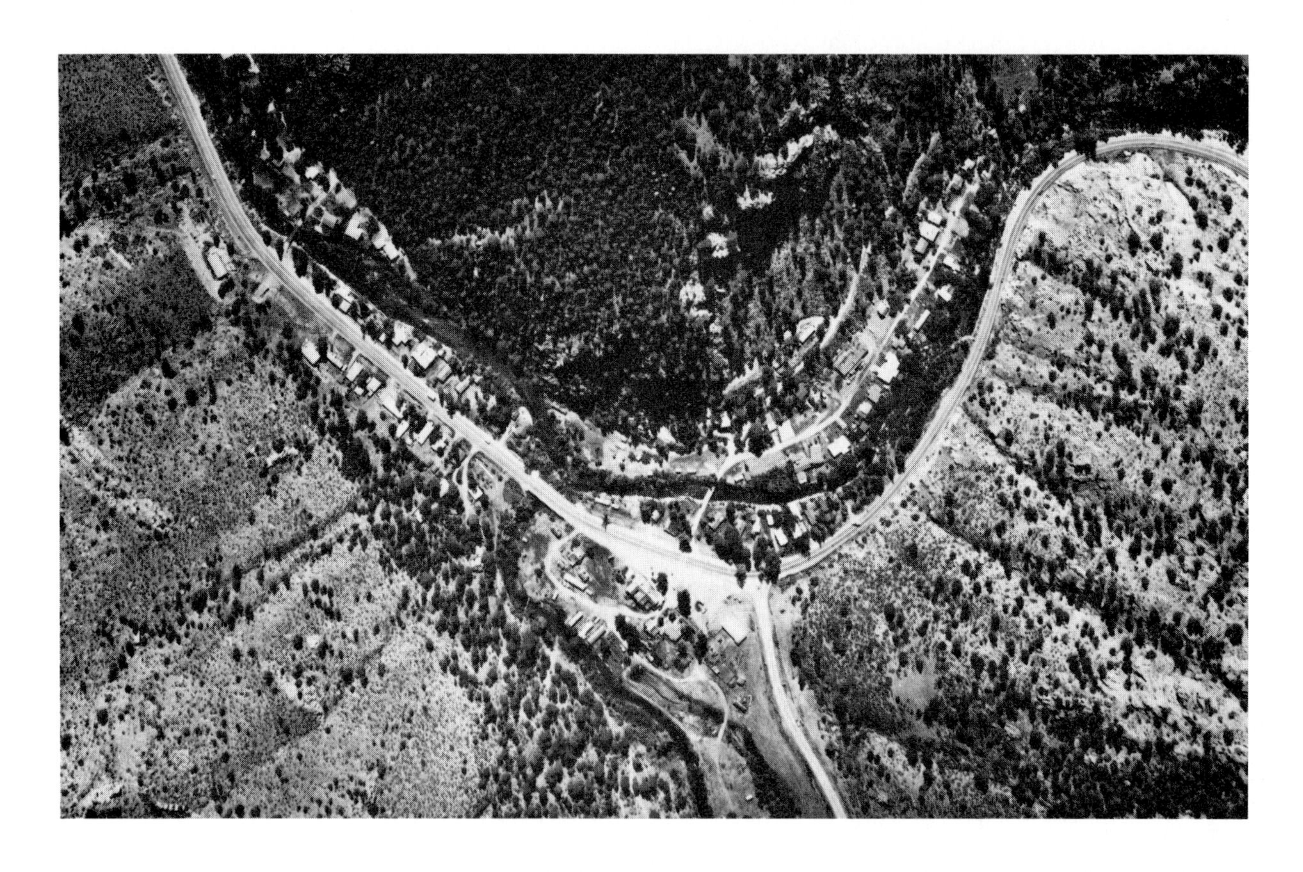

Fig. 11-16 On the evening and night of July 31 and August 1, 1976, 20.9 cm (8 in.) of rain fell in the canyon of the Big Thompson River, Colorado. The soil could not absorb that amount of water and the river flooded, reaching a maximum depth of almost 6 m (20 ft) and killing at least 139 persons. Such floods are estimated to occur every 100 to 500 years in that canyon. **A** and **B.** (opposite) Before and after photographs of the town of Drake, where at least 19 lost their lives. *Hogan* and *Olhausen* **C.** (right) A 2-m (6.5-ft) boulder moved by the flood punched a hole into the side of this house. The largest boulder moved by the floodwaters is estimated to weigh 230 metric tons. *Dave Tewksbury*

eleven persons perished. Damage approached $1 billion. No one knows exactly what happened, but it is thought that the highly fractured and porous bedrock probably was not adequately plugged with concrete, allowing water to flow from the reservoir to the core of the earthfill dam. Once the water reached the dam, it rapidly eroded the structure away.

STREAM DEPOSITION

The broad plains bordering many of the large rivers of the world have been tempting sites for settlement since the beginning of history. Both the Egyptian and Babylonian civilizations were essentially riparian, and the life of their people was bound to the river, whether the Nile or the Euphrates, the giver of life in an arid land.

Across the wide expanse of lowland bordering a river, a stream such as the Nile can spread its waters in time of flood. In fact, the annual flood was an event of such importance to the survival of Egypt that a whole pantheon of deities was given credit for the phenomenon. For centuries, the Nile floods laid down sediment—natural fertilizer—on farmlands adja-

Fig. 11-17 A. Hypothetical relationship between the rainfall associated with an actual storm and the discharge of a stream. The lag between peak rainfall and peak discharge occurs because of the time required for the ground to become saturated. **B.** Alteration of the storm-discharge curve after urbanization. Pavements and other impervious ground result in greater runoff and an earlier arrival of peak discharge, and floods may occur more frequently.

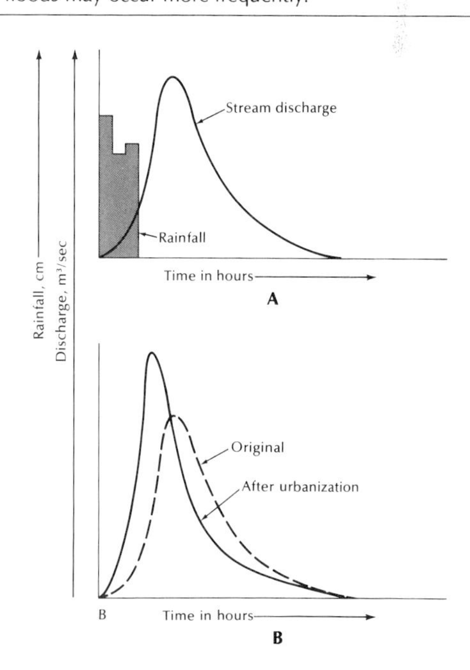

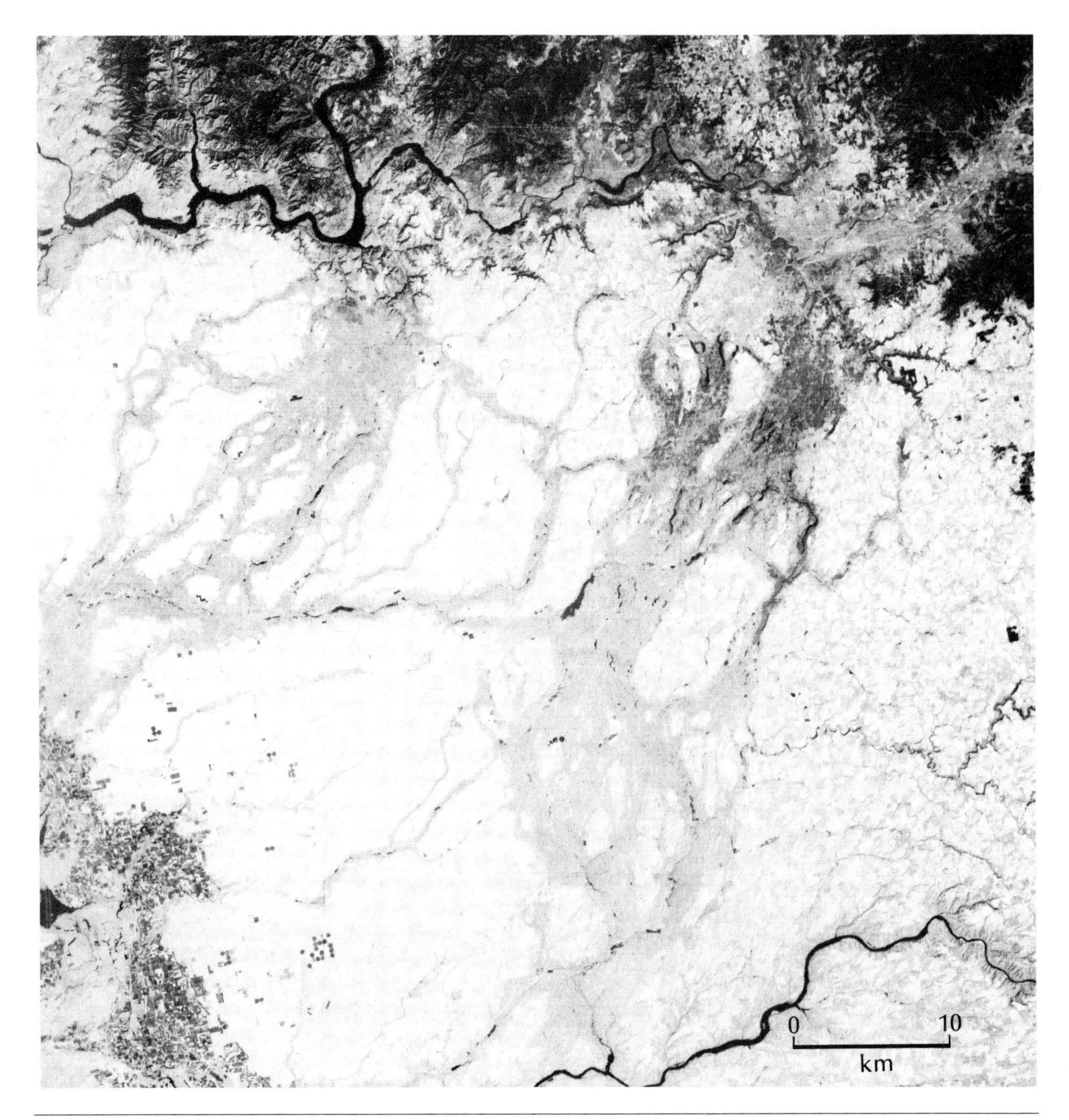

Fig. 11-18 Satellite photograph of the channeled scablands of eastern Washington, carved out by the ancient Spokane Flood. The path taken by the flood is shown by the gray braided pattern, whereas the present course of the Columbia River can be seen in the upper part of the photograph. *Eros Data Center, USGS*

Fig. 11-19 Artist's rendering of the Spokane Flood with the glacier in the background. The cataract in the foreground is Dry Falls and is 5.6 km (3.5 mi) wide and over 120 m (394 ft) high. *Washington State Park and Recreation Commission*

Fig. 11-20 Niagara Falls, as powerful as it is, has only a fraction of the power that was put forth by Dry Falls at the time of the Spokane Flood. The falls, as depicted in this painting by George Catlin, are about 50 m (164 ft) high, and the total width is 701 m (2300 ft). *Private Collection. Photo courtesy of Hirschl and Adler Galleries*

cent to it. But the building of the Aswan Dam in the mid-twentieth century cut off that natural process. Even along the coast, where the Nile meets the Mediterranean, a decline in fish harvests and increased erosion may be tied to the fact that the dam checks not only the floods but the nutrients and sediments that the floodwaters once carried to the sea. Of equal economic importance is the increased incidence of the parasite *Schistosoma,* which causes a debilitating, sometimes fatal disease in humans.

In the pages that follow, we will discuss the characteristic features of floodplains, mention some problems in floodplain maniuplation, and conclude with a description of sedimentation and landform patterns where rivers meet the sea.

Floodplain features From the nature of the term itself, we expect the surface of a floodplain to be covered with deposits from a flooding river. For example, the channel of the lower Mississippi River, from a short distance below its junction with the Ohio down to the Gulf of Mexico, lies wholly upon its own alluvium that it deposited.

Most of the more common floodplain features are along rivers that swing in large curving bends, or ***meanders,*** as they progress toward the sea (Fig. 11-22). The word "meander" comes from the name of a river in western Turkey, the Menderes, and the derivation of the word is from the Latin *maendere,* to wander. Meanders are one of the two most common river-channel patterns. If a meandering river cuts down rapidly, entrenched meanders are the result (Fig. 11-23).

Floodplains sometimes are bordered by low bluffs marking the outer limits of the band of swampy ground across which the river has been free to meander (Fig. 11-24). Most such rivers—and the lower Mississippi is typical—are confined by low embankments, or natural ***levees,*** which slope gently away from the river. A section of low-lying ground, the ***backswamp,*** commonly is found between the bluff and the natural levee. It is a boggy section on which the surface waters cannot flow back into the river because the slope of the natural levee is against them.

Natural levees are built by the river when it overtops its banks during a flood. Rather than surging violently over its banks at a single outlet, a typical flood moves away from the river toward the backswamp as a sullen, tawny, inexorably spreading tide of muddy water. The greatest check to the velocity of floodwaters comes when they leave the stream channel and first encounter the lake-like expanse of floodwaters that have inundated the backswamp. The bulk of the suspended sediment is deposited forthwith at the channel's edge, where the sudden drop in velocity occurs. The result, then, is that a narrow embankment, chiefly of fine sand and silt, builds up on either side of the river. Because their water content is less than the immediately adjacent, ill-drained backswamp, and the grain size of their sediment is larger than that of the swamp muck, the natural levees provide the only solid ground in such a saturated region as the lower Mississippi floodplain. For that reason, roads, settlements, and farms are clustered along the higher and firmer ground of the levee immediately adjacent to the river. The backswamp, with its intricate pattern of bayous, branching channels, and lakes is virtually uninhabited, with the exception perhaps of birds and muskrats and their hunters.

In some places, tributary rivers in the low areas between the natural levee and the bluff (see Fig. 11-24) run parallel to the mainstream for many kilometers before they are able to join it. Geologists have termed such rivers ***yazoos,*** after a stream of that type—the Yazoo River in Mississippi (Fig. 11-25).

In meandering rivers, the main current follows the outer side of a bend, where it is shunted by centrifugal force. There the current is strong, and the bottom of the channel is scoured more deeply than elsewhere. When the current leaves a bend, it normally does so on a tangent and may occupy a variety of positions before reaching the next bend. Upon entering the next bend, it crosses to the opposite bank. The ***crossings***, as they are called, are be-

Fig. 11-21 A. (above) The Teton Dam in southern Idaho during the time of failure. The remaining section of the dam is to the right of the falls. To the left is the spillway. *Robert Jensen* **B.** (below) Looking upstream a day after the dam failed. The debris adjacent to the stream below the dam was deposited by the floodwaters. *U.S. Army Corps of Engineers*

Fig. 11-22 River meandering across the plains of southern Wyoming. Meanders usually grow from the outside of the bends, and a cutoff course occurs when two outward-cutting bends meet. Oxbow lakes mark the course of the river cutoff. *Janet Robertson*

set by shoals and shallows, or sand bars.

If we draw two profiles of a river channel, the first between two bends, where the current is approximately in midstream, and the second in a bend, they will be quite different. In the first profile, the river flows in a broad, shallow, nearly flat-floored trough. But in the bend, the outside of the curve is deep and the inside is shallow. In other words, the cross-sectional pattern is wedge shaped, with the thicker part of the wedge on the outside (Fig. 11-26).

How meanders enlarge is a question that has been long debated. The most generally accepted theory is that the outside bank is undermined by deep scouring, especially when the river is in flood, thus over-steepening the bank and slicing away its foundations. Commonly, the bank fails by slumping into the stream through the removal of support at the base, rather than by being sawed horizontally by the river. The outside bank is called the ***cut bank.*** Much of the material derived through slumping is deposited on the inside of the next bend downstream as a ***point bar.*** Thus, meanders can migrate laterally, yet keep the same cross-sectional area, for generally the same amount of material that is taken from the cut bank of a bend has been added to the point bar of the next bend downstream (see Fig. 11-25).

Meanders migrate in the direction of the cut

bank. In places, the neck of land between the bends has been eroded through and a new, more direct channel—called a ***cutoff***—has been cut. The old channel then contains a crescent-shaped lake—an ***oxbow lake***—that eventually fills in with sediment. All told there have been about 20 such naturally occurring cutoffs on the lower Mississippi since 1765. About 15 have been made artificially by the Mississippi River Commission since 1932 to straighten the river's course, thereby increasing the gradient and thus the velocity, and as a consequence theoretically diminishing the flood hazard by improving the hydraulic efficiency of the channel. As we shall see later, however, straightening channels can have some disastrous side effects.

Historically, one of the more interesting cutoffs of the Mississippi occurred at Vicksburg in 1876. Before that date the river made a broadly sweeping curve past the city (Fig. 11-27). In 1876, however, the river formed a cutoff south of Vicksburg that isolated that town from the river. Ironically, in 1862, General Ulysses S. Grant tried unsuccessfully to construct an artificial cutoff at about the same place so that Union river traffic could by-pass Confederate guns at Vicksburg.

Let Mark Twain, in his role of steamboat pilot, have the last word on meandering streams. In *Life on the Mississippi* he writes about the river meanders and how the river is shortened when it cuts through the narrow neck of a meander. He grossly misuses the principle of uniformitarianism and is not very kind to science in general, but who will argue?

Therefore, the Mississippi between Cairo and New Orleans was 1215 miles long 176 years ago. It was 1180 after the cutoff of 1722. It was 1040 after the

Fig. 11-23 Entrenched meanders, locally known as goosenecks, of the San Juan River, Utah. *Tad Nichols*

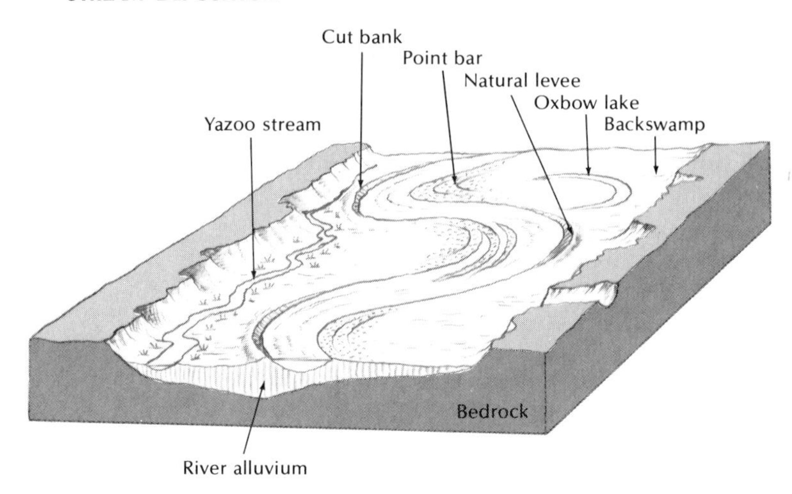

Fig. 11-24 Characteristic features of the floodplain of a meandering river.

American Bend cutoff. It has lost 67 miles since. Consequently its length is only 973 miles at present.

Now if I wanted to be one of those ponderous scientific people, and "let on" to prove what had occurred in the remote past by what had occurred in a given time in the recent past, or what will occur in the far future by what has occurred in late years, what an opportunity is here! Geology never had such a chance, nor such exact data to argue from! . . . Please observe:

In the space of 176 years the Lower Mississippi has shortened itself 242 miles. That is an average of a trifle over one mile and a third per year. Therefore, any calm person, who is not blind or idiotic, can see that in the Old Oolitic Silurian Period, just a million years ago next November, the Lower Mississippi River was upwards of 1,300,000 miles long, and stuck out over the Gulf of Mexico like a fishing rod. And by the same token any person can see that 742 years from now the Lower Mississippi will be only a mile and three-quarters long, and Cairo and New Orleans will have joined their streets together, and be plodding comfortably along under a single mayor and a mutual board of aldermen. There is something fascinating about science. One gets such wholesale returns of conjecture out of such a trifling investment of fact.

In places, meandering rivers are considered a real nuisance, especially if the lateral migration is at the expense of valuable land. Remedial measures to thwart such wayward rivers meanders vary, but one common one is to line the outside bend with that symbol of American affluence—the junked car.

If rivers do not have a meandering pattern, chances are that the pattern will be ***braided,*** the other major river pattern. Rather than flowing in a single, rather narrow, but deep channel, the river follows a braided pattern—a series of wide, shallow anastomosing channels that continually join and part from one another in a downstream direction (Fig. 11-28). The individual channels may change position hourly, daily, or seasonally—they are quite active.

Some rivers are braided and others meander mainly as a result of the type of sediment they carry. Meandering streams require a rather tough bank material to restrict the river to a single channel. The toughest material is cohesive silt and clay, and so it is that meandering rivers are those with a relatively high suspended load. In contrast, a stream can easily spread into many shallow channels if the bank material is loose and non-cohesive. The weakest bank materials are sand and gravel; hence, braided streams are characteristically bed-load streams.

Stream pattern is a part of the equilibrium of a stream, which is why Mackin refers to "prevailing channel characteristics" in his definition of a graded stream. A high-gradient, boulder-carrying braided stream can be at grade just as can a low-gradient, low-velocity meandering stream burdened with silt.

Deltas Herodotus, in the fifth century B.C., impressed by the branching pattern of the distributaries of the Nile, compared the form of the watery, muddy region between Cairo and Alexandria to the Greek letter *delta* Δ. The comparison is so apt that it has won acceptance in most of the languages of western Europe.

The Nile delta is a nearly ideal example of that particular landform, so much so that few others measure up (Fig. 11-29). A high-altitude photograph shows how the main channel of the river separates into a host of branching, lesser arms called, quite appropriately ***distributaries.*** Another feature of the Nile delta is the bordering bays and lakes, of which one, Abu Qir Bay, is a good example. Abu Qir was the site of the Battle of the Nile, in which the French fleet was destroyed by Nelson, thus ending Napoleon's hopes for an Eastern empire. Similar ***delta-flank depressions,*** as such water bodies are called, border many of the other deltas of the world. Well-known ones are Lake Ponchartrain by the Mississippi delta, the Zuider Zee (IJsselmeer) and marshes of Zeeland adjacent to the Rhine, and the lagoon surrounding Venice at the mouth of the Po.

Not all the world's rivers form deltas where they enter the sea. Two large ones that do not are the St. Lawrence and the Columbia—the St.

Fig. 11-25 Meandering streams that migrate laterally at a rapid pace leave the impression of point bars all across the landscape. Yazoo River, Mississippi. *Frank Beck*

Lawrence because it has little chance to pick up much sediment in the short run between Lake Ontario and the Gulf; the Columbia because it discharges directly into the open sea, where powerful waves and currents quickly redistribute the rivers's burden of sand and silt.

The best-developed deltas are most likely to be constructed where a river moves a large load of sediment into a relatively undisturbed body of water. Examples of such impressive accumulations of riverine deposits are the great deltas at the mouths of the Ganges-Bramaputra in India and East Bangladesh, the Indus in Pakistan, the Tigris-Euphrates in Iraq, the Niger in Nigeria, the Yangtze Kiang and Hwang Ho in China, the Mississippi in the United States, the Danube in Romania, and the Volga in the USSR. One of the most picturesque of the deltaic worlds is the Camargue, the land of cowboys and semi-wild cattle, at the mouth of the Rhône in southern France.

In North America, the Mississippi is by far the best known geologically, not only because of its long record of channel changes but because about 90,000 oil wells have been drilled in its sediments and because repeated geophysical surveys have been made up and down its length. All that information combined gives us a uniquely detailed, three-dimensional picture, not only of the 29,000 km² (11,194 mi²) of delta surface, but of the over 1.5-km (0.9-mi)-thick sediments below the waters of the Gulf of Mexico as well.

The great delta of the Mississippi is not a simple structure; it is a compound feature built up through a complexly overlapping pattern of ***subdeltas*** that have been constructed over the last 5000 years. From the older deltas, it can be seen that the river established its present course only very briefly before the arrival of the first European explorers, Cabeza de Vaca in 1528(?) or Hernando de Soto's followers in 1544.

More than 10,000 years ago, sea level was lower than it is at present as a result of the stockpiling of some oceanic waters on land in the form of glaciers. At the height of this glaciation, the Mississippi flowed in a channel entrenched in older deltaic sediments. It reached

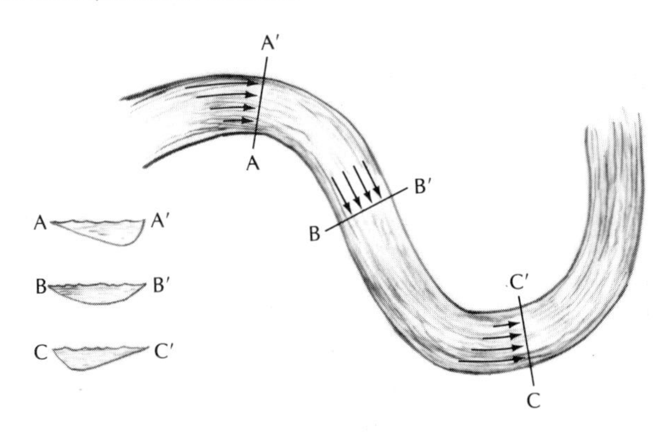

Fig. 11-26 Plan view of a meandering river, showing the change in channel shape with position along the river. The length of the arrows depicts relative river velocity.

the Gulf of Mexico at the head of a submarine canyon located south of New Orleans and southwest of the present birdsfoot delta, so named for its obvious shape. Melting glaciers returned water to the sea, the sea level rose, and the present delta began to take shape.

At Head of the Passes, the present-day river breaks up into three major and a number of minor distributaries (Fig. 11-30), and there the river depth is around 12 m (39 ft). The slope of the channel bottom in the distributaries is upstream against the gradient of the river surface because most deposition occurs at the mouth of the distributaries; there the river builds up shallow sand bars as it enters the Gulf of Mexico. Thus, the normal low-water depth at the mouth of most distributaries is as little as 3 to 4.5 m (10 to 15 ft)—without considering dredging and jetty construction—and in some of the minor, less frequently used channels, as little as 1 to 1.5 m (3 to 5 ft).

Natural levees border the channels of distributaries that are growing outward into the Gulf. The height of the natural levees increases gradually inland from the Gulf, until they are around 4 m (13 ft) high at New Orleans, a distance of 165 km (103 mi) from Head of the Passes.

This strange deltaic world—half water and half land, the abode of water birds of great variety, with a continuously changing pattern of lakes, swamps, marshes, and constantly shifting streams—is one of the most important environments for human beings. From the beginnings of historic time, their fertile soil, their network of waterways, and their position as a meeting ground between seafarers and those who make their living along the rivers of the world have made deltas tempting sites for coastal cities. Such a bewildering maze of channels, large and small, was made to order for pirates, as witness the successful operations for many years of the brothers Lafitte at Barataria in the Mississippi delta.

Quite a price is exacted, however, from a delta city in return for its communication advantages. Such a city is under constant threat of inundation by flood, and building foundations are insecure—Venice is an impressive illustration of what differential settling can do to structures built on delta mud. Even such a shallow excavation as a grave may fill with water (the vaulted sepulchers of New Orleans are an answer to that problem), and developing an unpolluted local water supply is difficult. But the greatest threat of all is that the river may change its course completely or that the channel may silt up. A good example of the latter fate is the ancient city of Ravenna; in the days of the Emperor Justinian (483–565) it was a leading seaport on the Adriatic; now it is about 10 km (6 mi) from the coast.

The low country bordering the mouths of the Rhine (the Randstad) provides an impressive example of the problems besetting the inhabitants of a delta. The Randstad is one of the most densely inhabited regions of the world and one where an immense volume of seaborne and riverborne trade moves through such ports as Rotterdam and Amsterdam. Those cities, already below sea level, throughout their entire existence have not only had to fight off overland armies—a hazard to which delta cities are peculiarly vulnerable—but have had to hold back the sea. The reasons for the problems faced by the people of the Rhine delta are complex, but they certainly include

Fig. 11-27 Major changes in the courses of the Mississippi and Yazoo rivers over a 16-year period.

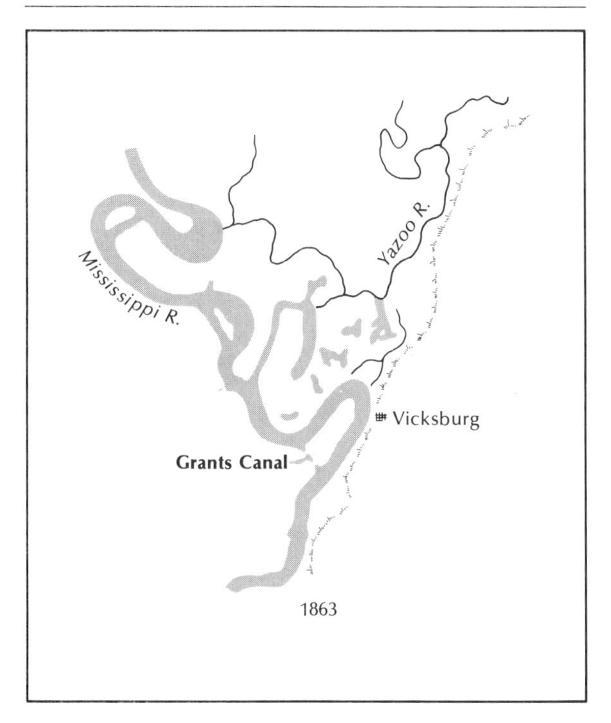

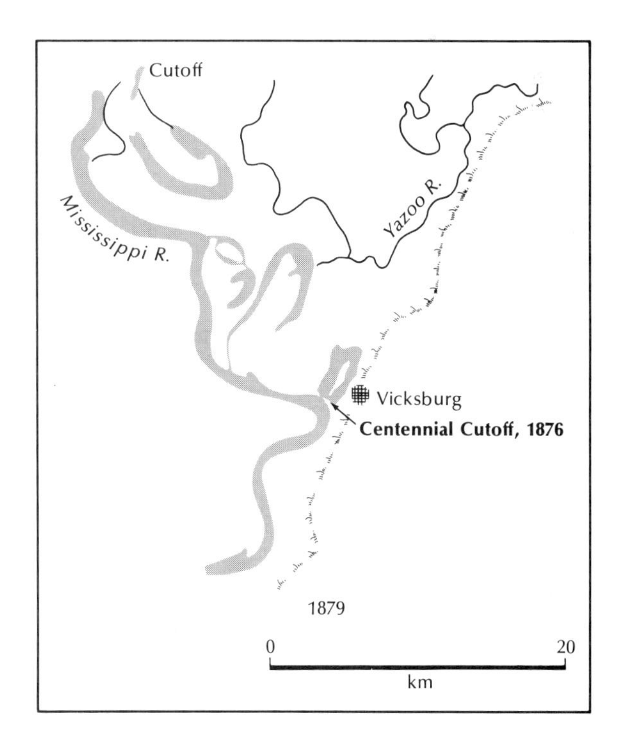

Fig. 11-28 Braided channels of the Muddy River near its junction with the McKinley River, Alaska. *Boston Museum of Science, Bradford Washburn*

the currently rising level of the sea, the compaction of the water-logged clays upon which the cities stand, and the apparently geologically active subsidence along their part of the North Sea coast.

Many of the same problems plague New Orleans, not the least being the problem of subsidence. Indications of a relative lowering of the land with respect to the sea are everywhere. Among them are the remains of Indian settlements far out on the bottom of Lake Ponchartrain, drowned cypress trees and inundated farmland around the lake's margin, and the sunken streets and graves of the deserted settlement of Balize near Head of the Passes. Its now silent shell-paved streets are buried under the marsh about 1 m (3 ft) below sea level.

There is little disagreement over the evidence of subsidence in many of the world's deltas, but there are strong differences of opinion as to the cause of subsidence. There are those who believe that the addition of as much as 1.8 million metric tons of sediment a day, as in the Mississippi delta, overloads the earth's crust, which subsides as a consequence. Others point out that the excess load is not so great as might appear at first, since the weight of the displaced sea water has to be considered, too.

Thus, the material added to the crust may have a specific gravity of about 1.8. How the lighter material displaces heavier sub-crustal material with a specific gravity of perhaps 3.3 is a tricky question to answer.

The answer probably is not a simple one, but certainly involves many of the factors operating in the Netherlands. The addition of an extra burden of sediment, as in the Mississippi delta, may not be adequate in itself to bring about a broad crustal downwarping, but if a load such as the Mississippi's accumulates in a region where subsidence is the dominant geological process, the weight certainly is contributory.

Stream terraces—a case for disequilibrium Extensive flat areas slope downstream more or less parallel to the slope of many streams (Fig. 11-31). If the "flats" are underlain by river deposits, the landform is called a ***river terrace***. In essence, such terraces are ancient floodplains, long abandoned and stranded high enough so that present-day floods no longer overtop them. River terraces and their associated deposits are important in deciphering the history of a river—a task, as we shall see, that is far from easy.

Two major kinds of terraces are recognized (Fig. 11-32). One is a ***cut terrace,*** so named because it consists of a thin veneer of gravel resting upon a fairly smooth surface cut from bedrock. How thin is thin? Geologists have their opinions, but most would say that the layer of gravel can amount to no more than the thickness moved during the deep scour that accompanies major floods. For small rivers, that means a depth of about 5 m (16 ft) or less, and for large rivers about 10 m (33 ft). The other kind of river terrace is a ***fill terrace.*** It too rests on bedrock, but the river deposits are relatively thick. It is difficult to generalize, but the deposits are thicker than the depth of scour and hence thicker than the gravels of cut terraces. (Geologists find such qualitative rules of thumb indispensable, and this is just one of many.)

We can learn something about former river behavior through the study of terraces and deposits. First, consider cut terraces. In some places they might be 1 km (0.6 mi) or more wide. In order to have developed such an extensive former floodplain, a river must have been in near-perfect equilibrium for a long time, swinging back and forth across the floodplain and nipping away at valley walls. Imagine a horizontal saw cutting through the land-

Fig. 11-29 Satellite photograph of the Nile Delta. *NASA*

Fig. 11-30 At Head of Passes, the Mississippi River divides into several distributaries, which lead out to the Gulf of Mexico. *Humble Oil and Refining Co.*

scape. Although the rates at which rivers cut through bedrock are hard to determine, we can estimate that surfaces of 1 km (0.6 mi) or more wide found in many parts of the American West may have required near-equilibrium conditions for hundreds of thousands of years. Something then changed the equilibrium of the river, so that it subsequently downcut, leaving the abandoned floodplain as a cut terrace.

Fill terraces have a more complicated history. Initially, a river flows at a low level in its valley, perhaps swinging back and forth on a floodplain. Aggradation then follows as river sediment is deposited on the floodplain rather than being carried farther downstream. Next, the floodplain becomes a terrace as the river downcuts. The sequence of events, therefore, is aggradation followed by degradation. Some valleys have many fill terraces inset one into another (Fig. 11-33)—indicating that the aggradation-degradation cycle has taken place many times.

It takes nothing short of a good detective to come up with a reason for terrace formation, and, like so many other geological solutions, the best one usually seems to be a list of possibilities. In general, geologists look for "downstream" or "upstream" causes. Downstream variation in base level always is a possibility because a river responds quickly to variations in sea level at its lower end, aggrading as the sea level rises and degrading as it falls. An upstream cause could be the reaction of a river to changes in the size and amount of load or in the volume of discharge—changes that might have resulted from a climatic change. The slope of a terrace, the characteristics of its deposits, and any remaining ancient channel patterns etched on its surface all have to be compared with similar features of the present-day river before any one hypothesis is preferred over others.

Tectonism also can affect river-terrace formation. For example, rivers usually entrench in areas undergoing uplift, but aggrade in areas undergoing tectonic flattening. Some thousands of years later, an uplifted area might be characterized by cut terraces and a flattened area by fill terraces. Furthermore, because faults might not be recognizable in the loose gravels, river terraces may be the only clue to earlier tectonic activity.

Artificial stream disequilibrium Rivers may be in or out of equilibrium naturally or they can be helped along by humans. The problem here is that people choose to live along rivers, but do not always build in the right places—suitable sites can only be chosen after long-term study of the behavior of the river. So when catastrophic events take place—catastrophic to landowners but run-of-the-mill events in the life of the river—the people affected want somebody to do something. In many places somebody did do something, and in many places the results have had unforeseen side effects.

Examples of river disturbances caused by humans abound. Because there are so many, we will cover only one fairly "innocent" one here, innocent at least at first glance. It should be pointed out that, as with any patient, before remedial measures are prescribed it is not a bad idea to try to find out how a river reacted to natural changes in the past before imposing new changes on it. As was emphasized earlier, however, many rivers, like many people, do not divulge their past histories readily.

An example is that of the meandering Blackwater River in Missouri. Local flooding had been a problem before 1910, and measures to control it seemed in order. The parameters of the original river in the headwaters reach were 54 km (34 mi) long, 1.7 m/km (9.3 ft/mi) gradient, with 1.8 meanders/km (3 meanders/mi). Bridges 15 to 30 m (49 to 98 ft) wide spanned the river. Because meandering rivers are slow moving, they cannot get rid of the rapidly increasing discharge that accompanies storms, and the result is flooding. The common remedial practice is to cut a straighter channel, which increases both stream gradient and velocity and disperses the floodwaters more quickly.

In 1910, the Blackwater was straightened and channelized. As Mark Twain points out so vividly, straightening also results in shortening. The river was shortened by about one-half at the expense of increasing the gradient to 3 m/km (17 ft/mi). The artificial channel was cut 9 m (30 ft) wide at the top, 1 m (3 ft) wide at the base, and 3.8 m (12 ft) deep, for a cross-sectional area of 19 m^2 (23 yd^2).

The floods were contained in the channel, but the latter grew quickly. Over the following 60 years, the river sector that enlarged the most measured 71 m (233 ft) across at the top, 13 m (43 ft) at the bottom and 12 m (39 ft) deep, for a cross-sectional area of 484 m^2 (579 yd^2). In any river system the tributary streams follow the lead of the main stream, and this drainage was no exception. The tributaries of the Blackwater downcut just as fast as the main stream, and gullies began to spread over the landscape.

Bridges were built across the straightened channel, but as the channel grew wider and deeper, the bridges had to be lengthened and the vertical pilings extended to avoid collapse. In spite of those efforts, many bridges could not withstand the stress and did collapse.

The effects of channel straightening were also felt downstream beyond the limits of the artificial channel as sediment derived from the enlarging channel was deposited. Deposition was so rapid that two successive generations of fence posts were buried in the flood debris.

The story of the Blackwater is an excellent

Fig. 11-31 A. (opposite, above) Flight of river terraces along the Madison River, Montana. *W.C. Bradley* **B.** (opposite, below) River terraces along the Rangitata River, Canterbury, New Zealand. Most of the terraces were formed during the time of the last glaciation. *New Zealand Geological Survey* **C.** (above) Rounded river gravel overlying light-colored bedrock in a terrace of the Awatere River, New Zealand. Finer-grained deposits, either of floodplain or aeolian origin, overlie the gravel. Before a flat area adjacent to a river can be called a river terrace, deposits such as these must be present. *Peter W. Birkeland*

small-scale example of what can happen when an attempt is made to regulate a river. Once the forces are set in motion, they are difficult to check. We now have enough case histories to intelligently predict the side effects of most regulatory measures. We can, then, only hope that the arguments of cost versus benefit can be properly laid before those individuals charged with making decisions.

Role of streams in landscape evolution

Today virtually everyone recognizes that valleys, even such imposing ones as Hells Canyon in Idaho, the Grand Canyon in Arizona (Fig. 11-34), or Kings Canyon in California—all of them more than 1 km (0.6 mi) deep—were cut by the narrow ribbon of turbid water in the stream at the bottom, barely visible hundreds of meters below the canyon rim.

This belief, so obvious now, was one disputed and was not fully accepted even by such illustrious figures as Sir Charles Lyell and Charles Darwin. As late as 1880, many geologists believed that although streams were capable of some downcutting, deeper and more impressive gorges, such as the Grand Canyon, resulted from a violent sundering of the earth's crust. The presence of such a river as the Colorado was purely fortuitous—instead of cutting the canyon it simply followed the course because it was the lowest and easiest route to follow.

Although a belief in a cataclysmic origin un-

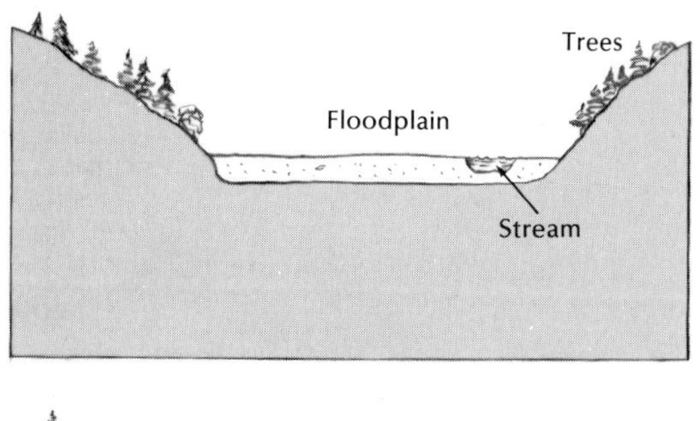

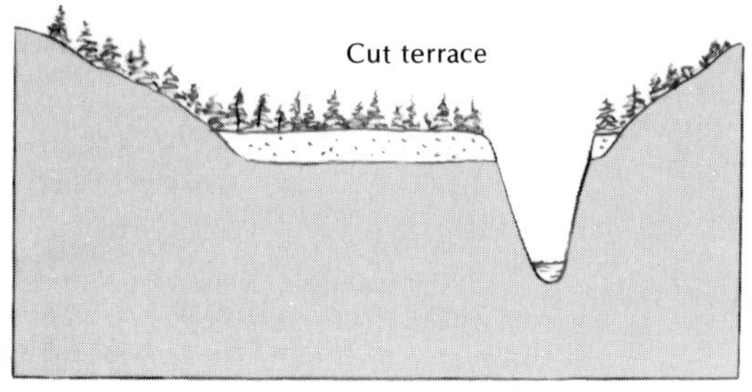

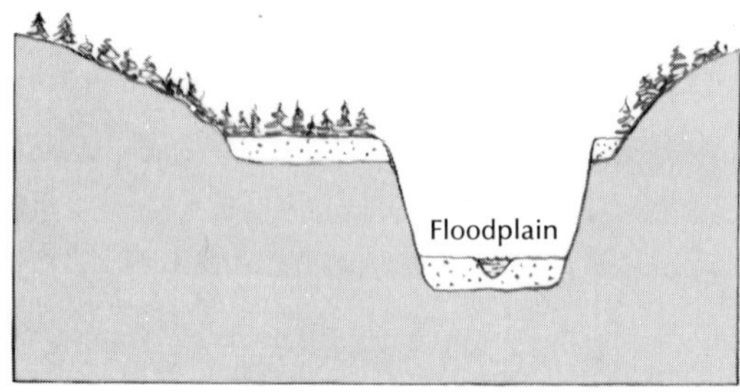

Formation of a cut terrace

Younger

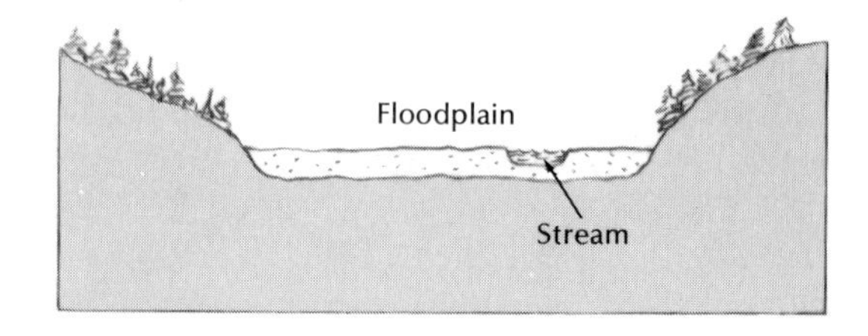

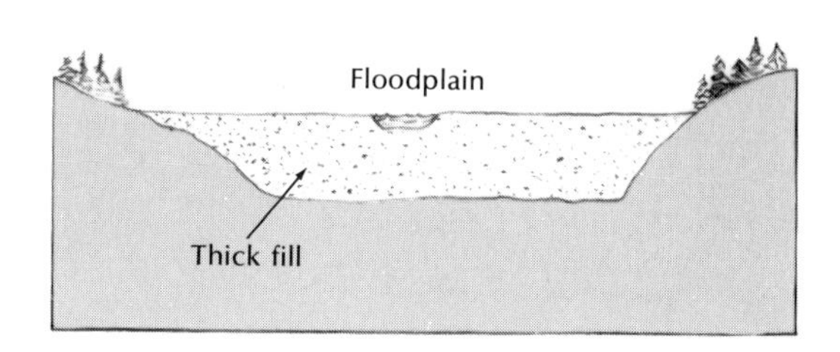

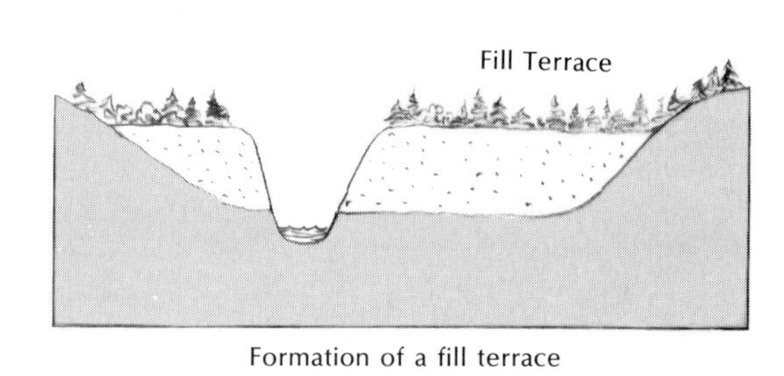

Formation of a fill terrace

Fig. 11-32 The formation of a cut terrace (left) and a fill terrace (right).

questionably has greater appeal to the imagination, it simply means that many geologists of a century ago had overlooked the succinct and eloquent statement made in 1802 by a Scottish mathematician and amateur geologist, John Playfair (1748–1819). Although his essay is now widely quoted, it is worth repeating because of the clarity of style—an attribute not always characteristic of scientific writing today—and because his choice of essentially simple, straightforward words made the point so long ago that streams do, in fact, excavate the valleys they occupy.

If indeed a river consisted of a single stream, without branches, running in a straight valley, it might be supposed that some great concussion, or some powerful torrent, had opened at once the channel by which its waters are conducted to the ocean; but when the usual form of a river is considered, the trunk divided into many branches, which rise at a great distance from one another, and these again subdivided into an infinity of smaller ramifications, it becomes strongly impressed upon the mind, that all these channels have been cut by the waters themselves; that they have been slowly dug out by the washing and erosion of the land; and that it is by the repeated touches of the same instrument that this curious assemblage of lines has been engraved so deeply on the surface of the globe.*

*From Illustrations of the Huttonian theory of the earth, by John Playfair, Edinburgh, 1802.

The Colorado River is responsible for cutting its channel to the bottom of the Grand Canyon, but, as noted earlier, it is not responsible for the entire excavation of the gorge, 21 km (13 mi) wide at the top. The widely flaring upper portion of the canyon, outside of the narrow slot actually cut by the Colorado, is the result of mass movement downslope, of weathering, and of slope wash. The river has served as a gigantic, constantly running conveyor belt, carrying away most of the debris supplied to it while continuing to cut downward.

The long-term effect of all rivers is to wear the land down to a nearly featureless plain near sea level where stream erosion virtually ceases. Sea level was called the ***base level of erosion*** years ago by Major John Wesley Powell (1834–1902), a pioneering geologist of the far western United States and the leader of the first party to explore the Grand Canyon of the Colorado (Fig. 11-35). In his thinking, the formation of such a plain included much more than the carving of a narrow stream channel down to sea level; it also involved the wearing down of all interstream areas until an entire region was nearly at sea level; in addition, intermediate stages in the process could be identified.

The effect of the slow wasting away of the area between streams is shown in the accompanying diagram (Fig. 11-36). A once broad upland, trenched by narrow canyons and reduced over many years to a nearly level plain not very far above sea level is called a ***peneplain*** from

Fig. 11-33 Three fill terraces, each underlain with stream deposits of a different age. The highest terrace is the oldest and the lowest is the youngest.

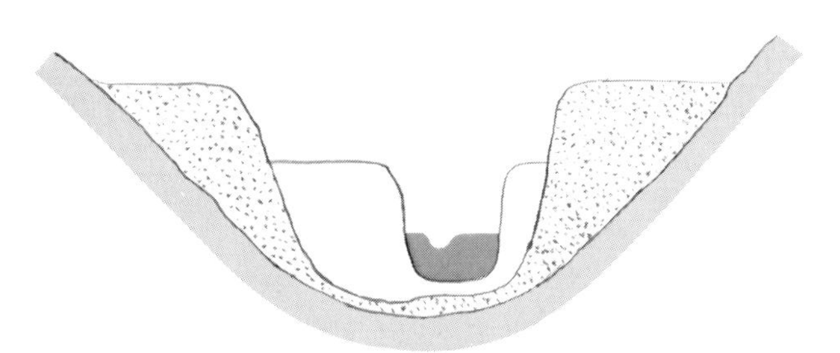

Fig. 11-34 (overleaf) The Grand Canyon at the foot of the Toroweap. The Colorado River was mainly responsible for cutting the canyon vertically, but its breadth is largely the result of slope processes that etched out the rocks in delicate relationship to their resistance to erosion. Rocks that are readily eroded form the gentle slopes, whereas more resistant rocks form the cliffs. *Drawing by William H. Holmes, from J.W. Powell, The exploration of the Colorado River, 1875*

the Latin *paene,* meaning almost, and the English "plain."

A peneplain, because it is a product of widespread degradation by streams and mass wasting, can never have a perfectly level surface, although it may come very close to it. The bedrock surface may not be worn down everywhere to the same level; portions underlain by more resistant rock may stand higher than their surroundings. Such isolated, residual hills, or even mountains, are called ***monadnocks,*** after Mount Monadnock in New Hampshire (Fig. 11-37).

Few incontestable examples of peneplains have been noted from round the world, although many partial peneplains have been described. Many expanses of nearly level plains of barren rock, such as the Hudson Bay region of Canada or much of the Scandinavian Peninsula, have complex histories, and each involves widespread glacial stripping.

Some geologists have proposed that a few of the flattish uplands in the western mountains of the United States and other areas might be uplifted remnants of former low-relief erosion surfaces. In such cases, erosional flattening would have preceded any mountain-building activity. Because such surfaces occur in high mountains that themselves are eroding rapidly, their days of preservation are numbered. In geology, however, numbers run large, and the high flatlands may remain part of the landscape for several million years to come.

There is less than general agreement among geologists about the way in which a combination of stream erosion and mass wasting operates to produce peneplains (see Fig. 11-36). One way is through ***downwasting,*** a process in which the steepness of slopes adjacent to a stream valley gradually diminishes through soil

creep, gravitational transfer, and the decomposition of rocks. The process might be regarded as a point of view based on the work of William Morris Davis (1850–1934), a pioneer American scientist. A different solution was advocated by an Austrian geologist, Walther Penck (1858–1945). According to Penck, the landscape evolves through ***backwasting,*** a process in which valley slopes retreat essentially parallel to themselves. An equilibrium slope is first established, as in Figure 11-36. Once equilibrium has been reached for that particular environment, the slope will continue to retreat. Eventually a graded, stripped rock surface develops at the base and widens as the slope retreats. Ultimately, all land above the level of the widening platform will be stripped off as the separate, retreating slopes meet, and an entire region will have been worn down to a base level of erosion with a very gentle gradient toward the sea.

That geologists still argue the merits of backwasting versus downwasting is testimony to the impact Davis and Penck had upon geological thought. However, a choice between the two explanations would be premature in view of our state of knowledge. The answer to the problem of the wearing away of the land between rivers is complex and probably involves elements of each explanation. However, the possibility seems strong that backwasting is more important in the arid lands of the world, whereas downwasting is more important in humid lands, where vegetation blankets the slopes, soil is deep, and mass movement dominates.

Regardless of the details, geologists today

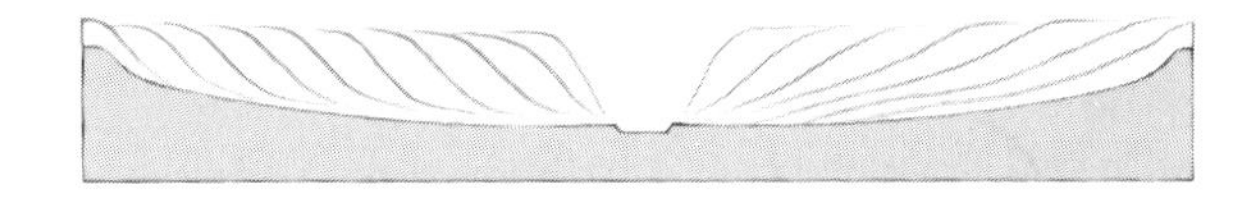

Fig. 11-36 The successive cross profiles of a valley, beginning with a narrow canyon and progressing toward a peneplain. On the right, the slopes are diminished by downwashing; on the left, the slope angles are parallel, and the valley walls retreat through backwasting.

Fig. 11-35 Major John Wesley Powell led the first expedition into the Grand Canyon in 1869. His own words describe well the mood of the small band as it moved into the chasm:

August 13. We are now ready to start on our way down the Great Unknown. Our boats, tied to a common stake, are chafing each other, as they are tossed by the fretful river. They ride high and buoyant, for their loads are lighter than we could desire. We have but a month's rations remaining. The flour has been resifted through the mosquito-net sieve; the spoiled bacon has been dried, and the worst of it boiled; the few pounds of dried apples have been spread in the sun, and reshrunken to their normal bulk; the sugar has all melted, and gone on its way down the river; but we have a large sack of coffee. The lighting of the boats has this advantage; they will ride the waves better, and we shall have but little to carry when we make a portage.

We are three quarters of a mile in the depths of the earth, and the great river shrinks into insignificance, as it dashes its angry waves against the walls and cliffs, that rise to the world above; they are but puny ripples, and we but pigmies, running up and down the sands, or lost among the boulders. . . .

With some eagerness, and some anxiety, and some misgiving, we enter the canyon below, and are carried along by the swift water through walls which rise from its very edge. They have the same structure as we noticed yesterday—tiers of irregular shelves below, and, above these, steep slopes to the foot of marble cliffs. We run six miles in a little more than half an hour, and emerge into a more open portion of the canyon, where high hills and ledges of rock intervene between the river and the distant walls. Just at the head of this open place the river runs across a dike; that is, a fissure in the rocks, open to depths below, has been filled with eruptive matter, and this, on cooling, was harder than the rocks through which the crevice was made, and, when these were washed away, the harder volcanic matter remained as a wall, and the river has cut a gateway through it several hundred feet high, and as many wide. As it crosses the wall, there is a fall below, and a bad rapid, filled with boulders of trap; so we stop to make a portage. Then we go, gliding by hills and ledges, with distant walls in view; sweeping past sharp angles of rock; stopping at a few points to examine rapids, which we find can be run, until we have made another five miles, when we land for dinner.

Then we let down the lines, over a long rapid, and start again. Once more the walls close in, and we find ourselves in a narrow gorge, the water again filling the channel, and very swift. With great care, and constant watchfulness, we proceed, making about four miles this afternoon, and camp in a cave.

A. Powell and his companions before the journey. *Library of Congress* **B.** Powell with Tau-gu, a Paiute Indian chief. *Smithsonian Institution* **C.** Powell sat in a captain's chair lashed to his boat. Making up the party were four other boats and nine men. Not all of them completed the journey; three men who climbed to the canyon's rim met their deaths at the hands of Indians, who feared them because they could not believe that the explorers had actually navigated the river. *J.K. Hillers, 1869, USGS*

Fig. 11-37 Mount Monadnock, New Hampshire, rises above a low-relief erosion surface that was subsequently modified by glaciation. *John S. Shelton*

generally agree that the running water of streams combined with mass wasting—given time—can wear away the highest mountains made of the most resistant rocks, until they are reduced to a nearly featureless plain (Fig. 11-38).

That so much of the land surface of the world is not so featureless indicates in itself the recency and the continuing nature of the forces of deformation affecting the rocks of the earth's crust. Add to that crustal unrest and the fluctuating level of the oceans during the ice ages, and no wonder we cannot convince the skeptics. To them we can only say "It could happen, given more time."

Low-relief erosion surfaces probably have

Fig. 11-38 A. (above) Flat erosional remnants form the summit plateau of the northern Wind River Mountains, Wyoming. *Janet Robertson* **B.** (left) Low-relief erosional surface cut in schists and subsequently dissected by streams, west of Dunedin, New Zealand. *W.C. Bradley*

Fig. 11-39 Hell's Canyon, cut by the Snake River in Idaho, averages 1680 m (5510 ft) in depth and is deeper than the Grand Canyon. Layers of dark basalt make up the walls–a marked contrast to the play of colors in the Grand Canyon. *Oregon State Highway Department*

been formed again and again through the long life of the earth. Some ancient and buried unconformities could be former peneplains graded to an unknown base level.

RATIONAL USE OF RIVERS

In the early days of our country, before the transcontinental railroad was completed and before the present web of highways was constructed, rivers provided the principal path of transportation for both goods and people. Now, of course, their overall importance is not as great, but certainly they are used in other ways. One of the most important uses is as a water supply and as a transport system for our monumental amount of waste products and sewage.

In an attempt to harness the forces of nature, we have built and propose to build a great many dams across our rivers. The water stored behind the dams is to be used for a number of purposes: production of electric power, a steady supply of water for irrigation and other purposes, and flood control, among others. At the same time, the number of rivers now preserved in national parks and monuments, or designated as Wild Rivers indicates that their aesthetic and recreational value is receiving more and more recognition.

That there is a conflict among users is clear from the even minimum attention given by the media. Choices must be made, but too often they are made by a small group of persons whose views do not necessarily represent those of the majority of those affected.

Even when plans for development of a river are disclosed, the proposers are often remarkably stubborn, refusing to listen to the opposition's arguments or to consider alternatives.

Luna Leopold, a prominent hydrologist, believes that a reason for such behavior is that although proposed benefits can be *quantitatively* stated, the "non-monetary values are described either in emotion-laden words or else are mentioned and thence forgotten." In an effort to remedy such situations, Leopold devised a chart to evaluate quantitatively some of the aesthetic factors of river sites. Disclaiming any personal bias, he applied his method to twelve river sites in Idaho in the vicinity of Hells Canyon of the Snake River, an area in which the Federal Power Commission wants to construct one or more hydroelectric dams.

After comparing the Hells Canyon site with other sites in Idaho capable of hydropower development, he found Hells Canyon the most worthy of preservation (Fig. 11-39). Leopold then compared the site to other river valleys that lie within national parks: (1) the Merced River in Yosemite, (2) the Grand Canyon of the Colorado River, (3) the Yellowstone River near Yellowstone Falls, and (4) the Snake River in Grand Teton National Park. His conclusion was that "Hells Canyon is clearly unique and comparable only to Grand Canyon of the Colorado River in these features." It remains to be seen whether his method will be followed in determining the rational use of our river resources.

The Colorado River is an example of the myriad of problems facing river planners. Although its headwaters are in the Rocky Mountains of Colorado, the river basin encompasses parts of seven western states and Mexico before the river empties into the Gulf of California (Fig. 11-40). Each of these states has been allotted a portion of the water, and a certain amount is guaranteed to Mexico. Cities in the basin—which is part of the Sunbelt—have grown rapidly in recent years, and the demand for water has risen greatly. Even Denver and Salt Lake City, which lie outside the watershed, tap the Colorado, diverting water through tunnels. As other cities in the area grow, agriculture and industrial needs will also increase. In addition, development of oil shale deposits in the northern part of the basin will require large amounts of water. Increased pollution, too, is a threat. Dams are controversial; and proposals for future dams will meet increasing opposition from conservation groups and residents, who argue that there should be no further interference, by developers, with the Colorado or other rivers. Such a conflict of interests is becoming more common and more political as resources become scarcer.

Fig. 11-40 Colorado River and the limits of its basin.

SUMMARY

1. Streams are the main agents that transport material from land to sea.
2. Water in a stream flows in a turbulent fashion, and turbulent flow aids greatly in the transportation of the stream's *solid load* (*bed load* and *suspended load*). The *dissolved load* quantitatively makes up an important percentage of the total load carried by a stream. It is derived from chemical weathering.
3. Stream velocity is determined by channel shape and roughness, discharge, and stream gradient. The greater the velocity, the greater the size of the material a stream is capable of moving.
4. Streams will reach an equilibrium in which the load added to the stream is balanced by the material carried by the stream. A stream that has reached equilibrium is called a *graded stream*.
5. Stream terraces reflect former stream disequilibrium during which the stream either *aggraded* or *degraded*. The two main kinds of stream terraces—*cut terraces* and *fill terraces*—each reflect a different stream history. Human intervention can bring about stream disequilibrium, and often the effect is detrimental to people and to structures along the river.
6. Floods are natural to all streams, and the identification of the extent of inundation for floods of various magnitudes is an impor-

tant part of land-use planning. More frequent flooding can be caused through building practices.

7. The two main kinds of stream patterns are largely related to the kind of load. *Braided* patterns reflect a high bed load, *meandering* patterns a suspended load.
8. High-relief landscapes, given sufficient time and tectonic stability, can evolve to a low-relief rolling surface. In a humid climate slope processes are mainly responsible for the lowering of the original landscape, whereas the role of the rivers is to transport material out of the drainage basin.

QUESTIONS

1. Discuss the factors that would contribute to (1) an increase in velocity downstream and (2) a decrease in velocity downstream in an imaginary river.
2. How would climatic change influence the amount and size of solid load, and the amount of dissolved load, in a particular river?
3. What are the factors that must be considered, regarding costs and benefits, before damming a river?
4. Would the recent geological history of a river be the same if it were lined with cut terraces or with fill terraces?
5. If low-relief erosion surfaces can be formed in both arid and humid climates, how could one determine the climatic conditions that prevailed during the formation of an ancient surface, some 100 million years old?

SELECTED REFERENCES

Baker, V. R., 1973, Paleohydrology and sedimentology of Lake Missoula flooding in eastern Washington, Geological Society of America Special Paper 144.

Bloom, A. L. 1978, Geomorphology, Prentice-Hall, Inc., Englewood Cliffs, New Jersey.

Bretz, J. H., 1969, The Lake Missoula floods and the channeled scabland, Journal of Geology, vol. 77, pp. 505–43.

Easterbrook, D. J., 1969, Principles of geomorphology, McGraw-Hill Book Co., New York.

Emerson, J. W., 1971, Channelization: a case study, Science, vol. 173, pp. 325–26.

Garner, H. F., 1974, The origin of landscapes, Oxford University Press, New York.

Leopold, L. B., 1969, Quantitative comparison of some aesthetic factors among rivers, U.S. Geological Survey Circular No. 620.

Leopold, L. B., Wolman, M. G., and Miller, J. P., 1964, Fluvial processes in geomorphology, W. H. Freeman and Co., San Francisco.

Mackin, J. H., 1948, Concept of the graded river, Geological Society of America Bulletin, vol. 59, pp. 561–88.

Morisawa, M., 1968, Streams, their dynamics and morphology, McGraw-Hill Book Co., New York.

National Research Council, 1968, Water and choice in the Colorado Basin, National Academy of Sciences Publication 1689, Washington, D.C.

Ritter, D. F. 1978, Process geomorphology, W. C. Brown, Dubuque, Iowa.

Ruhe, R. V., 1975, Geomorphology, Houghton Mifflin Company, Boston.

Schumm, S. A., 1977, The fluvial system, John Wiley and Sons, New York.

Thornbury, W. D., 1969, Principles of geomorphology, 2nd ed., John Wiley and Sons, New York.

Fig. 12-1 Sand dunes encroaching on the Sierra del Rosario Range, Sonora, Mexico. *Peter Kresan*

12

DESERT LANDFORMS AND DEPOSITS

The world's deserts, aside from the extreme Arctic, are the least familiar of land areas (Fig. 12-1). Perhaps their seeming mystery lies in their distance from lands such as western Europe and the Atlantic coast of North America, where modern Western civilization became industrialized. Had Western life remained centered on the Mediterranean, deserts would have been much closer to our daily lives because the limitations imposed by aridity bear heavily on such desert-bordering countries as Spain, Morocco, Algeria, Libya, Egypt, and Israel.

In earlier days, much of the southern shore of the Mediterranean was the granary of Rome, but with time once-flourishing cities, such as Leptus Magnus in Libya, are now stark ruins, half buried in the sand. One of the problems in studying deserts is that the boundaries are not fixed, but may change through the centuries. There is much evidence from ***paleobotany***—the study of fossil plants—that deserts are relatively late arrivals among the earth's landscapes. Deserts require a rather specialized set of circumstances for their formation and existence, and we shall inquire into what some of them are.

First we must agree on what constitutes a desert. Temperature is not the only factor; some are hot almost all of the time, others may have hot summers and cold winters, and some are cold throughout much of the year. Drought is their common characteristic; in a general way we might call those regions deserts where more water evaporates than actually falls as rain. Because drought is their prevailing characteristic, deserts notably are regions of sparse vegetation. Few are completely devoid of plants, but some almost are. Typically, desert plants are widely spaced. Their colors tend to be subdued and drab, blending with their surroundings; but in the spring, even the cactus blooms with colorful flowers. Their leaves may be small and leathery to reduce evaporation. In fact, some, such as the saguaro of Arizona or the barrel cactus of the Sonoran Desert, do not have true leaves. Other desert plants, such as the ubiquitous sage, may develop deep root system in proportion to the part of the plant that shows above ground. Plants with such adaptations as extensive deep roots and leathery leaves, which have a large water-holding capacity, are called ***xerophytes,*** from a combination of Greek words meaning dry and plant.

Fig. 12-2 Apollo 7 photograph of the arid Zagros Mountains, of Iran. Most geological features are well expressed because there is little vegetation and soil to hide them. The circular structures are salt domes, which can be preserved at the surface only because of the extreme dryness. Salt rock glaciers exist in some areas. *NASA*

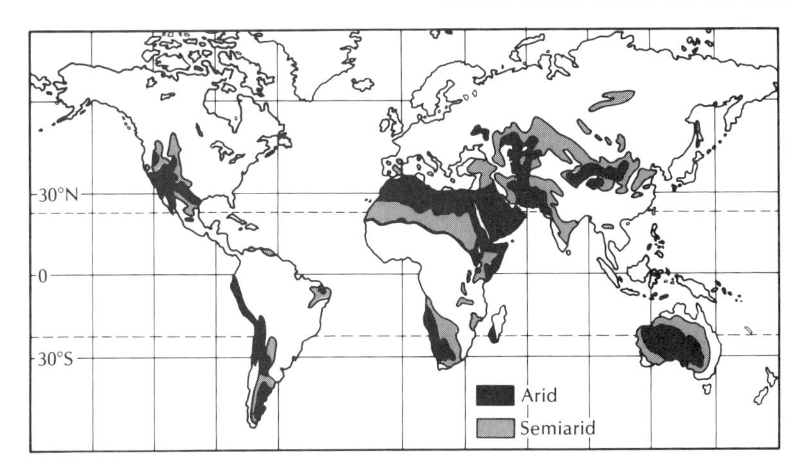

Fig. 12-3 Distribution of the earth's hot desert regions.

Deserts range from those that are completely arid and essentially barren expanses of rock and sand, devoid of almost all visible plants (Fig. 12-2), to those that support a nearly continuous cover of such plants as sagebrush and short grasses. A dry region with such a characteristic seasonal cover is best called a ***steppe,*** and commonly is marginal to more desolate areas.

Arid and semi-arid regions occupy one-third of the earth's land area and are concentrated in subtropical and in middle-latitude regions (Fig. 12-3). For example, there is nearly continuous desert from Cape Verde on the west coast of Africa, across the Sahara, the barren interior of Arabia, the desolate mountains of southern Iran, and on to the banks of the Indus in Pakistan. The preponderance of the dry areas of the earth—exclusive of the Arctic—are on either side of the equator, chiefly around latitude 30°, and they tend to occur on the western side of continents. Deserts also occur in polar regions, but such places are cold. They will be discussed at the end of this chapter.

Contrary to the popular image, most deserts are not vast shimmering seas of sand. Although many deserts are sand covered, the majority are not. They are more likely to be broad expanses of barren rock or stony ground with only shallow soil profiles. Ground colors are largely those of the original bedrock. For example, the bright red color that we associate with such places as Grand Canyon and Monument Valley comes in large part from coloring matter within the rocks themselves. In deserts, red colors can also be attributed to soils and to weathering.

It is typical of many desert regions, especially those in continental interiors, that streams originating within the desert often falter and disappear within the desert's boundaries, such a pattern of streams that do not reach the sea is called ***interior drainage.*** Some desert streams simply shrink and sink into the sand, such as the Amargosa River in Death Valley. Others may carry enough water to maintain a lake in a structural basin at the end of their course. Such lakes have no outlets and are almost universally salty or brackish—the Dead Sea, about 396 m (1300 ft) below sea level, into which the River Jordan flows, is a renowned example. A larger water body without an outlet is the Caspian Sea, and even though it is supplied by the mighty Volga, not enough water reaches it to overcome losses of evaporation and to allow the lake to spill over the low divide separating it from the Don River and the Black Sea. A third notable example is Lake Chad, which covers some 22,000 km^2 (8494 mi^2) in the southern Sahara.

In many arid parts of the world, where the water brought in by streams evaporates, desert

Fig. 12-4 Laguna Salada, a playa at the base of the Sierra los Cucapas Mountains, northeastern Baja, Mexico. *Peter Kresan*

lakes may be only short-lived seasonal affairs and may be completely dry for decades. Such ephemeral lakes are called ***playa lakes*** in the southwestern United States; they evaporate quickly, leaving a ***playa*** (Fig. 12-4). Some playas may be glaring expanses of shimmering salt, such as the Bonneville Salt Flats near Great Salt Lake in Utah, or they may be broad, dead-flat, clay-floored dry lakes.

Before going into how deserts develop, it might be appropriate to elaborate on some of the extreme conditions that characterize them. A temperature of 57°C (135°F) has been recorded in Algeria, and in Death Valley, California. No wonder people talk of frying eggs on rocks. Rock surfaces reach even higher temperatures, so that it is not surprising that many geologists attribute some rock-weathering features not only to high temperatures, but also to day-night, hot-cool cycles in temperature. Although the lowest recorded mean annual precipitation of 0.4 mm (0.016 in.) comes from the Sahara, the precipitation rate of some of the larger Saharan storms is 1 mm/minute (0.04 in./minute). In other areas, years go by without rainfall, the record being 15 years in Southwest Africa.

CONDITIONS CAUSING DESERTS

What special circumstances are responsible for the aridity of some parts of the earth's surface? If we omit the polar regions, there are three major types of arid regions: (1) Horse Latitude deserts, (2) rain shadow deserts, and (3) deserts produced by cold coastal currents in tropical and subtropical regions.

Most of the world's deserts are located about 30° north and south of the equator, as noted earlier, in the so-called Horse Latitudes (Fig. 12-5). Such low latitude deserts are near the equator, where the sun's rays strike the earth's surface more directly than at other places, resulting in more solar radiation to heat the air. The hot, moist air rises, cools, and loses its ability to hold water. The excess water falls as rain in the equatorial regions. The cooled air spreads northward and southward from the equator into the areas of the Horse Latitudes where it descends and is warmed, enabling it to hold much more water. The Horse Latitude deserts, then, are deserts because they are continually parched by warm, dry winds that evaporate any moisture present.

Fig. 12-5 Diagram of atmospheric circulation.

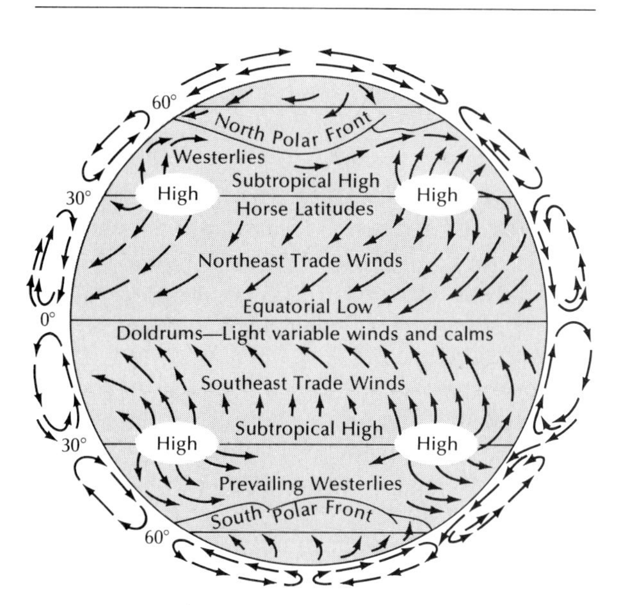

A second cause of aridity is the interposition of a mountain barrier in the path of a moisture-bearing air current. A striking example may be found in the western United States, where the desert stretches eastward in the so-called rain shadow of the Sierra Nevada of California. As moisture-laden air from the ocean rises along the west side of the Sierra Nevada, it is cooled, and rain falls. The now-dry, cooler air flows down the eastern slope, is warmed again, and evaporates any water in its path. In this mountain range, maximum rainfall actually occurs west of the divide.

The third type of deserts, formed because of cool coastal currents, are perhaps the least familiar to North Americans. Such deserts are found along the mid-latitude coasts of continents where the shores are washed by cold coastal currents, such as the Humboldt Current off the coast of Chile and Peru or the Benguela Current off the Namib Desert of southwest Africa, both of which run northward along the coasts.

The latter deserts are impressive because of the dramatic climatic contrasts encountered within extremely short distances. The desert of southern Peru and northern Chile, for example, is one of the driest lands on earth, yet its seaward margin is concealed in a virtually unbroken gray wall of fog. Winds blowing across the cold waters of the coastal current are chilled so that moisture is condensed to form a seemingly eternal blanket of fog that stands over the sea. Once the fog drifts landward, where air temperatures are higher, it burns off almost immediately, and the water-holding capacity of the air increases rather than decreases as it moves across the heated land.

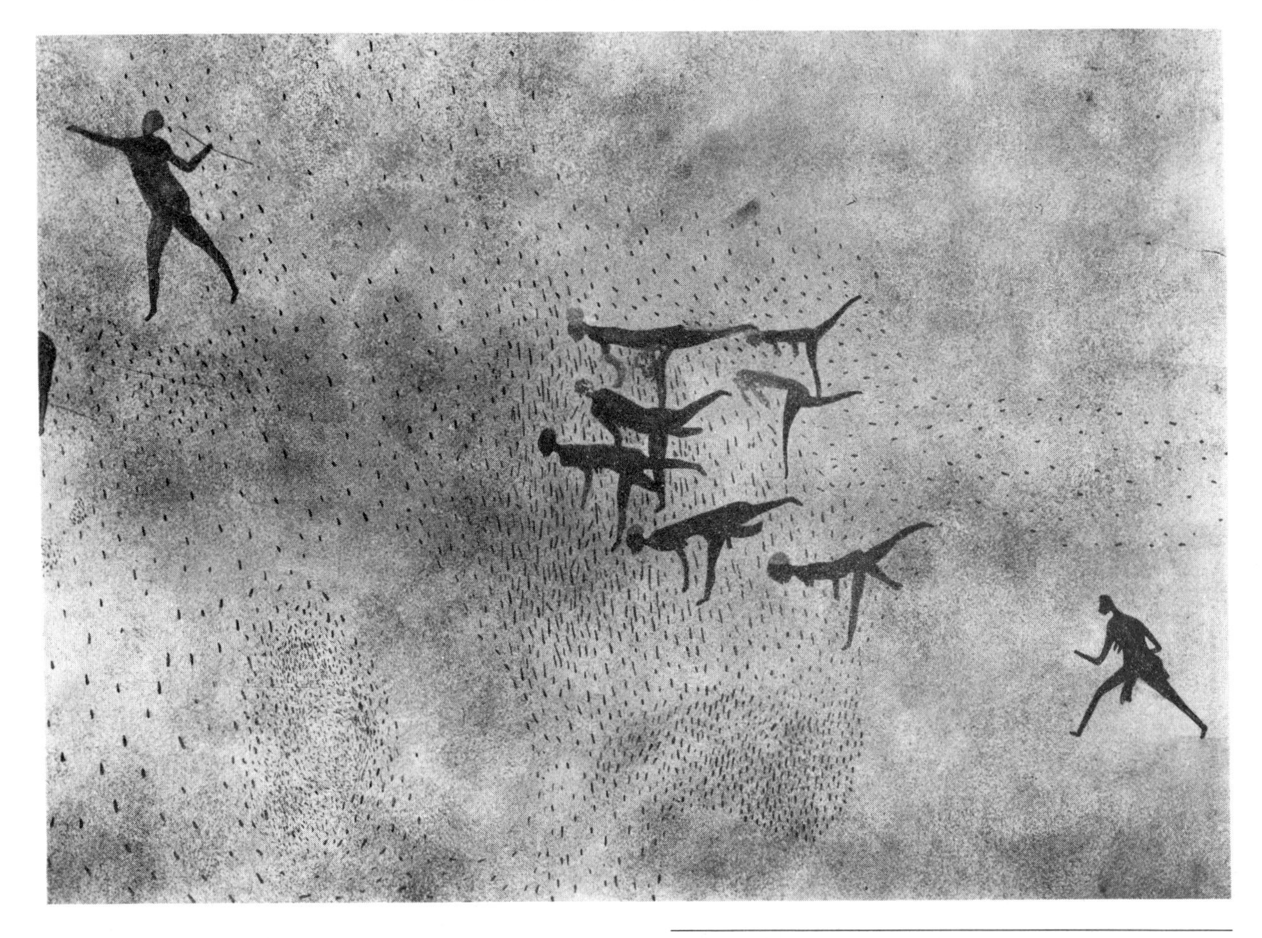

Fig. 12-6 A Stone Age rock painting in the African desert depicts people gathering grain. This and other similar paintings suggest a wetter climate in the past. *Eric Lessing, Magnum*

Few deserts, however, are the product of a single cause. For example, the Atacama Desert on the western slope of the Chilean Andes is not only affected by the cold Humboldt Current, it is also located in the Horse Latitudes. Furthermore, cold air sweeps down from the Andes, so that all three causes occur simultaneously, in one place.

In summary, we can say that the origins of deserts are complex, but are well worth trying to understand, not only for the intellectual challenge but also because so much of the future of the human race depends upon the utilization of arid lands. We now recognize that the activities of humans and their domestic animals can result in the migration of desert boundaries. Grazing and agricultural practices can so deplete the vegetative cover that the desert can expand into these overused areas. This is of much concern at present because such ***desertification,*** as it is called, commonly takes place in countries that can least afford to lose agricultural lands.

As we will see, the climate of the world has slowly changed in historic time. That change has been strikingly true of deserts, and much of the most compelling evidence comes from the Sahara and the Middle East. Artifacts, stone implements, and rock paintings of extraordinary subtlety and sophistication testify to the presence of early man in what today are desolate expanses of the Sahara (Fig. 12-6). Later, much of that region was the granary of Rome, and colonial cities of that day, as well as roads along which the legions marched, are now covered over by drifting sand. The expansion of the desert broke the slender thread of communication linking the cultures of the Mediterranean and the African worlds. Apparently thereafter the two had a vague and uncertain

Fig. 12-7 Geometrical patterns, created by piling up rocks coated with desert varnish, stretch across the desert in the Nazca Valley, southern Peru. The light color of the pattern is the exposed surface of the desert floor. The patterns date from about 100 B.C. to A.D. 700, but no one knows why they were made. The size of many of the designs suggests that they were meant to be visible only from the air, or, possibly, that they were monuments to important gods, and, hence, large. © *Marilyn Bridges*

awareness of each other over the centuries, but it was not until the introduction of the camel caravan that the land connection was reestablished. By that time the culture of each world had evolved along very different paths.

VARNISH AND PAVEMENT

In many parts of the desert, the land is simply barren rock outcrop or is strewn with rocks and boulders. Commonly, the rock is coated with ***desert varnish,*** a shiny, bluish-black coating peculiar to desert environments. It can coat entire stones or form patterns on high vertical cliffs. Early inhabitants of the Peruvian desert made patterns with such stones, the significance of which is still a mystery (Fig. 12-7). North American Indians scratched off the very thin coatings to make their picturesque petroglyphs (Fig. 12-8).

Chemical analyses indicate that the varnish is a mixture of iron and manganese oxides derived from weathering—from the rock on which they form, from the adjacent soil, or from airborne substances. Precipitation of the oxides could be inorganic, but recent research suggests that microorganisms played a role in their formation. Archeological and geological evidence indicates that several thousand years are required before desert varnish becomes a prominent feature.

Another unique feature of deserts is ***stone pavement***—a thin veneer of stones one to two stones thick. Looking closely at the surface, one can see that the stones form a tightly packed mosaic, as if put together by a mason—hence the term (Fig. 12-9). Typically, the layer of soil directly beneath the surface has much fewer stones.

Two hypotheses have been advanced to explain the origin of most stone pavements. One is wind erosion. Consider a sand and gravel deposit exposed at the surface with little vegetation for cover. Winds sweeping across the surface can pick up sands and finer-grained materials and carry them away, but the gravel will remain. If the process goes on long enough, the gravel and larger stones will form a tightly packed lag concentrate. The other hypothesis states that the stones move from the shallow depth to the surface. If the gravels were originally scattered throughout a somewhat clayey matrix, the shrinking, cracking, and swelling that accompanies drying and wetting cycles would move gravel toward the surface and there concentrate it into a pavement. Whatever their origin, stone pavements form an armor that protect the surface from further wind erosion. And, like desert varnish, they take a long time to form; commonly, those gravel surfaces with the best development of stone pavement also are well coated with varnish.

STREAM EROSION AND DEPOSITION

Paradoxical as it may seem, running water is the agent most responsible for sculpturing desert landforms, perhaps more so than in humid regions. It is indeed puzzling how the enormous volumes of rock that once filled desert canyons or shaped the mountains themselves were removed or worn down to bare rock plains in a land where there appear to be no streams.

It may be that not all landforms in all deserts were formed under the climatic regime we see today; they may be fossil landscapes, in a sense, survivals of erosional patterns carved during a more humid time. We shall see one result of a recent climatic shift a little farther on when we discuss the lakes that formed in desert regions during times of recent glaciation. Not all the erosion of desert landscapes, however, can be attributed to an earlier, wetter climate. Almost all deserts have some rainfall, even though 10 to 15 years may elapse between showers. So it could be that the infrequent storms, over a long enough period of time, provide enough water for erosion to take place.

The nature of runoff

Contrary to popular belief, cloudbursts are relatively rare in arid lands—they are much more common in areas where rainfall is greatest, for example, in the rainy tropics or the southern Atlantic coastal states. But a moderate rain in

Fig. 12-8 Desert varnish on sandstone in the Glen Canyon of the Colorado River. Shamans of the prehistoric Indian tribes inscribed petroglyphs by scraping away the dark coatings. Note the bighorn sheep, important prey of the Indians, and the self-portraits of the shamans with headdresses of bison horns. *Utah Museum of Natural History*

Fig. 12-9 Well-developed stone pavement in the Mojave Desert, California. *W.C. Bradley*

the desert can assume the proportions of a cloudburst in a more humid region and also do a very effective job of erosion. The reason is that little or no vegetation protects the slopes from the spattering effect of raindrops or from water running across the surface, rapidly cutting ravines and arroyos. ***Arroyo*** is a Spanish word for stream or brook, and in the southwestern United States it is used for the steep-sided, flat-floored river washes that are so common there (Fig. 12-10). Accelerated erosion in the late 1800s caused many of these features. It may be that a subtle climatic change or overgrazing by livestock recently introduced, or both, were responsible for the acceleration of erosion. ***Badlands,*** or slopes scored with great numbers of gullies, large and small, are also characteristic desert landscape elements caused by runoff in areas where the bedrock is mainly impermeable shale (Fig. 12-11).

In a sudden desert downpour, in a matter of moments, a dry, sandy arroyo bordered by low but steep cliffs is flooded with surging, mud-laden water. The stream swirls and churns violently forward, sweeping along a great mass of debris. Boulders of all sizes can be moved in the torrents. Such flash floods make the desert impassable until the arroyos drain. And they drain almost as rapidly as they fill, since there is no continuing source of water. After only a few hours beneath the desert sun, an arroyo floor earlier covered by 3 m (10 ft) of water may be dry sand again, with only an occasional pool of muddy water. Evidence of the event, however, will remain in the form of transported rocks and other debris.

Occasionally, the short-lived torrential floods overflow the low banks of dry washes, spreading a sheet of muddy, turbulent water over the desert floor. Such a sheet flood will pick up loose sediments and shift them around the landscape.

Rates of erosion in deserts vary with mean annual rainfall. In small drainage basins in the United States, the maximum amount of erosion seems to occur in semi-arid regions, with lower amounts in arid regions and in humid regions. Less erosion takes place in more arid regions because rainfall, the main cause of erosion, is scant, whereas in humid regions the stabilizing effect of vegetation lowers the rate of erosion (see Fig. 11-11).

Depositional landforms

Wherever erosion occurs, deposition takes place nearby. Deposition is especially prominent in arid regions because ordinary streams cannot escape beyond the confines of the desert. Their water sinks underground into the sandy stretches of their normally dry stream beds, it evaporates, or it may be taken up by fiercely competitive water-seeking plants, such as the mesquite and tamarisk, that line the banks of many desert watercourses. Much of the desert landscape is dominated by stream deposits, in large part because there is not enough running water to move debris out of a desert basin and on to the sea.

One of the most characteristic desert landforms is the ***alluvial fan*** (Fig. 12-12). It takes its name from its shape, which most approximates a portion of a cone that enlarges downslope from the point where a stream leaves the mountain front. Fans commonly are composed of accumulations of gravel and sand, and deposition on their surfaces is attributed to two factors, both of which reduce the velocity of the stream. One factor is that much of the water sinks into the porous, sandy subsurface layers of the fan, thereby decreasing the dis-

Fig. 12-10 Arroyo incised in a river valley in Nevada. The flat river bed and steep walls are typical of many arroyos. *Karl Birkeland*

Fig. 12-11 Badland topography, formed in shale, in the arid region of southern Israel. *Peter W. Birkeland*

charge, which helps to control the velocity. The other is that the main channel, upon leaving the mountain front and entering the fan, soon branches into a score of distributaries in a braided pattern. The increase in length of channel perimeter also helps to decrease the velocity and to bring about deposition.

Alluvial fans vary in size and degree of slope, but there is consistency in the variation. In a general way, the area of a fan is proportional to the area of the drainage basin from which the fan debris is derived; thus, small drainage basins produce small fans and large drainage basins, large fans. The slope of a fan is related to the parameters that control the slopes of streams. Fans derived from large drainage basins tend to have relatively low slopes, as do those constructed by streams with large discharge. Finally, sediment size helps to control the slope, for larger gravels are usually associated with fans having a steeper gradient.

Streams may be able to cross the fan in time of flood, but under normal conditions they sink into the sandy ground almost as soon as

One of the erupting volcanoes on Jupiter's satellite Io, as photographed from Voyager I at a distance of about 500,000 km. *NASA*

(Above) Computer-enhanced photograph of a Martian sunset over the region known as *Chryse Planitia,* as taken by Viking I. *NASA* (Below) The Great Red Spot of Jupiter, taken from Voyager I spacecraft; the distance from the top to bottom is about 24,000 km. The photo illustrates the turbulent features characteristic of this immense atmospheric storm center. *NASA*

(Above) Bachelor Mountain, Oregon, is one of many composite cones comprising the Cascade Range which have been built during the last few hundred thousand years. *Edwin E. Larson* (Below) Volcanic eruption on Heimay Island south of Iceland, January 1973. Showers of pyroclastic material and plumes of volcanic gases accompanied a lava flow which nearly destroyed the city. *Iceland Airways*

(Above) Incandescent pyroclastic material bombard the ash- and gas-shrouded fishing town on the island of Heimay. *Icelandic Airways* (Below) Molten toe of a ropy (pahoehoe) lava flow, in the Hawaiian Islands. *Glen Kaye, National Park Service*

(Above) Remarkable Rocks, Kangaroo Island, Australia. Weathering of a uniform early Paleozoic granite produced this extreme pitting. *W.C. Bradley* (Below) Ayers Rock, an erosional remnant of arkosic sandstone rises dramatically from the desert plain of central Australia. *Pan American World Airways*

Zion Narrows of the Virgin River, Zion National Park, Utah, is an example of deep canyon-cutting in Mesozoic sandstone in an arid climate. *Bill Ratcliffe*

Steep-sided valleys with waterfalls are characteristic of intensely glaciated alpine valleys such as this headwater of the Aare River, Switzerland. *Swissair*

(Above) Weathering and erosion over millions of years carved these formations in Monument Valley, Arizona, the vestiges of once-continuous sandstone beds. *Edwin E. Larson* (Below) Lake Powell, formed by the Glen Canyon Dam on the Colorado River, contrasts markedly with the surrounding arid landscape. *American Airlines*

Desert in bloom near Bahia San Louis Gonzaga, Baja California, Mexico. Because so little of the ground is protected by vegetation, infrequent torrential rains can cause a considerable amount of erosion in a short time. *Eliot Porter*

(Above) Detail of the topography in the ablation region of North Enilchek Glacier in the Tien Shan of central Asia. Note that the layering in the ice is near-vertical, no doubt due to deformation in the ice as it moves down-valley. (Below) In the lower parts of a glacier where ablation is dominant, the ice is commonly debris-laden and very rough as in Turamis Glacier in the Pamirs, Tadzhikskaya SSR. *Both photos, Nicolai Gridin*

(Right) The Lauterbrunnen, a U-shaped glacial valley in the Bernese Oberland, Switzerland. *Swiss National Tourist Office* (Below) The braided pattern of the Valannuker River, Iceland, photographed under the midnight sun. This stream heads at the terminus of an active glacier. *Barbara Jarvis*

Active erosion along the coast at Santa Cruz, California, produced this arch in bedrock and isolated a portion of the bedrock as a stack. *Orville Andrews*

Plunging breaker battering the south shore of Long Island, New York, during a storm. *Hal McKusick*

Green Pool, West Thumb area, Yellowstone National Park. Heated ground water, along with volcanic emanations, rises to the surface, forming geysers and pools. *William C. Bradley*

Travertine deposits at Mooney Falls, Havasu Canyon, Arizona. Carbonate-rich stream water fed by springs precipitates travertine along the course of the stream. *Edwin E. Larson*

Machu Piccu, in the Peruvian Andes, is the remnants of an Inca stronghold on the Urubamba River. The rock spires and towers were eroded from a Late Cretaceous-Tertiary pluton intruded into Precambrian and Paleozoic rocks. *Jeffrey House*

they leave the bedrock of the mountains. Visitors to a dry country almost always are surprised to see a stream waste away, growing thinner and thinner downstream and then vanishing completely—quite the reverse of humid-climate streams, in which volume commonly increases downstream. The water does not really vanish but percolates slowly through pore spaces in the fan. Far down the fan, the same water may be forced to the surface and seep out in springs—commonly marked in the desert by clumps of mesquite trees, one of the best guides to water in the arid Southwest.

If there is a one alluvial fan at the base of a desert range, there will probably be others—in fact, there probably will be one at the mouth of each principal canyon. Overlapping like palm fronds, the fans form a nearly continuous apron sloping away from the mountain front to the basin floor. The apron is called a ***bajada,*** from a Spanish word meaning a gradual descent (Fig. 12-13).

Many desert basins are closed ares with interior drainage so that rain either sinks into the ground or evaporates. Debris shed from the mountains fills in the basins, but the process is slow. In a humid region, where precipitation exceeds evaporation, such areas would be occupied by a lake, with waters that would rise until they spilled over the lowest point of the basin rim.

The Great Basin in the western United States is a vast desert basin, stretching from the Sierra Nevada eastward to the Wasatch Range in Utah and from the area north of the Colorado River into eastern Oregon and southern Idaho. Early pioneers often followed rivers as they moved westward across the area, only to find that the rivers drained into lakes with no outlet or that they simply vanished in the desert. Most lakes

Fig. 12-12 Alluvial fans in Death Valley, California. *John S. Shelton*

Fig. 12-13 Alluvial fans form bajadas on both sides of Death Valley, California. The view is toward the west side of the valley, where the largest fans have formed. Active stream channels are light in color, in contrast to the inactive surfaces darkened by desert varnish. The white material on the valley floor is precipitated salt. *USAF*

in a basin are saline and fluctuate in depth with climatic variations. Accounting for the salinity are salts brought to the lake as dissolved stream load and concentrated as pure water evaporates from the lake surface. Because everything soluble in desert rocks may be carried into such a lake, it is no wonder that some desert lakes are good sources of minerals. Saline lakes are likely to be especially prolific sources of such minerals as potash, potassium salts, and, in the eastern desert of California, boron compounds.

During a dry-climate cycle, some saline lakes may evaporate completely, leaving an almost blinding white residue of salts, of complex chlorides, sulfates, and carbonates. Called ***salt playas,*** such surfaces are most likely to be found at the end of a long, integrated drainage system. The jagged terrain of salt pinnacles on the floor of Death Valley (Fig. 12-14), nearly 90 m (295 ft) below sea level, was a fearsome stretch for the first party of emigrants to cross, in their battered wagons drawn by plodding oxen. That playa receives some of its water and salt from the Amargosa River, which name appropriately enough means bitter.

Clay playas are more likely to be found in smaller basins. Most of the year they are dry and their surface is baked as hard as a brick; in fact, they make ideal emergency landing fields. One such lake floor was converted into a gigantic multidirectional landing ground—24-km (15-mi)-long Rogers Dry Lake at Edwards Air Force Base in eastern California.

When a typical short-lived downpour is over, desert streams disappear as suddenly as they sprang into being. Then the arroyo bottoms quickly return to their seemingly unchanging state of dry, shifting sand enclosed between steep arroyo walls. The playa lake may endure a little longer, but ultimately its murky, dark-hued water evaporates or sinks underground. Before that happens, the suspended load of finely divided silt and clay particles is disseminated throughout the lake. Thus, when the lake has vanished, its floor is a dead-flat ex-

Fig. 12-14 The Devil's Golf Course, Death Valley, California, is composed of salts, complex chlorides, sulfates, and carbonates, which were concentrated in Lake Manley at the end of a long drainage system. The salts were precipitated as the lake disappeared. *Ansel Adams*

Fig. 12-15 Pediment cut from granite rock sloping away from steep mountain fronts in the southern Sinai Desert. This pediment has been dissected, and the underlying bedrock is exposed. In the same area is the famous sixth-century monastery of St. Catherine, as well as Mt. Sinai, a landmark since the time of Moses. *Ran Gerson*

panse of uniformly distributed, fine-grained, tan sediment. Very often, too, surface layers of playa clay shrink as they dry, cracking into thousands of small, polygonal blocks.

Erosional landforms

In addition to gullies, valleys, and canyons, there is in the desert another erosional landform, the ***pediment.*** Pediments are bedrock surfaces stripped bare, or with a thin veneer of gravel, that slope gently away from desert mountains toward the lower part of an intermontane basin (Fig. 12-15). The surface usually meets the mountain front at a sharp angle. From a distance the broad, encircling surfaces look like a uniformly sloping bajada rather than the product of long-continued degradation involving the removal of many thousands of cubic meters of bedrock (Fig. 12-16). In fact, there is no easy way to visually differentiate fans, which are areas of deposition, from pediments, which are areas of erosion.

Pediments were first described in this country in 1897 by an American exploring geologist, W. J. McGee. He was as surprised as anyone by their true nature:

> During the first expedition of the Bureau of American Ethnology [1894] it was noted with surprise that the horse shoes beat on planed granite or schist or other rocks in traversing plains 3 or 5 miles from mountains rising sharply from the same plains without intervening foothills; it was only after observing this phenomenon on both sides of different ranges and all around several buttes that the relation . . . was generalized.

Pediment formation has long been a puzzle to geologists, who have vigorously debated the topic since McGee first described them. Very briefly, there are two principal schools of thought on the subject. Some geologists think that a pediment is formed by lateral planation of streams; that is, erosion caused by a stream as it swings back and forth over a rock surface. The materials carried by the stream slowly cut away at the bedrock, and where the stream touches the mountain front, it causes it to retreat slightly. Other geologists believe that pediment surfaces weather and that the products of weathering are removed by water sweeping across the pediment during rainstorms. The mountain front also weathers, and various processes, including gravity and running water, later deliver this debris to the upper end of the pediment. Debris of a certain size will continue across the pediment; all other debris will remain at the mountain front until weathering reduces it to a size that can be transported. Between those two extreme views is one that says that both processes are important in pediment formation and that, depending on special local circumstances, one process will dominate.

Fig. 12-16 Pediments near Phoenix, Arizona. It is believed that the mountain fronts retreated parallel to their trends, leaving broad pediments behind. *W.C. Bradley*

The most recent hypothesis on the origin of pediments was advanced to explain those in the Mojave Desert in southeastern California. Evidence suggests that the Tertiary landscape, say eight million years ago, was one of soil-covered hills with fronts retreating in a parallel fashion (Fig. 12-17). A climatic change toward greater aridity a few million years ago increased the erosion rate and stripped the land of its soil cover. Bedrock pediments and bouldery mountain fronts are part of the stripped surface. The pediment is inherited from the past, so that one does not have to look to present arid conditions for its formation.

Certainly this hypothesis has much merit. Whether it explains the origin of all pediments, however, is doubtful. One problem in understanding many arid-land features is that in an area with scant precipitation there is little opportunity to observe formative processes if running water is the main formative agent. Hence, multiple working hypotheses must be utilized to infer the origin of many landscape features.

Erosional cycle in an arid region

Geologists have made a considerable effort to determine if landforms evolve through time, from one characteristic form to another. Some, for example, envisage the landscape in a humid climate as initially hilly, only to be converted into a peneplain after millions of years of erosion, dominantly backwasting. And, so the argument goes, perhaps an erosion cycle can also be deciphered from desert landforms.

As an example, we shall use the Southwest (Fig. 12-18). We start with a recently uplifted

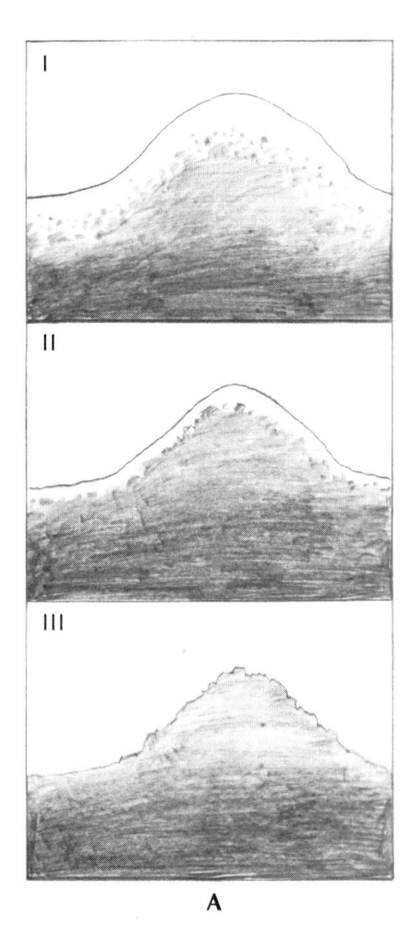

Fig. 12-17 **A.** Landform evolution in the Mojave Desert. Stage I is the soil-covered granitic landscape in Tertiary time. The bedrock weathers more toward the surface, and jointing controls the blocky weathering forms. Stages II and III represent progressively greater stripping of the landscape to the bedrock that predominates today. The low-lying flat areas would be our present-day pediments. **B.** Mountain front in the Mojave Desert made up of huge granitic boulders similar to those in Stage III of **A.** *W.C. Bradley*

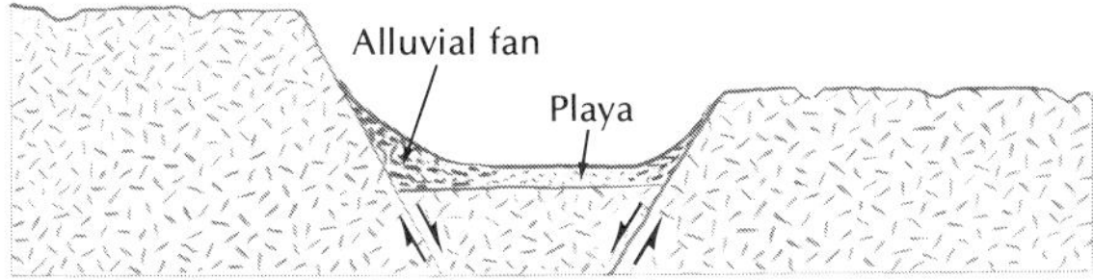

A Initial block faulting forms a basin-and-range topography. Undissected uplands may be present in the mountains.

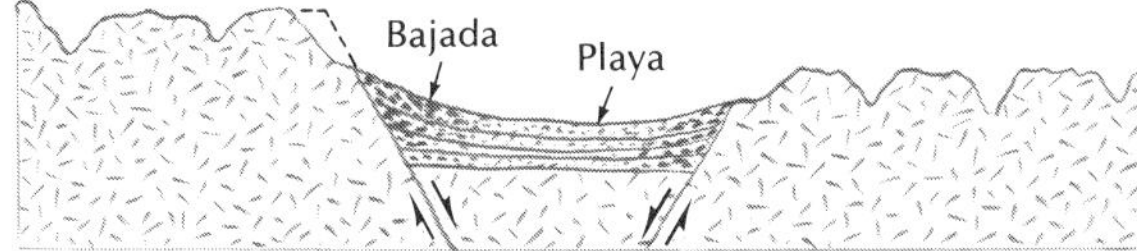

B The basin fills with sediments—fine grained playa sediments in the center and coarse-grained fan deposits around the margins. Drainage becomes through-going.

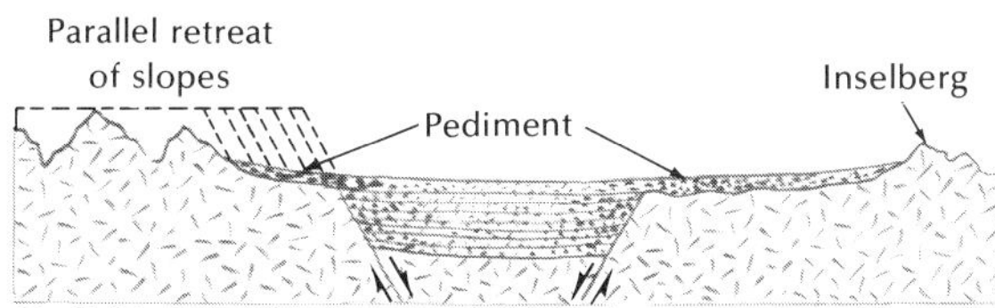

C With through-going drainage the local base level is relatively stable. Slopes retreat parallel to themselves, forming pediments with a thin gravel veneer.

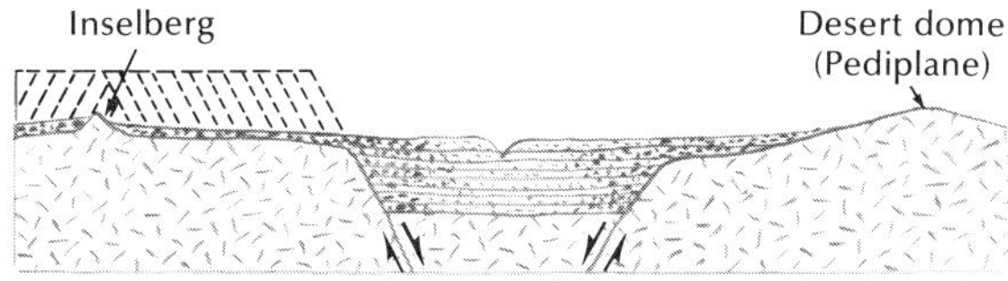

D Erosion continues, leaving a dome where an inselberg had been before.

Fig. 12-18 The erosional cycle in the semiarid to arid Southwest.

mountain range, bounded by normal faults. Narrow gullies cut its upland surface into broad, flat divides. Down-dropped blocks form basins that separate adjacent ranges. As weathering and erosion proceed, the mountains recede nearly parallel to their original slopes; flattening and rounding caused by soil creep in humid regions is not to be found. Also, debris is not carried out of the area, but accumulates in the basins. Because streams are few and flow short distances only, an integrated drainage system cannot develop. As a consequence, the valley floor is occupied by a playa. Alluvial fans lead out to the playa. The basin gradually fills with sediment, and drainage eventually can spill into the next basin, becoming "throughgoing."

Provided the area remains tectonically stable for a long period of time, and through-flowing drainage is established, slopes eventually can retreat from the original mountain front to form pediments. The pediment continues to expand at the expense of the mountain, until only mere fragments of the mountain—called ***inselbergs*** (German for island mountains)—remain. Finally, the inselbergs are consumed, and low bedrock domes are all that remain. These extensive coalescing pediments are called ***pediplanes*** and correspond to the peneplains of a humid region. Thus, it is believed that gently rolling terrain of very low relief is the end result of erosion in any region. Few areas remain tectonically stable long enough for such landscapes to develop, however.

Active faulting and climatic variation can alter the processes that form alluvial fans and pediments. In a very general way, at least in the Great Basin, alluvial fans are found in the north, where active faulting is taking place, and pediments occur in the south, where faulting processes have long been dormant. So, tectonic stability is essential to pediment formation.

Within most major alluvial fans, several different surfaces can be recognized (Fig. 12-19). A stream will incise the old fans, forming a new one at a lower level. This process can happen again and again. In such places as the west side of Death Valley, the incision may have been caused by the upward tilting of the mountain block. That tilting could also explain the drastically different sizes of fans on either side of the valley. The fans on the west side of the uptilted block are large, whereas those on the east side (the down-tilted part of the valley) are being buried by playa sediments, which keep the fans small. Fans, therefore, afford an important clue to tectonic history. Studies of fans thus become more relevant as desert terrain is increasingly encroached upon by humans, with some areas being considered as sites for nuclear power plants.

Before one insists that multi-fan surfaces indicate tectonic activity, climatic change also should be considered as a cause of such sur-

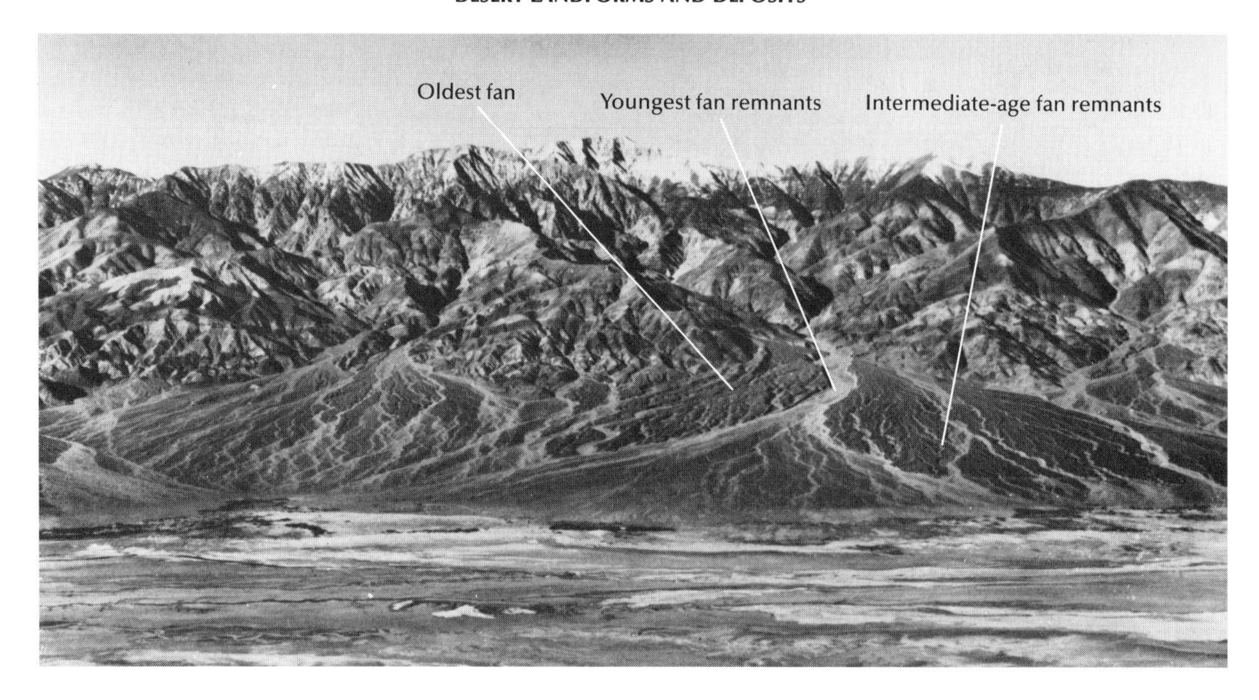

Fig. 12-19 View westward across Death Valley to the Panamint Range. Fans of several ages can be seen, the age depending upon position above the stream (the higher fans are older) and upon tonal differences caused by desert varnish. *H.E. Malde, USGS*

faces. Streams that build fans should respond to climatic changes by degrading at some times and aggrading at other times. For example, with a change to a wetter climate and greater discharge, would a stream degrade, only to be followed by aggradation during a drier climate? We know that climatic changes have taken place in the deserts, and we have attempted to read the changes in the fan deposits. But they are not easily interpreted, since, for example, little organic material, which can be used to establish past climates, is preserved in the deserts.

WIND

Wind erosion

Over many of the dry lands of the world the wind blows seemingly without restraint, adding a note of melancholy to an already desolate terrain. The drier part of Patagonia, of the far southeastern reaches of Argentina, is such a land. Other deserts are perhaps as windy on occasion, but in most of them times of extreme windiness alternate with times of calm.

When the wind blows in the desert, its erosional effectiveness is likely to be much greater than in humid regions. In the latter, the ground surface is protected by vegetation, which also acts as an extremely effective baffle to decrease wind velocity, and by a more tenacious mantle of weathered soil, which also may be damp throughout most of the year. The wind is very effective in transporting certain sizes of particles, but not others. It is that high degree of selectivity that makes the wind such an unusual agent of erosion. Obviously there is an upper limit to the size of particles that it reasonably can be expected to move. Very few boulders of the size that are transported by streams, waves, or glaciers will be moved by the wind—even if it is a tornado. Such immobility is true for loose material down to the size of large pebbles—although the smallest pebbles, about 4 mm (0.16 in.) in size, commonly are moved (Fig. 12-20).

Sand grains move along the ground in a series of hops—a motion known as ***saltation*** (Fig. 12-21). Grains may bounce off other grains on the surface, or they may stop abruptly, the force of their impact setting other grains in motion. The height to which the grains bound depends on the characteristics of the surface. Grains bound higher when bouncing off a hard surface, such as a pebbly desert pavement or a road, than from a loose sand surface, which absorbs much of the impact energy. Saltating sand particles seldom rise to heights greater than 1 m (3 ft) or so above the surface (Fig. 12-22).

Saltating grains have a marked effect on the shape of rock outcrops (Fig. 12-23) or gravel. In general, the upwind side of a gravel particle will be cut gradually to form a low-angle planar surface called a ***facet*** (Fig. 12-24). Facets form at approximately right angles to the wind, and attempts have been made to deduce ancient wind directions from faceted rocks, or ***ventifacts.*** But a stone can shift position, presenting a new face to the driving winds, and a new facet is cut. In fact, multi-faceted stones are more common than not. Stones can be shifted about by wind scouring near their bases, or by other means, including, especially, frost heave or animal disturbance.

Fig. 12-20 Gravel moved during a 1969 windstorm in Boulder, Colorado. These grains bounded down an asphalt road driven by winds exceeding 200 km/hr (120 mi/hr). *University of Colorado Daily*

Odd as it may seem, very fine-grained particles, such as silt or clay flakes, are not easily set into motion by the wind. In general, the same principles apply here that apply to streams. Silt and clay particles, which are small and strongly cohesive, cannot be picked up readily by a moving current of air because wind velocity diminishes close to the ground (Fig. 12-25). This is demonstrated when the wind blows full strength across the surface of a clay playa. Very little dust is stirred up, as a rule, and the hard-packed clay particles hold firm. Some of the looser sand and silt around the margin, however, may be picked up, especially if the playa surface is sun cracked. And if these sand grains saltate across the clay, particles of clay can be kicked into the air and carried aloft.

Once silt- or clay-size particles are picked up by the wind, however, they remain aloft much longer than sand does. There are no restraints imposed on their travel comparable to those of their waterborne equivalents, which are restricted by the drainage pattern of a stream or a glacier's course. Windborne dust swept from the fields of Colorado in the Dustbowl days of the 1930s was carried as far east as the New England states. In fact, some dust was blown far beyond, out over the North Atlantic.

There are scores of other examples of the efficacy of air currents in moving fine particles over vast distances. Volcanic eruptions of the explosive sort are the most telling because they constitute a point source for the volcanic dust. Also, because of the unique nature of the dust, it can be traced for great distances. In 1883, Krakatoa, near Java, hurled dust into the upper atmosphere to circle the earth several times, with an appreciable fall in regions as re-

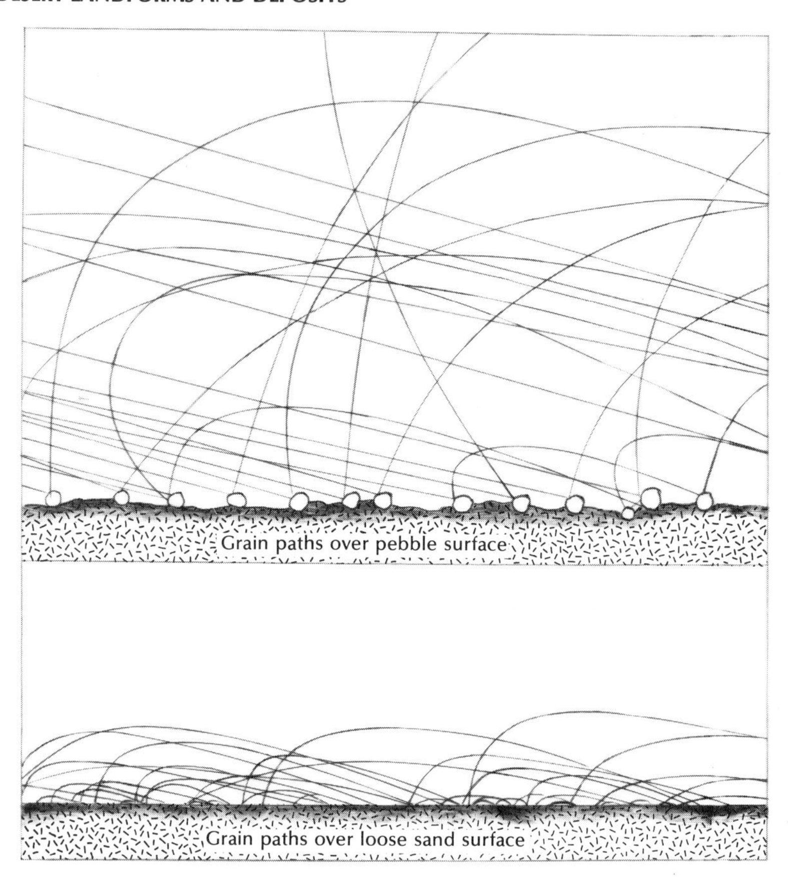

Fig. 12-21 Paths saltating sand grains take across two different surfaces.

mote from the source as western Europe. Icelandic eruptions have made appreciable deposits in Europe, too, as well as in eastern North America.

More typical, however, of the aspect of dust transport with which we are concerned here is the dust whirled high into the air by the turbulent winds of the Sahara and carried far and wide over the Mediterranean, southern Europe, and on occasion, as far as England, 3200 km (1988 mi) away. Even dust from central Asian deserts has been found on remote islands in the Pacific.

The amount of material in the air at any given time is surprising. In a dust storm of average violence, a cube of air 3 m (10 ft) on a side might well have 28 gm (1 oz) of dust suspended in it. Such an amount seems trifling, but if we increase the size of our cube of air until it is 1.6 km (1 mi) on each side and maintain the same saturation of dust, then the air current is supporting 4000 metric tons of solid material. Thus, a storm 500 to 600 km (311 to 373 mi) across might well be sweeping 100,000,000 metric tons of solids along with it.

Such facts are strong evidence that the wind can be an effective erosional agent and one of great significance in arid regions. In fact, it is the only agent that can transport material beyond the confines of a typical desert. Some dust from Africa, for example, has been identified in the Caribbean region.

Some of the closed basins in the desert excavated, or as earth scientists say, ***deflated,*** by the wind are very large (Fig. 12-26). A famous example is the great Saharan oasis of Kharga, west of the Nile, which is 190 km (118 mi) long by 19 to 80 km (12 to 50 mi) wide by 180 to 300 m (590 to 984 ft) deep. A series of longitudinal sand dunes trails off downwind from the oasis, its sandy floor almost certainly the source of supply. That is not to say that the wind hollowed out the whole basin; its origin, unquestionably, is more complex. Rain wash and streams would have eroded the friable sandstone and swept the loose sand down to the floor of the trough. There the sand remains until it is picked up by the wind to form dunes to the leeward of the basin.

Groundwater sets a lower limit to wind erosion. Once the desert surface is deflated to the level of the water table, so that the ground is

Fig. 12-22 A. (left) Erosion of this road marker is greatest close to the ground, where most of the sand grains are moving. **B.** (right) Telephone poles sheathed in metal to protect them from saltating sand grains. *D.C. Strong*

Fig. 12-23 A. (left) This granite outcrop in the Atacama Province of Chile has been undercut by the abrasive action of windblown sand. *K. Segerstrom, USGS* **B.** (above) Wind-driven sand etched this rock photographed near Fortification Rock, Arizona, in 1871, by T.H. O'Sullivan. *Library of Congress*

Fig. 12-24 Ventifact of granite in Wyoming, showing several faceted faces. *W.H. Bradley, USGS*

kept damp, the wind can no longer pick up loose material with the same ease, and deflation slows to a virtual halt. In Kharga a beneficial effect of deflation has been the appearance of springs around the margins of the great depression. Such springs are the source of water for the true oases, for with only a little ground water to draw on, date palms flourish, making a startlingly green contrast with their stark surroundings.

Another impressive closed basin in the Sahara is the Qattara, about 300 km (186 mi) along with a floor 134 m (440 ft) below sea level, a searingly forbidding quagmire of salt and shifting sand covering about 18,000 km^2 (6950 mi^2). Whatever the origin of the Qattara Depression, the wind almost certainly played a prominent role in enlarging it and in deepening it to the water table. Evaporation of the water through

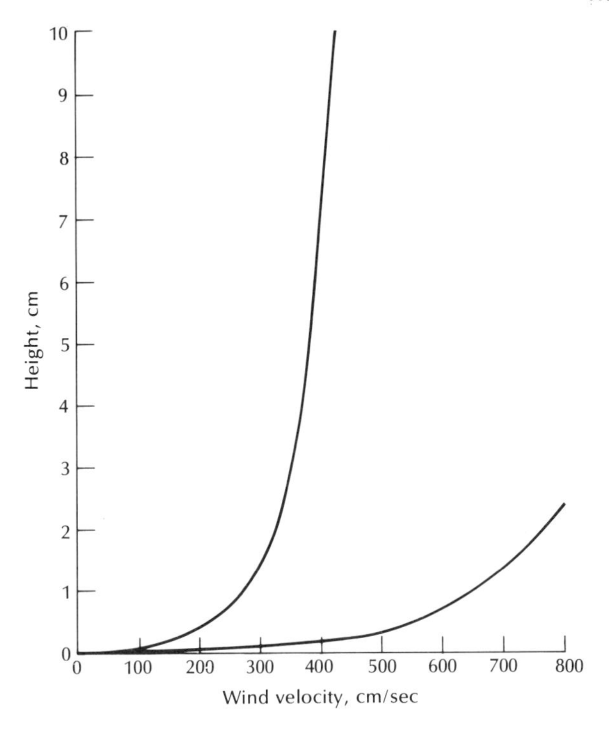

Fig. 12-25 Variation in wind velocity for two different events with proximity to the ground surface.

Fig. 12-26 Satellite photograph of deflation basins in the Sahara, cut in folded sedimentary rocks. Although basins can have many origins, those cut in desert bedrock often are best explained by the removal of material in an uphill direction. *Eros Data Center, USGS*

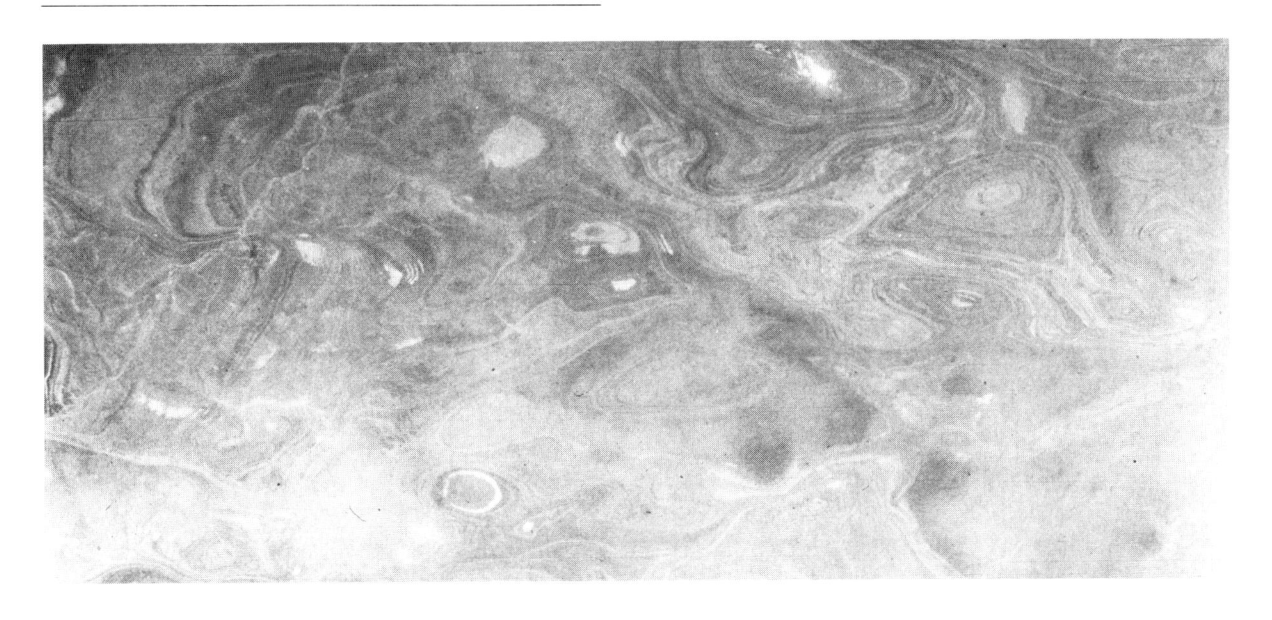

centuries has led to the accumulation of salt deposits. That immense, impassable saline trough acted as a barrier in World War II, making it impossible for Rommel's Afrika Krops to turn Montgomery's flank, and compelling them to attack the grouped British forces at El Alamein on the narrow neck of land separating the Qattara from the Mediterranean.

The Qattara Depression may yet have surprising utility if an unorthodox proposal should be acted upon, to lead Mediterranean water to it through canals and a tunnel and then to use the 460-m (1509-ft) drop from the rim of the depression to generate power from the water available in the sea. Gradually the depression would fill with water to form a concentrated saline lake the surface of which, it is estimated, would stabilize around 46 m (151 ft) below the level of the Mediterranean when the input of sea water balanced the loss of lake water through evaporation.

Fig. 12-27 (above) Cave houses dug in deposits of loess, Shansi Province, northern China. Loess is not only easily excavated, but it stands in a vertical face without failing. *H.E. Malde, USGS*

Fig. 12-28 (below) An elephant and graffiti carved into loess exposed in a steep road-cut in Atchison County, Missouri. Highway engineers like to work with loess because it usually holds a vertical cut well. *Ardel Rueff, Missouri Department of Natural Resources, Division of Geology and Land Survey*

Fig. 12-29 Great sand seas, or **ergs,** shown here in the eastern Rub' al-Khali, are typical of the desert of Saudi Arabia, as well as parts of the Sahara and other deserts of the world. In the foreground is the camp of a geophysical party exploring for petroleum. *Aramco*

Wind deposition

Loess Dust lofted out of the desert by winds may be transported scores of miles before it sifts down to accumulate in a tawny blanket of sediment. Such material in which silt-size particles predominate is called ***loess.***

The most renowned of such deposits are those of northeast China, and it was to them that Baron von Richthofen (1833–1905), a leading German geologist, gave the name of *löss* while on an exploring expedition to the outermost parts of the Russian and the Chinese empires (Fig. 12-27). The loess of China was transported by the wind out of the Gobi and across the barren Kalgan Range, on which the Great Wall was built. Deposited on the North China plain by the dust storms of centuries, loess lies deep in the valleys of that ancient land, often to thicknesses of hundreds of meters, although

Fig. 12-30 Avalanche tracks on the slip face of a miniature active sand dune. *W.C. Bradley*

it is much thinner on the crests of divides. The tan-colored silt gives the Yellow River its name, as well as the Yellow Sea, which is colored for hundreds of kilometers from shore by suspended dust particles. Loess is also found in other regions, where it is closely associated with glaciation (see Chap. 13). In the United States, dust derived from glacial-age floodplains forms extensive loess deposits, producing some of our more valuable, present-day farmland (Fig. 12-28).

Sand dunes When sand is plentiful in arid regions, winds move it and pile it up in characteristic heaps called sand dunes. Such features are not, however, limited to deserts. Many of the larger and more renowned of the world's dunes are along shorelines, such as those on the eastern shore of Lake Michigan, the length of Cape Cod, and the coast of Somalia. Dunes also border the sandy plains of some large rivers—the Volga is an outstanding example.

Few deserts are completely sand covered. Nevertheless, there are sandy areas in almost all deserts; the south-central part of Arabia and the western part of the Sahara are among the best known. Broad dune-covered areas in the Sahara and elsewhere are called ***ergs*** (sand seas) because they so resemble a wave-tossed sea (Fig. 12-29).

Sand dunes consist dominantly of sand-size grains, which bear testimony to the extraordinary sorting ability of the wind. Finer material such as silt may be blown far away, perhaps to form a loess deposit, whereas coarse rock fragments, such as pebbles and gravel, may lag behind the sand.

Dunes are neither stable nor permanent features of the landscape and may be continuously on the march. Usually they have a gentle side facing toward the wind and a steep side facing away from the wind. Wind-drifted sand blows up the gentler slope of a dune, and when it reaches the crest it may be carried a short distance over it—the tops of dunes sometimes seem to be smoking when the sand is driving across them. Behind the crest the sand drops out of the windstream to accumulate on the steeper slope, the ***slip face.*** When the sand is dry and well sorted, the inclination of the slip face may be as much as 34°. Should the slope steepen, the sand becomes unstable and shears along a slightly gentler plane, with the result that a small avalanche of dry sand glides to the base of the dune (Fig. 12-30). When new sand falls on the slip face, the slope steepens once more, and so the process repeats itself. The net result is a transfer of sand from the upwind to the downwind side of the dune. Thus, the dune slowly migrates—in a sense, rolling along over itself.

Dunes have a fascinating variety of shapes and patterns. The eventual shape is a function of many factors, for example, wind strength and direction, amount of sand, distance from the source of sand, and the presence of vegetation.

Where winds blow generally from one direction, as they often do along a sea coast, dunes are likely to have a more persistent geometry. For example, they may be aligned at right angles to the wind, in which case they are called ***transverse dunes*** (Fig. 12-31A). Such dunes may be quite short, and flourish where sand is abundant and the winds are strong. Typically, many coastal dunes are transverse.

A curious, aesthetically appealing dune is the ***barchan*** (Fig. 12-31B and 12-32), a sometimes perfectly proportioned crescent. Pointing downwind are the horns of the crescent and the steeper slip face that lies between them. Sand blows up the gentle slope of the crescent and slides down the slip face—a procedure typical of most dunes. Sand swept around the ends of barchans, however, tails off downwind to form the horns of the crescent. As the dune migrates, those points continue to pace its progress. Barchans, which are migratory and move across the desert at rates of 25 m (82 ft) or more per year, flourish in the trade wind deserts or in coastal deserts such as the Atacama of Chile and Peru. The area surrounding a barchan is likely to be barren bedrock, almost completely sandless. The wind is remarkably efficient, whisking up loose sand from the rocky floor of the desert between the dunes. Grains of sand bounce along much more readily over bare rock than they do over sand. Once they start to accumulate, as in a dune, their free-roving days are over—at least temporarily.

It is not unusual for one type of sand dune to grade laterally into another as conditions change. For example, transverse dunes can grade into barchans in a downwind direction as the supply of sand diminishes.

A ***parabolic dune*** is roughly similar in shape to a barchan (Fig. 12-31C; 12-33). Either U- or V-shaped, it forms where vegetation has partly stabilized the sand. Sand is blown from the center of the dune and deposited immediately downwind so that the nose migrates, but the arms remain anchored.

Other kinds of dunes occur in areas of ample sand supply, as a result of winds blowing, generally, in several dominant directions. ***Longitudinal dunes,*** for example, are linear ridges aligned generally in the direction of the prevailing wind, with slip faces on each side of the ridge (Fig. 12-31D and 12-34). Wind directions are thought to alternate from one side of the ridge to the other. Such dunes are com-

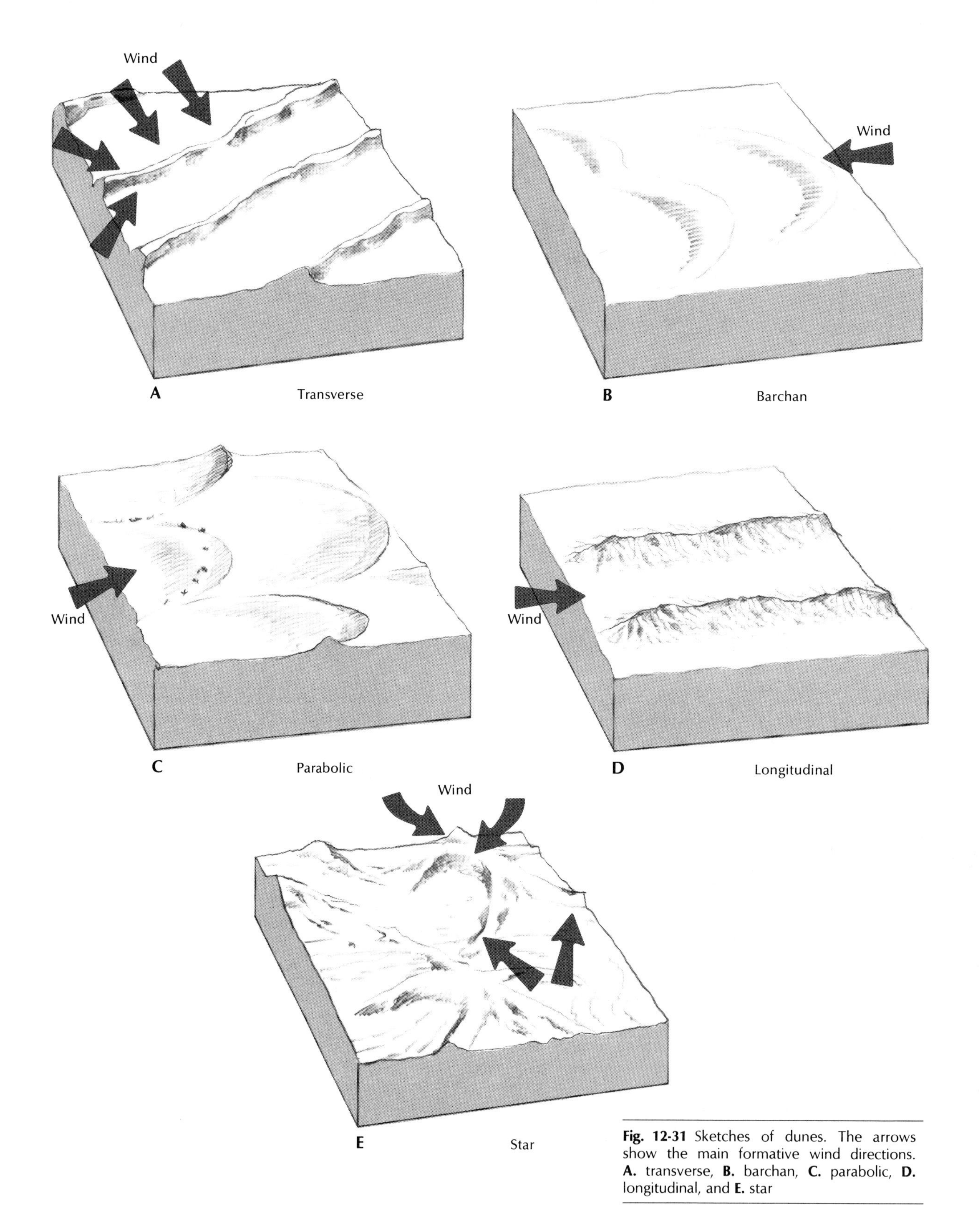

Fig. 12-31 Sketches of dunes. The arrows show the main formative wind directions. **A.** transverse, **B.** barchan, **C.** parabolic, **D.** longitudinal, and **E.** star

mon in the remote Great Sandy Desert of Australia, where ridges can be as long as 95 km (59 mi).

More complex are ***star dunes*** (Fig. 12-31E; see also Fig. 7-5), where winds blowing in many directions are involved, and the dune grows vertically rather than migrating laterally.

If it can be shown that dune forms became fixed in the past, as a result of anchoring by vegetation, they can be used to establish a history of wind directions. This information can then be incorporated into a climate model for some past period of time.

DESERT LAKES

Among the many distinctive features of such a dry and furrowed landscape as our own Southwest are desert lakes. They owe their existence, in large part, to the inability of desert streams to develop through-flowing courses. Where such streams are blocked, even though by no more than the advancing toe of an alluvial fan, their water is ponded and a lake results. Other lakes occur in troughs created when large crustal blocks dropped or downfaulted between adjacent mountains.

Fig. 12-32 Barchan dunes (foreground) in Saudi Arabia near Zalim. The wind moves sand grains from right to left. *Tad Nichols*

Fig. 12-33 A group of sparsely vegetated parabolic dunes in northern Arizona. The wind direction is from right to left. *Tad Nichols*

Some desert lakes, like Great Salt Lake with a surface area of about 3800 km^2 (1467 mi^2), are quite large. Others are little more than saline ponds.

Almost all desert lakes have brackish or saline water. The concentration of salt in Great Salt Lake ranges from as little as 14 per cent to as much as 27 per cent of the weight of the water, compared to 3 per cent in oceans. So Great Salt Lake is one of the world's most saline lakes—one in which swimmers have little problem floating. The Dead Sea, on the Israel-Jordan border, is another very saline lake.

Desert lakes are extremely sensitive indicators of climate. In a dry cycle, their water level falls as the result of evaporation, and drought-diminished streams cannot deliver enough water to counteract the loss of water in the

Fig. 12-34 The main linear feature here is a longitudinal dune, in which sand transport more or less parallels the orientation of the dune. Namib Desert, Southwest Africa. *Tad Nichols*

Fig. 12-35 Late Pleistocene lakes in the Great Basin, relative to present lakes. Arrows indicate direction of stream flow from one lake basin to another. In a few places, the lakes rose so high that water spilled out of the Basin.

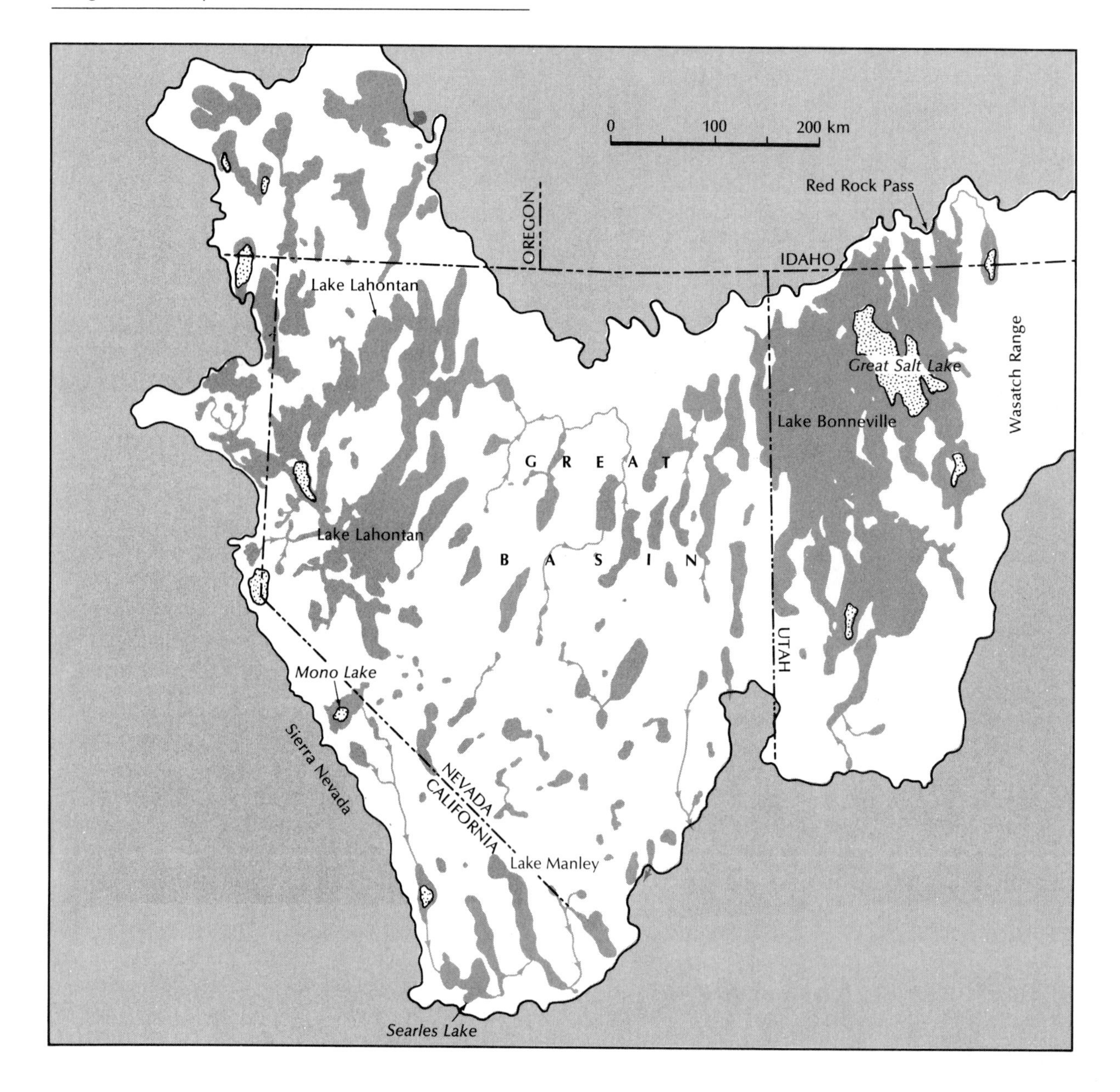

Fig. 12-36 Nineteenth-century etching of the Lake Bonneville shorelines on the northern end of the Oquirrh Range, Utah, *G.K. Gilbert, 1890, Lake Bonneville: U.S. Geological Survey Monograph 1.*

lake. An ever-widening band of salt borders the shore as the lake level drops. Most famous of such expanses is the Bonneville Salt Flat, adjacent to Great Salt Lake.

Ancient shorelines are another interesting feature of deserts—evidence that far larger lakes once existed (Fig. 12-35). The most redoubtable of the now-vanished inland seas in the United States was Lake Bonneville, the precursor of Great Salt Lake. Shorelines of this ancient lake can be seen on the higher slopes of the Wasatch Mountains, more than 300 m (984 ft) above the modern lake. Lake Bonneville covered an area of more than 51,000 km^2 (19,691 mi^2). At one time, Lake Bonneville had an outlet at Red Rock Pass, north to the Snake River, and thence to the Pacific by way of the Columbia River. At times the lake level dropped catastrophically, as shown by evidence of exceptionally large floods along the Snake River valley downstream from Red Rock Pass. One such flood approached depths of 100 m (328 ft) in some places.

A contemporary of Lake Bonneville was Lake Lahontan, located mostly in western Nevada not far from Reno. In that rugged mountainous area, all the intervening valleys were filled with long, narrow arms of the lake. Pyramid, Walker, and Winnemucca lakes are the chief remnants of Lake Lahontan.

Many desert lakes that were formerly higher, especially Lahontan and Bonneville, left their

Fig. 12-37 Tufa mounds in and along Mono Lake, at the eastern edge of the Sierra Nevada, California. Some mounds are still forming under water, whereas the lower ones at the lake edge have been exposed recently because some of the water has been diverted to Southern California. Still higher mounds formed during a higher lake level in the recent past. *The Mono Lake Committee, photo Jim Stroup*

Fig. 12-38 Lake sediments near the southern end of the Dead Sea, Israel, were deposited when the Dead Sea was much deeper, the result of a late Pleistocene climatic change. *Ran Gerson*

imprint on the landscape in an impressive array of wave-cut and wave-built landforms (Fig. 12-36). Among them are extremely well-preserved beaches, gravel bars, sea cliffs, deltas, and limy, tower-like deposits known as ***tufa*** (Fig. 12-37). The latter deposits are built up underwater by calcareous algae.

A remarkable set of lakes was briefly a part of the California landscape in the desert east of the Sierra Nevada. Individually they were far smaller than such giants as Lahontan and Bonneville and were part of an extensive network of lakes and streams. One series extended north from the site of modern Lake Arrowhead, in the San Bernardino Mountains, across one of the driest parts of North America, the Mojave Desert, to Death Valley. The lake, then perhaps 145 km (90 mi) long and 183 m (600 ft) deep, was named Lake Manley—in honor of Lewis Manley, who saved the first party of pioneers to reach Death Valley, where they ran out of supplies.

To the west of Death Valley, a similar set of lakes and streams led from Mono Lake, at the base of the Sierra Nevada, down the length of Owens Valley, to the basin of Searles Lake and on to Death Valley. Mineral deposits concentrated in Searles Lake are now being recovered from the dazzling white expanse of the saline playa.

Desert lakes occur in arid regions throughout the world. Among well-known examples are Lake Chad in Africa and Lake Eyre, in Australia, and Lop Nor, Lake Balkhash, and the Aral Sea in Central Asia. Not only were the Aral and Caspian seas larger in the recent geological past than they are today, but they were connected with one another as well as with the Black Sea. Many other seas, such as the Dead Sea, are rimmed by older shorelines that scar the barren slopes of the bordering desert hills, much like gigantic flights of steps, and lake sediments fill the basins (Fig. 12-38).

The obvious recency of these large lakes, coupled with the fact that in a few desert locations, such as along the flanks of the Sierra Nevada and the Wasatch Range in Utah, their shorelines cut deposits laid down by ancient glaciers, shows that the last "high stand" of the lakes generally coincided with the time of ice advance. Furthermore, the record is very clear that such events happened more than once, that all the high stands occurred during glacial times, and that the "low stands" (times of near desiccation) coincided with the times of glacier disappearance. Climatic change and the in-

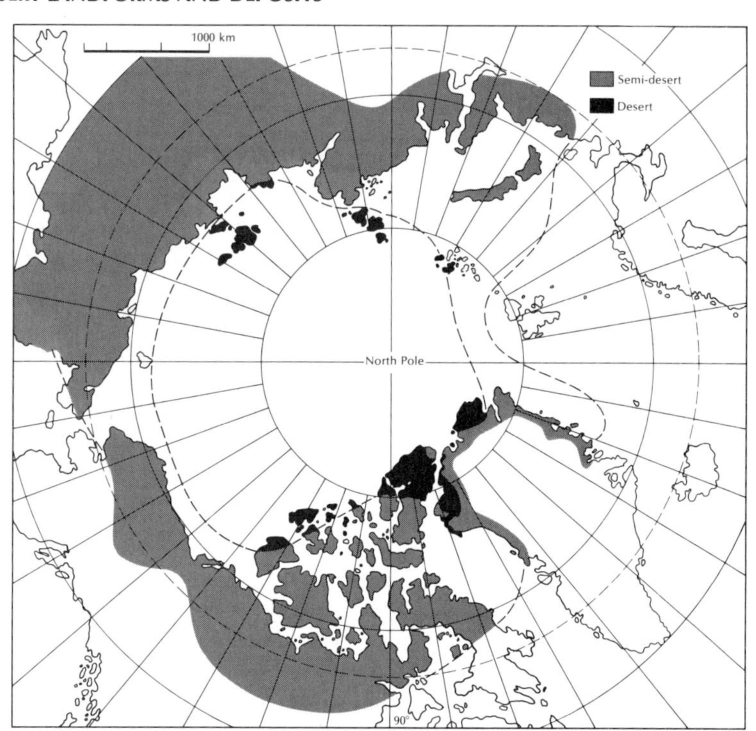

Fig. 12-39 Probable distribution of desert and semi-desert regions in the Arctic.

crease of meltwater appear to have caused lake expansion, and, as the glaciers of the world receded, the level of North America's desert lakes fell. Lakes in other climatic belts, such as those in Africa, did not rise and fall at the same time as our lakes—thus, lakes in these two areas are said to have been out of phase.

POLAR DESERTS

Another desert environment we are getting to know better is the polar desert—the cold dry areas that circumscribe the North Pole (Fig. 12-39) and that occur in the ice-free areas of Antarctica. The main differences between polar deserts and hot deserts is that the former receive most of their precipitation as snow and sleet and are characterized by permafrost (see Chap. 10).

Although some features of polar deserts are closely linked to the presence of permafrost, other features are quite similar to those of the hot deserts. So striking are some of the similarities that, were it not for the temperature, visitors might think they were in a hot desert. Rock weathers into weird forms in both cold and hot deserts, even though the processes of weathering might differ (Fig. 12-40). For example, salt weathering and wind erosion occur in both environments, and the little precipitation that does occur has some effect in both. Temperature alone, however, is not an effective weathering agent in polar deserts. Soils in both environments can be quite similar, with thin horizons and subsurface salt accumulations. Information on the rates of weathering and soil formation is somewhat scanty, but it seems that those processes are slower in the cold deserts. Ventifacts occur in both environments, and even salt lakes, the trademark of the hot deserts, are found in some polar deserts. The lakes in the Dry Valleys of Antarctica are an example.

Some hot-desert features, however, seldom

occur in polar deserts. Although in places there seems to be an adequate source of sand in the ubiquitous glacial outwash plains, for some reason the winds have not picked up the sand grains and shaped them into the sand seas so often characteristic of hot deserts. Although alluvial fans are seen in the polar deserts, pediments are not. Perhaps, as with so many other geological anomalies, we can call upon time to explain their absence. Many polar deserts only recently were covered by glaciers. Given more time, might pediments form? Nobody knows, and it is possible that processes associated with permafrost counteract pedimentation processes.

A FRAGILE LANDSCAPE

In many desert areas in the United States, population is burgeoning and with it problems in land use and abuse. In well-watered areas, vegetation usually grows rapidly after any disturb-

Fig. 12-40 Cavernously weathered granitic boulders. **A.** (above) in Death Valley National Monument, California. *R.M. Burke* **B.** (below) in Taylor Dry Valley, Antarctica. *T.L. Péwé*

Fig. 12-41 Erosion of a steep slope caused by the use of off-road vehicles near Gorman, California. Photographs taken in May 1976 (**A**) (above) and in March 1978 (**B**) (below) dramatically show the accelerated erosion. When the riding surface becomes too gullied, vehicle owners seek undisturbed land elsewhere. *H.G. Wilshire, USGS*

ance, helping to hide scars. In deserts, however, the works of humans remain in plain view for many years.

A recent threat to desert landscapes has been the damage caused by recreational vehicles. The ground is often disturbed to the extent that tire tracks can be seen crisscrossing the desert from miles away (Fig. 12-41). The tracks not only mar the landscape, but they also compact the soil and disturb the vegetation to the point that increased runoff, erosion, and gullying become a real problem. Motorcycles have made their way as far as the Arctic, where they are driven across the tundra—only time will tell how badly they are damaging that fragile terrain. We do know that deserts heal very slowly and that damage done today will undoubtedly affect generations to come. Some areas will never be restored. Even the Peruvian stone patterns (see Fig. 12-7) are threatened by off-road vehicles.

SUMMARY

1. Hot deserts occupy about one-third of the earth's land area and are characterized by rates of evaporation and plant transpiration that greatly exceed the precipitation rate. A sparse, or absent, vegetation cover plus the torrential but localized nature of the rainfall

help account for many desert features.

2. Many surfaces in the desert are characterized by *desert varnish,* which results from weathering. *Stone pavements* result from wind action or from the upward movement of stones from shallow depths.
3. The main depositional landform characteristic of hot deserts is an *alluvial fan,* formed from material deposited by a stream as it leaves the mountains and enters an intermontane basin.
4. The main erosional landform in hot deserts is a *pediment.* Its surface form is similar to that of a fan, but it is bedrock rather than deposited material. The origin of pediments is still debated.
5. Wind is a very effective agent in shaping some deserts. The sand it carries cuts away at stones and bedrock alike, and in some places it cuts and excavates bedrock basins.
6. Wind-blown sand can be deposited as a variety of dune forms.
7. Many deserts were wetter in the past, as shown by evidence that lakes existed in many present-day desert basins during times of glaciation.
8. Polar deserts are cold deserts and, with the exception of dune fields and pediments, have many features in common with hot deserts.

QUESTIONS

1. What conditions are responsible for deserts? Cite the evidence, within deserts or areas peripheral to deserts, that indicates that the boundaries of past deserts were different from the deserts of today.
2. Compare stream erosional and depositional processes downstream in a humid versus an arid environment.
3. How would you distinguish a pediment from an alluvial fan in the field? Would this distinction be important to groundwater exploration?
4. What factors are important in the formation of the main types of sand dunes?
5. What kinds of evidence could be used to suggest that desert lakes grew and dried up repeatedly, just as glaciers advanced and retreated repeatedly?

SELECTED REFERENCES

Bagnold, R. A., 1941, The physics of blown sand and desert dunes, Methuen and Co., London. (Repr. 1965, Halsted Press, New York.)

Bull, W. B., 1968, Alluvial fans, Journal of Geological Education, vol. 26, no. 3, pp. 101–6.

Cooke, R. U., and Warren, A., 1973, Geomorphology in deserts, University of California Press, Berkeley.

Glennie, K. W., 1970, Desert sedimentary environments, Elsevier Publishing Co., New York.

Hadley, Richard F., 1967, Pediments and pediment-forming processes, Journal of Geological Education, vol. 15, no. 2, pp. 83–89.

Leopold, A. S., and the Editors of Life, 1962, The desert, Time Inc., New York.

McKee, E. D., ed., 1979, A study of global sand seas, U.S. Geological Survey Professional Paper 1052.

Morrison, R. B., 1968, Pluvial lakes, pp. 873–83 *in* The encyclopedia of geomorphology, R. W. Fairbridge, ed., Reinhold Book Corp., New York.

Oberlander, T. M., 1974, Landscape inheritance and the pediment problem in the Mojave Desert of Southern California, American Journal of Science. vol. 274, pp. 849–75.

Ritter, D. F., 1978, Process geomorphology, W. C. Brown Co., Dubuque, Iowa.

Wilshire, H. G., 1977, Study results of 9 sites used by off-road vehicles that illustrate land modifications, U.S. Geological Survey Open-file Report 77-601.

13

GLACIERS AND EFFECTS OF GLACIATION

Scenically, the world is more indebted to glaciation than to any other process of erosion (Fig. 13-1). Without glaciation there would be few of the jagged peaks that stand in isolated splendor along the crest of many of the world's lofty mountain ranges.

The formation of steep cliffs in the valley heads and along the valley walls of the Alps, the Rocky Mountains, Alaska, and the Sierra Nevada is but a single aspect of glaciation. Because a glacier can also erode more deeply in some parts of its channel and less deeply in others, and because the material it deposits has an irregular surface, it forms lakes that add to the beauty of the landscape in alpine regions throughout the world, as well as in such lower-lying areas as the Great Lakes region, northeastern Canada, and Scandinavia. Unfortunately, however, the same lakes are mainly responsible for a summer flourish of insects. And the deep valleys whose outlines have been sharpened by the glacial file—such as Yosemite Valley, the Lauterbrunnenthal of Switzerland, and the Norwegian fiords—are spectacular.

Fig. 13-1 Extensively glaciated terrain in the region around Mount Blanc (4807 m; 15,771 ft), France. Characteristic of this area are the knife-like ridges called arêtes, and slender needle-like forms of many peaks, features that have challenged mountaineers for more than 200 years.

The Ice Age just ended has affected the lives of us all to a greater or a lesser degree. Soil and loose rocks were stripped from vast land areas, leaving barren rock behind. The load of stripped material was deposited toward the glacier margins. In addition, glacially produced silt and clay were blown off the floodplains of the world's major glaciated rivers and deposited across the landscape for many kilometers downwind. Such deposits constitute some of the world's finest agricultural lands.

Large lakes were created where none so extensive had existed before. Lake Agassiz located just west of the present Great Lakes was one of them, and its remnant, Lake Winnipeg, is a giant in its own right. The Great Lakes, also in large measure a product of glaciation, are still here. Great Salt Lake in Utah is a remnant of the much larger lake. And the rapid emptying of a glacier-dammed lake in the northwestern United States unleashed a prehistoric flood that could be the largest ever recorded.

Fig. 13-2 Enormous boulder resting on glacial ice, Baffin Island, Northwest Territories, Canada. Once the ice melts the boulder will be known as an *erratic. J.D. Ives*

Many of the areas blanketed by glacial ice were bowed down under its weight, and when the ice disappeared, the land rebounded. In Canada, uplifted wave-cut features show that the Hudson Bay region has risen 300 m (984 ft) or more (see Fig. 3-9). Historic records and various shoreline structures, such as ancient landing places, show that the land is still rising. It rises at the unusually high rate of about 2 m (6.5 ft) per century at the southern end of Hudson Bay and decreases to about zero in the vicinity of the southern Great Lakes.

Worldwide, the sea level reflects with the waxing and waning of the ice sheets. When the ice sheets expanded, tens of millions of cubic kilometers of water were withdrawn from the sea via the atmosphere and locked up on land as ice. Sea level was lowered as a consequence, perhaps by as much as 140 m (459 ft). That may not seem like much, but it was enough to alter profoundly world geography. Land areas now separated were then connected, and migrations of both people and animals occurred across these ***land bridges.*** Among the natural causeways were those that once connected Tasmania and Australia, Sri Lanka and India, New Guinea and Australia, and some of the islands of Indonesia. Most renowned of all was the land bridge that joined Alaska and Siberia, areas now separated by the 55-m (180-ft)-deep waters of the Bering Strait.

There are many reasons why geologists study glaciers and the effects and deposits of former glaciations. Perhaps the foremost reason is that they indicate rather dramatic past climatic changes. If we can determine when such changes occurred, we might be able to predict the frequency of climatic change. Another reason for studying glaciers is that glacial deposits often create special engineering problems. Further, monitoring the amount of ice on land helps us to determine if the rise and fall of sea level along some coasts is caused by worldwide changes in the volume of ice or by local tectonic changes.

THE GLACIAL THEORY

No wonder much of the evidence of past glaciation attracted the attention of observant men in Europe and in New England in centuries past. The lavish supply of boulders on New England farms was a source not only of wonderment to the early settlers but also of wearisome, backbreaking toil. So much labor was involved in clearing fields strewn with glacially transported stones that more than one young man was readily convinced that a life at sea could be no harsher—even on a New Bedford whaler.

For many, the presence of those strange stones, found far from their place of origin and very often completely out of harmony with their new environment—for example, granite blocks resting on a limestone terrain—could be explained as the work of that "vindictive affliction," the Great Flood of Noah.

In Great Britain, much of which was recently covered by glaciers, the widely scattered glacial deposits were called ***drift,*** a name betraying the belief that it originated as a deposit spread far and wide by an all-encompassing sea. Some even accepted the idea that icebergs and ice floes may have transported the ***erratic boulders*** (Fig. 13-2), as such out-of-place rocks are called, for British whalers working off the coast of Greenland had seen such debris embedded in sea ice there, as had explorers elsewhere in the Arctic.

Persuading English geologists that glaciers had scoured the inland surface and transported rocks the size of small houses for scores of kilometers was a difficult task, for there were no existing glaciers to serve as models. It is not surprising, therefore, that the most eloquent advocates of the notion that ice could perform prodigies of work in shaping mountains and excavating valleys were the Swiss. There are about 2000 glaciers in the Alps, and through the centuries their snouts have advanced or retreated, and alpine passes have been alternately ice free or ice blocked. Some villages occupied in medieval times are now buried by ice. The silver mine of Argentiere, active in the Middle Ages, is now covered by the glacier of Mont Blanc, near Chamonix, in the French Alps. Many alpine villagers must have been aware that when a glacier receded it left behind it a trail of barren, stony ground interrupted by low, rocky ridges, diversified by lakes and ponds, and strewn freely with rock fragments, large and small.

In the early nineteenth century, several European geologists became convinced that an ice sheet had covered much of northern Europe, an idea that aroused the curiosity of one of the leading Swiss naturalists of the day, Louis Agassiz (Fig. 13-3). He persuaded one of the geologists to take him on an expedition in the Alps. Agassiz became a believer, and when in 1837 he came to the United States and to a professorship at Harvard University, he spread the word far and wide of a "Great Ice Age" that had once refrigerated most of the Northern Hemisphere. A concept so novel aroused opposition, for some accepted the far more labored explanation that (1) the land sank; (2) the sea spread inland, and boulders and other detritus were rafted by icebergs far and wide across its waters; and (3) the land rose, thus shedding its oceanic waters, which left behind a residue of rocks and boulders scattered over the landscape.

Although Agassiz deserves full honor for carrying the word from Europe to North America, he was not the first to believe that glaciers had once advanced across the European landscape. The problem of discovering who had the original idea is one of the many difficulties confronting the science historian. So it was with the glacial theory. A number of remarkably perceptive persons had glimpsed the truth. As

Fig. 13-3 Louis Agassiz (1807-73). *Harvard University Archives*

early as 1802, erratic boulders in the Jura Mountains were recognized for what they were—glacially transported rocks. In Germany, a professor of forestry, Bernhardi, employing the same reasoning, wrote a paper in 1832 stating his belief that the ridges of drift and erratic boulders that are significant terrain elements of the northern plain were evidence that glaciers had advanced southward from lands far to the north. He suffered the familiar fate of a prophet in his own country in that few people paid any attention to him. Not until 1875 did German scientists accept the idea that their homeland had once been overridden by a sheet of ice.

DISTRIBUTION AND FORMATION OF GLACIERS

Glaciers today occupy about 15,000,000 km^2 (5,791,500 mi^2) of the earth's surface—about 10 per cent of the total land area. Most of the ice is locked up in two ice caps: Antarctica, which accounts for about 84 per cent of the ice in the world, and Greenland which accounts for 11 per cent. The rest of the ice is scattered around the world, generally in mountainous areas. Glaciers are especially prominent features in the ranges that parallel the coasts of Alaska, British Columbia, and Washington; in the Rocky Mountains of Canada and the northern United States; on many islands of the Arctic region, including Greenland; and in Scandinavia, the European Alps, the Southern Alps of New Zealand, and the Andes.

The amount of water locked up in glacial ice can be converted to depth of ocean water; thus we can appreciate its quantity and gain some idea of what might happen if all the ice on earth were to melt. It is no simple task to measure the amount of existing ice because the configuration of the bases of glaciers are not well known. The best estimates, however, are that if all the present ice were to melt, sea level would lie some 100 m (328 ft) above its present level. A glance at a map of the United States shows that many coastal cities would lie under 100 m of water, whereas many inland cities would become seaports (Fig. 13-4). There is no need for alarm, however, because the major ice caps are fairly stable, and any changes that might take place probably would do so gradually, over a thousand years or so. The glacial recession since the 1890s, for example, has resulted in a general sea level rise of 12 to 30 cm (5 to 12 in.)—a change perhaps not important to many of us, but certainly important to people living on low-lying coasts. In low-lying areas of Holland, for example, dikes have been built to keep the rising sea from submerging the land. It should be pointed out, however, that a rising sea level is not the only reason for the sea-water flooding of some areas. Other reasons are tectonic subsidence and compaction of the sediments that make up the land. More will be said of those processes later.

Change of snow to ice

Without snow, there would be no accumulation of extensive bodies of ice, and without ice, there would be no glaciers. True glaciers, however, must display evidence of flowage. Glaciers may develop in those parts of the world where the combination of sufficient winter snowfall and low summer temperatures allows some snow to remain the year around. Thus, glaciers are active today in high mountains over much of the globe, and in the farther reaches of the Northern and Southern hemispheres. On the high upper slopes of such equatorial mountains as Kilimanjaro in Africa, the summits of the Andes in South America, and the Carstenz Toppenz Range in New Guinea, ice fields lie at altitudes of 4800 to 5400 (15,744 to 17,712 ft). In mid-latitudes, as in the Sierra Nevada of California and the Swiss Alps, they lie close to 3000 m (9840 ft). Finally, they form near sea level in Antarctica (55°S) and at about 600 m (1968 ft) in lands bordering the Arctic Ocean.

More than cold temperatures are needed for snow to accumulate and remain from year to year. We can demonstrate that readily by pointing out the vast areas of the cold Arctic that are not covered with glaciers. Sufficient snowfall also is required to maintain glaciers, but the exact amount will vary from place to place. In maritime regions with high summer snowmelt, about 4 m (13 ft) of snowfall (on a water-equivalent basis) are required to maintain glaciers, whereas only a fraction of a meter is required in the polar deserts of the northern Arctic and Antarctica, where summer snowmelt is at a minimum.

Local environment is significant in determining the position of glaciers. In the Northern Hemisphere, north-facing slopes are shadier and receive less solar radiation than south-facing slopes; under the right condi-

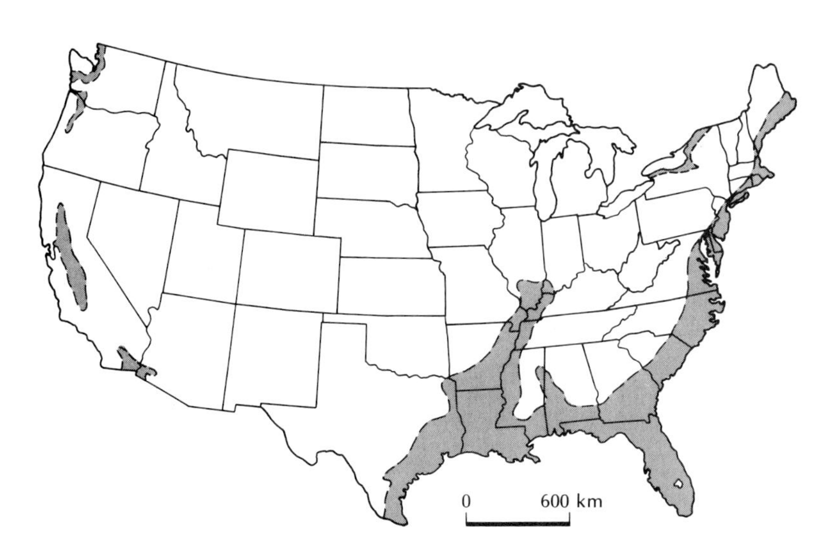

Fig. 13-4 Map showing the approximate location of the shoreline in the United States, if all the glaciers in the world were to melt.

Fig. 13-5 A. (above) Newly fallen snowflake. The specific gravity of new snow is low because much of its volume is occupied by air. **B.** (below) With time, snowflakes melt and refreeze into rounded snow particles. *E.R. LaChapelle*

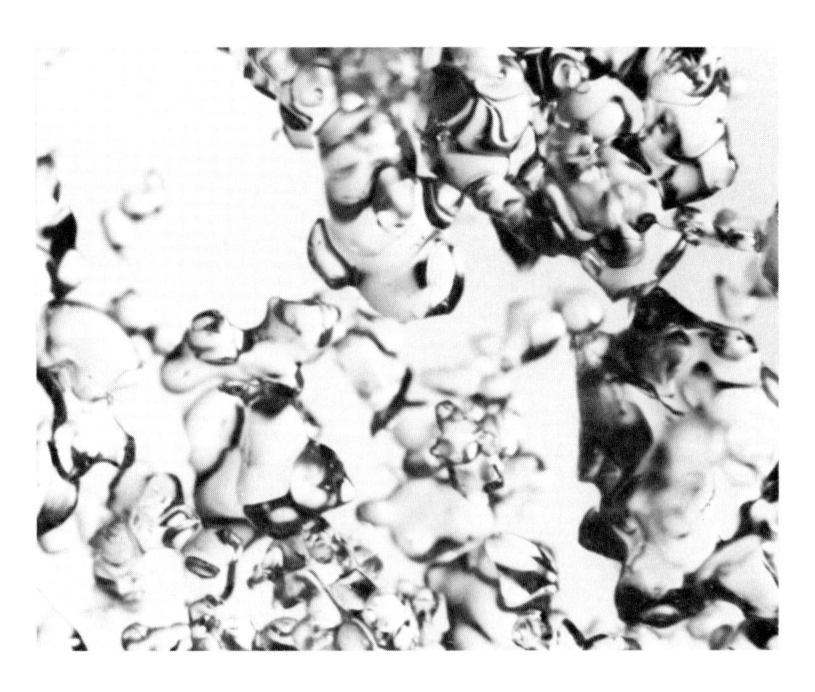

tions, ice can persist on northerly exposures, but not on southerly ones. In many North American mountain ranges, the western side may be the windward slope, where precipitation is greater and cloud cover more persistent. There glaciers commonly lie at lower altitudes than they do on the eastern side. Wind is also a determining factor in some places, since it redistributes winter snowfall. In parts of the Rocky Mountains of Colorado and Wyoming, especially, dry snow falling on the alpine tundra slopes is picked up by winds from the west and redeposited on the heads of east-facing valleys, eventually to be converted to glacier ice.

Before it becomes a glacier, snow must convert to ice, which is no simple process. Snow, strictly speaking, is not frozen water, as is ice, but frozen water vapor. In other words, it is water that has crystallized directly from water vapor in the atmosphere. Since snow is a crystalline substance, snowflakes grow in regular geometric patterns (Fig. 13-5A). Although they seem to show an infinite variety, it probably is not strictly true that no two snowflakes are alike. Crystalline water behaves like other mineral crystals in that there is an established crystal form for the compound—a variant of the hexagonal system—the same system of which quartz is a member. The specific gravity of snow is much less than that of water, so that 1 cm (0.4 in.) of snow is commonly equal to 1 mm (0.04 in.) of rain water.

After snowflakes lie on the ground for a short while, they change form (Fig. 13-5B). Individual flakes may ***sublimate*** (pass directly from a solid to a gaseous state) or they may melt and refreeze into granules of ice. That gritty, granular snow, with a texture much like coarse sand, is a familiar phenomenon in snowbanks that survive for most of the winter behind a building, in the shade of a forest, or in the lee of a cliff. Such granular, recrystallized snow is called ***firn*** (from a German adjective, of last year) in the German-speaking parts of Switzerland; ***névé*** (a word going back to the Latin stem of *nix* for snow) is used in French-speaking areas. Both terms are used in English as well.

Firn, which typically accumulates on the upper slopes of alpine mountains, goes through a gradual transition into glacier ice. Firn is usually white or grayish-white, and the spaces between the granules are filled with trapped air. At a depth equivalent perhaps to an accumulation of three to five years of firn, the pore spaces become smaller, or may even disappear, and the transition into blue glacier ice made up of interlocking ice crystals is complete (Fig. 13-6). The process is accompanied by an increase in the specific gravity from perhaps 0.1 in newly fallen snow up to 0.9 in solid ice. The change from firn to ice is aided by an increase in pressure resulting from the weight of the overlying snow and ice. As a glacier moves downvalley, the ice crystals recrystallize and may reach diameters as large as 7 to 10 cm (3 to 4 in.).

The conversion of snow to ice is an excellent example of present-day, rapid metamorphism. Snow falls on the ground and builds up sedimentary layers (Fig. 13-7). As temperature and pressure conditions change, the snow is metamorphosed into ice, which moves downvalley, developing many flow patterns that resemble patterns made by folding in metamorphic rocks. As the snow becomes ice, much of the glacier remains solid, just as rocks do during metamorphic change. If the formation of glaciers were not a solid-state change, the ice would melt and the glacier would vanish quickly.

Mechanism of glacier movement

Nobody doubts that glaciers flow. A common proof is that the rocks being deposited at the front of alpine glaciers are derived from the cliffs that flank the upper end of the glacier. Unusual evidence of such movement came from the disappearance of an early mountaineer in the European Alps. Efforts to find him failed, but his body appeared decades later, at the lower end of the glacier. In the section that follows we will look first at glacier flow and then try to explain how such motion comes about.

The cross-valley velocity of a glacier in an alpine valley is very similar to that of water in a stream. The flow is greatest in the center, and the least on the sides and bottom (Fig. 13-8). The diminishing velocity toward the valley sides and bottom is the result of friction between the ice and the bedrock. Average velocities vary from glacier to glacier, but most fall between 3 and 300 m (10 and 984 ft) per year.

Occasionally a glacier decouples from the rock floor and sides and advances downvalley at truly fantastic velocities, some approaching as much as 6000 m (19,680 ft) per year. Such movements are known as ***glacial surges*** and are usually characteristic of valley glaciers, although ice caps have also been known to surge. Surges generally take place after a glacier has become stagnant or is even receding; suddenly it moves several kilometers in a few months. For example, in 1953 the Kutiah Glacier in the Himalayas surged 11 km (7 mi) in three months. Various theories have been suggested to account for glacial surges. Some geologists have thought that earthquakes might shake great quantities of snow and ice down onto a glacier or that snowfall for a few years at the head of a glacier could increase. Others have suggested that an increase in the amount of meltwater percolating down through a glacier might help it to slide along its bed. A recent, credible explanation is based upon the formation of a block of stagnant ice at the end of a valley glacier. Such a block would act as a dam until the pressure of the flowing ice behind it forced it to give way.

Fig. 13-6 Interlocking crystals of glacier ice, as photographed under a microscope. *Chester Langway, Jr.*

Ice flows in complex ways. Generally, the lowest part of most glaciers moves by actually sliding over the rocks, as shown in Figure 13-8B, by the displacement of the base of the pipe from X′ to Y′. Evidence for such movement, which is called ***basal slip,*** are the polished and scratched rocks left behind when glaciers melt (Fig. 13-9). The horseshoe-like indentations in the polished bedrock are ***chattermarks*** (Fig. 13-9A), the long scratches ***striations*** (Fig. 13-9B). In contrast, the lower part of a glacier (Fig. 13-8B, between Y′ and Z) creeps downvalley through ***plastic flow*** because the ice crystals deform under the pressure of the overlying ice. The upper part of a glacier (above Z in Fig. 13-8B) consists of brittle ice since there is not enough ice and snow above it to cause it to deform plastically. Such ice rides piggyback on the lower ice, which is continually undergoing deformation. Deep flowage, combined with the valley-side friction and the low strength of the brittle ice, causes innumerable cracks or ***crevasses*** to form in the brittle ice (Fig. 13-10). Such crevasses can extend to 30 m (98 ft) or more in depth; at greater depths, plastic flow of the ice seals them off.

The above discussion is accurate for most glaciers, but it is not quite as accurate for those located in polar regions. There, temperatures are so low that glaciers are most likely frozen to their bedrock bases. Basal slip, therefore, is

Fig. 13-7 Sedimentary layers in glacial ice are well shown at the edges of this glacier in Vatnajökull, southeast Iceland. *J.D. Ives*

not so important in the motion of polar glaciers as it is in the motion of temperate-climate glaciers, which are not frozen to their bases.

The glacier budget

Glaciers continually gain and lose mass, and by a series of detailed measurements we can determine their gains and losses; in short, we can determine a budget. Although we can manipulate a household budget, a glacier cannot easily hide its surpluses or deficits. If, for example, it has a surplus year, the front may advance downvalley, whereas in a deficit year the front may retreat.

In order to understand a glacier's budget, we need to determine where the gains and losses are taking place. Every glacier has a fairly narrow zone on its surface known as the ***snowline*** (Fig. 13-11). That zone is identified late in the melting season; above it, some snow lingers from year to year and will accumulate, whereas below it, the snow from the previous winter (and some of the underlying ice) melts and is lost. Thus, there are distinct areas of ***accumulation*** and areas of wastage, or ***ablation,*** on glaciers (Fig. 13-12). In a very general way, the area of accumulation makes up about two-thirds of the total surface area of a valley glacier.

With the glacier budget in mind, we now can examine how a glacier advances and retreats. We will focus our attention on the front of a stable glacier, that is, one that is neither advancing nor retreating because the accumulation of snow is balanced by ablation (Fig. 13-13, position A). The glacier maintains the same form over the years, so that all losses in the ablation area are precisely balanced by the flow of ice into that area from up-valley. To illustrate the point, we might compare a glacier to a side of bacon being fed into a slicer. Although the bacon is continuously being shoved forward, it never advances beyond a fixed point because it is always being cut off by the oscillating blade. In contrast, if the snowline were to lower (Fig. 13-13, position B), the area of accumulation would enlarge. The amount of ablation would then be less than the amount gained, and the front would advance. Advance takes place until, for that snowline position, a balance between gains and losses is reached. Similarly, if the snowline rises, the opposite effects are seen—the area of wastage enlarges, more ice melts than can be compensated for by accumulation, and the front retreats until a balance with respect to the new snowline (Fig. 13-13, position C) is reached. If the snowline continues to rise until it no longer intersects the land surface, the glacier melts away.

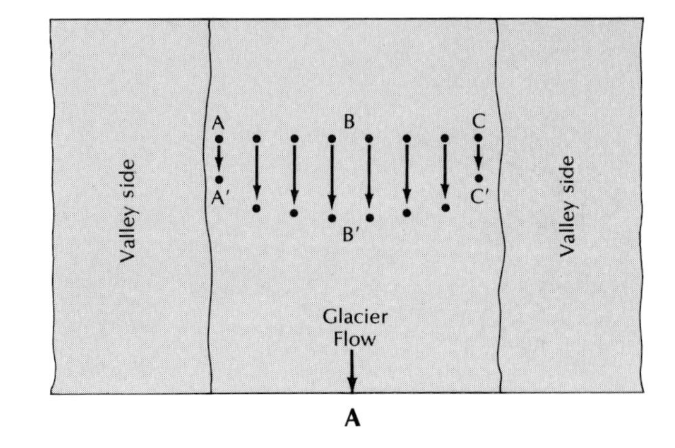

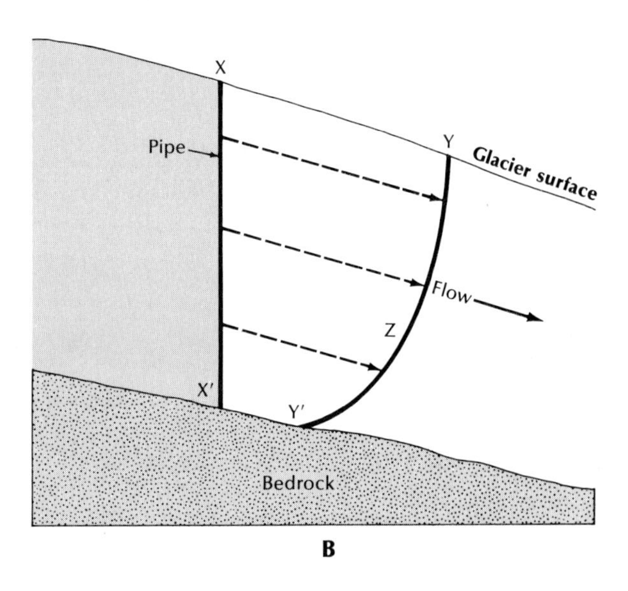

Fig. 13-8 A. Plan view of an alpine glacier, showing how velocity measurements are made. Stakes are set in the ice in a straight line (A-B-C). Eventually the movement of the ice will cause the line to curve (A'-B'-C'). The distances between the original and present positions of the stakes indicate the amount of displacement, from which the velocity is calculated. **B.** Vertical profile of an alpine glacier, showing the velocity distribution with depth. A vertical pipe extending down to bedrock is placed at X. After several years the pipe will bend (Y-Y'), its top moving from X to Y and its base from X' to Y'.

ALPINE GLACIATION

Although individual glaciers and ice fields are tremendously diverse, they can be placed in two broad categories: ***alpine glaciers*** and ***continental ice sheets.*** Alpine glaciers have their origin on mountain slopes and summits that rise above the snowline. They advance downslope

Fig. 13-9 A. (left) Glacially polished bedrock indented with chattermarks in the Sierra Nevada, California. *G.K. Gilbert, USGS* **B.** (below) Limestone bedrock striated by glacial action exposed in a quarry in Vermilion County, Illinois. *Illinois Geological Society*

Fig. 13-10 A. (right) Crevasses in a glacier on Mount Rainier, Washington. Irregular bedrock topography sets up complex flow patterns in the plastically deforming ice near the base of the glacier, causing the brittle uppermost ice to crack and form crevasses. *Peter W. Birkeland* **B.** (below) Crevasses as a tourist attraction, 1895. Their patterns change with the constant glacier movement, and in places they are covered by weak snow bridges—a hazard to mountaineers. Note the depth of this crevasse in Muir Glacier, Alaska. *Library of Congress*

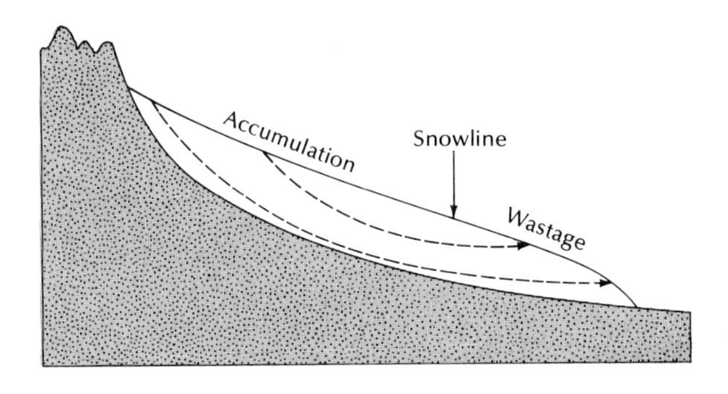

Fig. 13-11 Cross section of a valley glacier, showing the areas of accumulation and of wastage. The dashed lines are the approximate flow lines of the glacier particles. Note how they descend and then rise toward the surface in the wastage area. A snow particle in the accumulation area will follow such paths, convert to ice on the way, and melt when it reaches the surface.

under the pull of gravity, and more often than not take the path of least resistance by following a pre-existing stream valley. In contrast, continental ice sheets are irregular in shape, cover a large land area, and are not necessarily guided in their flow pattern by the underlying topography.

Continental ice sheets were much more important in the recent geological past than they are today. During the ice ages, North America as far south as the Ohio and Missouri rivers was buried under up to about 3 km (1.8 mi) of ice, as was most of northern Europe. Fortunately, such vast expanses of frozen water are gone, and surviving relics, such as the Greenland ice cap, although huge, are greatly subordinate to the vanished titans. Continental glaciers override the terrain and smooth out many of the irregularities they encounter. Alpine glaciers are more likely to accentuate irregularities in the landscape, making bold peaks even more jagged, and steepening the walls of already deep canyons.

Although alpine glaciers are conspicuous elements of the world's snow-capped mountain ranges, they, too, are much smaller than they were in the Ice Age. At that time, glaciers in the Alps advanced northward far beyond the foothills of the Alps onto the lowlands near Munich and southward into the low hills marginal to the Po Valley of Italy. Reduced though they may be, some of the surviving alpine glaciers are impressively large. Several of the Himalayan ice streams are 40 km (25 mi) long, and the Great Aletsch in Switzerland is 22 km (14 mi) long. The Seward glacier in Alaska with its tributaries approaches 80 km (50 mi) in length.

In the next section, we will discuss glacial erosion and deposition, emphasizing alpine glaciation. The discussion should help us to understand the effects of continental ice sheets on the landscape.

Glacial erosion

There appear to be two major ways in which alpine glaciers shape the land surface upon which they rest: ***glacial quarrying*** and ***glacial abrasion.*** In the first process, rocks are sprung, or pried, out of place in much the same way that they are in commercial rock quarries—except, of course, at a far slower rate. Abrasion takes place when the rock surface on which the ice rests is scoured or worn down—in much the same way that a wood surface may be sandpapered.

Quarrying is not completely understood, which is scarcely surprising because it is a process that takes place beneath the glacial ice in an environment ordinarily inaccessible to us.

The nearest access we have to that frigid, subglacial world is through the crevasses that extend down through the ice to the bottom of the glacier. Such deeply penetrating fractures are reasonably common in the accumulation area at the headward end of a glacier, but rare elsewhere. In 1899, in the Sierra Nevada of California, one of the leading topographers of that day, W. D. Johnson, had himself lowered on a line to a depth of 45 m (148 ft) into a crevasse—certainly no venture for the fainthearted—and saw that glacial ice indeed was capable of prying strongly jointed rocks loose from their foundations.

Glacial crevasses provide an avenue by which ***meltwater,*** supplied by melting of snow, névé, and surface layers of ice, streams down

Fig. 13-12 The snowline on these glaciers is marked by the transition from clean snow at their upper ends to streaked, "dirty" ice at their lower ends. The accumulation area lies above the snowline; the wastage area below it. Clyde Fiord on Baffin Island, Northwest Territories, Canada. *J.D. Ives*

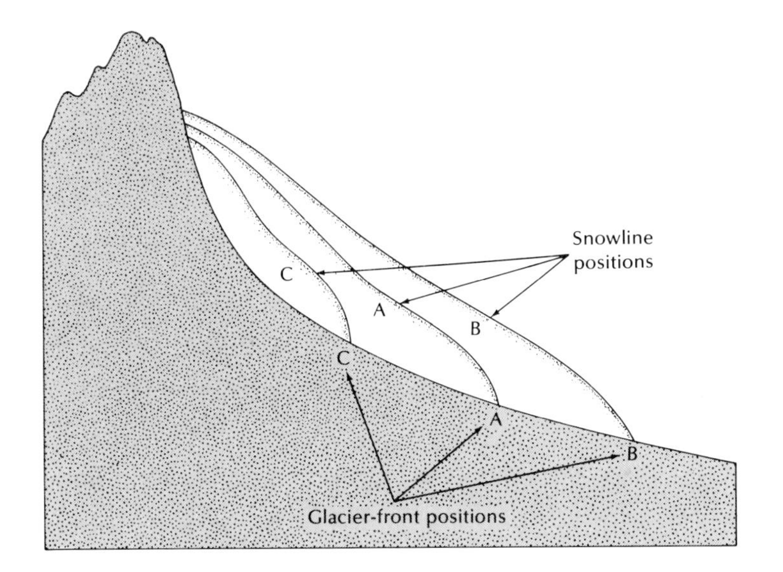

Fig. 13-13 Cross section of an alpine glacier, showing how its size is governed by the snowline. A change in size comes only through frontal advance or retreat because the upper end of the glacier is always anchored to the headwall.

into the inner recesses of a glacier. If meltwater penetrates to the rocky headwall or to the glacier floor, it percolates into the cracks and joints of the bedrock. As the water freezes it expands, exerting a tremendous pressure on the enclosing rocks, which may then be pried loose.

Should those rocks then be frozen into the glacier, they are carried along with the glacial ice as it moves downslope, and a new surface is bared for the process to be repeated. The more often glacial meltwater freezes in the subglacial rock joints, the more effective the process of rock quarrying will be.

Rocks from the valley walls, ranging from blocks the size of boxcars down to fragments the size of flour grains, are embedded in the ice as the result of glacial quarrying and mass wasting. The rocky debris is dragged along with the glacier, and as it moves downslope, it acts as an abrasive. When the embedded rock fragments are large, they may gouge out long grooves and scratches, called ***glacial striations,*** in the bedrock. Where the rock fragments are fine grained, they may polish the surface of the overridden rocks, much like a lapidarist's fine emery powder does. Visitors to the high parts of Yosemite National Park are impressed by the broad expanses of smoothly polished, shining granite, as fresh looking as though it had been given its bright sheen only yesterday (see Fig. 13-9). Abrasion works both ways; many of the rock fragments embedded in the ice develop scratches on their surfaces.

Landforms produced by glacial erosion

As the debris-laden ice quarries and grinds away the surface over which it moves, characteristic landforms are produced. Most commonly it smooths down irregularities, so that on a typically ice-abraded landscape rock knobs are rounded off and the lower ends of spurs and ridges are blunted or even worn away. Valleys are deepened and made more linear (Fig. 13-14). In the most striking examples, their sides are steepened until they are almost vertical, as in Yosemite (Fig. 13-15), or in the Lauterbrunnenthal in Switzerland.

Of the features characteristically resulting

from glacial quarrying, one of the most impressive is a ***cirque*** (Fig. 13-16). A cirque is a horseshoe-shaped, steep-walled, glaciated valley head; it is such a distinctive landscape element that a name for it exists in the language of every western European country in which it is found. A French word meaning circus, the term "cirque" is not completely appropriate in the glacial connotation because a true circus is a fully round figure. Actually, a glaciated valley head is a half-round feature and is more nearly comparable to the traditional form of a Greek theater. To return to the other names for a cirque: in German-speaking lands it is a *kar;* in Wales, a *cwm;* in Scotland, a *corrie;* and in Scandinavia, a *botn* or a *kjedel.* This brief linguistic excursion illustrates that geology is a truly international science; no single country has a monopoly on all the names for landforms.

The origin of cirques is still somewhat of a mystery. They seem to result partly from active plucking or quarrying at the head of a glacier, probably as the result of the downward percolation of water and frost-riving of the rocks in the headwall that towers above the glacier surface. Another process that seems important is the rotational movement of the ice, which gouges out closed basins in the floors of cirques. Should the glacier disappear, as so many have since the end of the Ice Age, then the cirque will be seen. The rock wall at the upper end of such a titanic box canyon may be a cliff 1000 m (3280 ft) high, and the base of the cliff is usually completely free of the long and sloping apron of talus blocks so common at the base of other cliffs. Commonly a lake occupies the closed basin on the cirque floor (Fig. 13-17).

Scores of clear, often dark blue, rock-basin lakes, called ***tarns,*** add immeasurably to the beauty of alpine scenery; many of them result from differential glacial scouring. Unlike streams, glaciers, as noted, dig deeply in some places and much less so in others. The floor of a glacial trough may consist of closed basins, in which water collects after the ice has melted, and of intervening barren ice-smoothed ridges. Such an irregular floor, sometimes deep beneath the ice, sometimes near the ice

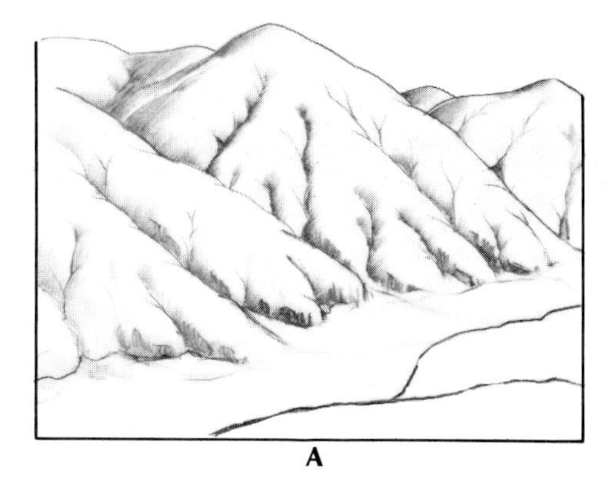

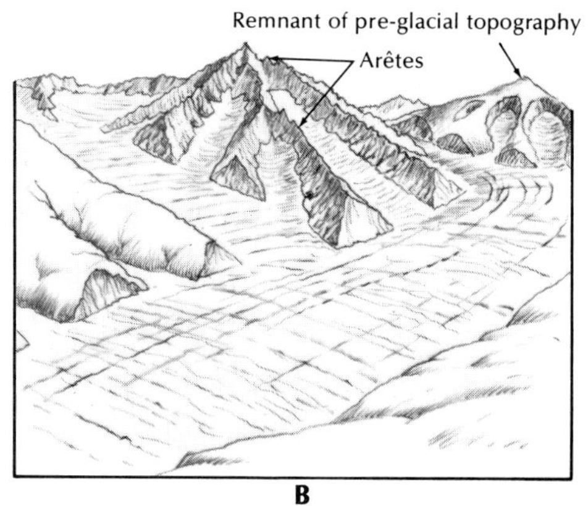

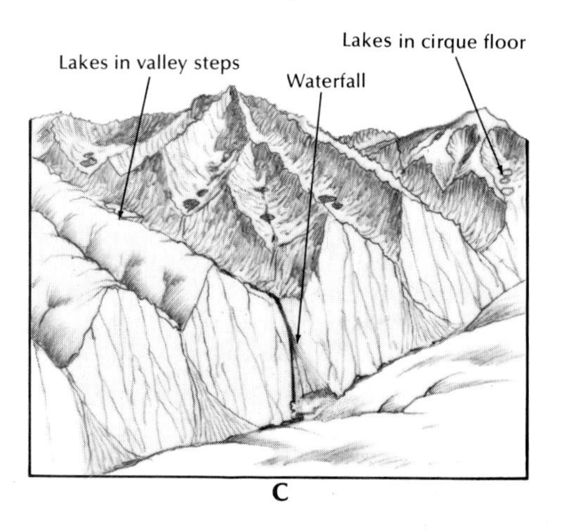

Fig. 13-14 The alteration of a stream valley by alpine glaciation. **A.** Terrain before glaciation with smooth, rounded hillslopes and a stream meandering through a V-shaped valley. **B.** Maximum extent of glaciation. Ridges and peaks close to the active glaciers are steepened and take on a jagged appearance. Not all the rounded mountain summits are consumed, however. **C.** After the ice melts, the characteristic glaciated landscape appears: U-shaped valley, hanging valleys, cirques, and lakes.

surface, can be excavated so long as there is a forward slope to the surface of the glacier. In the same way, lava flows are able to cross a corrugation of ridges and valleys.

Although glacial erosion is responsible for the extremely rugged alpine scenery, it is also responsible for the passes that allow for travel through such forbidding terrain. When two glaciers exist in parallel valleys close to one another, then the divide separating them may become narrowed through over-steepening until it is reduced to little more than a rock screen (see Figs. 13-1 and 13-14) called an ***arête*** (from the French word for ridge or fishbone). Similarly if two glaciers are on opposite sides of a divide, and flowing away from each other, the headwalls of the cirques may intersect until only the narrowest sort of rock partition separates the two cirques. The top of such a rock screen may be almost razor sharp and commonly is surmounted by jagged pinnacles or spires. If glacial quarrying continues actively, in very little time the intervening ridge may be stripped away completely. When that happens, the two glacial troughs intersect, and a sometimes strategically important transmontane pass or ***col*** may result. Some of the famous alpine passes, such as the St. Bernard, St. Gothard, and Simplon, are of that type, as are Berthoud Pass in the Rockies and Tioga Pass in the Sierra Nevada. In some glaciated mountains, glaciers radiate away from the summit area like spokes of a wheel. Should the glaciers continue to quarry actively at their upper ends, then the mountain may be whittled away until only a jagged, saw-toothed pinnacle called a ***horn*** survives. The Matterhorn is the world's most familiar example of such a glacially accentuated peak, and Mt. Assiniboine in Canada is well known to North Americans (Fig. 13-18).

Downvalley, a primary result of glacial erosion is the overdeepening of valley floors and the steepening of valley walls. The part of an alpine valley occupied by a glacier resembles the channel occupied by a river, but on a far grander scale, because the volume of water in the glacier is far greater.

The statement commonly is made that such a glacially occupied, ice-scoured valley has a ***U-shaped*** cross section (Fig. 13-19), in contrast to the ***V-shaped*** form of a stream valley. In reality, a cross section of a glaciated valley more nearly approximates a ***catenary*** (a curve produced by a wire or chain hanging from two points not in the same vertical line). Such curves are seen in the sweeping loops of the wire between the towers of a transmission line. If the points of suspension are close together, the curve is narrow and the slope is steep; if they are far apart, the curve is open and the side slopes are gradual. Glaciated valleys also commonly are characterized by a series of gigantic cliffs separated by flat stretches that resemble cyclopean steps. They owe their origin to the variabilities of glacial flow (and thus the ability to erode) and the relative resistance of the bedrock to glacial erosion.

Among the results of alpine glaciation are waterfalls that plunge over the rims of glaciated valleys and cascade in long streamers down the shining walls (Fig. 13-20). There are a number of explanations for the origin of such dramatic side streams. A widely accepted view is that they were unable to cut down rapidly enough to keep pace with the main canyon, which was being deepened by the glacier. Thus their valleys are called ***hanging valleys.***

Other spectacular landforms of glaciated terrain are the ***fiords*** that flank many high-latitude coasts, as in Alaska, the Canadian Arctic, Norway, Chile, and New Zealand (Fig. 13-21). Narrow troughs, they differ from land-based flat-floored glacial valleys (compare with Fig. 13-15) mainly in that they are submerged by the sea, but also in their truly remarkable depths. One in Norway, for example, is 1300 m (4264 ft) deep; another in Antarctica is more than 1900 m (6232 ft) deep. Not uncommonly, there are fairly shallow submerged rock barriers that lie 100 to 200 m (328 to 656 ft) below sea level at their mouths. Geologists have long speculated upon the origin of fiords, and the theories range from one of tectonic origin to one of glacial scouring and overdeepening of former stream valleys. Surely, many fiords owe much of their origin to glacial erosion and stand as testimony to the tremendous work a glacier can perform.

Glacial deposits and depositional landforms

Much of the debris carried by glaciers comes to rest beneath or along the periphery of the ice downvalley from the snowline. The sheer volume of debris brought down by the ice, as well as the size of many fragments, is more than the meltwater streaming away from a glacier can remove—at least not as quickly as it is supplied. Consequently, the sides and lower end of the glacier nearly always have a heavy rock burden (Fig. 13-22). The mass of ice-transported debris may accumulate as a hummocky, crescent-shaped, rocky ridge, looped around the snout of the glacier. A ***terminal moraine*** (from the Provençal French word *morena,* heap of earth), or *end moraine,* forms at the end of the glacial lobe. Ridges of debris that continue up-valley along the sides of the glacier form ***lateral moraines*** (Fig. 13-23). As a rule they are higher and bulkier than terminal moraines, and their crests slope forward with about the same inclination as the glacier surface. Should two glaciers join, then the lateral moraines that

Fig. 13-15 A. (below) Distant view of Yosemite Valley, California, looking east. Glacial erosion extensively modified its original sloping walls, creating a U-shaped valley. Yosemite's numerous vertical walls challenge rock climbers from around the world. *USGS* **B.** (opposite) Within Yosemite Valley. The lower end of the stream valley on the right was oversteepened by glacial erosion, producing a hanging valley and the picturesque waterfall. *Ansel Adams*

meet at their intersection may unite and continue together down the middle of the ice stream as a dark band of rocky debris known as a ***medial moraine***. In fact, there may be as many bands as there are unions of trunk and tributary glaciers, resulting in wonderfully banded strips of white ice interspersed with darker morainal layers (Fig. 13-24).

Lateral and terminal moraines form by several processes. Most of the material in lateral moraines is derived from the valley walls above the ice. Material that avalanches or slowly moves down the slopes is caught in the trough between the ice and the valley wall and forms a ridge as it is dragged downvalley by the moving ice. There are several opinions as to the origin of terminal moraines, but we will touch on just two here. One is that glaciers, as they advance over terrain covered by loose debris, push or bulldoze the material into a moraine. The other theory has to do with the melting of the ice at the lower end of a glacier. Such ice carries a fair amount of debris in it, and the latter material is left behind as the ice melts, although some may be carried away by the meltwater streams.

Fig. 13-16 Cirques at the heads of glaciated valleys in the Wind River Mountains, Wyoming. The cirques are cut into an old erosion surface, the dissected remnants of which now form rolling plateau-like uplands. Tarns or glacial lakes commonly occur in the cirque basin, but they also occur downvalley in bedrock-rimmed basins and are a characteristic feature of glaciated terrain. *Austin S. Post*

Fig. 13-17 Iceberg Lake, a tarn, in Glacier National Park, Montana. *Austin S. Post*

Fig. 13-19 U-shaped valley formed by a glacier flowing on granite, Wind River Mountains, Wyoming. *Peter W. Birkeland*

After the ice has disappeared end moraines and lateral moraines form a semi-circular ridge across the valley. For a time, morainal embankments may serve as natural earth-fill dams, and quite successfully, even though in a sense they are pointed the wrong way. They are concave upstream, instead of being arched convexly against the reservoir, as a well-designed dam is. Nonetheless, such morainal dams effectively impound the waters of some of the world's most scenic lakes. The best known probably are those bordering the Alps, such as Como, Maggiore, and Garda on the Italian side and Neuchâtel, Geneva, Luzerne, and Constance on the north. North America has such moraine-blocked water bodies, too; Lakes Mary and MacDonald in Glacier National Park are two of the most striking examples.

Fig. 13-18 (opposite) Mount Assiniboine, in the Canadian Rockies, is a classic example of a horn formed by glacial erosion. *Austin S. Post*

Material laid down directly by ice is called ***till*** and consists of poorly sorted debris in which the size of particles ranges from clay particles to huge boulders (Fig. 13-25). Till makes up the moraines, a term for the resulting landforms.

Fig. 13-20 (opposite) When Yosemite Valley (foreground) was being deepened by glacial erosion, a small sidestream unable to keep up with erosion was left hanging when the glaciers melted; thus Yosemite Falls was created. Photographed in 1866 by Carleton E. Watkins. *Metropolitan Museum of Art*

Fig. 13-21 View east along a fiord cut into a high mountain plateau. The waters are 1000 m (3280 ft) deep, and the valley walls are 1500 m (4920 ft) high. *Geodetic Institute, Copenhagen*

Fig. 13-22 Terminal moraine forming along the northeast margin of the Barnes Ice Cap, Baffin Island, Northwest Territories, Canada. *J.D. Ives*

Few geological processes leave such a jumble of debris, with, perhaps, the exception being desert and volcanic mud flows. In fact, in places it is difficult to determine if a deposit is a till or a mud flow. The material can be identified by examining the boulders it contains for striations and the bedrock on which it rests for polishing. Differentiation of the deposits is not merely an academic exercise, however, because assessment of geological hazards on the high, glaciated volcanoes of the Northwest, for example, demands that one be able to tell one deposit from another. One might feel rather foolish in identifying as a mud flow—and thus assigning to an area a high-hazard potential—what was actually a till laid down by a sluggish glacier 20,000 years ago. Yet such errors have been made by geologists.

Fig. 13-23 Terminal and lateral moraines, Mount Jacobsen, British Columbia, Canada. The moraines mark the former extent of the glacier, perhaps during the early 1900s. A change in climate caused the ice to retreat, leaving behind the moraines (a), a moraine-dammed lake (b), and ice-scoured bedrock (c). *Austin S. Post*

Fig. 13-24 Medial moraines formed by the joining of lateral moraines, Kaskawulsh Glacier, St. Elias Range, Yukon Territory, Canada. The number of medial moraines often indicates the number of tributary glaciers that feed a main glacier. *Austin S. Post*

Rock glaciers

In some areas with a continental climate, such as the Rocky Mountains of the United States, ice glaciers are not too common—probably the result of a combination of light winter snowfall and relatively high summer snowmelt. In their place we commonly see rock glaciers, tongue-shaped masses of rocky debris and ice slowly moving downvalley (Fig. 13-26). Because of the inherent problems of digging into a rock glacier, we have little data on their interiors. Some seem to consist of essentially clean glacier ice overlain by a thin mantle of rocky debris, whereas others probably contain rocks throughout cemented together by ice. Rock glaciers move downvalley just as ice glaciers do, through basal sliding and plastic flow. As they advance, their fronts are being oversteepened, and rock avalanches are common there.

Apparently special conditions are necessary for the formation of rock glaciers. One is that the altitude must be high enough for some snow to linger into the late summer. The other is that the surrounding cliffs must supply rocks rapidly enough to bury the summer snow and protect it from melting. In fact, the surface mantle of rock protects or insulates the underlying ice so well that rock glaciers persist at lower altitudes than those at which some present-day ice glaciers could survive.

Ice and rock glaciers have different sensitivities to climatic change. A warming climate during the last century has caused many ice glaciers to retreat. In contrast, some rock glaciers are not affected. Many of the larger ones are still advancing down valleys, knocking over trees in their path. We need not send out a rock-glacier alert, however, as their movement is very slow—about 1 m (3 ft) per year.

CONTINENTAL GLACIATION

Continental ice sheets, unlike alpine glaciers, did not move under the impetus of gravity in tongue-like masses through pre-existing valleys countersunk in the flanks of mountain ranges. Instead, during the ice ages, they sprawled broadly as huge disk-shaped masses of ice over much of northern Europe and northern North America, covering more than 20 million km² (7.7 million mi²) of land surface. In North America, the ice mass must have been a wall of nearly unimaginable size, extending almost 6400 km (3977 mi) across the entire width of Canada. That frozen tide spread southward into the United States roughly to the line of the pendant loop marking the courses of the Ohio and Missouri rivers today. In order to find a model for the vanished ice sheets we need to turn to the ice sheets of Greenland and Antarctica, both of which resemble gigantic domes.

The Greenland ice cap covers some 2 million km² (0.8 million mi²) and is shaped like an elongate dome that runs parallel to the trend of the island (Fig. 13-27). It reaches an altitude of 3.3 km (2 mi), just slightly lower than the highest peaks that flank the east coast, and extends to below sea level. The landscape along both coasts is quite irregular, and in places the ice overtops all the topographic barriers in its way. Such behavior is typical of a continental glacier—it flows in the direction of the surface slope, and if a mountain range is in the way, it simply rides over it. If a mountain is too high to be overridden, the ice flows around it. In general, the ice from the central part of the island moves down through deep, narrow fiords to the sea. There the ice may break off, a process known as ***calving*** (Fig. 13-28) and drift away as

Fig. 13-25 Glacial till in the Sierra Nevada, California. Typical of many alpine areas, the till is very bouldery, and the boulders occur in a wide range of sizes.

icebergs into the sea lanes of the North Atlantic (Fig. 13-29). Such an iceberg sank the *Titanic* in 1912, with a loss of 1489 lives. The same icebergs carry entrained glacial debris, dumping it far out at sea as they melt. Such material rains down as sediment on the ocean floors.

The Greenland ice cap is similar to alpine glaciers in that there is a central area of accumulation surrounded by an area of wastage. As shown in Figure 13-27, ice flow in the central area is inward, whereas in the surrounding area the flow lines intersect the ice surface. Unlike glaciers located farther from the sea, however, a larger proportion of the loss is through calving rather than through surface melting.

Fig. 13-26 Rock glacier on Mount Sopris, Colorado. Rockfalls from the high cliffs continually feed debris to the glacier. The slow movement downvalley results in transverse furrows and ridges on its surface. Rock avalanches are common on the front of the glacier, which is steepened by constant pressure from behind. *Edwin E. Larson*

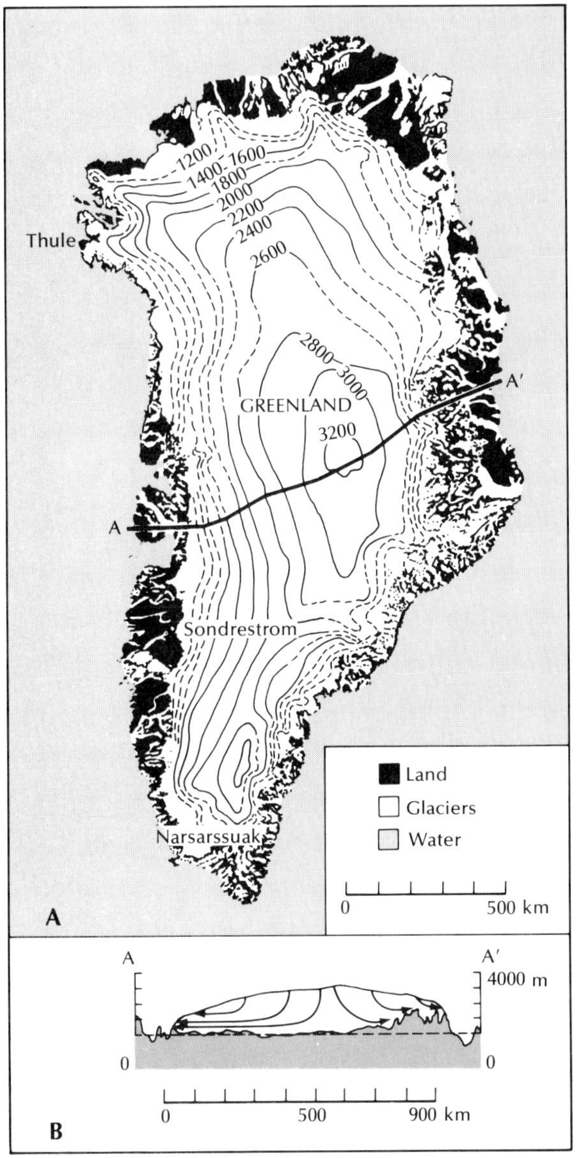

Fig. 13-27 **A.** Map of Greenland, showing the areas of land and ice, with contours depicting areas of equal altitude on the ice. **B.** Cross section of Greenland along line A-A′ of the map. Arrows depict approximate flow lines of the glacial ice.

The Antarctic Ice Sheet is more than seven times the size of the Greenland sheet, but it has much in common with it (Fig. 13-30). For example, the Antarctic sheet also extends below sea level, and in places it flows against the "topographic grain" to reach the sea. It differs from the Greenland sheet in that its outer edge is mostly grounded on the continental shelf, and, because of the extremely cold climate and the general absence of ice melting, most of the wastage is through calving. Thus, an ice advance could only occur if the sea level were to drop, exposing a larger expanse of continental shelf. Parts of the ice sheet receive so little precipitation (less than 5 cm, or 2 in., per year) that the flow of ice is extremely slow and the area is a desert, but of the cold variety. Indeed, landforms and soil and weathering features in many areas of Antarctica closely resemble those found in hot deserts.

Today, interest in the Antarctic Ice Sheet is high. Because the amount of water locked up in it is so great, small changes in its volume will strongly affect the sea level, worldwide. Most likely, the changes will take place very slowly, but some earth scientists feel that parts of the sheet could surge, just as alpine glaciers can. If that happened over a short enough time period and involved large masses of calving ice, the effect surely could be catastrophic in many coastal areas.

Enough data are now available to reconstruct the last major continental glaciers with some accuracy (Figs. 13-31 and 13-32). The generating centers for much of the North American ice appear to have been Baffin Island, the Labrador Peninsula, the country west of Hudson Bay, and the islands north of the Canadian mainland. Ice from those centers coalesced into one enormous ice sheet, which spread to the south and to the west. At times it joined the Cordilleran Ice Sheet, which was centered over the mountains of western Canada.

Both geologists and archeologists are interested in the placement and timing of the joinings, because it was a major migration route for human beings and animals from Alaska southward. The Cordilleran Ice Sheet also sent a thick tongue of ice southward into the Puget

Fig. 13-28 A. (above) Icebergs calving from Miles Glacier, Alaska. *Bradford Washburn* **B.** (opposite) Glacier flowing into the sea at Glacier Bay, Alaska. *George Orser*

Lowland of Washington, covering areas where the cities of Bellingham and Seattle now stand. In Europe, the ice sheets spread outward from the backbone of the Scandinavian Peninsula, and to a lesser degree from Britain and Ireland, traveling south beyond the present-day cities of Warsaw and Kiev, almost to the gates of Volgograd.

Hills and mountains that were overriden by those great masses of ice are more likely to be rounded off than to be surmounted by the spires, castellated divides, and horns so typical of alpine glaciation. Mountains of surprising height have been buried beneath the ice of continental glaciers. Among those in the northeastern United States are the Catskills (1280 m; 4198 ft), the Adirondacks (1615 m; 5297 ft), and the Presidential Range of New England, the highest point of which is Mount Washington, with an altitude of 1917 m (6288 ft).

Landforms of continental glaciation

Landforms associated with continental glaciation are complex, and, although many are similar to those produced by alpine glaciation, they occur on a much larger scale. Cirques, of course, are not present because the ice does not flow from a jagged mountain top, but instead flows from the central portion of an enormous ice dome.

The central part of the ice sheets, such as in northeastern Canada, scraped and scoured the landscape. We find, therefore, vast stretches of smoothed bedrock, stripped of its soil, in which all minute bedrock features, such as joint patterns, have been etched into relief (Fig. 13-33). Weaker portions of the bedrock were scooped out, forming closed depressions that now hold lakes, such as the uncountable ones that dot Labrador. Telltale striations in the bedrock give us the only evidence of the direction in which the ice flowed during the latter stages of the ice sheet.

In contrast to the scrubbed interior areas covered by the ice sheets, large areas of till were deposited around its margin, leaving an undulating terrain known as ***ground moraine*** (Fig. 13-34). Indeed, many of our major northern cities are built on till. Where it is sandy and bouldery, till makes a good base for buildings and freeways, but where it is rich in clay, it makes for very slippery foundations. Tills and associated deposits often change properties over short distances laterally; thus geologists find it difficult to predict the kind of surface materials that might be encountered during a

Fig. 13-29 Between April 1978 (above) and May 1980 (below) Columbia Glacier, Alaska, retreated about 0.8 km, exposing Heather Island (lower right) for the first time in recorded history. As the glacier continues to retreat, some 208 km^3 (50 mi^3) of icebergs will be discharged into Prince William Sound over the next 30 to 50 years. This is causing some alarm because oil tankers cross the sound in the journey from Valdez, the southern terminal of the Trans-Alaska Pipeline, to the lower 48 states. *USGS, Project Office–Glaciology*

Fig. 13-30 Portion of the Antarctic Ice Sheet near McMurdo, looking northward. In the foreground are dry valleys, with the tongue-shaped Victoria Upper Glacier. The mountains in the middle ground are part of the Transantarctic Mountains; beyond is the Mackey Glacier, which drains part of the Antarctic Ice Sheet. *USGS*

construction project. Extensive field mapping and drill-hole data are essential to builders working in glaciated areas.

In places, blocks of ice may have been entrapped in the till that, upon melting, produced a depression called a ***kettle*** (Fig. 13-35). It is no wonder then, that glaciated areas abound with lakes.

One major landscape feature of North America—the Great Lakes—resulted from the advances and retreats of the last major glaciation, called the Wisconsin Glaciation. In part the lake basins are ice-gouged, and their floors lie far below sea level; the bottom of Lake Superior is 213 m (699 ft) below sea level and that of Lake Michigan 104 m (341 ft). They are partially

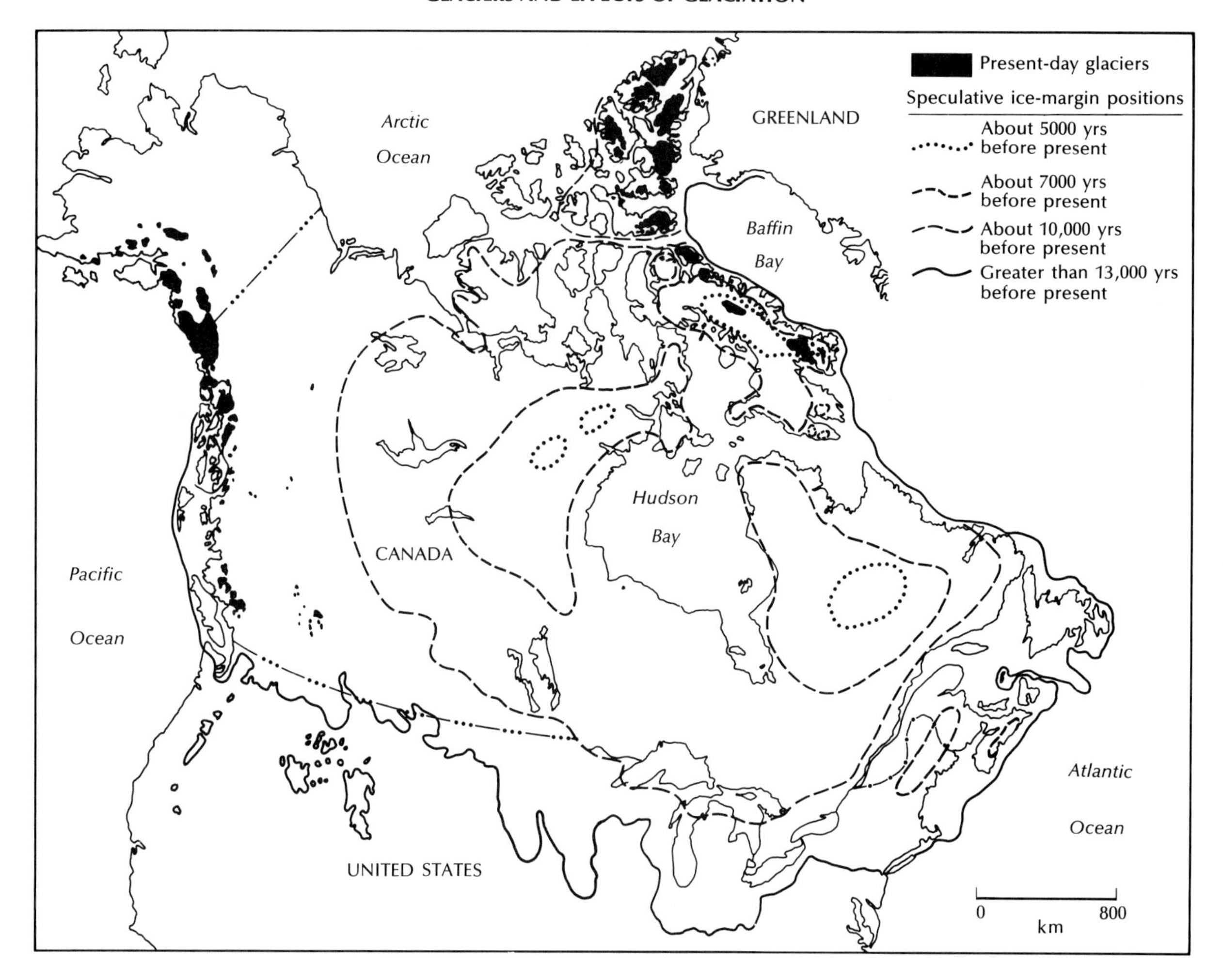

Fig. 13-31 Area covered by major continental glaciers in North America. The lines denote various positions of the ice front at different times (in years before present).

blocked by moraines, especially around the south end of Lake Michigan, with a pendulous, lobate form that is a close counterpart to the moraine-outlined glacial lobe the lake usurped. The Great Lakes had a different pattern during the last part of the Ice Age than they have today; for one thing they were dammed to the north by the retreating wall of the receding ice sheet; for another, their levels were higher then and their outlets were quite different. One outlet was via the Mohawk Valley and the Hudson; another was down the course of the Illinois River to the Mississippi and thence to the Gulf of Mexico, rather than to the Atlantic by way of the Gulf of St. Lawrence, the present outlet.

Some glacial forms on ground moraine are not random heaps of till, but show a regular geometry. Among the shaped features are swarms of curious elliptical, rounded low hills, resembling a whale or the bowl of a teaspoon turned upside down. Such hills are called

drumlins, from an Irish Gaelic word *druim* which means the ridge of a hill (Fig. 13-36). Of those curious features, certainly the most renowned is Bunker Hill, although there are many others in the Boston region, including some of the islands in the harbor. Drumlins vary widely in size and shape, but few are more than 1 km (0.6 mi) long or 30 m (98 ft) high. In general, they appear to contain a high percentage of clay. Although details of their origin are uncertain (no one ever saw a drumlin being made), there is little doubt that they originated beneath moving ice. We can say that with some certainty because they occur in groups, parallel to one another and to the known directions of ice transport. Proposals for their origin range from erosion of pre-existing till or bedrock to a purely depositional landform feature related to the mechanics of till deposition.

Eskers, from the Gaelic, are elongate, narrow, sinuous ridges of stratified sediment that commonly wander across the countryside, much like a canal levee. Generally their crests are rounded, their side slopes are moderately steep, and their longitudinal slope is gentle (Fig. 13-37). Like conventional streams, they may meander; occasionally they are joined by tributaries, but unlike ordinary streams they may climb up hill slopes, especially where they cross low ridges through passes. Seldom do their crests stand much more than 30 m (98 ft) above their surroundings. Some may be as long as 500 km (311 mi), although most are a great deal shorter. The consensus today is that eskers probably are deposits made by streams flowing in ice tunnels at the bottom of a glacier.

Along the terminal portion of the former continental ice sheets are crescentically looped morainal ridges that are characteristic of landscapes in the North American mid-continent (Fig. 13-38). The height of such ridges rarely exceeds 30 m (98 ft), and in some places wide gaps may appear in the ridge, either because the ice at that point was not loaded with debris or perhaps because the moraine has been

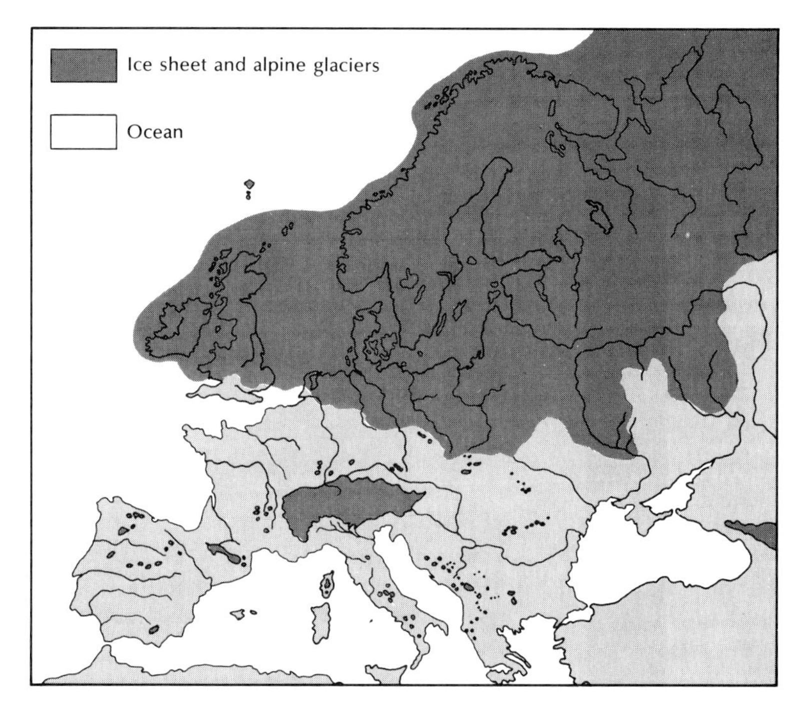

Fig. 13-32 Limits of glaciation in Europe. A continental ice sheet spread south from Scandinavia and the British Isles into northern Europe and Russia. South of the ice sheet, alpine glaciers covered many parts of the Alps, but the two main ice masses did not join. Other local glaciers also are shown. The coastline shown is the present one. In glacial times, however, sea level was lower, and the coastline was seaward of its present position.

Fig. 13-33 Glacial scouring has been such in this granite terrain that joints and other structural features show up very clearly. Melville Peninsula, Northwest Territories, Canada. *Department of Energy, Mines, and Resources, Canada*

Fig. 13-34 Hummocky ground moraine near Cypress Hills, Saskatchewan, Canada. *Geological Survey of Canada*

eroded away. Such massive moraines formed while the ice front was stationary, following either advances or retreats of the front.

Most of the rocks in till are locally derived; that is, they have been carried no great distance from their source. Thus, if the surrounding countryside is chiefly limestone, the till will be mostly limestone fragments, large and small. Interspersed with them, however, may be a number of far-traveled rocks. If the latter have a distinctive composition, they may be traceable all the way back to their source, in which case they are called ***indicators***. Many are known to have traveled 500 km (311 mi); none apparently can be traced more than 1200 km (746 mi). Fragments of native copper from the

Fig. 13-35 Closed depression in till in the Matapedia Valley, Québec, that probably formed when a block of glacial ice melted. *Geological Survey of Canada*

southern shore of Lake Superior are found as far away from their origin as southern Iowa and southern Illinois. Other more challenging examples of glacial transport are diamonds, some as large as small pebbles, found as far south as southern Ohio and Indiana and having a presumed source north of the Great Lakes. Technically the diamonds are not indicators because their actual source is not known, or if it is, it is a remarkably well-kept and presumably profitable secret.

Deposits closely associated with glaciation

Two characteristic deposits are closely associated with proximity to glaciers, be they alpine

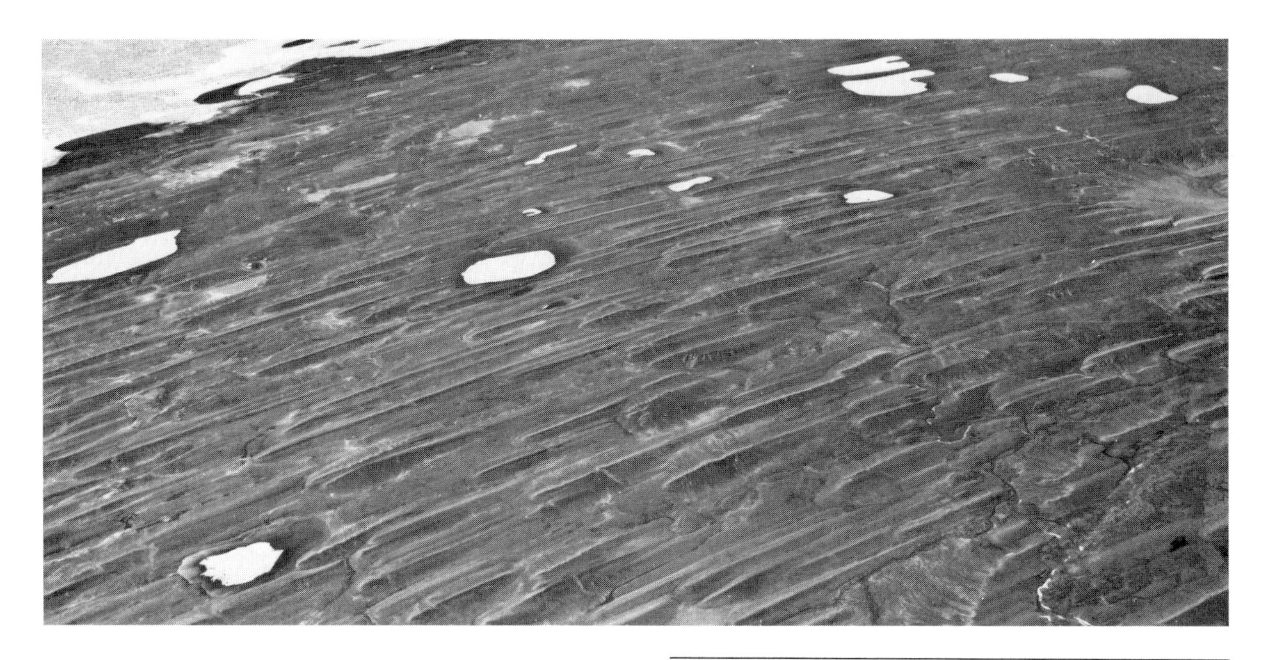

Fig. 13-36 Well-developed drumlin field, Stefansson Island, Northwest Territories, Canada. The ice flowed to the right. *National Air Photo Library, Canada*

Fig. 13-37 This esker, near Fort Ripley, Minnesota, consists of gravel and sand probably deposited in a winding tunnel at the base of an ice sheet. When the ice melted, a winding ridge was left behind. *W.S. Cooper*

or continental. Streams draining off the front of the ice are heavily loaded with debris and build up their beds to form vast floodplains, or ***outwash plains*** (Fig. 13-39). Those featureless plains have been graded and regraded endlessly by ever-shifting streams. Thus, the material is fairly well sorted and, when formed, is unvegetated. Because the material is so fresh, so uniformly textured, and in the mid-continent so fine grained, it makes fertile soil. In fact, much of the best farmland there is located on glacial outwash plains. Outwash associated with alpine glaciers, however, commonly is very bouldery, and the soils that form from it are not fertile.

Because the glacial mill grinds so exceeding fine, the surface of an outwash plain can be covered with ***rock flour,*** a fine silt that winds can pick up and transport. With strong winds sweeping across an open, unprotected, silt- and sand-covered plain, great clouds of dust are picked up and swept for scores of miles beyond the barren floodplain (Fig. 13-40). Deposited, the fine, wind-transported glacial debris, called loess (see Chap. 12), may blanket much of the neighboring countryside, sometimes to depths of 30 m (98 ft). Loess was broadcast over the length and breadth of the Mississippi Valley as well as across the lowlands of Central Europe and far down into the Danubian plain of Hungary. It is always thickest near its floodplain source, and systematically thins downwind. The deposits are responsible for some exceptionally fertile soils the world over.

Fig. 13-38 Major moraine systems in the mid-continent east of the Rocky Mountains. All were formed during the last major glaciation, the Wisconsin, older moraines having been muted by subsequent erosion or buried by younger deposits.

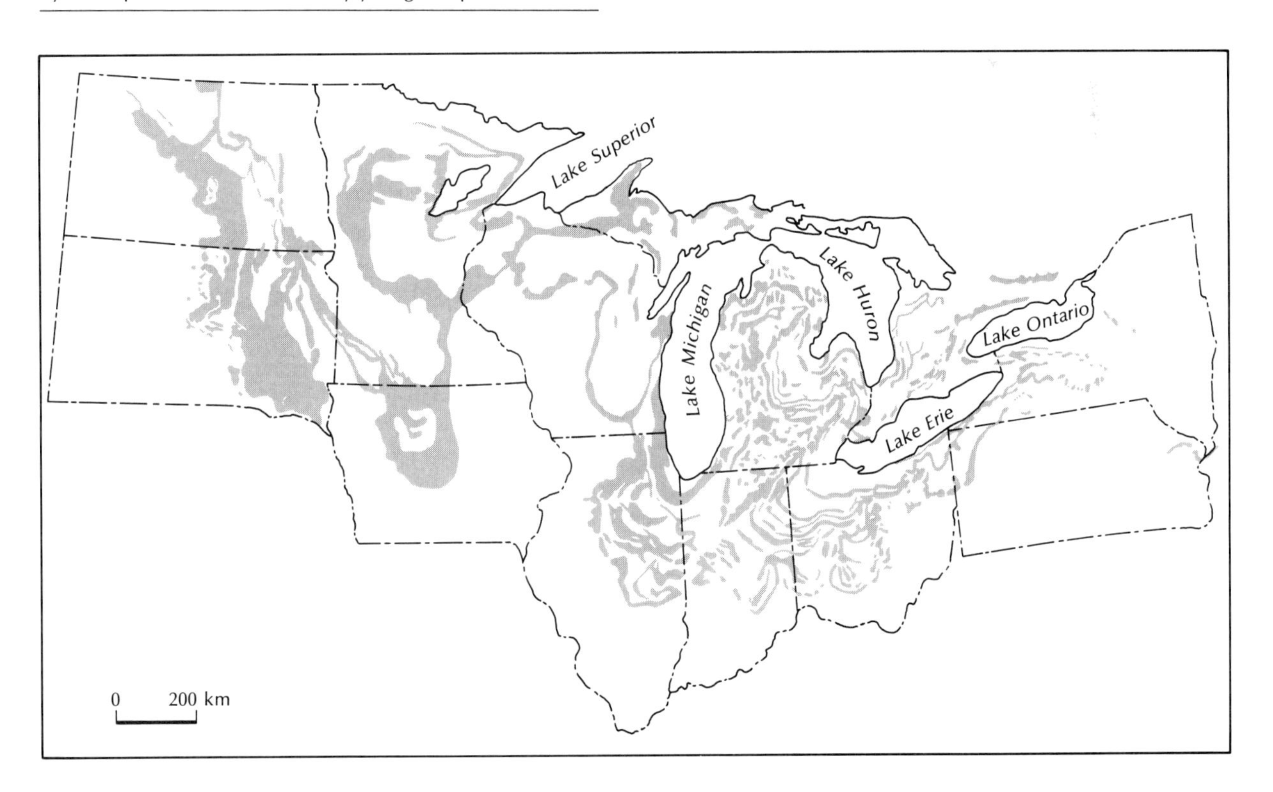

Fig. 13-39 Glacial outwash in Ekalugad Fiord, Baffin Island, Canada. The light-colored surface is recent glacial outwash, colonies of lichens have darkened the outwash surface (center) deposited about 4000 to 5000 years ago. Note the patterned ground of the latter surface. *J.D. Ives*

Fig. 13-40 Clouds of silt rise from the floodplains of the Knik River and tributaries, Alaska. *W.C. Bradley*

MULTIPLE NATURE OF GLACIATION

When the glacial hypothesis was advanced more than a century ago, people spoke of a Great Ice Age. The idea was generally held that in some mysterious way the ice advanced across the northern lands, lingered a while, and then withdrew. Today we know that the Ice Age was a vastly complex event, involving multiple advances and withdrawals of continental ice sheets and alpine glaciers.

A more formal term for the geological time during which most of the recent large-scale glaciers advanced and retreated is the ***Pleistocene Epoch.*** It began about 2 million years ago and ended about 10,000 years ago. The Pleistocene is not synonomous with the time of glaciations, however, because there is evidence that the last period of glacial advances and retreats began before 2 million years ago in such places as Alaska and Antarctica.

A map of the United States shows the limits of tills of different ages (see Fig. 13-41). But how do we really know, though, that there was not more than one advance, and how do we determine how many actually took place?

Most geologists interpret the evidence available in the mid-continent to indicate that there were at least four major advances of the ice during the Pleistocene, separated by three interglacial phases when the ice withdrew, perhaps completely. Evidence for that complex succession is based in part on (1) the way in which moraines and other deposits of a later

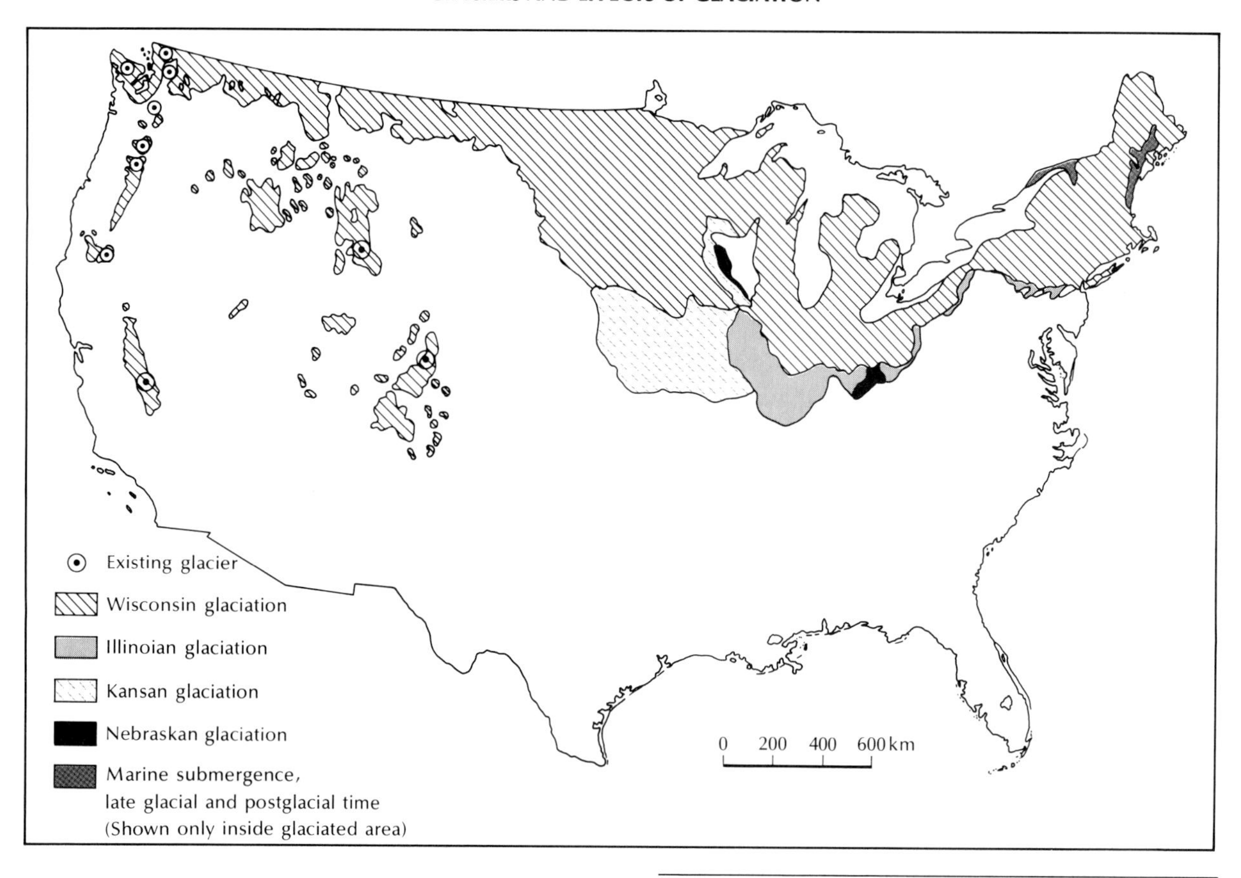

Fig. 13-41 Glacial geology of the United States.

glacial stage may overlap those of an earlier advance and (2) the degree of weathering of the glacial deposits. The original constructional pattern of older moraines and other glacial landforms will be blurred or perhaps even obliterated by subsequent erosion. Older deposits have developed a strong soil profile and younger ones a weak profile. Weathering may have progressed to depths of 3 m (10 ft) or more in older glacial deposits, and some boulders, even though they appear solid and intact, are so deeply decayed that they can be sliced through with a bulldozer blade. Some weathered glacial material is known by the eminently descriptive word ***gumbotil.*** Gumbotil was originally defined as a gray, thoroughly leached clayey soil that characteristically is sticky when wet, but hard and firm when dry. When stepped in wet in the field, it literally grabs your boots off. In places, such soils formed in interglacial periods and were subsequently buried by the till of the next glacial advance; their presence constitutes fairly firm evidence for extensive retreats of the ice for long periods of time (Fig. 13-42).

The glaciations and interglaciations of the central part of the United States generally were named after the localities or areas in which deposits or soils were well exposed for study. Names for the glaciations were taken from states, but those for the interglaciations came from Sangamon County, Illinois; Yarmouth,

Iowa; and Afton Junction, Iowa. The major glaciations and interglaciations are (youngest at top)

Glaciation	*Interglaciation*
Wisconsin	
	Sangamon
Illinoian	
	Yarmouth
Kansan	
	Aftonian
Nebraskan	

Evidence from deep-sea sediments is causing geologists to doubt that only four major glaciations have taken place; there may have been many more (see Fig. 15-20).

It is difficult to determine exactly when the various mid-continent glaciations occurred. The Wisconsin commonly is considered to have started about 75,000 years ago and lasted until about 10,000 years ago. The only other "dated" glaciation is the Kansan, which occurred at about the same time as a 600,000-year-old volcanic ashfall deposit that resulted from an enormous volcanic explosion in what is now Yellowstone National Park.

Alpine glaciation seems to have occurred more or less in concert with the continental glaciations and interglaciations, as shown by weathering and topographic features, as well as by absolute dates on the various tills. Major canyons in the high mountain ranges were glaciated, and bulky moraines commonly are found at their mouths (Fig. 13-43). In a very general way, the preservation of alpine moraines is related to the age of the till, just as it is in the mid-continent. Moraines dating from the Wisconsin glaciation are relatively well preserved, Illinoian moraines, less well, and older moraines have been worn down completely through erosion. In addition to the moraines that mark the maximum extents of the ice, smaller moraines are found far upvalley, in an area adjacent to the cirques. They record slight climatic changes over the last 5000 years, but enough to have caused cirque glaciers to advance and retreat several times.

The beginning of the last period of glaciation seems to have varied from place to place. The Antarctic Ice Sheet reached a substantial size about 10 million years ago, or even longer ago. Extensive glaciation also took place 10 million years ago in Alaska. In contrast, the oldest glaciation recognized so far in the western United States may be about 3 million years old, or even younger. In like manner, the end of the last glaciation varies from place to place. Glaciers have all but disappeared from many of our western mountains outside of the Pacific Northwest, so that we are now in an interglacial period. Such examples point out an important lesson for glacial geologists—climatic change does not take place in all areas at the

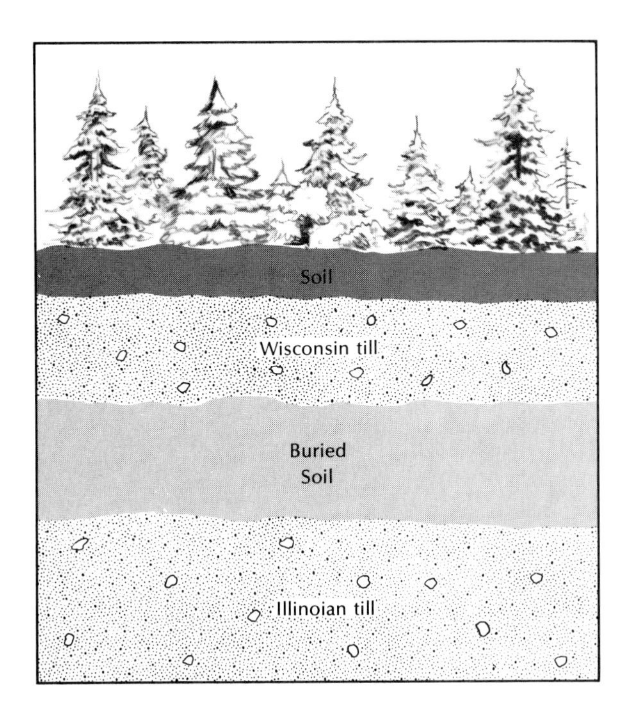

Fig. 13-42 The kind of field evidence necessary to prove multiple glaciation. At the surface is the younger (Wisconsin) till. Since the retreat of ice some 15,000 years ago, a weak soil has formed on its surface. Beneath it is a soil that formed on older (Illinoian) till. The buried soil is deeper, richer in clay, and more weathered, representing a surface exposed to weathering and soil formation between the Illinoian and Wisconsin glaciations for more than 15,000 years.

Fig. 13-43 Moraines at the mouth of Bloody Canyon, in the Sierra Nevada, California. The older moraines formed when the ice took a path to the right, as viewed downvalley, after leaving the mountain front. In contrast, the ice that laid down the younger moraines moved directly out of the canyon and deposited a lateral moraine across the upper end of the valley enclosed by the older moraines. *Mary Gillam*

same time. Thus, we would not expect climatic change to be synchronous worldwide. In geological terminology, the change is ***time-transgressive,*** occurring in one place first, and in another place later—depending on local circumstances.

Finally, the climatic changes that produced the glaciers were sufficient to greatly change the environment in areas where glaciers did not form. One such place is the western American desert, which incorporates parts of California, Nevada, and Arizona. We know, for example, that during the Wisconsin glaciation trees grew at altitudes several hundred meters below their present lower limit. The evidence comes from careful examination of tree fragments in abandoned rat nests found where it is too hot and dry today for trees to exist. The consensus is that rats do not travel far overland on a hot day; thus, in those prehistoric times, the trees must have gradually migrated down to the altitudes at which the nests were found. Other evidence for climatic change is the abundant evidence that lakes of all sizes dotted the western basins at that time (see Chap. 12).

POSTGLACIAL CLIMATIC CHANGES

Enough is now known to tell us that the climate since the end of the Pleistocene has not always been the same. Most of the information comes from non-climatic sources because the length of time during which climatic observations have been made is far too short. Reliable instrumental records date only from the mid-nineteenth century; in fact, the earliest rainfall record in the modern sense was made in Padua, starting in 1725, and the earliest temperature record in Florence, in 1655. As a result, some of the scattered information seems authentic, some of it interpretive, and some of it frankly speculative. Yet we are working in the realm of human history, and historic records have survived many climatic hazards, such as drought, floods, crop failure, blocking of alpine passes by long-enduring accumulations of ice and snow, and successions of unusually cold winters. Accounts of those events often were written down, others have to be inferred from such occurrences as the discovery of village sites far up a slope laid bare through the evaporation of lake waters. Conversely, the original site of lake dwellings may now be concealed by a rising lake level. A good example of a historical record that reflects changing climatic patterns is the annual flooding of the Nile, for water levels have been recorded since A.D. 641.

By means of such techniques as carbon-14 dating, tree-ring dating, measurement of the sizes of lichens on rocks, and historical records, it is possible to determine the approximate age of mountain moraines and thus a glacier's pattern of recent oscillation (Fig. 13-44). Interpreting the evidence, earth scientists have determined that three major advances have occurred over the past 5000 years—the most recent advance is referred to as "The Little Ice Age" (Fig. 13-45). In many places, the advance that culminated in the last several centuries was the most extensive.

For the first 50 years of the twentieth century the rate of recession for most, although not all, Northern Hemisphere glaciers was rapid. During that half-century, a significant warming of the atmosphere, accompanied by shifts of marine currents and fish populations, migrations of land animals, and changes in annual temperatures and precipitation was seen. Since about 1950, however, slightly cooler and moister climates have slowed the retreat of some glaciers, and hastened the advance of others. Understanding the long-range effects of those subtle changes is important because they come at a time when the world's population is expanding rapidly, and the need for increasing food production is greater than ever.

Piecing together fragmentary evidence is a largely unfinished task, although a vast store of information has been accumulated, chiefly from Europe, where a long and often turbulent record can be found. We know little, however, of the postglacial climatic changes in lands that are more remote or in areas with radically different climates—such as the monsoonal tropics.

The doomed Norse colony in Greenland is an example of the impact of climatic change. The colony was founded in A.D. 984 and perished around 1410. In its early history, the Arctic seas were iceless and perhaps calmer, and Viking ships could make passage where today ice floes and stormy seas bar the way. The colonists raised cattle and hay, built permanent habitations, and the settlement flourished to such an extent that it had its own bishop. With a climatic change that brought the Greenland ice southward again, with the pressure of the Eskimos at their gates, with a succession of crop failures, with the rise of permafrost in the ground—so that even such shallow excavations as graves were no longer possible—and with the perils of the ocean crossing too great for the frail vessels of that day, the colony and all its inhabitants perished.

Other examples, almost without number,

might be cited of the impact of changing climates on the lives of human beings, and thus on the course of history. A powerful description of the effect of a prehistoric climatic change of the magnitude of some of those that occurred in historic time is the following passage from Palle Lauring's *The Land of Tollund Man, The Pre-history and Archaeology of Denmark:*

The change which set in altered not only living conditions but the country itself. It became windy, rainy, and foggy, and there was a fall in temperature. The change is evident in the form of a clear stratum in bogs, which indicates that the rain turned to torrents. Grey mists swept like a veil across the land, the cattle congregating miserably round the houses. The winters set in with drifting snow, frost, and yet more snow. Wondering, men advanced through cold and death-like forests, where the snow stifled every sound and where oak tree branches were weighed down by it. Cattle froze to death; wolves failed to find food; belts and sounds were overlaid with ice, rendering navigation impossible for months on end; corn would often be destroyed by frost and water, and harvest would mean waiting for the air to dry while the ears blackened and rotted. Gone were the days when young women sang as they went about clad only in corded skirts, golden-brown from the sun.

Where the present trend is leading, no person can say. Will the earth's atmosphere generally continue to warm and present ice melt, or will the air chill once more and massive glaciers start their march again?

OTHER ICE AGES

A single catastrophic deep-freeze of the earth would not fit very well with our ideas of uniformitarianism, and, when we look back through the geological record, we do indeed find evidence for at least two periods of extensive glaciation preceding the last one. One occurred about 260 million years ago, or in the Permian Period. The evidence for it consists largely of glacial deposits known as ***tillites,*** which are unsorted mixtures of sand, gravel, boulders, and clay that have been consolidated. Some of the boulders show striations, and many of the tillites rest on striated rock surfaces. Other quite extensive tillites are marine in origin. In such deposits, boulders and cobbles from melting sea ice appear to have dropped into much finer-grained bottom sediments, deforming their horizontal layers. The presence of a number of distinct tillite layers indicates that the Permian glaciation, like the Pleistocene, had stages and substages.

There are several puzzling things about the Permian glaciation. One is that it seems to have been limited to the Southern Hemisphere. Another is that the directions of the striations indicate that a number of Southern Hemisphere continents were attached at that time and later drifted apart. The latter idea has been a difficult one for geologists to accept, and it will be looked into further in Chapter 20.

The oldest ice age we know of occurred about 600 million years ago at the end of the Precambrian Era, but before the extensive fossil record of the Cambrian Period. Evidence for it, too, consists of tillites, many of which were marine, indicating a widespread distribution of sea ice. In contrast to the Permian ice age, this age was not limited to one hemisphere.

Other glaciations may have occurred, but they are not recorded in preserved rocks. Of all the sediments that accumulate on land, glacial remnants by their very nature might be considered least likely to survive. Certainly deposits left by alpine glaciers on mountain slopes are most vulnerable to erosion. Continental glacial deposits may not be too permanent, either.

CAUSES OF GLACIATION

Naturally, a phenomenon as challenging as an ice age has brought forth many attempts at explanation, and each explanation has its adherents and its detractors. At least 50 hypotheses must have been advanced over the years, but most of them contain an inconsistency that in the end disproves them.

Rather than worry about pre-Pleistocene glaciations, we will concentrate on the origin of

Fig. 13-44 Young moraines front the Isfallsglaciären in Swedish Lapland. The numbers refer to the ages of the moraines, based on historical records for the 1910 moraine and lichen sizes and carbon-14 dating for the older ones. *G.H. Denton*

Pleistocene glaciation because more data are available on it. Among the things to be kept in mind about that glaciation, before we consider its explanation, are the following:

1. The Pleistocene glaciation was a multiple rather than a single event, and at least four major advances of ice commonly are recognized in North America and Europe.
2. The glaciation appears to have been synchronous on both sides of the Atlantic, and apparently Northern Hemisphere glaciers advanced or receded at the same time as Southern Hemisphere glaciers did. In other words, the entire earth seems to have responded to the same climatic pulses.
3. The advances were not of equal size, nor were the interglacial times of equal length.
4. Glacial buildup seems to have been a slow process, but the retreat was rapid, verging on rates that might be termed catastrophic. For example, it has been estimated that the

North American continental ice sheet of Wisconsin age took some 20,000 years or more to attain maximum size, but it disappeared in only 7000 years.

5. Times of glacial advance appear also to have been times of lowered temperature, as demonstrated by the nature and the fossil content of cores of sediments from the bottom of the sea and by evidence of a lowering of the regional snowline on the high mountains of the world. Furthermore, in areas such as Arctic Canada, much more precipitation is required to initiate an ice sheet than occurs at present. With those observations in mind, let us turn our attention to some of the major theories of glaciation.

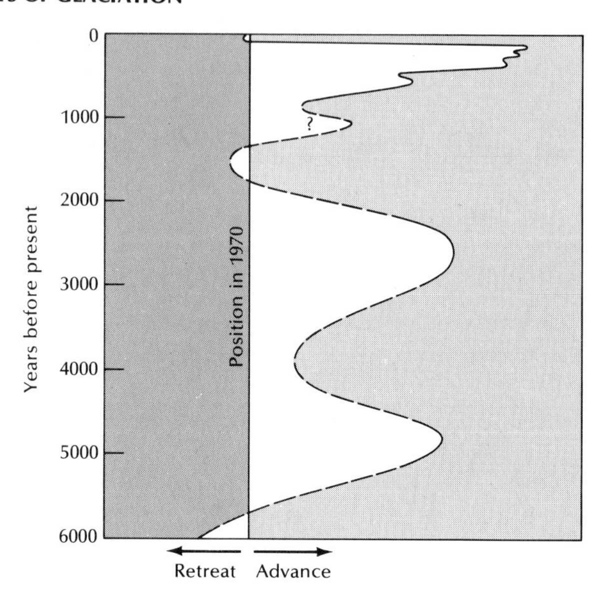

Fig. 13-45 Worldwide average position of mountain glaciers over the past 6000 years, relative to their position in 1970.

Terrestrial causes

The presence of widespread land areas in polar positions and the subsequent generation of glaciers and ice shelves are said to be major prerequisites for cooling the earth and setting the stage for a glaciation. In support of that theory, Antarctica appears to have moved into a polar position at about the time of the first evidence for its last major glaciation. High mountains also seem to have been necessary, because the land surface must intersect the regional snowline. Most of the world's major mountain ranges were elevated to their present heights during or just prior to the Pleistocene, including such lofty summits as those of the Himalayas, Andes, Caucasus, and Alps. Not only was the Pleistocene a time of unusual crustal activity, but the north-south pattern of significant mountain ranges in North and South America became firmly established, and since those ranges lie athwart the general planetary circulation of the atmosphere, they may have had a significant effect on the growth and dispersal of glaciers. Neither theory explains multiple glaciation, however: only a heretic would call for mountain ranges to rise and fall like yo-yos to produce glaciations and interglaciations.

Before we go on we should mention that although one theory might explain the initiation of an ice sheet, another might beter explain its continued growth. For example, in whatever way the North American ice sheet began, it could subsequently have been able to propagate itself. The continental ice advanced on fronts thousands of kilometers long, and ice moved for hundreds of kilometers across low-lying terrain. Once the glacier reached such massive proportions, much of the snow needed to nourish it could have accumulated near its margin rather than at its center. The great temperature contrast between a glacier and its surroundings makes it operate meteorologically much like a permanent polar front, while the ice itself acts much like a rising mountain mass, to alter the climate. Thus, with more rapid accumulation of snow close to the margins of the ice sheet, ***ice domes*** (parts of a glacier at which surfaces stand higher than adjacent areas) could accumulate, and from them the ice could spread out, rather than from a remote northerly center in far-off Canada.

Atmospheric causes

Other hypotheses cite changes in the composition of the atmosphere to explain the events of the last glaciation. One hypothesis says that climatic variations extensive enough to trigger glacial epochs may be caused by variations in the amount of carbon dioxide (CO_2) in the atmosphere. A marked increase of CO_2 would produce a so-called greenhouse ***effect,*** with a general rise in temperature. Energy from the sun reaches the earth's surface because our atmosphere is transparent to much of it. Solar energy radiates from the earth back into space as heat (the infrared portion of the spectrum). But the CO_2 as well as the water vapor and ozone contained by the atmosphere is partially opaque to infrared radiation; thus solar heat is kept close to the earth. The more CO_2, the warmer the atmosphere. In fact, the warming trend that started about 1900 is considered by some to have resulted from the enormous quantities of CO_2 added to the atmosphere through the burning of coal and oil once the Industrial Revolution hit its stride. Although the theory explains how the earth could heat and cool at various times, no one has really explained how the amounts of CO_2 in the atmosphere could vary enough to produce climatic changes of glacial-interglacial proportions.

Another theory involves the dust in the atmosphere. Dispersed particles can block part of the sun's rays from the earth's surface, thus cooling it down. Objections to such a theory are that major volcanic eruptions, such as that of Krakatoa in 1882, did not affect the weather enough to bring about a major change in the budget of glaciers or to initiate glaciers where none had previously existed. Some geologists, indeed, are turning the argument around, suggesting that increased volcanism could result from crustal stresses induced by the loading of ice on land. We see how geological ideas can change, for what was a popular cause at one time may well become, as more data are obtained, nothing more than an effect at another time. The "dust" argument, however, is far from settled.

Astronomical causes

Astronomical explanations for the origin of glaciation make much of the fact that the eccentricity of the earth's orbit changes slowly with respect to the sun. The inclination of the earth's axis also changes slowly—part of the year the earth is tipped toward the sun, part of the year it is tipped away from it. The Northern Hemisphere is inclined away from the sun in the winter when the earth and sun are a little closer to each other than they are in the summer. The result is that winter temperatures in the Northern Hemisphere are slightly higher than they are in the Southern. A third variable is that the earth's inclination to the plane of its orbit may change slightly over millennia. Because the various eccentricities in the relation of the earth to the sun are not systematically linked, there will be variations over the years in the distribution of solar energy at any given place on the earth's surface, but not in the total amount received from the sun.

Arguments can be marshaled both in favor of and against the effect of astronomical changes on ice ages. In favor of it is the evidence in deep-sea sediments and in marine-terrace sequences that temperature variations and glacially controlled ocean levels both seem to coincide with the predicted temperature variations. The correlation appears to apply, at least, to the last glaciation and part of the preceding interglaciation. However, Northern and Southern Hemisphere glaciation should have been out of phase, when in fact they seem to have been in phase.

Another astronomical explanation involves variation in the amount of incoming solar radiation. We know that variation does occur, but does it happen with such a frequency to trigger major glaciations? No one has answers to the question.

By no means have we exhausted the explanations put forth, but we have presented the most current ones. Much additional work needs to be done to establish the validity of any glacial hypothesis. Nonetheless, theories are important in provoking debate, in forcing scientists to marshal arguments pro and con, and in encouraging them to seek new evidence. Such an effort might even prevent us from initiating an irreversible climatic trend and starting a new ice age.

SUMMARY

1. Glaciers form when winter snowfall does not melt entirely, but instead builds up year by year until the mass, partly converted to ice, flows under its own weight.
2. The mechanisms of glacier movement are (1) slippage along a bedrock base and (2) the internal deformation of the ice crystals.
3. The *glacier budget* is determined by the annual accumulation and the annual loss, or *ablation,* of ice. If the two are balanced, the glacier, although flowing, will maintain a stationary front. If accumulation exceeds ablation, the glacier advances. If ablation exceeds accumulation, the glacier retreats.
4. Glacial erosion is responsible for much of the world's scenic mountain terrain. *Cirques* form at the valley heads; the valley lower down is U-shaped. Also, glaciers gouge out bedrock depressions that can fill with water to become lakes—the legacy of glacial erosion either in mountainous terrain or in flat shield areas overrun by continental glaciers.
5. *Till* is the common sediment deposited by glaciers. It is found heaped into *moraines* around the former periphery of the glacier or as *ground moraine* where it is deposited at the base of a glacier. Deposits closely associated with glaciation are glacial *outwash,* along rivers that drained glaciers, and *loess,* a fine silt blown from the glacial-age floodplains.
6. Multiple glaciation took place during the Pleistocene Epoch, with no less than four major glaciations recognized on land. The main mountain glaciations seem to coincide with the major continental glaciations, and the times of glaciations in both the Northern and Southern hemispheres may have been the same.
7. The causes of glaciation and its multiple character are not known. Many hypotheses have been advanced, but all have major drawbacks in explaining the Pleistocene glaciation.

QUESTIONS

1. Compare the formation and movement of a glacier in a dry polar area with one in a more temperate region.
2. What major erosional landforms are associated with alpine glaciation and with continental glaciation?
3. How are lateral, medial, and terminal moraines formed?
4. What is the evidence for multiple glaciation, and how can we decipher environmental conditions between times of glaciation?
5. Compare the geological causes of glaciation with the astronomical causes.

SELECTED REFERENCES

Andrews, J. T., 1975, Glacial systems, Duxbury Press, North Scituate, Massachusetts.

Denton, G. H., and Hughes, T. J., eds., 1980. The last great ice sheets, Wiley-Interscience, New York.

Denton, G. H., and Porter, S. C., 1970, Neoglaciation, Scientific American, vol. 222, no. 6, pp. 100–10.

Flint, R. F., 1971, Glacial and Quaternary geology, John Wiley and Sons, New York.

LaChapelle, E. R., 1969, Field guide to snow crystals, University of Washington Press, Seattle.

SELECTED REFERENCES

Paterson, W. S. B., 1969, The physics of glaciers, Pergamon Press, New York.

Post, A. S., and LaChapelle, E. R., 1971, Glacier ice, University of Washington Press, Seattle.

Roberts, W. O., and Lansford, H. 1979, The climate mandate, W. H. Freeman and Company, San Francisco.

Sharp, R. P., 1960, Glaciers, University of Oregon Press, Eugene.

Sugden, D. E., and John, B. S., 1976, Glacier and landscape, a geomorphological approach, Edward Arnold, London.

Wright, H. E., Jr., and Frey, D. G., eds., 1965, The Quaternary of the United States, Princeton University Press, Princeton, New Jersey.

14

THE SHORE

One of the most visible and dramatic interfaces on earth occurs where the land and the ocean meet. It is an area, unique both physically and biologically, that has fascinated people for thousands of years (Fig. 14-1).

WAVES

Waves can be almost hypnotic in their effect. Although one wave may look like another, no two are the same. Their rhythm depends not only upon the local wind for the shorter, steeper waves, but also upon distant fierce winds that start the long, even-spaced ridges of a ground swell moving outward from a storm center half a world away.

Before we go on, a few terms relating to waves should be introduced: ***Wave length*** is the horizontal distance separating two equivalent wave phases, such as two crests or two troughs. The ***velocity*** is the distance traveled by a wave in a unit of time, which can be related to its other physical properties by a number of simple concepts. The ***period*** of a wave is the length of time required for two crests or two troughs to pass a fixed point. The ***frequency*** is the number of periods that occur within an interval of time—say a minute.

Fig. 14-1 The sea attacks the rocky coast at Schoodic Point, Maine. *John Steenstra*

As we watch the endless procession of waves, it is difficult to believe that it is the form of the wave that moves forward through the water and not the water itself. That statement may not appear to make sense at first, but watch a bottle bobbing on the surface of a bay. Waves pass under it repeatedly, but other than slowly drifting with the current the bottle moves relatively little. An analogy is the rippling motion wind makes as it blows across a wheat field. Waves follow one another across the stalks of wheat, and yet the wheat does not pile up in a heap on the far side of the field. Instead, the motion in the grain results from the nodding of the individual stalks each time a wave passes.

As long ago as 1802 it was known that water particles within a wave do not move forward with the advancing wave itself, but instead follow a circular orbit (Fig. 14-2). Detailed studies,

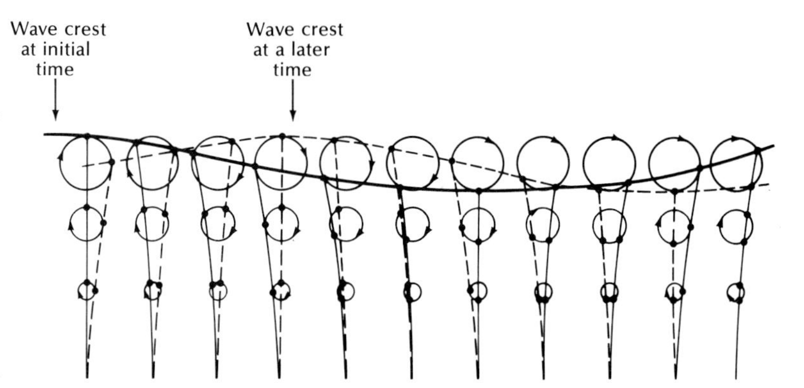

Fig. 14-2 Cross section of an ocean wave, showing the paths the water particles follow. The wave profiles and positions of the water particles are shown at two moments which are one-quarter of a period apart in time. Notice that the orbit of the water particles diminishes with depth. The nearly vertical lines show how grass would be bent as a wave form passes. The stalks are vertical beneath a crest or a trough.

of course, have been made since that time, but the basic principles of water motion in a ***wave of oscillation*** are the same as those recognized in the early nineteenth century. As the wave crest approaches, the water in the preceding trough moves toward the advancing crest. Then, progressively, the particles move upward, forward with the crest, downward, and then seaward again in preparation for the passage of the next wave crest.

The same cross section also shows how rapidly wave motion diminishes with depth. The diameter of the circles decreases in a geometric ratio with increasing depth. For practical purposes, wave motion ceases when the water depth is approximately equal to one-half the wave length.

Some of the effects of that orbital motion are familiar to every surfboarder, and others can test the effects by simply wading or swimming into the ocean a short distance. You will be conscious that the water is running strongly seaward toward the oncoming wave. As the wave surges shoreward, however, the water will sweep you strongly toward the beach. A common escape route, to avoid being caught in the force of a breaking wave, is to dive to the deeper water in either the trough or beneath the next crest, where motion is less. Most of us are aware of the backward and forward pulse of the sea in the breaker zone, but not too many people realize that the motion is only part of the orbital path described by water particles within a wave.

There appear to be finite limits to the possible size of waves. Among the largest waves observed was one that rose 34 m (112 ft) high when it was sighted off the stern of the U.S.S. *Ramapo* in 1933 during a gale in the North Pacific. Wave lengths are likely to be shorter than most people imagine, but they are impressively long at times (Fig. 14-3). One of the largest swells ever reported had a wave length of 792 m (2598 ft), with a period of 22.5 sec and a velocity of 126 km/hr (78 mph). The figures are formidable when one considers the enormous mass of water involved.

How such volumes of water are set in motion is a fair question. Almost everyone knows that the wind driving across the surface of the sea is the primary cause. Yet, how is it then, that on completely windless days a tremendous surf may pound some exposed coast? Or that in a violent gale the wind may hammer the sea flat into a turbulent mass of dark water streaked to the horizon with foam?

For large waves to form in deep water, the following requirements must be met. First, there must be a strong wind to set large masses of water moving. Second, the wind must be of fairly long duration—more than just a sudden gust is needed. Third, the water must be deep, at least deep enough to round out the full circular pattern—waves 9 m (30 ft) high are not

Fig. 14-3 The Coast Guard cutter *Ponchartrain* wallows in the trough of a following sea in the North Atlantic. *U.S. Coast Guard*

Fig. 14-4 Stormy seas are characteristic of the oceans off Antarctica. *W.R. Curtsinger, National Science Foundation*

likely to occur in a water basin only 3 m (10 ft) deep. Fourth, the distance that wind friction can operate on waves—called ***fetch***—is important. When waves have a long, uninterrupted run, they can reinforce one another. The ripples crossing a small pond are a good example. They are small on the upwind side of the pond, yet they may have grown to fair dimensions by the time they reach the downwind shore.

It is not surprising, then, that some of the largest seas are those driven before the strong westerly winds south of Cape Horn. Around the margins of Antarctica, an unbroken sweep of ocean encircles the earth; waves in those seas may be said to have unlimited fetch.

To return to an earlier question: How, if waves are formed by the wind, can they travel shoreward in an endless, rhythmically advancing succession on a dead calm day? The answer is that such waves, to which the name of ***swell*** is given, may have originated in gales thousands of miles away. Waves that break on the exposed coast of Cornwall may have had their start in the far-distant reaches of the South Atlantic. Waves crashing on the west coast of the United States may have been born in the Antarctic (Fig. 14-4), whereas the waves the surfers ride in Hawaii may have come from the Arctic.

A swell is made up of ***long-period waves,*** which outrun the more randomly distributed ***short-period waves*** characteristic of a storm. Short-period waves are left behind, and the

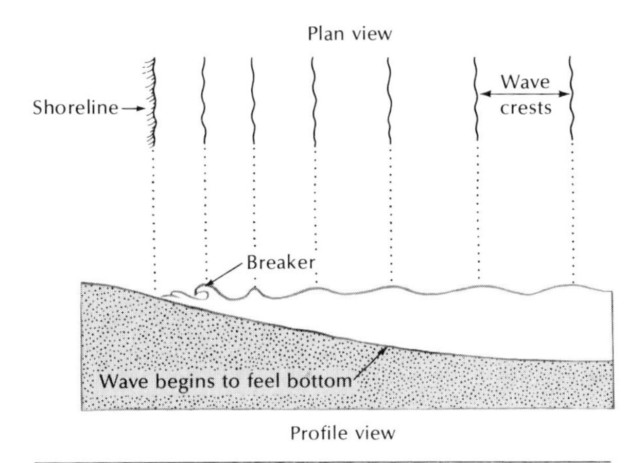

Fig. 14-5 How waves shorten, increase in height, steepen, and break as they advance into shallow water.

more uniformly spaced swell far outdistances the gale winds localized around some cyclonic center.

Formation of surf

The formation of surf is a complex phenomenon. The endlessly changing pattern of breaking waves—and their variations with the tide, with wind and calm, with storm, and with the lulls between—has inspired generations of painters, photographers, writers, and ordinary daydreamers. Few manifestations of the natural world are more dynamic or make us more

Fig. 14-6 Surfer riding a plunging breaker. *Aaron Chang, Surfer Magazine*

conscious of the force of moving water than a strongly running surf.

As waves move from deep water shoreward, they begin to feel bottom when the depth of water is equivalent to about one-half the wave length. As they move into shallower water near shore, their wave length shortens and their height increases, relatively (Fig. 14-5). Their velocity is also reduced, and if one looks at their crests along a pier or breakwater, they do indeed seem to rise up out of the sea. In a general way, waves of oscillation break when the stillwater depth is roughly equal to 1.3 times the height of the wave. There appear to be at least two major causes leading to the formation of breakers. The first is an increase in the circular velocity of particles in a wave that has moved into shallowing water, so that the velocity of the particles at the wave crest exceeds the decreasing velocity of the wave form. The second is that in the shallow depths near shore there is not enough water to complete the wave form. That is especially clear in the so-called ***plunging breaker,*** in which a wave curls over in a beautifully molded half cylinder, topples, and crashes with a thunderous roar (Fig. 14-6). The entrained air is violently compressed, and converts the entire roller into a froth of foam-whitened water.

Spilling breakers are those in which the crest foams over and cascades down the front of the advancing wave, without actually toppling over. Such breakers ordinarily diminish in height as they move landward, and they may advance in nearly parallel rows over a broad stretch of beach.

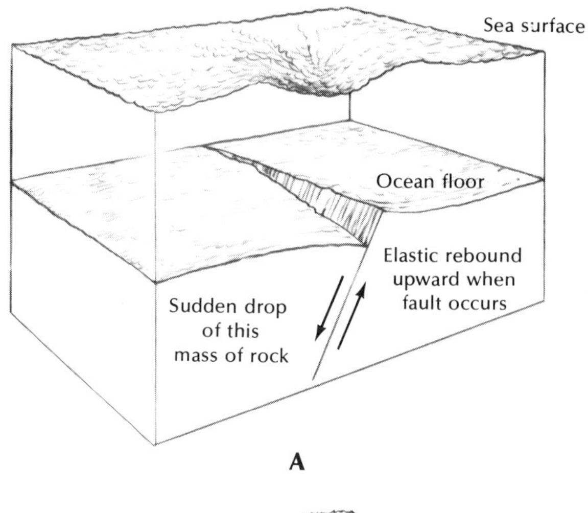

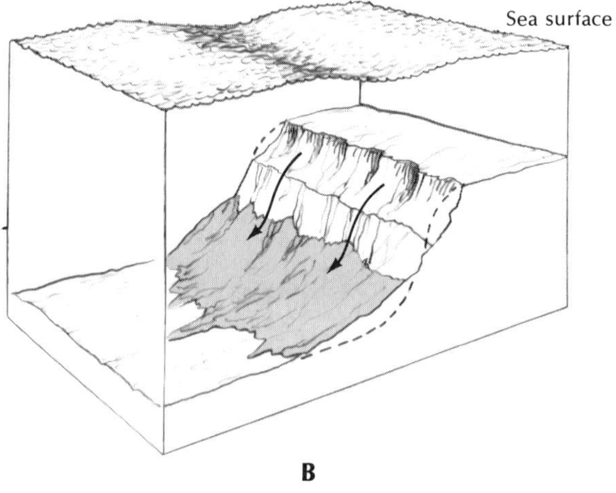

Fig. 14-7 Two common ways in which tsunami form. **A.** A sudden drop of the ocean floor along a fault causes the water surface to drop, and a wave is generated. **B.** An earthquake triggers a submarine landslide of loose sediment, which displaces water and sets up a tsunami.

Tsunami

One of the most destructive waves on earth is the ***tsunami*** or ***seismic sea wave.*** In the past it was called a tidal wave, but the term has been abandoned by earth scientists because the wave is not related to the tides. It is generated in response to disturbances on the ocean floor, such as fault displacement or earthquake-caused landslides (Fig. 14-7). American oceanographers chose the word ***tsunami*** (Japanese for harbor wave) to designate such waves. Certainly the destructiveness of such waves is enhanced in the progressively narrowing reaches of a bay.

The waves are barely noticeable in the open ocean, for they may be no more than 1 m (3 ft) high, with a wave length approaching 200 km (124 mi), and a period in the order of 15 minutes. Such waves move at the fastest natural velocities on earth, however, exceeding 600 km/hr (373) mph).

Fig. 14-8 Boats washed into Kodiak, Alaska, by a tsunami triggered by the Alaska Earthquake of 1964. A section of the harbor and partly submerged breakwater can be seen in the upper left. *USGS, photo U.S. Navy*

Only when this oceanic ripple hits a coast is violence unleashed (Fig. 14-8). Like most waves, it too feels bottom, increases in height, and breaks. Willard Bascom has kept track of the damage of some tsunami, and here records the effects of one generated by a colossal landslide in the Aleutian Trench. The date was April 1, 1946:

Later, over a period of years, I traveled to many Pacific shores asking about the effects of that tsunami. Remarkably often points facing into the waves and bays facing away from them were hardest hit. For example, Taiohai village at the head of a narrow south-facing bay in the Marquesas Islands four thousand miles from the earthquake epicenter was demolished. Hilo, Hawaii, only half as far from the disturbance and whose offshore topography seems precisely suited to funnel tsunamis toward the town, fared worse. There the captain of a ship standing off the port watched with astonishment as the city was destroyed by waves that passed unnoticed under the

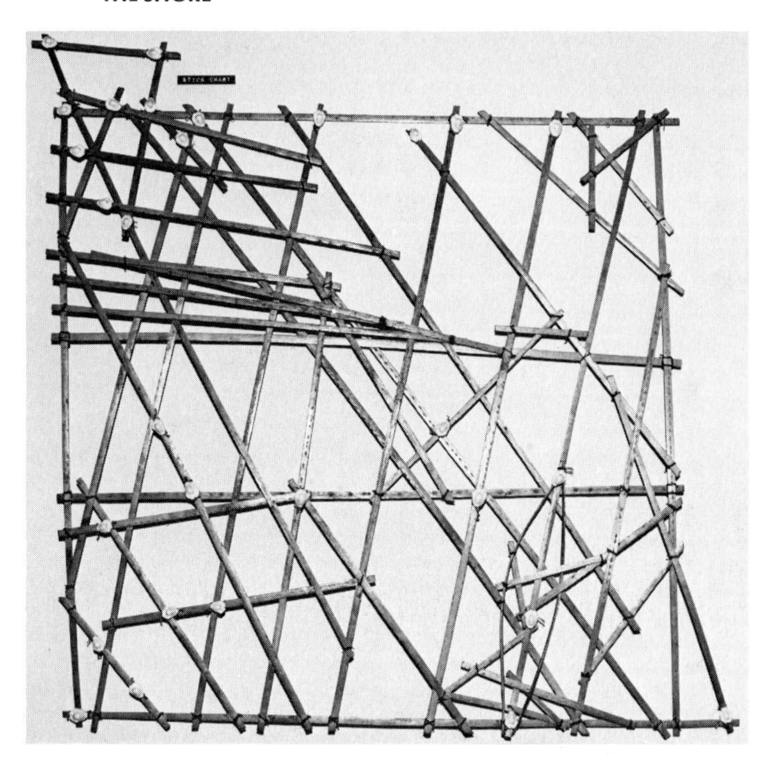

Fig. 14-9 Stick chart used by Polynesians in navigating between islands in the Pacific. The shells represent islands and the stick pattern wave orientation and refraction. *Janet Robertson, Geological Society of America*

ship. Another ship, the *Brigham Victory,* was unloading lumber at Hilo when the tsunami struck. The ship survived with considerable damage but the pier and its buildings were destroyed. One hundred and seventy-three persons died and $25 million in property damage was done by the waves at Hilo that morning.

But the truly great waves of April 1 struck at Scotch Cap, Alaska, only a few hundred miles from the tsunami's source, where five men were on duty in a lighthouse that marked Unimak Pass. The lighthouse building was a substantial two-story reinforced concrete structure with its foundation thirty-two feet above mean sea level. None of the men survived to tell that story but a breaking wave over one hundred feet high must have demolished the building at about 2:40 A.M. The next day Coast Guard aircraft, investigating the loss of radio contact, were astonished to discover only a trace of the lighthouse foundation. Nearby a small block of concrete one hundred and three feet above the water had been wiped clean of the radio tower it once supported.

Today, a tsunami warning system set up around the Pacific helps alert people so such catastrophes can be avoided.

Wave refraction

As wave fronts approach land they commonly are bent, or refracted, so that they almost always parallel the shoreline closely, no matter what their configuration. Refraction occurs because of variations in the velocity of a wave as it reaches shallower water: different parts of a wave touch bottom at different times, slowing its forward progress and changing its direction. In Polynesia, seafarers use wave orientation and refraction to navigate between islands, fashioning charts from sticks and shells (Fig. 14-9). By observing wave patterns, they chart accurate courses between the islands, which are visible only at very close range.

Wave refraction explains the variation in the intensity with which waves break on an indented coast (Fig. 14-10). Land promontories

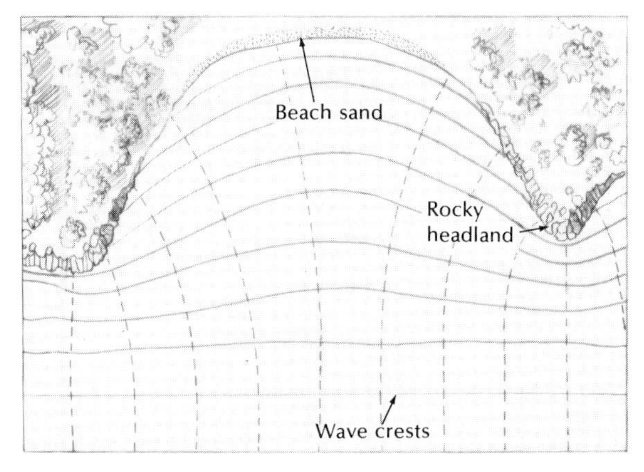

Fig. 14-10 A. How waves are refracted. Wave crests crowd together as they move into the shallow water around the rocky headlands; in the deeper waters of the bay, however, they are more widely spaced. Wave energy, equally distributed between the parallel dashed orthogonal lines in deep water, is concentrated on the headlands where the lines converge but is diminished in the bay, where they diverge.

Fig. 14-10 B. (below) Wave refraction on a coast with rocky headlands and small islands. Eyre Peninsula, South Australia. *South Australia Lands Department*

are eroded because the wave energy is focused there by refraction. In contrast, the energy is diffused through an adjacent bay, as any mariner knows. Because of diffusion, wave energy decreases in bays, which are commonly the site of deposition of sediments eroded from the promontories.

WAVE EROSION

It is in the shore zone, where breakers are able to bring their full force to bear against the land, that most of the erosive work of the sea occurs Fig. 14-11). Because there is an upper and lower limit to each wave, the work of waves

Fig. 14-11 Storm waves from the Southern Ocean have eroded coastal cliffs on Banks Penninsula, New Zealand, plucking out the large boulders that now rest on the beach. *Peter W. Birkeland*

might be likened to that of an enormous horizontal saw. Their upper effective limit is the maximum height they reach at high tide during a storm. That height will differ among coasts, depending upon how the land faces the sea and how strong is the gale driving the waves. Some storm-driven waves have been known to reach heights of 65 m (213 ft) or so.

The lower limit of wave erosion is much less certainly known. Estimates vary widely and possibly reflect the bias of the estimator. Observational evidence seems to indicate that most waves can generate currents capable of moving sand at depths of not much more than 12 m (39 ft), although ripples and other signs of disturbance of the sea floor are often seen at depths of 107 m (351 ft) below sea level and even deeper. Gravel and cobbles are not moved by wave currents much below 6 m (20 ft). The lower effective limit of wave transportation and erosion is called ***wave base***; obviously, its depth differs on different coasts, just as the upper limit does.

Waves erode mostly by abrasion, just as streams do. In times of calm, very little erosion takes place; in fact, if there is a source of sediment, deposition commonly occurs. During storms, however, when the waves are highest and most capable of carrying such abrasive agents as sand, gravel, and even cobbles, their erosive power is great. Since waves cannot erode above their maximum height, their action is restricted to a narrow zone of the shore. On a cliffed shoreline the base of the cliff is undercut, and, if the rocks are weak because of jointing or poor consolidation, landslides will occur. In such cases, the landslides are actually the eroding agent of the upper parts of the cliff.

The shore zone is one of the most dynamically active erosional areas on earth. During high tide, it is submerged. When the tide is low, and especially if the tidal range is great, a broad expanse of the shore zone may be exposed, and acted upon by rain and wind. In the Arctic, the shore may be modified by sea ice, and in other regions the alternation of wet and dry periods may contribute to the process of erosion.

The rate of wave erosion varies from place to place. Some coasts show little change in a generation; on other coasts, erosion can be quite rapid, as shown by undercut roads, patios, and steps built down the face of sea cliffs. Rates in excess of 1 m (3 ft) per year are not uncommon (Fig. 14-12). The rate is a function of the strength of rock making up the cliff and the energy of the waves. Cliffs along the English Channel have retreated as much as 12 to 27 m (39 to 89 ft) overnight during a single storm.

Fig. 14-12 The cliff on this beach near Montara, California, retreated 51 m (167 ft) between 1866 and 1971, an average rate of about 0.5 m (1.6 ft) per year. *K.R. Lajoie, USGS*

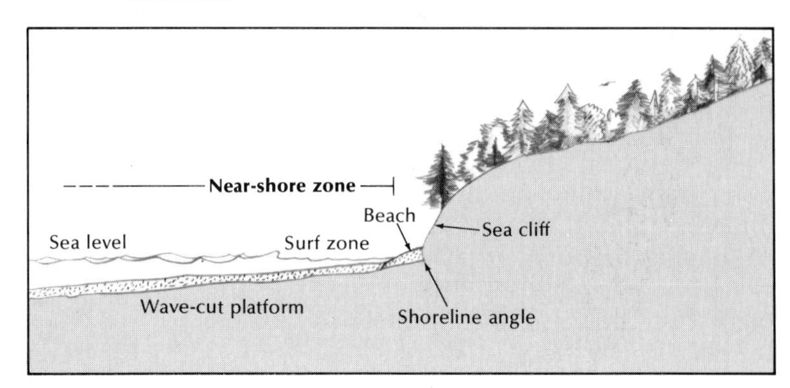

Fig. 14-13 Cross section, showing characteristic landforms along a steep coast.

Because wave erosion has an upper and a lower limit and because it works horizontally, it produces landforms that differ from those made by downward-cutting streams and glaciers (Fig. 14-13).

The most noticeable coastal landform is a ***sea cliff*** (Figs. 14-13 and 14-14). Its height depends upon a variety of factors, including the energy of the attacking waves, the slope of the land surface, and the resistance of its rocks. Obviously, a headland of granite is likely to be much more resistant than one of chalk. The imposing height of the chalk cliffs near Dover results from the erosional weakness of that rock as well as the fury of the sea's attack in a North Atlantic gale (Fig. 14-15).

As a sea cliff recedes before the onslaught of the waves, a planed-off rock bench called a ***wave-cut platform,*** is cut at its base (see Fig. 14-13). Sometimes it will be bare, abraded rock, interrupted, perhaps, by tide pools and occasionally by unreduced remnants of the cliff, known as ***stacks*** (see Figs. 14-14 and 14-15). Where the shoreward portion of the platform is mantled with sand, it is called a ***beach.*** Seaward of the platform there may be an accumulation of wave- and current-transported material, which constitutes a ***marine-built terrace.*** Whether or not the last feature is present depends in part upon the energy of waves and currents, especially the longshore current. Should the current be strong enough, it may sweep the sediment into the mouth of the next bay down the coast instead of depositing it at the base of the cliff from which it came.

Beaches are the land's first line of defense against wave erosion, particularly by hurricane and other storm waves. Much of the force of storm waves is expended upon the gradual slope of the beach, and even if the beach is removed, it will usually be rebuilt by deposition in calmer weather if a source of sediment exists. Sea cliffs do not have such renewal ability; once parts of them are eroded and removed, they are gone forever. There are protective as well as recreational reasons, then, for close observation and study of our beaches in order to preserve them.

BEACH DEPOSITION AND EQUILIBRIUM

Wave and current erosion and transport make some portions of a coast retreat and other portions advance. Material that is eroded from a

Fig. 14-14 Near-vertical sea cliffs on the Daisei Islands, located off the west coast of Korea, testify to the high rate of beach erosion and shoreline retreat. The stacks of vertical rock remnants detached from the cliffs have not yet been consumed by the sea. Note the contrast in topography between the sea cliff and the hilly terrain behind it, which was formed by subaerial processes. In places, the upper parts of the valleys have been truncated by sea-cliff retreat. *Department of Defense*

Fig. 14-15 Vertical cliffs cut in chalk along the southern coast of England near Dover. *Aerofilms, Ltd.*

point or headland may be deposited by a longshore current in a nearby bay or marsh. In the long run, erosion and deposition may attain a rough equilibrium, or one may outpace the other. In Britain, for example, in the 35 years immediately preceding 1911, it is estimated that 1900 hectares (4692 acres) were lost to the sea, while 14,345 hectares (35,444 acres) were gained, the latter chiefly as salt marshes, sand spits, beaches, and bars. On the other hand, the island of Heligoland, in the North Sea off the German coast, in A.D. 800 had a 200-km

(124-mi) long shoreline, and was a much-fought-over piece of terrain in the Viking era. By 1900, the one-time independent dukedom had been reduced to a strongly fortified rock only 5 km (3 mi) around, enclosed by sea walls, breakwaters, and other defenses.

How are beaches formed? Beach sands are a product of weathering and erosion that generally takes place far inland, and are brought to the coast by streams. Some sand, however, is derived from erosion of the bedrock platform or of the sea cliff itself. In shallow waters, so-called rivers of sand move parallel to the shore and provide some of the raw materials for beach building. Waves, even though refracted, break against the beach face at a slight angle, but recede in a straight path. The resulting pattern of water movement produces the ***longshore current,*** which carries sand particles along with it (Fig. 14-16).

An equilibrium is commonly reached along a beach between the amount of sand available and the variations in wave energy. As we shall see, that equilibrium has many attributes common to stream equilibrium. Even if some of the sand is deposited in submarine valleys and shunted to deeper parts of the ocean floor, it is replaced by sand brought down to the sea by streams or by sea-cliff erosion, and so equilibrium is maintained.

Permanent loss of beaches is becoming increasingly common, and ironically enough, such loss is usually caused by our meddling with the beach equilibrium. Sand accumulates

Fig. 14-16 Waves striking a gravelly beach near Banks Peninsula, New Zealand. The angle at which the waves meet the shore indicates that longshore transport of sand is away from the viewer. *Peter W. Birkeland*

Fig. 14-17 At Santa Barbara, California, a breakwater constructed to protect small boats not only cut off the supply of sand to the coast beyond, but interrupted longshore currents, resulting in the formation of a spit (just beyond the moored boats), which threatens to close off the harbor. Equilibrium has been restored, but at the cost of dredging the spit and transporting the sand to the eroding coast on the far side of the harbor. *John S. Shelton*

behind every dam we build, and thus never replenishes beach sand moving along the coast. In the construction of beach resorts, if much of the available sand is removed to provide flat surfaces for building foundations, the result can be accelerated beach erosion.

Jetties and breakwaters, however, are probably most disruptive of beach equilibrium, because they act as dams. They disrupt the longshore current and sand transport, causing damage that is extremely expensive, and sometimes impossible, to correct. For example, sand deposited by the longshore current on the upcurrent side of a jetty or breakwater will accumulate there until the beach becomes excessively wide. Meanwhile, on the other side, the current, deprived of its supply of sand, removes and carries away the pre-existing sand until the beach disappears. Furthermore, some sand may be swept around the structure, filling the harbor or marina it was built to protect in the first place (Fig. 14-17). At Port Hueneme, in California, a 915-m (3001-ft) jetty now channels sand directly into Hueneme Submarine Canyon, so it is lost forever.

Unfortunately we seldom recognize that our own activity is destroying our precious beaches. Instead we blame the ocean and attempt to conserve beach property by fighting the ocean rather than working with it. In addition to jetties and breakwaters, seawalls are built in an attempt to control erosion. Seawalls are seldom effective, however, because they provide a vertical battering surface rather than a gradual slope, as on a natural beach. The excessive wave erosion that can result, not only prevents accumulation of a new beach, but also quickly undermines the structure itself. Commonly, too, seawalls are not built high enough to exclude damaging storm waves, and the water thât does break over them is kept from returning to the ocean. Another often-tried remedy to prevent beach erosion is the ***groin,*** a short wall built at right angles to the shore. The structure does indeed trap sand on its upcurrent side, but below the groin, prop-

Fig. 14-18 Groins along the Anna Maria Key south of Tampa, Florida. Sand transport is toward the viewer. *Coastal Engineering Archives, University of Florida*

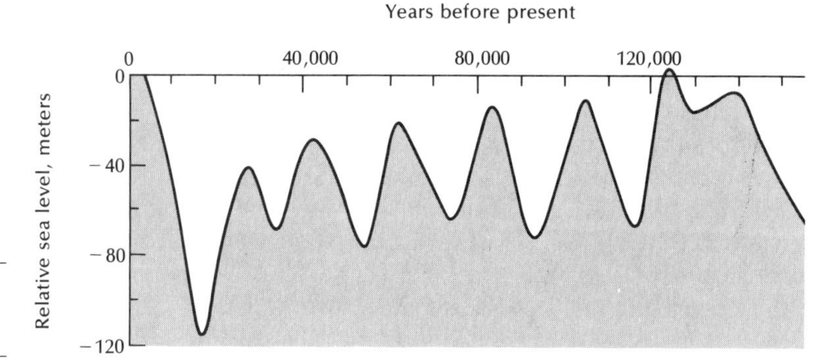

Fig. 14-19 Worldwide sea-level fluctuations based on studies of marine terraces in Barbados.

erty owners may find themselves without a beach, as the undernourished longshore current rapidly removes the sand. They may even erect groins on their own property. Thus groin-building has a natural tendency to proliferate and, in fact, parts of the Atlantic shore bristle with them (Fig. 14-18).

Beach engineering has progressed to the point where suitable and effective structures have been designed, but usually they cost more money than the individual property owner wishes to pay, and in the absence of any clarification of legal responsibility, local governments are often unwilling to take on the added tax burden.

One of the remedies that has proved useful is an artificial beach, made by dumping a large quantity of imported sand. If the sand grains are not the correct size, however, the artificial beach may disappear faster than the original beach. If there are cobbles in the imported sand, as happened when the beach was renewed at Oceanside, California, the sand vanishes, leaving a very stony shore area. Although stone-strewn beaches may not be suitable for recreational purposes, they do help to prevent further erosion. Off Long Branch, New Jersey, on the other hand, 459,000 m^3 (600,370 yd^3) of sand was dumped too far offshore. Beyond the reach of the longshore current, the sandpile has remained essentially intact.

But where conditions are appropriate and the supply of sand is sufficient, artificial beaches can work quite well. Dune fields, too, can be rebuilt by adding sand and stabilizing it with a plant cover. Determining where an artificial beach is feasible presents a problem, however, since it is difficult to measure rates of erosion and other factors in a zone as active as the shore. Finding sand is less difficult—usually a suitable source is nearby—an up-coast harbor filling with unwanted sand or an ancient beach a short distance offshore or even inland. But with the increasing use of the world's shorelines, sand to replenish them will become a problem of economic interest. For example, a proposal once was made to import sand from the Grand Bahama Banks for the eroded beach along 160 km (99 mi) of the

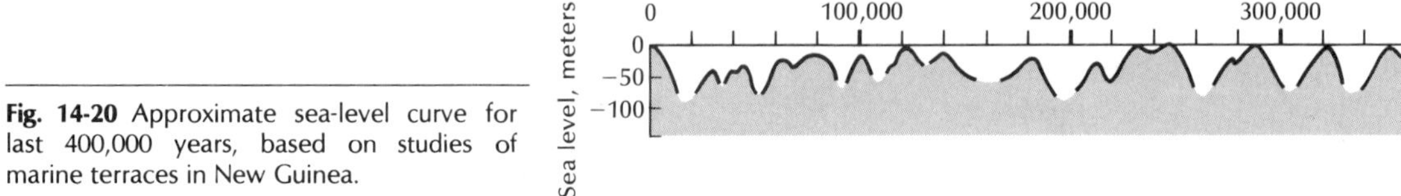

Fig. 14-20 Approximate sea-level curve for last 400,000 years, based on studies of marine terraces in New Guinea.

southeast coast of Florida. The project would involve the design and construction of a special hopper dredge, as well as a payment to the government of the Bahamas.

The extent of beach erosion problems makes it obvious that control cannot long be left to individual landowners. For reasons of economy alone, as well as the large-scale research, planning, and organization required, such operations will increasingly fall to local, state, and federal agencies. The situation may lead to more public ownership of beach property, an effect that will make the beaches accessible to more people for recreation.

SEA LEVEL FLUCTUATIONS

Along some coasts, crustal instability has caused the shore to be uplifted or depressed relative to sea level. In addition, there have been actual changes in sea level itself, and by their very nature the latter have worldwide effects. Seldom, in fact, has sea level remained stable over the past million years or so, although it is generally agreed that sea level has changed little during the last 5000 years (Fig. 14-19).

The worldwide relative stability in sea level coincides roughly with the appearance of the maritime civilizations around the shores of the Mediterranean Sea and the Persian Gulf, so that the ancient harbors of the Egyptians, Persians, Phoenicians, and Minoans correspond roughly to the sea level of today. Over 5000 years ago, sea level was lower, but fluctuating, and we have to look back some 120,000 years for the last time it was close to its present position (Fig. 14-20). However, during the last 400,000 years, sea level was generally lower than it is today.

Present-day sea level is far from stationary. Beginning around 1850 it started to rise, as shown by tide gauges in seaports all over the world. C. A. M. King has compiled data from around the world on the annual rate of present-day rise—that for the Atlantic coast of North America is 4 to 7 mm (0.16 to 0.28 in.), Italy 1.7 mm (0.07 in.), Holland 1.5 mm (0.06 in.), and western North Africa 0.6 mm (0.02 in.). There are various reasons for that nonuniform rise from port to port, but nobody doubts that the sea level is rising.

The important question is what accounts for

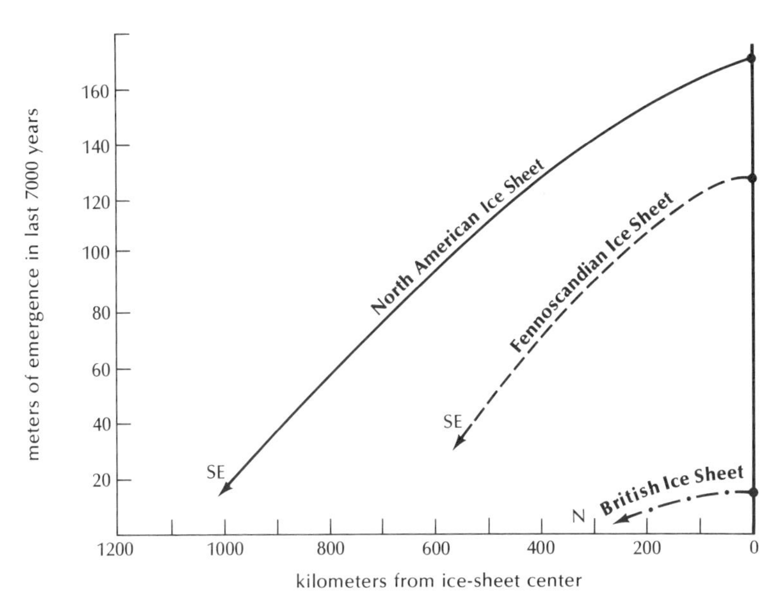

Fig. 14-21 Hypothetical cross sections across terrain occupied by former ice sheets show how the amount of emergence (marked by shoreline features in the field) is related to the size of the ice sheet (the North American ice sheet being the largest, the British the smallest), and to the distance from the center of the sheet.

Fig. 14-22 Elevated beach deposits near Fort Severn, Canada, on the southwest shore of Hudson Bay, clearly demonstrate that the land has been rising since the disappearance of Pleistocene ice. *Department of Energy, Mines, and Resources, Canada*

the fluctuations in sea level? The current rise is probably related to the worldwide recession of glaciers and the return of their waters to the sea. Thus, over the last 120,000 years (Fig. 14-19), periods of relatively low sea level correlate with major advances of continental glaciers and periods of relatively high sea level with major recessions, all during the Wisconsin glaciation (Chap. 13). The similarity between Figures 14-20 and 15-20 suggests a very strong glacial-interglacial influence on sea level even further back in time. Such changes are called ***eustatic.*** The fluctuation cycle can also allow us to estimate the extent of glaciation, to determine ice volumes, weights, and ice-cap thicknesses during the Pleistocene.

Variations in sea level can have far-reaching effects. A rise over several generations can wreak havoc, especially along flat coasts.

EXAMPLES OF COASTAL DEVELOPMENT

The formation of coasts is complex, for not only do present-day processes operate on coasts, but many coastal landforms carry a legacy of the past. Three major kinds of coasts will be described below.

Emergent coasts

It is quite clear that the land along some coasts is rising. The best evidence for such change is found in areas that were once covered by the Pleistocene glaciers. The sheer weight of the

Fig. 14-23 Prominent marine terrace along the California coast, north of San Simeon. *John S. Shelton*

Fig. 14-24 Marine terraces of coral reef material near Finschhafen, New Guinea. The highest terrace, about 260 m (853 ft) above sea level, was formed about 140,000 years ago; this coast, therefore, has been uplifted that amount because the highest terrace formed when sea level was close to its present position. *Department of Defense*

ice pushed the earth's crust down; when the ice melted, the crust rebounded. So in areas that were under or near ice caps, prehistoric shoreline features are found far above present sea level. The patterns of shoreline uplift are related to the size of the ice cap; the larger the ice cap, the greater the uplift (Fig. 14-21). Those glaciated areas are still emerging (Fig. 14-22). For example, the Gulf of Bothnia is rising at the rate of 0.9 cm (0.35 in.) and Hudson Bay at 2 cm (0.8 in.) per year. It is estimated that the latter areas still have a considerable amount of rebounding to do, so that we can expect more harbors to become increasingly shallow or eventually to go dry.

Other coasts far from centers of glaciation are also rising. Again, the evidence is found in old marine beaches and wave-cut platforms—collectively called ***marine terraces***—and sea cliffs standing far above sea level. Such landforms certainly were pushed up by tectonic movements because, as we have mentioned, there is no evidence that sea level has risen much above its present level for the past million years. Classic examples of marine terraces are found along the California coast (Fig. 14-23). Perhaps the most spectacular flight of terraces are the coral reef terraces of New Guinea (Fig. 14-24). Also, part of the New Guinea coast is the most rapidly rising one in the world—some 3 m (10 ft) in a thousand years. This means that beach deposits of the last intergalciation (about 120,000 years ago) are now more than 300 m (984) ft) above sea level.

Submerged coasts

Other coasts of the world show distinct signs of submergence. A familiar example is the coast of the Netherlands, which is in a deltaic area of rapid subsidence and thus is sinking at a rate of nearly 10 cm (4 in.) per century. No wonder that with a total rate of sinking of around 20 cm (8 in.) per century—about one-half from subsidence and one-half from rising sea level—that, beginning with the great floods of medieval times, the Dutch have been compelled to construct an extraordinary complex of dikes and coastal defenses upon which their survival depends.

Typical submerged coasts are those in which the sea extends inland, sometimes for long distances, in embayments. Should the indentations have been shaped by stream erosion before their invasion by salt water, the coast is known as a ***ria coast,*** from the name applied to the southern shore of the Bay of Biscay. The Gulf of Maine, the north coast of South Island, New Zealand (Fig. 14-25), and the southern coast of Brazil are ria coasts.

How a particular coast becomes embayed may be difficult to determine. Perhaps the land subsided, in which case the sea invaded or "drowned" pre-existing river valleys. Or it could be that the land remained stationary, and the postglacial rise of sea level flooded low-lying parts of it. In either case the effect is the same.

Some of the progressive changes a ria coast is likely to undergo as a result of its modification by the waves and currents of the sea are shown in the accompanying set of diagrams (Fig. 14-26).

In the first diagram, the sea has come to rest

Fig. 14-25 Drowned stream-carved valleys form the Marlborough Sounds, South Island, New Zealand, a classic example of a ria coast. *D.L. Homer, New Zealand Geological Survey*

Fig. 14-26 The possible successive stages in the development of a submerged coastline.

upon a landmass the surface of which has been shaped by stream erosion. Former ridges now extend seaward as headlands, and the sea may reach inland as an embayment or estuary, perhaps much as Chesapeake Bay does (Fig. 14-27).

The second and third diagrams (Fig. 14-26) show some of the changes that might be anticipated with time. Because most of the erosional energy of the sea is concentrated on the headlands, they soon terminate in sea cliffs. If there is a longshore current, debris from the sea cliffs or the rivers can pile up in a curved embankment called a ***spit*** that trails downcurrent from the land. In places, spits grow almost long enough to close off a bay, and they are then called ***bars*** (Figs. 14-26 and 14-28). In other places, the beach sand moving along the coast piles up in the low-energy zone behind near-shore islands and eventually are connected to the mainland by a strip of sand, called a ***tombolo.*** Streams entering the bay deposit sediment when their velocity is abruptly checked, building deltas out into the relatively still water. Thus, the outer exposed parts of the coast are pushed back by erosion while the innermost parts are built out by deposition.

In the later history of such a coast, as shown in the last two diagrams of Figure 14-26, the headlands may retreat inland until the coast is more or less straight, and an equilibrium may be reached between sea erosion and deposition of sediments from terrestrial erosion. If very little sediment is being added, the headlands may recede as far as the innermost bay head. Then the coast will have lost its original indented character and will be cliffed along much of its length. What irregularities there are will very largely reflect the relative resistance of the rocks cropping out along the cliff face.

Fig. 14-27 Chesapeake Bay is a fine example of a landscape shaped mainly by subaerial processes, but later drowned by the rising sea. *NASA*

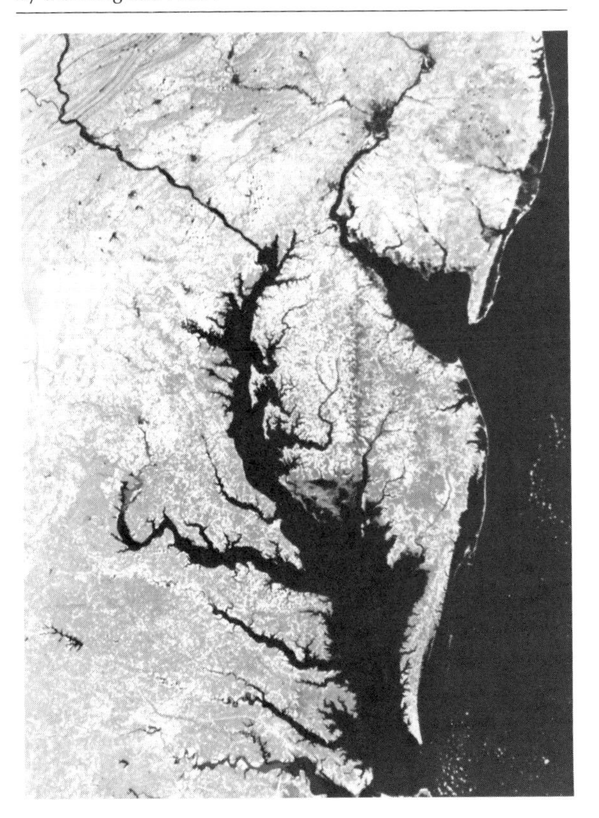

Plains coasts

Some coasts of the world with a low gradient and an adequate sand supply are fringed by ***barrier islands,*** long, narrow islands of sand separated from the coast by narrow bodies of waters called ***lagoons.*** Some 282 barrier islands extend, like linked sausages, from Long Island, New York, around Florida and the Gulf of Mexico to the southern end of Texas (Fig. 14-29). Cities such as Atlantic City and Galveston, and many resorts, have been built on these islands, which lie from 3 to 30 km (1.9 to 19 mi) offshore, and range from 2 to 5 km (1.2 to 3 mi) in width and 10 to 100 km (6 to 62 mi) in length. Most of them rise no more than 6 m (20 ft) above sea level. A comparable chain of sandy barriers stretches along the low coasts of the Netherlands, Friesland, and Denmark around the southeastern margin of the North Sea.

During the Civil War, the sandy bars and islands along the east coast were of strategic importance in amphibious landing and minor naval operations. Gaining control of the offshore islands and the passes between them was of the utmost importance to the Union Navy and proved to be a formidable task. The shallow, unlighted passes (or openings) through the barrier islands, with their endlessly shifting channels, were a boon to the blockade runners, whose shallow-water craft were more maneuverable than the vessels of the blockading fleet.

Barrier islands have two different types of origin. One is the growth and ultimate breaching of a spit (Fig. 14-30). The other, which is probably the way most barrier islands originate, involves a coast undergoing slow submergence (or a rise in sea level) so that the beach, or dune ridges, becomes an island and the low area behind it becomes a brackish lagoon (Fig. 14-31).

The stability of any barrier island depends on

Fig. 14-28 A bar nearly closes the opening into Bolinas lagoon, north of San Francisco, California. A tidal channel cuts through the bar and several such channels meander across the mud flats behind the bar. The position of the channel across the bar is unusual, because the longshore currents move the sand away from the viewer. *Aero Photographers*

several factors. Abundant sand supply is crucial. Waves carry sand in from the ocean floor, and longshore currents sweep it along the coast. During storms, large waves wash over parts of the island, depositing additional sand. In time, winds may pile up sand dunes, which further increase the height of the island. Storms can carve inlets to the lagoons, and, although currents may keep some of the inlets open, others will disappear as sediment fills them in. Depending upon the rate of submergence, the amount of wave energy, and the

Fig. 14-29 Barrier islands fringing the outer coast of North Carolina, photographed on the Apollo 9 space mission, 1969. *NASA*

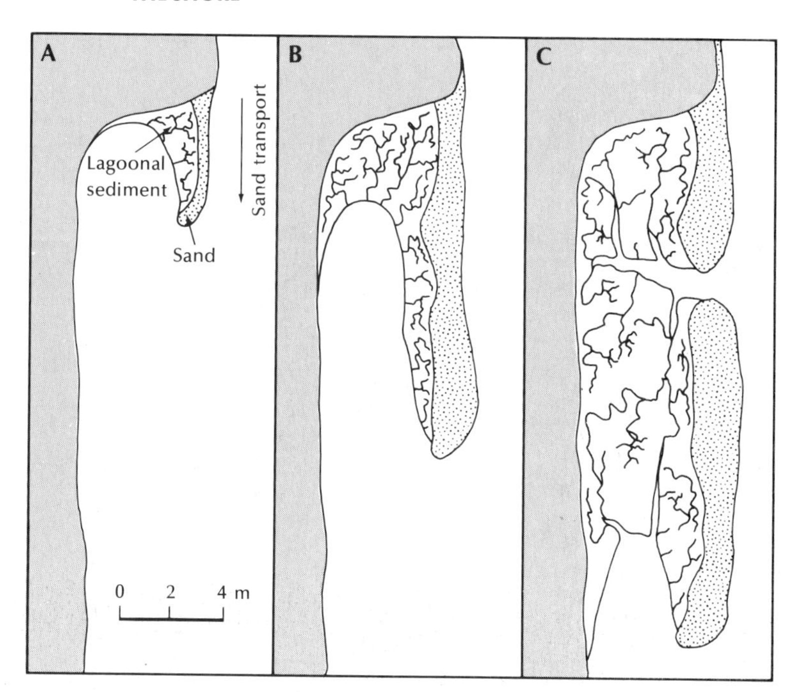

Fig. 14-30 Formation of a barrier island from a spit. In (A) and (B), the sandy spit grows and fine-grained lagoonal sediments fill in behind it. In (C), the spit has been breached by a tidal channel, and the lower end of the spit has become a barrier island.

rate of sediment supply, a barrier island can maintain its general shape and position, be pushed landward, or grow seaward.

With the development of our barrier islands for recreational and other uses, we have learned more about their dynamic nature, but not without cost in lives and dollars. About one-fourth of the barrier islands along the Atlantic and Gulf coasts are developed and urbanized, about one-third have been put aside for recreational use or as wildlife refuges. The rest—more than 100—are privately owned and undeveloped. Throughout recent geological time, these islands have acted as buffer zones, absorbing the impact of winter storms and hurricanes, shifting and changing shape with time.

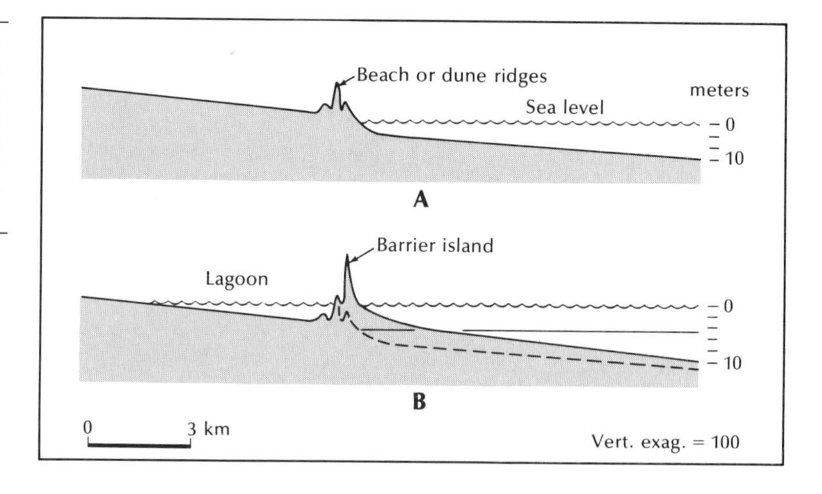

Fig. 14-31 Formation of barrier islands by submergence. A. Waves and wind build a rampart of beach and dune ridges. B. With a submerging coast or rising sea level, the ridges are separated from the land, forming a barrier island and a lagoon. Under appropriate conditions they can maintain themselves by growing upward and seaward.

Fig. 14-32 Resort hotels built at the water's edge, Fenwick Island, Maryland. © *Robert Dolan*

Yet people often build houses, hotels, and other buildings virtually at the water's edge with little thought for protection against exceptional storms (Figs. 14-32 and 14-33). They also try to control the flow of sediment or to retain the naturally shifting sands by building groins, but this can erode away other parts of the shore. The sea level is still rising along the barrier islands—25 cm (1 ft) from 1900 to 1974—bringing the edge of the ocean still closer to the already precariously placed buildings. A study of a section of the mid-Atlantic coast indicates an average rate of erosion of 1.5 m/yr (5 ft/yr).

What can be done to prevent further erosion and to compensate for the effects of erosion in developed areas? One answer would be to add sand to the beaches, perhaps by dredging offshore deposits. This manuever would be expensive, and possibly never ending, and the offshore changes might lead to some unforeseen damage. Another answer to save life and to prevent property damage would be to move threatened buildings back from the shore.

To prevent man-induced erosion in areas being developed, better geological information concerning cause and effect could be incorporated into development schemes. Thus, restricted development could be limited to inland areas, and more fragile shorelines protected, perhaps through public ownership as wild areas.

Fig. 14-33 The mid-Atlantic coast was struck by the most severe winter storm of record on March 7, 1962. Most of the damage occurred on low-lying barrier islands, this one being Fire Island. The storm caused over $500 million in damages between Long Island and Cape Lookout, and left 32 dead. *United Press International*

SUMMARY

1. Waves cut away at the edges of land, and the depth of their most effective erosion is only one-half the wave length. Water particles move in an orbital fashion as waves pass by, but the orbital motion is not completed in shallow waters; there the waves break in the surf zone.
2. *Tsunami* commonly are earthquake-generated waves that reach exceptional heights as they approach land; hence they are among the most destructive waves on earth.
3. Waves are refracted as they approach coasts, and most of their erosive energy is concentrated on headlands. Bays between headlands commonly are sites of deposition of material derived from the headlands. On steep coasts, wave erosion will form a *sea cliff* and a *wave-cut platform*. Beach materials (sand and gravel) cover the platform.
4. Wave refraction produces a *longshore current* that moves beach material along the coast. In places, an equilibrium is set up between the wave energy and the amount of material moving along the coast. Interference by man has caused problems by altering coastal erosion and deposition.
5. Sea level during the Pleistocene was low during times of glaciation and high during times of the interglaciations. At no time during the Pleistocene was it much higher than it is at present.
6. Three major types of coasts are recognized: *emergent* coasts, *submerged* coasts, and *plains* coasts.

QUESTIONS

1. Explain the movement of sand along a beach. What would be the consequence of building a jetty into deep water at the mouth of a river?
2. What recommendations would you make for the development of a barrier island?
3. If sea level remains constant, what factors would influence the distance that erosion will push the coastline back?
4. What combination of landforms and deposits are used to distinguish an emergent coast from a submerged coast?
5. How can it be shown that a particular coast is subsiding tectonically rather than being submerged by a rise in sea level related to the melting of glaciers?

SELECTED REFERENCES

Bascom, W., 1964, Waves and beaches, Anchor Books, Doubleday and Co., New York.

Dolan, R., Lins, H., and Stewart, J., 1980, Geographical analysis of Fenwick Island, Maryland, a middle Atlantic Coast barrier island, U.S. Geological Survey Professional Paper 1177-A.

Kaufman, W. and Pilkey, O., 1979, The beaches are moving, Anchor Press/Doubleday, Garden City, New York.

King, C. A. M., 1972, Beaches and coasts, Edward Arnold Publishers, London.

Shepard, F. P., 1973, Submarine geology, Harper & Row, Publishers, New York.

Fig. 15-1 The research submersible *Alvin* scours the floor of the Cayman Trench, at a depth of 3659 m (12,000 ft) south of Grand Cayman Island in the Caribbean Sea. *Emory Kristof* © *National Geographic Society*

15

THE SEA

The sea is the most alien of all the world's environments, and therefore the most mysterious. In the United States, a country of east-west continental extent, our interests are largely directed inward upon the affairs of the land, though there was a time when the sea played a significant role in American life. This was the time of whaling ships, when whale oil was used for illumination, and of long voyages under sail, as chronicled by Herman Melville and Richard Henry Dana. Crossing the ocean now can be a matter of hours rather than weeks and months.

The sea is almost the last great scientific frontier on earth, although it covers 71 per cent of the earth's surface. We know more about any land area, however remote, than we do of the sea floor directly off our shores.

Geologists were long unable to study the deep oceans—not because they had no direct interest in them, for they did. Many sedimentary rocks, for example, are marine, and in order to understand them better we need to know more about the marine environment. In the last three decades, however, new technology has made possible study of the deep oceans to a much greater extent, and a better understanding of the sea and the sea floor is emerging (Fig. 15-1). The theory of plate tectonics (see Chap. 20) is based in part on oceanic features, and the study of marine organisms sensitive to variations in temperature and in sea-water composition has given a fairly lucid picture of the timing of the Pleistocene glaciations and interglaciations. Recent deep-sea dives have revealed life forms previously unknown (Fig. 15-2).

Economic reasons also help explain part of our present interest in the ocean, or better, in ocean sediments. Several billion dollars worth of oil and gas are extracted annually from shelf sediments, with about one-quarter coming from shelves off the United States. Although environmental problems caused by offshore drilling are numerous, the solutions to many of them are in sight, and deposits on the oceans' shelves probably will increase in importance as a fuel source. More industries may look to the sea floor for concentrations in ancient submerged beaches of such valuable heavy minerals as tin, gold, or diamonds; but exploration costs are high, and profitable concentrations

Fig. 15-2 Such animals as these typically live in shallow waters, near land, so you can imagine the surprise when *Alvin* scientists found them in deep waters, close by hot springs, in a mid-ocean ridge near the Galapagos Islands. **A.** Tube worms. *Jack Donnelly, Woods Hole Oceanographic Institution.* **B.** Worms nicknamed "spaghetti," possibly acorn worms. *James Childress, University of California at Santa Barbara / Woods Hole Oceanographic Institution*

may not be common. Much is made of the presence of manganese nodules in the deep sea, but the depth of the water makes mining difficult, and the manganese content of the nodules is not always so high as it is in terrestrial ore bodies.

OCEAN FLOOR FEATURES

The ocean floor is far from being flat (Fig. 15-3). On a small scale, it probably has as many irregularities as do land areas. On a large scale, however, its features are fairly regular and predictable. The origin of many oceanic features will be discussed here, but the origin of those that are related to plate tectonics are given in Chapter 20.

Continental shelves

As we leave the beach and proceed seaward, we find a relatively shallow platform that sur-

Fig. 15-3 Relief map of the rugged ocean floor in the vicinity of Australia and Antarctica. A major ocean ridge lies between the two continents.

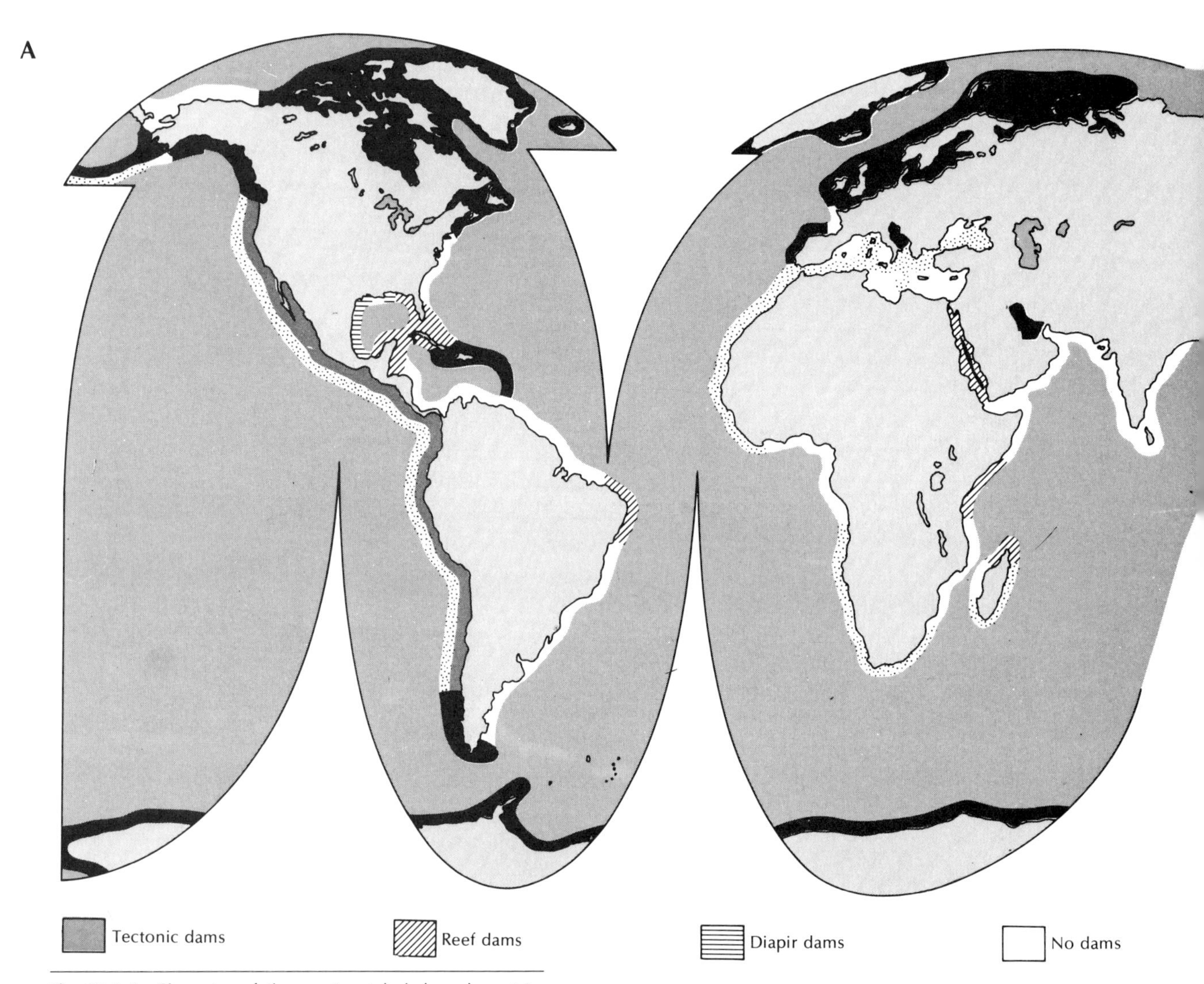

Fig. 15-4 A. Character of the continental shelves, by origin. **B.** Generalized cross sections across continental shelves of different origin.

rounds almost all the continents, the ***continental shelf***. Although it makes up only 7.5 per cent of the ocean floor, it is equal in area to 18 per cent of the earth's land area. The shelf has an average width of 80 km (50 mi), and can be up to 1500 km (932 mi) wide. It approximates a smooth, gradually sloping platform, but often has an irregular surface with depressions, hills, and terraces, showing as much as 100 m (328 ft) of relief. In formerly glaciated regions, glacial erosion and deposition give the shelf considerable relief. And, in areas of deep fiords, the

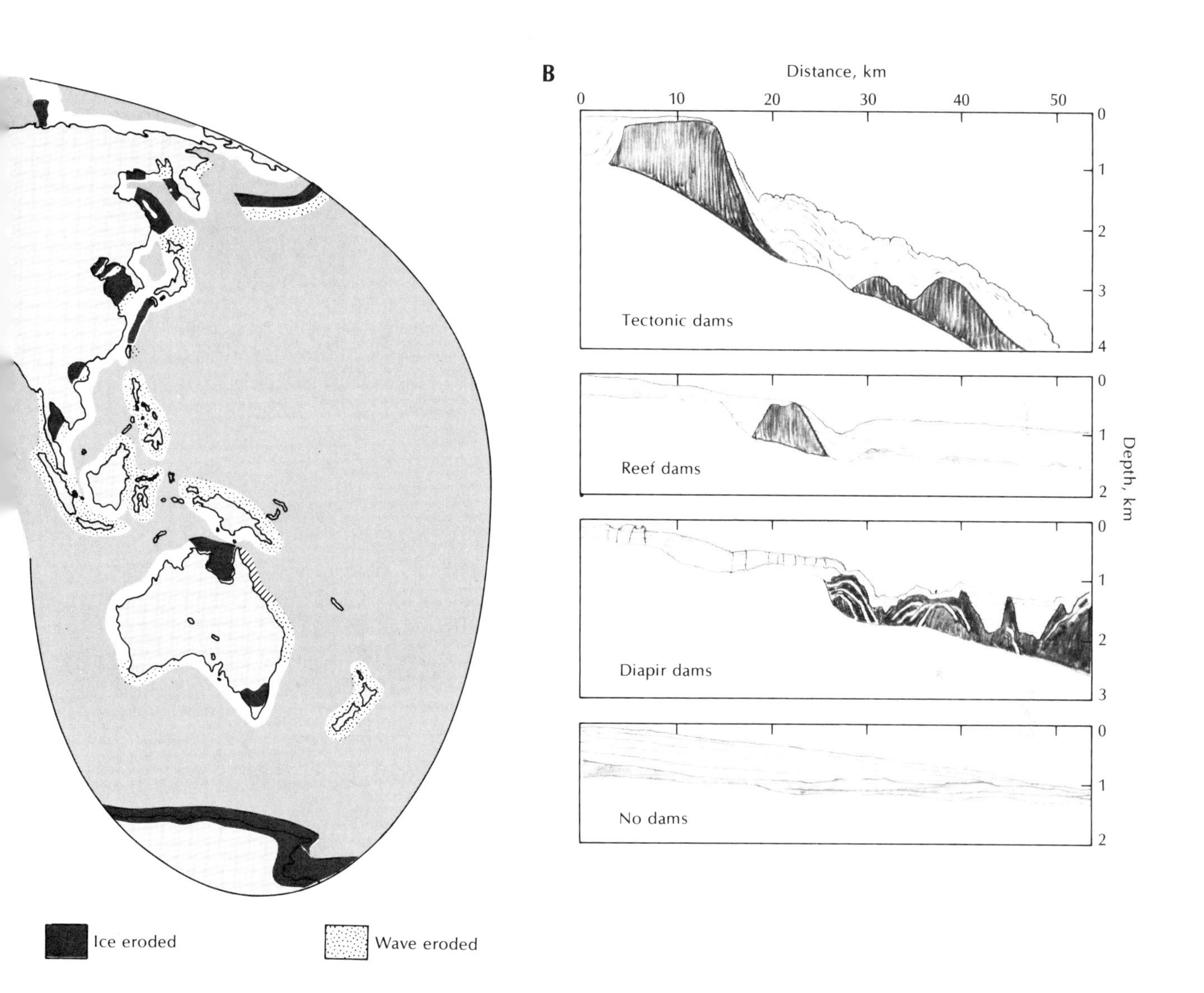

deepest portion of the shelf might even be the landward part. The outer edge of the shelf, called the ***shelf break,*** is characterized by a sudden steepening of the slope. The depth of the shelf break ranges from 20 to 550 m (66 to 1804 ft) and averages 133 m (436 ft).

Most continental shelves are composed of sediments and sedimentary rocks that are on the average of 2 km (1 mi) thick. The form of many shelves is related to the position and origin of various natural dams (Fig. 15-4). Common around the Pacific Ocean are tectonic dams, formed by long blocks of rocks that have been broken or folded, with some blocks pushed upward. The basins thus formed behind the dams are then filled by sediments from the continents. Sometimes parts of the dam rise above the surface of the sea as islands, such as the Farallons near San Francisco and the Channel Islands off southern California. At other places, chiefly the east coast of the United States, such dams are very ancient and sediments have not only filled the basins behind them, but have covered the dams as well. In such cases, and where there are no dams, the shelf sediments form a great wedge, oftentimes thickening in their seaward direction.

Off the coast in the western Gulf of Mexico, the dams are giant domes of salt, called ***diapirs,*** which have pushed up through overlying sedimentary rocks. In tropical waters, dams are often formed by outlying reefs of coral and marine microscopic plants (algae). That type of dam is also seen on shelves off the eastern Gulf coast and southeastern United States.

During the Pleistocene, when part of the ocean's water lay in great ice sheets on the land, extensive areas of the continental shelves were exposed to subaerial and glacial erosion. In places, the erosion exposed the tops of the dams, revealing igneous and metamorphic rocks that have not been covered to this day. Climate during the Ice Age fluctuated between colder and warmer. Those fluctuations, of course, were reflected in changing levels of the sea. When temperatures remained constant for any length of time, the waves cut cliffs and terraces in the exposed continental shelves. Such wave-cut features are now submerged, and some of them have since been buried by more recent sediments. Others, however, still persist and can be detected by echo-sounders. They provide us with clues to the earth's recent history. The fluctuating sea level also resulted in a complex pattern of sediment distribution on the shelf. In fact, in time, some of the Ice-Age sand and gravel beach deposits might become major sources of building materials, as more readily accessible sources on land are used up.

Continental slopes

The ***continental slopes*** extend downward from the shelf break to the deep-ocean floor. It should not be thought that the gradients of the slopes and shelves are very great, a conclusion that might be drawn from looking at typical drawings of cross sections of ocean basins, because such drawings have high vertical exaggeration. The average slope of the shelves is 0°07′, whereas that of the continental slopes is 4°17′ for the first 1825 m (3986 ft). Like the shelves, the slopes are not gradually descending, flat expanses, but have basins, hills, valleys, and canyons. In general, the slopes are a quite rugged transition zone, about 16 to 32 km (10 to 20 mi) wide, which connects the two main levels of the earth's surface—sea level, or close to it, and ocean bottom, 3660 m (12,000 ft) below sea level.

Submarine canyons

One of the most characteristic features of the continental slopes is the numerous valleys, or canyons, that have been cut into them. Some are caused by landslides and other gravity mass movements in the unstable materials of the slopes, and other follow breaks in the underlying rocks. Those valleys are among the most impressive features of the ocean floor and the most difficult to explain.

Submarine canyons are found throughout the world. Of those that can be recognized, none can be considered typical because so few have been thoroughly studied. Probably the best known is La Jolla Submarine Canyon and Fan Valley, with its major tributary, Scripps Canyon. Those canyons are located just off the coast of La Jolla, California, the site of Scripps Institution of Oceanography, and within a few miles of the Naval Undersea Research and Development Center. The canyons are of interest, also, as the location of the first dives by a submersible in the United States when the *Trieste* explored them in 1960. In addition, they head very close to shore, making their shallower portions, down to 60 m (197 ft), easily accessible to small boats and scuba-diving scientists.

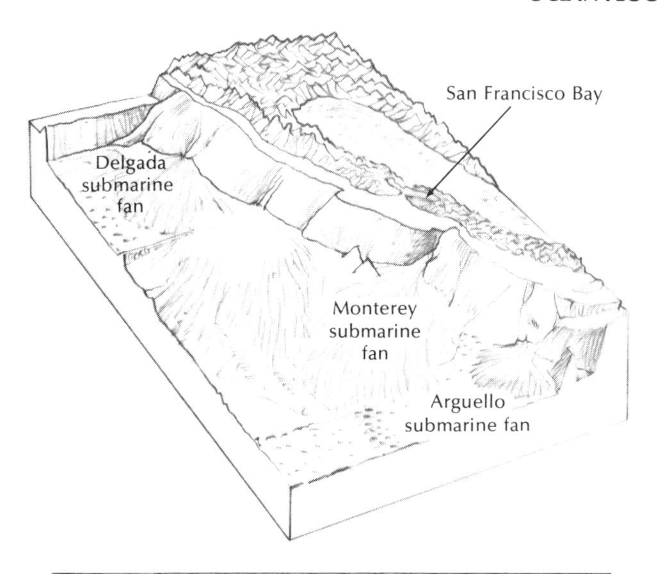

Fig. 15-5 Diagram of the coast and sea floor off central California, showing several submarine fans.

Deeper portions have been studied using Jacques Cousteau's *Diving Saucer.*

From the beach, a flat, gently sloping, sand-covered terrace extends seaward. About 213 m (699 ft) from shore, at a depth of 12 m (39 ft), the bottom suddenly drops away in the precipitous 24-m (79-ft) headwall of La Jolla Canyon. At the base of the cliff is the wide, bowl-shaped head of the canyon. Seaward, the valley widens and then narrows again until it is a rock-walled gorge, its steep sides covered with a lush growth of marine plants and animals. The bedrock of the canyon floor shows through its normal sandy covering, in some places; in others, there are thick mats of seaweed and other organic material. In places, progress down the canyon is by giant steps or terraces, caused by a slumping in the loose bottom sediments.

Partway down, Scripps Canyon joins La Jolla Canyon, and a glance through the submarine gloom reveals that its walls have been so undercut at their bases that they actually overhang as much as 6 m (20 ft). The height of the walls decreases gradually, and the valley widens until it is a little more than 1 km (0.6 mi) wide with an entrenched channel wandering across it. The walls are of semi-consolidated clay rather than hard rock, and there are natural levees on either side of the valley. We have entered the La Jolla Fan Valley, which is cut not in the La Jolla Terrace as was the canyon, but in its own fan-like deposits. Eventually it evens out until it merges with the sea floor. Such ***submarine fans,*** formed by sediment channeled down the canyon, are not uncommon, and if a number of canyons or other submarine valleys are spilling sediment out onto the ocean floor, a marine equivalent of an alluvial plain is formed (Fig. 15-5). That sediment-laden boundary of coalescent fans between the continental slope and the deep-ocean floor is called the ***continental rise.***

Larger canyons exist. One off the California coast—known as the Monterey Canyon—rivals the Grand Canyon of the Colorado River in its vertical dimension (Fig. 15-6). The Hudson Canyon is the most famous canyon off North America's east coast. It heads off Long Island, New York and trends south-eastward in a fairly straight line. It is about 915 m (3001 ft) deep. For world-record length, we look to the deep canyon off the mouth of the Congo River, for it extends 240 km (149 mi) offshore.

The origin of submarine canyons has been the subject of debate by geologists for decades, and it is not necessary here to go into the many theories that have been proposed, only to be discarded. Much work remains to be done before the solution can be found, but current thought seems to favor subaerial erosion as a starting point for many canyons. Such erosion would have commenced with the lowering of sea level during Pleistocene glaciations. Yet subaerial erosion could not carve out canyons as continuous features onto the deep-sea floor. If a canyon were to result solely from such erosion, sea level would have to drop to the level of the floor of the canyon. Past glaciations have removed some water from the oceans, but hardly enough to expose entire canyons.

The one exception is the Mediterranean Sea, where the submarine canyons seem to be subaerial. Six million years ago the Mediterra-

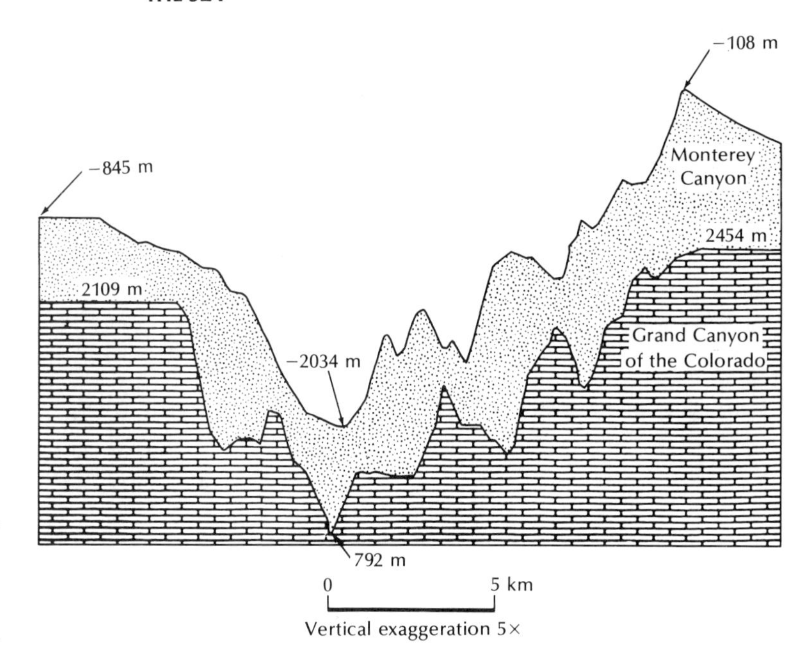

Fig. 15-6 Transverse profile across Monterey Canyon, off the California coast, compared to profile of the Grand Canyon of the Colorado River.

nean was isolated from the Atlantic Ocean. The sea quickly dried up, leaving a desert basin some 3000 m (9840 ft) below sea level. Rivers carved gorges into the sides of the basin, and the subaerial gorges became submarine canyons when the oceanic waters returned.

The consensus, then, is that in order to understand the origin of most submarine canyons, we need to know more about the processes of erosion in submarine regions. We know that strong bottom currents exist throughout the ocean and that they have been observed in canyons as well. Such currents can transport sand-size sediment. Marine organisms no doubt play their part in the mechanical and chemical disintegration of canyon walls, some of which are hard bedrock. Density currents of debris-laden water, set into motion by the slumping of oversteepened and unconsolidated material near the heads of canyons, are another force commonly called upon as an erosional agent. They are called ***turbidity currents,*** and their role in the erosion of submarine canyons is accepted by some, but not by others.

Turbidity currents, as their name implies, are clouded or muddy streams of moving water. They were first described scientifically in 1840 by Forel, a Swiss engineer, who noticed that the muddy waters of the Rhône disappeared where the river flowed into the clear waters of Lake Geneva, and reasoned, quite correctly, that the colder water of the Rhône, laden with glacial silt, was heavier than the water of the lake. Therefore, the river sank through the lake and flowed as an undercurrent along its bottom.

Much of the same thing happens where the muddy Colorado River flows into Lake Mead, which is backed up several kilometers behind Hoover Dam. A muddy current makes its way along the bottom of the lake to deposit sediment all the way to the power-house intake towers. The effectiveness of such a current apparently was not anticipated at the time the dam was designed, when it was assumed that much of the river sediment would be deposited as a delta at the upper end of the lake.

Such currents present two major problems: (1) the abrasive effect of the sediment particles on the turbines and (2) sediment transport across the entire length of the lake, which might seriously shorten its life expectancy since it will fill more rapidly than planned. A model of the lake was prepared in order to study the circulation of the silt-laden bottom current, and photographs show how the turbidity current glides along the reservoir floor below the less dense and clearer water of its upper levels.

That turbidity currents may exist in the sea was suggested strongly on the afternoon of November 18, 1929, when a severe earthquake shook the Grand Banks off the Newfoundland coast. Apparently the shock was violent enough to trigger what may have been a submarine landslide or slump, which was soon converted into a roiled cloud of suspended sediment that swirled down the continental slope. Such an event normally would go unnoticed were it not for the unique circumstance that the path followed by the earthquake-induced current was directly across one of the greatest concentrations of submarine cables in the world—one of the communication networks linking North America with Europe.

One after another the cables were broken that afternoon and night, and, because the time as well as the point of rupture could be determined electrically, the pattern of failure was determined (Fig. 15-7). It turned out that the cables broke progressively downslope, and when they were fished up for splicing, many of them were found to be chaotically snarled and jumbled, as though they had been thrust aside by a giant hand.

From the distance separating the individual breaks and from the time of failure it was estimated that, early on, the current had a velocity of about 20 m/sec (66 ft/sec) or more, but that the velocity gradually decreased with time until the last cable, located about 600 km (373 mi) from the epicenter, broke some 13 hours after the earthquake.

Whether the cable-breaking pulse traveling across the sea floor was a turbidity current or not probably never will be known. About all we do know from that episode and a few other episodes is that occasionally currents set in motion in the depths of the sea, for short periods, are capable of doing an immense amount of work and of shifting great volumes of sediment.

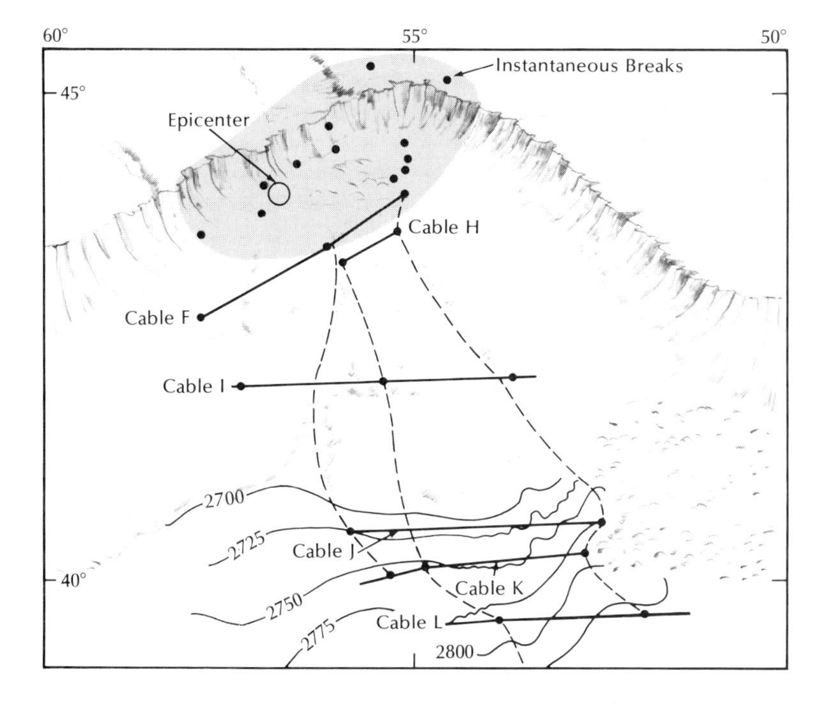

Fig. 15-7 The setting of the Grand Banks earthquake of 1929, showing the area in which the cables broke simultaneously, and the positions of parts of other cables that broke as the turbidity current progressed downslope. The dark circles are points of breakage, and the dashed lines are paths that parts of the turbidity current might have taken. The numbers on the contour lines in the lower part of the diagram represent water depth in fathoms (1 fathom = 1.8 m or 6 ft).

THE ABYSS, OR DEEP OCEAN

At the bottom of the continental slope in the depths of the sea, the ***abyss*** floor stretches from continental margin to continental margin. Some parts of the floor of the abyss are a dead-level plain and those so-called **abyssal plains** leave an unwavering smooth line on the fathometer, kilometer after kilometer. Some appear to be sediment covered, as is a broad expanse in the northwest Atlantic Ocean south of Newfoundland. Others, such as one in the Indian Ocean, seem to be the nearly level surfaces of vast sub-ocean floods of basalt, perhaps comparable to the Deccan lava plateau on land in nearby India. Hills, known as ***abyssal hills*** commonly rise above the plain, but remain well below sea level.

Other parts of the sea floor are more varied, and include such diverse forms as broad rises and plateaus that may have quite gentle slopes. Ridges are narrower features, with steeper marginal slopes and narrow crests. Many of them are submarine mountain ranges. Outstanding among them is the **Mid-Atlantic Ridge,** which stretches down the full length of the ocean and is practically equidistant from the continents on either side. Where the ridge pierces the surface of the sea its lofty peaks are the foundations of such islands as Ascension, the Azores, Iceland, St. Helena, and Tristan da Cunha. The highest of the latter is Pico in the Azores, which towers 2300 m (7544) ft) above sea level and stands on a base that rises 6100 m (20,008 ft) above the ocean floor.

A curious feature of the Mid-Atlantic Ridge is the **median valley,** an essentially continuous central depression, or **rift,** with a nearly level floor and steep sides that follow the crest. The deepest part of the rift lies some 1000 to 1300 m (3280 to 4264 ft) below the adjacent ocean floor. In some ways the trough resembles the rift valleys of eastern Africa.

Almost equally curious are the great ***trenches*** countersunk below the general level of the ocean floor. Like many mountain ranges, they are long, narrow features (Fig. 15-8). The more prominent ones are in the Pacific, and in general they are clustered along the Asian margin.

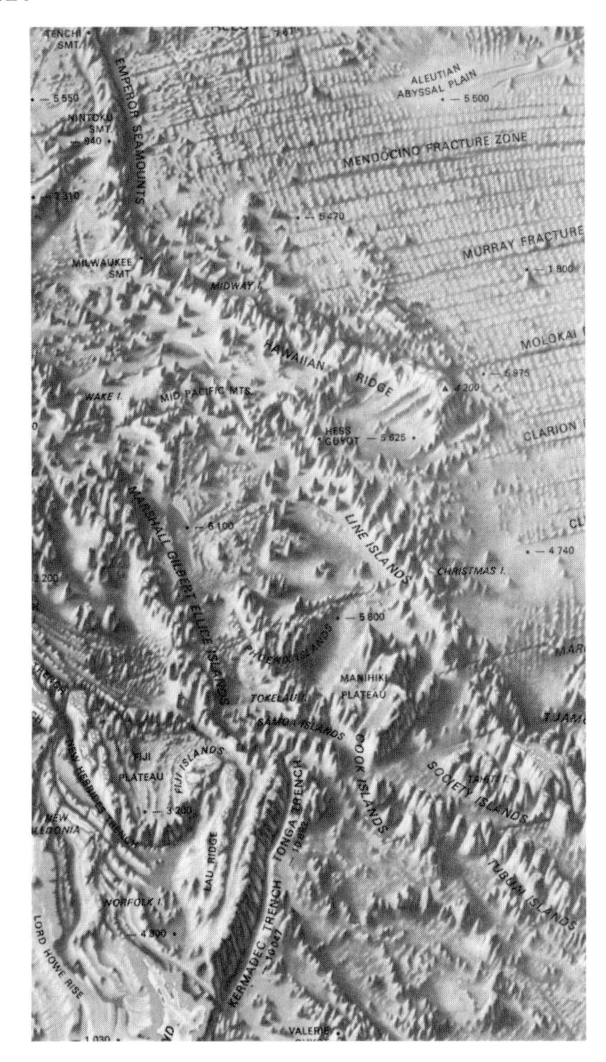

Fig. 15-8 Features of the ocean bottom in the western Pacific. Deep trenches are clustered along the Asian margin of the ocean.

Except for the trenches off the coast of Chile and Peru, such submarine wrinkles are associated with the chains of volcanic islands that festoon the sea between the trenches and the adjacent mainland, following roughly the same pattern as the **island arc.**

No deeper spots on earth exist than those trenches. The Mariana Trench off the Mariana Islands at 10,910 m (35,785 ft) and the Tonga Trench north of the Tonga Islands at 10,770 m

(35,326 ft) are the two deepest known. It is interesting to note that adventurers, in their efforts to get as far from sea level as possible, have always gone up—to the top of Mount Everest—when in fact the points farthest from sea level are down, deep under the sea.

The extent and pattern of the deep, elongate trenches on the ocean floor, especially in the Pacific, were recognized during World War II. With naval vessels continually crossing normally unfrequented waters, an enormous amount of fathometer records was accumulated. When the records were compiled, the great system of trenches paralleling the Mariana Islands—one of the world's immense submarine mountain chains—became clearly apparent. The work of surveying those hidden features has continued and slowly we are learning more about their shape, if not their origin.

Another discovery was that not all oceanic volcanoes are associated with trenches, island arcs, and ridges. Most such volcanoes, in fact, are not high enough to rise above the surface of the water. They are known as ***seamounts*** and have the familiar shape of land volcanoes of the intermediate type, with steep sides and a small summit area (Fig. 15-8). Generally they stand about 1 km (0.6 mi) above the ocean floor. The number of such submerged volcanoes is almost unbelievable. In 1964, the oceanographer Henry Menard estimated that there are 10,000 in the Pacific Basin alone, which makes them one of the ocean floor's most prominent features. Although some are isolated cones, many others occur in clusters and linear arrangements. All of those investigated thus far are of basaltic composition (Fig. 15-9).

Much more puzzling than seamounts are the features called ***guyots*** (Fig. 15-8 and 15-10). They, too, are volcanic mountains, but their truncated summits form a nearly level submarine plateau. Their strange name honors Arnold Guyot (1807–1884), a Swiss geographer and associate of Louis Agassiz, who came to the United States more than a century ago to teach at Princeton University. Guyots commonly rise to within 910 to 1520 m (2985 to 4986 ft) of sea level. Their surfaces typically consist of barren, planed-off volcanic rock, but many are covered with rounded boulders, which may have been shaped when the platform stood closer to sea level. Some guyots have been photographed, and rock samples have been dredged from them. Fossil coral belonging to the Cretaceous Period (more than 70 million years ago) has been found attached to some of the samples. In one guyot in the Pacific near Bikini, fossils of *Globigerina,* a minute, single-celled, surface-dwelling organism that lived during the Eocene Epoch (about 50 million years ago) had sifted down into cracks in the truncated summit.

How were the guyots' summits planed off to such remarkable uniformity, at such great depths? If we argue that they were beveled by wave action—which seems the most logical solution—then we are caught in the dilemma of believing either (1) that sea level has risen in the not-very-remote geological past or (2) that the guyots were carried down to their present depth by regional subsidence. Although the latter explanation is the more likely one, it raises many questions—such as the likelihood of great expanses of the ocean floor sinking and the mechanics of transferring huge quantities of material at subcrustal depths. Plate tectonics (see Chap. 20) can be used to explain the location of guyots far below sea level.

Before we leave the subject of submarine volcanoes, we should note their enormous size. We commonly conceive of terrestrial volcanoes as large features, but they are dwarfed by the size of many of their submarine cousins (Fig. 15-11).

CORAL ATOLLS

The origin of coral ***atolls,*** ring-shaped islands made up of the skeletons of marine animals, has been another subject of intense discussion among geologists (Fig. 15-12). Their appeal reaches back across the centuries, when the first Western European seafarers made their way into the distant reaches of the tropical seas. That is the world made famous by Herman Melville, by Robert Louis Stevenson, by the stirring events arising out of the conten-

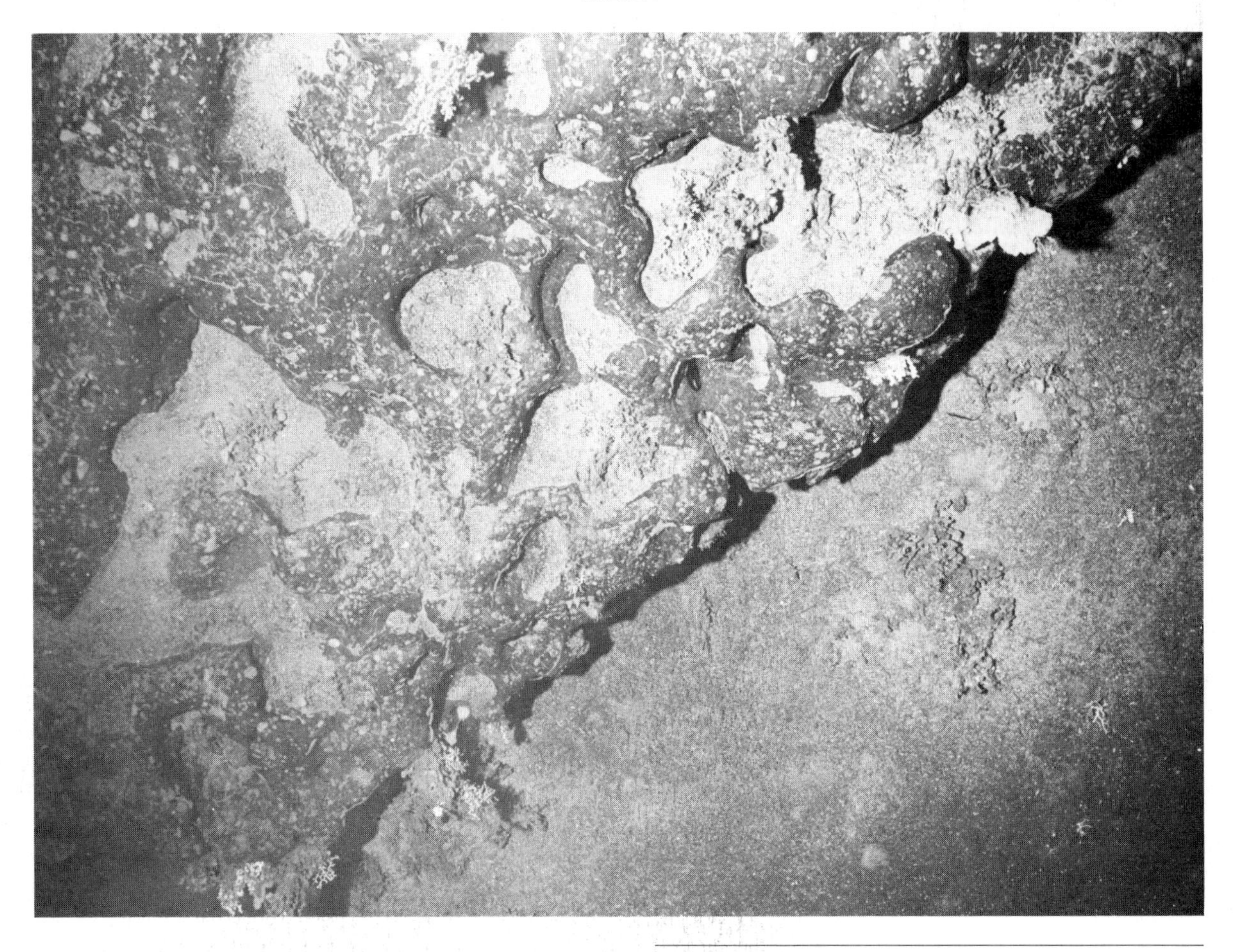

Fig. 15-9 Pitted lava with attached coral and bryozoan colonies forming the surface of the Kelvin Seamount, western North Atlantic, at a depth of 903 m (2962 ft). *Bedford Institute of Oceanography*

tious voyage of H.M.S. *Bounty,* and, in our time, by the naval conflict that surged back and forth across the vast domain of the Pacific in the years 1942–45. For those who are interested, the modern world of those far-distant islands is effectively described in the novels and essays of James Michener:

Scattered over a thousand miles of ocean in the eastern tropical Pacific, below the Equator, lies a vast collection of coral islands extending in a general northwesterly, southeasterly direction across ten degrees of latitude. Seventy-eight atolls, surf-battered dikes of coral, enclosing lagoons, make up this barrier to the steady westward roll of the sea. Some of the lagoons are scarcely more than salt-water ponds; others, like those of Rangiroa and Fakarava, are as much as fifty miles long by twenty or thirty across. The *motu,* or islets, composing the land, are threaded at wide intervals on the encircling reef. The smaller ones are frequented by sea fowl which nest in the pandanus trees and among the fronds of scattered coconut palms. Others, enchantingly green and restful to sea-weary eyes, follow the curve of the reef for many miles, sloping away over the arc of the world until they are lost to view. But whatever their extent, one feature is common to all: they are mere fringes of land seldom more than a quarter of a mile in width, and rising only a few feet above the sea which seems always on the point of overwhelming them.

A coral atoll, circular in form, subtended a shallow lagoon. On the outer edge giant green combers of the Pacific thundered in majestic fury. Inside, the water was blue and calm. Along the shore of the lagoon palm trees bent their towering heads as the wind directed, and after a thousand more years brown men in frail canoes came to the atoll and decided it should be their home.

The world contains certain patterns of beauty that impress the mind forever. They might be termed the sovereign sights and most men will agree as to what they are: the Pyramids at dawn, the Grand Tetons at dusk, a Rembrandt self-portrait, the Arctic wastes. The list need not be long, but to be inclusive it must contain a coral atoll with its placid lagoon, the terrifyingly brilliant sands and the outer reef shooting great spires of spindrift a hundred feet into the air. Such a sight is one of the incomparable visual images of the world.

Probably relatively few people realize that coral reefs once excited the scientific world as greatly as they fired the thoughts of the more romantically inclined literary figures of the nineteenth century. Few environments provide a setting more unlike that of the Western world nor a way of life more different than that of the sea-roving Polynesians.

Early in the nineteenth century the more expansionist powers fitted out a number of exploring expeditions, and in many cases the expeditions included a naturalist among the ship's company. The status of such persons in that nautical world was never high, but we can be grateful today that a few dedicated men blazed the scientific trail for us. Charles Darwin (1809–82) will always stand chief among them. As a young man of twenty-two he was assigned as a naturalist to H.M.S. *Beagle*, a 10-gun brig of about only 230 metric tons, scarcely 30 m (98

Fig. 15-10 Perspective diagram of Bikini Atoll and Sylvania Guyot adjoining it on the west.

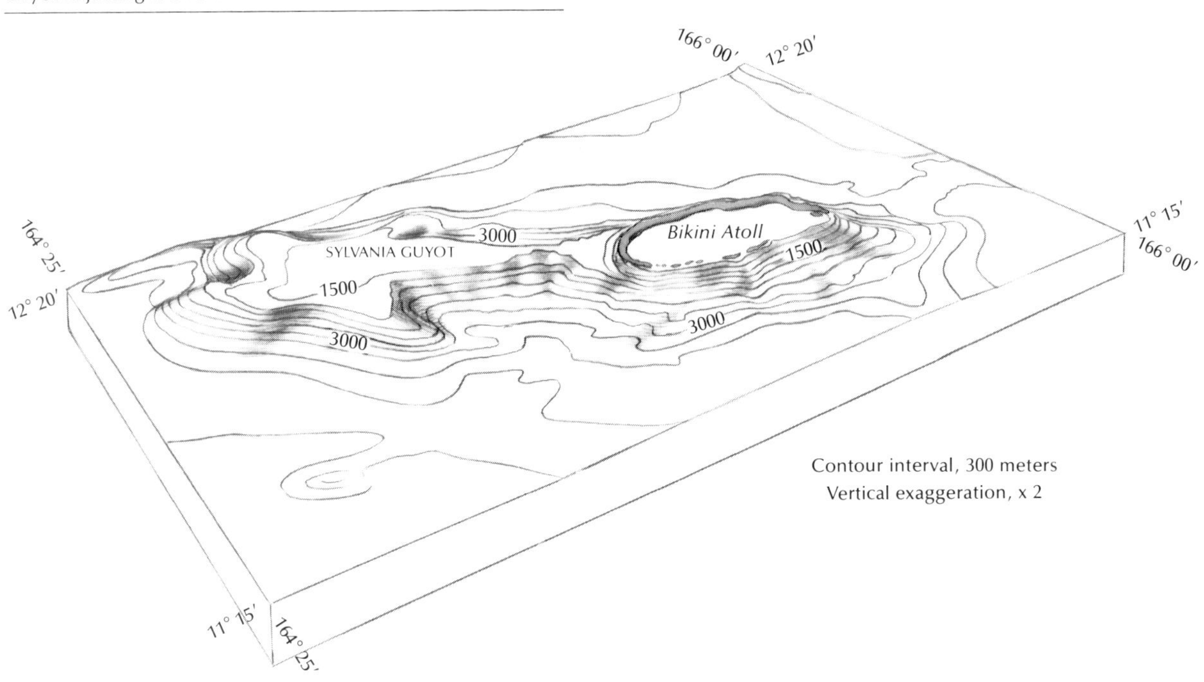

ft) long, carrying 74 persons. The ship circumnavigated the world, and Darwin used both his geological and zoological observations on that voyage in formulating his theory of evolution published in his *Origin of Species*.

Few people are aware that Darwin's scientific training had been in geology as much as in any other branch of science, and that on the voyage of the *Beagle* he made great numbers of observations on rocks, volcanic features, and fossils. Among the many wonders he beheld, few aroused his interest more than the fairyland world below the sea created by the corals. Though he actually saw only a few reefs, Darwin had the insight to recognize that there were three major kinds: (1) fringing reefs, (2) barrier reefs, and (3) atolls; and that the three were related to each other in a logical and gradational sequence.

Darwin believed that the succession from one reef type to another could be achieved by the upgrowth of coral from a sinking foundation, such as a subsiding volcano. As long as the rate of coral growth was more rapid than the rate of sinking, Darwin argued, the progression would be from a fringing reef through the barrier stage, and, with the disappearance through subsidence and erosion of the central island, only a reef-enclosed lagoon, or atoll, would survive (Figs. 15-13 and 15-14).

Fringing and barrier reefs are not as difficult to explain—even such imposing ones as the Great Barrier Reef of Australia—as are the atolls. Some atolls are very large—Kwajelein in the Marshall Islands of the South Pacific is 120 km (75 mi) long and averages as much as 24 km (15 mi) across. Most are far smaller. Nearly all true atolls are low—few have a freeboard of much more than 3 m (10 ft). The reef ring consists of solid coral and over it the sea sweeps freely at high tide, with telling effect during storms. The rest of the time the reef may be just barely awash. In the ring are the reef islands—in a sense they are like sand dunes—which are composed of thousands upon thousands of small fragments of ground-up coral, algae, pieces of shell, and the like. On these are rooted the pandanus and the coco palm—lifegivers of the Polynesian world.

In that two-dimensional world, one is always aware of the encircling sea. No place on earth could be more vulnerable to the unbridled sav-

Fig. 15-11 The sizes of various volcanoes, on land and at sea.

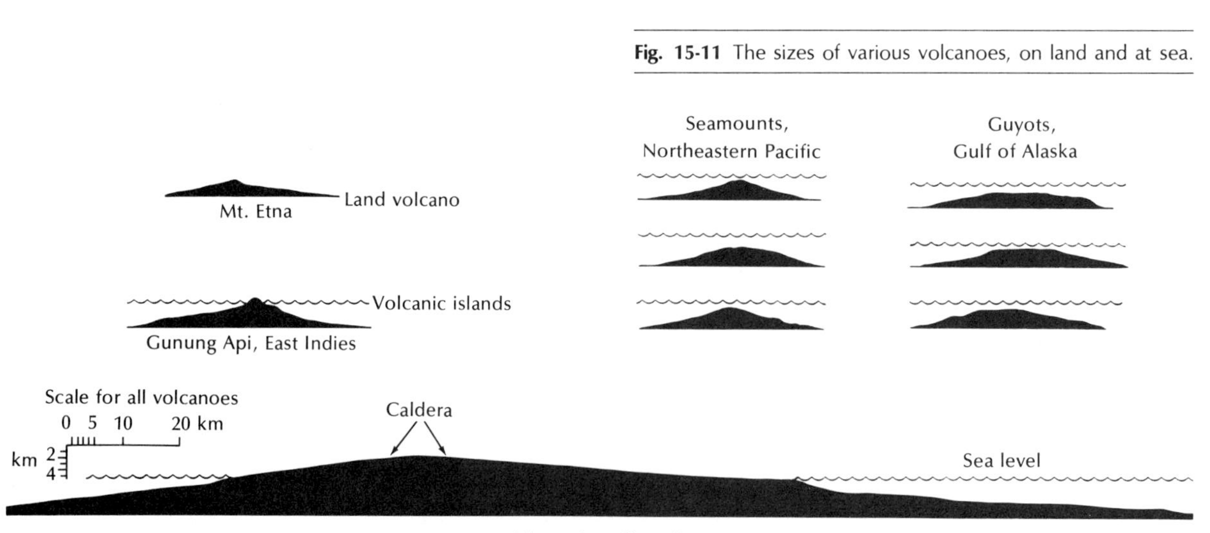

Fig. 15-12 Atolls of the Tuamotu Archipelago in the southern Pacific Ocean. *NASA*

Fig. 15-13 (Overleaf) Barrier reef encircling a volcanic island, an intermediate stage in the formation of an atoll. *Office of Naval Research*

agery of a hurricane (Fig. 15-15), its impact on such a defenseless world vividly described by Nordhoff and Hall in their novel *Hurricane:*

The very land itself had all but disappeared beneath the seas that swept the islet from the north. In the swiftly changing light, now bright, now dim, only the trees could be seen, and the church with its gleaming white walls. It gave the impression of sinking slowly, as the seas, a fathom and more deep, swept around it, meeting beyond in great bursts of spray. Then it would appear to rise a little, as though buoyant, to meet the onset of the next wave. And all this while, above the tumult of the hurricane, we could hear, at times, the faint clanging of the bell, tolled by the wind. Its tones, remarkably sweet and clear, reached us as faintly, almost, as imagined sound. It was the voice of that night, in so far as things human were concerned, and a desolate voice it was!

By the middle of the afternoon, we knew that we had seen the worst, and at nightfall the hurricane left us, moving southward like the monster it was, in search of new lands to lay waste. The stars came out in a cloudless sky, and when the moon, one night past the full, rose, its mild light revealed as pitiable a scene of desolation as could have been found the world over. One might well have said that Manukura had ceased to exist. The village islet, certainly, had been destroyed as a habitation for man. From our mound of rocks we looked down upon . . . upon . . . what shall I call that moon-blanched corpse of an island? It bore no resemblance to the place we had known. Nothing remained to show where the village had been. The sea had half devoured the land itself, and what had been one islet was now two, divided by a channel swept clean to the reef bed, and a full fifty yards wide.

Research since Darwin's time generally has upheld his ideas on the formation of atolls. If, for example, the reef foundation was sinking as the reef built up, then atolls should be underlain by hundreds of meters of coral, all of which grew in shallow waters. Darwin realized that drilling a hole through a reef would be the surest test of his theory, as he suggested in 1881 in a letter to Alexander Agassiz, an American oceanographer and the son of Louis Agassiz.

Several atolls were drilled prior to World War II, but results regarding their origin were not conclusive. But, as part of the environmental studies made in the Marshall Islands in connection with atom-bomb testing, the U.S. Navy drilled a series of deep holes. A total of five were put down on Bikini in 1947, three shallow and two deep, the latter to 404 and 767 m (1325 and 2516 ft). Even the deepest was coralline material all the way. Of three deep holes drilled in 1951–52 on nearby Eniwetok, the two deeper, 1266 and 1389 m (4152 and 4456 ft), went completely through a 1200-m (3936-ft) cap of shallow-water reef limestone and bottomed in a typical mid-Pacific volcanic rock, basalt. The age of the fossils in the basal limestone is Eocene, which meant that Eniwetok atoll is at the top of a coralline accumulation that has grown upward during the last 60 million years. The rate of subsidence does not appear to have been constant, but ranged between perhaps 15 m (49 ft) and 51 m (167 ft) per million years, slowing down later in the history of the reef.

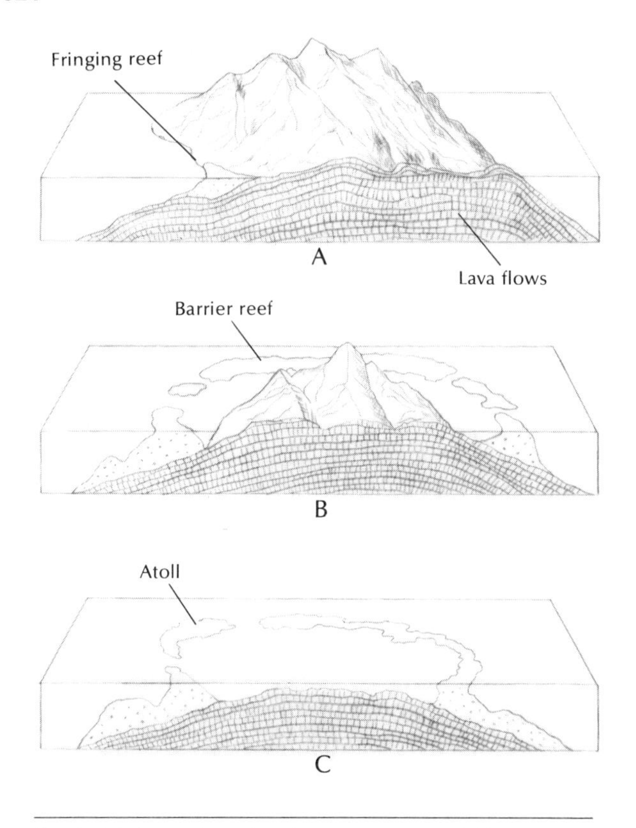

Fig. 15-14 Succession of reef types around a sinking volcanic island, from a fringing reef to a barrier reef to an atoll.

Thus, some of the events ushering in the nuclear age, a little more than a century after Darwin saw his first atoll, serve to vindicate his altogether remarkable insight. There can be little doubt that at least two of the great atolls of the western Pacific, in the Caroline and Marshall Islands, resulted from the upgrowth of reef organisms on a slowly sinking foundation.

The same nuclear age, however, has brought difficult times to the atolls and their inhabitants. Because of their isolation and low populations, atolls were favored locations for testing atomic weapons, and now some are being studied as possible depositories of nuclear wastes. Again, one of nature's most delicate and beautiful ecosystems is being threatened.

Another complicating factor in atoll formation, or for that matter in the formation of reefs in general, is that of changing sea level with glaciations. In a period of glaciation, when the sea level was at least 100 m (328 ft) below the present level, the reef would be exposed to weathering and battering by waves. In an interglaciation however, sea level rises and reefs build upward. If that cycle is run many times, as has happened over at least the past million years, the formation of these islands becomes quite complicated, especially the formation of the top 100 m (328 ft) or so.

Many other perplexing features have been observed, as increasing numbers of oceanographic vessels cross the seas of the world and as the technology of marine surveying improves. Great escarpments, several of them more than 1600 km (994 mi) long, interrupt the sea floor in places. Four of them, roughly equally spaced, strike almost due west from the coasts of California and Mexico into the Pacific, at least to the longitude of Hawaii. Their relationship to geological structures on land is difficult to determine. Rather than being a continuation of the prominent land fracture, the San Andreas fault, the two off the California coast intersect the fault very nearly at right angles. One of the escarpments off the coast of Mexico possibly may be a seaward continuation of the line of Mexican volcanoes aligned along the nineteenth parallel. That volcanic system includes not only the lofty cones of Popocatépetl and Ixtaccihuatl outside Mexico City but also the recently active volcanoes of Parícutin and of the island San Benedicto off the coast.

OCEAN SEDIMENTS

A great variety of sediments blankets the ocean floor. Most people are familiar with the sand

Fig. 15-15 A storm in the late 1970s sent water and fragments of the coral reef across Majuro Atoll, Marshall Islands, Micronesia. Damage was severe, but there were no deaths. *Judy Knape*

and gravel of beaches. Seaward of the beach, silt and clay mantle the continental shelf, but what of the material that covers the deep floor of the sea?

Many of the descriptions of sediment found on the floor of the abyss are taken from the reports of the H.M.S. *Challenger* expedition of a century ago (Fig. 15-16). The *Challenger* left England in December 1872, making an epic oceanographic study of the Atlantic, Pacific, and Southern oceans before returning to England in May 1876. Six scientists gathered data to better understand the physical, chemical, and biological properties of both the ocean water and the sediments that blanket the ocean floor. Today we know that the pattern of the sea floor, although it may resemble the one worked out by the men of the *Challenger* in broad outlines (Fig. 15-17), is far more complex.

Fig. 15-16 H.M.S. *Challenger* passing by an iceberg.

Fig. 15-17 **A.** Distribution of sediment types on the sea floor. **B.** Sediment thickness. **C.** Rate of sediment accumulation.

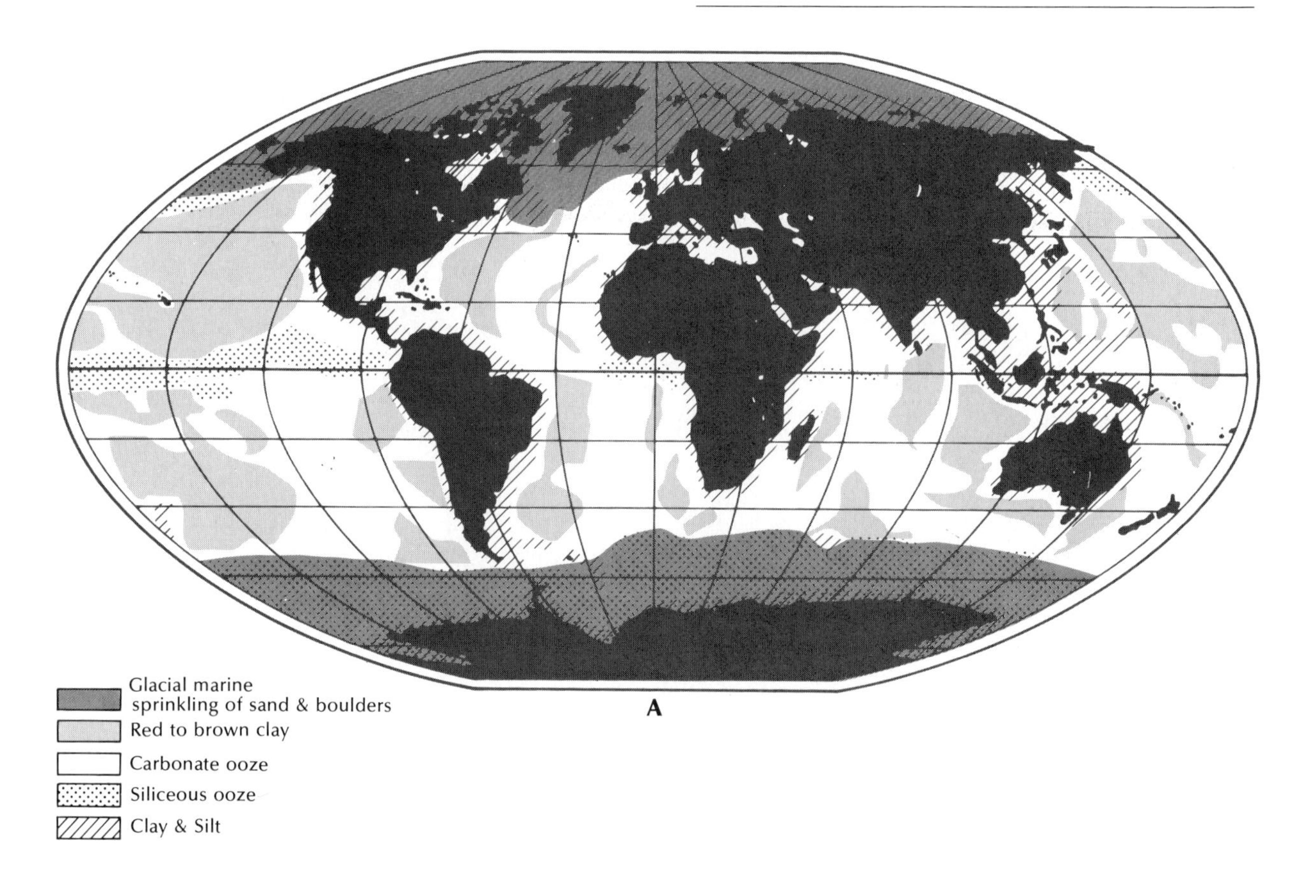

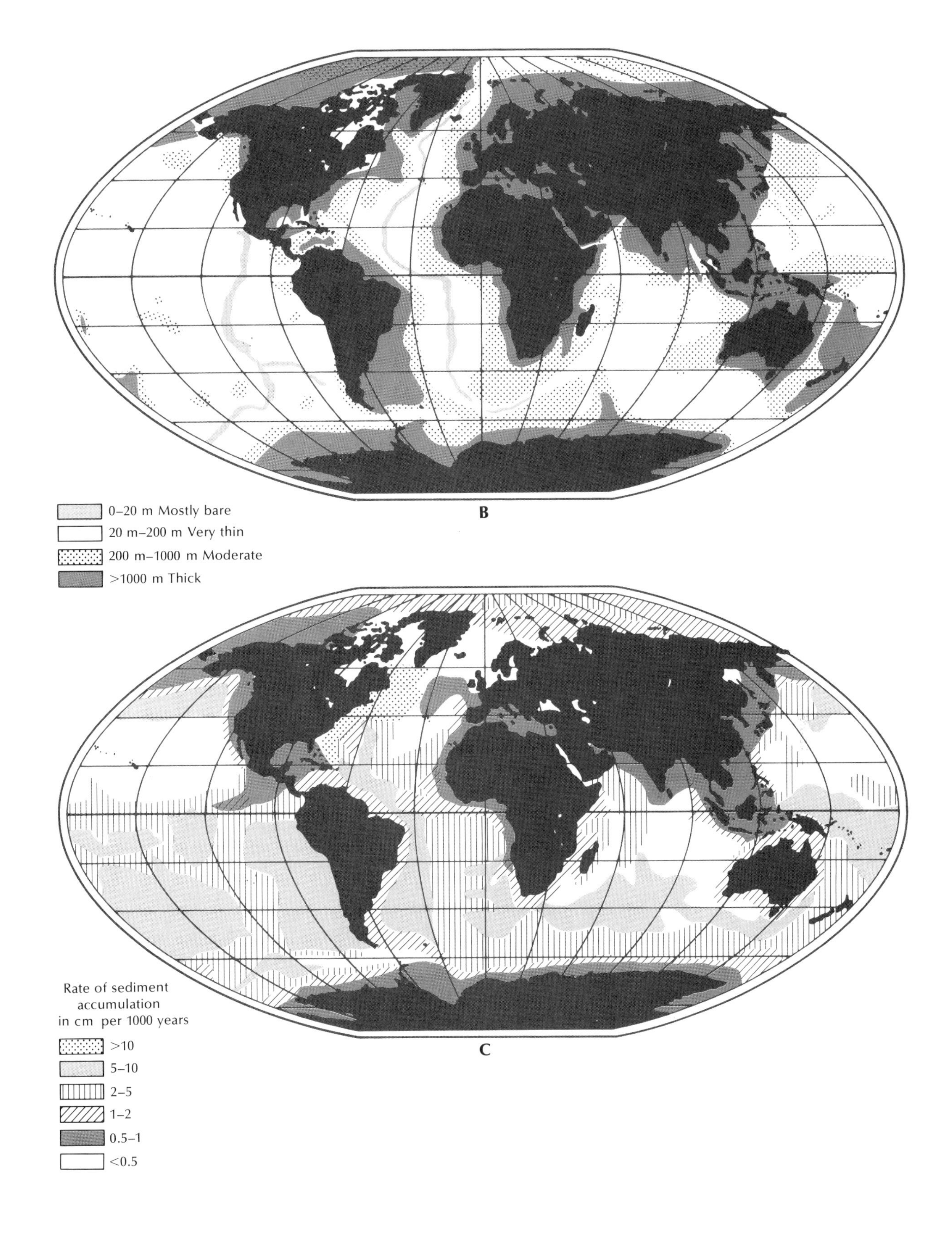
0–20 m Mostly bare
20 m–200 m Very thin
200 m–1000 m Moderate
>1000 m Thick
B
Rate of sediment accumulation in cm per 1000 years
>10
5–10
2–5
1–2
0.5–1
<0.5
C

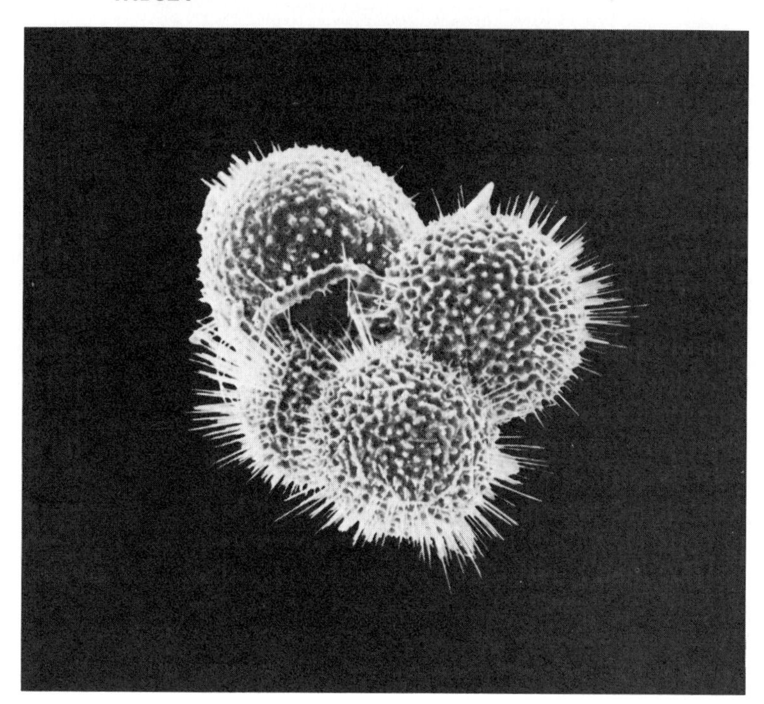

Fig. 15-18 A. Scanning electron photomicrograph of the Foraminifer *Globergina,* as it appears floating in the ocean waters. Once the shell is shed, the spines are lost, and as a fossil it looks similar to other Foraminifera. *American Museum of Natural History*

In brief, the *Challenger* men demonstrated that generally on the deep sea floor between the lower part of the continental slope and a depth of around 4 km (2.5 mi) the most abundant sediment is ***ooze.*** That wonderfully descriptive term gives an impression of what organic flour is like. Photographs of ooze, as well as verbal descriptions of it by bathyscaphe divers, suggest that it looks like an ivory-colored blanket. The slightest disturbance sends a dust-like cloud swirling up through the dark water, making us realize how powdery it must be. Ooze does not form in place on the sea floor, but accumulates as the result of a gentle, unceasing "snowfall" of the remains of microscopic, free-floating surface organisms. When they die, they sift down from the sunlit surface to the lightless floor of the sea. It is only because the terrestrial supply of sediment is so slight that such organic deposits can build up.

Not all oozes are the same, however; their composition varies systematically across the ocean floor. Much of the abyssal plain beneath tropical and warm-temperate seas, in depths of less than about 4000 or 4500 m (13,120 or 14,760 ft), is carpeted with microscopic shells of **Foraminifera,** a single-celled creature that secretes calcite (Fig. 15-18A). Like so many single-celled creatures, they do not die, but reproduce themselves by division. That is, one organism divides to make two, each of which grows a new shell, and the original shell, now vacated, is cast off to sink to the bottom of the sea to form **calcareous oozes.** As the shell drifts downward to great depths, however, generally below 4000 to 4500 m (13,120 to 14,760 ft), the temperature and chemical conditions of the water are such that the carbonate dissolves, and calcareous ooze does not form.

In other parts of the ocean, ***siliceous oozes,*** which consist mainly of ***radiolaria*** (the coarse-meshed animals in Fig. 15-18B) and ***diatom*** remains, predominate. They are minute animals and plants, respectively, that secrete a silica cell. The siliceous oozes are found in areas of either high productivity of such organisms combined with low productivity of the carbonate forms, or in places where the water depths are so great that the carbonate forms dissolve completely. Although siliceous forms also dissolve in sea water, their rate of dissolution is less than that of carbonate forms—thus, they can survive as fossils where carbonate forms cannot.

Siliceous oozes consisting mainly of diatoms are abundant in a broad band in the colder seas marginal to both the Arctic and Antarctic. Those cold-water-inhabiting plants thrive in the frigid seas in uncountable multitudes, and in a sense compose the true pastures of the sea. It is curious that the largest creatures on earth, the Arctic whales, should depend for sustenance on the humble diatom, among the smallest of living things.

Like many land flora, diatoms grow during part of the year, attain maturity, and then decline. Often when they die, they appear to perish in hordes. The water in such protected inlets as fiords then turns green and soupy looking, and oars and boats are immersed in such a flood of organic matter that they become green and slimy and acquire a distinct fragrance.

A unique sedimentary accumulation is the monotonous soft red to brown clay that carpets the deeper parts of the oceans. In terms of area, it is the most widespread of all deposits of the earth. Curiously, it is almost wholly in-

Fig. 15-18 B. Middle Eocene ooze (45 million years old) from a Deep Sea Drilling Project core in the western Indian Ocean. Most of the fossils are radiolaria, but a few are Foraminifera (F) and sponge spicules (S). *Scripps Institute of Oceanography*

organic, save for such exotic souvenirs of marine life as the ear bones of whales and the teeth of sharks—both relatively insoluble in sea water. The fact that some of the sharks' teeth are those of species long dead, and known to us only as fossils, is evidence of the extraordinarily slow rate at which the clay accumulates.

The clay is extremely fine grained, and much of it comes from land. Transported to the sea by rivers, waves, or the wind, it was then carried far and wide by ocean currents. It sifted down, particle by particle, through the lightless depths to accumulate gradually on the bottom of the sea. Volcanic and cosmic dust also contribute to the carpet of clay, but they prob-

Fig. 15-19 Boulders on the sea floor at a depth of 3562 m (11,683 ft) in the Drake Passage between South America and Antarctica. *Smithsonian Oceanographic Sorting Center*

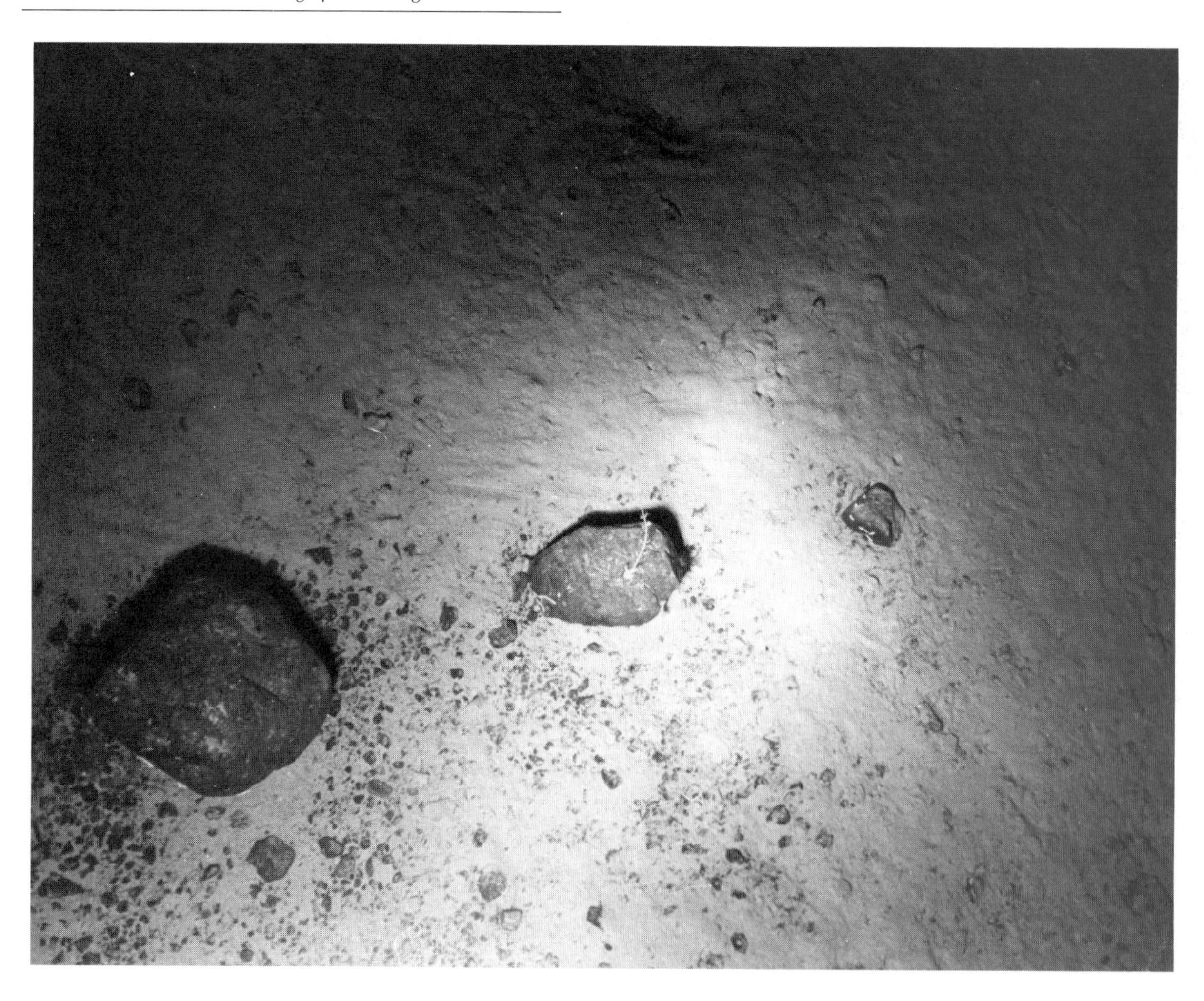

Fig. 15-20 Iceberg with boulders carried piggyback fashion. As the iceberg rotates or melts, the boulders and sands rain down on deep sea floor thousands of meters below.

ably only make up a small proportion of the whole.

As one might expect, the thickness of sediments varies from place to place on the ocean floor (see Fig. 15-17B), as does the rate of deposition (see Fig. 15-17C). One reason the thickness varies is that the ocean floor varies in age from very young over the oceanic ridges, to progressively older proceeding away from the ridges (see Chap. 20). The rate of deposition varies with proximity to continents, erosion rate on the continents, and the biological productivity for the organic oozes. Add to that the prevailing ocean currents and the solution of the carbonate or silicate remains as they settle to the deep bottom, and one can see that the rate is difficult to calculate.

In places, relatively coarse-grained sediments are found on the abyssal plains. For example, it is not uncommon to find broad, fanlike expanses of sand spreading out across the ocean floor from the continental margins. Such sands are found in regions where ooze normally would be found, and often contain the remains of organisms which, to the best of our knowledge, live in shallow waters near the margins of the ocean basin. It is likely that many such sediments were deposited by turbidity currents moving off the continental slope or down submarine canyons (see Fig. 15-5).

Perhaps more anomalous is the presence of land-derived boulders far from the continental shores (Fig. 15-19). No ocean current could have moved rocks of such size by lateral transport into their present position. Rather, they seem to have fallen straight down to the ocean floor, after having been carried far offshore by icebergs (Fig. 15-20). Concentrations of such stones are found around glaciated areas and areas of pack ice, and the resulting sediment is described as ***glacial marine*** (see Fig. 15-17A).

GLACIATIONS AND INTERGLACIATIONS—THE STORY FROM THE OCEAN DEEP

Recent intensive study of cores from the floor of the sea has caused scientists to alter their ideas on the numbers and timing of Pleistocene glaciations. How was their thinking changed? A sample core is a historical record of events in earth history, with the youngest events recorded near the top of the core and the older ones at the base. Further, many scientists think that the deep oceans were repositories of sediment throughout the Pleistocene, and even farther back in time, and that no erosion occurred in such submarine areas. Thus, the record preserved by the sea is nearly complete.

What do ocean sediments tell us about glacial periods? For one thing, some cores show

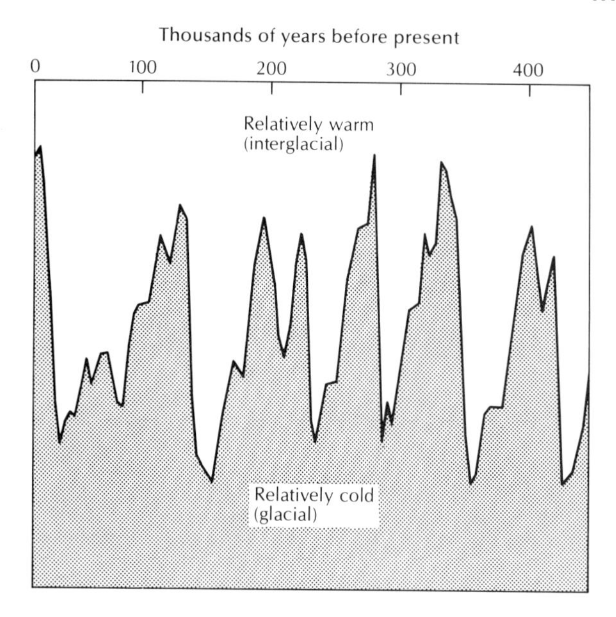

Fig. 15-21 Chemical data on fossil shells of foraminifera can be used to help date the time of glacials and interglacials in the past. Here the data from a core brought up from the floor of the Caribbean Sea are interpreted.

glacial-marine deposits interbedded with normal marine deposits. In other cores, the Foraminifera population changes in number, in species, and in chemical composition with depth. Such marine life seems to be sensitive to either the changes in ocean water temperatures, or changes in water composition that accompany climatic change, and the stockpiling of water on land as ice.

When data are carefully analyzed, one can see a procession of alternating warm and cold intervals that somehow must be related to glaciation and interglaciations on land (Fig. 15-21). The revelation of more glacial periods than had hitherto been recognized on land was fairly revolutionary. From Figure 15-21, for example, one can postulate six major glaciations in little more than the past 400,000 years. How are the six to be correlated to the traditional four recognized in the land record? Nobody has come up with an answer acceptable to the majority of earth scientists, but surely some of the land record has been lost by erosion. The other interesting aspect of the curve in Figure 15-21 is its sawtooth shape, which can be interpreted to mean that ice sheets take a long time to build up—say some 50,000 years or more—but that their disappearance by melting verges on the catastrophic—it is all over in as little as 10,000 years.

Much work still needs to be done. We have to learn to read the deep-sea sediment record better. We also need to correlate more carefully the oceanic record with the glacial record on land, which will require more thorough field work. Also, from studying the relationship between warm and cold intervals and periods of glaciation, as shown in Figure 15-21, we might well be able to predict future glaciations, the climatic conditions signaling their onset, or the catastrophic warming conditions that accompany their disappearance. The diagram also reveals that the interglacials are of relatively short duration. Could we be nearing the end of a warm period?

SUMMARY

1. The ocean floor consists of several major topographic features. The *continental shelf* forms the relatively shallow, gently sloping platform found off the coast of most continents. The shelf gives way seaward to the *continental slope,* which extends down to the ocean floor. Submarine canyons, some larger than any on land, are cut into both the shelf and the slope in places. Major features of the deep ocean floor are ridges and deep trenches, both of which form prominent topographic lineations and have their origin in plate tectonic activity (see Chap. 20).
2. Volcanoes are a prominent feature of the ocean. Some rise to the surface as islands, others are submerged and are called *seamounts,* and some submerged ones have flat tops and are called *guyots.* The flat tops of the guyots are generally ascribed to erosion when the sea level was lower than at present. Plate tectonics (see Chap. 20) can

be used to explain their present position far below sea level. *Atolls* consist mainly of the buildup of coral limestone deposits on slowly subsiding volcanoes.

3. Deep ocean sediment consists mainly of the fine-grained *oozes* of calcareous or siliceous shell grains (the remains of minute animals and plants) and clay. In places, nodules of ore minerals are present. In other places, gravel clasts rest upon the fine-grained ocean bed far from shore. They probably were carried far offshore on icebergs and deposited when the iceberg melted.
4. The chemical and biological properties of the oozes vary with depth in ocean cores, and the variations provide important clues as to the number of glaciations and interglaciations in the Pleistocene. There are more such periods recorded in the ocean sediments than in the record of terrestrial glacial tills, as the latter are presently interpreted.

QUESTIONS

1. Describe the processes that helped shape the continental shelves.
2. What are the arguments against submarine fans being submerged subaerial features?
3. What are several processes that would reduce the height of the volcanic portion of an island, relative to sea level, in progressively going from an island with a fringing reef to an atoll?
4. What are the major factors influencing the rate of deposition of terrestrially derived sediments in a deep ocean basin?
5. Why are we looking to the oceans for data on past glaciations, and what kinds of data are used?

SELECTED REFERENCES

Heezen, B. C., and C. D. Hollister, 1971, The face of the deep, Oxford University Press, New York.

Menard, H. W., 1964, Marine geology of the Pacific, McGraw-Hill Book Co., New York.

The Ocean, a Scientific American book (1969), W. H. Freeman and Co., San Francisco.

Thurman, H. V., 1975, Introductory oceanography, C. E. Merrill Publishing Co., Columbus, Ohio.

van Andel, T., 1977, Tales of an old ocean, Stanford Alumni Association, Stanford, California.

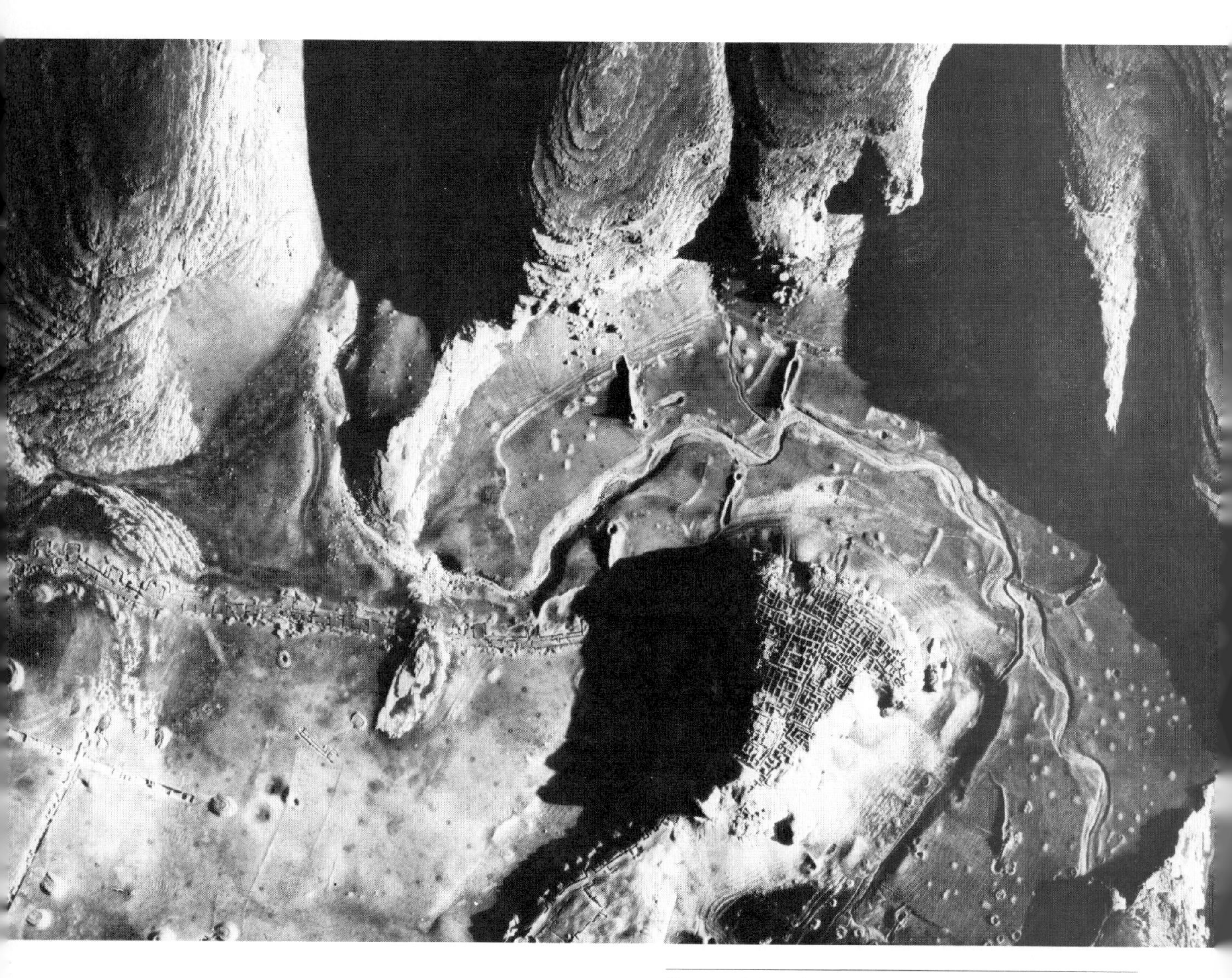

Fig. 16-1 Ruins of an ancient city along a dry river bed in Iran. The circular mounds mark the tops of the vertical shafts of **kanats,** ancient water transportation systems. *The Oriental Institute, University of Chicago*

16

GROUNDWATER

> In Xanadu did Kubla Khan
> A stately pleasure-dome decree
> Where Alph, the sacred river, ran
> Through caverns measureless to man
> Down to a sunless sea.

Coleridge's verse, quoted above, reflects the remarkable image that many people have of water within the earth. Some are prone to speak glibly of underground rivers flowing for miles beneath the parched surface of some of the world's most absolute deserts, and to many of us springs are nearly as mysterious as they were to people long ago.

Springs played a leading role in Greek and Roman mythology. An example is the spring of Arethusa, which appears on the island of Sicily in the ancient harbor of Syracuse. The river god Alpheus, in pursuit of the wood nymph Arethusa, flowed as an underground river all the way from Greece to Sicily, where he finally caught the elusive spirit and changed her into a spring.

Comparable beliefs were held in those early days. Generally, there were two leading schools of thought. One held that springs drew their water from the sea—how the salt was eliminated and how the water was elevated to the great heights it reached in mountain springs remained unanswered questions. The other belief was that springs and streams had their own origin within subterranean caverns, large enough perhaps to have atmospheres of their own from which water condensed as a sort of rain to feed them. Aristotle (384–322 B.C.) was content with that idea, since rainfall was obviously inadequate:

> The air surrounding the earth is turned into water by the cold of the heavens and falls as rain, [so] the air which penetrates and passes into the crust of the earth also becomes transformed into water owing to the cold which it encounters there. The water coming out of the earth unites with the rain water to produce rivers. The rainfall alone is quite insufficient to supply the rivers of the world with water. The ocean into which the rivers run does not overflow because, while some of the water is evaporated, the rest of it changes back into the air or into one of the other elements.

Seneca (3 B.C.–A.D. 65) gave the seeming death blow to any concept so preposterous as one that water in the ground had anything to do with rain:

> Rainfall cannot possibly be the source of springs because it penetrates only a few feet into the Earth whereas springs are fed from deep down. . . . As a diligent digger among my vines I can affirm my observation that no rain is ever so heavy as to wet the ground to a depth of more than ten feet.

It was not until the mid-seventeenth century that two Frenchmen, Pierre Perrault (1608–80) and Edme Mariotte (1620–84), demonstrated that there was a relationship between rainfall and the discharge of springs. The approximate relationship of evaporation from the sea to the amount of rainfall and runoff was worked out by the astronomer Edmund Halley—known to us for the comet named after him. According to the geologist Oscar Meinzer (1876–1948):

> Perrault made measurements of the rainfall during three years, and he roughly estimated the area of the drainage basin of the Seine River above a point in Burgundy and of the run-off from this same basin. Thus he computed that the quantity of water that fell on the basin as rain or snow was about six times the quantity discharged by the river. Crude as was his work, he nevertheless demonstrated the fallacy of the age-old assumption of the inadequacy of the rainfall to account for the discharge of springs and streams. Perrault also exposed water and other liquids to evaporation and made observations on the relative amount of water thus lost. He also made investigations of capillarity, established the approximate limits of capillarity in sand, and showed that water absorbed by capillarity cannot form accumulations of free water at higher levels.
>
> Mariotte computed the discharge of the Seine at Paris by measuring its width, depth, and velocity at approximately its mean stage, making the velocity measurements by the float method. He essentially verified Perrault's results. In his publications, which appeared after his death in 1684, he defended vigorously the infiltration theory and created much of the modern thought on the subject . . . he maintained that the water derived from rain and snow penetrates into the pores of the earth and accumulates in wells; that this water percolates downward till it reaches impermeable rock and thence percolates laterally; and that it is sufficient in quantity to supply the springs. He demonstrated that the rain water penetrates into the earth, and used for this purpose the cellar of the Paris Observatory, the percolation through the cover of which compared with the amount of rainfall. He also showed that the flow of springs increases in rainy weather and diminishes

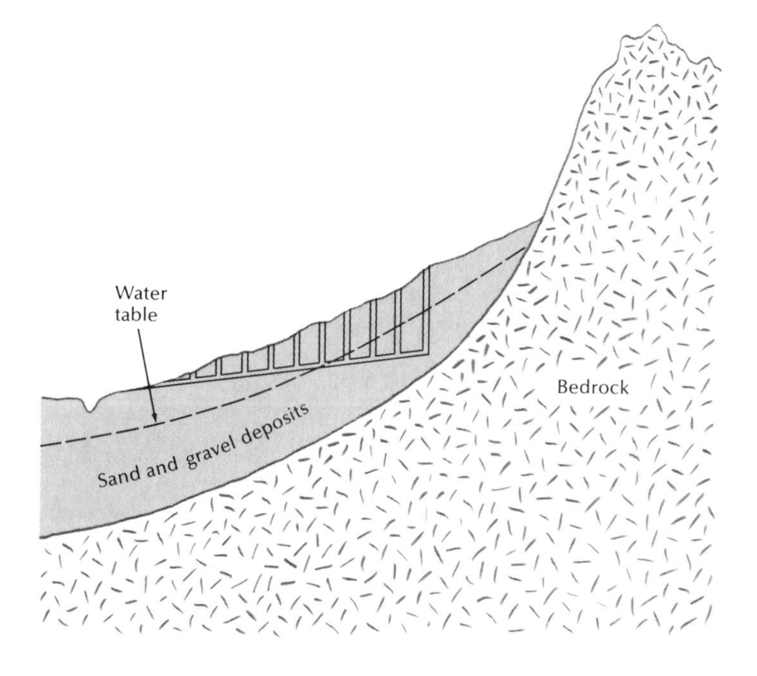

Fig. 16-2 Cross-sectional diagram of a kanat system in Iran. The vertical shafts and near-horizontal tunnels intersect the water table in the alluvium flanking the mountains, and the tunnels guide the water to the surface, where it is used for irrigation.

in time of drought, and explained that the more constant springs are supplied from the larger underground reservoirs.

In ancient times, considerable use was made of water from the ground. Wells were the center of village life for centuries and still are over much of the world. Not only were wells the focal point of village life, they were absolutely essential to the survival of a walled city or castle. It would be a foolhardy baron who attempted to hold off a siege without an intramural source of water.

The demand for water created by large concentrations of people in urban centers has resulted in a more extensive development of underground sources of water than most people realize. Almost everyone knows of the heroic measures the Romans took to conduct water to their cities by building imposing, valley-spanning aqueducts. Oddly enough, the Romans knew little about water in the ground. They placed their chief dependence on springs and streams with the result that they went prodigious distances to the Apennines for water when they had a perfectly adequate supply almost directly underfoot had they dug for it.

But other ancient people did dig for water, and their underground pursuit led to the construction of remarkable burrow-like excavations. The chief example are the ***kanats*** of ancient Persia, now Iran (Figs. 16-1 and 16-2). They center largely around Teheran and, for the most part, are dug in the gravels of the great apron of alluvial fans at the base of the Elburz Mountains. The kanats are long passages that serve as collection galleries in the porous gravels of the fan. The longest known one is 80 km (48 mi) long, and individual tunnels may be as deep as 300 m (984 ft) below the surface. In the old days they were truly multipurpose structures because they served as a source of drinking water and as a means of sewage disposal. In general, a kanat follows a water-bearing layer of sand or gravel within the fan and every few hundred yards is connected with the surface by a shaft sunk during construction.

In Egypt, a remarkable water-collecting tunnel system was built around 500 B.C. Constructed in the Nubian sandstone, it gathered water that was probably introduced into the rock as seepage from the Nile. All told, the tunnel system has a length of 160 km (99 mi) or so, although no one can say with certainty because it has almost entirely caved in. Water still escapes from the tunnel entrance, which was once thought to be a spring. Actually, that extraordinary system was so important that the temple of Ammon was dedicated in its honor.

Not only did people of long ago drive tunnels to intercept water in the ground, but they drilled wells to surprising depths. An outstanding achievement among dug wells was one at Orvieto, in Italy, which was sunk to a depth of 61 m (200 ft) in 1540. Two spiral staircases lined the walls, one above the other, with one being used by descending, the other by ascending, water-bearing donkeys.

Wells were drilled to great depths, as deep as 1500 m (4920 ft) in China, for example. Deep wells drilled at Artois in France in the twelfth century and at Modena, in the Po Valley of Italy, flowed water at the surface, exciting great interest, since they were the first true artesian wells of medieval times.

In this chapter we will discuss the origin and occurrence of groundwater, the ways to prospect for it, and the problems of contamination in underground reservoirs, and such features as hot springs and geysers, which are a groundwater phenomenon as well as an energy resource (see Chapt. 21).

ORIGIN OF GROUNDWATER

Nearly all the water in the ground comes from precipitation that has soaked into the earth. Additionally, some water is included with marine sediments when they are deposited, and some reaches the upper levels of the crust when it is carried there by igneous intrusions, volcanoes, and hot springs. In practical terms, however, the latter sources provide only a minor part of the total budget of usable water in the ground.

Many things happen to water that falls as rain or snow, as we saw in the discussion of the hydrologic cycle (see Fig. 11-2). For the United

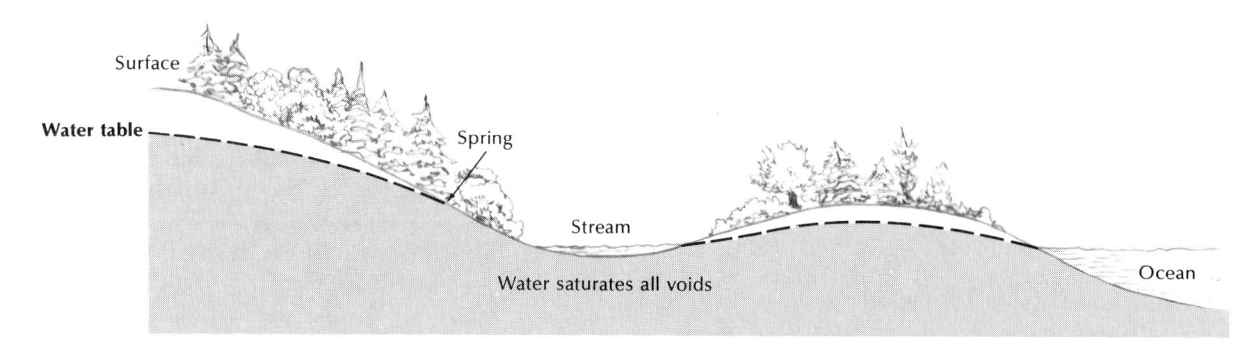

Fig. 16-3 Position of the water table in a humid region.

States, earth scientists Luna Leopold and Gordon Wolman have estimated that water is divided up as follows: There is an average of 76 cm (30 in.) precipitation each year. Of that amount, approximately 53 cm (21 in.) is returned directly to the atmosphere by evaporation and transpiration. Only 23 cm (9 in.) runs off in streams directly to the sea, and of the total runoff, nearly 40 per cent escapes by the Mississippi River—an impressive fraction of the continental supply.

Where does groundwater come from, then, if the budget balances as closely as the above figures indicate? The answer is that although the amount of water entering the ground by infiltration is slight—perhaps as little as 0.25 mm (0.01 in.) per year in some places, more in others—with the passage of many millennia, great quantities of water slowly accumulate in the ground. It is that vast reservoir built up gradually over thousands of years that we draw on today—unfortunately, in some areas, more rapidly than it is replenished. Where the latter situation exists, for example, in the Tucson-Phoenix area in Arizona and in parts of the Great Valley in California, the groundwater can be considered a non-renewable resource that will be depleted some time in the future.

The total amount of groundwater in the world has been estimated at some 8.4 million km^3 (1.8 million mi^3). A glance back at Figure 11-2 indicates that the amount is more than that in the rivers and lakes at any one time, but less than that in the oceans or glaciers. It is that great readily available abundance that makes groundwater so important to the development of farms, communities, and industries.

Fig. 16-4 Position of the water table with respect to influent and effluent streams. Arrows depict the flow of water beneath the water table.

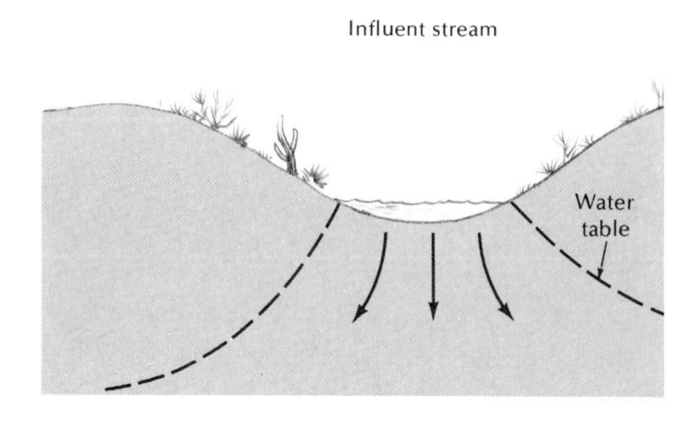

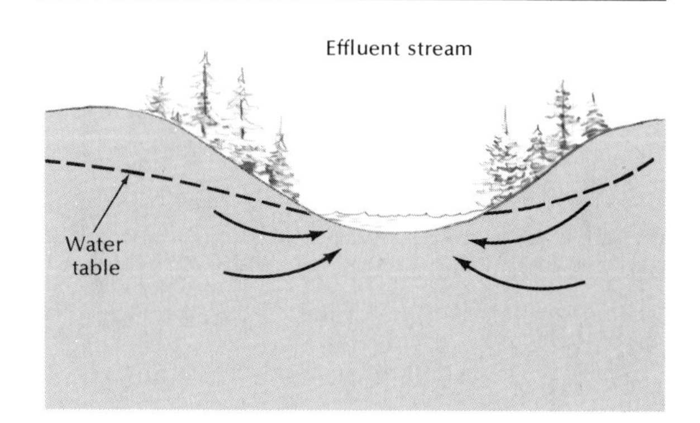

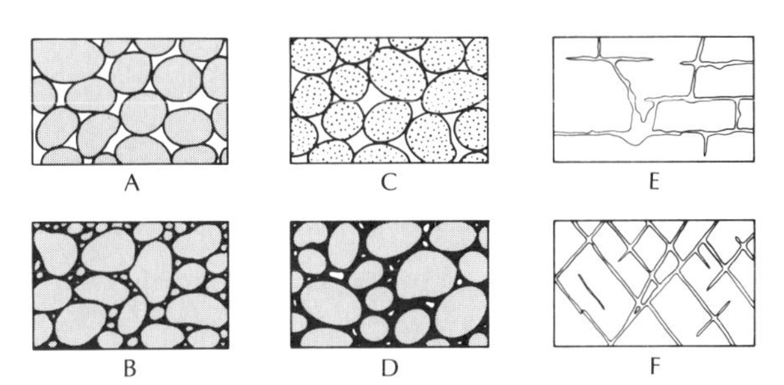

Fig. 16-5 Degrees of rock porosity. **A.** Highly porous, well-sorted sandstone. **B.** Poorly sorted sandstone of low porosity. **C.** Highly porous, well-sorted pebble conglomerate made up of porous sandstone pebbles. **D.** Well-sorted sandstone of low porosity (cementing material fills much of the void space). **E.** Limestone rendered porous through solution. Some of the voids could eventually grow into large underground caverns. **F.** Dense igneous rock rendered porous by joints.

OCCURRENCE AND MOVEMENT OF GROUNDWATER

Probably many of us have only seen groundwater standing in shallow wells—very often green, scummy, and unappetizing. If we were to determine its level and then to compare that with the level at which water stands in nearby wells, we would quickly discover that in many regions the water surface is at the same level. The surface at which water stands in wells is called the ***water table.*** All the voids, or openings, in rocks below the water table are filled with water, or are saturated; this area is called the saturated zone. Above the water table, in the zone of ***aeration*** or ***vadose zone*** (from the Latin, shallow), the pore spaces in the ground may range from completely dry to partially full. In many places, the water moves downward through the zone of aeration to the water table. Such movement, however, is quite slow.

Actually, the water table is rarely dead level. Instead it is more likely to be a blurred replica of the ground surface (Fig. 16-3), rising under hills and sinking under valleys. It intersects the surface at lakes and streams and also at springs. Sometimes it adds water to a stream, especially if the stream is at a low elevation (Fig. 16-4). This situation is common in humid regions, and the stream is then called an ***effluent stream.*** If the stream flows above the water table and thus adds to the supply of water in the ground, it is an ***influent stream.*** The latter are common in arid regions. The position of the water table related to an effluent stream is more or less stable. That beneath an influent stream, however, is apt to fluctuate. It will intersect the surface when the rivers are flowing, but will drop below the surface if they run dry.

A closer look at the zone of aeration shows that it is actually made up of three zones: (1) the ***zone of soil moisture,*** (2) the ***intermediate zone,*** and (3) the zone encompassing the ***capillary fringe.***

The zone of soil moisture is the portion of the profile most familiar to us. It is the ground layer that becomes wet after a rain or when a lawn is watered. When completely saturated, it may become a quagmire; at other times, it may be bone dry and dusty.

Commonly there is a lower margin to the surface zone of soil moisture. It may be only a few centimeters down, or several meters. When we dig into the ground it generally becomes drier, until the soil is no longer moist—as Seneca believed it to be. Typically, though, in that intermediate belt the water percolates slowly downward through openings until it reaches the water table. How well or how rapidly it percolates depends largely on the ***porosity*** and ***permeability*** of the ground. What those two terms mean will be described a few paragraphs later.

Extending a short distance upward from the water table, the capillary fringe is comprised of thread-like extensions of water that have migrated upward in the minute passageways between individual soil grains. The movement occurs in about the same way that kerosene climbs the wick of a kerosene lamp or that water rises in a very narrow glass tube. The fringe is usually less than 4 m (13 ft) thick.

Contrary to what many people think, the zone of high water content under the water table does not continue indefinitely downward. In other words, drilling a well to greater depths will not necessarily increase the flow of water. With increasing depth, the pore spaces in the rocks close up, and their water-bearing capac-

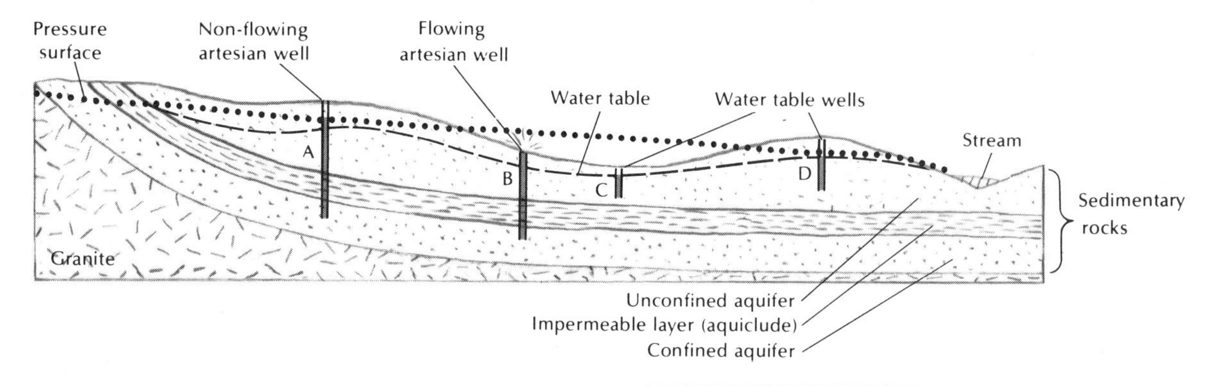

Fig. 16-6 The two main kinds of aquifers, the geological setting necessary for artesian wells, and an aquiclude.

ity diminishes until they may become completely dry. The upper levels of deep mines may require constant pumping to keep them from flooding, whereas the lower levels may be so dry that water has to be brought down for use in drilling.

Porosity

Porosity is of the greatest importance in controlling the movement of water in the ground. We are familiar with the general meaning of the word when we think of a porous substance as one that contains many holes. Actually, porosity is expressed as the percentage of the total volume of the rock that is occupied by openings. If one-half the volume of a rock is taken up by openings, the material has a porosity of 50 per cent; if only one-quarter is, then its porosity is 25 per cent, and so on.

Many factors determine the porosity of a rock. In clastic sedimentary rocks, the packing arrangement is important, but so, too, is the degree of sorting, the porosity of the clasts, and the amount of cementing material in the void spaces. Limestone itself may be quite dense, but masses of it are porous because dissolution of the rock can result in large cavities and thus in high porosity. The porosity of igneous and metamorphic rocks is largely determined by joint frequency, because the rock itself is so dense (1 per cent, or less, porosity).

It is important to note that grain size does not influence porosity in clastic sediments. BB's or basketballs, if packed in the same manner, would have identical porosities. In fact, relatively fine-grained materials, such as silt, may have higher porosities than such seemingly open material as gravel. Among the most highly porous materials are newly deposited muds, such as those of the Mississippi Delta. They may reach the incredible value of 80 to 90 per cent porosity—they contain so much water that individual particles scarcely touch one another. Quicksand is another example. The porosity of most materials, however, is less than 15 to 20 per cent. Various degrees of rock porosity are shown in Figure 16-5.

Permeability

Permeability is a measure of the capability of a rock to transmit a liquid through it. Therefore the actual size of the openings in a rock is much more impórtant than the percentage of open space. For example, a silt or clay may have a higher porosity than a gravel, but since the void spaces are so small the permeability is less. Large void spaces obviously are more permeable than small ones. Also important are connections between the openings. If the pore spaces in a rock do not communicate, the water will not flow, no matter how large those spaces may be.

Aquifers

Not all rocks are equally permeable, nor do they all have the same capacity to hold water. A permeable, highly porous sandstone layer, for example, may not only be able to hold much more water than its enclosing rocks, but it also may provide a route along which groundwater moves with relative freedom. Such a layer that readily yields water to a well is called an ***aquifer.*** In contrast, a layer that is too impermeable or too tight to accept water, such as one high in silt and clay, is appropriately called an ***aquiclude.***

There are two common kinds of aquifers (Fig. 16-6). One is an ***unconfined aquifer,*** which may be nothing more than a surficial layer of permeable sand or gravel. The other is a ***confined aquifer,*** a layer of permeable sandstone between layers of impermeable rock. Typically, a sandstone aquifer may crop out in a band paralleling a mountain front, dip below the adjacent plain, and flatten as it extends farther away from the mountain. Such an aquifer, called the Dakota Sandstone, dips east of the Rocky Mountains and under the Great Plains in the Dakotas and Colorado (Fig. 16-7). The first

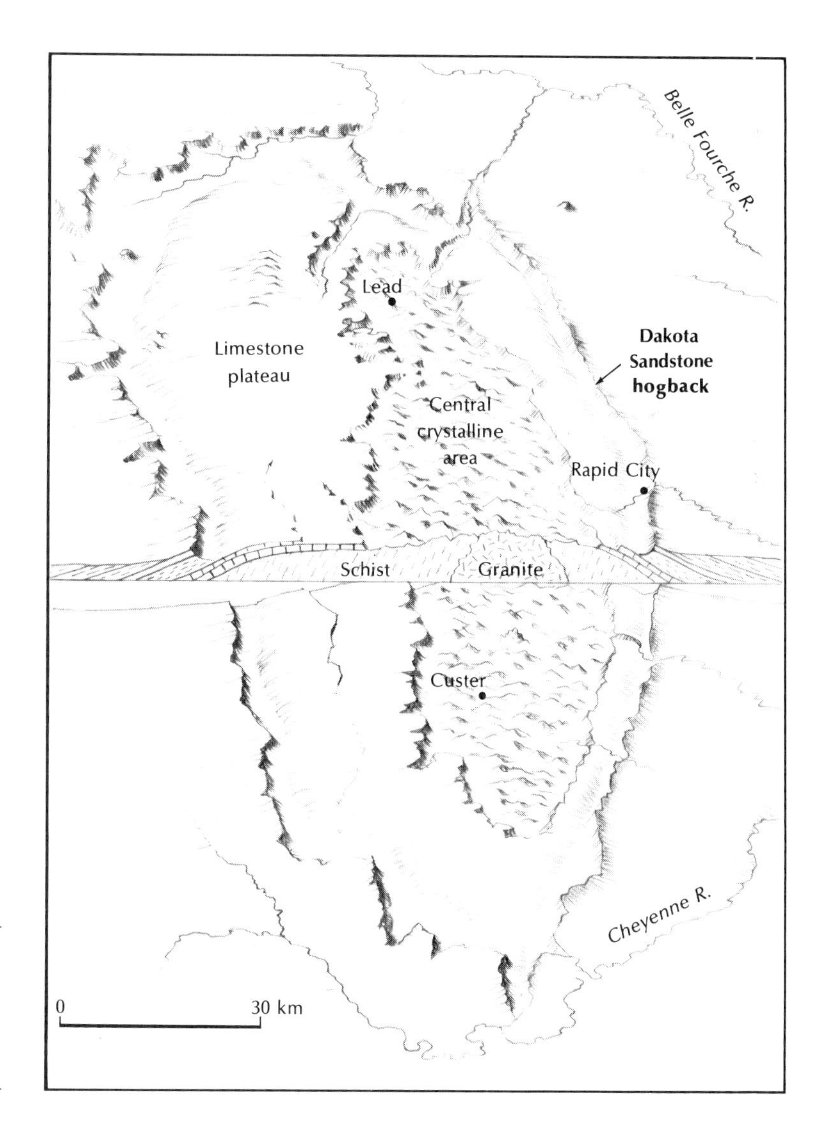

Fig. 16-7 Diagram of the Black Hills and the surrounding plains. Water that falls in the mountains enters the Dakota Sandstone, a major aquifer, and slowly travels along that confined aquifer to great depths beneath the dry plains.

wells were drilled into that sandstone in the 1880s, and since then it has yielded a prodigious quantity of water.

Water wells

Wells are constructed to tap such underground supplies of water as those described. In an unconfined aquifer, the water level in the well lies at the water table (see Fig. 16-6, wells C and D). The water level varies, however, as a result of precipitation patterns as well as pumping practices. Those who have lived on a ranch dependent on a well for irrigation water, for example, are fully conscious of the fact that when the well is pumped the water level in it drops. A short time after pumping ceases, the water rises, although not always to the same level, should the amount of water removed be exceptionally large.

How much does a single well affect the water table of an entire district? Do the water levels in adjacent wells rise or fall in concert? The answers to such questions have been established through observation in many localities over many years. If a well is pumped heavily, and water is taken out of the ground faster than it can be replenished, then the water table is pulled down in the form of an inverted cone centering on the well, known as a ***cone of depression.*** Obviously, the water level in nearby wells will be affected more drastically than that in more distant ones. Studies show that the effect of an individual well may be seen in others over distances of as much as 0.4 km (0.25 mi). Hard pumping in many wells will make the rims of individual cones overlap, until the water level of an entire basin is lowered.

Some wells that tap confined aquifers flow at the surface of the ground (Fig. 16-8). They are called ***artesian wells,*** from the Roman province of Artesium, now Artois, in southern France. To almost everyone, the term artesian means a well that flows freely. Yet in practice the word has a more restricted use, and is now applied to a well in which the water is under pressure because a confined aquifer has been penetrated.

Whether or not water reaches the top of the ground depends on the relationship of the ***pressure surface*** and the shape of the terrain (see Fig. 16-6). The pressure surface is the level to which water rises in a confined or unconfined aquifer. Theoretically, in a confined system it equals the highest point in the aquifer. However, the pressure surface does not coincide with that point because energy is lost through friction as the water moves through the aquifer; hence, the pressure surface lies at a lower level and declines in altitude away from the recharge area.

Fig. 16-8 Artesian well flowing in Montana. *J.R. Stacy, USGS*

Artesian wells can be flowing or non-flowing. They flow where the pressure surface is above ground (Fig. 16-6, well B). In contrast, if the pressure surface is below ground, water must be pumped (Fig. 16-6, well A). The pressure in a confined aquifer, however, may cause the water to rise to a higher level than that of the water table in an adjacent, unconfined aquifer.

Should a large number of wells tap an artesian reservoir, the pressure will drop, and the flow will ultimately diminish. Such is the case with the Dakota Sandstone and the other aquifers associated with it. When, 40 to 70 years ago, water poured out of the ground under pressure high enough in some places to operate waterwheels, today, after the drilling of about 10,000 wells, pressure has dropped to the point where many wells must be pumped, and in flowing wells the yield has greatly diminished. This will have an enormous impact on our agriculture in the near future.

It might be imagined that under ideal conditions the amount of water withdrawn from an aquifer would be replenished quickly. But replenishment takes place slowly, whereas pumping goes on at a rapid pace. The result is that the water table is lowered and wells must be drilled to greater depths at greater cost.

The problem of overdraft is especially critical in the Southwest, where groundwater is crucial to the billion-dollar agricultural industry as well as to other industries. In parts of California and Arizona, for example, maximum water-table decline approaches or exceeds 5 m (16 ft) per year. Obviously this drawing off of water cannot go on forever, and other sources of water are being sought. In California, for example, an aqueduct has been constructed to transfer surface water from the northern part of the state, where supplies are still more plentiful, to the southern counties. This has led to much political haggling.

Coastal areas have their own peculiar problems when it comes to pumping overdrafts. Groundwater beneath the land is fresh, but that beneath the ocean is salty. Fresh water, since it is less dense than salt water, rests on the latter, and the contact between the two dips beneath the land. If fresh groundwater is removed faster than it is replenished, salty groundwater moves inland.

A case in point is an area on Long Island, New York, where salt-water contamination of local wells has been a problem for many years. A good freshwater aquifer underlies the island, and before extensive groundwater development took place, salty groundwater did not enter the aquifer (Fig. 16-9). In earlier days, much of the fresh groundwater was returned to the aquifer via cesspools. But as the population and the number of cesspools increased, contamination became such a problem that sewage was dumped into the sea. By that time, water was being removed from the system more quickly than it was replenished, and the interface between fresh water and salt water moved inland, and deep wells that once

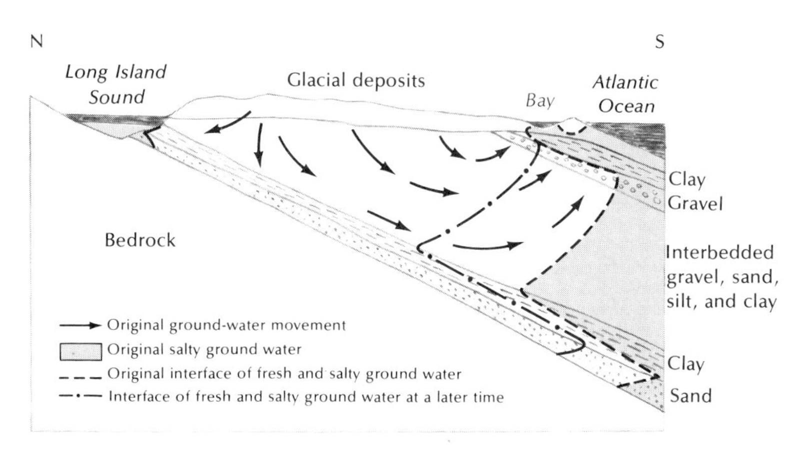

Fig. 16-9 Groundwater conditions on Long Island, New York.

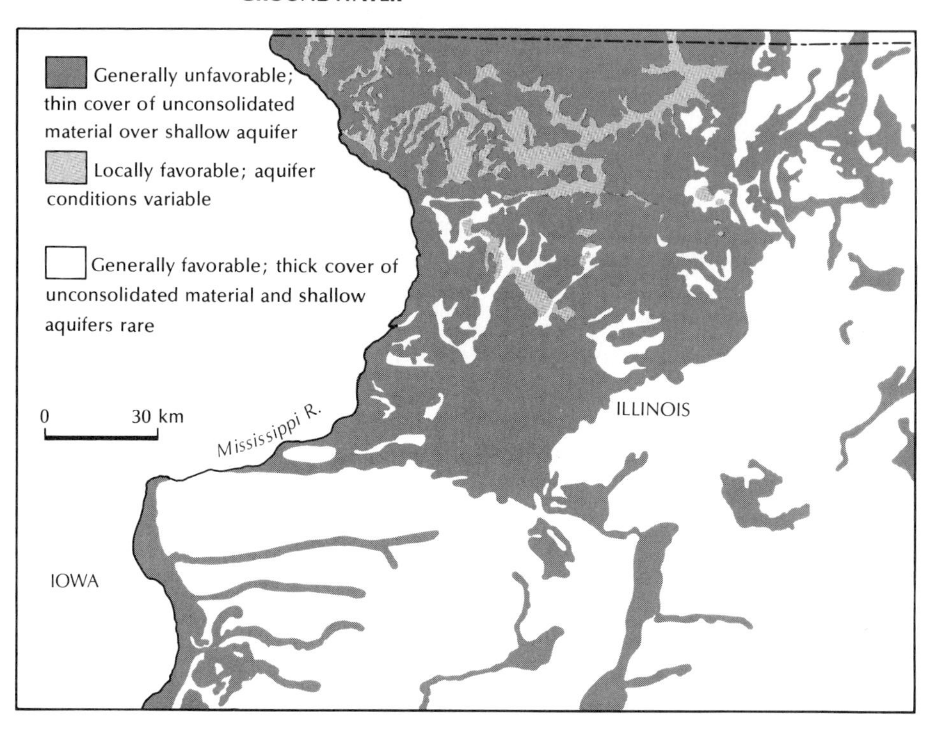

Fig. 16-10 Map of northern Illinois indicating the chances of success in locating a suitable site for sanitary landfill.

tapped fresh water have become brackish.

If the situation is allowed to continue, a very important fresh-water aquifer will be destroyed. What can be done? Perhaps one of the simplest solutions would be to reclaim the sewage and to pump water back underground. In that way the fresh water–salt water interface would be forced seaward, away from the domestic water wells.

Extensive withdrawal of groundwater can have other deleterious side effects—for example, the land may subside. Sinking occurs because unconsolidated sediments compact when ground water contained in the sedimentary pores is removed. In some areas the amounts of subsidence are large indeed: up to 2 m (6.5 ft) in the Houston-Galveston area, 1 m (3 ft) at Las Vegas, and 8 m (26 ft) in the southern portion of the Great Valley of California. The areas involved measure from hundreds to thousands of square kilometers.

Such subsidence creates enormous problems. Coastal areas suffer from the encroachment by the sea, and levees have to be built to keep it back. Perhaps a larger problem is that water-transport systems that require gravity flow, such as canals and sewers, can be disrupted by the changes in gradient that accompany subsidence. Californians are especially aware of the problem because the very expensive California Aqueduct, which brings water to the southern part of the state, crosses subsiding areas. Another serious problem is that compacting sediments can exert enough stress on the well casings to cause their rupture. One way to counteract such subsidence is to recharge the aquifer with water.

CONTAMINATING GROUNDWATER

Pollutants are a by-product of our society that are not easy to dispose of. For example, many pollutants have been dumped or piped into our streams and oceans with serious consequences. Because so many areas rely heavily on groundwater, we must guard against contaminating our aquifers. Recharge areas for aquifers, especially, should be kept from possible contamination.

A common practice today is to dispose of refuse in sanitary landfills (garbage dumps). What makes the operation "sanitary" is that periodically the refuse is bulldozed over and buried beneath compacted earth. Such landfills can become unsanitary, however, if the water draining through them picks up undesirable substances in solution and subsequently contaminates the water supply. Hence, communities should site their dumps only after the geology and the groundwater conditions of their area have been thoroughly studied. Where sufficient information on the subsurface is available, regional maps have been prepared to aid planners in siting landfills (Fig. 16-10).

In places it is feasible to dispose of liquid waste by pumping it to great depths. Obviously the geology of an area must be studied with great care to avoid any possibility of contaminating present or future aquifers or disturbing areas in which earthquakes are a possibility. In one case (see Chap. 18), deep pumping of radioactive wastes temporarily made Denver, Colorado, a seismically active region.

PROSPECTING FOR GROUNDWATER

"Prospecting" for water is not so strange as it may sound, since groundwater is essentially a mineable resource. Radiometric dating has shown that thousands of years may be required for water to accumulate in underground reservoirs. Furthermore, in places, the permanent lowering of water tables with continued pumping indicates that the annual rainfall is not sufficient to replenish the water. Yet the demand for groundwater increases, although much water is wasted, and the need to prospect for new water supplies is becoming more and more evident to urban planners, irrigation scientists, hydrologists, and others.

The search for groundwater has produced its colorful prospectors. It has long been customary to use "water witching" to determine the presence of water. The water witch, or dowser, walks back and forth over the land, holding two ends of a forked stick, or divining rod, keeping it horizontal (Fig. 16-11). When the stick, through some magical power, dips sharply downward of its own accord, the dowser announces that this is the place to dig. More often than not the patently unscientific method brings in a good well, at least often enough in humid regions to perpetuate the belief in water witching, even among those who should know better. One reason it is successful is that in many places in the world there is a plentiful supply of groundwater, so the placement of a well is not a critical factor. Another is that some water witches are very observant. They have noticed that the water table is closer to the surface in valleys than on hilltops, or that certain types of vegetation indicate the presence of water. Even though scientists condemn the practice, it is estimated that about 25,000 water witches are still at work in the United States.

Much more accurate methods, however, are now used to find the vast quantities of water required by people everywhere. The developing countries especially—many of which are in arid regions where water must be found before agricultural production can be increased—require reliable and accurate prospecting.

There are several direct methods of searching for water. Mapping the rock units in an area will show potential aquifers, as will records of existing wells. Geophysical methods used in petroleum exploration can also be used to help locate aquifers (for example, the seismic methods described in Chap. 18).

Prospecting for ground water in sedimentary rocks is not too difficult because the positions of the rocks at depth are predictable. Prospecting in the glaciated part of the mid-continent, however, is much more difficult. In places, glacial tills of low permeability are interlayered with permeable stream deposits consisting of sands and gravels. The latter are good groundwater sources, but they vary in thickness over short distances and do not form extensive blanket deposits as some older sandstones do (e.g., the Dakota Sandstone, Fig. 16-7). It is very difficult to predict where they will occur. In places, the gravels of pre-glacial valleys buried beneath the fills are sought as groundwater reservoirs.

Once aquifers are located we can call upon computers to help us decide how to develop the resource most efficiently. Computer models for an entire groundwater system can simulate the flow of the water, the conditions at each well, the amount of water being removed, and the actual water-table level at any time. They can also predict the effects of future withdrawals and help to plan the distribution of wells and the timing of water removals. In some areas, efficient management may require the collection and storage of rainfall runoff in huge groundwater reservoirs to prevent excessive loss through surface evaporation. In others it may be economically feasible to recharge a groundwater system by pumping into it water brought from a distance.

GEOLOGICAL ROLE OF GROUNDWATER

Water in the ground does work of geological significance, comparable in many ways to the more visible achievements of rivers, glaciers, lakes, and the sea on the earth's surface.

A cementing agent

Among the more significant accomplishments of groundwater is that of providing the means by which the various natural cements, such as calcite ($CaCO_3$), silica (SiO_2), and iron oxide (Fe_2O_3), are introduced into the pore spaces of unconsolidated sediments. Such cements are reasonably soluble substances and may be dissolved from rock or soil layers by water when it starts its journey underground. Later, when the saturation is sufficiently high, and the temperature and the pressure are right, those substances may precipitate out of solution. Gradually, as they are deposited on the surface of individual grains, much as scale is deposited in a teakettle or a hot-water heater, they bind the grains together. In that way, pore spaces are drastically reduced in size until finally they may become sealed off almost completely, and thus become less permeable.

Underground caverns

Groundwater plays a unique role in areas underlain by rocks that are soluble in water. Limestone, marble, and such evaporite deposits as gypsum and salt are examples of these rocks, and when they dissolve and slowly waste away, large underground caves are formed. The Carlsbad Caverns in New Mexico, Mammoth Cave in Kentucky, and the caves decorated by Stone Age peoples near Lascaux in southern France are renowned. And there are scores of others in many parts of the world. Their dark, silent recesses have intrigued explorers since the earliest days, and even today there are few countries or states without active speleological groups within their borders.

The origin of limestone caves has long been debated and is far from settled. The crux of the debate is whether the solution responsible for the removal of thousands upon thousands of cubic meters of solid rock in some of the larger caverns occurred above or below the water table. Two factors are being debated. On the one hand, in sections of caverns above the water table the leading present-day process appears to be deposition rather than solution. At least deposition is the process in stalactite and stalagmite formation. On the other hand, water below the water table is often already saturated with lime; it cannot pick up any more, and thus solution stops. A continuing supply of circulating water with a low lime content is called for, seemingly an unlikely situation in a region underlain dominantly by limestone.

A further complication is that many caverns include deposits of clay, silt, and even gravel, which has led some geologists to conclude that those caves were eroded, at least in part, by subterranean streams. Such rivers are fairly common in limestone terrains, such as those in Indiana and Kentucky.

A theory that appears to apply to the Carlsbad Caverns is that the caves were formed by solution at a time when the water table stood higher than it does now. As a result of canyon-cutting by nearby streams, the water table was lowered and passageways made by solution along joints and bedding planes were opened to the air. It was possible then for such distinctive features of the cave world as stalactites, stalagmites, columns, and ribbons and sheets of travertine ($CaCO_3$) to be formed.

Few geological phenomena arouse more curiosity than the strange, eerie structures made by dripping water in underground caverns. Most familiar of those to visitors to the great number of national, state, and privately controlled caves are the icicle-like pendants of ***travertine*** hanging down from the cave roof (Fig. 16-12A). They are ***stalactites,*** and they normally form where dripping water seeps from the rocks above the cave. When the water reaches the air of the cave, some of the carbon dioxide contained in solution escapes, and calcium carbonate ($CaCO_3$) is precipitated. Also, if some of the water evaporates, a residue of calcium carbonate is deposited. Because the drops of water that hang suspended momentarily from the cave roof are likely always to be about the same size, the tiny rings of travertine they leave nearly always will have the same diameter. Gradually, a series of rings forms a long pendant, customarily with a narrow tube extending through its full length. Seldom, though, is such perfection achieved. The tube may become plugged, the amount of water may vary, or new holes may break out along the sides of the stalactite, rather than at the tip. All such changes lead to a great variation in form.

Fig. 16-11 Seventeenth-century water witch at work. From Pierre de Vallemont, *La physique occulte,* Paris, 1663. *Rare Book Division, New York Public Library*

Stalagmites (Fig. 16-12B) are deposits built upward from the cave floor, and, characteristically, they grow below stalactites. When a drop of water falls from the tip of a stalactite it may lose some of its carbon dioxide, or its dissolved lime may become concentrated through evaporation, and more lime is deposited. Thus a counterpart accumulation of lime gradually builds up from the floor of the cave to oppose the stalactite growing downward from the roof. Stalagmites, unlike stalactites, do not have a central tube, and, because they are built up by the saturated water that spatters over their surface, they usually are thicker and more diversified in shape. With two such structures growing toward one another, the stalactite and stalagmite may eventually meet and fuse to form a column.

Other cave deposits may take on a wide variety of shapes—fluted, columnar, or sheet-like masses—that are often enhanced by indirect lighting in commercially developed caves.

Karst

The landscape that may develop in a region underlain by limestone or other soluble rock differs in a multitude of ways from one underlain by less soluble rocks. Parts of China display such a landscape well (Fig. 16-13). In the United States, there are many karst areas—the major ones being in Pennsylvania, Maryland, Virginia, Indiana, Kentucky, Tennessee, Florida, Missouri, New Mexico, and Texas (Fig. 16-14). Puerto Rico is also known for well-expressed karst topography (Fig. 16-15). But perhaps the best-known region of that type is the Karst, the portion of Yugoslavia bordering the Adriatic, the Dalmatian Coast. It is one of the picturesque coasts in the world, with the sea penetrating far inland in long, fiord-like inlets. They are separated by barren, whitish limestone ridges and islands that contrast vividly with the sea. ***Karst,*** in fact, is the name given to similar

Fig. 16-12 A. (opposite) Stalactites hanging from the roof of a cave, Lehman Caves, Nevada. *Hal Roth* **B.** (below) Stalagmites growing upward from the floor of a cave, Carlsbad Caverns National Park, New Mexico. The stalagmite in the center has joined a stalactite to form a column. *National Park Service*

Fig. 16-13 Karst topography in Kwangsi Province, China, creates a surrealistic landscape. A probable explanation for the strange towers is that the limestone terrain may have been lowered by erosion, leaving the former caves exposed as valleys and the solid rock between the caves as steep hills. *H.E. Malde, USGS*

topography formed by the solution of rocks everywhere.

Dalmatia is one of the historic coasts of Europe. The once-forested slopes of the now barren hills of what was then known as Illyricum provided timbers for the galleys of Rome and later for the wide-ranging vessels of the Venetian Maritime Republic. Today it is a harsh, stony land, and it is difficult to visualize the widespread forests and soils that once mantled its slopes before destruction by overcutting, overgrazing, and subsequent active erosion.

The region has one of the heavier rainfalls of Europe, yet surface streams are absent. Limestone—if joints and other fractures abound—is so permeable that rainwater sinks rapidly into the ground. A stream will flow for short distances, disappear underground, and then reappear several kilometers away as a river emerging from a giant spring. Such a limestone terrain is pocked with large numbers of closed depressions, some large, some small. Commonly, the depressions are floored with red clay, and that thin accumulation of reddish soil is likely to be all that is available for agriculture. In Yugoslavia, the larger depressions may be several kilometers across—large enough to shelter a village and its surrounding patchwork of fields. The origin of the large depressions is uncertain. Most are probably caused by removal of material by solution.

Smaller, closed depressions in Yugoslavia and elsewhere are almost certainly caused by solution. Some of them extend downward into the earth through near-vertical shafts, which commonly lead to deep caverns. In North America, such solution pits are called ***sinkholes,*** and some may hold small lakes if they are floored with a layer of clay (Fig. 16-16). Should the clay seal be broken, then the lake will drain away through solution channels into the underlying limestone. In places, rivers end in such sinkholes and disappear.

Sometimes sinkholes serve as natural wells. Their steep sides extend downward for scores or even hundreds of meters, until they intercept the water table, which stands as a pool of green water at the bottom. Renowned examples of such formation are the *cenotes* of Yucatan. Mexico's Yucatán Peninsula is a nearly level limestone plain, without surface streams because the rainwater sinks almost immediately into the ground. When the peninsula was the site of the Mayan Empire, the dense agglomerations of people at such cities as Chichén-Itzá depended upon so slender a supply of water as the dank fluid at the bottom of a limestone sink. No wonder that, to preserve it, a maiden burdened down with bangles and ornaments was ceremoniously hurled into the cenote in order to assure a continued supply.

Sinkhole formation is still an active process, as some unfortunate landowners have discovered. It is a geological hazard common to many areas in Florida, Texas and in other parts of the country where it causes damage to houses, building foundations, and other structures (Fig. 16-17) and makes the maintenance of sta-

Fig. 16-14 Distribution of karst areas in the United States.

Fig. 16-15 Karst topography formed in limestone terrain, Puerto Rico. *USGS*

ble road beds for highways very difficult. Research is now being directed at finding ways to determine which areas are prone to sinkhole formation, so that future damage can be avoided. One method of study is aerial photography with remote-sensing devices. Data thus obtained may reveal thermal and vegetation patterns that could indicate the presence of caverns that could collapse.

Geysers and hot springs

By far the most spectacular manifestation of groundwater is its appearance at the surface in the form of geysers and hot springs. They are the leading attraction of Yellowstone National Park, and few have not heard of Old Faithful or seen it run through its repertoire (Fig. 16-18). Yellowstone is not the only geyser area in the world; in fact, the extensive one of Iceland gives its name to this sort of aqueous outburst, since all are named for a large Icelandic spring, *geysir*. Another large and touristically attractive geyser region is the Rotorua region of North Island in New Zealand. It is currently being developed as a source of thermal power.

Although the actual process that goes on in an erupting geyser is something of a mystery, a generally held view is that groundwater percolating downward in a geyser area comes in contact at depth with a heat source. That source may be cooling igneous rocks or steam, or other hot gases given off by magma. Even though the water at the bottom of a tube may be heated to 100°C (212°F), it does not boil because the boiling point increases with an increase in pressure. We are more familiar with the opposite effect—lowering of the boiling point in the thin air of high mountain tops to the point where potatoes, for example, do not cook through.

Thus, the temperature at the bottom of a column of water may rise above the boiling point at normal atmospheric pressure. However, nothing is likely to happen until all the water is heated to the top of the column, perhaps until it begins to surge, or spill, over the rim. Should enough water drain off, then the pressure throughout the column is reduced, with the result that the superheated water near the bottom flashes into steam. That action is enough to propel the whole column of water upward,

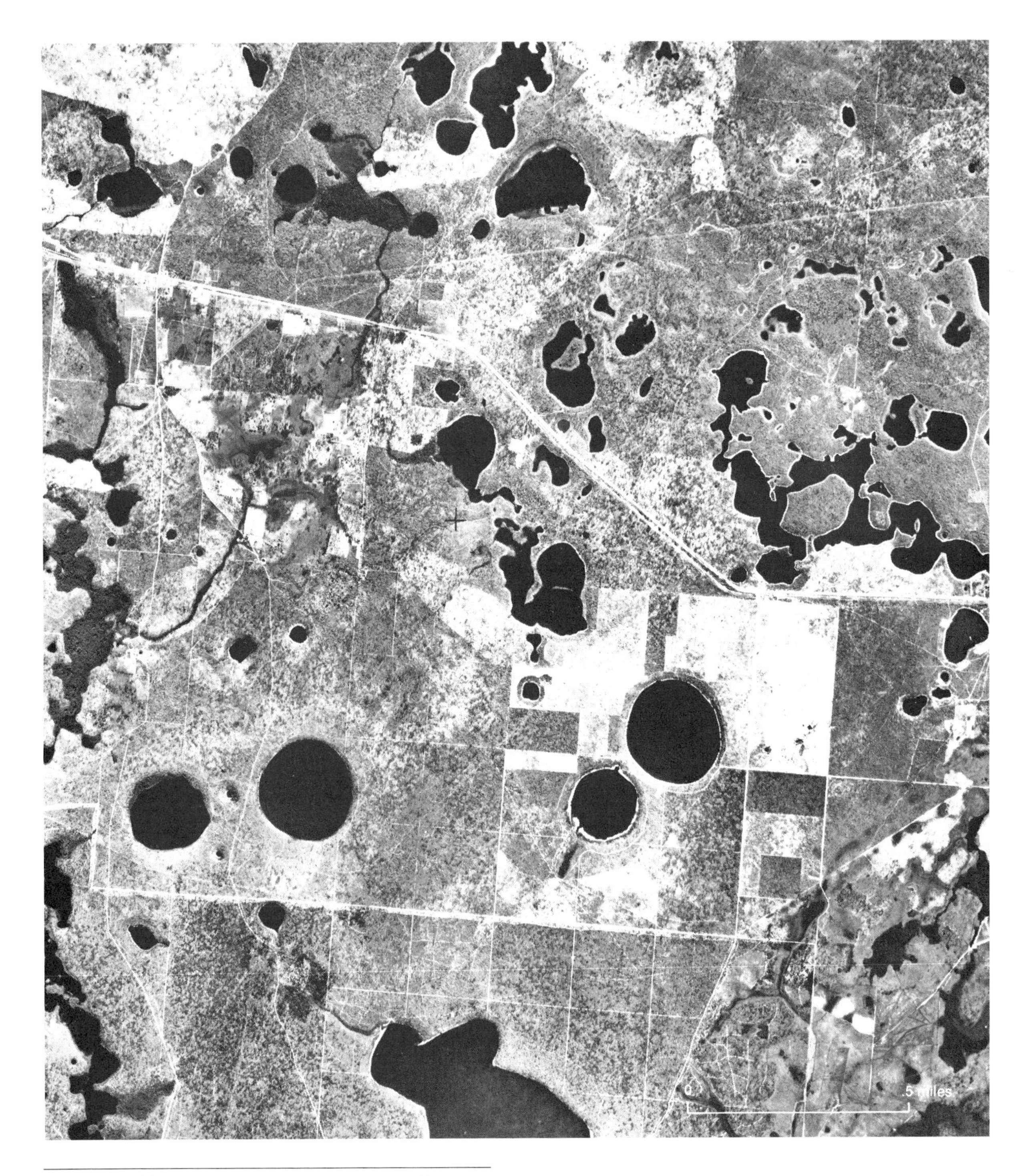

Fig. 16-16 Sinkholes are common features in the limestone terrain of central Florida, and because of the high water table, many became lakes. *USGS*

Fig. 16-17 "Wink Sink," a recently formed sinkhole in northwestern Texas, is the result of the collapse of underlying rock into a cavern formed by solution of a subsurface salt layer. The depression, which formed in only 48 hours, is 110 m (361 ft) across and 34 m (112 ft) deep. Note the broken oil pipeline. *R.W. Baumgardner, Jr.*

and since a similar pressure reduction and a near-instantaneous conversion to steam occurs throughout its length, a mixture of hot water and steam is hurled skyward—in Old Faithful for about 46 m (150 ft). The details of how the subterranean geyser reservoir is filled after being blown clear, and how some geysers achieve their remarkable periodicity, have become the center of much of the debate.

The castellated rims, platforms, and particolored deposits surrounding the geysers and hot springs of Yellowstone are especially interesting features to park visitors. In general, there are two kinds of hot-water deposits. Those deposited directly from mineral-rich geyser water often are composed of silica—supplied in part from the underlying igneous rock—and are called ***siliceous sinter.*** They are likely to be grayish and to consist of amorphous silica, very much like opal. Limy deposits, made by calcareous algae that can survive in the temperatures of hot springs and pools, are called ***travertine*** (Fig. 16-19).

Hot springs are more widely distributed over the face of the earth than geysers. There are more than one thousand in the United States,

Fig. 16-18 (Opposite) Steam rising from Old Faithful Geyser, Yellowstone National Park. *Janet Robertson*

Fig. 16-19 Travertine terraces at Mammoth Hot Springs, Yellowstone National Park. Photographed by W.H. Jackson in 1864. *The Metropolitan Museum of Art, Rogers Fund, 1974*

and most of them are located in the montane parts of the far West. Fundamentally, hot springs result when groundwater comes into contact with a source of heat in the earth's crust. Typically, the source may be igneous rocks that have not yet lost all their initial heat. Or it may be ***juvenile water;*** that is, water freed for the first time by cooling igneous bodies at depth.

The energy of geysers and hot springs in places is harnessed for geothermal power (see Chap. 21).

SUMMARY

1. Rainwater percolates beneath the surface of the ground, eventually filling the pores to become *groundwater.* The latter is an extremely valuable source of water.
2. The top of the groundwater zone is the *water table,* and its configuration generally parallels that of the land surface. Where the water table intersects the land surface, water flows; springs are a common example.

3. *Permeability* of rock is the main property controlling the rate at which groundwater flows from one point to another. Rock units that are permeable and hold sufficient quantities of water are called *aquifers*. Porous sandstone beds make some of the best aquifers.
4. Before aquifers are extensively used, they should be studied to avoid certain problems: too-rapid lowering of the water table, land subsidence accompanying excessive withdrawal, encroachment of salty groundwater in coastal areas, and pollution of groundwater.
5. The geological role of groundwater ranges from the cementation of sand deposits into sandstone to the formation of underground caverns and the stalactites and stalagmites, for which many are so famous, to the formation of the peculiar topography called *karst*.

QUESTIONS

1. How do porosity and permeability influence the flow of groundwater?
2. Compare the suitability of the following rock types for groundwater reservoirs: sandstone, shale, limestone, granite, and firmly cemented conglomerate.
3. What factors must be considered in storing toxic chemical wastes in ponds so that the groundwater is not contaminated?
4. How does karst topography form?
5. Are there any similarities between the eruption of a geyser and the eruption of a rhyolitic nueé ardente?

SELECTED REFERENCES

Davis, S. N., and De Wiest, R. J. M., 1966, Hydrogeology, John Wiley and Sons, New York.

Jennings, J. N., 1971, Karst, Australian National University Press, Canberra.

Leopold, L. B., 1974, Water, a primer, W. H. Freeman and Co., San Francisco.

Meinzer, O. E., 1939, Ground water in the United States; a summary, U. S. Geological Survey, Water-Supply Paper 836-D.

Monroe, W. H., 1976, The karst landforms of Puerto Rico, U.S. Geological Survey Professional Paper 899.

Poland, J. F., and Davis, G. H., 1969, Land subsidence due to withdrawal of fuels, *in* Reviews in Engineering Geology, D. J. Varnes and G. Kersch, eds.: Geological Society of America, Boulder, Colorado.

Rinehart, J. S., 1980, Geysers and geothermal energy, Springer-Verlag New York, Inc., New York.

Todd, D. K. 1959, Ground water hydrology, John Wiley and Sons, New York.

Fig. 17-1 Sedimentary strata deformed on a large scale; Murdafil, Iran. *Aerofilms, Ltd.*

17

DEFORMATION OF ROCKS AND MOUNTAIN BUILDING

When the crust of the earth is subjected to deformational stress, it buckles and breaks and the rocks become permanently deformed. In some places, originally horizontal strata are tilted and folded (Fig. 17-1), and in others the rocks are cracked and often offset along faults (Fig. 17-2). The area of geology that deals with the deformation of the earth, especially that recorded in ancient rocks, is called ***structural geology***.

Geologists are now aware that the rocks in the outer part of the earth are still being bent, broken, uplifted, and depressed at countless places in response to forces acting deep beneath the surface. The fact that the earth is in dynamic readjustment is brought home to us forcibly every time there is a large earthquake—a tangible indication that stresses in the earth have built to the point at which rocks fracture and suddenly shift.

The rates of deformation are slow, however, so that geologists, who have been making observations for only a short moment of earth history, see today what appear to be only minor modifications in the rocks at the earth's surface. Yet, given long periods of geological time, those nearly imperceptible small-scale changes can accumulate, eventually to bring about deformation on a grand scale. Surface rocks can be broken and separated along faults by as much as tens or even hundreds of kilometers; sedimentary rocks can be crenulated into great and small folds that resemble those of a gigantic rumpled tablecloth; mountain chains of spectacular proportions, like the Himalayas, can be pushed up where none existed before; and continents can be rent asunder and ocean basins formed.

The realization that forces working inside the earth can bring about such drastic changes was a long time dawning. Leonardo da Vinci (1452–1519), the Renaissance genius, observed fossil shells preserved in the rocks of the Tuscan hills of Italy and very early reached the conclusion that the Apennine Mountains had once been covered by the sea. The same observation was made later by Nicolaus Steno (1630–87), in Florence, who made the additional observation that since sedimentary rocks were deposited essentially horizontally, those that deviated

Fig. 17-2 Nearly 20 feet high, this prominent fault scarp developed in the foothills along the east base of the Sierra Nevada Range, near Lone Pine, California, during the so-called Owen's Valley earthquake on March 26, 1872. Many consider this to have been the largest historic earthquake in California. *David B. Slemmons*

from the horizontal must have been tilted by some force after the sediments had been laid down (Fig. 17-3).

The idea that deformation of rocks is accomplished by small deformational increments over long periods of geological time is a uniformitarian view. Charles Lyell, as you recall, was the champion of that idea in the early nineteenth century, and in fact he provided one of the examples that demonstrated the painful slowness of the process. Because he was able to visit many lands, he acquired a world outlook conspicuously lacking in many of his Victorian contemporaries. And few spots are more favorable than the shores of the Mediterranean he visited in providing an understanding of the rates at which deformation proceeds. Buildings had been constructed along the shores of that nearly tideless sea for millennia and the more durable ones were made of stone, which had withstood the ravages of time and weather in the relatively dry climate remarkably well. Those built close to the shore served as unusually sensitive recorders of changes in the sea level.

Lyell noted that all three of the surviving columns of the Temple of Jupiter Serapis, not far from Naples (Fig. 17-4), had lines encircling them about 7 m (23 ft) above sea level. Below the line, each column is riddled with small holes bored by shallow-water marine shellfish through a band almost 3 m (10 ft) wide. He concluded that the uppermost line on each column represented a former sea level. And the holes bored by the marine shellfish further supported his belief that the temple was once submerged to that level.

The historical record bears out Lyell's conclusion. Apparently the temple was built around the second century B.C., and it must have started to subside shortly thereafter, because a new floor was built on a fill of about 45 cm (17.5 in.) covering the original mosaic floor, and false bases were built around the columns. At an unknown later date, a fall of volcanic ash buried the court of the temple to a depth of about 3 m (10 ft). Continued subsidence of the land eventually allowed the sea to invade the entire structure. The 3-m (10-ft) band on each of the columns marked by borings of marine organisms represents the depth of water between sea level and what was then the mud- and ash-covered floor of the Mediterranean.

When the land started to rise again is not known. In medieval times the sea extended inland to the bluff behind the temple. Certainly the rise was well under way by A.D. 1503, for in that year, according to Lyell, Ferdinand and Isabella of Spain granted to the University of Puzzuoli "a portion of land, 'where the sea is drying up' (*che va seccando el mare*)." Also, according to Lyell, the main rise of the land occurred at the time of the historically destructive eruption of Monte Nuovo in A.D. 1638. Then "the sea abandoned a considerable part of the shore, so that fish were taken by the inhabitants; and, among other things Falconi mentions that he saw two springs *in the newly discovered ruins*" (Lyell).

That example, based on observations he made in 1828, demonstrated to Lyell that much of the deformation of the earth goes on relatively slowly, and is a far cry from the catastrophes most of his contemporaries relied upon to account for the more striking elements of the landscape. His book, *The Principles of Geology* (1830), was the first textbook of geology, and it had a profound influence on the Victorian world of letters. Tennyson, with many other literary lights of the period, was excited by Lyell's discoveries, as the poet's emotion-charged passages indicate:

> There rolls the deep where grew the tree
> O earth, what changes hast thou seen!
> There where the long street roars hath been
> The stillness of the central sea.
> The hills are shadows, and they flow
> From form to form, and nothing stands;
> They melt like mist, the solid lands,
> Like clouds they shape themselves and go.
>
> *In Memoriam*, 1850

The vertical fall and rise of the land surface Lyell documented near Naples is minor, however, when compared to the subsidence now affecting much of the Netherlands. The coast of Holland is protected by dikes, and much of the country, including the cities of Rotterdam and Amsterdam, is below sea level. Few realize how unremitting and laborious a struggle must be waged to hold back the sea, because the coastal part of Holland is sinking at a rate of about 21 cm (8 in.) per century. That makes the achievement of the Dutch all the more remarkable.

The hills along the coast near Los Angeles, California, provide another illustration of uplift of a segment of land from the sea. A person approaching the Palos Verdes Hills, the bold headland that partially shields the harbor of San Pedro in southern California, is impressed by the promontory's seaward slope, which rises above the sea like a cyclopean stairway. Wave-cut terraces separated from one another by steep cliffs rise in thirteen steps to 396 m (1299 ft) above the sea. They are evidence that the coast in that area has been elevated verti-

Fig. 17-3 These tilted, folded strata between Rawlins and Laramie, Wyoming, were once nearly horizontal. *J.R. Balsley, USGS*

Fig. 17-4 Temple of Jupiter Serapis, near Naples, Italy, showing the zone of marine shellfish borings made in the columns during submergence.

cally so recently geologically that fossil seashells preserved on the flat terraces are identical to the shells of marine organisms living today in the ocean.

Another even more impressive set of uplifted wave-cut platforms exists marginal to the coast of Peru. Some of them are 16 to 24 km (10 to 15 mi) wide and are littered with marine shells appearing as if they had lived in the sea only yesterday.

And as a final example, high on the flanks of the world's loftiest mountain, Mount Everest, are found water-deposited rocks containing marine fossils that lived in the sea 60 million years ago.

Rates of movement

What do we actually know about the rate at which the earth's crust is being deformed? The answer is "not a great deal," because accurate surveying records go back only about two centuries, and in many cases, the best-surveyed areas are among the more stable parts of the earth. The Coast and Geodetic Survey ran levels in 1906, 1924, and 1944 across Cajon Pass

Fig. 17-5 Plastic deformation in limestone along the Rio Extorax, Querétaro, Mexico. The beds have been thickened through tight chevron folding or thinned through stretching. *K. Segerstrom, USGS*

between Victorville and San Fernando in the Transverse Ranges of southern California. The measurements show that the area around the pass where it crosses the San Gabriel–San Bernadino Mountains rose as a gentle arch during those 38 years by about 20 cm (8 in.) or at the rate of about 53 cm (21 in.) per century. As James Gilluly of the U.S. Geological Survey pointed out, though the rates may appear modest indeed, with sufficient time—and time is not lacking in geology—such a rate yields an uplift of 122 m (400 ft) in 25,000 years for the San Gabriel–San Bernadino Mountains. Mount Everest, rising at a comparable rate, could have reached its present height in 2 million years. Deformation of the earth need not be a uniform, continuous process, but can occur in abrupt jumps. For example, in 1899, during a severe earthquake in the area centering around Yakutat Bay, Alaska, a large portion of the earth's surface was uplifted as much as 14.4 m (47 ft). And during the 1964 Alaskan earthquake, a maximum uplift of 11.9 m (39 ft) was recorded in the region of Patton Bay.

These examples of the rate of earth movement have all been in a vertical direction, either sinking or rising, or, in the case of the Temple of Jupiter Serapis, doing both. Horizontal shifts have been recorded, too. In the San Francisco earthquake of 1906, roads, fences, and so forth were offset along the San Andreas Fault by as much as 6.4 m (21 ft) almost instantaneously. Repeated surveys also show a gradual creep of points on one side of the fault past corresponding ones on the other side, without any discernible break of the ground surface. Recent geodetic and geophysical measurements along the San Andreas Fault indicate a steady shift of about 2 cm (0.79 in.) per year of one fault block laterally past the other. And as we shall see in Chapter 20, the slow drift of continents across the face of the earth and the opening of ocean basins proceed at rates of about the same magnitude.

STRUCTURAL RESPONSE OF ROCKS

Rocks in the outer portion of the earth, in response to large- and small-scale forces, are subjected to compression, extension, shearing (lateral slippage), and torsion (twisting). Some rocks fracture, others break and slide past one another, still others fold with varying degrees of ease. How any particular body of rock responds structurally to stress depends on a number of factors—confining pressure, temperature, the time span of stress application, and the amount of reactive pore-space fluids present. In general, at the low confining pressures and temperatures within a few kilometers of the earth's surface the rocks are brittle, and respond by jointing (fracturing with no movement of blocks) and faulting (fracturing with slippage between adjacent blocks). At higher confining pressures and temperatures at greater depths rocks tend to become plastic, ultimately folding and flowing like toothpaste from a tube (Fig. 17-5). Plasticity also increases with the time over which stresses are applied and with the amount of reactive pore-space fluids contained within the rock.

At depths, where temperatures are relatively high (particularly if fluids are present), the rocks undergoing folding and flowage commonly recrystallize (metamorphoze).

When we consider all these factors, it becomes evident that jointing, faulting, and some types of folding are characteristic of rocks a few kilometers in depth, whereas plastic folding and flowage are characteristic of rocks between about 3 and 30 km in depth.

ANCIENT DEFORMATION RECORDED IN ROCKS

One step in the study of an area's structural geology is to describe the geometry and complexity of the folds, fractures, and faults (Fig. 17-6). When the structural features are small, they can easily be investigated by directly observing them. Some structures are so large, however, that they can only be studied piecemeal, and then they can only be investigated by making and studying geological maps.

The kinds of rocks that best lend themselves to structural study are sediments. Most strata were deposited horizontally, and deviations from that position or offsets in the continuity

Fig. 17-6 Steeply tilted strata adjacent to a fault in the Dinosaur National Monument, Utah. The Green River is in the foreground. *Philip Hyde*

of beds indicate deformation. Moreover, strata that possess distinctive rock properties can be traced easily for long distances, so that deformation can be studied over wide areas.

If a sedimentary layer is tilted at an angle to horizontal it is said to be a ***dipping*** bed: the amount of ***dip*** is the angle between the bed and the horizontal surface as determined by a level.

The line of intersection made by the dipping stratum and the horizontal surface is called the ***strike*** of the bed. The dip is measured in a plane that is perpendicular to the strike.

Perhaps the best way to visualize the dip and strike of an inclined stratum is to imagine a single dipping bed of sandstone that projects out of the calm sea (Fig. 17-7). The intersection of the sea surface (a horizontal surface) with the bed is the strike, and the dip is the amount of tip of the bed in a plane perpendicular to the sea-level trace, measured downward from the horizontal. A marble placed on the surface of the tilted bed would roll down the dip direction of the bed: the line of dip thus is directional.

The two ends of the strike line point in opposite compass directions. That is, if one end of the line points northeast, the other end points southwest. By convention, we only record the end of the strike line that makes an acute angle with the true north direction. A strike line measured and denoted by N 30 E, would be one in which one end points to a position 30° east of north, and the other, 30° west of south. On a map the geologist would draw a short segment with that orientation, as shown in Figure 17-8.

The dip can be represented on a map by a short line drawn in the direction of dip, at right angles to the strike line and attached to it. Consider the dip of such a bed to be 40° and the dip direction southeast. In a field notebook, the dip and the strike direction are written as follows: N30E40SE, and the map representation is that shown in Figure 17-8.

Vertically dipping beds are denoted on the map as —+—. Horizontal beds, of course, lack both a dip and a definable strike and field geologists use the symbol ⊕ to denote them on a map. If a bed is rotated past 90°, that is, is ***overturned,*** the geologist measures the dip and strike as before, but denotes it on the map with the symbol ⌓.

By measuring the dips and strikes of sedimentary strata and plotting them on a map it is generally possible to determine the large-scale fold geometry and offsets or abrupt changes in the geometry that might indicate faulting.

Fig. 17-7 Strike-and-dip of a tilted bed. The strike is represented by the line of intersection of the sea surface with the dipping bed. The dip is perpendicular to the strike and represents the path a marble would take as it rolls down the bed. *William Estavillo*

Folds

Some structural processes result in the warping or folding of the rocks near the earth's surface. In sedimentary rocks, the fold structures generally are more easily determinable, and they will appear as small- to large-scale wave-like crenulations in the strata (Fig. 17-9).

The folds resembling wave crests are called ***anticlines***—from the Greek, to be inclined against itself, and those resembling wave troughs are called ***synclines***—also from the Greek, to lean together (Figs. 17-9 and 17-10). The sides of such folds are called ***limbs,*** or ***flanks,*** and a line drawn along the points of maximum curvature of each bed is called the ***axis.*** It is apparent that in all open and some asymmetrical anticlines, the limbs dip away from the axis, whereas in all open and some asymmetrical synclines they dip toward it (Fig. 17-11). The axis of most anticlines would correspond to the ridge line of a roof, for example, just as the axis of most synclines would correspond to the keel of a ship.

A plane connecting the axial lines in successive beds in a fold is called the ***axial plane.*** It can be flat or curved, and vertical to flat lying (Fig. 17-12), depending on the straightness of the axis and the difference in dip angle between the two flanks. Small-scale recumbent folds are shown in Figure 17-13. Large-scale recumbent anticlinal folds, largely containing deformed metamorphic and igneous rocks in their centers, have been called ***nappes*** and are conspicuously developed in the higher reaches of the Alps.

Deep mines and oil wells show that in deformed sedimentary rocks the geometry of a fold deep underground may be very different from its geometry at the surface. Such structures are called ***disharmonic folds.*** Trying to second-guess what a fold at the surface is doing at depth is the sort of gamble that makes the work of a geologist such an intellectually stimulating challenge.

Folds in which the axes are horizontal are said to be ***non-plunging,*** and they could be represented by a piece of tin roofing placed on a table (Fig. 17-14). In most cases, the axes dip at an angle to the horizontal and the folds plunge into the ground (Figs. 17-15 and 17-16). Many folds plunge in one direction at one end of the fold and in the other direction at the other end. They are said to be ***doubly-plunging folds*** (Figs. 17-17 and 17-18).

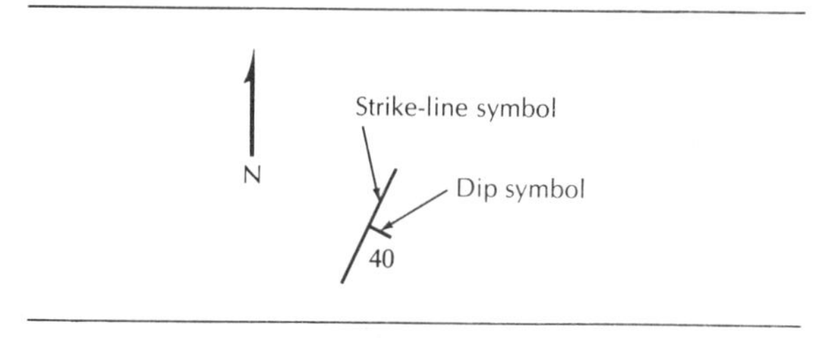

Fig. 17-8 Map symbol for a bed striking N 30 E and dipping 40° toward the southeast.

Careful study of the preceding diagrams and photographs of folds should bring out the following relationships more clearly than many paragraphs would:

1. Anticlines plunge in the direction at which their sides converge (come together).
2. Synclines plunge in the directions at which their sides diverge (spread apart).
3. Thus, in anticlines it can be seen that the rocks become progressively older toward the axis; in synclines, they are younger.
4. The dip measured on the axis also determines the angle at which the fold plunges. Note also that the strike of the beds where they cross the axis is at right angles to the axis in both the anticline and the syncline.

Anticlines that lack a well-defined elongation and that plunge from a point nearly equally in all directions are called ***domes*** (Fig. 17-19), and the corresponding synclinal structures are called ***basins.***

One other fold to be described is the ***monocline*** (from the Greek word for "one inclination"), a one-limbed structure with horizontal strata on either side of steeper-dipping strata (Fig. 17-20).

Fig. 17-9 Large-scale crenulations in strata; anticlinal and synclinal folds of the Grande Chartreuse north of Grenoble in the French Alps. *Swissair-Photo*

Joints

Nearly all rocks visible on the earth's surface are cut by cracks and fractures (Fig. 17-21). They are so commonplace that few people grant them more than casual notice. A minor number of the cracks, or ***joints,*** as they are called, are the result of the cracking of thin sheets of igneous rocks as they crystallize and cool. When igneous rocks are broken into a close-spaced set of prisms, the pattern is called *columnar jointing* (see Chap. 5).

Most joints are a result of deformation of the rocks, although the exact details are not at all clear. Many occur in closely spaced subparallel alignment—called a ***joint set***—and not too uncommonly, three such sets are mutually perpendicular (Fig. 17-22). Jointing is less likely in

Fig. 17-10 Anticlines and synclines near the Sullivan River in the southern Rockies of British Columbia. *Geological Survey of Canada*

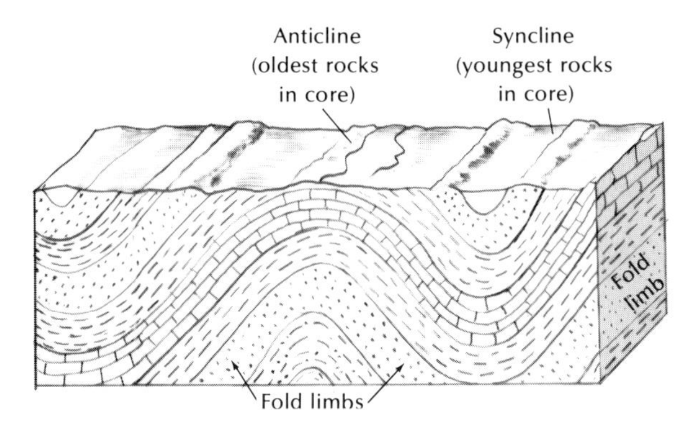

Fig. 17-11 An open symmetrical anticline and syncline. As a result of erosion, as shown, the oldest rocks will be found in the core of an anticline and the youngest in the core of a syncline.

Fig. 17-12 Different types of anticlinal folds, showing in each case the relation of the axial plane and the axis to the fold geometry.

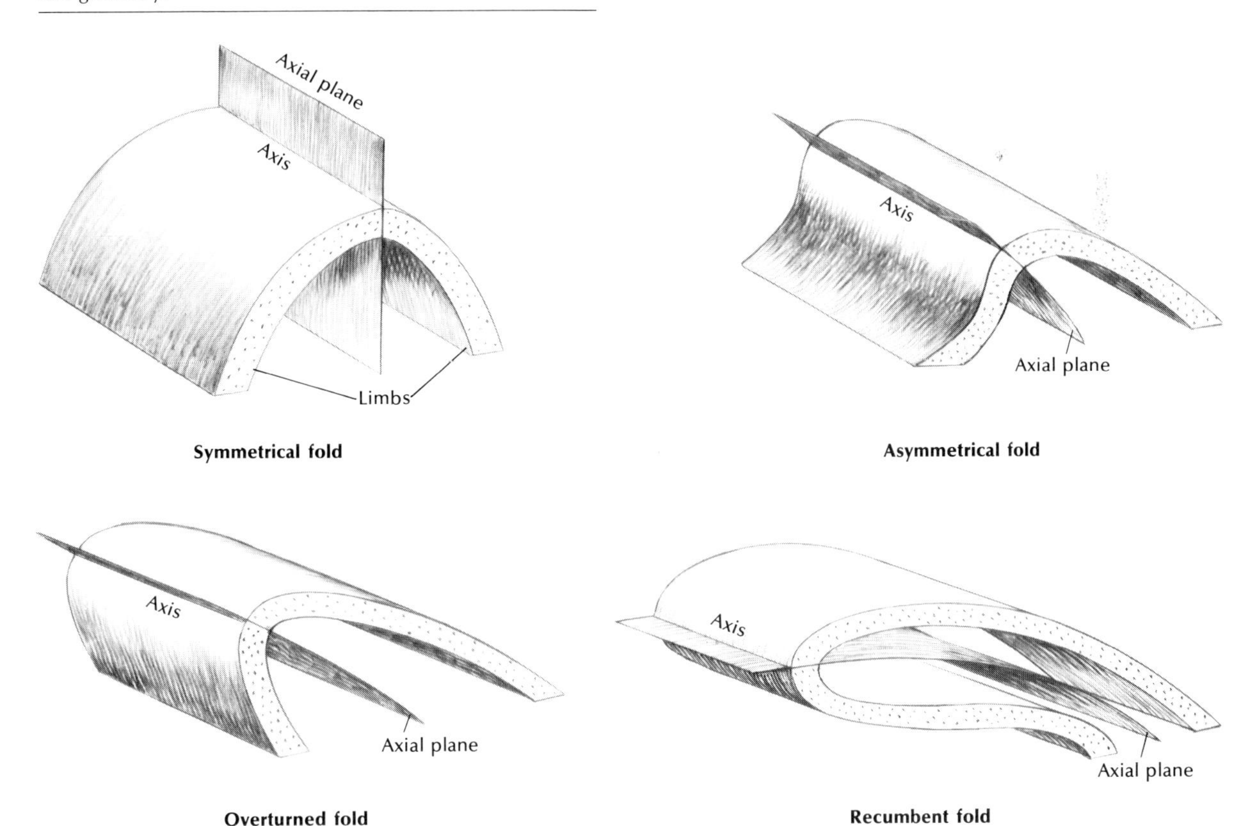

Fig. 17-13 Recumbent folds in slate folded about 350 million years ago. *Geological Survey of Canada*

Fig. 17-14 Symmetrical folds with horizontal axes (**A**) before erosion and (**B**) after erosion. Note how the resistant layers form parallel ridges after erosion.

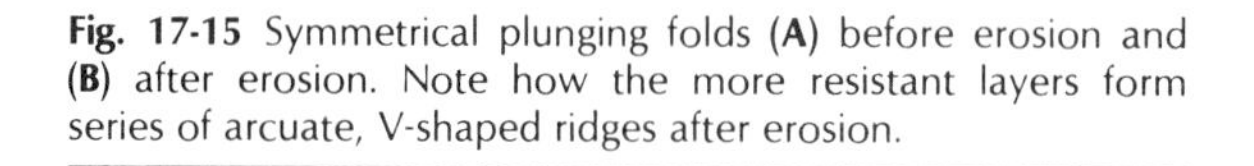

Fig. 17-15 Symmetrical plunging folds (**A**) before erosion and (**B**) after erosion. Note how the more resistant layers form series of arcuate, V-shaped ridges after erosion.

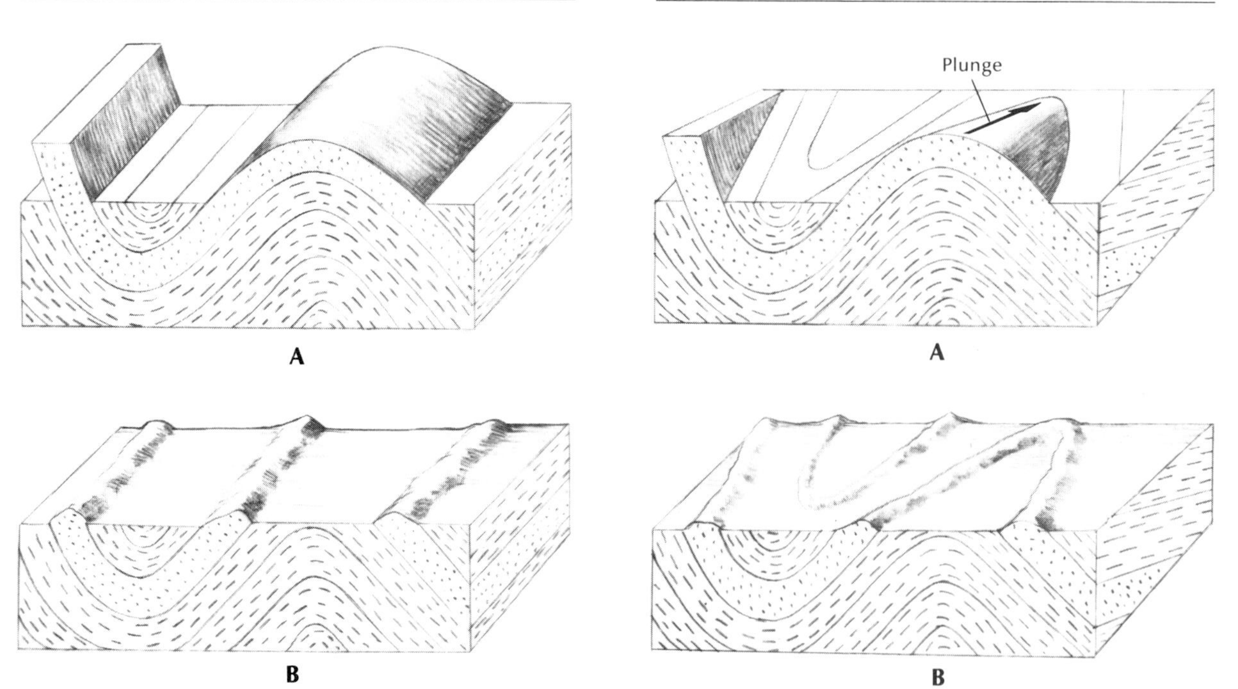

laminated or foliated rocks than it is in the texturally massive varieties.

FAULTS

Faults are fractures along which movement has occurred. In some cases, the offset amounts to a centimeter or less; in others, it may amount to as much as 200 km (124 mi).

Because fault movement involves the sliding of one block past another, many fault surfaces are smoothly polished and grooved (Fig. 17-23). These features, called ***slickensides,*** provide one clue to the direction of fault slippage, but they record only the most recent event. Earlier events may have involved slippage in totally different directions, but the slickensides associated with them, will have been obliterated by the most recent slippage.

Often, too, the rocks adjacent to a fault will have been pulverized or ground to bits, forming a clayey, soft material called ***fault gouge.*** In some instances, the rocks in the fault zone may be broken and sheared, creating a ***fault-zone breccia.***

If fault movement occurs close to the earth's surface, the surface itself may be broken and offset. The resulting low linear cliff is called a ***fault scarp*** (Fig. 17-24). Quite a number of such low, fault-induced cliffs have appeared sud-

Fig. 17-16 Plunging anticline in sediments along the eastern flank of the Front Range, near Fort Collins, Colorado. Note the similarity of the eroded pattern of this fold to the diagram in Figure 17-15B. *Edwin E. Larson*

denly in widely scattered parts of the world where earthquakes are especially prevalent. Such historically recent scarps are especially common in Japan, New Zealand, and India and in the contiguous United States in California and Nevada.

The tremendous eastern wall of the Sierra Nevada in the Mount Whitney sector also is often cited as an example of a fault scarp (see Fig. 6-16). Although it is true that the eastern front of the range is the result of fault action, the escarpment is not an exposed fault plane. Steep as the mountain front appears when viewed full face, when looked at from the side the average inclination of most of the ridges can be seen to be about 25°, whereas the exposed faults within the range have average dips of 60 or 70°. It is apparent that the face of the range is a fault-controlled erosional landform. The high part of the Sierra Nevada has been uplifted along faults, but weathering and erosion have reduced the original steep fault slope from nearly 70° back to about 25°.

Since fault planes have all the geometrical attributes of a dipping stratum, we can describe their orientation in space by means of strike-and-dip notation. For example, a particular fault plane might trend N 45 W and dip 52° toward the southwest.

Many faults, because they provide avenues for circulation of underground fluids, have been mineralized by ore-bearing solutions. Miners have long been aware of that fact, and numerous tunnels and shafts have been dug to reach the riches trapped in the fault zones. A miner working a mineralized fault zone where the fault dips at a moderate angle literally stands on the block on one side of the fault, with the face of the other block hanging above his head. The block beneath his feet is called the ***footwall*** and that above his head the ***hanging wall*** (Fig. 17-25). In the description of faults that follows we will use those two common terms.

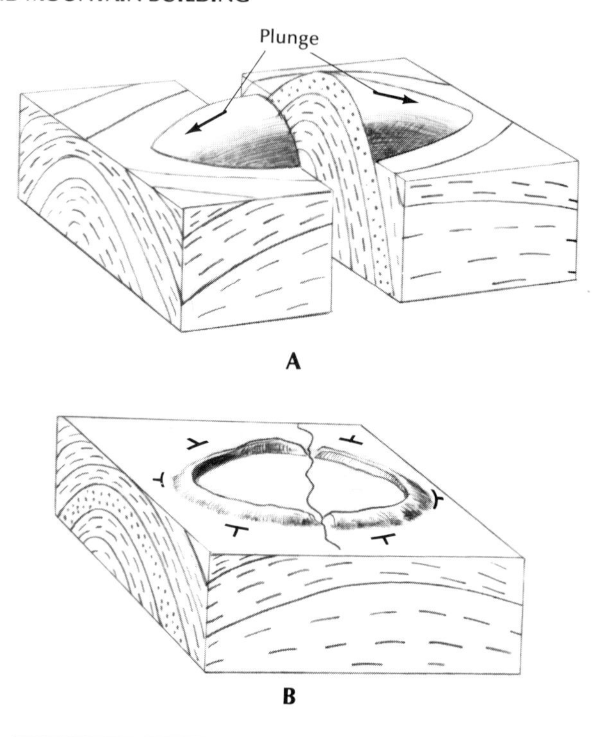

Fig. 17-17 Doubly plunging anticline **(A)** before erosion and **(B)** after erosion. Dip and strike symbols show attitudes of the layers.

Apparent relative movement

In most cases it is almost impossible to tell the exact direction of movement of the two blocks on either side of a fault from observing the rocks alone. In Figure 17-26, for example, the final pattern (Fig. 17-26D), which is all we see, could have resulted from (1) slippage of block **a** upward, in relation to block **b**—as seen in Figure 17-26B, (2) the horizontal slippage of block **a** relatively past block **b** (Fig. 17-26C), or (3) slippage at an oblique angle, involving some vertical and some horizontal displacement. In most cases, then, we can determine only an apparent relative movement.

Known directions of relative displacement—normal, reverse, strike-slip, and overthrust faults

In those cases where a line, as for example a fold axis, or volume of small dimensions is cut by a fault, the exact nature of the relative displacement can be determined.

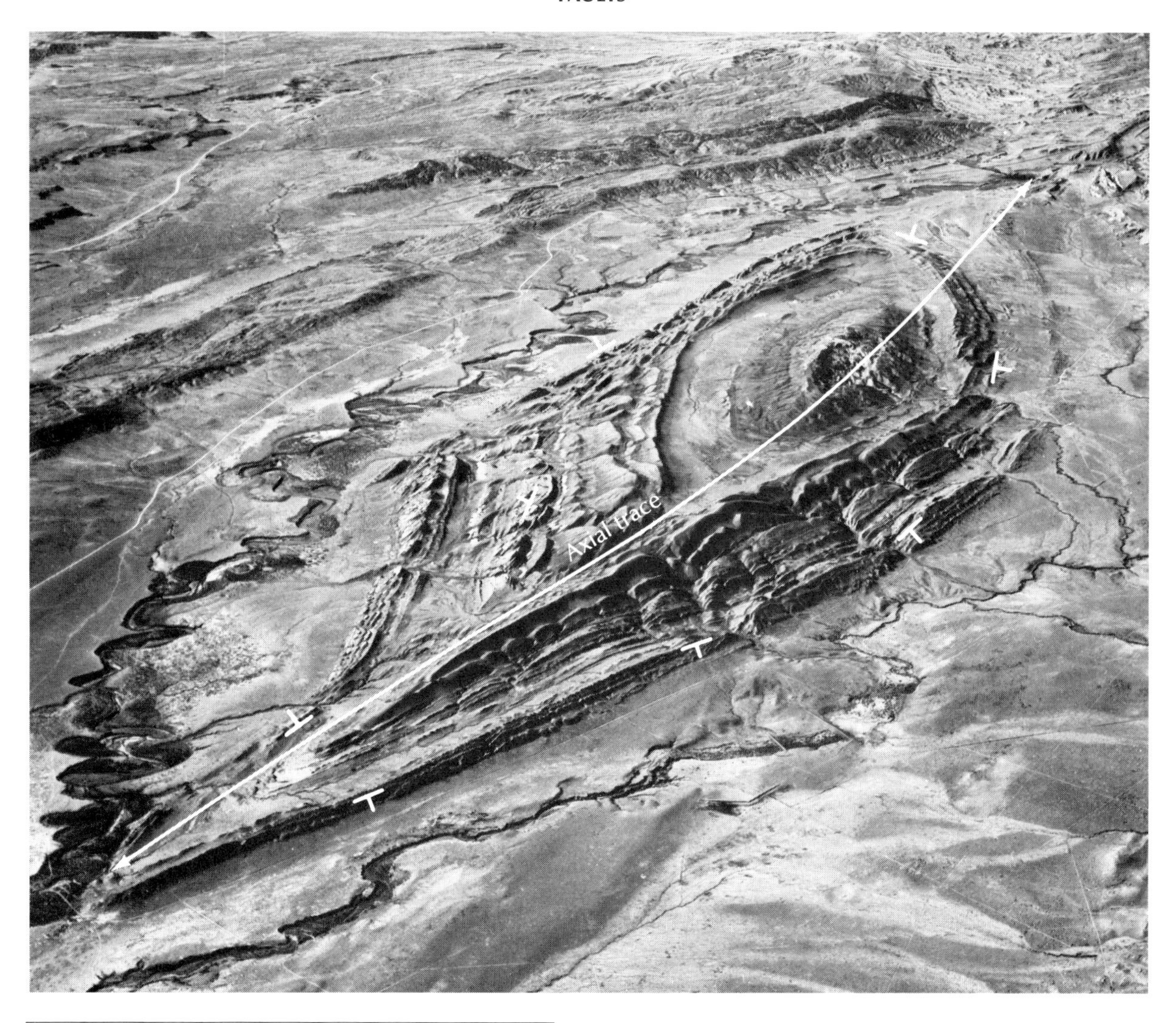

Fig. 17-18 Little Maverick Dome, northwest of Riverton, Wyoming. A doubly plunging anticline, or elongate dome, showing the trace of the doubly plunging axis and appropriate dip-and-strike symbols. *John S. Shelton*

The term used to denote the actual relative displacement of the once adjacent points, measured in the plane of the fault, is ***slip.*** Figure 17-27 illustrates the three kinds of slip. ***Dip slip*** (A) is movement on the fault parallel to the dip; ***strike slip*** (B) is a measure of displacement parallel to the fault strike; and ***oblique slip*** (C) is movement at an angle to the dip and strike of the fault. A fault in which the displacement is primarily dip slip and in which the footwall has moved up relative to the hanging wall is called a ***normal fault*** (Fig. 17-28). If the displacement is primarily dip slip, but the footwall has moved down relative to the hanging wall, the fault is called a ***reverse*** or ***thrust*** fault (Fig. 17-28C). Depending on the dip angle of the fault,

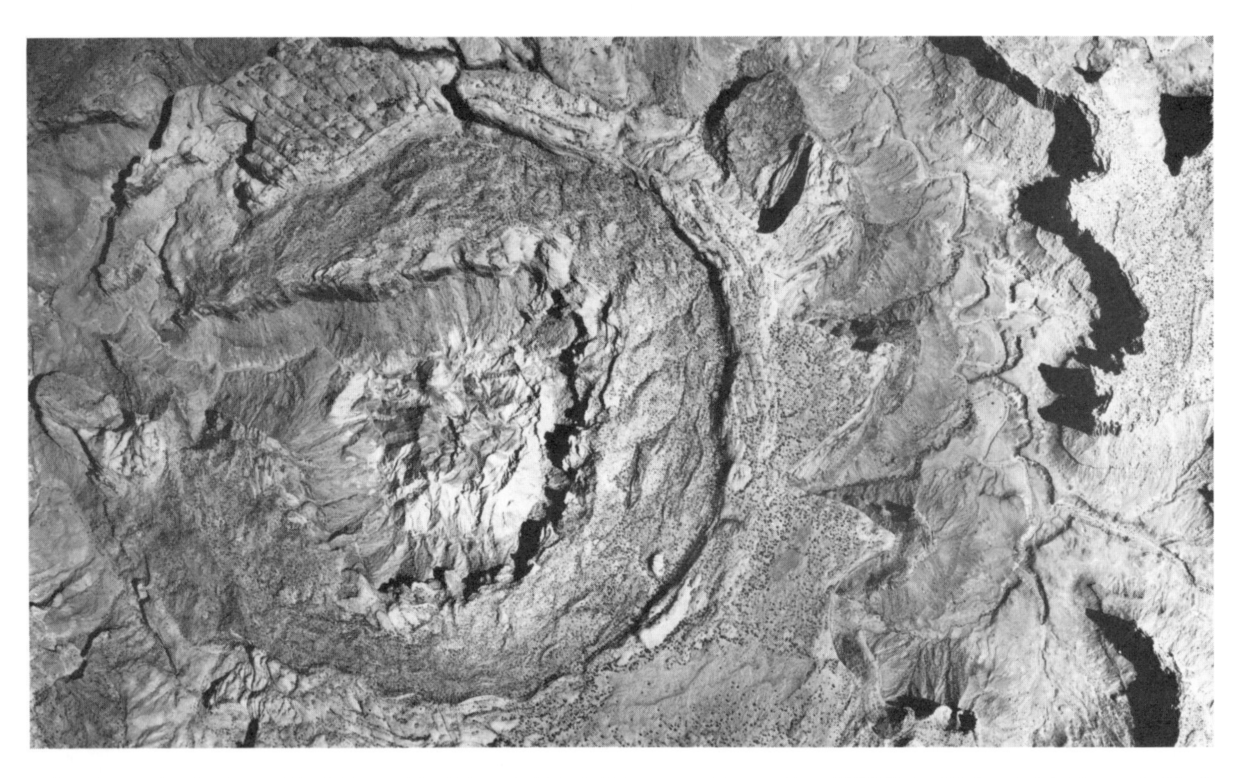

Fig. 17-19 Upheaval dome, Canyonlands National Park, Utah; an almost perfect example of a dome structure. *Eros Data Center, USGS*

Fig. 17-20 Looking south along the monocline that forms the west flank of Raplee Anticline, Colorado Plateau, southern Utah. The upland beds to the left are part of Monument Upwarp. In the foreground is the San Juan River. *Don L. Baars*

it can be described as a low- or high-angle fault: if less than 45°, ***low angle***; if more than 45°, ***high angle.*** If the dip angle on the thrust fault plane is less than 10°, it is called an ***overthrust fault.*** If the slip is primarily parallel to the strike of the fault it is called a ***strike-slip fault*** (Fig. 17-28D).

High-angle normal and reverse faults Many faults show displacements that are primarily dip slip in nature (Fig. 17-29). Along some of the larger faults of that type, displacements up to 2 km (1 mi) or more have occurred.

If a block rises relatively above the blocks on either side of it the fault-bounded upland block is called a ***horst*** (from the German word meaning, among other things, a ridge). Relatively down-dropped linear blocks between horsts are called ***graben,*** after the German word

Fig. 17-21 Closely spaced joints in nearly horizontal sandstone strata, eastern Utah. Differential erosion has emphasized the jointing pattern. *J.R. Balsley, USGS*

Fig. 17-22 Three sets of joints nearly at right angles (two are perpendicular to the rock face, and the third is parallel to it), in granitic rocks on the east face of Mount McAdie in the Sierra Nevada, California. *Tom Ross*

Fig. 17-23 Slickensided fault surface. The grooves indicate that the last fault motion was parallel to the dip of the fault. Sumter county, Alabama. *W.H. Monroe, USGS*

for "trough or trench" (Fig. 17-30). A renowned example is the Rhine graben, which is followed by the Rhine River from Basel to Mainz. There the valley is linear and trench-like with walls that are fault scarps marginal to highlands—the Schwarzwald to the east in Germany and the Vosges to the west in France.

An impressive set of graben is visible as the nearly continuous line of fault-bordered troughs that extend the length of Africa and part of the Middle East, from Mozambique to the Dead Sea. Those great down-dropped segments of the earth's crust are far too large to be referred to as trenches, and for that reason the frequently used term ***rift valley*** is appropriate.

One of the better-known rift valleys holds Lake Tanganyika, with a length of 672 km (418 mi) and a width from 32 to 64 km (20 to 40 mi). The water surface is 771 m (2530 ft) above sea

Fig. 17-24 Surface rupture resulting from an earthquake on the Motagua Fault, Guatemala, in 1976. *R.C. Bucknam, USGS*

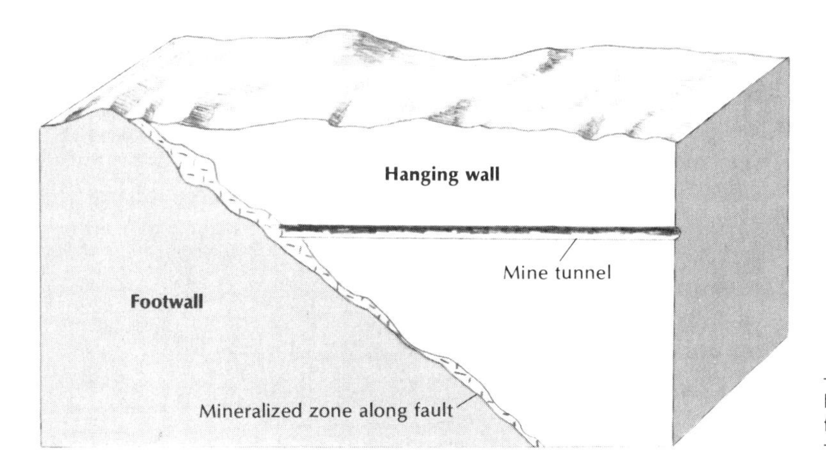

Fig. 17-25 Relation of a hanging wall and a footwall along a fault.

Fig. 17-26 Two possible displacements on a fault leading to identical offset of dipping stratum. **A.** Unfaulted dipping bed, with incipient fault plane shown. **B** and **C.** Two types of fault motion that could produce offset. **D.** Faulted dipping bed as observed in field. The relatively uplifted block in **B** has been eroded.

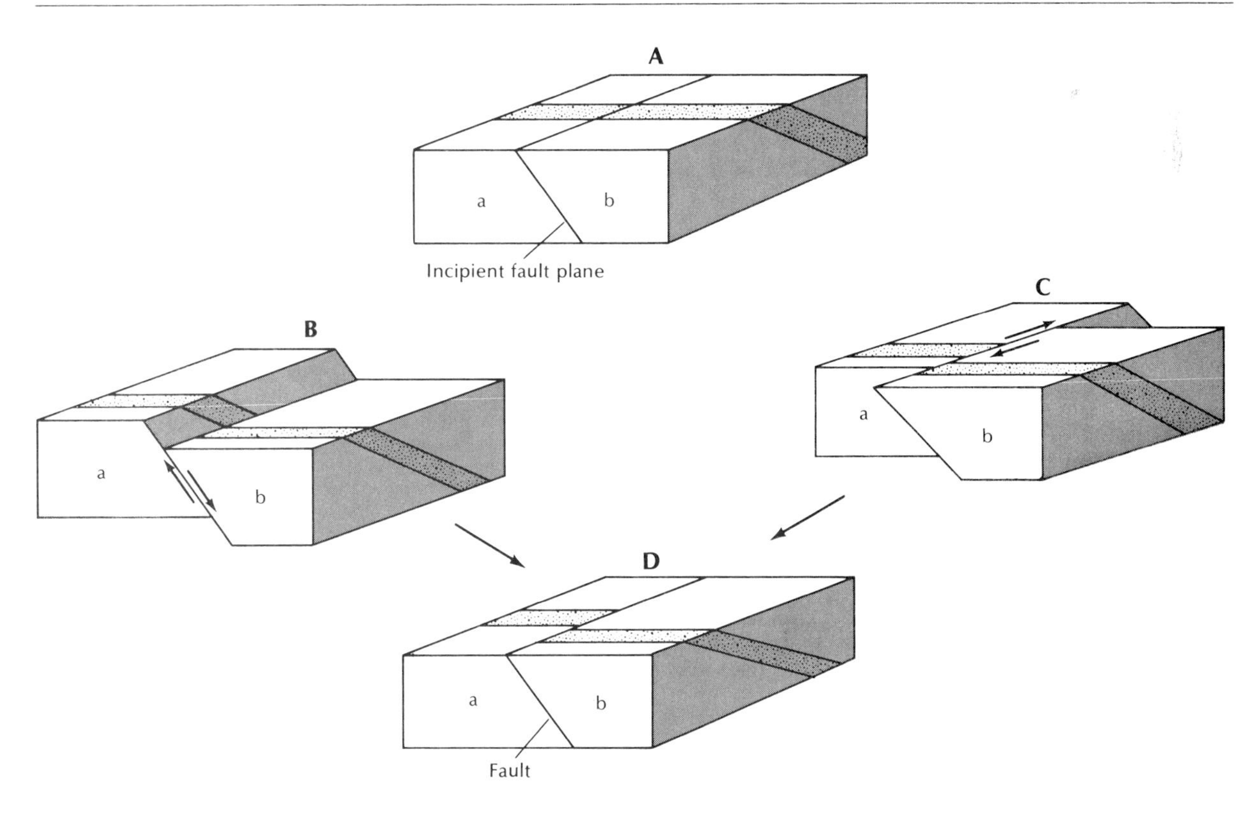

level, but the bottom is 506 m (1660 ft) below sea level. The impressive Lake Albert Nyanza is set in a trough that has an eastern wall 305 to 457 m (1000 to 1499 ft) high and a mountainous western slope rising to 2439 m (8000 ft).

Strike-slip faults In most strike-slip faults, which primarily involve displacement parallel to a fault strike, the dip of the fault plane is nearly vertical. If you place two bricks side by side on a table and move them backward and forward, you will have reconstructed the movement of a typical strike-slip fault.

Most of the world's faults that show the greatest amount of movement, measurable in places up to scores of kilometers, are strike slip in nature, as, for example, the San Andreas Fault.

Since the 1906 San Francisco earthquake, the San Andreas Fault and others like it have been

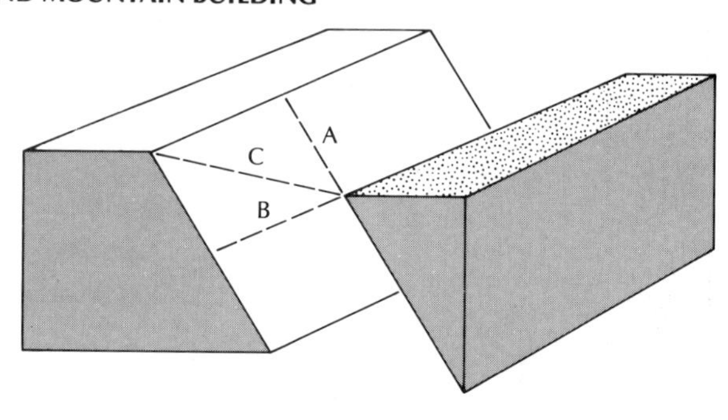

Fig. 17-27 The types of slip on a fault plane: Dip slip (**A**), strike slip (**B**), and oblique slip (**C**).

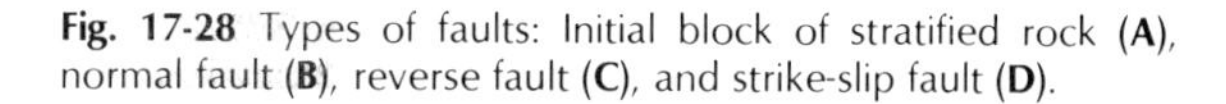

Fig. 17-28 Types of faults: Initial block of stratified rock (**A**), normal fault (**B**), reverse fault (**C**), and strike-slip fault (**D**).

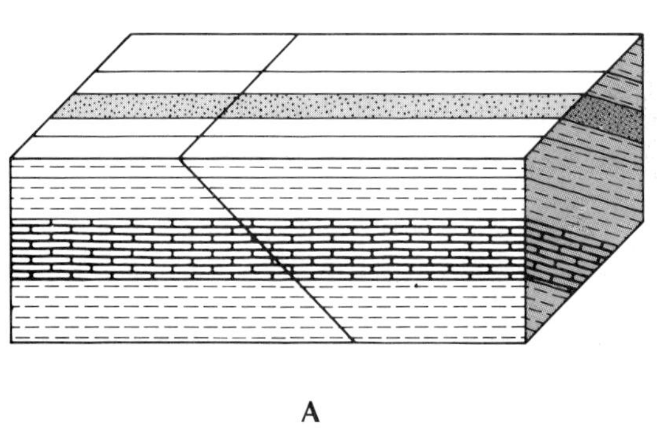

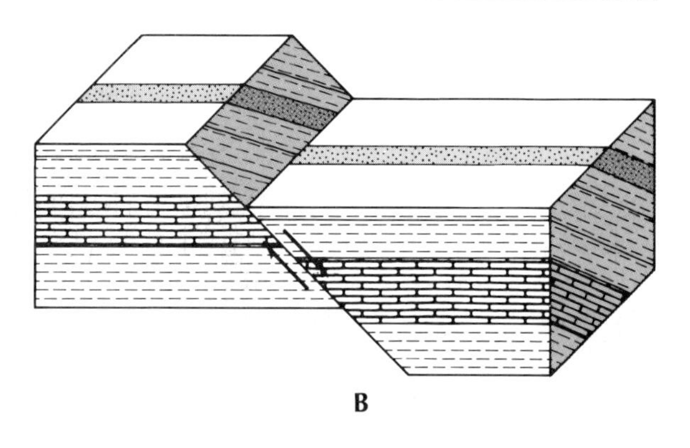

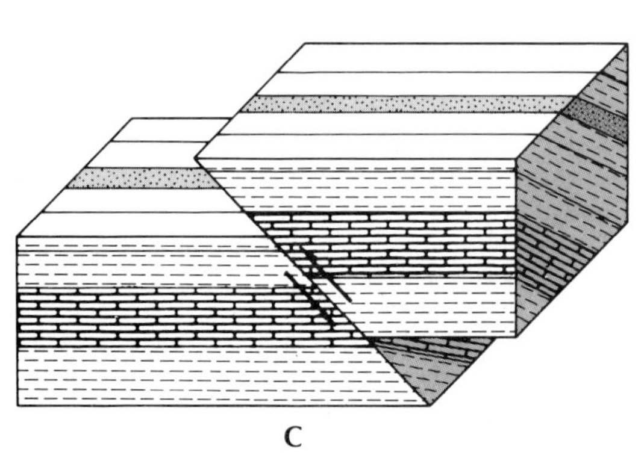

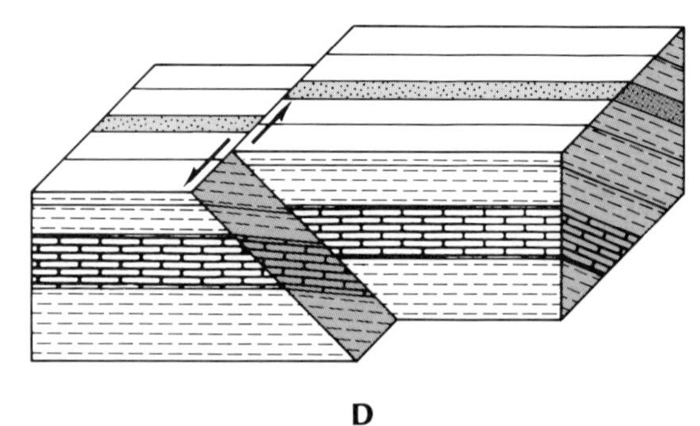

Fig. 17-29 Surface effects resulting from movement on a thrust fault during the Meckering earthquake in Western Australia in 1968. *West Australian Newspapers, Ltd.*

studied in more detail, and such faults have been recognized in parts of the world as distant from one another as California, Canada, Scotland, Switzerland, the Philippines, Japan, and New Zealand. With such a wide range, it is not surprising that a widely varied terminology has been applied to strike-slip faults. British geologists are likely to call them ***transcurrent faults,*** or ***wrench faults.*** In the United States, they are commonly called ***lateral faults***; thus, we can speak of ***left-lateral*** faults for those in which the side opposite a person facing the fault has apparently been displaced to the left, and ***right-lateral*** if the movement has apparently been to the right (Fig. 17-31).

From the statements made thus far, it should be apparent that the San Andreas Fault is one of the world's more renowned examples of a strike-slip fault. The map (Fig. 17-32), showing the trace of the fault, demonstrates its great length, roughly 960 km (597 mi) from where it cuts the coast north of San Francisco until its several branches disappear beneath the waters of the Gulf of California. The map also shows that it is not a single fracture at its southern end, but rather a complex system of faults.

Theoretically, in a fault system as extensive as the San Andreas, it would seem reasonable that the distance at which rocks have been offset on either side of the fault would be easy to determine, but actually it is very difficult. Part of the problem arises from the fact that the strike of the rocks cut by the San Andreas is nearly the same as that of the fault itself, so that intersections are not sharply defined. Also, ancient displacements along the fault are concealed beneath more recent deposits, which in a sense have covered over old scars. A tentative estimate by the geologists M. L. Hill and T. W. Dibblee would place the amount of movement along the fault complex at 192 km (119 mi) in the last 60 million years and a possible 560 km (348 mi) in the last 120 million years—roughly the age of the oldest rocks marginal to the fault. Another well-documented example of a strike-slip fault along which considerable lateral movement has occurred is the Great Glen Fault in Scotland. Almost everyone who has ever looked at a map of Scotland is likely to wonder about the nearly continuous vale that extends across the middle all the way from the North Sea to the Atlantic Ocean. The valley is also the site of Loch Ness, where the legendary sea serpent has been sighted. Fortunately, the rocks and structures on either side of the Great Glen Fault can be matched up with some confidence, and according to the British geologist W. Q. Kennedy, they indicate a strike-slip movement of about 104 km (65 mi), with the northern part of the Scottish Highlands being displaced relatively southwestward.

Overthrust faults Geologists struggling to interpret the perplexing relationships of the Northwest Highlands of Scotland early in the nineteenth century—around the time of the War of 1812—were puzzled to find sandstone, shale, and limestone interbedded in what appeared to be a normal sequence with gneiss and schist. Even such an astute observer as Charles Lyell accepted it as a conformable succession of seemingly related rocks. Fortunately for Lyell's reputation, no one in that day was really certain of the true nature of metamorphic rocks.

However, with continued field work throughout the nineteenth century, the complacent state of mind brought about by that misconception was increasingly disturbed by (1) the awareness that metamorphic rocks are not normally interbedded with sedimentary rocks, (2) the fact that some of the contacts separating different rock units are not normal depositional surfaces, but are faults, and (3) the discovery of rocks containing fossils of earlier geological periods, which were found resting on rocks containing fossils known to have lived later in geological time.

In a region as renowned for disputation as Scotland, it is no wonder that the "Highland Controversy" is one of the more celebrated intellectual conflicts in the history of geology. By 1861, the view gained acceptance that perhaps those aberrant geological relationships were the result of the gliding of great sheets of rock over one another in a fashion similar to what had been found, in 1849, to be part of the internal structure of the Alps. As a result of nearly a century of effort, the large-scale, almost horizontal displacement of many miles' extent was demonstrated to the satisfaction of most. The presence of at least three rock slices was established in the Scottish Highlands, and they were shown to have moved from the southeast toward the northwest, carrying regionally metamorphosed schists over younger unmetamorphosed sedimentary rocks.

Such faults, as described earlier, are called ***overthrust faults,*** and in typical examples, the displacement may be very large: 16 to 24 km (10 to 15 mi) are not uncommon—and in some cases may amount to 50 to 60 km (31 to 37 mi).

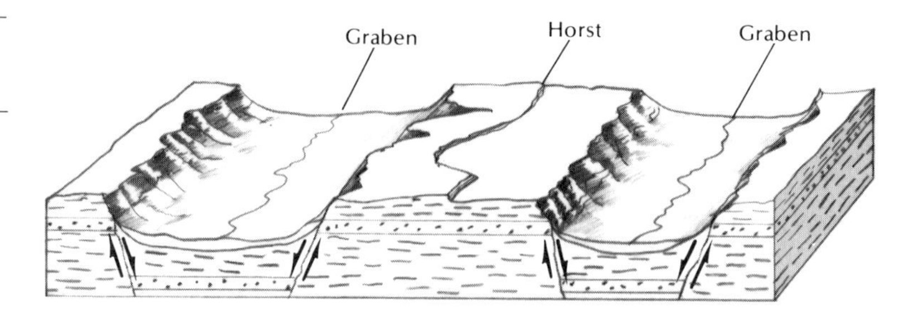

Fig. 17-30 Structural and physiographic relations of a horst and an adjacent graben, features found in many parts of the world.

Fig. 17-31 The trace of the right-lateral San Andreas Fault, Carrizo Plain, California. *A.M. Bassett*

In the United States, the pioneer work in establishing the existence of thrust faults was done around 1900, in large part through the efforts of Bailey Willis, who for many years was a professor of geology at Stanford University. Willis had been doing field work in Glacier National Park, where he showed that the eastern wall of the Rocky Mountains in the park area consists of sedimentary rocks that are more than 500 million years old resting on rocks with ages ranging from 60 to 130 million years. The differences in those unlike strata are further accentuated by the fact that the older sedimentary rocks are much more resistant to erosion and form the castellated ridges and steep cliffs scored by deep canyons that make the park landscape so renowned. The younger rocks, however, are less resistant and are responsible for the gently rolling, undulating landscape of the high plains of Montana to the east. The explanation Willis came up with was that the older sediments slid over the top of the younger sequence along a large overthrust fault (Fig. 17-33).

One of the remarkable features of the great thrust fault of Glacier Park—to which the especially appropriate name of Lewis Overthrust was given to commemorate the pioneer exploratory efforts of Meriwether Lewis in 1803–06—is the outlying peak, Chief Mountain (Fig. 17-34). That mountain is a remnant of the hanging wall of the fault, isolated by erosion, and thus in a sense it is a "mountain without roots." Such an erosionally isolated fragment of a thrust sheet is called a ***klippe*** (from the German word for cliff; pl., *klippen*). Conversely, erosion may cut through the overthrust plate of such a fault, exposing the rocks beneath (Fig. 17-35). A ***fenster,*** from the German word for window, is the term given to such an occurrence.

Many thrust faults are not regular geometric planes, but have complexly curved surfaces; in places, the dip may be 10° or less, in others it may steepen to 45° or so (see, for instance, Fig. 17-35). Some even have been folded or bent after formation as the deformational processes continued. Whether or not they have been folded, they appear to have very gentle dips near the leading edge of what may once have resembled a tongue-like lobe, and steeper dips in the root zone to the rear.

Having established the geometry of those great over-riding rock plates that have produced such large-scale displacement—somewhat like the telescoping of a collapsible drinking cup—geologists are now ready to attack the central problem of determining the mechanism responsible for their existence. Were they shoved from behind? Are they the result of gravitational gliding of great masses of rock down gentle slopes? Neither possibility can be advocated with certainty. Recent investigations, however, have shown that if the pore spaces within the rocks are filled with water under abnormally high pressure, so that a buoyant effect results, less force than had formerly been thought necessary is required to overcome friction and set such a mass in motion. Large overthrust sheets, then, may be able to slide downhill on relatively gentle slopes.

Many overthrust sheets are found flanking some of the larger complex mountain chains of the world (Fig. 17-36). It appears probable that, as mountain building progressed, the central part of the range was uplifted to some degree, thereby providing the slope necessary for overthrust slippage—and the blocks of sediments slid, like large single landslide blocks, downslope for many kilometers or even tens of kilometers. Quite commonly, the rocks in the upper plate (hanging wall), especially if they are relatively weak sediments, will be folded complexly and cut by numerous smaller thrust faults. Such an overthrust sheet, composed primarily of rumpled sediments, is called a ***décollement,*** and is common in the Jura Mountains marginal to the Alps (Fig. 17-37) and in parts of the Appalachian Mountains in the eastern United States.

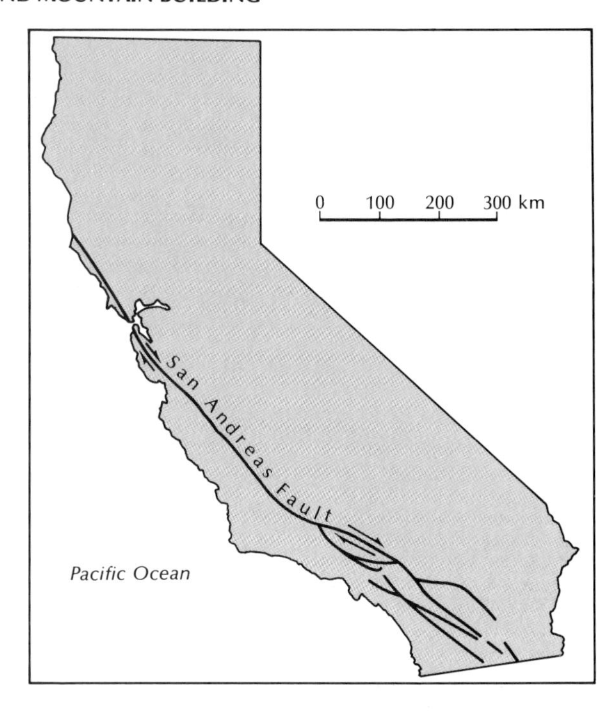

Fig. 17-32 Map of California, showing the trace of the San Andreas Fault. Note how the fault splays and splits at its southern end.

MOUNTAINS AND MOUNTAIN BUILDING

The description of such structural features as folds, joints, and faults is only a prelude to a discussion of mountains and mountain-building processes.

Of all the landforms and structural features on the earth's surface, none, surely, is closer to the heart of geologists than mountains (Fig. 17-38). The greatest variety of rocks is visible in their valleys and on their ridges and peaks. In the long, linear mountain belts, metamorphic

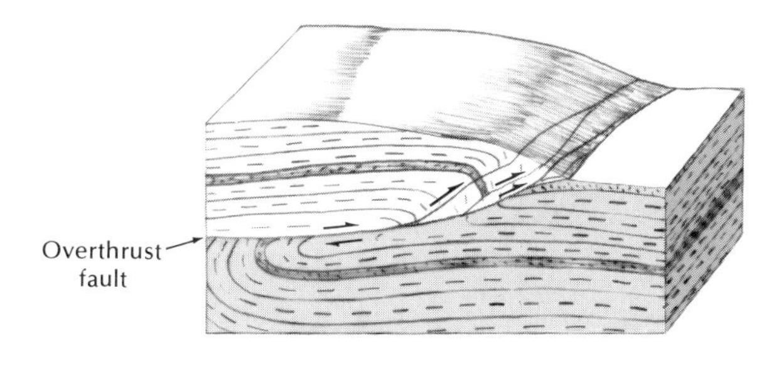

Fig. 17-33 Cross section of an overthrust fault.

Fig. 17-34 Looking northwest toward Chief Mountain with Cable Mountain directly to its left, Glacier National Park, Montana. Chief Mountain, composed of sediments over 500 million years old, sits on top of rocks ranging from 60 to 130 million years old (forming the lower, gentler slopes) and represents a remnant of an overthrust sheet that moved for many kilometers over the Lewis Overthrust fault. *Doug Erskine, Glacier National Park*

Fig. 17-35 Diagram showing the existing situation along the Lewis Overthrust Fault (A) soon after movement along it ceased and (B) after erosion brought about partial dissection. The erosional remnant, or klippe, is isolated from the thrust sheet and the window, or fenster, is eroded through the sheet.

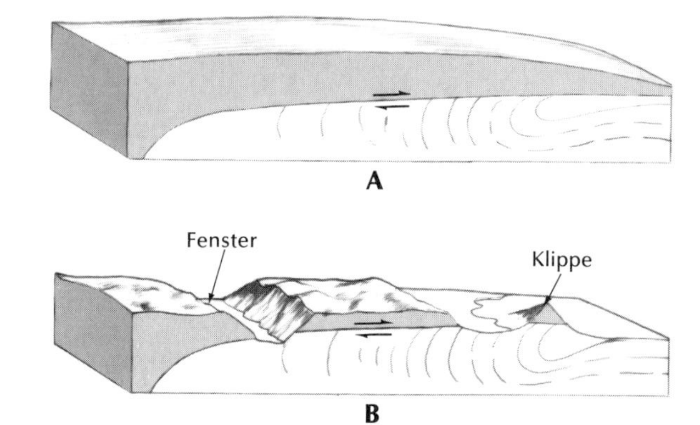

Fig. 17-36 The Keystone Overthrust Fault, in the Spring Mountains west of Las Vegas, Nevada, separates dark-colored Cambrian limestone, above, from light-colored Jurassic sandstone below. *John S. Shelton*

and plutonic rocks commonly are found in close association in the axial portions and outward from that, sedimentary rocks are broken, overthrust, and complexly folded. Dramatic erosional landforms are common in mountains: landslides and other kinds of mass movement have their maximum development; streams are more powerful because their gradients are steeper; frost-riving efficiently subdues the higher alpine summits; and last, in the high mountains glaciers make the montane scenery a source of joy and inspiration to many dwellers of the lowlands.

That mountains can be a place of solace and of beauty is a relatively new point of view. Mountains were greatly feared by travelers in medieval times, and rightly so. Roads crossing them were few, and almost all were rough and tedious. Inns were incredibly crude by our standards and distances between them, in terms of travel time, great. Dangers from landslides, cold, snow, and armed highwaymen were very real. Certainly, few people would have climbed a mountain just for the sake of climbing it.

By the eighteenth century a change set in. Not only had climbers started Alpine ascents, but curiosity about the nature of the mountain world was awakening. One of the leaders in that inquiry was a Swiss, Horace Benedict de Saussure (1740–99), who, in addition to making the first ascent of Mont Blanc, wrote a four-volume work, *Voyages dans les Alpes,* which contains, among descriptions of birds, flowers and trees, a first account of the complex structure of the rocks of those famous mountains.

Mountains have been studied scientifically through the years since the mid-eighteenth century, until today we know vastly more than our predecessors about their rocks and structure. Yet we still do not understand clearly how mountains are formed and what processes operate within the earth's crust to raise some of them—such as the Andes and Himalayas—to their imposing heights.

Every mountain and mountain range is unique however it is judged. Therefore, like many other natural phenomena, mountains are difficult to classify. We are prone to assign to rigid, pigeonhole categories features that may have had more than one kind of origin or that tend to merge with another. The short classification below serves to differentiate the more distinctive types. Yet it is not so arbitrary as to be inflexible.

Volcanic mountains are built up of an accumulation of volcanic eruptives, such as ash, pumice, bombs, and lava flows.

Fault-block mountains owe their elevation to differential movement along faults, so that some parts of the crust are raised and others are lowered relative to one another.

Oceanic ridges and rises form a nearly continuous elongate mountain system throughout the world's ocean basins.

Folded and complex mountains generally consist of igneous and strongly deformed sedimentary and metamorphic rocks. Normally, they occur in great elongate belts and require, on the average, at least several hundred million years to form.

Volcanic mountains

Some of the world's best-loved and most scenic peaks are volcanoes. Among the more familiar are Fujiyama, Mount Rainier, Mount Etna, Mauna Loa, and lofty Andean summits such as El Misti and Cotopaxi. Mauna Loa, which, counting the submerged as well as the visible part, rises about 9150 m (30,012 ft) from a base 145 km (90 mi) in diameter on the sea floor, is the world's largest isolated mountain mass.

Volcanoes constitute a fairly high percentage of the world's mountains. Add to them the extinct or mostly dormant centers, such as the Cascade volcanoes of the western United States (Fig. 17-39), and the total number is large. Could the waters of the sea be rolled away, we should be doubly impressed because the peaks of many of the volcanic islands would loom far above the surrounding abyssal plain, and we would also see for the first time numerous submarine volcanoes.

However, volcanic mountains differ fundamentally from the others in our classification, since they are accumulations of material piled

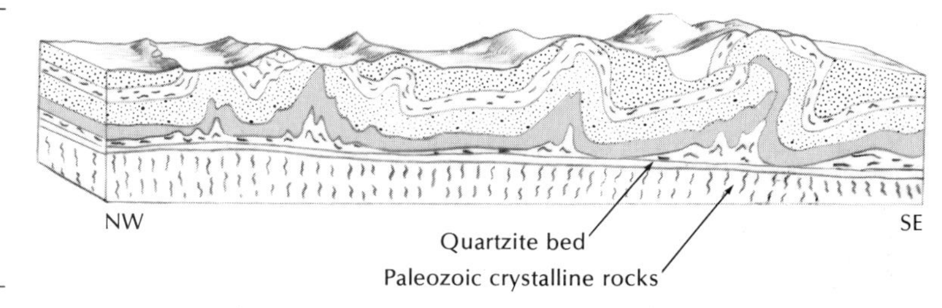

Fig. 17-37 Décollement in the Jura Mountains, western Europe. The lowest, nearly vertical structure consists of Paleozoic crystalline rocks (mostly metamorphics). Directly above is a thin bed of flat-lying Triassic quartzite, then a structurally weak unit consisting largely of layers of shale and salt along which most of the sliding has occurred.

up on the earth's surface. In Chapter 6 we discussed the great diversity of form volcanic mountains may show; there is no need to repeat the discussion here. The chief point to be emphasized is that volcanoes are heaps of pyroclastic material or lava or both. Although they may be grouped in clusters, or even in chains, as in the Cascades, they do not form the long, nearly continuous ridges so characteristic of such folded or complex mountains as the Alps or the Himalayas. Characteristically, volcanoes rise as conical or dome-like mountains above their surroundings.

Fault-block mountains

We have already discussed some fault-block mountains in connection with the description of normal and reverse faulting. They are conspicuously developed in the Basin and Range province of the United States from Colorado westward to California and from Idaho southward to Arizona.

To the geologists who accompanied the early exploring expeditions, those mountains were a challenge. Obviously, their structure was wholly unlike that of the central Appalachians, with nearly continuous ridges broken at long intervals by water gaps such as those of the Cumberland, Potomac, and Susquehanna rivers. By the 1840s, the geological structure of the Pennsylvania portion of the Appalachians had been deciphered; it was shown that the internal arrangement of the mountains was a succession of synclines and anticlines with crests and troughs that resembled a train of waves.

Although men looked for a similar geometry in the Great Basin, its discovery eluded the nineteenth-century geologists attached to the various railway surveying parties or to military expeditions. Instead of being relatively simple, like the central Appalachians, the internal structure of many of the Great Basin ranges is wonderfully complicated, with intricate patterns of thrust faults, folded stratified rocks, and complex hierarchies of igneous intrusions. Certainly there was no wave-like progression of folded strata.

It remained for one of the United States' greatest geologists, Grove Karl Gilbert (1843–1918), to solve the riddle of the origin of the Great Basin ranges. It was a difficult feat indeed, for scientifically, next to nothing was known of the remote, arid Southwest. There was not even an adequate map to show the location and extent of many of the desert ranges. As a young man of twenty-eight, Gilbert accompanied an exploring party of the U.S. Army Corps of Engineers. The difficulties imposed by the hostile terrain were severe enough, but Gilbert, as a civilian, also had to contend with the arbitrary decisions governing the movements and route of the militarily oriented, topographic surveying department. The accomplishment of so much in so short a time, while he was still an untried and largely self-taught geologist, earned him a place among the pioneers of scientific exploration.

In simple terms, Gilbert saw that the unusual topographic form of the generally straight-margined mountains, separated from one another by broad, gravel-floored basins, was most plausibly explained by the presence of one or more faults along their margins, faults by which they had been uplifted. Such features,

Fig. 17-38 Mt. Gardner, Ellsworth Mountains, Antarctica.

bounded at one or both lateral margins by large, high-angle normal or reverse faults, are called ***fault-block mountains*** (Figs. 17-40 and 17-41). For the time (1872), his was a bold generalization, but investigations of those and other mountains of the Far West, as well as of comparable ones around the globe in recent years since then, support Gilbert's prescience in making a sound generalization from the limited data available to him.

As is often true of scientific discoveries, later work demonstrates that an originally simple concept becomes complicated as more and more information comes to light. No one as yet has devised a completely satisfactory explanation of the origin of the fault-bordered mountain ranges in the Basin and Range province. It appears that geologists generally agree on the nature of the boundary faults—they are relatively straight along their strike; their dip is steep, perhaps 60° to 70°, and they resemble normal faults more than other types. There is evidence that the fault surface of some mountain-front faults decreases in dip, downward, and, several kilometers beneath the surface, incline at angles of only 40° to 50° or less (Fig. 17-41).

Clearly, vertical movements of the earth's crust have been responsible for the origin of fault-bordered mountains and troughs. There is no evidence of compression or of shortening of the earth's crust, the process that commonly is believed necessary, at least in part, for the formation of such structures as anticlines, synclines, and thrust faults. However, there is evidence of extension in the Basin and Range country across its entire east-west width—about 700 km (435 mi). If one were to construct a model of the numerous horsts and graben, and to slide the blocks back along their inclined fault planes to their original unbroken condition, there would be a reduction in the east-west dimension—that is, at right angles to the fault trends—equivalent to about 64 km (40 mi). That amount of extension has occurred in the region during the last 25 million years.

In addition to east-west stretching, there is abundant geological evidence that the area, in general, has been uplifted over the same time period, particularly the last 10 million years or so. No one really knows what could bring about both the uplift and stretching of a large segment of continental crust. It has been suggested that the East Pacific Rise, a zone where two plates are pulling apart (diverging), does not die out where it butts into the North American plate near the southern tip of Baja California, but continues northward beneath the continent (Fig. 17-42). The extension and uplift of the basins and ranges, then, are thought to represent cracking and rifting of the crustal blocks above the land-locked rise system. But a drawback to the theory is the extremely wide zone of extension. Most earth scientists think that the zone of stretching associated with a diverging plate boundary should be concentrated in a narrow band right along its crest—as in the rift systems in the Red Sea area and eastern Africa.

Moreover, many earth scientists do not believe that the East Pacific Rise extends beneath the continent. They believe either that the rise ceases to exist where it abuts Baja California or that it has been offset northward along the San Andreas Fault system. They point to the small segment of a zone of divergence found in offshore Oregon and Washington as the other end of the severed rise system.

There is one bit of geophysical data that probably has some bearing on the origin of the Basin and Range province. The entire western part of the United States, including all the areas that have been structurally and volcanically active during the last 30 million years, is underlain by an anomalous upper mantle that extends 100 km (62 mi) or so downward, below the base of the crust. The mantle is anomalous because it is less dense than it should be. Some geophysicists, therefore, have speculated that in some way the upper mantle has changed chemically to produce a relatively less-dense (and, accordingly, more-voluminous) material. In response to the increase in volume in the mantle, the crust above has been uplifted and stretched. What might cause such chemical change is unknown at present.

Fault-block mountains larger than those in the Basin and Range province (commonly ris-

Fig. 17-39 A chain of volcanic mountains form the Cascade Range in California, Oregon, and Washington. Mount Jefferson is closest, and in the background, from left to right, are Mounts St. Helens (before the 1980 eruption), Rainier, Hood, and Adams. *John S. Shelton*

ing to elevations over 3800 m; 12,464 ft) are found in New Mexico, Colorado, and Wyoming. In fact, most of the large, generally north-south to northwest-southeast trending ranges in those three states, which include the Sangre de Cristo, Front, Gore, Sawatch, Owl Creek, Wind River, and Big Horn ranges, owe their existence, in part, to block faulting that has occurred along one or both sides of the mountain blocks. In many cases, the mountain block has been raised so high that erosion has stripped off most of the overlying Paleozoic

Fig. 17-40 Looking northwestward across the Paradise Range, Nevada, a mountain block that has been relatively uplifted along high-angle faults that more or less parallel the margin of the range. *J.R. Balsley, USGS*

and Mesozoic strata to expose the Precambrian crystalline rocks beneath—rocks that were at least 7625 m (25,010 ft) below sea level only about 70 million years ago. Oil-well drilling and field studies have shown that most of the faults are almost vertical at depth, but, as they are traced upward, they flare outward into high-angle thrust faults. In some instances, the thrust fault flares outward to such an extent that it becomes a low-angle thrust or overthrust fault. It is almost as if the mountain block expanded as it rose along the frontal faults, much in the way a cork expands as it is pulled from a wine bottle.

Whatever the cause of those mountains, the rocks of the upper mantle and the overlying crust have been pushed upward for distances as great as 13 km (8 mi). Vertical movements of crustal blocks of such large magnitudes are hard to fit into any existing theory of mountain building. In fact, the existence of such mountains so far from the active plate margins is particularly hard to fit into the plate-tectonic model.

Oceanic ridges and rises

The world's longest mountain chain lies mostly beneath the ocean's surface. As described in Chapter 3, the ridge-and-rise system stretches, almost unbroken, nearly 64,000 km (39,770 mi) throughout the world's ocean basins. The range, basically unlike any on the continents, appears to be caused by a broad, linear rise of the ocean floor. It is cut by abundant normal faults that parallel the ridge axis, and commonly it is offset, for distances up to 1000 km (621 mi) and more, along fracture zones that cross it at nearly right angles.

The ridge system is the site of many shallow-focus earthquakes that occur directly beneath the ridge axis and along parts of the cross-cutting fracture zones. Volcanic activity is common along the ridge axis, and many earth scientists feel that the elevation of the ridge is due in part to the accumulation of basaltic lavas and the injection of multitudes of dikes at shallow depths. Some believe, however, that the elevation of the linear welt is caused by a volume increase as a result of serpentinization of olivine-rich rocks along a narrow zone beneath the sea floor.

It has become apparent that oceanic ridges and rises are located astride the zone where two plates of lithosphere are being pulled apart—the zones of divergence. Such oceanic mountain chains, then, are closely associated in some way to the origin and movement of lithospheric plates. We will leave further discussion of the subcrustal process responsible for such features to the chapter on continental drift and plate tectonics (Chap. 20).

Folded and complex mountains

When geologists think of mountain-building processes, they usually have in mind those that produce ***alpine chains,*** the complex linear mountain belts found on all continents (Fig. 17-43). Because of their complexity, no two alpine chains are exactly alike, either in their features or in their history of development. Much, if not most, of the thought on the origin of mountains has been given to alpine chains.

Alpine chains are essentially continuous, extending for several hundred up to 1600 km (994 mi) or more in length, with most of the following features:

1. A core of structurally complex, dynamothermally metamorphosed rocks (generally sedimentary or sedimentary and volcanic in origin), which are locally intruded by and intimately associated with granitic plutonic rocks, particularly batholiths.
2. Elongate belts of thick sections of relatively unmetamorphosed sedimentary rocks, usu-

Fig. 17-41 Typical fault block mountains of the Basin and Range province, western United States. Many of the range-border faults decrease in dip, downward, as shown. Arrows at the base of the block indicate the directions of extension required to form these mountains.

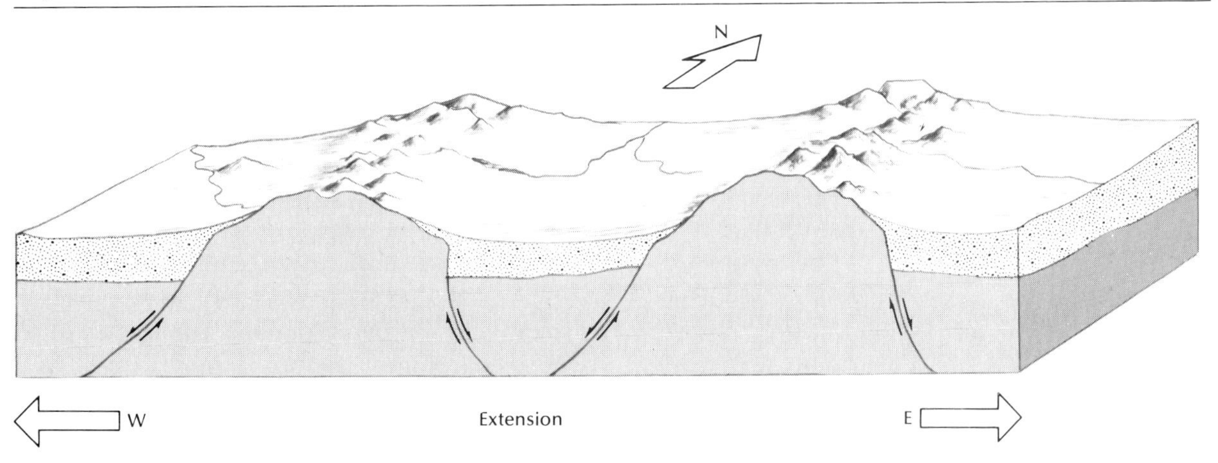

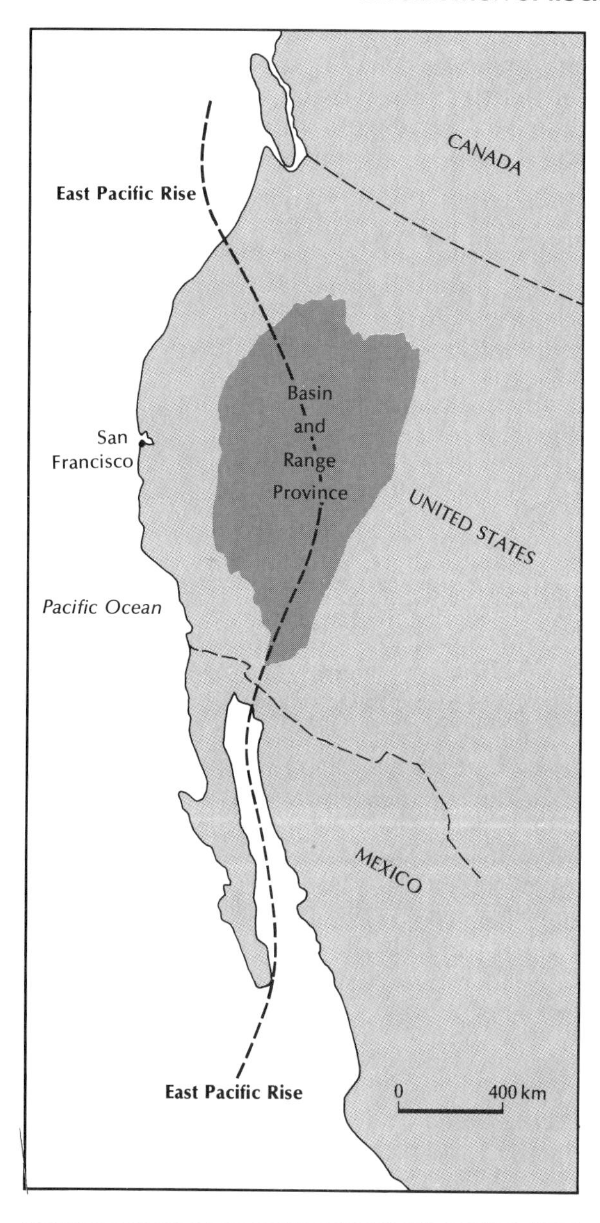

Fig. 17-42 Western North America, showing the general area of Basin and Range province and the possible location, at depth, of the East Pacific Rise divergence zone.

ally on one side, but in places on both sides of the core, which have been complexly folded, overturned, and thrust faulted. Décollement structures are common. The degree of folding and faulting decreases with distance from the mountain core until, at a sufficiently great distance, the rocks are essentially flat lying. The thickness of the sedimentary section decreases progressively from the core region outward.

Most alpine mountains, just after they are formed, apparently rise high above sea level to form linear topographic prominences, such as the Alps and Himalayas. Even after erosion has essentially reduced them to low-relief features once again, if the main geological elements just outlined are recognizable, it is still possible to describe the feature as an alpine mountain chain. The Appalachians in eastern North America certainly do not have the surface grandeur of many mountain ranges. Yet they possess all the features just enumerated, in a nearly classical manner, and can be spoken of as an alpine chain, even though the last mountain-building activity there occurred about 225 million years ago (Figs. 17-44 and 17-45).

Commonly, but not always, alpine chains are formed where thick sections of sedimentary rocks have accumulated. That relationship was commented on in 1859 by James Hall (1811–98), a young and energetic member of the pioneering New York Geological Survey, at a time when much of upstate New York was still a frontier. He found that Paleozoic strata underlying the Mississippi Valley states, such as Iowa and Illinois, were much thinner than those of identical age underlying the Allegheny Plateau of New York and Pennsylvania. From that observation, as well as from the knowledge he acquired through reading, he reasoned that many of the world's mountains had somehow been formed in regions with abnormal thicknesses of sedimentary rocks.

A contemporary of Hall's, James Dwight Dana (1813–95), had done his stint of wandering as a geologist attached to the vessels of the U.S. Exploring Expedition (1838–42). It was a round-the-world investigation that deserves to stand with the much better-known voyages of the *Beagle* and the *Challenger.* Dana was greatly impressed by Hall's discovery of the abnormal thickness of sedimentary rocks exposed in the Appalachians and, recognizing the rather unusual circumstances under which such sediments accumulated, coined the name *geosynclinal,* now shortened to ***geosyncline,*** for the immense trough in which they were laid down (Fig. 17-46).

Although the two men agreed on the physical relationships of the rocks and structures in the Appalachians, they differed strongly on how they got that way. Hall believed that the earth's crust was bowed down as a consequence of load imposed upon it by a localized accumulation of sediment. The more sediment that was deposited in such a wedge, the deeper the crust would be bowed downward. Obviously, Hall reasoned, such a situation made for a rather uncertain equilibrium. The

Fig. 17-43 Looking across the rugged, alpine terrain of central Switzerland. In the foreground is Mt. Pilatus, to its right the Bürgenstock, and in the background, the Lake of Lucerne. *Swiss National Tourist Office*

comparatively weak material of the sediments could be crumpled readily and thrown into contorted folds by inward movement of the vise-like margins of the trough. Fracturing of the infolded sediments could also provide avenues through which magma could invade shallower zones in the earth's crust or possibly even reach the surface.

Dana, on the other hand, did not believe that the weight of such accumulated sediments was capable, in itself, of bending the crust downward. Without the knowledge that seismology has given us today of varying densities of the material in the earth's crust and mantle, he recognized that only with great difficulty could relatively light, water-soaked, unconsolidated or recently consolidated sediments displace higher density subcrustal material. Dana believed that the earth was slowly contracting through time and that the contraction of the interior (very likely, he believed, through loss of heat) caused a geosyncline to form and eventually led to the folds and faults in the

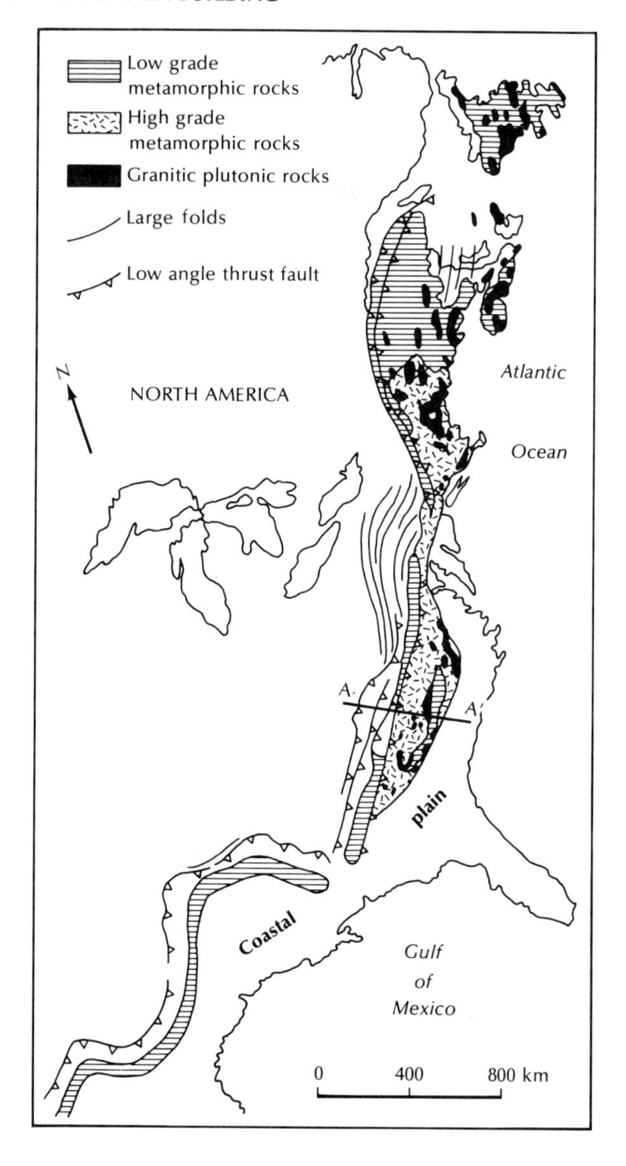

Fig. 17-44 (right) Main structural features of the Appalachian Mountains.

Fig. 17-45 (below) Cross section A-A' (see Fig. 17-44) across the Appalachian Mountain belt. In the axial portion are high-grade metamorphic and plutonic rocks; outward from the axis, the degree of metamorphism and the intensity of deformation decrease.

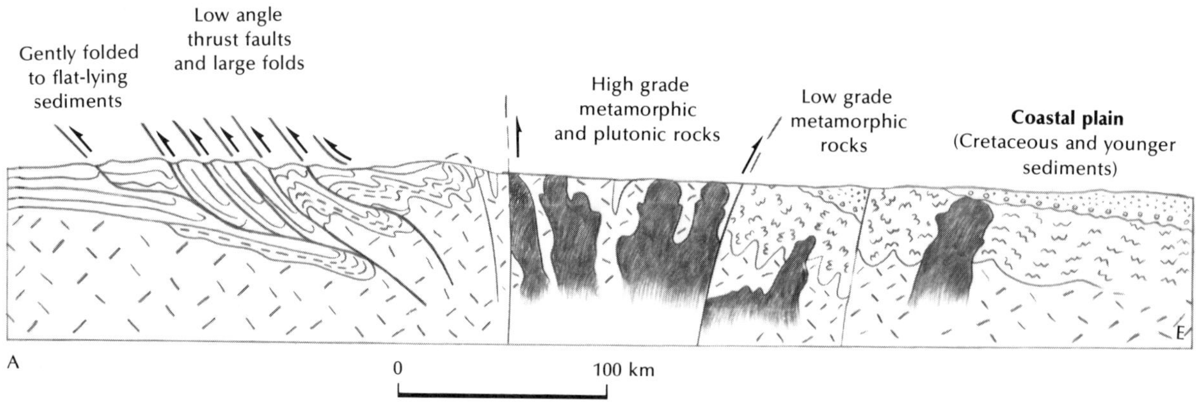

thickened geosynclinal sedimentary section, much like the wrinkles on the skin of a dried apple. Ultimately a mountain chain like the Appalachians would be formed.

In the century since then, both points of view have continued to find their adherents, but as we have learned more and more about mountains and the physical properties of the earth, we have been forced to search for more sophisticated theories. Plate tectonics, for example, can more fully explain all the features of alpine mountains and other geological features of the earth as well.

To become more familiar with the way an alpine mountain chain develops, let us look in more detail at the Appalachian system. Erosion during the last 225 million years since the last stages of its development has been both a blessing and a bane. Because of erosional stripping we can now look deep into the innards of the range to see, more or less, what transpired far below the surface. Unfortunately, the upper levels of rocks, which undoubtedly were different from those below, have been stripped away, removing the record of the processes occurring near the surface in the geosyncline. There is general agreement, however, that the Appalachians progressed through an evolutionary sequence somewhat as follows:

1. A lens of sediment, perhaps 9150 to 12,200 m (30,012 to 40,016 ft) thick, 3200 km (1988 mi) long, and 480 km (298 mi) broad, accumulated slowly during the later part of Precambrian time and throughout the Paleozoic Era—roughly from around 650 million to 225 million years ago. From all the evidence, the source of most of the sediment was east of the present-day Appalachians. Sometimes the source land stood relatively high, and streams emptying into the geosynclinal sea carried coarse material, such as boulders, gravel, and sand. At other times the source land may have stood lower. One certainty from the record of the geosynclinal strata is that there was considerable volcanic activity in the region. The volcanoes probably made their appearance as volcanic island arcs that occurred from time to time in the geosynclinal trough. Some became sufficiently high to project above the sea that covered the geosyncline, only to stagnate and to be planed off by subsequent erosion, their debris added to the continuous rain of sediments filling the trough. The evidence of the physical and the paleontological record is that the water was generally less than 2000 m (6560 ft) deep during the 425 million years the trough was in existence. During part of its history, the time in which the coal beds of West Virginia and Pennsylvania were being laid down, some of the geosyncline was a broad marshland above sea level. At other times, it may have been an expansive delta, perhaps resembling the Mississippi Delta of today. The trough continued to downwarp during its development and the sediments continued to be carried in to fill it.

   As time passed, much more happened than a continuous, but nearly imperceptible sinking of the geosynclinal floor. From time to time the rocks were compressed and deformed. These intervals are recorded by unconformities in the rocks and by such structures as anticlines, synclines, and faults. There were occasional volcanic episodes and alternating incursions and retreats of the sea.
2. Sporadically, during the development of the range, rocks in some parts of the deeper central portion of the basin were heated to temperatures high enough to bring about regional metamorphism, and slates and

Fig. 17-46 Hypothetical cross section through the Appalachian geosyncline as it filled with sediments, showing the thickening of strata toward the axial portion. The maximum thickness of sediments amounted to about 12,200 m (40,000 ft).

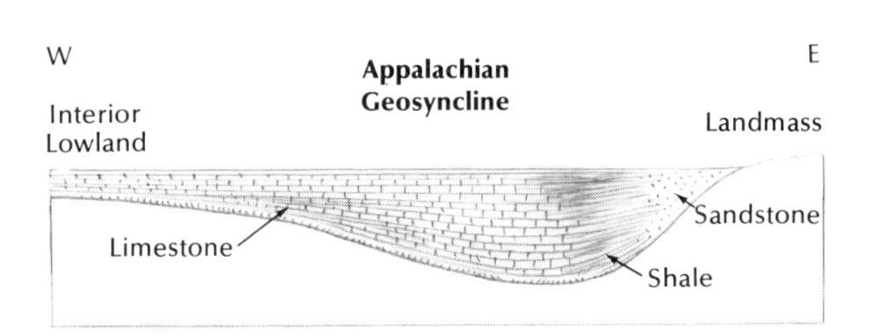

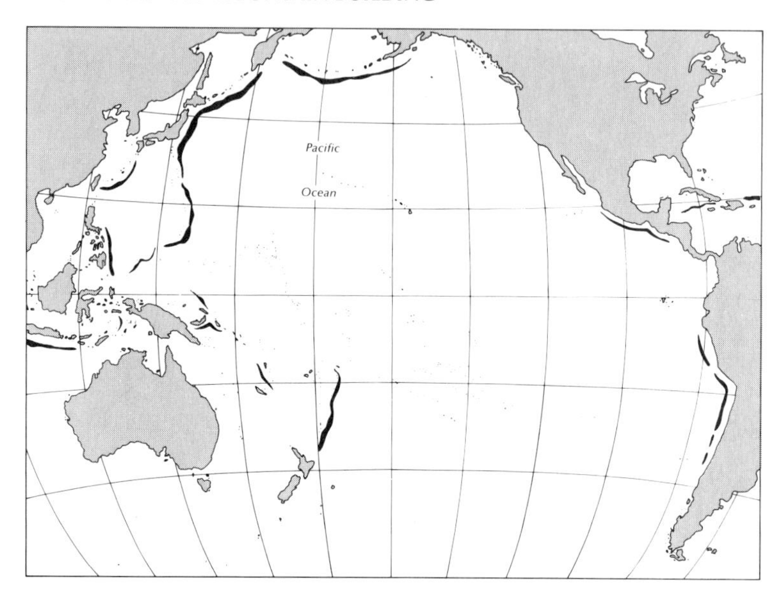

Fig. 17-47 Deep ocean trenches, shown in black, demarcate zones of convergence. Note that most border the Pacific Ocean basin.

schists were formed from the sedimentary and volcanic rocks. In some portions of the geosyncline, temperatures became high enough for melting to occur, and the bodies of granitic magma that formed worked their way upward to fairly shallow levels in the crust.

3. The long depositional episode ended at the close of the Paleozoic Era, about 225 million years ago. Then, the thousands of meters of accumulated strata were further metamorphosed and thrown into folds or were broken by great, low-angle thrust faults—the latter chiefly in the southern part of the range.

   Undoubtedly at that time a mountain range of considerable stature must have stood in bold relief to the lower-lying countryside around it.

4. After some minor episodes of deformation and volcanism, the mountains eventually were worn away, to be replaced by an erosional surface that had been beveled across the rocks and structures of the former chain.

Essentially, the same general sequence of events occurred in other parts of the world, at different times, as for example, in western North America during the Paleozoic and Mesozoic eras. Commonly, the rates of development in the other ranges were different, or, because of local geological situations, the number of steps varied. Each alpine range, then, is the product of a unique set of events.

Modern counterparts of ancient geosynclines and plate tectonism One might ask where today are geosynclines being developed and filled with sediments and where are alpine mountains being uplifted? The problem, of course, is in recognizing—mainly from features exposed at the surface—that mountain building is going on at depth.

What are the outward signs of metamorphism at depth? How does one determine that magmas of batholithic proportions have formed and are slowly rising? The absence of volcanic activity is no sure sign, since batholiths vent at the surface only after they have reached rather shallow levels. Conversely, volcanism may not be a valid indicator, since we know of many volcanoes and large plutons that have developed in response to processes other than alpine mountain building.

Will an alpine chain be formed wherever sediments accumulate to great thicknesses? The answer is a tentative no. In the region of the Gulf Coast of the United States, near the delta of the Mississippi River, sediments have piled up to a thickness of 12,000 to 15,000 m

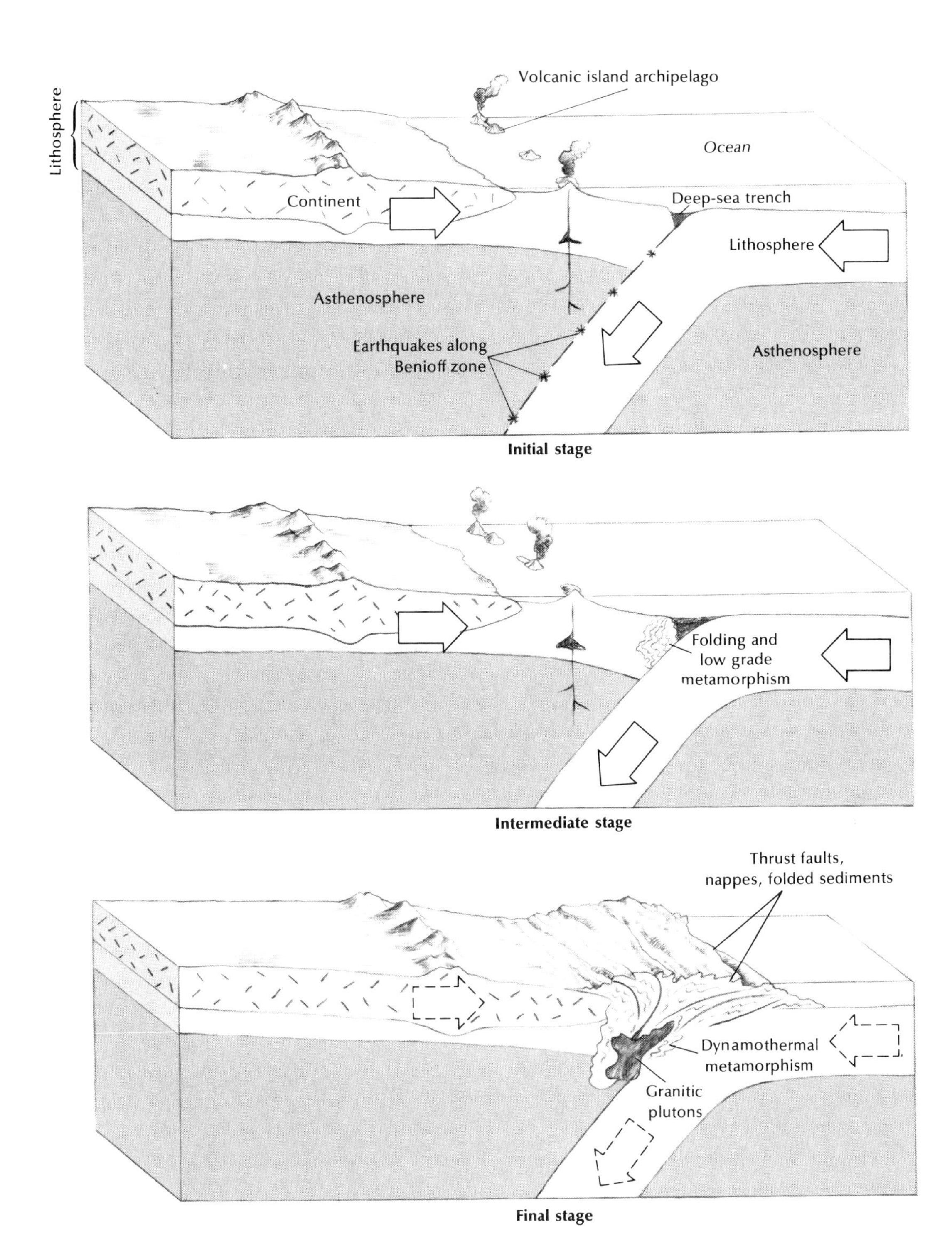

Fig. 17-48 Postulated sequential development of an alpine mountain system along a zone of convergence, according to the plate tectonic theory.

(39,360 to 49,200 ft)—a thickness equivalent to that found in many earlier alpine geosynclines. Yet there are few geologists foolhardy enough to say that the north side of the Gulf of Mexico is the site of a developing alpine chain. The great thickness of the sediments seems related mostly to the dumping of large amounts of clastic debris in the restricted Gulf waters by one of the world's largest rivers—and probably has little to do with mountain building. Moreover, sedimentation in the region has not been associated with volcanism, seismic activity, or any of the other aspects of mountain building we might expect to occur. Deep drilling for oil in the area has shown that the sediments are still largely unaltered, metamorphically and structurally.

In other parts of the world—Australia and the USSR, for example—there are found old sequences for essentially unaltered and undeformed sedimentary rock as thick as those on the Gulf Coast. Certainly from the point of view of thickness alone, the deposits are geosynclinal in character; yet mountain building has not occurred in those areas.

Theory of alpine mountain building Several theories have been advanced to explain the development of alpine mountains, but most have proved restricted in scope and have not met the test of time. Probably the best explanation of alpine features is the plate tectonics theory, which proposes that the likely place to expect mountain building is along the boundary where two plates converge (Fig. 17-47). Such a place is characterized by active andesitic volcanism, seismic activity down to 700 km (435 mi) in many arcs, and linear troughs deep enough to accumulate great thicknesses of sediments derived from the erosion of nearby continents and from volcanic islands. Such areas, of course, correspond to the island archipelagoes that border the Pacific Ocean today. Sediments that fill the geosynclines can be carried down to great depths in the active zone of convergence, where they can be heated, dynamothermally metamorphosed, melted to produce batholithic magmas, and folded and faulted. A diagram of how the process might unfold with time is given in Figure 17-48.

One island-arc segment, that of Japan, has had a long and complex geological history. The rocks exposed in the islands are complexly deformed, show signs of having been dynamothermally metamorphosed, contain numerous granitic stocks and batholiths interspersed with the metamorphic rocks, and include considerable thicknesses of sediments derived both from the Asian mainland and from the eruption and erosion of andesitic volcanoes. Many geologists point, therefore, to Japan as a site of active mountain building through the process of plate convergence.

SUMMARY

1. Rocks in the outer part of the earth are currently being bent, broken, uplifted, and down-dropped in response to the action of internal forces.
2. Evidence that deformation occurred in the geological past is seen in *jointed, folded,* and *faulted* rocks. Folding produces *synclines* and *anticlines,* which may be *symmetric, asymmetric,* or *overturned. Faults* are classified on the basis of known directions of relative displacement (*normal fault, reverse* or *thrust fault, strike-slip fault,* and *overthrust fault*).
3. There are many kinds of mountains: volcanic, fault-block, oceanic ridges and rises, folded and complex. Most difficult to understand are the generally linear *alpine chains,* such as the Appalachian Mountains. They are complex in structure and require up to hundreds of millions of years to form. Most contain a core of dynamothermally metamorphic and associated batholithic rocks flanked on one or both sides by sedimentary rocks. The degree of deformation in the strata progressively lessens from the core outward.
4. Most geologists believe that mountains are formed at converging plate boundaries, but modern counterparts of the ancient geosynclines are difficult to identify. They possibly occur in the ocean deeps and associated island-arc systems.

QUESTIONS

1. What is some obvious evidence that the earth's outer portion has been structurally deformed?
2. Discuss the factors that influence how a body of rock will respond to deforming forces.
3. Describe the various kinds of folds. How do domes, anticlines, and monoclines differ?
4. Compare joints with faults.
5. How does the nomenclature of faults differ when only the apparent relative movement is known, as compared to when the relative displacement directions are known?
6. Describe lateral faults in general and the San Andreas Fault in particular.
7. In what type of plate-tectonic boundary do you think overthrust sheets would be most likely, and why?
8. Discuss the various kinds of mountains, giving examples and outlining the principal features that differentiate one kind from another.
9. What are the proposed steps in the development of an alpine mountain range?
10. After the erosion of the superstructure of an alpine range, do you think that there would still be features that would indicate its former existence?

SELECTED REFERENCES

Billings, M. P., 1960, Diastrophism and mountain building, Geological Society of America Bulletin, vol. 71, pp. 363–98.

Billings, M. P., 1972, Structural geology, 3rd ed., Prentice-Hall, Englewood Cliffs, New Jersey.

Clark, S. P., Jr., 1971, Geologic structures, Chap. 2, pp. 9–25 *in* Structure of the earth, Prentice-Hall, Englewood Cliffs, New Jersey.

Collett, L. W., 1927, repr. 1974, The structure of the Alps, R. E. Krieger Pub. Co., Huntington, N.Y.

Curtis, B. F., ed., 1975, Cenozoic history of the southern Rocky Mountains, Geological Society of America, Memoir 144.

de Sitter, L. U., 1964, Structural geology 2nd ed., McGraw-Hill Book Co., New York.

Dewey, J. F., and Bird, J. M., 1970, Mountain belts and the new global tectonics, Journal of Geophysical Research, vol. 75, pp. 262–65.

Dott, R. H. Jr., and Batten, R. L., 1976, Evolution of the Earth, McGraw-Hill Book Co., New York.

Eardley, A. J., 1951, Structural geology of North America, Harper and Bros., New York.

Gilluly, J., 1949, The distribution of mountain building in geologic time, Geological Society of America Bulletin, vol. 60, pp. 561–90.

Gilluly, J., 1970, Crustal deformation of the western United States, *in* The megatectonics of continents and oceans, H. Johnson and B. C. Smith, eds., Rutgers University Press, New Brunswick, New Jersey, pp. 47–73.

Griggs, D. T., 1939, A theory of mountain building, American Journal of Science, vol. 237, pp. 611–50.

Hill, M. L., 1947, Classification of faults, American Association of Petroleum Geologists Bulletin, vol. 31, pp. 1669–73.

Hill, M. L., and Dibblee, T. W., Jr., 1953, San Andreas, Garlock, and Big Pine faults, California, Geological Society of America Bulletin, vol. 64, pp. 443–58.

Kay, M., 1951, North American geosynclines, Geological Society of America, Memoir 48.

Kennedy, G. C., 1959, The origin of continents, mountain ranges, and ocean basins, American Scientist, vol. 47, pp. 491–504.

King, P. B., 1959, The evolution of North America, Princeton University Press, Princeton, New Jersey.

Murray, G. E., 1961, Geology of the Atlantic and Gulf coastal province of North America, Harper and Brothers, New York.

Oakeshott, G. B., 1966, San Andreas fault: geologic and earthquake history, 1966, Mineral Information Service, October, pp. 159–65, California Division of Mines and Geology, Sacramento.

Rubey, W. W., and Hubbert, M. K., 1959, Role of fluid pressure in mechanics of overthrust faulting, Geological Society of America Bulletin, vol. 70, pp. 167–206.

Spencer, E. W., 1969, Introduction to the structure of the earth, McGraw-Hill Book Co., New York.

Umbgrove, J. H. F., 1947, The pulse of the earth, Martinus Nijhoff, The Hague, Netherlands.

18

EARTHQUAKES AND THE EARTH'S INTERIOR

Many of history's major catastrophes have been related to earthquake activity. It has been estimated that over the last 4000 years about 15 million people have died as a result of earthquakes. Each year, more than a million earth tremors occur around the world, and of those, about 50 are intense enough to cause significant loss of property and lives. About 10 each year can be considered major events capable of causing untold death and destruction within a moment's time in heavily populated areas.

The most sinister aspect of an earthquake is that it strikes with little or no warning. At one moment the landscape is serene; at the next it gyrates and dances.

CASE HISTORIES OF EARTHQUAKES

Perhaps the best way to become acquainted with the way in which earthquakes affect the earth and human beings is to recount a few of the numerous case histories.

Fig. 18-1 Damage sustained by the church of Monte di Buia, Italy, during the Friuli earthquake of May 6, 1976, which killed nearly 1000 people in northern Italy. *Courtesy of James Stratta*

United States

As yet, the United States has been fortunate in that there has been relatively small loss of life and property resulting from earthquake activity. Yet in the last 200 years, there have been several tremors of sufficient intensity to have caused immense damage had they occurred in an urban area.

San Francisco In the spring of 1906, San Francisco was the queen city of the California coast and the gateway to the Orient. Her business districts hummed by day and the Barbary Coast by night.

Early in the morning of April 18, 1906, when the schooner *John A. Campbell* was running on a southeast course with Point Reyes in northern California due east 233 km (145 mi), the vessel shuddered, almost as if she had run aground. The startled crew could scarcely believe their senses because the chart showed a depth of 2400 fathoms (14,389 ft) at their position. Although the crew had no way of knowing it, they were only a few among the many whose daily routine was disturbed, or even ended forever, by the events set in motion at 5:12 A.M. throughout a region covering about 975 thousand km^2 (376 thousand mi^2) surrounding San Francisco. That disaster of more than seventy years ago is still a most instructive example today because it devastated an essentially modern city. The population of the United States is becoming increasingly urbanized, and many of the problems faced at that time by the people of San Francisco are exactly the ones that might confront Civil Defense agencies today, with the additional burden of immense traffic jams, unimaginable a generation ago.

San Francisco in 1906 in some ways resembled the city of today, but in others it was very different. Cobblestone streets lined waterfront areas known as the Barbary Coast, and the air was pungent with the aroma of roasting coffee, and the beery blasts emanating from the dark, block-long sawdust-floored saloons. The delicate tracery of the upper yards and rigging of the Cape Horners rose above the pier sheds, and the waters of the bay were whitened by the paddle wheels of ferries and Sacramento River boats.

Much of that colorful world was obliterated in a series of violent shocks in the early morning hours (Fig. 18-2). Had the earthquake struck later when people were up and about, the casualty list would have been much larger. The number of people who died is not known, but it may have been as high as 700. Many transient residents, in such places as sailors' boarding houses, simply vanished, and, since even the sketchy pre-1906 records disappeared in flames, many of the permanent inhabitants could not be traced. Almost all the destruction was attributed to the fire that followed the tremor (Fig. 18-3). Fire was indeed the leading destroyer, but earthquake damage was not negligible, perhaps averaging as much as 25 per cent of the total damage.

Newer buildings in the San Francisco of 1906 looked much like those in the older downtown sections of American cities today. By 1906, riveted steel frames were coming into wide use for taller structures. Exterior walls were more commonly faced with masonry than they are today, and reinforced brick was widely used for smaller commercial buildings. In that day the Victorian influence prevailed, and most buildings were encrusted with ornamentation and gingerbread—real earthquake hazards. San Francisco was unusual in one regard, and that to its sorrow, in that it was one of the larger cities of the world in which the buildings were constructed principally of wood. Typically, the residential section consisted of block-long rows of wooden multi-storied houses or apartments on narrow lots. The fire was essentially unstoppable, once it reached those buildings.

The San Francisco earthquake emphasized that the kind of ground on which buildings are constructed is crucial in determining the extent and nature of structural damage. Those founded on solid rock showed slight damage when compared with virtually identical structures built on waterlogged or unconsolidated ground. Damage was especially severe, for example, in downtown San Francisco, where an area about 20 blocks square had been built on ground reclaimed from the bay after the Gold Rush of 1849. There, on a sludgy foundation made up of sunken ships, water-soaked refuse, bottles and bodies, all buried under poorly consolidated mud and silt, the most severe damage occurred.

Fires from a variety of causes broke out at many points almost immediately after the strongest shocks. At first there was little awareness that fire was the true enemy. People either gawked at the blazes or attacked them on a piecemeal basis—and not too successfully, because the alarming discovery was made almost at once that there was no water for the fire hoses. Most water lines beneath the streets had ruptured, and water flow dropped to a pathetic trickle.

Fire started near the waterfront and swept inland across the broken city. Should you ever visit San Francisco, try to visualize the swath,

Fig. 18-2 Earthquake damage in San Francisco, California, April 18, 1906. Although the quake caused much damage, the fire that subsequently swept unchecked through the city caused most of the property loss. *W.C. Mendenhall, USGS*

eighteen blocks deep, swept by fire from the Embarcadero at the waterfront inland to Van Ness Avenue, the first wide street where a fire line could be held. Elsewhere vain attempts were made to check the advancing flames by dynamiting whole rows of buildings to keep the fire from leaping from roof to roof as a crown fire does in the forest. Unfortunately, the dynamiters had little expertise, and as often as not the explosions hurled burning material far and wide, causing new fires in buildings as far as a block away.

Among those who witnessed the San Francisco earthquake and fire were a young boy,

Fig. 18-3 Sacramento Street, San Francisco, just after the 1906 earthquake. Note how the brick fronts of the buildings spilled across the streets. The people are watching one of the great fires resulting from the quake. *Arnold Genthe, California Palace of the Legion of Honor*

John Ohrwall, and his parents. His father was so shocked by the catastrophe that he salvaged his belongings as best he could and moved the family about 160 km (99 mi) away, to just south of Hollister; far enough, he thought, to be free from subsequent quakes. As fate would have it, he traveled southward, paralleling the San Andreas Fault, and built his new home next to the fault. John Ohrwall eventually went into the wine business and became superintendent of the Almaden Winery. Unbeknownst to him, one of the winery buildings was constructed exactly on the fault line; now, each year, scientists flock to the Almaden Winery to view and measure the cracking and splitting of the walls resulting from continued earth movement. Fate has not been so cruel as it might have been, for along that section of the San Andreas Fault, the fault walls do not stick for long periods of time. As a result, that section of the fault is relatively free from strong earthquakes—at least for the present.

From both a technological and a scientific point of view, the San Francisco quake was a particularly significant event. Immediately following the quake, the California State Earthquake Commission, appointed by the governor, but supported by the Carnegie Institution, conducted an exhaustive investigation. Hundreds of people were interviewed and evidence was collected from every damaged area.

Sudden slippage along the San Andreas Fault was the primary cause for the earthquake, which was demonstrated beyond any reasonable doubt (Fig. 18-4). Not only was faulting established as the mechanism, it was shown unequivocally that the displacement was strike-slip. Before 1906 the importance of that type of fault movement was only dimly appreciated.

But there for all to see was a nearly continuous trail of furrowed ground, fractured barns, and other features that stretched for hundreds of kilometers, with all the evidence showing that the western side of the San Andreas Fault moved horizontally northward with respect to the eastern side, which moved southward. The fault has a length of perhaps 966 km (600 mi) on land; during the 1906 quake, slippage occurred along about one-half of its length—from Point Arena north of San Francisco to San Juan Bautista to the southeast.

The maximum offset of 6.4 m (21 ft) was near Tomales Bay north of the city; elsewhere in the neighborhood of San Francisco it held rather consistently at around 4.6 m (15 ft). Most observers noted that the maximum displacement was in rather soft ground, which appeared to have "lurched," thereby amplifying the offset. A separation of about 5 m (16 ft) appears to be closer to the true maximum offset.

In the study authorized by the California Earthquake Commission the offset was remeasured by topographic survey stations in and around San Francisco. On the basis of the results, the geologist Harry Fielding Reid formulated a theory for the generation of earthquakes. He realized that there are blocks of the earth's crust that slide past one another along faults, in response to large-scale forces. As long as the blocks on either side of the fault are free to slide, slippage will be continuous and smooth. If, however, the walls on opposite sides of the fault become "locked" together because of irregularities along the fault surface, slippage will occur in fits and starts. Long time spans during which no movement occurs would be followed by short moments of rapid motion. Reid thought that during the times when the walls were locked the rocks adjacent to the fault would be progressively bent and would accumulate ***elastic strain energy.*** When the crust failed and rapid movement occurred, the strained rocks would snap back (rebound)—much like a springboard does after a swimmer has dived into a pool—and an earthquake would occur. The larger the amount of strain energy released, the larger the quake. Reid's explanation, which is diagrammed in Figure 18-5, is called the ***elastic-rebound*** theory, and today is still the accepted model for the generation of a quake.

New Madrid, Missouri, 1811–12 Although the San Francisco earthquake caused a great deal of damage, it was not the most severe quake to jolt the U.S. mainland in the last 200 years. The one with that distinction occurred in the unlikely location of New Madrid, Missouri, about eight years after the Louisiana Purchase.

The New Madrid quake occurred at the time of the War of 1812, when the region was still barely opened to settlement, and life on the Mississippi was harsh. The tremor was felt over an area of around 2,500,000 km^2 (965,240 mi^2)—from the Canadian border to the Gulf of Mexico and from the Rocky Mountains to the Atlantic Ocean (Fig. 18-6). It originated in the central part of the Mississippi Valley, very near Cairo, Illinois, where the Mississippi and Ohio rivers join. The spot was unlikely—a vast lowland plain, remote from any actively growing mountain range. By far the largest number of historic earthquakes in the United States have been concentrated in the mountainous parts of the Far West.

Although aftershocks continued for a year following the earthquake, the main effects resulted from three major shocks, beginning very early in the morning of December 16 and continuing into December 18, 1811, and again on January 23 and February 7, 1812. Were shocks of a similar magnitude to occur today, most of the cities of the central Mississippi Valley would be flattened, and newspapers, news magazines, and TV news programs would carry photographs of little else.

According to the American seismologist Myron L. Fuller, whose investigation a century later resulted in a detailed picture of the extent to which the landscape of the Mississippi Valley was altered by the cataclysm:

> The ground rose and fell as earth waves, like the long, low swell of the sea, passed across its surface, tilting the trees until their branches interlocked and opening the soil in deep cracks as the surface was bent. Landslides swept down the steeper bluffs and

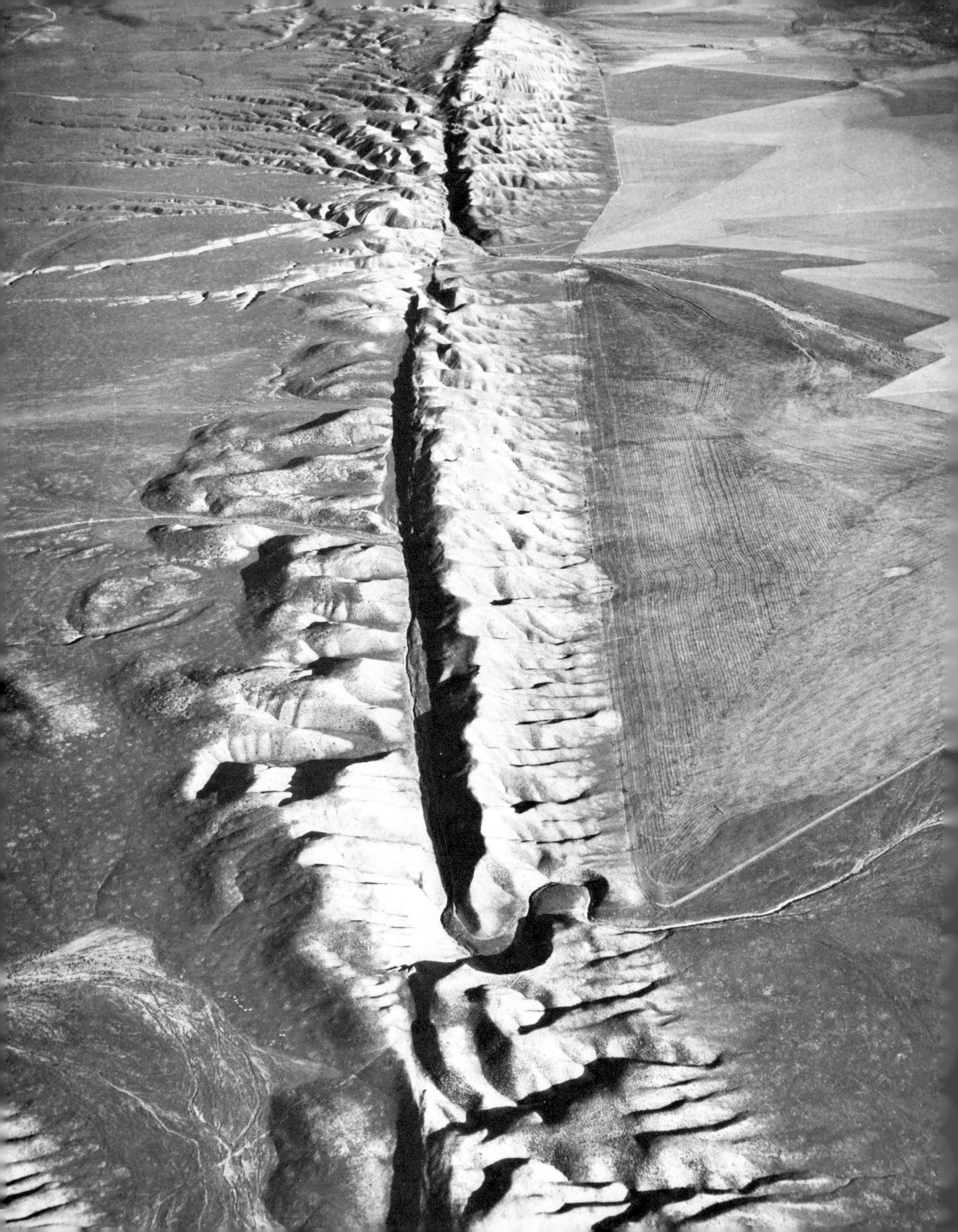

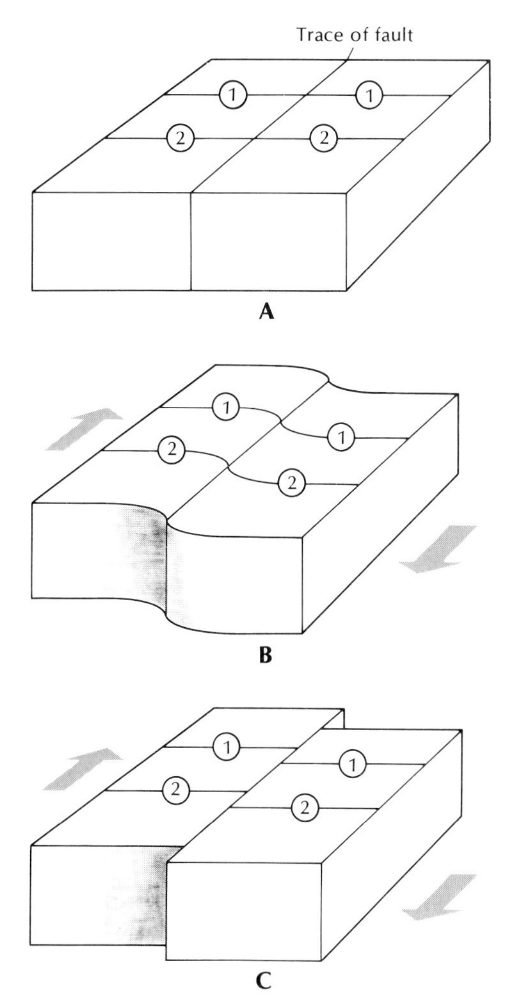

Fig. 18-5 Three stages in the movement of two blocks of the earth's crust past one another, as postulated in Reid's theory of elastic rebound. **A.** Undeformed blocks with imaginary straight lines (1, 2) extending across the trace of the fault. **B.** The two blocks move as indicated by the arrows, but the fault is "locked," resulting in bent lines (1, 2) near the fault trace. **C.** The blocks continue to move, leading to sudden slippage along the fault, and an earthquake. The lines are again straight, but offset along the fault trace.

Fig. 18-4 The San Andreas Fault is one of the major faults in the United States, running nearly two-thirds the length of California. Almost without exception, the trace of the fault, here shown in the Carrizo Plains, is easily discernible. *William A. Garnett*

hillsides; considerable areas were uplifted, and still larger areas sunk and became covered with water emerging from below through fissures or little "craterlets" or accumulating from the obstruction of the surface drainage. On the Mississippi great waves were created, which overwhelmed many boats and washed others high upon the shore, the return current breaking off thousands of trees and carrying them out into the river, sand bars and points of islands gave way, and whole islands disappeared.

Fortunately, the history of that earthquake was better documented than might have been expected, since there were a number of capable observers in the region at the time, including the renowned naturalist John James Audubon. The region was visited by Sir Charles Lyell at the time of his American tour in 1846, and he left a wealth of observations made when the evidence was still fresh.

There were three truly noteworthy geological effects of the earthquake: (1) the appearance of low cliffs cutting across country—very possibly faults scarps—some of which produced waterfalls around 2 m (6.5 ft) high where they intersected the Mississippi River; (2) the elevation of long, low, arch-like ridges or domes, along which former swamp areas were uplifted perhaps 3 to 6 m (10 to 20 ft). One of the ridges was 24 km (15 mi) long; (3) the sudden appearance of the so-called "sunken ground," which is a broadly depressed part of the Mississippi floodplain extending along the river 240 km (149 mi), and also the site of two very large lakes that came into being—St. Francis and Reelfoot. The latter, now a bird sanctuary, is an imposing sight; the gray trunks of cypress trees drowned more than a century ago stand in somber dark waters, which, in places, are up to 6 m (20 ft) deep.

The New Madrid quake was unusual in that (1) it consisted of three main shocks of extraordinary magnitude occurring over a period of nearly two months, and (2) it occurred in an area normally not thought of as earthquake prone. Because of the threat of future large-scale earthquakes the region around New Madrid has been studied intensively in recent years. The data indicate a relatively wide structurally disturbed belt, containing a number of

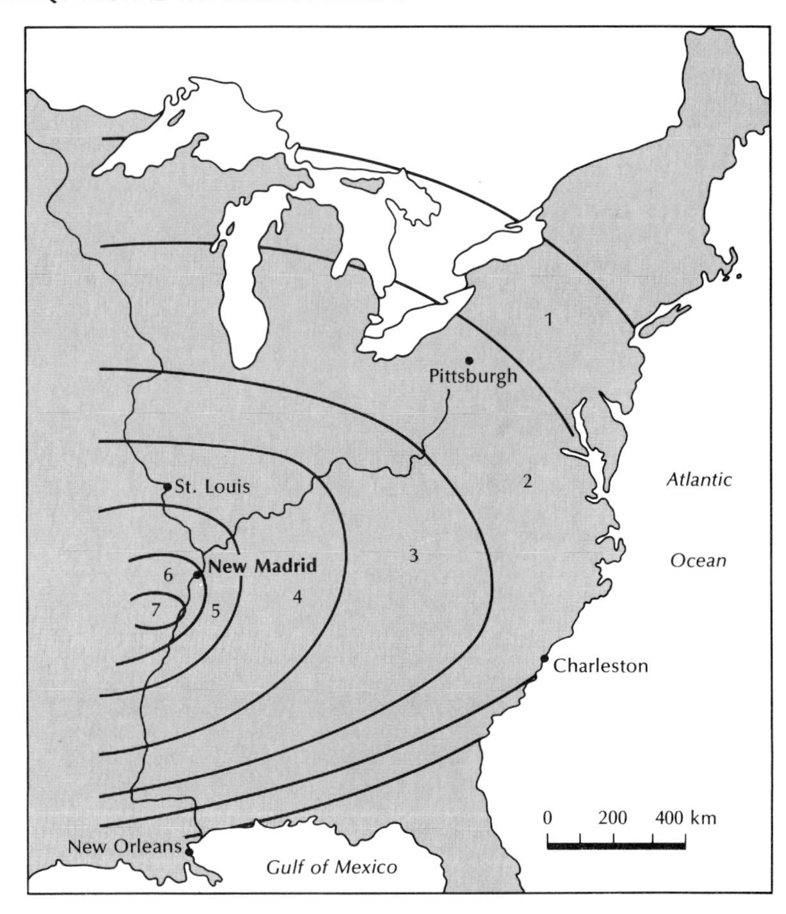

Fig. 18-6 The New Madrid earthquake: generalized zones of equal intensity east of the Mississippi. Intensity increases from moderate in Zone 1 to total damage in Zone 7. The shock would have been felt over an area of about 2,500,000 km² (965,250 mi²).

faults, that trends northeast-southwest (Fig. 18-7). It is a zone of tectonic weakness, which can suddenly break and shift, as it did in 1811 and 1812, in response to accumulated strain. It is thought that the belt is relatively old, having formed at least 500 million years ago, and deep, extending to depths of 40 km (25 mi). Still in question is the source of stress in the region. Some think that it is associated with movements of the earth's tectonic plates; others have suggested that it is related to loading along the southern margin of North America as a result of sedimentation by the Mississippi River during the last few million years.

Japan

The islands of Japan are even more beset by earthquakes than is the West Coast of the United States. Since 1700, approximately 200,000 people have died from quakes in Japan; since 1900 there have been more than 25 earthquakes equal in intensity to the San Francisco quake of 1906.

About 100 million people live on the islands, which together make up a land area about equal to that of California. Of necessity, people are crowded into restricted urban areas, a factor that surely has increased the probability of earthquake damage. Many of the tremors occur just offshore, commonly along the landward side of the deep trench that skirts the islands. Some of the earthquakes are associated with rapid changes in submarine topography, either as a result of offset along faults or slumping of loosely consolidated sediments that have accumulated in bays or other near-shore areas. Not uncommonly, after such sea-bottom changes, tsunami (seismic sea waves) are generated, which devastate coastal population centers. In 1869 an earthquake in the central Pacific sent a wall of water 34 m (111 ft) high against the Japanese coast north of Tokyo, resulting in the death of 27,000 people.

China

China, too, has been beset by earthquakes. As recently as 1976, a large quake struck in northeastern China producing extensive damage, and, although reports are only fragmentary, the death of about 600,000 people.

Undoubtedly the largest loss of life from earthquake activity ever recorded occurred during the 1556 tremor in China's Shansi region. At least one million people perished as a result of the quake, most from the collapse of buildings, but many from the famine and pestilence that followed when the countryside lay in ruin.

Europe and the Middle East

Southern Europe is not free from earthquakes, particularly around the Mediterranean Sea. Italy, for example, has been struck by six large quakes since 1780, which have caused the death of 150,000 people. As recently as November 1980, a relatively large quake struck southern Italy, devastating many small villages east and south of Naples, and killing about 4000 people.

One of the best known of the earth tremors

Fig. 18-7 Structurally disturbed belt near the apex of the Mississippi Embayment. Recent small earthquakes, shown by circles, fall mostly within the limits of the belt.

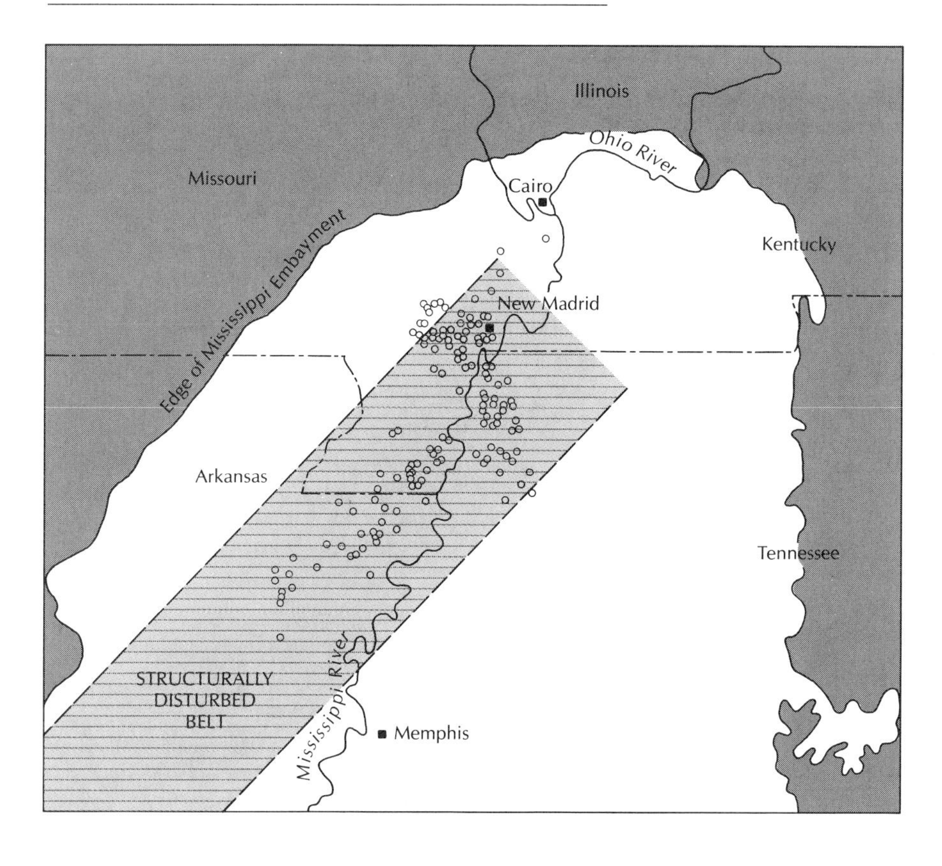

Fig. 18-8 Contemporary engraving depicting the destruction wrought by the Lisbon earthquake, 1755. *Prints Division, New York Public Library*

that have struck southern Europe occurred in the ancient city of Lisbon, Portugal, on the banks of the Tagus River, on All Saints' Day, November 1, 1755 (Fig. 18-8). There were three major shocks, at 9:40 A.M., at 10:00 A.M., and at noon. Many of the people who were at Mass perished in the collapse of the multitude of medieval churches, the spires of which had been so notable a feature of Lisbon.

It is no wonder that buildings, made as they were of loosely bonded masonry and huddled along narrow, irregular streets collapsed, in one thunderous crash, perhaps not greatly different from the succinct description given by Voltaire in *Candide:*

When they had recovered a little, they walked towards Lisbon, hoping to get something to eat after their ordeal, as Candide still had some money left; but they had scarcely entered the city when suddenly the earth shook violently under their feet. The sea rose boiling in the harbour and broke up all the craft anchored there; the city burst into flames, and ashes covered the streets and squares; the houses came crashing down, roofs piling up on foundations, and even the foundations were smashed to pieces. Thirty thousand inhabitants of both sexes and all ages were crushed to death under the ruins.

A startling phenomenon to the survivors at Lisbon, as well as to observers at other places along the Portuguese coast, was the sudden

appearance of a great wave (seismic sea wave) set in motion by a submarine shock off the coast.

In Lisbon, the wave reached a height of about 6 m (20 ft) as it swept up the Tagus River. Ships and boats were smashed together and sank, and much of the wave's fury was concentrated on a newly constructed marble pier, the *Cais de Pedra,* which was black with people fleeing the burning city. Their drowning gave rise to the legend that the pier and all the people on it had been engulfed by a great fissure that opened and closed over them, so that they were never seen again. Even Lyell believed, on the basis of a visit to Lisbon, that such a burial had occurred. But it is more likely that the quay broke up under the earthquake shock and the attack of sea waves.

The distance to which the shocks of the Lisbon earthquake were felt throughout Europe was one of its remarkable features. They were strong throughout the Spanish peninsula and well into France, and they also caused severe damage in Morocco (although some believe that the destruction there was due to a separate shock of closer origin.

The Lisbon earthquake had a profound effect on European thinking. It led Rousseau to point out in his *tout est bien* philosophy that earthquakes are all for the best; it is the evils of civilization that are bad. If we were not cooped up in cities, earthquakes would not kill us. Voltaire plainly ridiculed such unthinking optimism. As we have seen, Candide and his insufferable companion, Dr. Pangloss, were caught in that earthquake. For Dr. Pangloss it proved to be a time to test his belief that every horror encountered on earth was a necessary event in the pre-established harmony of the universe, the best of all possible worlds.

Spiritually, the Lisbon disaster was a challenge to the theologians of the time. Why did one of the most devout and justly renowned cities of the world suffer such a fate at the hands of a loving God—so that the good should perish in such multitudes along with the evil? According to T. D. Kendrick's study, *The Lisbon Earthquake,* which is concerned primarily with the related themes of eighteenth-century earthquake theology and the end of optimism, a widely held belief with regard to the earthquake was the following:

> The reason why God had overthrown Lisbon was not only because He intended to shock the whole of Christendom into a state of penitent obedience to Him by the staggering destruction of such a celebrated and wealthy city, one that was perhaps, thanks to its maritime trade, the best-known city in Europe; but also because Portugal was a kingdom under the special and principal care of Heaven, so that according to the rules of the divine discipline, the Portuguese for their own good and as a result of the heavenly priority that was their due, were singled out for the honour of being the first to be punished and those who were punished most severely.

Politically, the Lisbon earthquake was a prophetic event because it provided the setting for a modern dictator, the Marquez de Pombal. Although he was ultimately to seize all the trappings of power, to reduce the king to ineffectiveness, to expel the Jesuits, and to execute his opponents, during the disaster Pombal unquestionably preserved order when chaos would have prevailed and never faltered in his determination to rebuild a greater Lisbon. The modernization of the city, the cutting of boulevards, and the restoration of its ancient glories—if not its liberties—were essentially Pombal's doing.

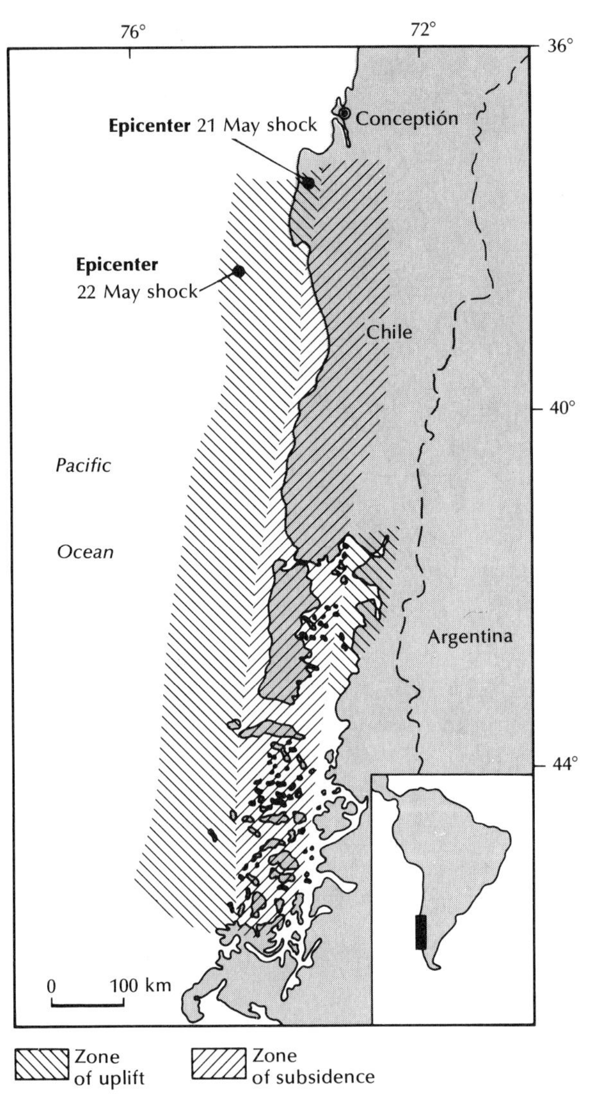

Fig. 18-9 Epicenters of the Chilean earthquake (May 21 and 22, 1960) and the area affected by uplift and subsidence accompanying the earthquake. Incredibly large forces changed the elevation of a considerable area of the earth's crust.

Destructive earthquakes are also prevalent in North Africa and the Middle East. In 1978, a very strong quake jolted a remote region in the salt flats of eastern Iran, flattening the city of Tabas, and 40 nearby villages, and killing about 25,000 people.

In 1980, Algeria was struck by a devastating quake, which wiped out about 20,000 inhabitants in and around the city of Al-Asnan. This city had been rebuilt after a 1954 earthquake, but unfortunately the new multistoried buildings had not been adequately reinforced. After the 1980 quake, the Algerian Cabinet discussed the possibility of moving Al-Asnan to a new location less prone to earthquakes rather than rebuilding the broken city.

South America

The west coast of South America is also subject to violent earth tremors. Charles Darwin, during his voyage on the H.M.S. *Beagle,* was in Chile in 1835 at about the time that the region near Concepción was struck by an earthquake of moderate severity. Unaccustomed to such activity, Darwin wrote an account of the event, noting that some parts of the earth had been uplifted by as much as 3 m (10 ft) after the quake.

On May 21, 1960, a particularly devastating series of earth tremors shook Chile. The first of the tremors occurred on Saturday and caused widespread damage in and around Concepción; they continued all the next day (Fig. 18-9). At 2:45 P.M., Sunday, an unusually strong shock was felt, and many persons throughout southern Chile left their homes and stood about in the streets. They were still standing there at 3:15 P.M. when the main shock rocked the entire region.

Pierre St. Amand, a well-known seismologist, in his discussion of the Chile quake, provided a lucid account of the events associated with the first large shock.

> The motion of the ground during the main shock was as if one were at sea in a small boat in a heavy swell. The ground rose and fell slowly with a smooth, rolling motion, smaller oscillations being superimposed on larger ones. In Concepción, cars and trucks parked by the side of the road rolled to and fro over a distance of half a meter when they bobbed up and down in response to the movement of the ground. The tops of the trees waved and tossed as in a tempest. Some already damaged buildings fell. The earthquake itself was silent; not a sound came from the earth. The period of vibration was of the order to ten to twenty seconds or more. The shaking lasted fully three and a half minutes and was followed for the next hour by other shocks, all having a slow, rolling motion. . . . In the Region of the Lakes . . . the movement began smoothly and continued for some two minutes, just as in other localities, when suddenly, a loud subterranean noise was heard followed by a sharp jarring motion and a more rapid, less regular vibration of the earth. Similar reports were obtained at other points to the east of the Lakes, and it seems from these that another earthquake took place here . . . while the ground was still shaking from the first shock.

Aftershocks continued to be felt for months. St. Amand recorded 119 shocks, from the main quake in May 1960, until June 1961. Although none was as severe as the main quake, 32 of the aftershocks were very strong. Because the active area was so large, 160 km (99 mi) wide by almost 1600 km (994 mi) long, and included a wide variety of land features and surface conditions, most of the possible earthquake effects were shown by the Chilean earthquake, with one notable exception. Even though a thorough search was made, both on the ground and from the air, in no place were large offsets of the surface along a fault trace found.

Other land surface effects, however, were numerous. Landslides were common; the earthquake in many places triggered the sudden movement of unstable material. Ground cracks were caused by the settling of fill and the subsidence of areas underlain by soil that had liquefied. Liquefaction also caused huge earth flows, probably because of the presence of quick clays. In the harbor of Puerto Montt a ship was caught in a current of sand and mud that flowed into the bay; thereby it had the unique distinction of being the first ship to have gone aground in a landslide. The ship was later converted into a small, but unusual hotel.

In the area of the most intense shaking, some trees were snapped off, while others were uprooted, and fell. In some cases, the broken branches formed a circular pile of debris on the ground around the trunk.

Flooding by sea water resulted from subsidence of the land. Some areas, however, were elevated 1.5 to 2 m (5 to 6.5 ft) above sea level. The rivers, too, flooded. In places the severe shaking compacted the poorly consolidated or

Fig. 18-10 Adobe buildings in Huaraz, Peru, destroyed by the 1970 earthquake. The epicenter of main shock was about 150 km (93 mi) west of Huaraz. *USGS*

unconsolidated material of the river banks, and they were lowered considerably. The earthquake actually shook water out of the ground, so that the rivers were unusually full. Almost continuous rain added to the flood waters.

Two days after the main shock, the volcano Puyehue in the Andes east of Concepción erupted and continued active for several weeks. Steam and ash issued from a fissure about 300 m (984 ft) long and from several smaller openings. The last stage of the eruption was characterized by a number of flows of viscous lava. St. Amand wrote

> The local newspapers, in an unparalleled burst of enthusiasm, reported that 12 volcanoes had exploded and that two new ones had been formed. Lava was reported to be flowing down the sides of several of the volcanoes, and towns were said to have been buried. Because of the bad weather and poor visibility in the central valley, the inhabitants of the valley towns all believed that the volcanoes were, indeed, erupting and were concerned by the situation for a period of several weeks. Newspapers and news magazines all over the world repeated and enlarged on these stories.

The Chilean Region of the Lakes lies inland from the coast, but is still well within the area affected by the earthquake. The lakes exhibited a feature known as a ***seiche***. The motion of the quake set up a wave that oscillated back and forth in each lake basin. The water sloshed, much like the familiar wave in a bathtub. Seiches in the Chilean lakes were small compared to those in Hebgen Lake, Montana, during the 1959 earthquake there. Hebgen Lake is a dammed lake. Of it the seismologist John Hodgson wrote

> An eyewitness standing in the moonlight on Hebgen Dam and looking down its sloping face could not see the surface of the water, so far had it receded. Then with a roar it returned, climbing up the face of the dam until it overflowed the top, and poured over it for a matter of minutes. Then the water receded again, to become invisible in the moonlit night. The fluctuation was repeated over and over, with a period of about seventeen minutes; only the first four oscillations poured water over the top of the dam, but appreciable motion was still noted after eleven hours.

In the Chilean quake, a similar secession of water was observed in the ocean itself. Shortly after the main shock, the coastal waters receded well below the lowest low-tide line. The inhabitants of the coast knew from experience

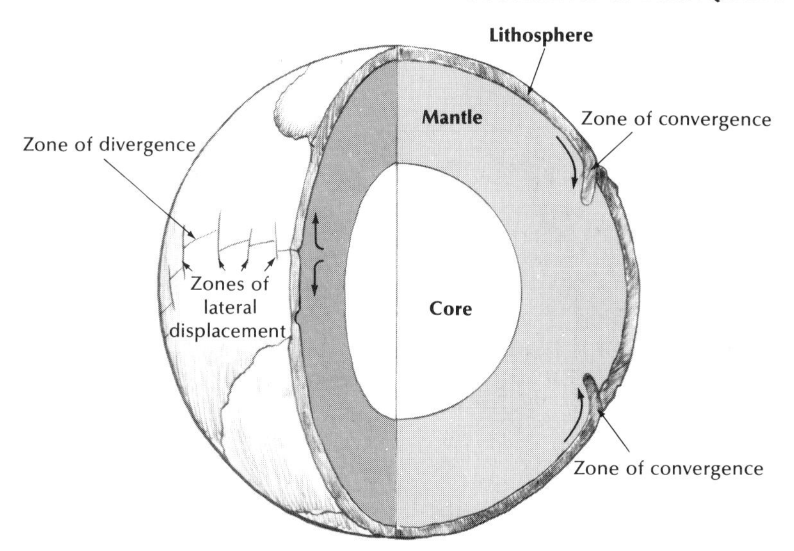

Fig. 18-11 The earth, showing the zones where plates converge, diverge, and slide laterally.

what to expect and fled from their homes to nearby hills to wait and watch for the tsunami. They had from 15 to 30 minutes to complete their evacuation, when the sea returned in a mighty wave that reached 6 m (20 ft) in height at some places and extended as much as 3 km (2 mi) inland. The waves continued for the rest of the afternoon, generally diminishing in height, although the highest wave is reported to have been the third or fourth, rather than the first one.

Leveling surveys run sometime after the 1960 quake indicated that two strips parallel to the coast, each about 1500 km (932 mi) long by 150 km (93 mi) wide, had changed in elevation; one went up while the other went down (Fig. 18-9). The amount varied locally, but it appears that, on the average, uplift amounted to about 1 m (3 ft) and subsidence to about 1.5 m (5 ft).

Although the Chilean quake of 1960 was of greater magnitude, the one that struck parts of Peru on May 31, 1970, caused ten times the destruction and loss of life. The quake was centered about 25 km (15.5 mi) off the coast and caused extensive damage on land. About 30,000 people lost their lives, largely as a result of the collapse of buildings (Fig. 18-10). The quake also set in motion a chain of events that led to a catastrophic rock avalanche and the death of thousands more in the towns of Yungay and Ranrahica (see Chap. 10).

OCCURRENCE OF EARTHQUAKES

Earthquakes occur essentially all over the world. Their distribution, however, is not haphazard; they tend to occur in linear belts or zones of seismic activity—zones that earth scientists now believe mark the boundaries of rigid plates (see Chap. 20). Earth tremors are generated at the plate boundaries where the plates converge, diverge, or slide past one another (Fig. 18-11). Away from the plate edges, earthquake activity is generally absent. The San Andreas Fault in California marks one such plate boundary between the Pacific Plate on the west and the North American Plate on the east.

Is is noteworthy that a belt of seismic activity (mostly related to a zone of convergence) almost completely girdles the Pacific Ocean basin and is essentially coincident with the zone of active andesitic volcanism that also rings the Pacific (see Chap. 6). Most earth scientists believe that both phenomena—seismicity and volcanism—are the result of large-scale forces within the earth that push the plates together.

Numerous, often large quakes occur in an east-west swath across southern Europe and Asia. Those quakes also apparently mark a zone of convergence of two main plates and perhaps a number of smaller ones.

Several earthquake belts occur within the ocean basins. Those that are coincident with the oceanic ridges and rises, such as the Mid-Atlantic Ridge or the East Pacific Rise, mark zones of divergence; that is, zones where the plates are moving apart. Volcanic activity occurs along such zones as well, but the lavas tend to be basaltic rather than andesitic. One land-locked zone of divergence with both earthquake and volcanic activity runs along the east side of Africa. It is marked topographically by nearly continuous elongate basins—the African Rift valleys.

Earthquakes also occur in zones where two plates slide laterally, one past the other. Such boundaries are relatively abundant in the ocean basins and are represented by the fracture zones that cross the oceanic ridges and rises.

Earthquakes rarely originate at the earth's surface; generally, the fault movement starts at some depth below ground. The point of initiation of rupture or slippage is called the ***focus,*** whereas the point on the earth's surface vertically above the focus is called the ***epicenter*** (Fig. 18-12). Once a rupture starts, it propagates along the fault at several kilometers per second. The total length of the rupture associated with the 1906 San Francisco earthquake was about 432 km (268 mi). Most quakes, however, originate from much smaller ruptures.

The focus of an earthquake can be as deep as 700 km (435 mi). Arbitrarily, quakes that originate at depths less than 70 km (43 mi) are called ***shallow focus***; between 70 and 300 km (43 and 186 mi), ***intermediate focus***; and between 300 and 700 km, ***deep focus.*** Most earthquakes taking place today fall into the category of shallow focus.

Along the plane of rupture, strain energy is converted and released in two different ways, as heat and as seismic wave motion. About one-half of the released energy goes into heating the rocks directly adjacent to the fault

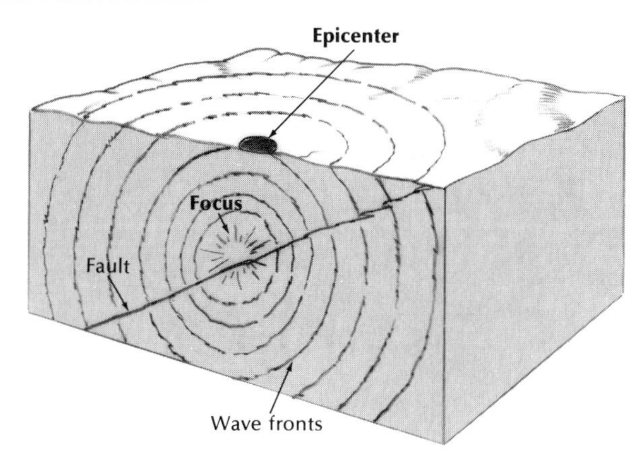

Fig. 18-12 Relation of the focus to the epicenter of an earthquake. Also shown are wave fronts of seismic waves radiating out from the focus.

zone. The seismic vibrations, on the other hand, travel away from the focus in all directions, much as ripples radiate from the point at which a pebble is dropped into a quiet pond. As the waves travel outward, the energy they carry is distributed over a progressively larger perimeter, and the intensity of wave motions diminishes.

Some seismic waves travel through the interior of the earth and are called ***body waves;*** others travel at and close to the surface of the earth and accordingly are called ***surface waves.*** It is the motion of surface waves that we feel during an earthquake. Such waves can be set in motion either directly by fault movement at or close to the surface or, indirectly, by body waves that emanate from the focus and strike the ground surface near the epicenter directly above. The violence and complexity of the shaking at the earth's surface is related to the magnitude of the released strain energy, the proximity of the focus, the duration of the quake, and the physical nature of the surface materials. Unconsolidated mud will develop much larger waves than will crystalline igneous or metamorphic rocks. We shall return to body waves later in the chapter, when we discuss

Fig. 18-13 Housing development, near San Francisco, sprawls across the trace of the active San Andreas Fault, which extends from the sag pond, lower left, to the upper center, in the distance. *Marshall Moxon*

what those waves tell us about the interior of the earth.

PEOPLE AND EARTHQUAKES

What can be done to modify the destructive effect of earthquakes on people and the structures they build? Certainly, it is inconceivable that urban areas will be moved to regions that are less active, seismically. Managua, Nicaragua has been destroyed at least four times by earthquakes, the last as recent as 1972; yet its residents continue to rebuild and to try to forget the past. California is hit by hundreds of quakes each year, and still its population continues to burgeon (Fig. 18-13). And there is still the possibility, even in the seismically quiet regions of the world, of a rare, but nonetheless destructive quake, as for example the one in New Madrid, Missouri.

At present, there are two approaches to the problem of possible earthquakes. One is concerned with establishing an awareness of the effects of quakes and with enforcing building codes; the other is concerned with predicting and controlling earthquakes themselves.

It has become apparent that through careful selection of building sites and the establishment of minimum construction requirements, the effects of earthquakes can be lessened.

In major earthquakes, buildings will be most severely damaged if they stand on filled, unconsolidated, or water-soaked ground and least damaged if they stand on solid rock. Selection of an actual building site is a factor over which we can exercise some control, yet, it is surprising how little thought has been given to the problem in most regions of high earthquake hazard. Fortunately, as we become more aware of the physical world around us, the matter of choosing more stable building sites is receiving increased attention.

The element over which we have the most direct control is the type of construction of the buildings we inhabit. It is probably most important to construct a building that will vibrate as a single unit. It helps, too, if it is not too rigidly structured. Fortunately, most wooden-framed houses fit those requirements, especially if the mudsills are bolted to a concrete foundation and if the studs are strengthened by angle-bracing at the corners. Also important is the use of reinforcing rods, run the length of

Fig. 18-14 Earthquake damage to freeway under construction, San Fernando earthquake, 1971. Geological hazards in California have led to public safety legislation. *Newhall Signal, California Divisions of Mines and Geology*

brick chimneys and tied to a concrete slab at the chimney base. Among the more memorable sights after the 1933 Long Beach earthquake were the hundreds of brick chimneys snapped off at the roof line.

Among the more vulnerable buildings are those with roofs extending over a comparatively broad span, such as supermarkets, bowling alleys, and auditoriums. If the roof of such a structure is arched, it exerts an outward thrust against the bearing walls, which may be hammered down, as it were, until they collapse.

It is necessary to consider the geological structure of an area when determining the placement of major transportation routes, aqueducts, pipelines, water-storage facilities, power plants, and the like. Geologists now know the major fault traces in most areas and can generally tell which ones are active at present or are likely to have been active recently. Most municipalities construct their vital industries with that knowledge in mind. For example, in the selection of possible building sites for nuclear power plants, those that would otherwise meet specifications may fail to qualify because of the presence of faults that appear to have been recently active.

Regardless of how carefully building sites are selected and how well structures are built, a major earthquake in a populated area will cause property damage and loss of life (Fig. 18-14). If only we could predict earthquakes, population centers could be evacuated, thereby reducing loss of life, and at least some preparations could be made, thereby reducing property damage.

Such goals may seem fanciful and beyond attainment. Yet earth scientists, particularly in those countries where earthquakes are prevalent and resources are available (the United States, Japan, the USSR, and China), are trying to reach those goals. ***Earthquake prediction*** has assumed a high priority in all of science and is another example of how science can be used to benefit people. In China today, earthquakes have been designated the "number-one naturally occurring enemy of the people," and work in the field of earth science has been directed singularly toward reliable prediction.

One type of general long-range seismic forecasting is based on analyzing the distribution and intensity of earthquakes that have occurred in the past. From such an analysis, seismic-risk maps can be generated that indicate the likelihood of earthquakes of a certain size for any geographic area (Fig. 18-15). Such maps, however, do not show the recurrence rates of large earthquakes. Both western California and the area surrounding Charleston, South Carolina, are shown as zones of potentially severe quakes. However, only one severe

Fig. 18-15 Seismic risk map of the United States, based on effective peak accelerations (given in terms of the acceleration of gravity, g) expected from ground shaking in earthquakes. Statistically the accelerations shown have only a 10 per cent chance of being exceeded in 50 years. The g ranges are negligible risk, <0.049 g; low risk, 0.05–0.099 g; moderate risk, 0.10–0.199 g; high risk, 0.20–0.399 g; severe risk, 0.40–0.599 g.

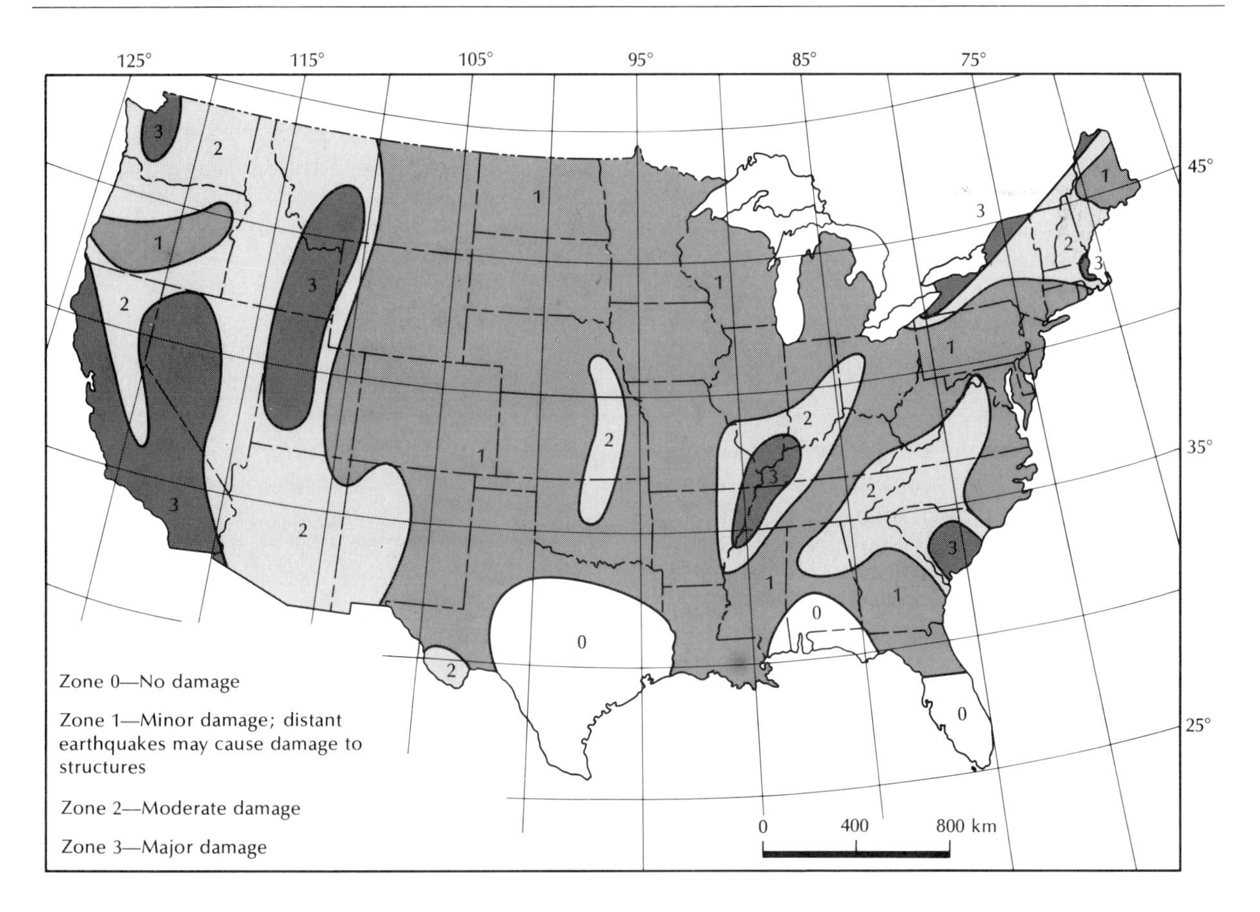

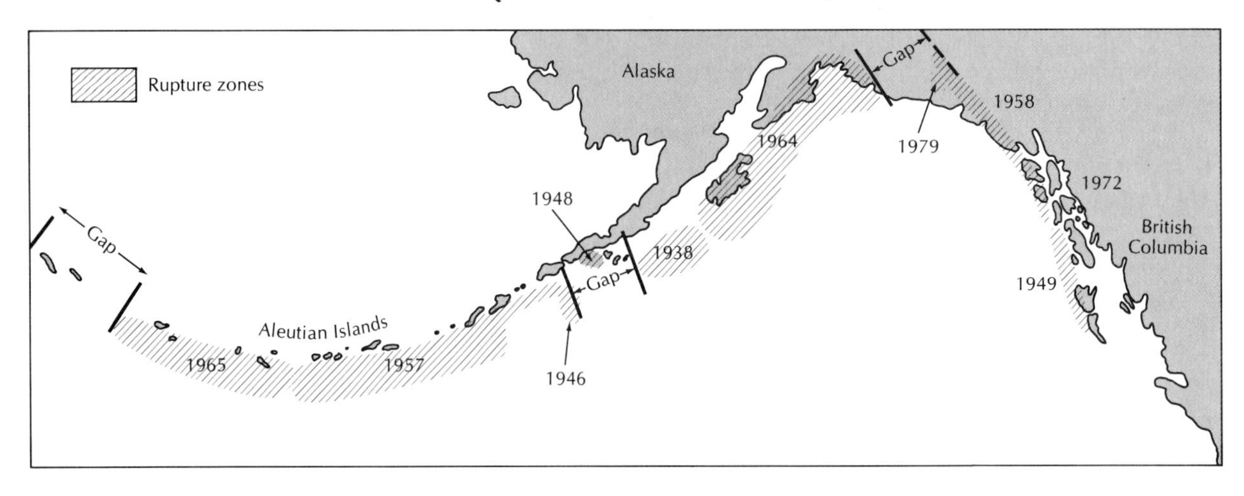

Fig. 18-16 Rupture zones of large, shallow earthquakes (1930 to 1979) and seismic gaps in southern Alaska, the Aleutian Islands, and British Columbia.

quake has been recorded near Charleston, and that in 1886, whereas many have been recorded over the years in western California. A part of eastern Massachusetts is ranked Zone 3 because of the occurrence, in 1755, of a single large earthquake, in the area of Cape Ann, about 100 km (62 mi) from downtown Boston. Studies of small-scale features in some Zone 3 areas have made possible the determination of general recurrence rates of large quakes. For example, the average length of time between large earthquakes along the San Andreas Fault, near Los Angeles, is about 200 years. Some quakes were only 50 years apart, although others were up to 250 years apart. The last great earthquake there, in 1857, was one of the largest seismic events observed in California. A quake of the same intensity would cause great destruction and loss of life in present-day Los Angeles.

From the recurrence rates of large earthquakes between 1784 and 1980, a major earthquake is predicted near the end of the Alaskan Peninsula sometime between 1980 and 2000.

Another long-range prediction technique, generally applied to regions and seismic belts in which large quakes have commonly occurred, is to identify the points of origin of recent large quakes (***rupture zones***) and associated aftershocks (Fig. 18-16), which are plotted on a map. The technique is based on the premise that strain uniformly accumulates along the length of any seismic belt, but is released only over small segments of the belt. Segments of a belt that have not undergone strain release recently are called ***seismic gaps*** and are considered prime regions of potential seismic activity. Once such gaps are identified, they can be closely monitored.

One of the principal detailed monitoring techniques involves determining how much strain has accumulated along a locked fault plane and at what rate. Earth scientists look for changes, such as tilt or strain, in the relative and actual positions of points on the earth's surface—either by geodetic or other measurement. Or they look for changes in the physical properties of the rocks, which are affected by strain buildup, along the fault. Such changes may be determined by measuring the local

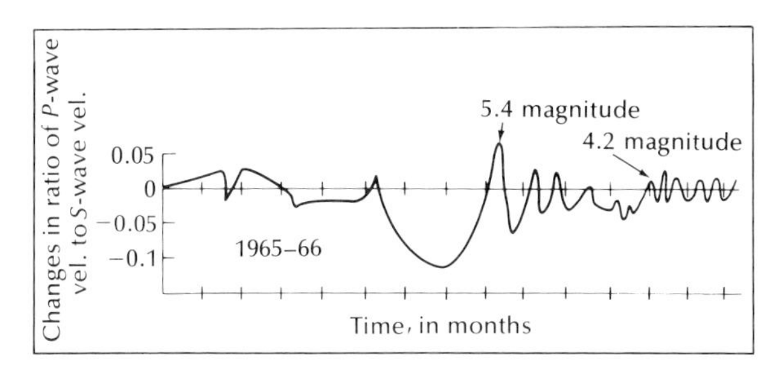

Fig. 18-17 Two earthquakes in the Garm district, USSR (1956–66), were preceded by drops in the ratio of *P*-wave to *S*-wave velocity.

magnetic field, the electrical conductivity of the rocks, or the velocities of seismic body waves (Fig. 18-17).

Commonly, the water levels in wells and rates of groundwater flow are anomalously affected during strain buildup. And it has been shown, in some wells and along some faults, that the concentration of the radioactive isotope radon (one of the daughter products of uranium) increases dramatically just prior to strain release.

Scientists also set up sensitive seismic instruments along a fault zone and "listen" to the crackling of the earth as strain builds before fault slippage occurs. As the time of slippage approaches, the number of micro-earthquakes per observational interval generally increases. It may be an increase in micro-activity that accounts for the common observation that dogs and horses become increasingly nervous and jittery just before a large quake. Recently, in fact, different animal groups have been observed to see if their behavior changes just prior to an earthquake. In California, for example, groups of animals, chiefly mammals, are housed near the San Andreas Fault, where they are continuously observed. Unusual behavior immediately preceding a shock is closely examined and recorded. No clear-cut results have been obtained yet, but some researchers are convinced that this line of study eventually may prove to be one of the best ways to predict earthquakes.

Many predictive techniques have been and are being tested, but no event, or set of events, has been found to signal the advent of an earthquake. Hindsight, however, seems to suggest that some earthquakes might have been predicted, whereas others could not have been. For example, in 1975, about 90,000 people were evacuted from the Chinese city of Haicheng after a large earthquake had been predicted. Shortly thereafter, a very large quake did occur. Undoubtedly, thousands of lives had been saved by the timely prediction. In 1976, however, more than 500,000 people perished in the city of Tangshan in an earthquake that was preceded by few if any signals.

In 1979, the prospect of accurate earthquake prediction was further dampened by the occurrence of a moderate quake in a well-monitored area about 100 km (62 mi) south of San Francisco, along the Calaveras Fault, a major branch of the San Andreas Fault. No precursor signals had been recorded for this quake. In view of the discouraging results so far obtained, it is clear that our approach to earthquake prediction must be studied further, and modified accordingly, and that other types of prediction will have to be determined and explored.

If and when we acquire the ability to predict accurately an earthquake of a certain magnitude, another problem will arise—relating to how citizens in the affected area respond to the prediction. For example, what if a potentially disastrous earthquake were predicted for the vicinity of San Francisco on or about a certain date? Many would panic and flee the danger area. Accidents and loss of life could occur, and who would be responsible? What if the alarm were false, and after a week or so people returned to find their houses ransacked, pets lost or dead, crucial jobs interrupted? Who would shoulder the blame? What if, after they returned to the city, an earthquake did hit, causing much destruction? How would the surviving citizens react? What would be the effect on the economy of a city such as Los Angeles if a large earthquake were predicted sometime within the next two years?

Those and other questions have led some

scientists to conclude that the only way to deal with earthquakes is to control them. But is there any evidence that we can influence their behavior? The answer appears to be "yes," and the approach to controlling them centers around a seemingly unlikely substance—water. About 20 years ago, a deep well was drilled into the Precambrian rocks underlying the Rocky Mountain Arsenal near Denver, Colorado. The purpose of the well was to dispose of radioactive waste fluids. Beginning in 1962, and continuing for three years thereafter, many thousands of gallons of waste fluids were pumped down the well. Shortly after pumping began, earthquakes began to occur in the Denver area. Officials at the arsenal denied any connection between the pumping and the earthquakes, but a Denver geologist, David Evans, noted that the frequency of quakes was directly related to both the amount of waste-water pumped in during any period and the pressures used to pump it in (Fig. 18-18). In 1965, pumping was stopped and earthquake activity slowly ceased.

Most earth scientists have concluded that the Denver earthquakes were the result of a release of tectonic strain in the Precambrian rocks adjacent to a fault. Before pumping, the walls were locked, and no faulting or quakes were recorded. By forcing waste-water down the hole, water pressure at depth was generated, the pressure being sufficient to force the walls of the fault apart, thereby effectively lubricating it. The fault "unlocked" and slippage occurred.

Such incidents have been documented elsewhere, leading geologists to speculate that pumped water might be used to control fault movement. Take the San Andreas Fault as an example. Along a segment of the fault, three equally spaced wells could be drilled down along the fault plane to a depth of 4600 to 6100 m (15,093 to 20,014 ft). Groundwater would be pumped out of the two wells at either end of the fault segment, drying up those parts of the fault and effectively locking them. Then enough water could be pumped down the center well to lubricate the fault and permit movement. Supposedly a small-magnitude quake

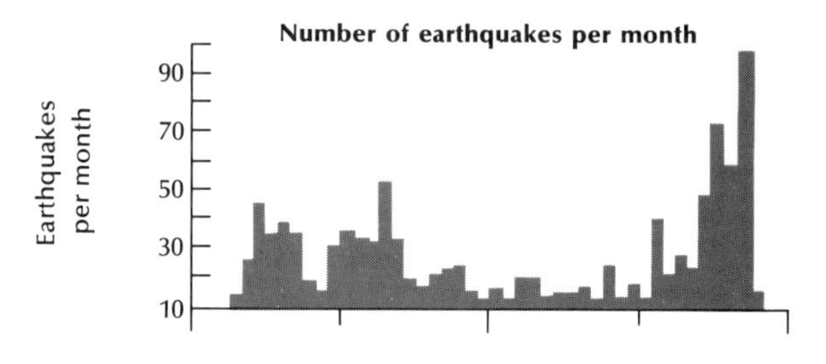

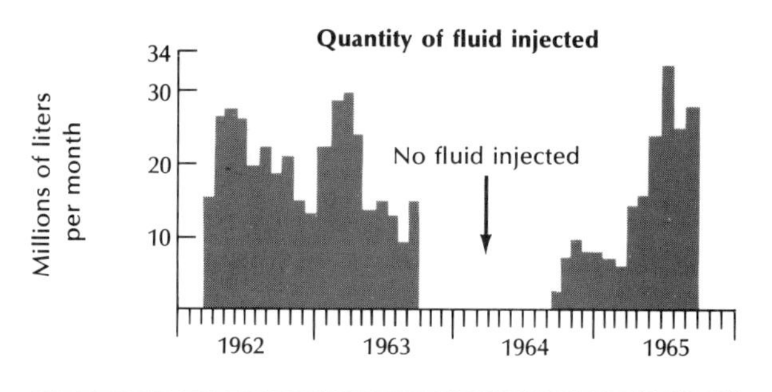

Fig. 18-18 Correlation between quantity of waste-water pumped into a deep well and the number of earthquakes near Denver, Colorado.

would result, and afterward that section of fault would be temporarily free from strain. The process of drying out and lubricating would be repeated up and down the fault until strain energy had been released along its entire length.

Many unknowns surround such an approach to earthquake control. Even if it eventually worked, large earthquakes could be triggered during the early stages of strain release. At the moment, no federal, state, or local agency feels ready to accept the responsibility of risking this method of control.

The same could be said about earthquakes as is said of the weather: Everyone talks about them, but no one does anything about them. But perhaps in the near future that will change.

MEASURING THE HEARTBEAT OF THE EARTH—SEISMOLOGY

There is another aspect of the earthquake story, one that is rather undramatic and has to do with the systematic study of earthquakes and seismic waves. Through it we can hope to learn a great deal about the nature of earthquakes around the world and about the material through which body waves pass.

Earthquakes, simply because they are dramatic natural events, have invited speculation as to their origin from the earliest times. Greek philosophers devised many explanations that today read like mythology. For example, Aristotle, who lived in the fourth century B.C., and whose interpretation of natural phenomena once had the force of dogma, believed that earthquakes resulted from the escape of air trapped deep within the earth.

The 1755 Lisbon earthquake prompted the first serious effort to study earthquakes scientifically. By 1700, from observations of swinging chandeliers, John Mitchell, an English physicist, surmised that earthquakes produced wave motion in the rocky crust of the earth. In 1859, Robert Mallet (1810–81), an English engineer who had been thinking about earthquakes in theoretical terms, had an opportunity to study the results of a real one when he visited Italy and saw the damage done to hill towns in the Apennines east of Naples by a destructive quake in 1857. Although he believed, incorrectly, that earthquakes were explosive in origin and were related to volcanic activity, he laid the foundations of observational ***seismology.*** The word seismology, a common one today; comes from the Greek *seismos,* earthquake. By observing such phenomena as the directions buildings and monuments fell and the nature of cracks in the ground, he worked out a rough method for determining the source direction of an earthquake. Mallet was among the first to try an experimental approach to earthquake study, setting off explosions and measuring the travel times of the resulting waves. He also started an earthquake catalog and through it achieved some understanding of the geographic distribution of earthquakes.

The seismograph

The founder of modern seismology was an English mining engineer, John Milne, who went to Japan in 1875 as one of a group of men who wanted to bring the technology and educational systems of the West to the Japanese people. By 1880, he had sufficient exposure to earthquakes, to emerge as the first full-fledged seismologist. Through his efforts the seismograph, the instrument used to record the vibrations set up by an earthquake, was transformed from a scientific curiosity into a precision instrument.

The instrument was perfected over the many years since the rudimentary one was contrived by L. Palmieri, in Italy, in 1855. The problem that had to be solved initially was how to measure the vibrations from an earthquake at the earth's surface when the instrument itself is attached to the surface and is shaking. The solution was both simple and elegant. A large mass was suspended from a thin filament (a pendulum), thereby virtually isolating the object from the earth's surface—it was, in essence, floating in space (Fig. 18-19). When earthquake waves set the ground into motion at the recording site, the mass, because of its large inertia, tended to remain at rest while the ground beneath it and the point of filament suspension shimmied and shook. A stylus, or pen point, was attached to the mass so that it could trace the motion of the ground as it moved to and fro.

The first primitive seismographs consisted of a single, vertically suspended pendulum, and the record was traced in smoothed sand or on a stationary piece of smoked paper (Fig. 18-19). Since the stylus traced its path back and forth past the same spot on the ground, each earth tremor produced a bewildering array of lines. But the problem was overcome when the recording paper was attached to a plate that moved beneath the pen at a known rate; thus it could not write over its previous trace (Fig. 18-20). In most modern seismographs, the pen records on a drum that turns at precisely four revolutions per hour. As the drum revolves it slowly spirals ahead; it is, in essence, a spiraling clock. The recording pen also makes a tic mark on the record after each minute, so that the onset of any motion to the hour, minute, and second can easily be read off. For ease of comparison of their records, all stations around the world have agreed to use the same time base; when the paper on their various drums is changed each day, it is reset and synchronized with Greenwich Civil Time.

A somewhat more complicated problem occurred with the early seismographs. During a quake the ground moves in three dimensions; that is, the motion at any one moment can be simultaneously east-west, north-south, and up-down. A single pendulum suspended vertically cannot be used to obtain a readable record of the three-dimensional movement. In the late 1800s, however, Milne, who had long struggled with the problem, hit upon the idea of using three separate seismographs at each station, each one capable of recording one of the dimensions (Fig. 18-21). In order to measure the horizontal movements he mounted two pendulums in horizontal positions, but at right angles to each other, and to measure vertical movement, he retained the pendulum suspension in a vertical position. His breakthrough made possible quantitative seismology. By 1882, Milne-Shaw seismographs had been installed at 50 cooperating stations around the world. Scientists throughout the world were then able to measure and compare records of the most complicated earth movements. It is no wonder that Milne is considered the founder of seismology.

To be sure, other problems, such as amplification of signals and damping of vibrations set up in the seismograph by a quake, had to be overcome as the seismographs of the nineteenth century gave way to more modern instruments. Several of the recent improvements in design are a result of research in electronics and experimental physics. For instance, in one seismograph today, relative motion between the pendulum and earth is sensed, amplified, and recorded electromagnetically.

Today's precision seismographs are relatively easy to use, maintain, and standardize (Fig. 18-22). Most are installed at permanent stations, but portable models are becoming increasingly popular. The latter are set up to record quakes at some strategic location for a few days, weeks, or months, and then are moved elsewhere. As you may recall, portable seismographs, taken to the moon on several of the Apollo missions, relayed to the earth re-

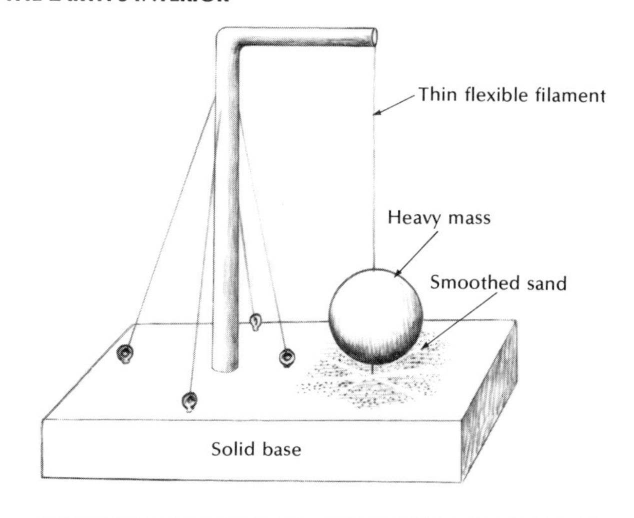

Fig. 18-19 A simple model of a seismograph in which a heavy mass is suspended from an arm on a thin wire. During an earthquake, the base (solidly attached to the earth), moves as does the arm, but the mass, because of its inertia, is relatively unaffected.

Fig. 18-20 In a refinement of the seismograph design, a rotating drum is placed beneath the stylus. After Fig. 23-14A (p. 394) from THE EARTH SCIENCES, 2nd Edition, by Arthur N. Strahler. Copyright © 1963, 1971 by Arthur N. Strahler. Reprinted by permission of Harper & Row, Publishers, Inc.

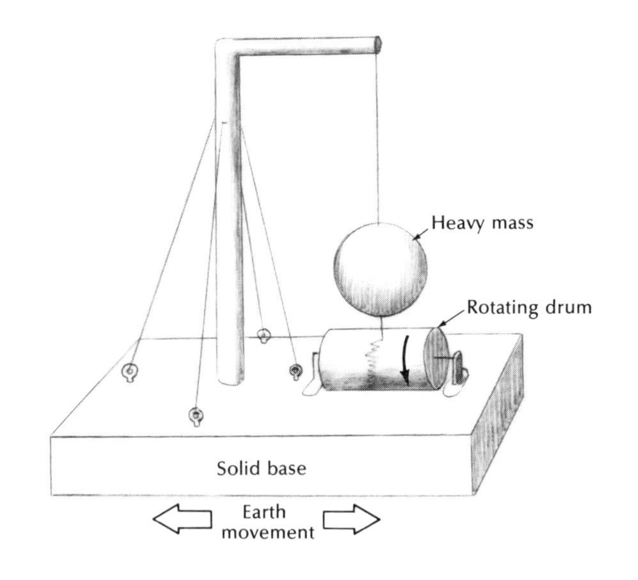

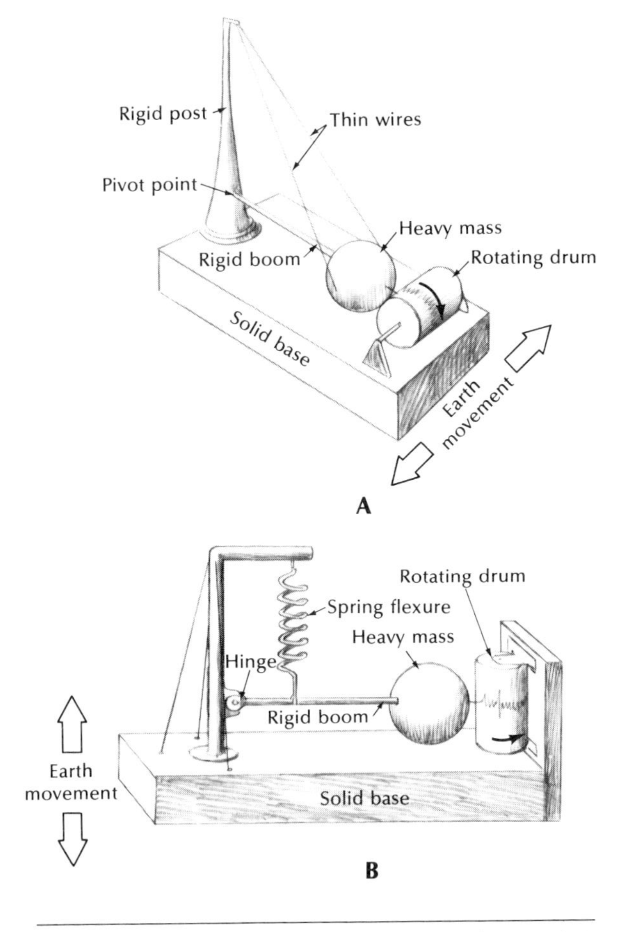

Fig. 18-21 Basic design of modern seismographs. **A.** Horizontal-type pendulum seismograph for measurement of N-S or E-W motions. **B.** Hinged pendulum seismograph for measurement of vertical motion. After Figs. 23-15 and 23-14B (p. 401) from THE EARTH SCIENCES, 2nd Edition, by Arthur N. Strahler. Copyright © 1963, 1971 by Arthur N. Strahler. Reprinted by permission of Harper & Row, Publishers, Inc.

cordings of "moonquakes" for a long period of time. Some seismographs are built specifically to monitor the intensely strong vibrations created close to the epicenter of an earthquake; others to measure incredibly weak seismic rustlings. Most seismographs can be tuned to pick up long or short seismic waves, as desired. Some special instruments can record the exceedingly long wave length seismic vibrations associated with large quakes, which require about an hour for the passage of one wave length.

The number of operating seismographs stands at more than 1000 today. Seismograph stations are not uniformly distributed around the world: countries that have the most earthquakes generally also operate more seismographs. Since the early 1960s, about 120 of the stations have agreed to operate according to a specified procedure and to transmit their records to a central collection agency at Golden, Colorado, maintained by the Earthquake Information Center of the U.S. Geological Survey. Most theoretical seismologists today use these records and do not themselves operate seismograph stations. The copying and distribution center for the world-wide network of stations is at Boulder, Colorado, and functions as a part of the National Oceanic and Atmospheric Agency (NOAA).

Seismology has come a long way in the last 100 years. As a result, we have begun to learn not only a great deal about earthquakes, but about the interior of the earth as well.

Waves and wiggles Most of the time the pens on a set of seismographs are relatively quiet. They record only low-level background vibrations that arise from a variety of causes: electronic instrumental noise, traffic vibrations near the site, wind gusts, and even the pounding of heavy surf on a distant coastline (Fig. 18-23).

From time to time, however, the seismographs will be shaken by waves emanating from an earthquake somewhere in the world, and the pens will record the onset, duration, and nature of the incoming wave motion (Fig. 18-24). For a quake that is relatively close by, the instrument generally will record three major wave pulses arriving in succession at the station. The first to arrive, generally a relatively low-amplitude wave, is called the ***primary*** or ***P wave.*** The second pulse, the ***secondary*** or ***S*** *wave,* is distinguished by sudden onset and is normally of greater amplitude than the *P* wave. And the *S* wave is followed by a wave train of higher amplitude and longer wave length, the ***long*** or ***L wave***. Depending on where the quake is located relative to the recording station, each of the three seismographs (E-W, N-S, and

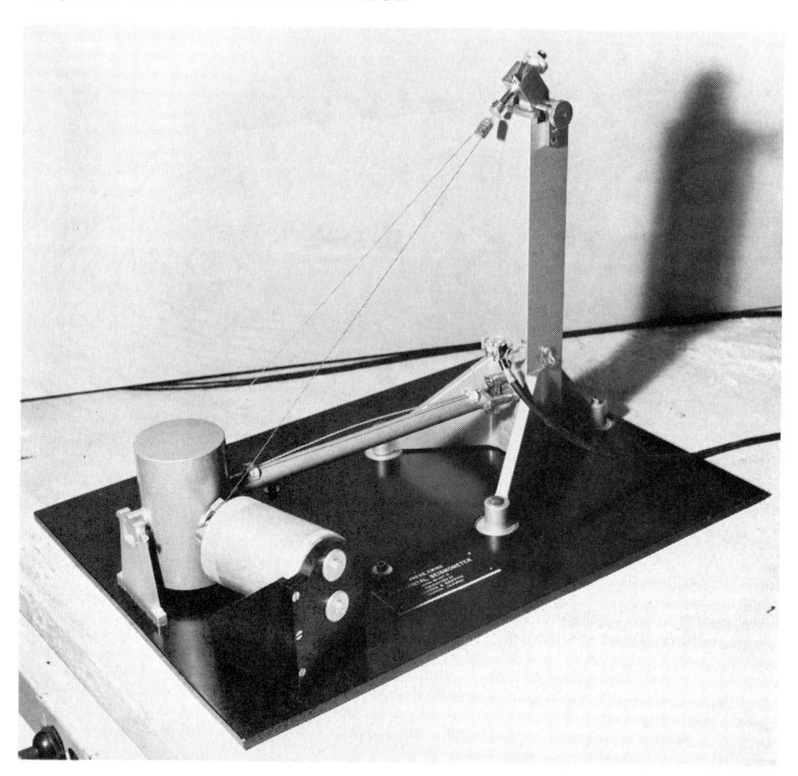

Fig. 18-22 Modern seismometer used to record either the east-west or north-south horizontal ground motion associated with seismic waves. *Lamont-Doherty Geological Observatory*

vertical) will show some variation in character in each of the three wave trains.

The *P* and *S* waves travel through the earth and accordingly are called ***body waves,*** whereas the *L* waves follow paths in the outer layers of the earth and are called ***surface waves.***

The nature of motion for each type of wave has been studied theoretically, and in the laboratory. It has been found that *P* waves travel through a solid as an accordion-like, push-pull sort of oscillation (Fig. 18-25). As the wave travels, the rock at any one place is at one moment compressed and at the next expanded. That type of compressional wave motion is familiar, since it is characteristic of sound waves as they move through the air and into our ears. Sound, as we hear it, is simply the recording of the to-and-fro motion of our eardrums in response to the push-pull waves moving through air. Sound travels in a similar way (but more quickly) in water. By means of sonar (the beaming and receiving of sound signals underwater), a ship's personnel can locate nearby ships and pinpoint obstacles and gain a knowledge of the topography of the ocean bottom over which the ship is traveling.

Compressional sound waves travel fastest in a solid. That is because the atoms are tightly bonded to each other, and thus are highly elastic. And anything that will increase the degree of elasticity will cause the *P* waves to be transmitted faster. For instance, when a rock specimen is compressed in a hydraulic press, the atoms in the rock are pushed closer together, thereby increasing its elasticity and increasing the speed with which *P* waves travel through it. Therefore, other things being equal, deep inside the earth, where pressures are extremely high, *P* waves travel more quickly that they do near the earth's surface. Increasing the temperature of a rock specimen, on the other hand, has a tendency to *decrease* elasticity and will correspondingly reduce the speed of *P* waves. Obviously, at any point inside the

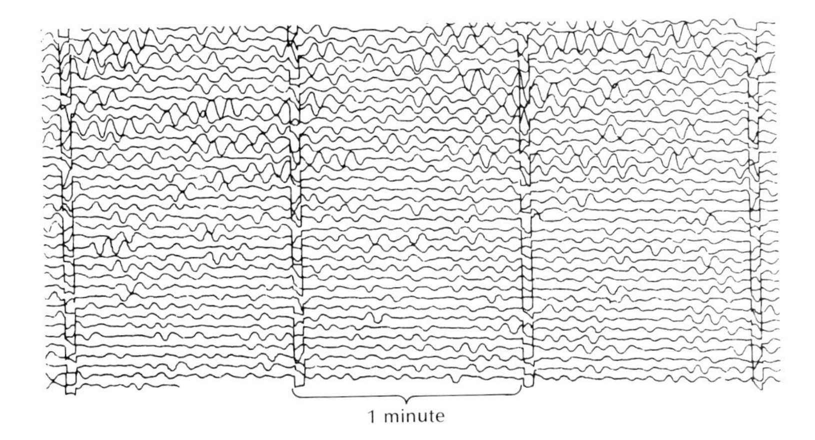

Fig. 18-23 Enlargement of a seismogram record showing low-level background vibrations that arise from a variety of causes. One-minute interval marks are transcribed on the record for precise determination of the onset of seismic signals.

earth, the elasticity will reflect the rock type, confining pressure, and temperature.

The *S* wave always travels more slowly than the *P* wave, and the reader would be correct in assuming that the difference in arrival time is due to a difference in the manner in which the *S* wave travels. It has been shown that in a solid, another type of oscillation is possible besides push-pull motion; seismic energy also can be transmitted by a shaking motion of the material transverse to the direction of passage of the wave (Fig. 18-26). In that mode, the rock particle at any one point moves back and forth sideways, while the *S* wave travels on in a snake-like fashion. It is much like the motion made by shaking a piece of a rope; the waves travel in *S*-shaped patterns as the rope slides from side to side. Rocks exhibit a lower degree of elasticity to such motion; therefore, the *S* wave is slower than the *P* wave. But, of course, increasing confining pressure will increase the elasticity of the rock, and the *S* wave will increase in speed. At depth in the earth, the *S* wave is expected, therefore, to move more quickly that it moves near the surface. And an increase in temperature will decrease its elasticity and slow the wave. Fluids and gases lack the type of elasticity that will permit the passage of *S* waves. Therefore such waves cannot travel through water, air, or any other fluid, such as magma.

The *L* waves, the slowest of the three, travel

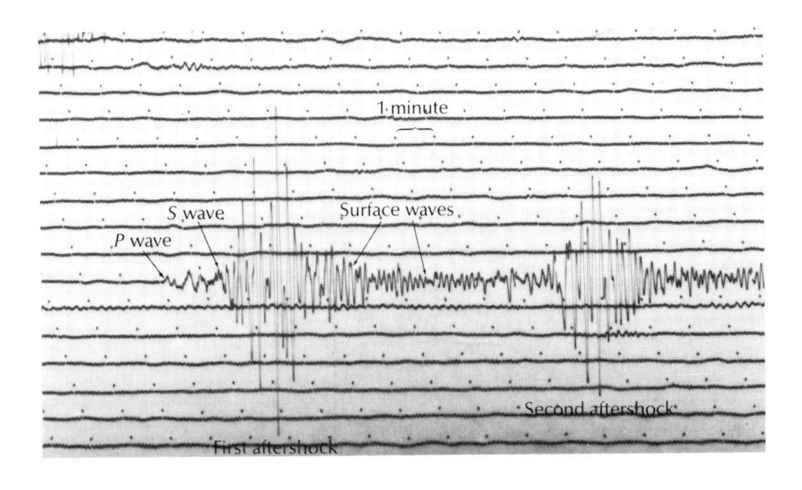

Fig. 18-24 Actual seismogram of north-south motions of two aftershocks recorded in Fairbanks, Alaska, a few days following a major earthquake in 1964. The onset of *P* and *S* waves are indicated for the first event. The *S* and *L* waves of the first shock obscure the onset of the *P* waves of the second event. The foci of the events were about 800 km (497 mi) south of the seismograph station.

in a rather complex way close to the surface of the earth. Because their passage is restricted to the outer levels of the earth, their speed is relatively constant.

All earthquakes generate all three types of waves. How they appear on the record at any one seismograph station will depend on the size and nature of the rupture at the point of origin or focus, the distance from the focus to the seismograph station, the depth of the focus of the quake, and the kind of material through which the waves travel before reaching the station.

Seismologists can gain insight into how waves travel by plotting the arrival time of the waves generated by a single earthquake at stations located at increasing distances from the quake epicenter (see Fig. 18-27).

If many stations are used, it is possible to construct a travel-time curve (Fig. 18-28). As you can see, the plot for each wave is decidedly different. The *L*-wave plot is almost a straight line, whereas the *P* and *S* waves plots are curved. Of the two body waves, the plot of the *P* waves bends more sharply than does that of the *S* waves. Moreover, from about 10,000 km or 6214 mi (equal to about 102° of earth circumference), to about 16,000 km or 9942 mi (equal to about 143°) from the epicenter, no *P* and *S* waves are directly recorded. It is as if seismographs in that portion of the globe were in the shadow of some seismic obstacle. That zone is called the ***shadow zone.*** At arc distances beyond 143°, seismic waves are again received, but their travel time is much longer than expected, as if they had been slowed during their passage through the central part of the earth.

Both *P* and *S* waves begin at the same time. However, a *P* wave is about twice as fast as an *S* wave, and the farther the two waves travel, the wider becomes the gap separating the wave fronts (Fig. 18-27 and 18-28). The difference in speed between *P* and *S* waves is indeed fortunate, for it provides us with a means of determining the distance a seismograph station is from the earthquake focus. For example, in the seismograph record in Figure 18-29, the *S* wave arrived about 8 minutes after the *P* wave. The travel-time curve (see Fig. 18-28) in-

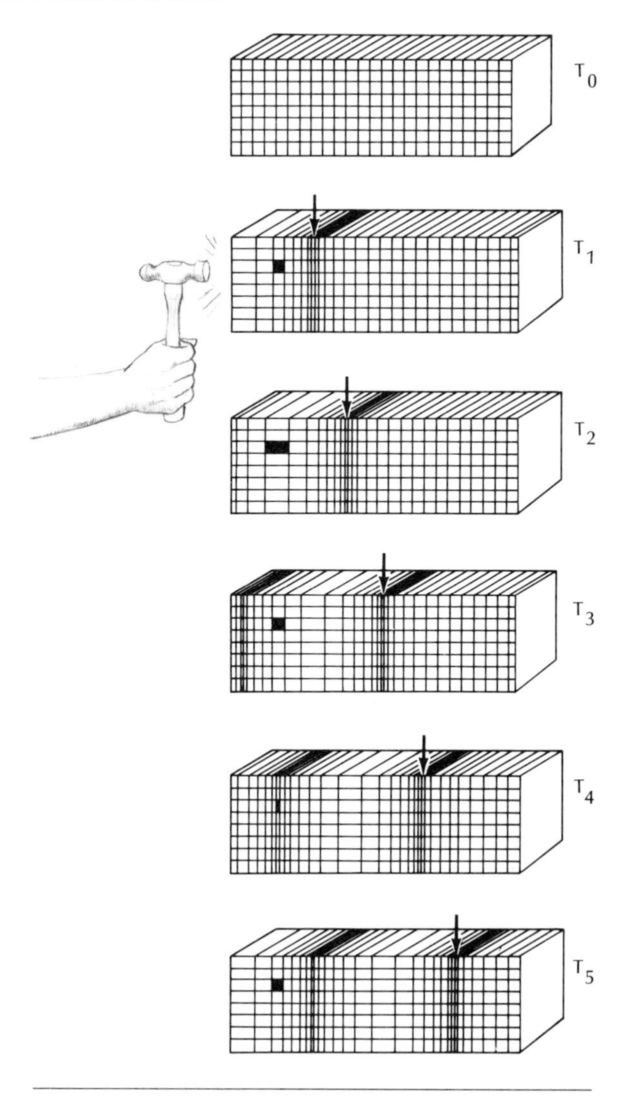

Fig. 18-25 Successive stages in the small-scale deformation of a rock by a *P* wave. As the sequence progresses, a wave of maximum compression (marked by the arrows) passes through the rock. As wave after wave pass through the block, the darkened square (representing any small volume of the rock) shakes back and forth, undergoing alternate compression and expansion.

dicates that the difference in arrival times is equivalent to about 8800 km (5468 mi). Now, seismologists cannot tell from which direction the quake came, so they cannot pinpoint the epicenter. They can only say on the basis of one set of travel-time data that the epicenter

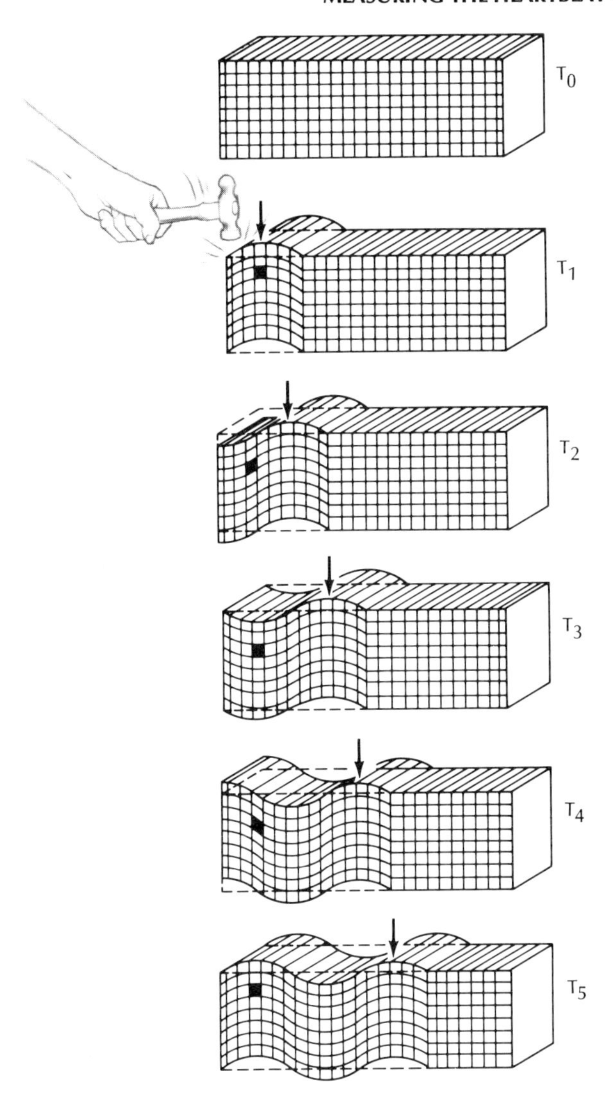

Fig. 18-26 Successive stages in the small-scale deformation of a rock by an *S* wave. As the sequence progresses, the *S*-wave crest (marked by the arrows) passes through the block, and the rock shakes sideways. Any particular volume (darkened block) undergoes cyclic changes in shape in response to the passage of the waves.

lies at a distance of about 8800 km from the station. A circle of that radius drawn around the seismograph station on a map will contain all the possible points of origin (Fig. 18-30). If the same quake is recorded at another station, however, it is possible to narrow down the area of the quake origin. Seismologists at the second station can similarly determine the distance to the epicenter and record a circle of the appropriate radius, with their station at the center. The circle will intersect that drawn around the first station at either one or two points. If information is available from a third seismograph station, a single intersection will be determined and thereby the earthquake epicenter can be located (Fig. 18-30).

In practice, the three circles will not usually intersect at a single point. For one thing, many earthquakes do not start at a point but rather along a fault line. The 1906 San Francisco quake, for example, originated along a fracture hundreds of kilometers long. For another, some quakes do not originate near the surface of the earth, but at depths as great as 700 km (435 mi). The travel-time curves given in Figure 18-28 are fine for shallow-focus quakes, but different curves must be used for deep-focus quakes. The existence of the latter was first suggested by H. H. Turner, a seismologist, in 1922, and during the following decade their reality was widely accepted only after much debate and discussion. Part of the evidence for their existence is the different sort of seismogram they write, compared to that of shallow-focus earthquakes. For one thing, the *L* waves are either lacking or ambiguous. Also, the *P* waves arrive sooner than they would have had they started near the surface. After all, they have the advantage of a considerable head start in their race toward a seismograph station. As we have already noted, deep-focus quakes commonly originate in zones of convergence, where two plates come together. Most earthquakes originate around the periphery of the Pacific Ocean, but others come from beneath the Himalayas, the Mediterranean Sea, Indonesia, and the Caribbean. The nature of rupture at a depth of several hundred kilometers is not exactly clear, since at those depths it might be expected that the rocks are so hot that they would tend to flow. Yet the quakes do occur, and the amount of energy released during any one event is just about the same as that from a near-surface quake. Today the focus of most quakes is located by the use of

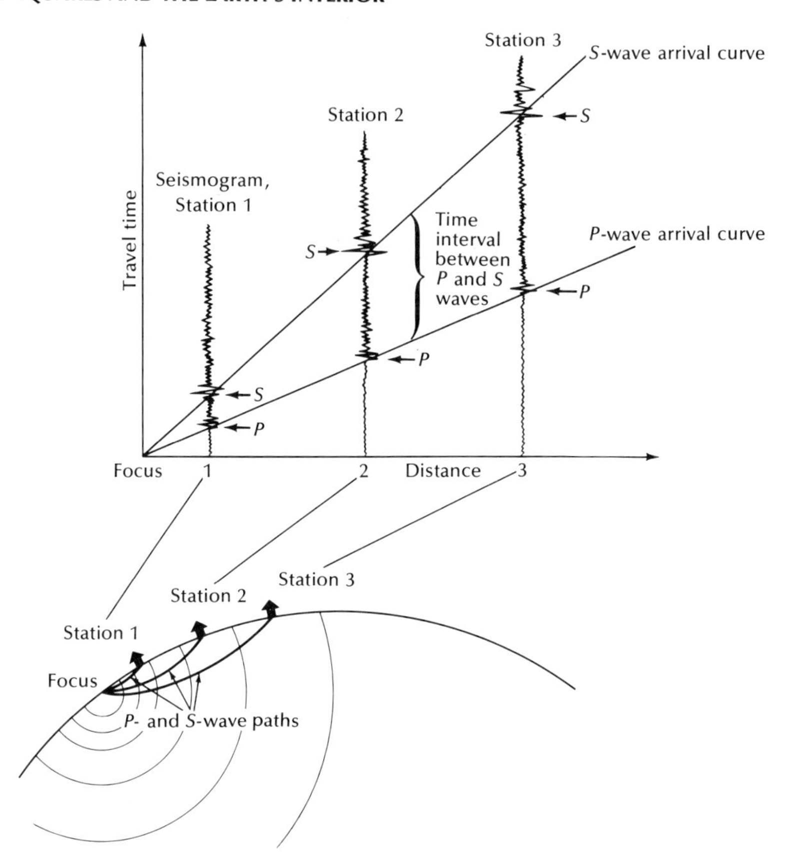

Fig. 18-27 Increase in the arrival time of *P* and *S* waves with distance from the epicenter. Note that the time interval between the arrivals also increases. As shown, the *P* wave is faster than the *S* wave.

high-speed computers. The records from as many stations as possible are used in the calculation. After the focus is roughly determined, by the use of travel-time curves as outlined above, the position of the focus is continuously moved (figuratively) in small steps laterally and vertically, and the travel times recalculated to all stations. The best estimate of the position of the focus is obtained when the calculated travel times to all stations best fit with the actual travel times measured directly from the records.

First Motions

The pattern of seismic waves also indicates the type of the relative fault movement—strike-slip, dip-slip or oblique-slip—that occurred during the initiation of the seismic waves. This is best understood by looking again at Figure 18-25. Note that in response to a hammer-blow—that is, a push—a compression wave moves away from the point of initiation. The first part of the wave signal received at any point will be a zone of compression. However, the wave could just as easily have been started by pulling the block sharply to the left; then, the first part of the wave signal would be a zone of rarefaction, or expansion. On the seismograph, the pen would initially move up or move down to start its wiggly trace depending on whether the first part of the *P* wave was a zone of compression or a zone of expansion, respectively (Fig. 18-31). The first motions received on the three different seismographs at any station can be used to determine the orientation of the fault and the relative motion of the sliding blocks during earthquake initation.

Intensity versus magnitude

As soon as the earth starts to rupture, *P* and *S* waves begin their outward journey, carrying the vibrational energy initiated by the seismic disturbances in all directions. The size of the waves (their amplitude—which is measured from the crest of one wave to the trough of another and divided by 2) at any one point varies with the size of the event. Small ruptures produce small waves, whereas large ruptures produce large ones. It is possible to obtain some idea of the magnitude of an earthquake at its source by measuring the amplitudes of the *P* and *S* waves as they are recorded on a seismograph. Of course, the amplitude diminishes as the waves travel away from the disturbance and as the vibrational energy is dispersed. Therefore, it is necessary to make a correction that takes into account the distance of the recording station from the seismic event.

Fig. 18-28 Travel-time curves for earthquakes originating at a depth of less than 100 km (62 mi). Notice how the *P*- and *S*-wave curves bend, indicating an increase in speed with distance.

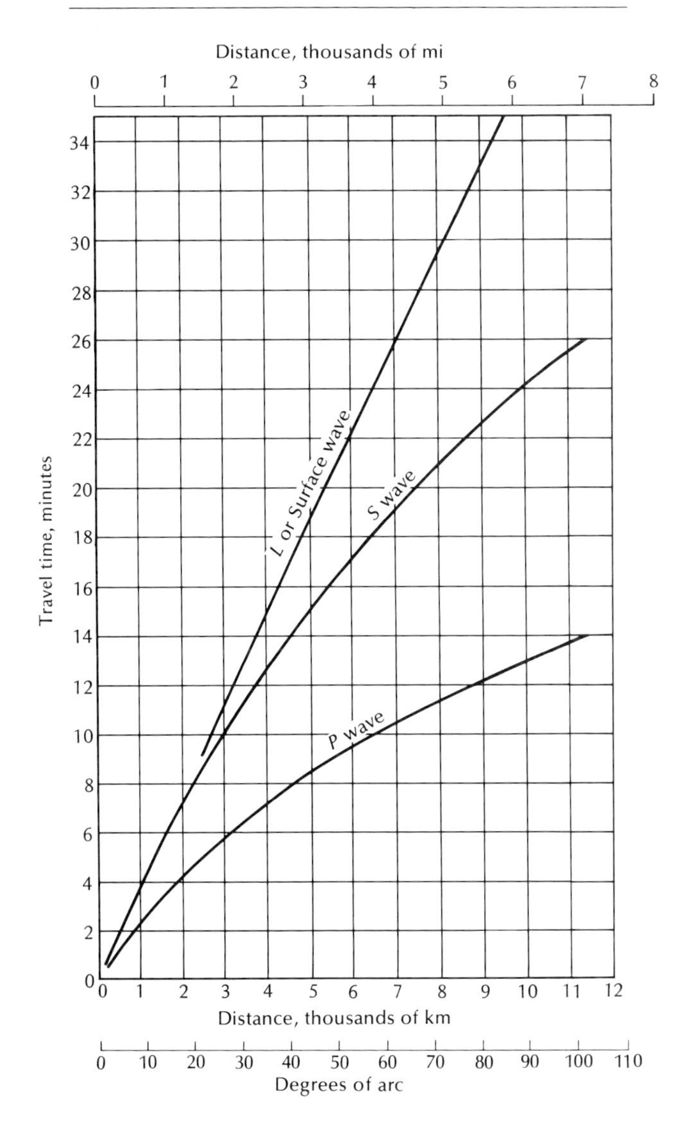

A scale of earthquake magnitudes based on that line of reasoning was first devised by the seismologist Charles Francis Richter in 1932 and bears his name. It enables seismologists at a distant recording station to get a pretty good idea of the amount of strain energy released during an earthquake (Table 18-1). The scale is roughly exponential, so that a difference of only one division corresponds to a difference in source energy of about 30 times. During a magnitude 4 event, for example, the energy released is not just twice that released during a magnitude 2 event, but about 900 times as great (Fig. 18-32). The largest earthquakes ever recorded have measured about 8.9 on the ***Richter scale:*** they correspond to the energy produced by detonation of about 100 million metric tons of TNT. In contrast, a magnitude 1 quake would be equivalent to the energy released by the detonation of less than 0.5 kg (1 lb) of TNT. You might wonder why no quake larger than 8.9 has ever been recorded. It appears that there is a limit to how much strain energy can be stored before the rocks fail and slippage occurs.

Each year, many hundreds of thousands of magnitude 1 and 2 quakes occur; only about one larger than magnitude 8 occurs every five to ten years (Table 18-1). Any quake larger than magnitude 6 can be considered a major event and capable of considerable damage.

The extent of damage done by an earthquake depends (as discussed earlier) on a variety of factors, such as proximity of the focus, duration, amount of energy released (Richter magnitude), type of rock underlying the site, type of structures, and density of population. The

1 minute
P arrival
S arrival

Fig. 18-29 Portion of a seismogram from Berkeley, California, recording an earthquake on September 20, 1968. Both the *P* and *S* waves are detectable, with the *S* wave arriving almost exactly 8 minutes after the *P* wave.

Fig. 18-30 From the time lag between *P* and *S* waves recorded at three separate stations, the approximate epicenter of an earthquake can be determined. The distance to the event is calculated, and a circle of that radius is drawn around each of the three stations. The common intersection marks the approximate epicenter.

Table 18-1. Earthquake intensities, magnitudes, statistics, and energies

Modified Mercalli Scale: Characteristic effects of shallow shocks in populated areas	*Approximate Richter magnitude*	*Number of earthquakes per year*	*Energy (Approximately equivalent, metric tons of TNT)*
Damage nearly total	≥8.0	0.1–0.2	>5,668,750
Great damage	≥7.4	4	907,000
Serious damage, rails bent	7.0–7.3	15	181,400
Considerable damage to buildings	6.2–6.9	100	27,210
Slight damage to buildings	5.5–6.1	500	907
Felt by all	4.9–5.4	1,400	136
Felt by many	4.3–4.8	4,800	27
Felt by some	3.5–4.2	30,000	1.8
Not felt, but recorded	2.0–3.4	800,000	6×10^{-3} to 0.8

Data from B. Gutenberg

amount of damage for any particular quake will vary as those factors vary. An earthquake scale—based on the damage wrought by a quake—is called an ***intensity scale,*** and it ranges from a minimum number, corresponding to a barely discernible event, to a maximum number, corresponding to total collapse of buildings and great loss of life (Table 18-1). The ***Modified Mercalli Scale*** is perhaps the most commonly used intensity scale today. It was formulated by the Italian seismologist Giuseppe Mercalli (1850–1913) in 1902 and modified in 1931 by American seismologists Harry O. Wood and Frank Neumann.

Seismic waves and the earth's interior

Our knowledge of the interior of the earth is restricted to observations of rocks once buried deep below the surface, to cuttings from drill

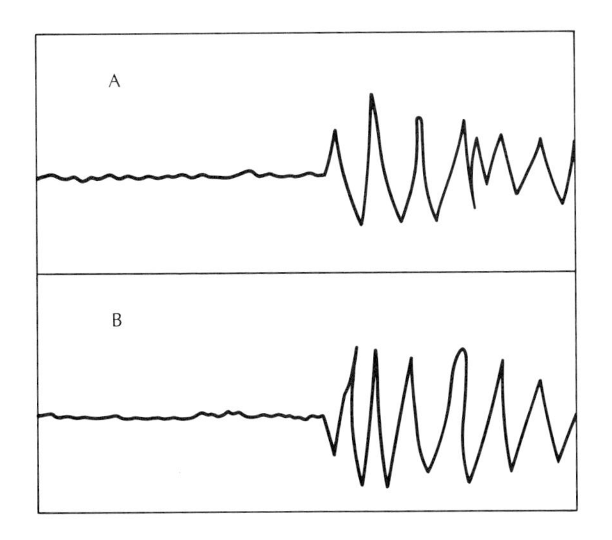

Fig. 18-31 Two different responses (first motions) of a seismograph, depending on whether the first portion of the incoming *P* wave is (**A**) compressional (due to a push) or (**B**) expansional (due to a pull).

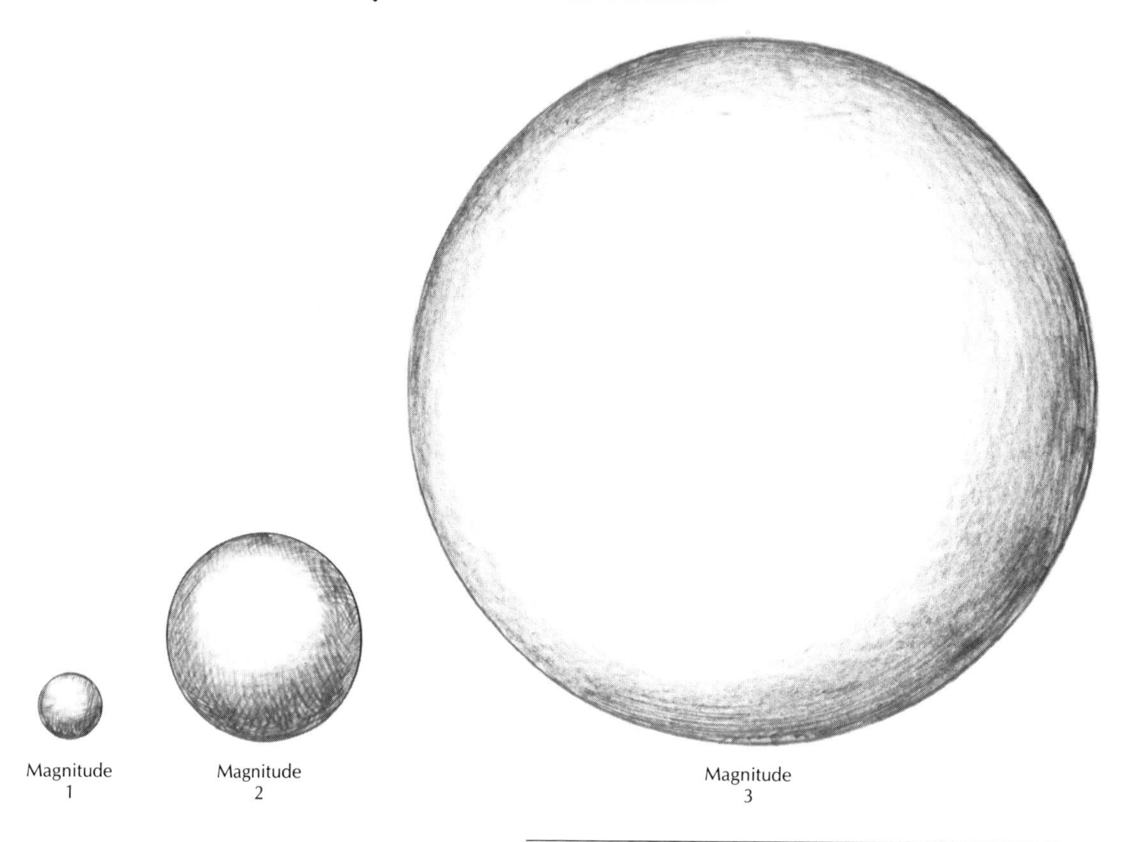

Fig. 18-32 Relation of Richter magnitude to energy released, for three magnitudes. The volume of the spheres is roughly proportional to the amount of energy released by earthquakes of the given magnitudes.

holes thousands of meters deep, and to rock fragments ripped from the walls of volcanic conduits and carried to the surface by eruptions of lava.

Fortunately, there are means by which we can make realistic inferences about the inaccessible parts of the earth. Of all the means, seismology has proved the most useful. From the analysis of countless earthquake records, seismologists and other earth scientists have put together a model of the structure and composition of the earth. Of course, we know that it is only a first approximation and that it will be modified somewhat as time passes and as new data become available. Most earth scientists feel so confident about the general features of the model, however, that they would expect few surprises even if at this moment we could drill a hole to the center of the earth.

Seismology has been so useful largely because body waves—both *P* and *S* waves—radiate outward in all directions from an earthquake. Many traverse the deeper parts of the earth as they make their way toward distant seismograph stations, and their seismographic tracings carry information about the materials through which they pass. Before trying to decipher the results, let us first look at some of the properties of body waves.

Body waves, like all traveling waves, are sub-

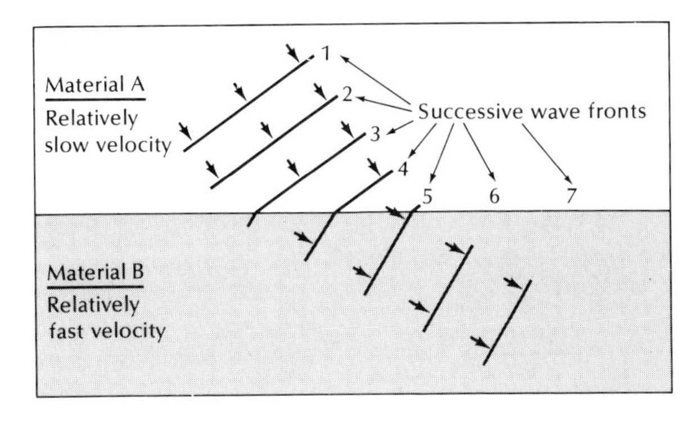

Fig. 18-33 Refraction of a light wave as it moves from one material (A) to another (B). The wave travels more quickly in (B) than in (A); therefore when it strikes the interface and is in both materials, it is bent, or refracted.

Fig. 18-34 Refraction and partial reflection of an incident light wave as it encounters a surface of discontinuity (sharp break) between one material (A) and another (B). The angle at which the ray is reflected from the interface (θ) equals the angle of impingement (ϕ) of the incident ray.

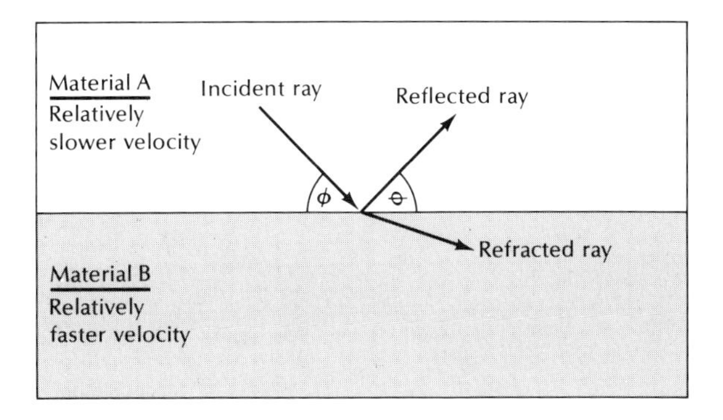

ject to refraction (bending) and reflection. In a homogeneous material at constant temperature and pressure, a seismic body wave will travel at a constant velocity—which depends on the density and elastic properties of the material. If the wave passes into a material of different density or elastic properties, its velocity generally will change, and refraction of the wave will generally occur (Fig. 18-33).

For example, you are aware that light rays, as they pass from the air into the glassy material of a pair of eyeglasses and back again, are refracted, or bent, in such a way that the rays are focused upon the retina of the eye.

Reflection of waves will occur when they encounter a sudden velocity change (discontinuity) in the material through which they are passing. As an example, light rays are reflected when they encounter a highly polished surface. Reflection surfaces generally are not totally reflective, particularly to seismic waves. Some of the wave energy passes through the discontinuity and into the material beyond, where it continues (Fig. 18-34).

If the rocks through which a wave is traveling increase in temperature, the wave will slow down; if there is an increase in rock confining pressure, the wave will speed up. In a fluid, in which all rigidity has been lost, the *S* waves cannot propagate at all, and the *P* waves are greatly slowed.

As we mentioned earlier, the waves that arrive directly at progressively more distant stations travel at increasingly greater speeds. That can mean only that those waves go deeper into the earth and move through materials that permit progressively faster wave propagation. The implication is that the earth's physical properties change progressively with depth so that seismic waves move faster and faster. Figure 18-35 shows how seismic waves would behave as they pass through a three-layered body.

By studying the details of travel-time and distance data collected from earthquake records from all over the world, seismologists have been able to put together a consistent picture of how body-wave velocities vary with depth (Fig. 18-36).

The results are intriguing, for they show that sharp velocity discontinuities exist within the earth; that *S* waves cannot be propagated between 2900 and 5000 km (1802 and 3107 mi), and that the increase in *P*- and *S*-wave velocities in the outer 700 km (435 mi) of the earth is far from uniform.

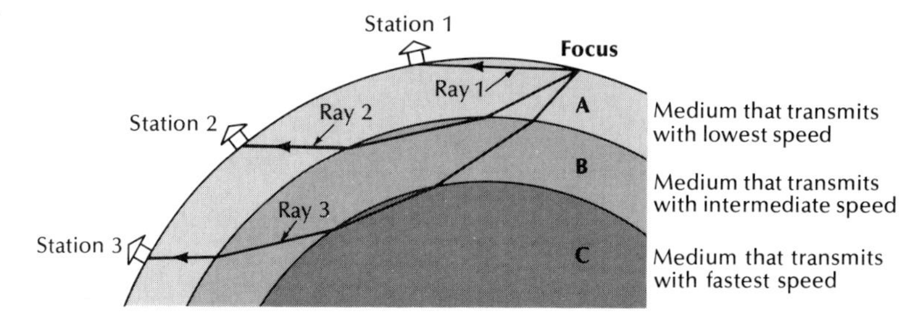

Fig. 18-35 Transmission of three earthquake waves in a hypothetical three-layered earth. Ray 1 travels only in layer A, and is the slowest. Ray 2 travels both in layers A and B, moving more quickly through B, the intermediate-speed layer. Ray 3 travels through all three layers, and moves with the highest average velocity through layer C. Wherever a wave encounters a velocity discontinuity, it bends.

In general, the velocities of both *P* and *S* waves increase down to 2900 km (1802 mi). However, *S*-wave velocities are lower than expected between about 100 and 250 km (62 and 155 mi), which has led to the designation of the region as the ***low-velocity zone.***

At 2900 km (1802 mi), the *P*-wave velocity drops drastically to a value close to its value near the earth's surface, and the *S* wave disappears entirely. Such behavior is exactly what would be expected if at 2900 km the rock changed from solid to fluid. The change in rock properties appears to be very sharp and has been designated the ***core-mantle boundary.*** The solid mantle is thought to consist of iron- and magnesium-silicate minerals, and the liquid core principally of iron. Within the core and at about 5100 km (3169 mi), the *P* wave velocity appears to increase abruptly, and there is even a hint of an *S* wave. Seismologists think that is caused by a change from a fluid ***outer core*** to a solid ***inner core.***

The existence of the sharp velocity discontinuity at the core-mantle boundary also provides an explanation of the so-called "seismic shadow zone" mentioned earlier (Fig. 18-37). Earthquake waves can move through the mantle to a depth of 2900 km (1802 mi), but deeper waves strike the core-mantle boundary, and some of the energy is sharply reflected back to the surface. But some energy moves into the core and is greatly refracted as it does so. Because of the extreme refraction at the core-mantle boundary, no direct *P* waves reach the earth's surface between 102° and 143° arc distance from the epicenter. Only *P* waves can traverse the fluid part of the core, but because they do so at reduced velocity, the time necessary for waves to travel through the central part of the earth is increased.

One other major seismic discontinuity lies near the outer margin of the earth. It is the ***Mohorovičić discontinuity,*** named for Andrija Mohorovičić (1857–1936), the Yugoslavian seismologist who first suggested its presence in 1909. The name has since been shortened to the ***Moho*** or ***M discontinuity,*** which has enabled many a seismologist to discuss it at a cocktail

Fig. 18-36 How *P*- and *S*-wave velocities change with depth. The variation in speeds is represented by the variable thickness of the lines. Note the lack of *S* waves and the reduced velocity of *P* waves in the fluid outer core.

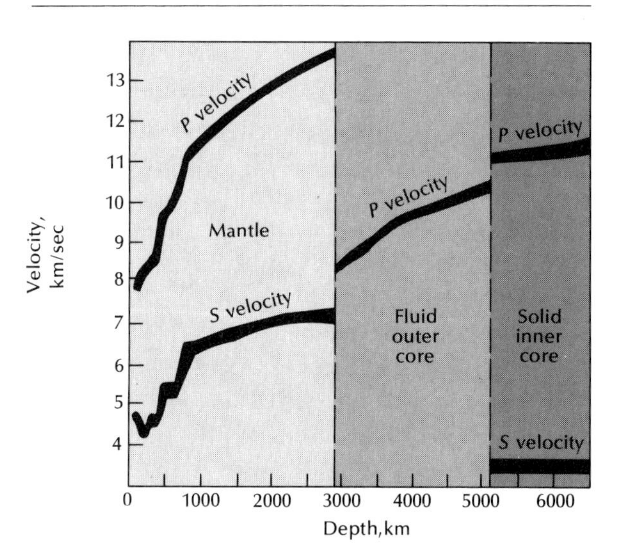

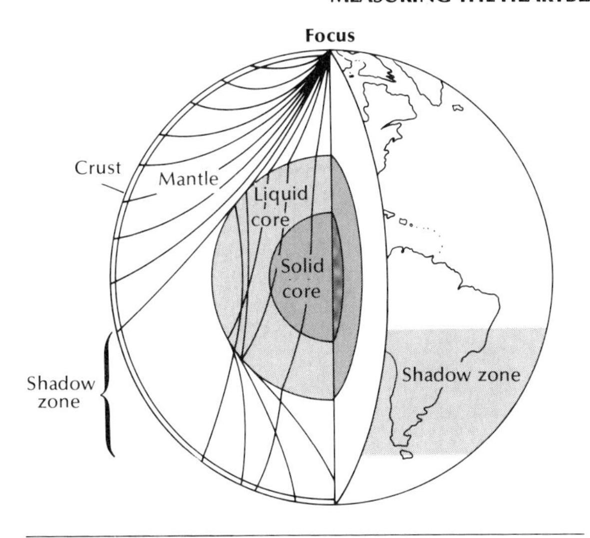

Fig. 18-37 The shadow zone for an earthquake originating at the North Pole. As shown in the cutout, *P* and *S* waves travel normally until they intersect the liquid core. Then the *P* waves slow down abruptly and bend toward the center of earth (accounting for the shadow zone), and the *S* waves disappear entirely. Beyond 143° of arc from the focus, *P* waves are again received directly.

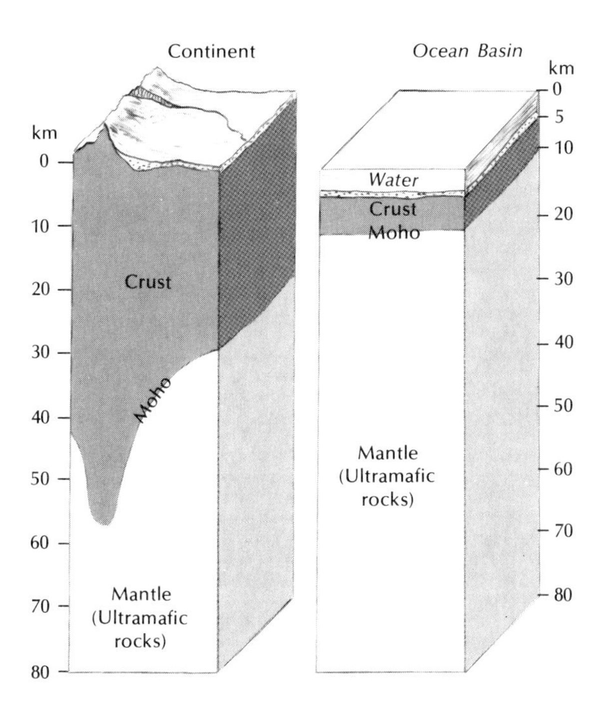

party rather than to remain tongue tied. The Moho separates rocks above, which have *P*-wave velocities of 6 to 7 km/sec (3.7 to 4.3 mi/sec) from those below which have *P*-wave velocities of about 8 km/sec (5 mi/sec). By tradition, rocks above that seismic-velocity break are called the ***crust*** and those below it, the ***mantle.*** Under the continents the crust generally varies from 30 to 60 km (19 to 37 mi) in thickness, whereas under the ocean basins it is about 6 to 8 km (4 to 5 mi) thick (Fig. 18-38).

Recently, on the basis of its seismic properties, as well as its mechanical strength, the outer 100 km (62 mi) of the earth have been designated the ***lithosphere,*** the layer extending from 100 to 250 km (62 to 155 mi) the ***asthenosphere*** (Fig. 18-39), and the layer below 250 km as the ***mesosphere.*** The lithosphere (meaning rock sphere) therefore includes both the crust and the upper part of the mantle. It is cooler, more brittle, and stronger than the asthenosphere. The large plates of the earth, which appear to slide to produce the effects described by the plate tectonic model, consist entirely of lithospheric material.

The asthenosphere (meaning sphere of weakness) corresponds to the low-velocity zone, mentioned previously, in which *S* waves slow down greatly. There, it is thought, the effects of increasing temperature predominate over the effects of increasing confining pressure. Thus, the rocks are less rigid. The presence of a relatively weak zone is thought by many to be absolutely necessary to the mobility of the lithospheric plates above. In fact, many scientists have concluded that temperatures are so high in the asthenosphere that the

Fig. 18-38 Exaggerated schematic view of the relationships between crust and mantle beneath the continents and ocean basins. The continental crust, which has a relatively low density (2.7 g/cm^3), floats isostatically in a high-density mantle and projects downward as much as 60 km (37 mi) beneath some high mountains. The oceanic crust, with a density of about 3 g/cm^3, is relatively thin. After Fig. 23-27 (p. 401) from THE EARTH SCIENCES, 2nd Edition, by Arthur N. Strahler. Copyright © 1963, 1971 by Arthur N. Strahler, Reprinted by permission of Harper & Row, Publishers, Inc.

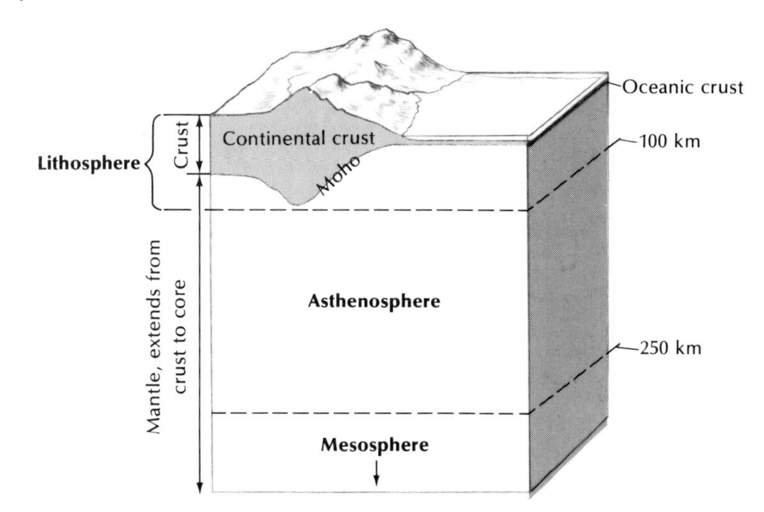

Fig. 18-39 The threefold division of the outer part of the earth, based on its physical properties. The lithosphere, which includes both the crust and the upper part of the mantle, is strong and relatively brittle; the asthenosphere is relatively weak and capable of flowing; and the mesosphere is strong.

rock is partially melted (up to 10 per cent) and that the molten part is the source of much of the iron- and magnesium-rich lava that erupts from volcanoes at the earth's surface.

The mesosphere (middle sphere) represents the mantle zone beneath the asthenosphere, where the pressure again becomes high enough to cause the rock to be strong.

Seismologists have also been able to tell us how the inside of the earth varies in density. Making such determinations involves combining (1) data from laboratory experiments, (2) theoretical considerations that relate density and body-wave velocities, and (3) a knowledge of the general distribution of mass in the earth (see Chap. 3). The result, as shown in Figure 18-40, clearly indicates that density increases with depth, paralleling the steps in the body-wave velocity distribution. The greatest jump in density occurs at the core-mantle boundary. At the center of the earth, rocks apparently have specific gravities as high as 12 or 13; that is, they are four to five times more dense than are most rocks at the surface.

What kinds of rocks could account for the variations in body-wave velocities and density? The elements making up the rocks must be those that are seemingly abundant within our solar system. Such a conclusion is based on our knowledge of the elements occurring in nearby stars (particularly the sun) and in meteorites that fall to the earth's surface. In its passage around the sun the earth is nearly continuously hit with meteoritic debris. Most of the meteorites are small and are consumed as they pass through our oxygen-rich atmo-

Fig. 18-40 Curves depicting the estimated density of the earth. Notice the sharp increase at the mantle-core boundary. The density of rocks at the earth's center is thought to be about four to five times that of surface rocks.

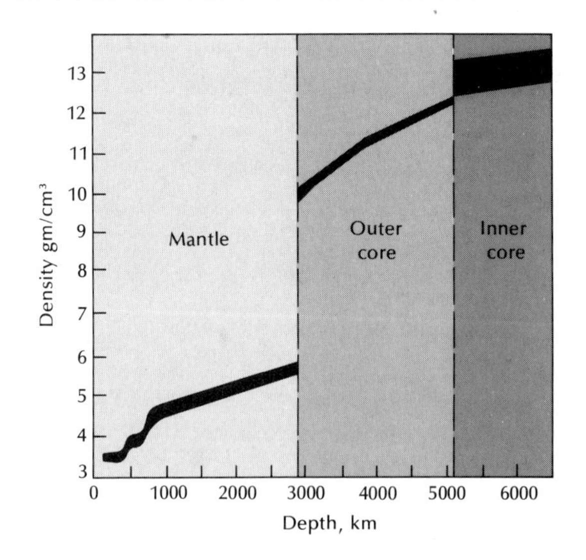

Table 18-2. Average composition of stony meteorites and the hypothetical composition of the earth

Element	*Stony meteorites, weight per cent*	*Earth, weight per cent*
Oxygen (O)	33.24	29.5
Iron (Fe)	27.24	34.6
Silicon (Si)	17.10	15.2
Magnesium (Mg)	14.29	12.7
Sulfur (S)	1.93	1.93
Nickel (Ni)	1.64	2.39
Calcium (Ca)	1.27	1.13
Aluminum (Al)	1.22	1.09
Sodium (NA)	0.64	0.57
Chromium (Cr)	0.29	0.26
Manganese (Mn)	0.26	0.22
Phosphorus (P)	0.11	0.10
Cobalt (Co)	0.09	0.13
Potassium (K)	0.08	0.07
Titanium (Ti)	0.06	0.05

sphere. But some are large enough to resist total destruction and they fall far and wide over the earth. Most of those that are recovered can be placed into two main groups: the ***iron meteorites*** and the ***stony meteorites.*** Iron meteorites consist primarily of iron-nickel alloys, whereas stony meteorites consist of silicate minerals and of lesser amounts of iron-nickel alloys and sulfides. Stony meterorites have a relative abundance of non-volatile elements—such as magnesium, silicon, aluminum, calcium, and iron—in about the same proportion as in the sun and other stars. Earth scientists have concluded, therefore, that the relative abundances of those elements in an average stony meteorite is a good representation of the average composition of the planetary bodies in our solar system, including the earth (Table 18-2).

The relatively dense core of the earth, which accounts for about 30 per cent of its mass, is thought to be most probably composed of iron, with smaller amounts of silicon, nickel, and sulfur. The mantle, on the other hand, accounting for about 67 per cent of the earth's mass, appears to be made of silicates—mostly those of iron and magnesium, with only small quanitities of aluminum, calcium, and sodium. In the outer part of the mantle the rock is chiefly olivine, with lesser amounts of magnetite and pyroxene. That conclusion is confirmed by analysis of nodules and clasts found in the basaltic lavas and kimberlites that were ripped from the mantle and carried to the surface during volcanic activity.

In the outer 700 km (435 mi) of the earth, the physical state of the iron-magnesium silicates in the mantle varies, apparently in response to variations in temperature and pressure (Fig. 18-41). We have already noted that in the asthenosphere temperatures appear to be high enough and confining pressure low enough so that partial melting of the rocks may take place. Below 250 km (155 mi), as increasing confining pressure becomes the dominant factor, the material solidifies, and in the next 450 km (280 mi), it undergoes at least two changes in physical state into more-dense crystalline forms of the iron-magnesium silicates. What appears to happen, and this has been verified by laboratory experiments in which pressures equivalent to those at a depth of 600 km (373 mi) are generated, is that as pressure continues to increase with depth the less-dense crystalline

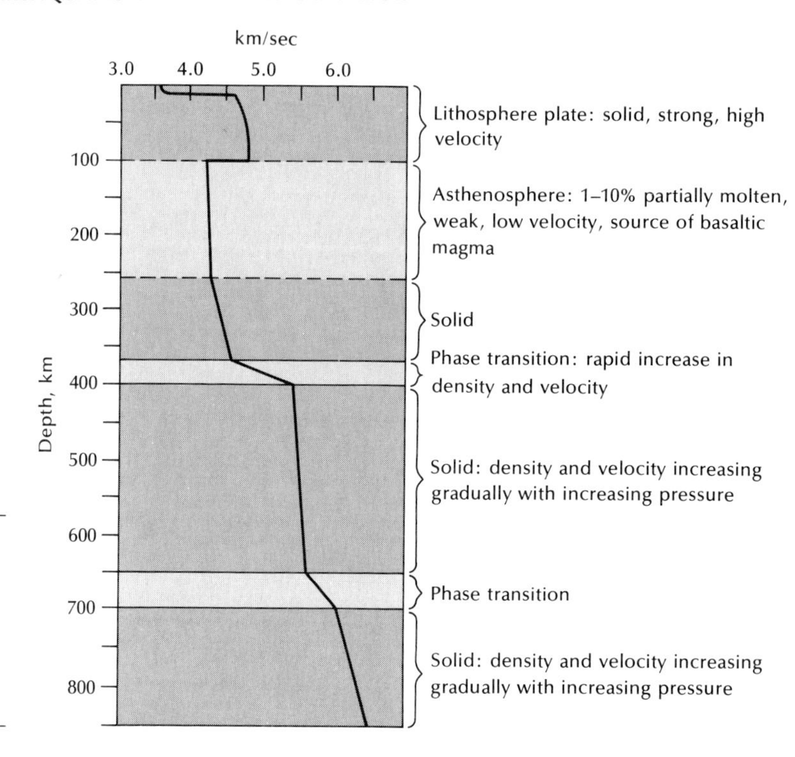

Fig. 18-41 Estimated nature of the outer part of the earth based on variations of the *P*-wave velocity with depth. The velocity break at 100 km (62 mi) represents the boundary between the lithosphere and asthenosphere. Increases in velocity near 400 km (249 mi) and 700 km (435 mi) apparently represent changes of state.

substances, which were stable near the outer part of the mantle, can no longer withstand the pressure and change to more-compact, denser forms. One such ***change of state*** occurs at a depth just above 400 km (249 mi) and another between 650 and 700 km (404 and 435 mi). Each change of state is associated with an increase in body-wave velocity. Below 700 km, no further changes are evident, and the mantle undergoes continued compaction, with continuing increases in body-wave velocities, down to the core-mantle boundary. Evidently, below 700 km, the rocks are in the densest possible state. Variations in temperature and pres-

Fig. 18-42 Estimated variation in temperature and pressure with depth inside the earth. Note that initially the temperature increases rapidly, but quickly tapers off.

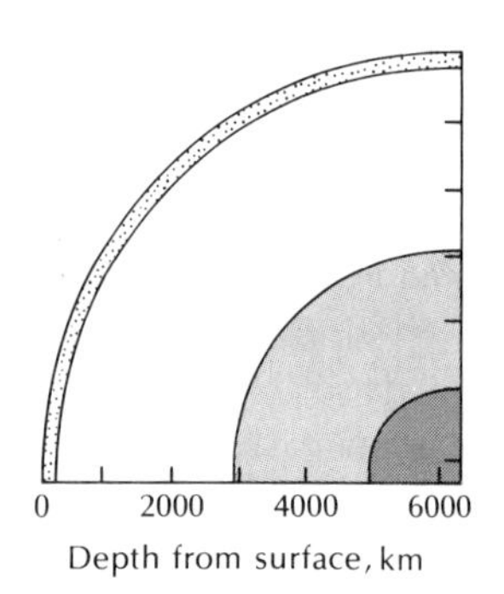

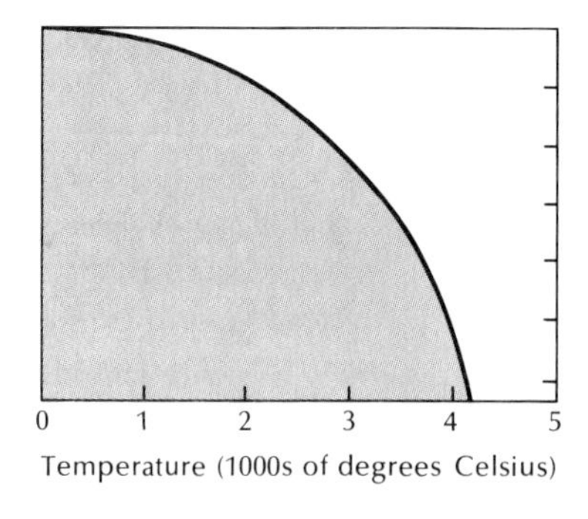

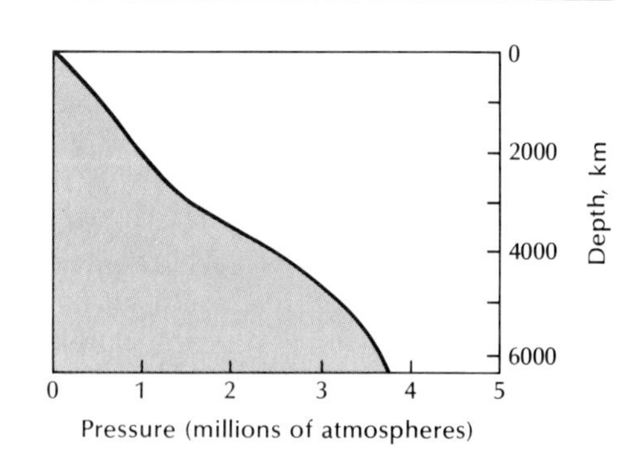

sure with depth are given in Figure 18-42. One atmosphere of pressure equals about 1 kg/cm^2 (14.7 lb/in.^{2}), the air pressure at sea level.

The earth's crust, as we know from direct observation, consists of diverse rocks of the lighter silicate minerals rich in calcium, sodium, aluminum, and potassium. In the ocean basins those silicates are mostly plagioclase, pyroxene, and olivine. In contrast, the continents are composed of a complex of the relatively light granitic, metamorphic, and sedimentary rocks.

In many ways the realm beneath our feet is as mysterious as interstellar space. It is true that we have come a long way toward understanding it, but the goal of full knowledge is still far distant. Just as it has been so up to now, wiggles on seismographs will undoubtedly prove to be the principal way by which we progress toward that goal.

SUMMARY

1. The San Francisco earthquake of 1906 resulted from slippage along the San Andreas Fault, which amounted to about 5 m (16 ft). Fire subsequent to the quake caused most of the destruction. In trying to explain the earthquake, H. F. Reid postulated the *elastic-rebound theory.*

   One of the largest quakes ever felt in the United States centered in New Madrid, Missouri, in 1811–12. The death toll and property damage were low because of a low population density at that time.
2. Japan, China, southern Europe, and South America also have earthquake-prone areas and experience a number of severe and destructive quakes.
3. *Shallow-focus* quakes are defined as those down to 70 km (43 mi); *intermediate-focus* quakes, between 70 and 300 km (43 and 186 mi); and *deep-focus* quakes, between 300 and 700 km (186 and 435 mi). Earthquakes occur in linear belts; many scientists feel that such belts delineate the edges of tectonic plates, which are converging, diverging, and sliding laterally. Shallow-, intermediate-, and deep-focus earthquakes are associated with convergent zones, whereas shallow-focus quakes are associated only with zones of divergence and lateral translation.
4. Amelioration of earthquake damage is being approached in two ways: through *earthquake prediction* and *earthquake control.* At present, prediction appears more feasible than control.
5. The first waves to arrive on a seismograph are *P waves,* then *S waves,* and finally *L waves. P* waves are *compressional* waves and can travel through solid or liquid. *S* waves are *transverse* waves that can only travel through a solid. *L* waves are the slowest and travel at or close to the earth's surface. Since *P* waves travel faster than *S* waves, it is possible to estimate the distance to a seismic disturbance by the separation in the arrival times of these two waves.
6. The energy released during an earthquake is measured on the *Richter magnitude scale.* The largest magnitude ever recorded is about 8.9. Nearly a million quakes occur yearly; of them only about ten are above magnitude 7. A quake of magnitude 8 or larger occurs every 5 to 10 years. Earthquake damage depends on several factors—magnitude, proximity, duration of quake, type of ground, and type of building.
7. Seismic data indicate that the earth is a layered body. The *crust* extends down to the *Moho discontinuity,* which occurs as deep as 70 km (43 mi) beneath the continents and 8 km (5 mi) beneath the oceans. Below the crust, the *mantle* is composed of ferromagnesian silicates and extends down to 2900 km (1802 mi). Below that is the *core,* which is liquid in character down to 5100 km (3169 mi) and solid down to the center of the earth.
8. On the basis of seismic properties, the outer 100 km (62 mi) of the earth has been designated the *lithosphere* and the layer between 100 and 250 km (62 and 156 mi) the *asthenosphere.* The asthenosphere, corresponding to the *low-velocity zone,* is relatively weak and is thought to be necessary to the mobility of the lithospheric plates above. The *mesosphere* is the mantle zone

beneath the asthenosphere, another zone with strong rocks.

9. Seismic speeds and laboratory data indicate that two *changes of state* occur in the outer part of the mantle in response to increased pressure. One occurs at about 400 km (248 mi) and the other at about 650 to 700 km (404 to 435 mi). Each is associated with an increase in body-wave speed.

QUESTIONS

1. Discuss the global distribution of earthquakes.
2. What factors make some earthquakes more devastating than others?
3. What is the elastic-rebound theory of earthquake generation?
4. What is being done to reduce earthquake threat?
5. Describe a simple seismograph and how it developed through time.
6. What are the principal waves generated by an earthquake? How do they differ?
7. How is the intensity of an earthquake measured? The magnitude?
8. Describe how the velocities of the *P* and *S* waves vary with depth in the earth.
9. Compare the lithosphere and the asthenosphere with the crust and the outer mantle.
10. Describe the variation with depth within the earth of temperature, pressure, and density.

SELECTED REFERENCES

Anderson, D. L., 1962, The plastic layer of the earth's mantle, Scientfiic American, vol. 207, pp. 52–59.

Baranzangi, M., and Dorman, J., 1969, World seismicity maps compiled from ESSA, Coast and Geodetic Survey, Epicenter Data, 1961–67, Seismological Society of America Bulletin 59, pp. 309–80.

Båth, M., 1973, Introduction to seismology, John Wiley and Sons, New York.

Bolt, B. A., 1978, Earthquakes—a primer, W. H. Freeman and Co., San Francisco.

Bronson, W., 1959, The earth shook, the sky burned, Doubleday and Co., Garden City, New York.

Bullen, K. E., 1954, Seismology, Methuen and Co., London.

Byerly, P., 1954, Seismology, Prentice-Hall, Englewood Cliffs, New Jersey.

CDMG Note, 1979, How earthquakes are measured, California Division of Mines and Geology, Note 23, pp. 35–37.

Davison, C., 1936, Great earthquakes, Thomas Murby and Co., London.

Eiby, G. A., 1957, Earthquakes, Frederick Muller, London.

Evans, D. M., 1966, Man-make earthquakes in Denver, Geotimes, May–June, pp. 11–18.

Fuller, M. L., 1914, The New Madrid earthquake, U.S. Geological Survey, Bulletin 494.

Gutenberg, B., and Richter, C. F., 1949, Seismicity of the earth, Princeton University Press, Princeton, New Jersey.

Hamilton, R. M. 1980, Quakes along the Mississippi, Natural History, vol 89, pp. 70–74.

Heacock, J. G., ed., 1971, The structure and physical properties of the earth's crust, Geophysical Monograph 14, American Geophysical Union, Washington, D.C.

Hodgson, John H., 1964, Earthquakes and earth structure, Prentice-Hall, Englewood Cliffs, New Jersey.

Howell, B. F., 1959, Introduction to geophysics, McGraw-Hill Book Co., New York.

Iocopi, R., 1964, Earthquake country, a Sunset Book, Lane Book Company, Menlo Park, California.

Kendrick, T. D., 1956, The Lisbon earthquake, Methuen and Co., London.

Lawson, A. C., and others, 1908. The California earthquake of April 18, 1906, Report of the State Earthquake Commission, Caregie Institute of Washington.

Leet, L. D., 1948, Causes of catastrophe, Whittlesey House, McGraw-Hill Book Co., New York.

Lovering, J. F., 1958, The nature of the Mohorovičić discontinuity, Trans. American Geophysical Union, vol. 39, pp. 947–55.

McCann, W. R., and others, 1980, Yakataga gap, Alaska: seismic history and earthquake potential, Science, vol. 207, pp. 1309–15.

Oakeshott, G. B., and others, 1955, Earthquakes in Kern County, California, during 1952, California State Division of Mines, Bulletin 171.

Plafker, G., Dricksen, G. E., and Concha, J. F., 1971, Geological aspects of the May 31, 1970 Peru earthquake, Bulletin of the Seismological Society of America, vol. 61, pp. 543–78.

Plafker, G., and Savage, J. C., 1970, Mechanisms of the Chilean earthquake of May 21 and May 22, 1960, Geological Society of America, vol. 81, pp. 1001–1030.

Poldervaart, A., and others, 1955, Crust of the earth, Geological Society of America, Special Paper 62.

Reid, H. F., 1914, The Lisbon earthquake of November 1, 1755, Seismological Society of America Bulletin, vol. 4, pp. 53–80.

Richter, C. F., 1958, Elementary seismology, W. H. Freeman and Co., San Francisco.

Robertson, E. C., ed., 1972, The nature of the solid earth, NcGraw-Hill Book Co., New York.

Saint Amand, P., 1961, Los Terremotos de Mayos—Chile, 1960, Technical Article 14, U.S. Naval Ordinance Test Station, China Lake, California.

Shepard, F. P., 1933, Depth changes in Sagami Bay during the great Japanese earthquake, Journal of Geology, vol. 41, pp. 527–36.

Sutherland, M., 1959, The damndest finest ruins, Coward-McCann Book Co., New York.

Verhoogen, J., 1956, Temperatures within the earth, *in* Physics and chemistry of the earth, vol. 1, Pergamon Press, New York.

Witkind, J. S., 1962, The night the earth shook; a guide to the Madison River Canyon earthquakes area. Department of Agriculture, U.S. Forest Service Misc. Publication 907.

Wyllie, P. J., 1971, The dynamic earth: textbook in geosciences, John Wiley and Sons, New York.

Wyllie, P. J., 1975, The earth's mantle, Scientific American, vol. 232, no. 3, pp. 50–63.

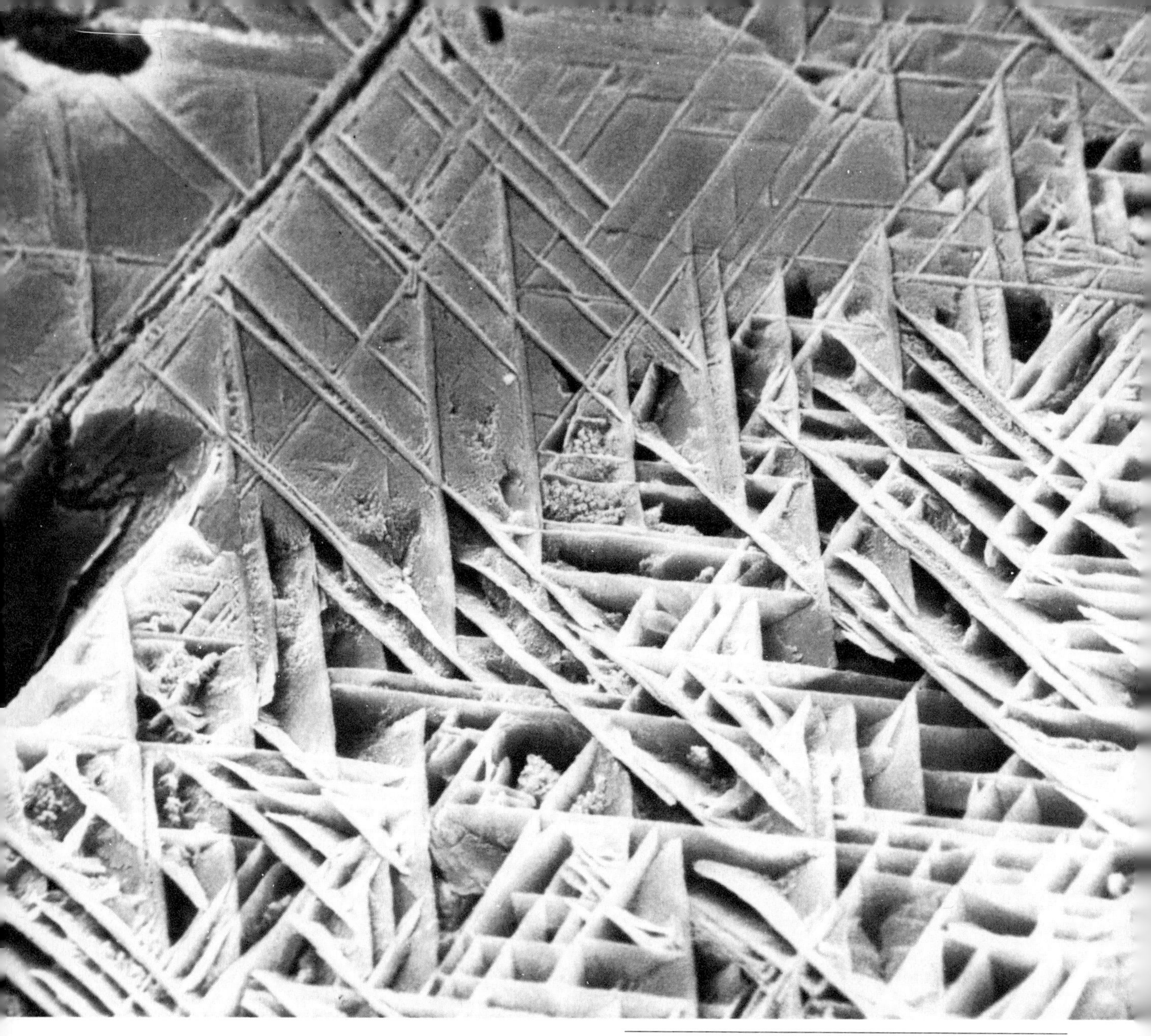

Fig. 19-1 The surface of a magnetic grain composed of an intergrowth of magnetite (interlaminar areas) and an iron-titanium mineral (laminar areas), as photographed under the scanning electron microscope at a magnification of about 1225 times. The grain was etched in hydrochloric acid to bring out the relief. *Rich Hoblitt* and *Edwin E. Larson*

19

THE MAGNETIC FIELD OF THE EARTH

The ancient Greeks divided all matter on earth into four categories: fire, water, air, and earth. They were aware that the last of those, earth, had several recognizable forms, among them a mineral called ***lodestone,*** which could attract other pieces of lodestone or iron. Today, we know lodestone as magnettite (Fe_3O_4), one of the common oxides of iron found at the earth's surface (Fig. 19-1). The attractive force of magnetite was so mysterious that the Greek philosopher Thales of Meletus (640–546 B.C.) finally attributed a soul to it to explain its strange properties.

By the early part of the Middle Ages, it was found that if a piece of lodestone or a piece of iron that had been rubbed on it was suspended so that it could freely turn, it always came to rest in the same position (Fig. 19-2). It was not until around A.D. 1100, however, that a magnetic compass, which could be used by travelers on land and sea for finding their way, was invented. The inventor, whose name is not recorded, was probably Chinese. Another century, apparently, had to pass before Europeans became aware of the design and use of a compass (Fig. 19-3).

Many curious properties were attributed throughout ancient times and the Middle Ages to the magnet. It was supposed to give comfort and to increase one's grace; it could stop hemorrhages and cure toothaches; and it was considered useful in reconciling husbands and wives.

In 1269, Petrus Perigrinus de Maricourt, a French crusader, not only described a compass in detail, but fashioned a sphere out of lodestone to investigate the nature of the magnetic field surrounding it by means of a small magnetic compass needle. Although not sophisticated, his study was nonetheless the first scientific investigation of any magnetic substance.

A much more detailed investigation of the properties of magnets was carried out by William Gilbert (1540–1603), an English physician and physicist, who published his results in 1600 in a book called *De Magnete*. He found that magnetic substances possess poles, and that ***unlike poles*** attract whereas ***like poles*** repel each other. He also noted that a compass needle comes to rest in only one orientation at any one location on the earth's surface (Fig.

Fig. 19-2 Lodestone (magnetite) is strongly magnetic. The magnetic field (lines of force) surrounding the samples shown here are made evident by a sprinkling of iron filings. *Janet Robertson*

19-4). He was aware that the earth possesses a magnetic field, and even went so far as to propose that the entire earth acts as if it were a gigantic spherical magnet.

In the nineteenth century, physicists began to explore the mysteries of electricity, a phenomenon thought to be totally different from magnetism. It came as a shock, therefore, when in 1819 the Danish physicist Hans Christian Oersted (1777–1851) announced that a wire carrying an electric current deflects a compass needle. Oersted's discovery was but the beginning of a brilliant series of studies by physicists André Ampère (1775–1836), James Maxwell (1831–79), and Michael Faraday (1791–1867), which showed that electricity and magnetism are but two sides of the same coin: that is, when an electric current flows, which involves the movement of electrons, a magnetic field is produced; conversely, where a strong magnetic field intersects a substance capable of conducting electricity, an electric current is generated. It was their work that clarified the relationship between the two phenomena.

Fig. 19-3 An early magnetic compass was made by placing a fragment of lodestone on a tiny wooden raft. Such an assemblage will rotate until its magnetic field aligns with that of the earth. Stars mark the poles of the lodestone. From *Magnes,* 1643. *New York Public Library*

THE EARTH'S MAGNETIC FIELD

The earth's magnetic field is an invisible force field that everywhere permeates the earth and extends outward from the earth's surface for hundreds of thousands of kilometers; even the moon 384,000 km (238,618 mi) away, is, as the astronauts have found, not beyond its effects. The best way to visualize the main elements of the earth's field is to place a bar magnet, which has two magnetic poles, on a piece of paper at the center of a circle representing a cross section of the earth (Fig. 19-5). Invisible lines of force, the number being determined by the strength of the magnet, converge at one magnetic pole and diverge at the other. The use of the north and south magnetic pole designation is by convention of physics only. Each line of force is continuous, so that one emerging from the one pole will bend through space and eventually re-enter the magnet at the other pole, as shown. It is apparent from the illustration that the lines are essentially parallel to the "earth's surface" at the equator, but that their angle of intersection with the surface increases toward the poles. At a spot above the earth's north magnetic pole, the field lines point vertically downward. A similar angular relationship is shown by the lines that emerge south of the equator. The lines are more concentrated near the poles than they are near the equator; the result is that the field is about twice as strong at the poles as it is at the equator. Since the earth is three dimensional, lines of force like those shown in cross section would be found to completely encircle the earth.

Fig. 19-4 William Gilbert's diagram of the earth with small magnets showing inclination of field. From *De Magnete,* 1600. *New York Public Library*

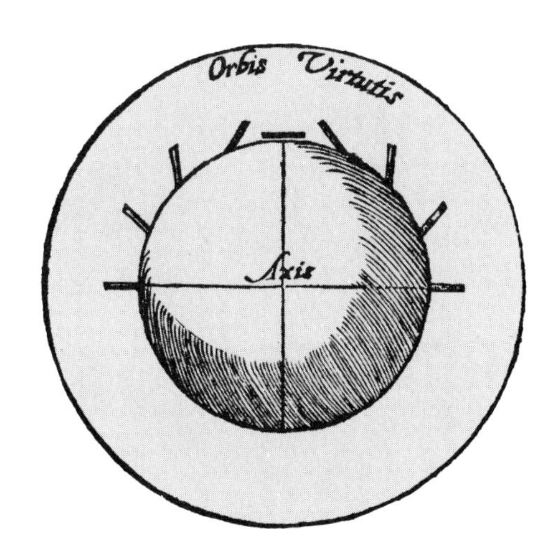

In order to map the longitudinal magnetic field lines just described, we can use a small magnetic compass, moving it about from place to place along the heavy line circumscribing the bar magnet. The compass needle will align parallel with the field lines at any particular point, and a unique orientation of the needle will result (Fig. 19-5). The north end of the needle will always point toward the north, but the angle that the needle makes with the heavy line representing the earth's surface will change with location. One can actually "see" the field lines by sprinkling iron filings or finely divided magnetite on a piece of paper. The tiny fragments will be pulled into crude, curved lines that approximate those of Figure 19-6.

The simple ***dipolar field*** we have just studied consists of two unlike poles called ***dipoles,*** and is very similar to the main field of the earth as Gilbert realized back in 1600. It is no wonder he believed that there was the equivalent of a large bar magnet within the earth.

The earth's dipolar field is generated within the metallic iron core so far below the surface that we can only speculate on what might cause it. Certainly there is no deeply buried permanent magnet, since the central regions of the earth are much too hot for any known, likely material to retain its magnetic properties. However, as you will recall, the outer core most likely is composed primarily of liquid iron. Being metallic, it can conduct electric currents, and being fluid, it can flow. Geophysicists now believe that in response to local heating the liquid core flows convectively while the earth, including the core, spins on its axis. If, for any number of reasons, a weak magnetic field is initially present, electric currents will be created during the flow, which will, in turn, create new magnetic fields in a complex interaction of electromagnetic coupling.

Once formed, the new magnetic field will be amplified and constrained to lie more or less along the axis of rotation of the earth; it is therefore dipolar. The mechanism can be crudely modeled in the laboratory, and has been termed a ***self-exciting dynamo*** (Fig. 19-7). Since the dynamo action is controlled in part by the rate and scale of movement in the liquid iron core, the field will vary in intensity, and to some degree in direction, as changes in the fluid motions in the core occur. Thus, although the field appears relatively stable from day to day, over hundreds of years it will vary. In fact, the field can vary greatly in intensity, even dropping to zero. When it regenerates the magnetic poles may even be reversed.

Today, the magnetic poles do not coincide with the rotational (geographic) poles: the north magnetic pole lies at 78½°N latitude, 69°W longitude (Fig. 19-8). This situation, which has prevailed since at least the early 1830s when the German mathematician and physicist K. F. Gauss (1777–1855) first described the earth's magnetic field in detail, is probably a momentary aberration (in geological time), and it is thought that, over the next 5000 years or so, the magnetic poles will slowly shift until their average positions match those of the rotational poles.

Although the earth's field is predominantly a long-lasting dipole, minor fluctuating (transient) fields also exist, causing local distortions of the dipolar field. Some of these distortions are the result of a flow of electric current induced by solar radiation in the upper atmosphere. Intensities associated with them are small—generally less than 2 per cent of the average dipolar intensity.

Other magnetic irregularities apparently are

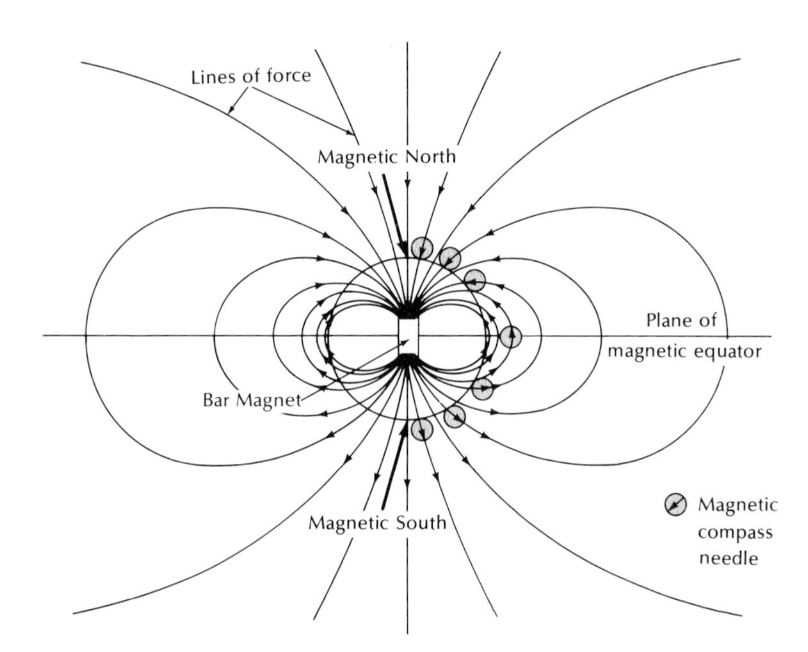

Fig. 19-5 Field lines around a bar magnet, shown in cross section through the bar's axis. Arrows on the field lines indicate the direction of the magnetic field; the magnetic compass needle is parallel to them. Note that the inclination of the compass needle is horizontal at the magnetic equator and vertical at the magnetic poles.

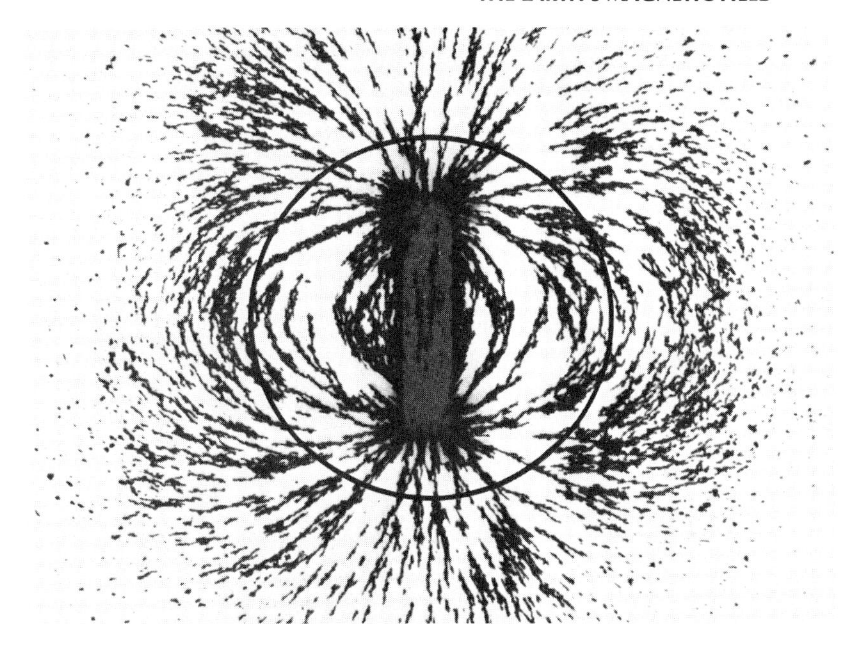

Fig. 19-6 Iron filings sprinkled over a bar magnet enable us to observe a magnetic field similar to that of the earth. Note the similarity between this figure and Figure 19-5. *Janet Robertson*

the result of fields generated in the outer part of the liquid core. Today there are about six magnetic zones over the face of the earth in which the field is greater or smaller than would be expected. Because these zones are not geometrically related, or paired, they are called ***non-dipole centers.*** They can possess intensities of as much as 10 to 20 per cent of the dipolar intensity, and their presence, therefore, noticeably warps the dipolar field, causing local deflection of magnetic directions (Fig. 19-9). A curious aspect of the non-dipole centers is that, with time, some vary in intensity, some drift westward at about $^{1}/_{5}°$ of longitude per year, circling the globe in 1500 to 2000 years, and some do both. Because these centers

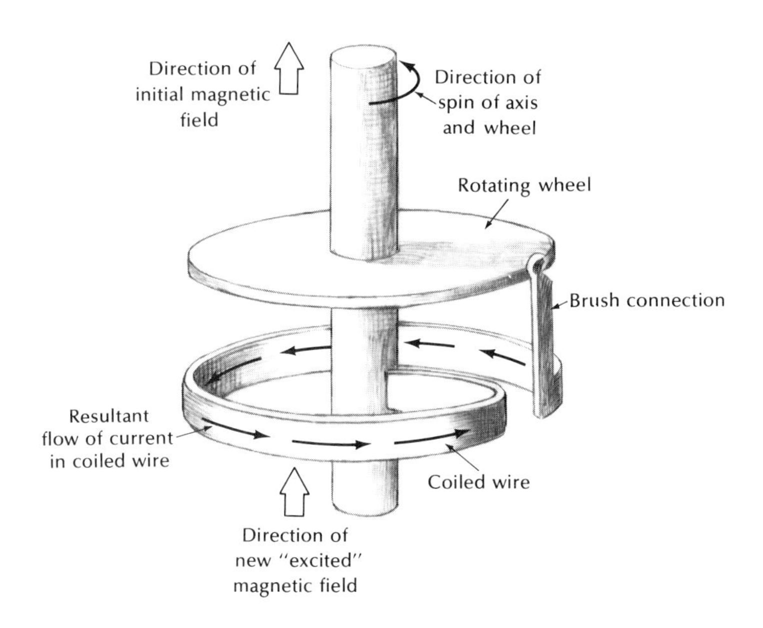

Fig. 19-7 Model of a simple, self-exciting dynamo. If the dynamo is to function, the wheel must be turning and an initial magnetic field must be present. The resultant electric currents produce a new "excited" magnetic field, which is sufficient to keep the dynamo functioning even if the initial magnetic field is removed.

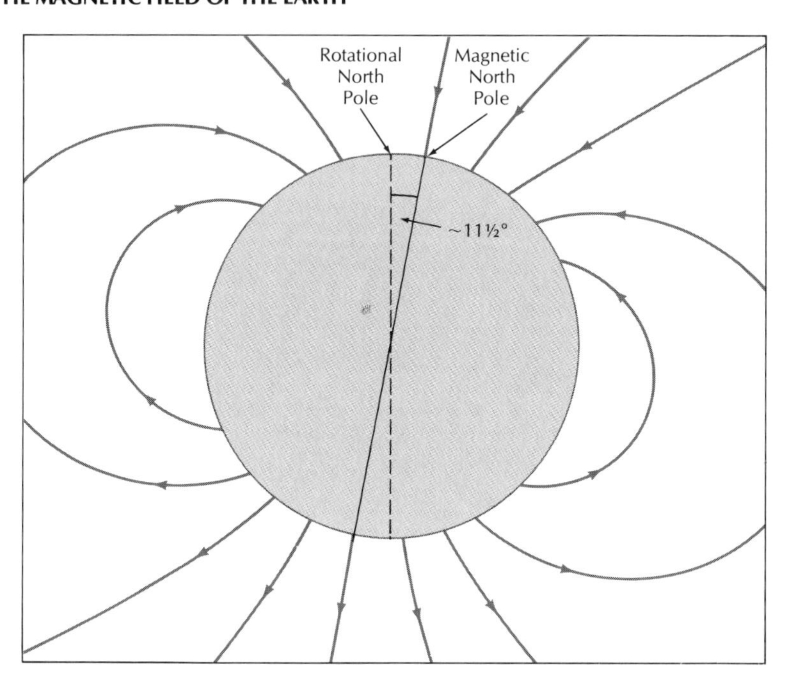

Fig. 19-8 Present-day angular offset of the dipolar axis of the earth's magnetic field with respect to the rotational axis.

change and drift with time, the local field measured at any point on earth will change slowly in magnetic direction and intensity from year to year. Over several hundred years, directional changes might amount to 25° or more.

Variations in the earth's magnetic field appear to be of relatively short duration, and will eventually average out. If the average positions of the magnetic poles for an ancient geological time can be determined through a study of fossil permanent magnetism (***paleomagnetism***), then the positions of the geographic poles at the time the rocks were formed can also be determined. Such information is particularly important to geophysicists who study how the continents have moved and shifted with respect to the geographic poles.

ROCKS, MAGNETISM, AND PALEOMAGNETISM

Almost all rocks contain some minerals that can become permanently magnetized. Magnetite (Fe_3O_4) is the mineral chiefly responsible for most of the magnetism in rocks, which can be measured by means of a sensitive instrument called a ***magnetometer.*** Magnetite will retain its permanent magnetization up to a critical temperature—580°C (1076°F)—known as its ***Curie temperature,*** which takes its name from the noted French chemist Pierre Curie. Above 580°C, the mineral will be essentially nonmagnetic. Thus when a lava solidifies and cools, it will remain unmagnetized until its temperature drops to 580°C, the Curie temperature of magnetite. As the temperature continues to decline, the lava will acquire a permanent magnetization—called ***remanent magnetization***—parallel to the earth's magnetic field lines at that locality and at that time. Since the permanent magnetization is the result of thermal cooling, as described, it is called ***thermoremanent magnetization,*** or ***TRM,*** which is figuratively frozen into the lava so that, regardless of later changes in the earth's magnetic field, the magnetization in the lava will record the initial field direction acquired during cooling. Inasmuch as the magnetic field direction (declination, inclination) for the earth's dipolar field is unique for any latitude, it is possible, after measurement of the remanence direction, to determine where the north magnetic pole was at the time a rock crystallized.

Generally, lava cools and acquires its remanence in only a few weeks, so that the magnetic directions recorded are only for that geological moment. In order to determine average dipolar directions, paleomagnetic measurements must be made on many different flows from the same area. Then the best estimate of the average north paleomagnetic pole can be made, which should be a close approximation of the paleogeographic pole. Such information can be used to reconstruct the drift paths of the continents through time.

Rocks can acquire a permanent magnetization in ways other than through cooling of their magnetic minerals. For example, it is possible for small magnetic clasts that accumulate during sedimentation to orient themselves like tiny compass needles as they settle through the waters in a lake or an ocean. Thus sedimentary rocks, too, can acquire a remanent magnetization at the time they form. This type of magnetization is called ***depositional remanent magnetization,*** or ***DRM.*** Generally, it is much weaker than TRM, but it can be measured easily with a sensitive magnetometer. In the paleomagnetic study of a sedimentary rock, many samples are taken from a section of strata, each sample representing a somewhat different time, as in the study of lava flows, and the average north paleomagnetic pole is calculated.

After initial rock formation, chemical processes can also lead to the formation of magnetic iron oxides, particularly hematite, and the development of a ***chemical remanent magnetization*** (***CRM***), which parallels the earth's field at the time the magnetic minerals were chemically formed. Chemical remanent magnetization is common in some sedimentary rocks, especially red beds. Dating the time of the chemical alteration that led to the acquisition of CRM is difficult, however.

How stable is remanent magnetization? Rocks vary, but many of the more stable kinds can retain the record of the original field direction for many hundreds of millions and even billions of years. In the Beartooth Range of Montana, there are basaltic dikes with TRM that has been frozen in for more than 2.5 bil-

Fig. 19-9 Map of the United States, showing how the magnetic declination varies at latitude 40°, across the continent due to local warping of the main field by non-dipole centers.

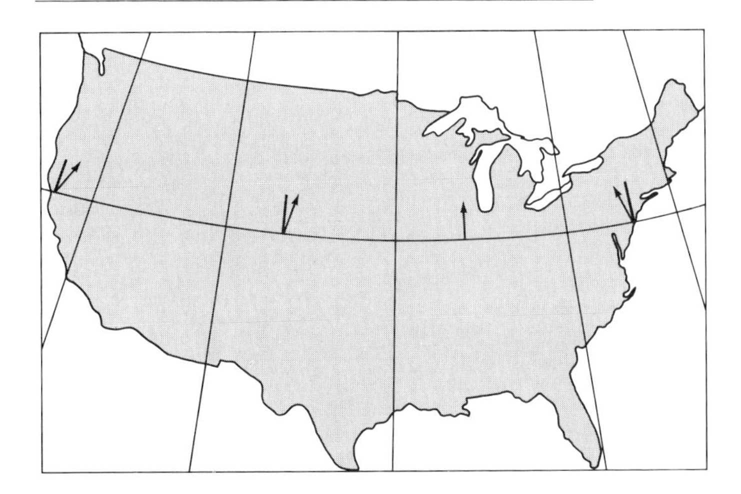

lion years. Its presence is one more proof that the earth has possessed a magnetic field—and therefore the liquid iron core necessary to generate the field—for more than 2.5 billion years.

REVERSALS OF THE EARTH'S MAGNETIC FIELD

From a study of the permanent magnetization in rocks, it is now known that episodically during the earth's history the magnetic field has *reversed*—that is, the magnetic poles have shifted by 180°. Such a field is said to be reversely oriented, and the rocks formed at those times will record ***reversed remanent magnetic directions.*** The last major event of reversed orientation of the field lasted until about 690,000 years ago. Since then, the field has been dominantly *normal* (parallel to today's field), except for minor fluctuations. Through radiometric dating of lava samples from around the world, it has been possible to establish the sequence of polarity changes in detail for the last few million years (Fig. 19-10). The sequence is called the ***magnetic polarity time sequence*** and, as Figure 19-10 shows, the earth's field has reversed about every 200,000 to 300,000 years. The longest polarity interval during this time span has lasted nearly 700,000 years; the shortest only about 10,000 years

The transition from a normal polarity to a reversed polarity, or vice versa, does not take place instantaneously. Rather it takes about 1000 to 5000 years—a short time, geologically. As the field changes over, it remains at a very low intensity. No one knows how the low magnetic field affects animals and plants during changes of polarity, but experiments on small organisms have indicated that normal life cycles could be adversely affected. In fact, some geologists have suggested that some species may have become extinct as a result of the reversals in polarity of the earth's magnetic

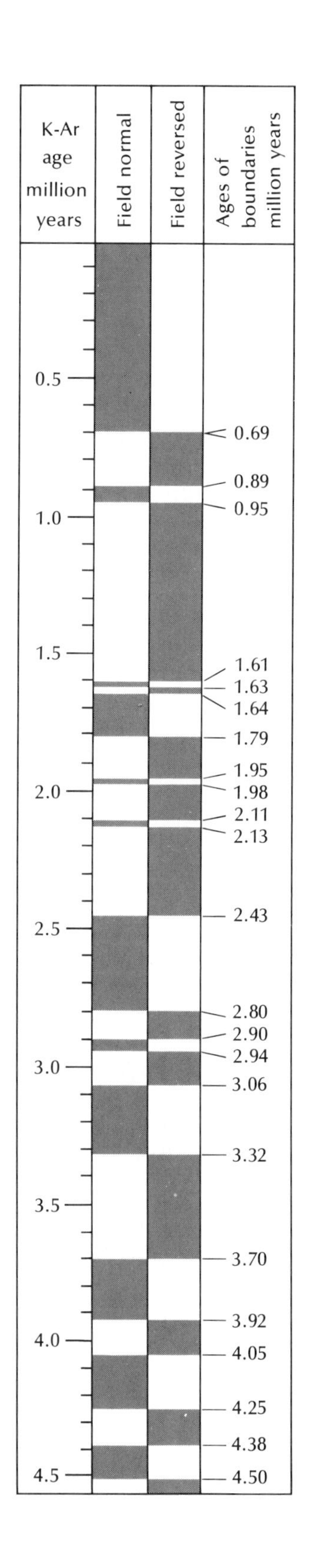

Fig. 19-10 Magnetic polarity time sequence back to about 4.6 million years ago, based on potassium-argon dating of lava flows with known paleomagnetic directions. Note that the field has changed back and forth between normal and reverse polarity.

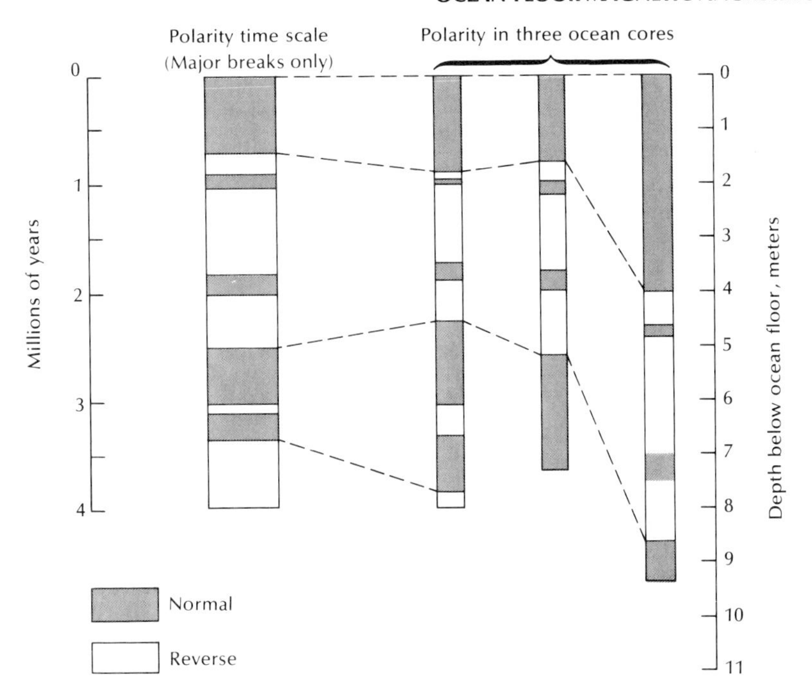

Fig. 19-11 Magnetism in deep-sea sediments. As the field changes from normal to reverse and back again, the accumulating sediments acquire corresponding magnetic directions. Hence, core samples will contain a record of the changes, which can be correlated with the magnetic polarity time sequence.

field. Further experiments will be necessary, however, to evaluate those suggestions.

It is interesting to note that in the last 150 years the earth's magnetic field has decreased in intensity by about 7 per cent. It is possible that the field of today is headed for another reversal in the relatively near future. At the present rate of decay, however, another 2000 years would be required. The observed decrease, of course, may only be part of a short-term fluctuation, completely unrelated to reversal.

During the last 10 to 15 years, oceanographic vessels have been sampling cores of the upper sedimentary layers of the ocean floors. Those sediments, deposited during the last several million years, contain a record, in their remanent magnetization, of magnetic polarity changes, just as the rocks on land do. It has been possible to correlate the reversals of magnetization recorded in the sea-floor cores with the magnetic polarity time sequence and thereby to date them throughout their length (Fig. 19-11). In that way, rates of sedimentation have been established for different parts of the ocean basins, and fossils contained in the cores have been accurately dated.

OCEAN-FLOOR MAGNETIC ANOMALIES

In the course of oceanographic studies, many ships towed magnetometers to measure the earth's magnetic field as they criss-crossed the oceans. After data had been recorded for a few years, it became apparent that the intensity of the field varied in a systematic, but oscillatory way. In some places it was slightly stronger than the average, and in others it was slightly weaker. When plotted on a map the anomalously strong and weak regions were found to occur in alternating linear belts paralleling the axes of the oceanic ridges. Detailed magnetic surveys were then made from oceanographic vessels, to see how the pattern varied across the oceanic ridges and rises. The results of one such survey over the Mid-Atlantic Ridge south of Iceland is shown in Figure 19-12. Over oceanic ridges it was found that not only were magnetic anomalies linear, but the pattern on both sides of the ridge was bilaterally symmetrical, outward from a higher-than-average linear anomaly that ran along the ridge axis. Whatever the pattern on one side of the ridge, its mirror image was recognizable on the other side.

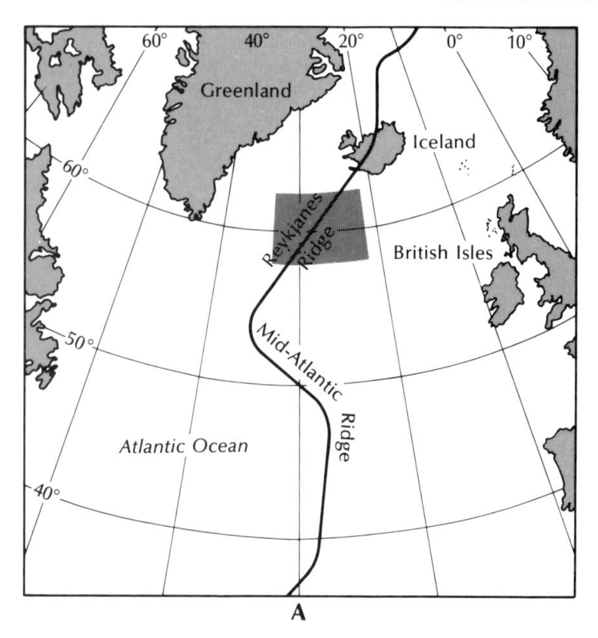

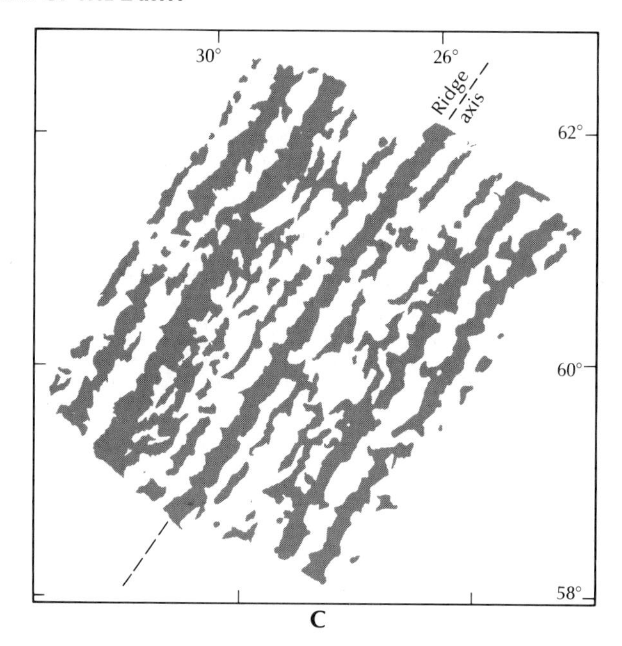

In 1963, two English geophysicists, F. J. Vine and D. H. Matthews, provided a plausible explanation of the mystery of the linear anomalies. They concluded that parts of the ocean floor beneath the sedimentary layers must be magnetized in a normal direction—that is, parallel to the field of today, whereas other parts must be reversely magnetized. The higher-than-average field intensities, they reasoned, corresponded to the normally magnetized stripes, and the lower-than-average anomalies corresponded to the reversely magnetized stripes (Fig. 19-13). The explanation was based on the idea that the fossil magnetism that paralleled the present-day field (***normal direction***) would add to the total intensity of the field and that the fossil magnetism opposite to today's field (***reversely magnetized***) would detract a little from it (Fig. 19-14). The bilateral symmetry could only mean that for each normal or reversely magnetized stripe on one side of the ridge, there was a nearly identical counterpart on the other side. That is, paired anomaly stripes existed, parallel to and on either side of the ridge axis.

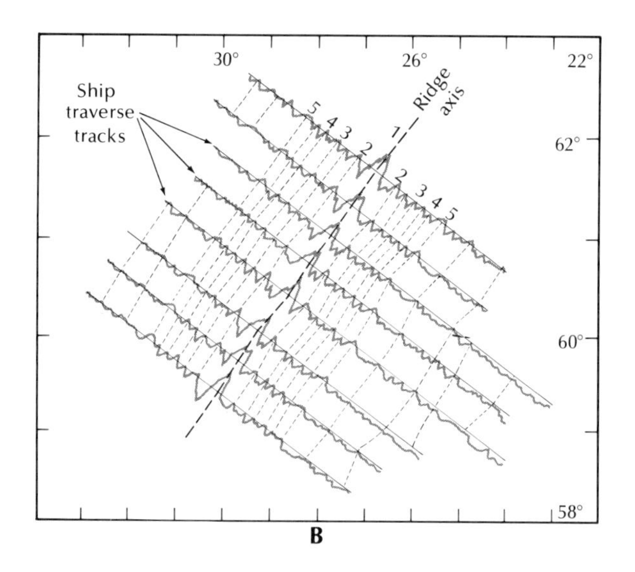

Fig. 19-12 Magnetic anomalies mapped in the area of Reykjanes Ridge, south of Iceland (**A**). In (**B**) is shown the relative variation in the earth's magnetic field determined along selected ship traverses across the ridge, and in (**C**) the inferred areal distribution of the variations. The black areas represent higher-than-average field density and the white areas, lower-than-average. There is a bilateral symmetry to the anomalies, such that those on one side of the ridge are mirrored on the other—as indicated by the numbers in (**B**).

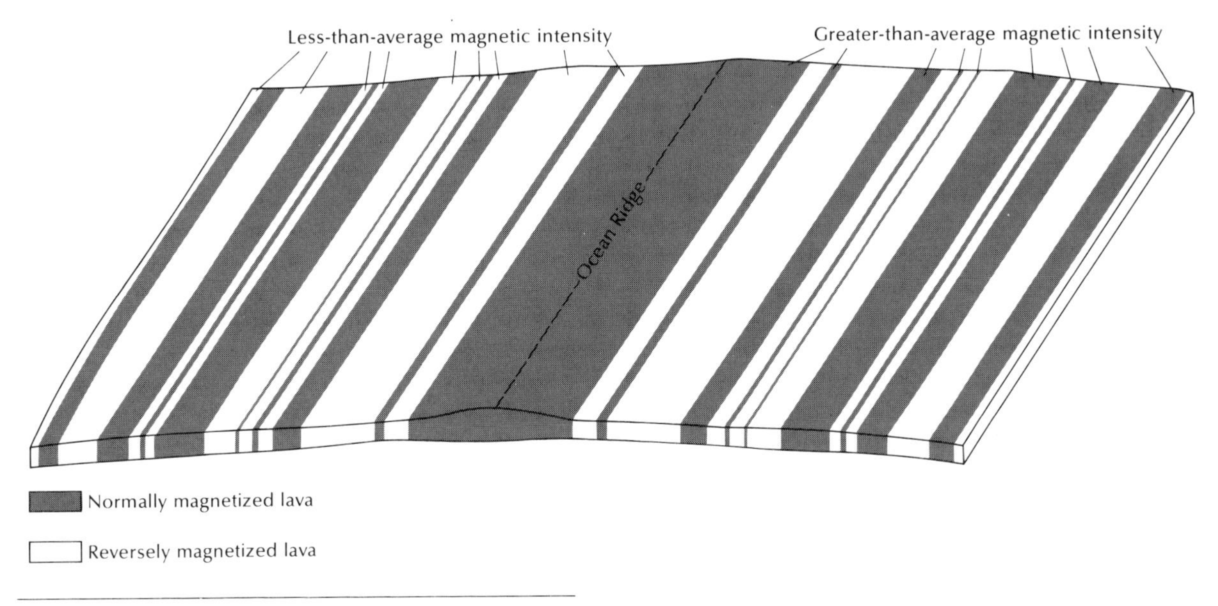

Fig. 19-13 An idealized magnetic anomaly pattern in the area of an ocean ridge and the magnetic polarity of ocean-floor lavas that could account for the pattern.

Fig. 19-14 The magnitude of intensity measured by a ship's magnetometer depends on both the field intensity of the earth and the direction and intensity of magnetization in the crust beneath the ship. When the two are parallel **(A)** the intensity is enhanced; when the two are opposite **(B)** the intensity is decreased.

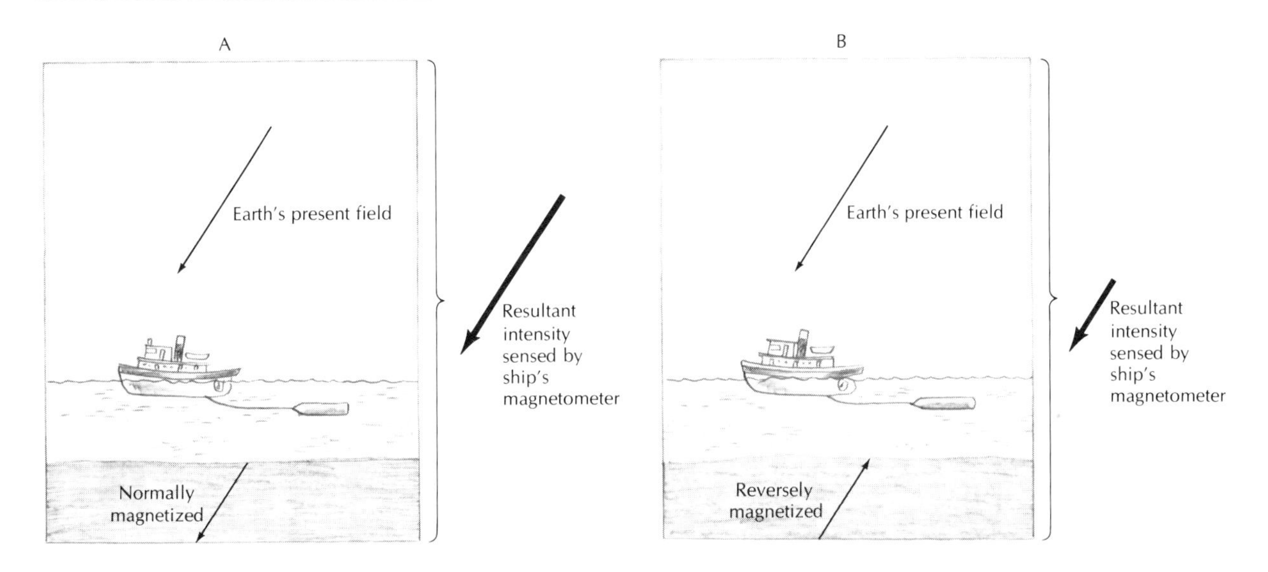

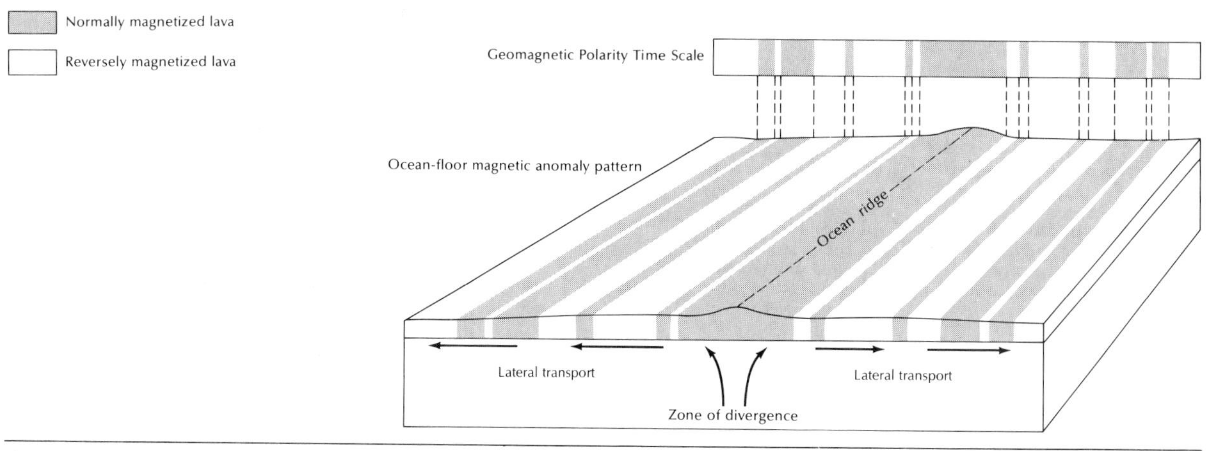

Fig. 19-15 Comparison of the linear magnetic anomaly pattern with the magnetic polarity time sequence, from which it is inferred that the ocean floor is spreading apart. Apparently, lava flows and dikes are added to the region of the ridge axis, where they cool and become magnetized parallel to the field direction at that time. As divergence continues, the field changes polarity, new material is added, and older material is carried away from the axis in both directions.

When the widths of the magnetic stripes on one side of a ridge axis were plotted against the land-based magnetic polarity time sequence, their relative positions matched nearly exactly, right up to the present (Fig. 19-15). The youngest subsediments occurred along the ridge axes, becoming progressively older outward in both directions. At the time, the discovery was an exciting one, for it indicated that new material was being added continually to the crust along the ridge axis, and that, subsequently, it was being split and moved laterally away as if on two outward-moving conveyor belts. Dating of the magnetic stripes provided further good evidence that the oceanic ridges and rises are zones of divergence and that the ocean basins are young features, continually being formed.

Since the magnetic polarity time sequence could be dated, it was possible to date the oceanic linear magnetic anomalies and to determine rates of spreading along the zones of divergence. They proved to be 2 to 6 cm (0.8 to 2.3 in.) per year.

In the last few years, extensive deep drilling in the oceans has verified Vine and Matthews's theory. The character of linear magnetic anomalies has provided key insights into how ocean basins form and has enabled the precise tracking of the movements of lithospheric plates and attached continental masses.

Certainly, the knowledge from magnetic studies of plate motions and divergent zones has become fundamentally significant in the development of the plate tectonic theory.

SUMMARY

1. The properties of a *magnetic field* were first investigated by William Gilbert in 1600. Magnetic materials possess *poles,* which if alike, repel each other and if unlike, attract each other. Later experiments demonstrated that electricity and magnetism are related.
2. The earth's magnetic field is *dipolar.* The field lines parallel the earth's surface at the magnetic equator and are perpendicular to the surface at the magnetic poles. At intermediate latitudes, the field lines are inclined to the surface; the higher the latitude, the steeper the angle.
3. The earth's magnetic field is generated inside the liquid iron core, in the manner of a *self-exciting dynamo,* through electromagnetic coupling.
4. Essentially all rocks contain magnetic min-

erals (most commonly *magnetite*) and can acquire a *permanent magnetization* (*remanent magnetization*) parallel to the earth's field at the time they form; for example, lavas become magnetized as they cool below 580°C (1076°F), the *Curie temperature* of their magnetic minerals, and sediments can acquire a remanent magnetization through alignment of magnetic clasts as they settle through water.

5. Periodically, the earth's magnetic field has reversed. Radiometric dating of lava samples has established a succession of dated reversals (*magnetic polarity time sequence*). The reversals recorded in sea-floor cores can be correlated with that sequence, thereby providing a method of dating them throughout their length and of determining sedimentation rates.
6. Oceanic magnetometer surveys reveal a magnetic anomaly pattern of stripes of higher-than-average field intensities alternating with stripes of lower-than-average intensity. The stripes parallel the oceanic ridge axes. Higher-than-average stripes correspond to normally magnetized ocean crust, the lower-than-normal stripes to reversely magnetized crust. The pattern can be correlated with the magnetic polarity time sequence to show that the oceanic ridges and rises are zones of divergence, and to provide a means of determining the rate of sea-floor spreading.

QUESTIONS

1. What is the simplified relation between electricity and magnetism?
2. Describe, in words and by means of a drawing, the nature of a dipolar field. How well does the main field of the earth match your description?
3. What is thought to be necessary for the generation of the earth's main field? Would you expect a strong field on the moon, on Mars, or on Venus?
4. Describe the nature and origin of short-term irregularities in the earth's magnetic field.
5. What is remanent magnetism, how is it acquired, and how is it used to provide information concerning the geological history of the earth?
6. What are geomagnetic reversals? What has been their pattern of occurrence during the last few million years?
7. Explain the occurrence of oceanic linear magnetic anomalies. How are they related to the magnetic polarity time sequence?
8. How has the magnetic polarity sequence been used to provide crucial information concerning geological processes in the oceans?

SELECTED REFERENCES

Cox, A., ed., 1973, Plate tectonics and geomagnetic reversals, W. H. Freeman and Co., San Francisco.

Doell, R. R., Cox, A., and Dalrymple, G. B., 1964, Reversals of the earth's magnetic field, Science, vol. 144, pp. 1537–43.

Heirtzler, J. R., Dickson, G. O., Herron, E. M., Pitman, W. C., and Le Pichon, X., 1963, Marine magnetic anomalies, geomagnetic field reversals, and motions of the ocean floor and continents, Journal of Geophysical Research, vol. 73, p. 2119.

Heirtzler, J. R., Le Pichon, X., and Bacon, J. G., 1966, Magnetic anomalies over the Reykjanes ridge, Deep-Sea Research, vol. 13; pp. 427–44.

Irving, E., 1964, Paleomagnetism, John Wiley and Sons, New York.

Nagata, T., 1961, Rock magnetism, Marizen, Tokyo.

Parasnis, D. S., 1961, Magnetism, Harper and Brothers, New York.

Phillips, O. M., 1968, The heart of the earth, Freeman, Cooper, and Co., San Francisco.

Takeuchi, H., Uyeda, S., and Kanamori, H., 1970, Debate about the earth, rev. ed., Freeman, Cooper, and Co., San Francisco.

Fig. 20-1 Continental drift in action. The coastlines of the Arabian Peninsula and the Somali Republic in northeast Africa fit so well that there is little doubt that at one time the two regions were joined and are now moving apart from one another. *NASA*

20

CONTINENTAL DRIFT AND PLATE TECTONICS

Earth science is in the midst of a revolution that is probably as profound as that which rocked biology in the mid-1800s, when Darwin presented his paper on evolution, or which rattled physics in the early 1900s, when non-Newtonian quantum mechanics was introduced. The revolution is a basic change in the way geologists view the earth and its internal workings. In the 1700s and 1800s, geologists were concerned mainly with the continents and knew little about the oceans. The continents were viewed as fairly static entities, which moved slowly up and down. It was thought that they might have grown slightly through marginal accretion, but by and large they were laterally stationary. In the many subfields of geology, scientists worked primarily on the problems related to their areas of interest, and were isolated from research and events in other areas, but much of this changed with the revolution.

The new philosophy, the theory of ***plate tectonics,*** largely came about from a study of the ocean basins. Within the context of this theory, the earth is viewed as an active planet powered by internal forces that produce a continually changing, interrelated set of geographic and geological phenomena. The idea, from which the name stems, is that the outer skin of the earth is broken into a relatively small number of blocks, or plates, that move laterally in response to internal forces. Many geological phenomena (volcanism, earthquakes, and mountain building, including plutonism and metamorphism) and features (ocean ridges, island arcs, deep-sea trenches, and mineral deposits) are related to the plate-margin interactions. Through the differential movement of plates, continents can "drift," relative to one another, a result that has greatly affected biological evolution (Fig. 20-1).

The theory of plate tectonics has unified geological thought. It ties so many facets of geology together that many earth scientists think that it must be close to the truth, and many researchers have set out to reinterpret geological history in terms of plate tectonics. Probably more than 98 per cent of all active earth scientists subscribe to the theory.

Regardless of its popularity, however, we must realize that this theory is only a more modern way of looking at the earth. It appears

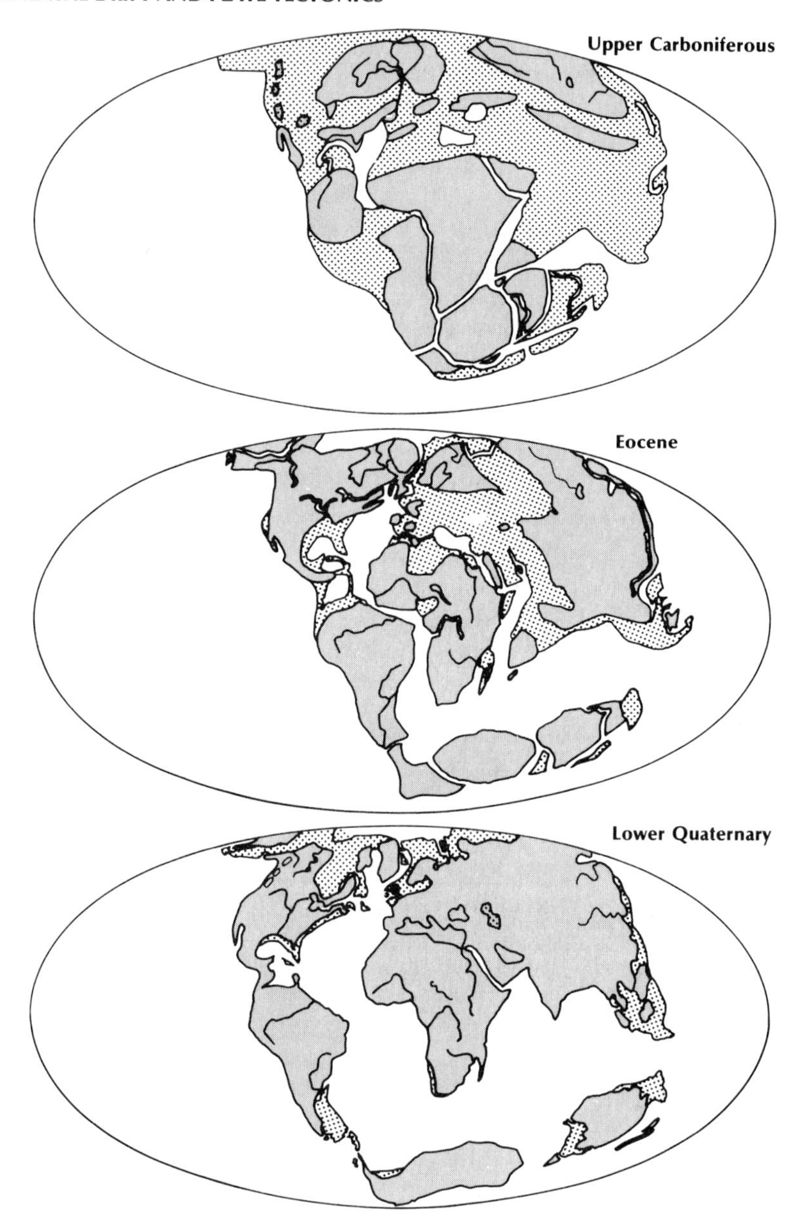

Fig. 20-2 Breaking up of the world's continents according to Wegener. He called the original giant continent *Pangaea*. The white areas are deep ocean, the darkened areas land, and the stippled areas shallowly submerged continental shelf.

to be an improvement over the old way, but it probably will not be the last way. What is perceived as revolutionary thought today may well become dogma tomorrow, or it may be abandoned. The theory has already been modified, and other modifications will undoubtedly be necessary. The real test will come with time.

In the following pages, we will first discuss the introduction and acceptance of the theory of plate tectonics, which is itself a fascinating documentation of the way science progresses. Then we will list the principal tenets of the plate tectonic theory, describing how they provide a basis for understanding earth processes and features. Lastly, we will assess the theory as it stands today.

CONTINENTAL DRIFT

The idea that one great land mass may have once existed on the earth, a land mass that subsequently broke into drifting continental-size fragments, is certainly not a new one. However, most of the earliest recorded suggestions of continental drift were casual remarks made after a cursory examination of a world map and had little substance to back them up.

As early as 1620, Sir Francis Bacon, in his book *Novanum Organum,* commented that the west coast of Africa and the east coast of South America were so strikingly similar that it could not be by accident alone. Soon after, in 1658, the Frenchman François Placet also remarked that the similarity of those coastlines suggested that the continents had been joined together at one time, and he conjectured that their separation may have occurred during the biblical flood. About 200 years later, Antonio Snider-Pelligrini wrote in *La Création et ses mystères dévoilés* that the continents bordering the Atlantic must have once been contiguous; he based his conclusion on slightly more substantial evidence—the similarity in fossil plants in the coal beds of North America and of Europe. He even went so far as to reconstruct the supercontinent. By 1908, two scientists, F. B. Taylor and H. B. Baker, had reached the same conclusion from their studies of the distribution of mountain chains around the world.

But all those suggestions were made in such an offhand manner that they evoked little support and even less interest in the subject of continental drift. Furthermore, such ideas ran contrary to the dogma of geology, which by 1900 was developing into a full-fledged science. Most geologists believed the continents to be permanent and fixed and regarded the weakly supported suggestion that continents could slip and slide about on the earth's surface as pure geological fantasy. Little did they realize that a storm of controversy regarding that very issue would break not too far in the future and would last for more than 50 years.

Alfred Wegener, in 1910, was an instructor of the newly developing field of meteorology in a school in Germany. During the course of his work that year, he too was struck by the similarity of the shapes of the coastlines of South America and Africa, just as were others before him. Yet it was only a fleeting idea, and Wegener pushed the thought from his mind. Then, in 1911, he came across an article on the similarity of fossils on the two continents that rekindled his interest. Unlike the earlier advocates of continental drift, Wegener became obsessed with the idea and was tireless in his efforts to uncover any evidence, from whatever source, that might indicate that the land masses around the Atlantic Ocean might have been contiguous. He finally published his speculations, with supporting evidence, in 1915. Wegener was convinced that at one time all the continents were part of one giant continental mass that he called ***Pangaea*** (Fig. 20-2). At some time during the Jurassic period, Pangaea split along many cracks, and the fragments began to drift apart. The response to the publication initially was mild, but by 1924, he had a full-fledged fight on his hands.

R. T. Chamberlin, an American earth scientist, wrote soon after: "Can we call geology a science when there exists such differences of opinion on fundamental matters as to make it possible for such a theory as this to run wild." Even in 1944, the American geologist Bailey Willis maintained that the theory as propounded by Wegener was nothing but a fairy tale and should be minimized because of its deleterious effect on students.

Wegener, in his synthesis, had attempted to bring together as many pieces of corroborating data as he could. He included evidence from physical geology, geodesy, geophysics, paleontology, zoology, and paleoclimatology. He was not an expert in any of those fields and chose, unfortunately, some poor and even inaccurate examples to bolster his argument. Seizing upon his errors, the opposition ridiculed Wegener's hypothesis mercilessly. Leading geologists ranted and railed against the evils of the drift theory with an almost religious fervor. Most of them prefaced their remarks with pleas for open-mindedness on the subject, but it was plain to see that their own minds were tightly closed. Yet in their own

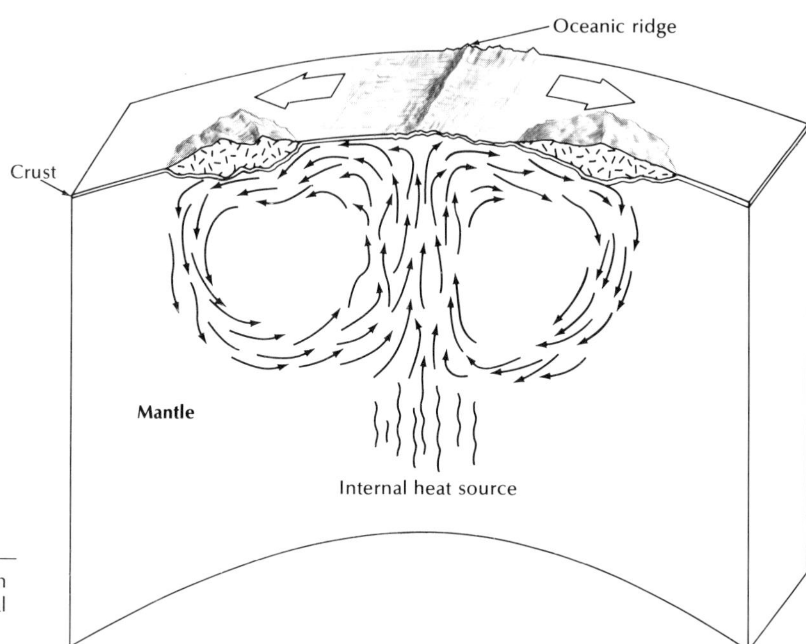

Fig. 20-3 Hypothetical thermal currents in the mantle and their relation to continental drift.

eyes they saw themselves as guardians of the truth, defending their science against a whimsical wave of idiotic speculation.

One of the main points the opposition made was that there appeared to be no mechanism for continental movement to occur. Wegener had outlined two possible mechanisms—one related to gravitational attraction between the continents, the moon, and the sun, and the other related to the centrifugal pull on the spinning earth—but almost everyone agreed that the forces produced in both instances would be woefully inadequate. Since a workable mechanism was lacking, most earth scientists thought that the hypothesis of continental drift could be discounted.

Wegener stood up well to the fusillade of criticism and continued to look for more and better evidence to support his idea, up to the time of his death at the age of fifty. In 1930, while on his fourth expedition to Greenland, he set out from the northernmost base in central Greenland bound for the west coast and was never seen again.

Although most earth scientists were outspokenly against Wegener's hypothesis, a few farsighted individuals accepted the idea, and tried to accumulate data that would support it. Perhaps the most capable disciple was the South African geologist Alexander du Toit. He was aware that the fate of the theory would depend largely on the quality of the data used as evidence. So he pioneered geological studies in both Africa and Brazil to ferret out undeniable proofs of continental drift. In the course of his work, du Toit came to believe that not one but two supercontinents existed before drift took place. He named them ***Gondwana*** and ***Laurasia.*** The former supposedly consisted of all the continents now in the Southern Hemisphere, and the latter of North America, Greenland, Europe, and Asia. An oceanic waterway he called the ***Tethys Sea*** separated the two land masses.

The great Scottish geologist Arthur Holmes (1890–1965) also was fired by Wegener's idea and came to accept it completely. Holmes in 1927 suggested that perhaps some sort of ***thermal convection*** in the earth's mantle was responsible for the movement of continents (Fig.

20-3). So at last, the advocates of the drift theory could point to a mechanism powerful enough to move continents. No one knows even today whether the mantle can flow convectively or not; the majority of earth scientists believe that it can, however.

As the controversy continued, the number of adherents to Wegener's hypothesis slowly increased. In time, many influential geologists, such as Felix Vening-Meinesz (1877–1966), S. Warren Carey, and Sir Edward Bullard, were listed among its supporters and the theory became impossible to ignore. Let us review some of the evidence that led more and more geologists to accept the theory.

Evidence of continental drift

Fit of continental coastlines One of the strongest pieces of evidence was how well coastlines fit together. The earliest reconstructions, however, were made on maps, which, as two-dimensional representations of a three-dimensional globe, distort earth features to some degree. So the first "fits of continents" were not so exact as hoped for (Fig. 20-4). In fact, as the opposition pointed out, the continents did not appear to fit together very well. Some of the poor fit, of course, resulted from the land masses being matched at the coastlines. Later, earth scientists realized that shoreline processes can greatly modify coastal margins and conceived of the idea of fitting the continents at a line *below* sea level, a line that more truly represented the shape of the continent. In 1955, Carey joined South America and Africa together at the 2000-m (6560-ft) isobath, carefully projecting his reconstruction onto a map. The fit he obtained was so excellent that he remarked that the argument that South America and Africa do not fit together closely should never again be used against the drift theory. In 1965, Bullard and two associates showed that by using a computer program they could obtain an even better fit of the continents around the Atlantic. They chose the 1000-m (3280-ft) isobath as their depth-of-fit line and were amazed at how slight the gaps and overlaps between the fitted continents were (Fig. 20-5).

Fig. 20-4 Fit of South America and Africa at the 183-m (100-fathom) isobath, as proposed by Wegener. An isobath is a line of equal topographic elevation below sea level. Although there are gaps between the continental margins, the fit is relatively good.

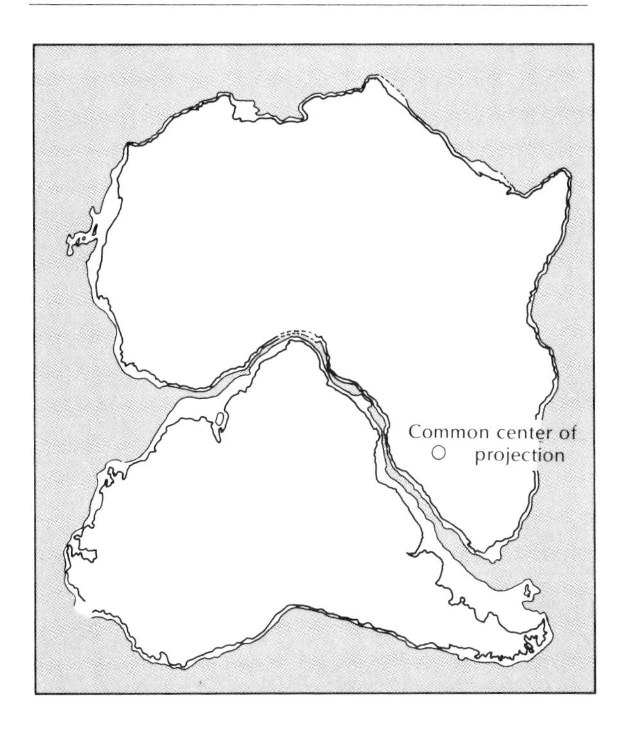

Stratigraphic and structural similarities If the continents seemed to dovetail, did geologically significant features also match across the continental boundaries? Such an idea might be termed the "torn newspaper" concept. If a page of newsprint is ripped into several pieces and reassembled, then not only should the pieces fit back together, but the lines of print should continue across the torn boundaries as well.

One geological feature that could be matched up is the truncated rock record on two or more continents. In Figure 20-6 two generalized stratigraphic sections are shown, one representing the rock record in southeast Brazil and the other in southwest Africa. Both should be closely similar if the conjectured fit of the continents has any merit—and they are

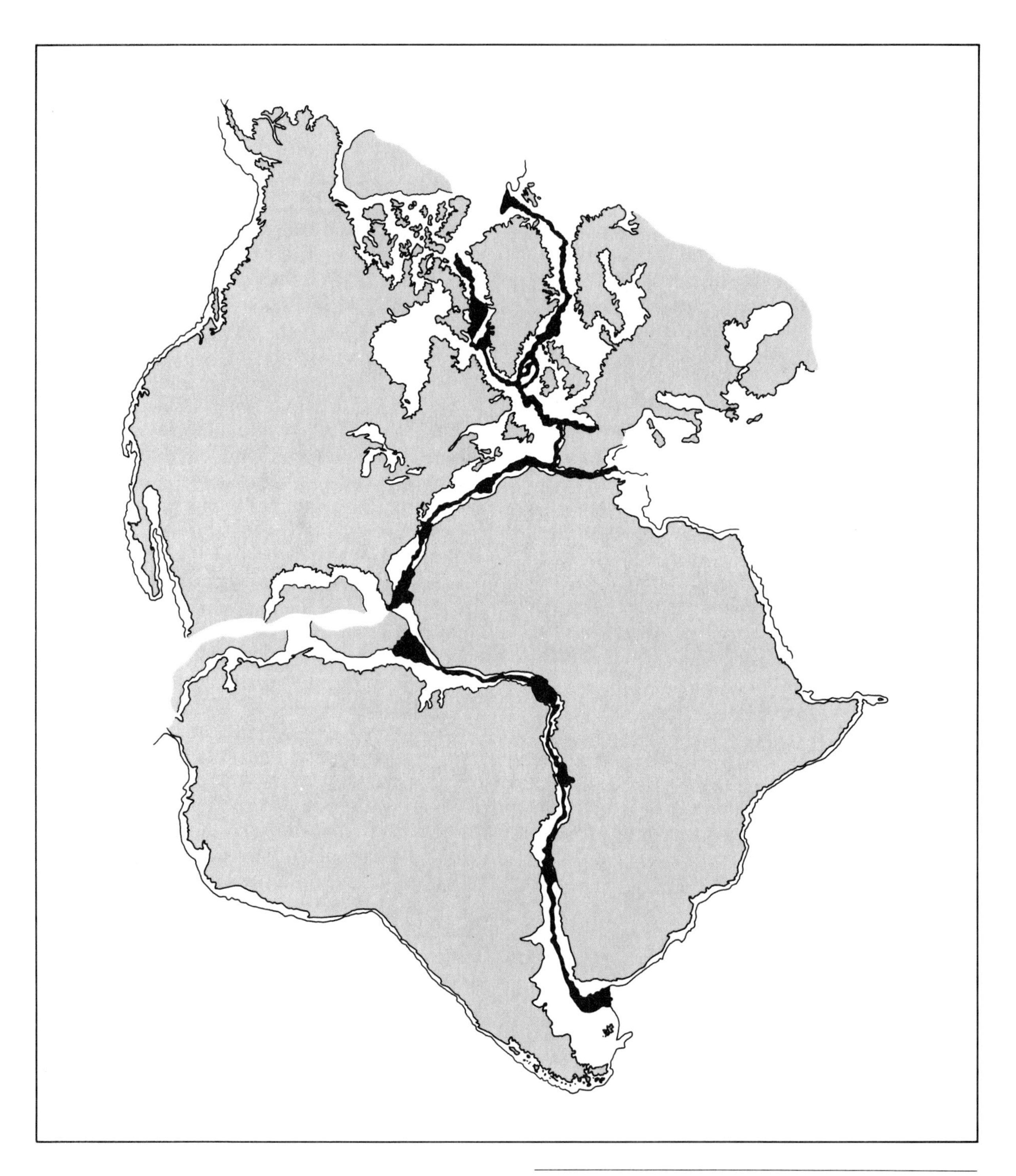

Fig. 20-5 Computer fit of all continents bordering the Atlantic. The thickness of the black line represents the amount of overlap.

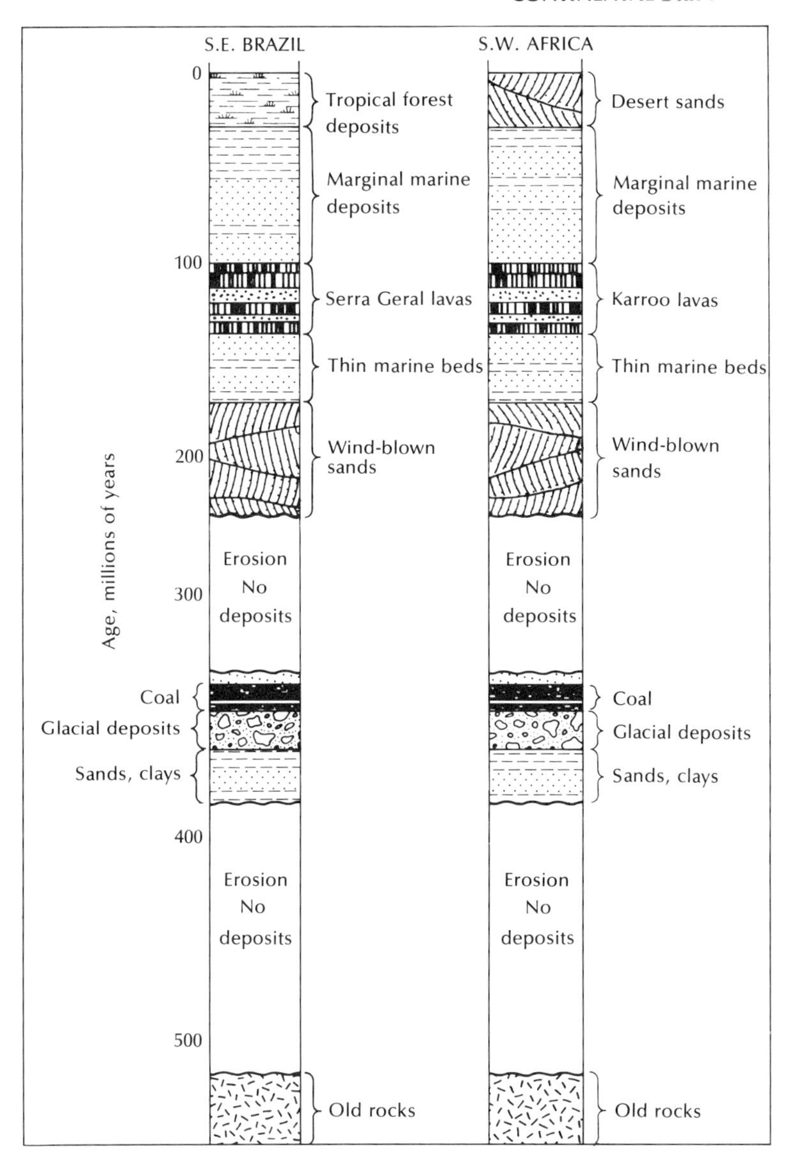

Fig. 20-6 Generalized columnar section of rocks from southeast Brazil and southwestern Africa. Notice the great similarity in sections up to about 100 million years ago.

nearly exact in detail. Such data cannot be used to prove beyond a doubt that drift has occurred, but they are telling evidence.

In the late 1960s, Patrick Mason Hurley, an American geochronologist, found that in the vicinity of Ghana in West Africa there is a distinct boundary between rocks of two different ages: one more than 2000 million years and the other about 600 million years. The boundary trends southwestward to the west coastline of the continent. Hurley arranged for a collaboration between the geochronology laboratories at the Massachusetts Institute of Technology and the University of São Paulo, Brazil, to see if the same boundary could be found in South America, on the east coast of Brazil. Indeed, a boundary between 2000- and 600-million-year-old rocks was found, which appeared to be an extension of the African trend when the two continents were fitted together at their coastlines (Fig. 20-7). Hurley was thus convinced of the reality of continental drift.

Segments of mountain chains in Europe, Greenland, and North America all are about the same age. One segment runs through Scandinavia, Scotland, and Northern Ireland, one through eastern Greenland, and one through eastern North America. Are they all parts of different ranges that developed almost simultaneously, or are they parts of one extensive range that has since been split apart by continental drift? If the continents are reassembled in the manner suggested by Bullard, then most of the mountain segments fit together in a nearly continuous chain. Again, the evidence supports the idea of drift, but by itself is not conclusive.

Paleoclimatology All the continents in the Southern Hemisphere were affected by glaciation during the Pennsylvanian and Permian periods as shown in Figure 20-8A. But many glacial deposits and polished and scoured bedrock surfaces are located close to the equator, where continental glaciation is very unlikely under any circumstances. Moreover, from the alignment of glacial grooves and the orientation of chatter marks left on the bedrock, it was possible to determine the direction of flow of the ice that created them: more often than not, the indications were that the ice had flowed from the sea toward land. That, too, was curious since geologists know of no glaciers today or during the recent ice ages that were formed in the oceans and that flowed onto the land. Just the opposite situation occurred.

If one reconstructs Pangaea (or Gondwana), however, it is apparent that all the glaciated regions fit together into one large area unbroken by seaways (Fig. 20-8B). Then the flow-direction data fit together in an understandable manner.

Such evidence, which was also presented by

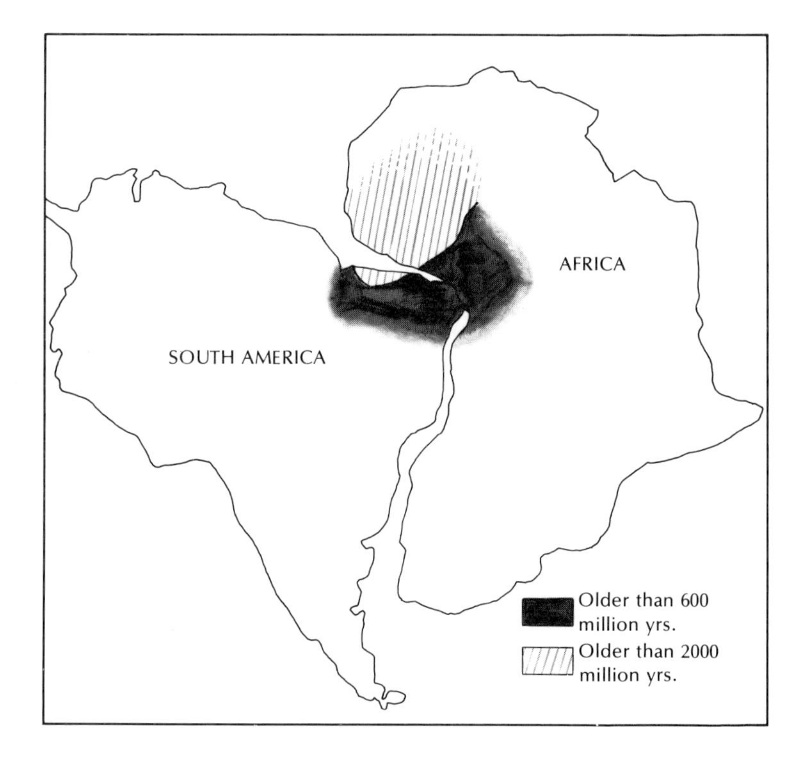

Fig. 20-7 Matching the boundary between rocks of different ages in Africa and South America. The age boundary shown extends from the vicinity of Accra, Ghana, to that of São Luiz, Brazil.

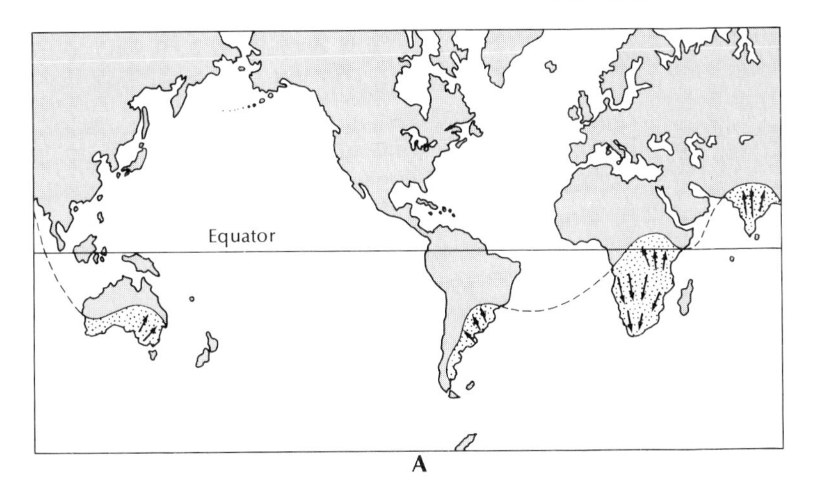

Fig. 20-8 Distribution of late Paleozoic glacial deposits as seen (**A**) on a present-day map and (**B**) on a map in which continents have been reassembled according to Wegener. Directions of glacial flow are indicated by arrows. In the present-day map, the glaciers appear to originate in the oceans; in the reconstruction, glacial directions form reasonable patterns.

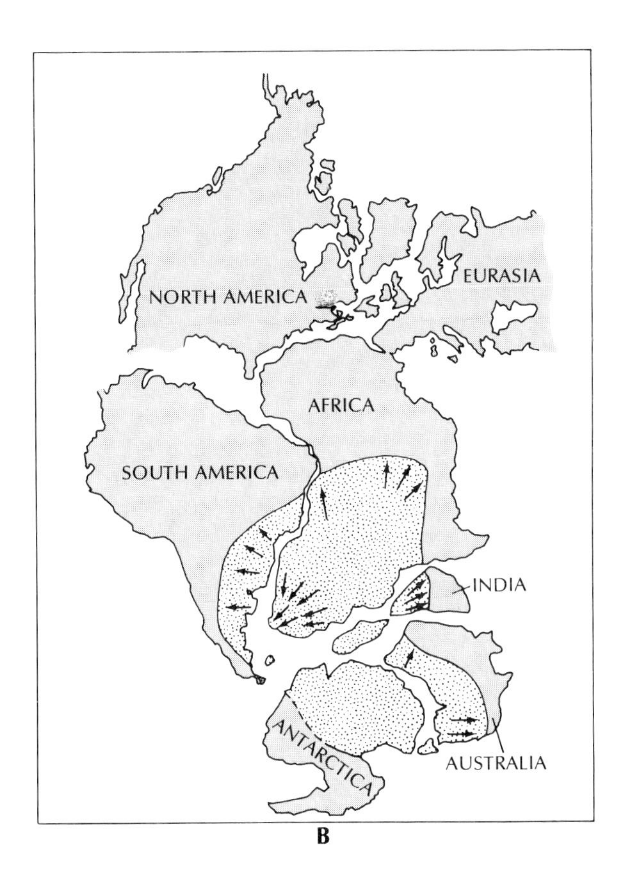

Wegener, is not in itself ironclad proof of the reality of continental drift, but when it is added to evidence already presented, the case is measurably strengthened.

From a study of the rocks on the different continents, it is possible to obtain an idea of the climates that existed when the rocks were laid down. The distribution of the mapped paleoclimates make a disjointed, checkerboard pattern on the continents as they now stand. When the continents are pieced back together, however, the patterns appear continuous and reasonably explicable.

Fossils If all the continents once had been one, animals and plants that lived on it should have been able to migrate freely across the land, restricted only by the availability of natural habitats and impassable features, such as mountain ranges. Fossils found on the various continents today, then, would show a similarity in ancestry, dating back to the time when the continents were contiguous. Wegener and his followers attempted to point out such similarities, noting, for example, that 64 per cent of Carboniferous and 34 per cent of Triassic reptile groups are the same on all southern continents. But the data available on the subject in the early 1900s were sketchy. Even in 1943, when George Gaylord Simpson, one of the world's leading paleontologists, reviewed the situation he concluded that the paleontological literature was too full of mistaken identities, unsubstantiated claims, internal inconsistencies, and shaky conclusions to be of any use in determining whether drift had occurred or not.

As more and better paleontological data have been presented, Simpson's objections have begun to fade. Similarities in fossil forms have been shown to exist between the continents. As recently as 1969, fossil land-restricted amphibia and reptiles of Triassic age, discovered on the isolated frozen continent of Antarctica, were determined to be exactly like Triassic fossils found in Africa, Madagascar, and Australia.

Paleomagnetism In the 1940s and 1950s, paleomagnetism was a growing branch of the field of geophysics. One of the leaders in that field was the renowned English geophysicist Stanley K. Runcorn. He and his associates measured the fossil magnetism in a great many rock samples that ranged in age from late Precambrian to Recent. From the paleomagnetic data they calculated an average north pole position for the rocks of each geological age, in the manner outlined in Chapter 19. On the well-documented assumption that the earth's dipolar axis is coincident with the rotational axis when field directions are averaged over long periods of time, they concluded that the calculated paleomagnetic poles corresponded to the rotational north poles. When the poles for European rocks were plotted on a present-day map, it was found that they fell on a curved path that ran from south of the equator to the present north pole (Fig. 20-9). The younger the pole, the closer to the present-day geographic pole it occurred. The plot, called an ***apparent polar wander curve,*** seemed to indicate that, with regard to Europe, either the north pole had moved or wandered or that the European continent had shifted and rotated in position with time. Today, for a variety of reasons, most paleomagnecists favor the latter explanation.

The significant point is that when Runcorn and his associates plotted the paleomagnetic poles for North American rocks covering the same time span, they obtained a similarly shaped apparent polar wander curve, but it was displaced about 40° toward the west from the European curve (Fig. 20-9). The similarity in shape between the two curves seemed to indicate that both Europe and North America had undergone similar changes in latitude and rotation relative to the pole. The displacement between the two curves Runcorn attributed to continental drift. When North America was "moved" back to adjacency with Europe, the two curves became nearly coincident over most of their paths from the Precambrian through the Triassic. From that point in time onward, the paths diverged. Runcorn concluded that drift must have been initiated in the Triassic and that it had been going on ever since.

Runcorn published his results in 1962, at about the time when the theory of continental drift was still unacceptable to most earth scientists. The paleomagnetic evidence appeared so incontestable that it swayed many scientists who had been fence-sitting up to that time. Ad-

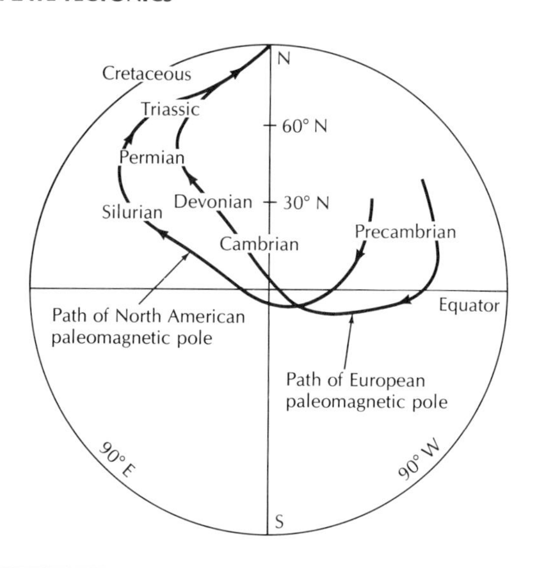

Fig. 20-9 Two apparent polar wander curves from the Precambrian Era to the present, one based on data from North American rocks, and one on data from European rocks. The paths are similar in shape but not coincident in position, suggesting that North America and Europe have been drifting since the Triassic Period.

ditional paleomagnetic studies of rocks of all ages from all the continents have confirmed the first results.

Linear magnetic anomalies We have already discussed the significance of oceanic magnetic anomalies (Fig. 20-10) in Chapter 19, but we should reiterate that their existence not only has confirmed that the ocean basins were cracking open along giant fractures but provided insight into the way the continental masses actually drifted apart (Fig. 20-11). When the anomalies in various parts of the ocean basins were identified and correlated, it was possible in many cases to determine when continental fragmentation began and what drift paths the continents took. The name given to the phenomenon of continental drift by an opening of the ocean basins was ***sea-floor spreading.***

Fig. 20-10 Pattern of bilaterally symmetric linear magnetic anomalies along the Mid-Atlantic Ridge. The areas of higher-than-background intensity are in black; those of lower-than-normal intensity are in white. Normally and reversely magnetized rocks beneath the ridge are youngest along the axis and increase in age laterally away from the axis.

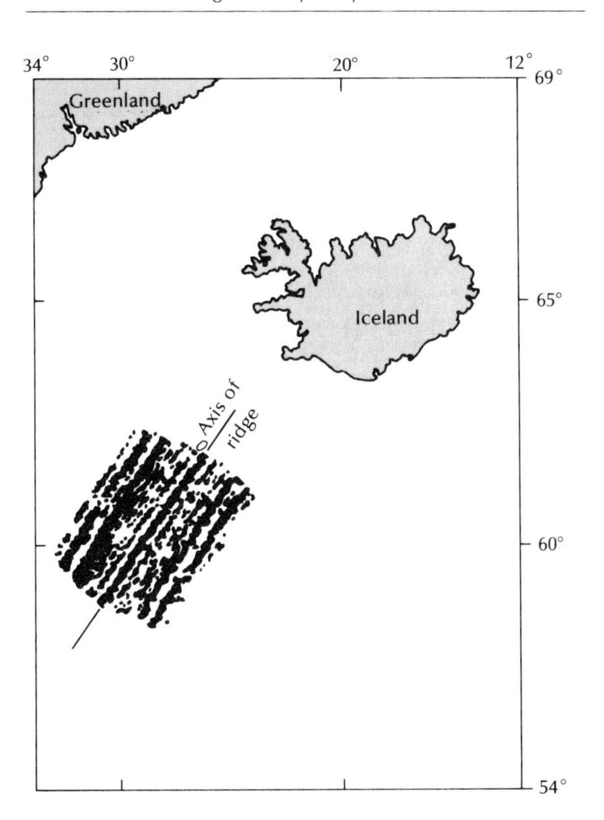

Deep drilling in the oceans If the oceanic ridges and rises are actually zones of divergence along which spreading occurs, then the ocean floor should become progressively older outward from the ridge crests. With the launching of the *Glomar Challenger,* the ocean-going drilling ship, it was possible to test that hypothesis. One of the first lines of drill sites chosen was in the South Atlantic Ocean, at about 30° south latitude, just south of Rio de Janeiro. Eight sites were drilled, each progressively outward from the Mid-Atlantic Ridge. Drilling at each site was halted when the drill bit had penetrated the sedimentary sequence and had begun to bite into the hard basaltic rocks beneath it. The sediments recovered from each coring site were described in detail, and fossils contained within them were used to date each core throughout its length. The age of the sediments directly overlying the basaltic rocks was taken as a rough estimate of the age of the basement rocks beneath.

The results (Fig. 20-12) are consistent and, much to the elation of the advocates of drift, show a continuous increase in age outward from the ridge axis. The slope of the curve, essentially a straight line, indicates uniform spreading from the Mid-Atlantic Ridge at a rate of about 2 cm (0.79 in.) per year—just about the rate determinable from the linear magnetic anomaly pattern in that part of the South Atlantic.

Continued deep-sea drilling turned up another significant fact, which at first seemed astonishing;—in none of the oceans were basaltic rocks encountered beneath the sediments that were older than about 150 million years. The ocean basins are all geologically young features! Yet the continents contained rocks 2500 to 3000 million years old or even older. Somehow the continents have floated about, like flotsam and jetsam, and have lasted for billions

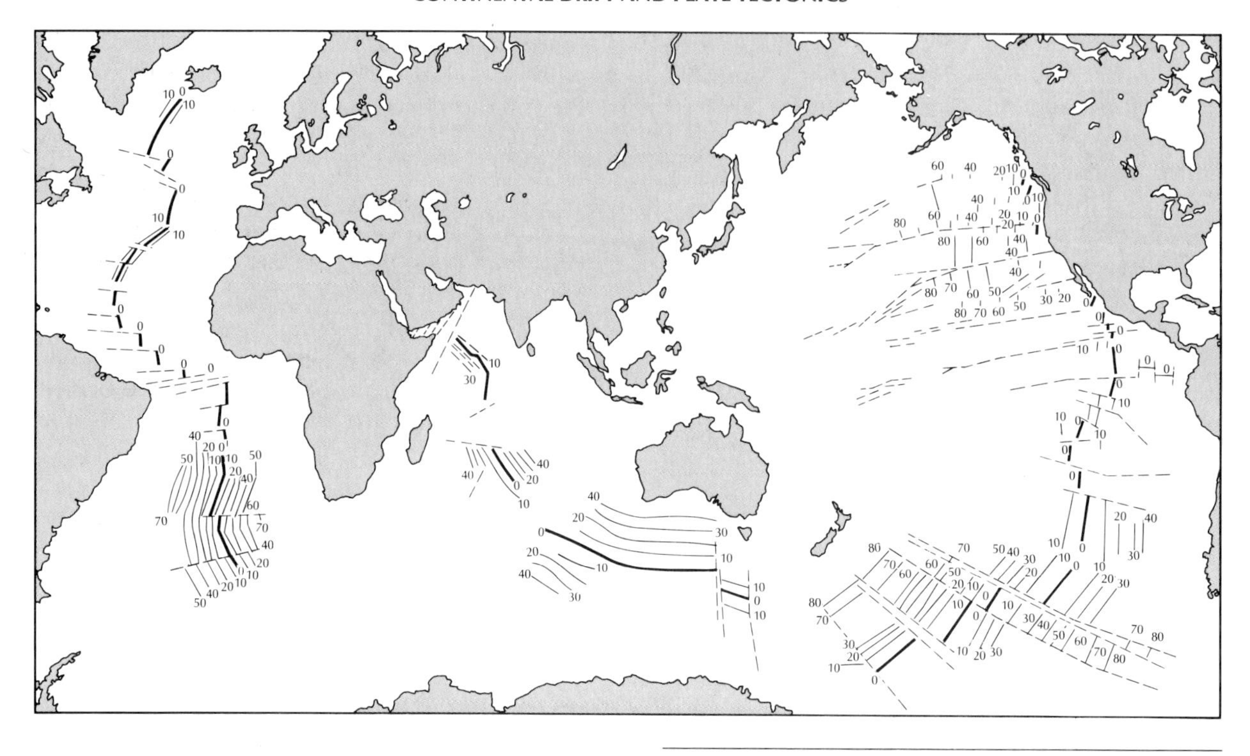

Fig. 20-11 Lines of equal age (in millions of years) on the ocean floor as determined from the linear magnetic anomaly pattern. The youngest rocks are found along oceanic ridges and rises, where the earth is splitting apart.

of years, whereas the oceans appear to be in a continual state of renewal. If there is no old oceanic crust, how did it disappear? That question will be pondered a little later in the chapter.

Continental Drift in the 1960s

By the 1960s the idea of continental drift was supported so strongly that most earth scientists considered it to be a certainty. It is unfortunate that Wegener did not live long enough to witness that complete turnabout in geological thinking.

With the acceptance of the continental drift theory, earth scientists turned their attentions during the 1960s and 70s toward a unified theory that could account not only for the drifting of continents, but for all the other internally generated phenomena, such as mountain building, genesis of magma, and metamorphism as well. Through the development of the plate tectonic theory, we have advanced a long way toward that goal. Yet there is still a long way to go. It is an exciting time in geology, for almost each day brings new facts to light that require modifications in the way we look at the earth and its formation.

Steps toward a unified theory Investigations of the ocean basins have indicated that new crust is continually being generated along the oceanic ridges and rises, but what happens to it after it is formed? If the crust is not being consumed at a rate equal to its generation then the earth would have to be expanding to accommodate it. And some geologists suggest that this is so. However, others have pointed out that if the rate of expansion in the geological past had been anywhere near the rate of sea-floor spreading observable in the recent past, the earth would have been an extremely small body initially, small enough to make that idea unrealistic. Moreover, the generation of new crustal material is not uniform, and if the crust is not consumed, then the earth's surface should become distorted. But its shape is a fairly regular ellipsoid of rotation.

Most geologists, therefore, have come to accept the idea that crustal material is being destroyed somewhere. In the early 1960s, the American geologist Harry Hess suggested that the crust was likely to be consumed at the island arcs of the world, an idea that was championed by the American oceanographer Robert Dietz. Island arcs, as you recall, are linear features and the sites of a variety of geological phenomena (Fig. 20-13). The islands themselves generally consist of a string of volcanoes that erupt andesitic lavas and pyroclastics. Oceanward from the chain of volcanoes is a linear trench. Earthquakes are common to island arcs, occurring down to depths of about 700 km (438 mi). When the foci of all the quakes are plotted on a cross section at right angles to the arc system, they define an inclined seismic zone, called the ***Benioff zone,*** that begins at the surface near the trench and dips beneath the island arc at a steep angle, commonly about 45° (Fig. 20-14). The zone was named after Hugo Benioff (1899–1968), an American seismologist who pioneered studies into the distribution of earthquakes in continental margins. Studies of first motions recorded on seismographs reveal that, in most cases, the Benioff zone can be likened to a large thrust-fault zone along which the oceanward block pushes under, or underthrusts, the continental-ward block.

New global tectonics—A forerunner hypothesis In the late 1960s, Jack Oliver and Bryan Isacks, two U.S. geophysicists, decided to investigate the possibility of the disappearance of crustal material at island arcs by studying the behavior of seismic waves generated in the Tonga Island arc system in the southwest Pacific. They set up seismographs on islands both east and west of the deep Tonga trench and found that the nature of the seismic waves, es-

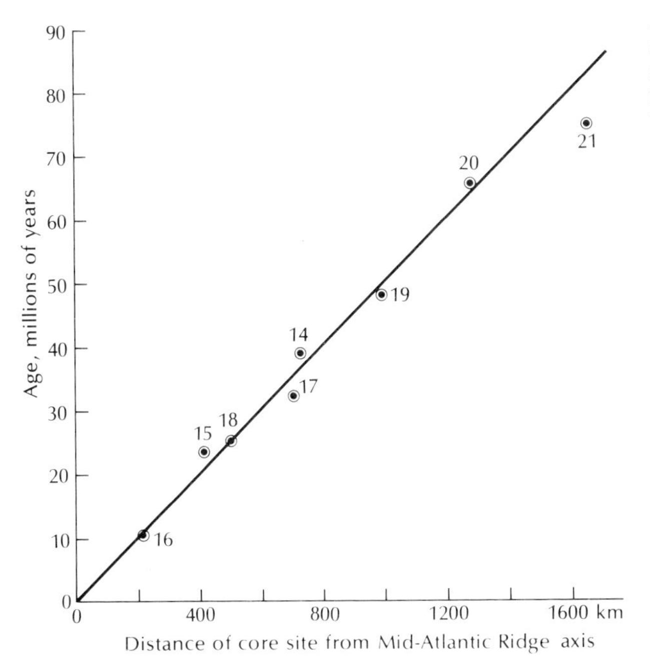

Fig. 20-12 The age, determined paleontologically, of ocean sediments immediately above basaltic rock, plotted as a function of distance from the Mid-Atlantic Ridge axis.

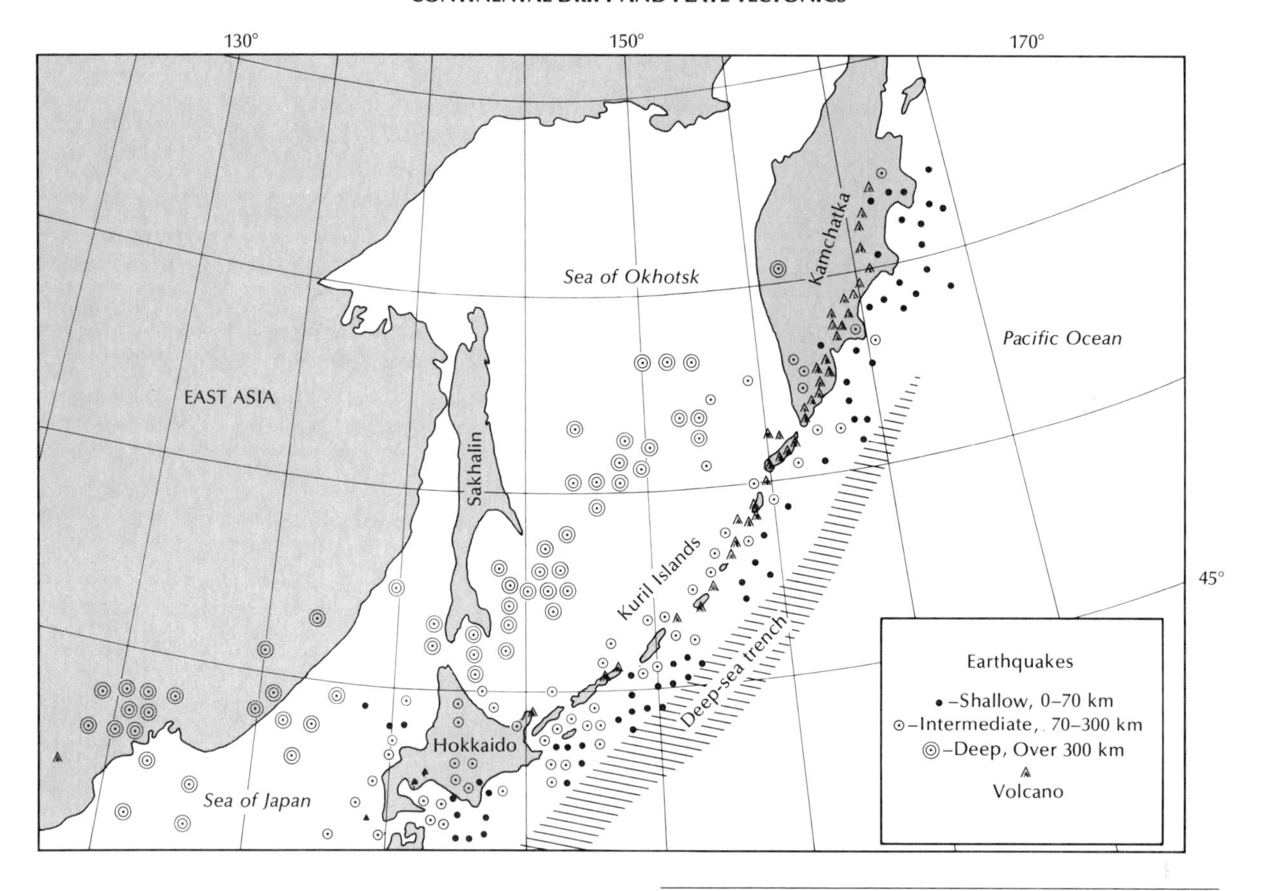

Fig. 20-13 Kamchatka Peninsula and the Kuril island arc, showing the relation of earthquakes, volcanoes, and a deep-sea trench.

pecially the *S* waves, were different, depending on the location of the stations. The differences, Oliver and Isacks reasoned, are the result of differences in the rocks beneath the island arc through which the waves travel. Their interpretation of the character of that rock substrate is shown in Figure 20-15.

Apparently a lobe of high-strength (brittle) material about 100 km (62 mi) thick parallels the Benioff zone and extends downward nearly 500 km (313 mi). They concluded that the brittle material, which they called the ***lithosphere,*** is being pushed down through a layer of lower-strength material, which they called the ***asthenosphere.*** The latter extends from a depth of about 100 km down to about 250 km (155 mi) and corresponds to the zone around the world called the ***low-velocity layer*** (see Chap. 18). Beneath the asthenosphere, the mantle again becomes stronger, to form what they called the ***mesosphere.***

In 1968, Isacks, Oliver, and another geophysicist, Lynn Sykes, incorporated their findings into a hypothesis, which they termed ***new global tectonics.*** They postulated that new lithospheric material being formed at the ocean rises, once formed, moves laterally away from the ridge axes and toward the island arcs (Fig. 20-16). They went on to speculate that in the vicinity of the trench off an island arc the lithosphere bends sharply downward and sinks or is pushed into the softer asthenosphere below. It is much like a giant treadmill, with upwelling and generation of new lithospheric material

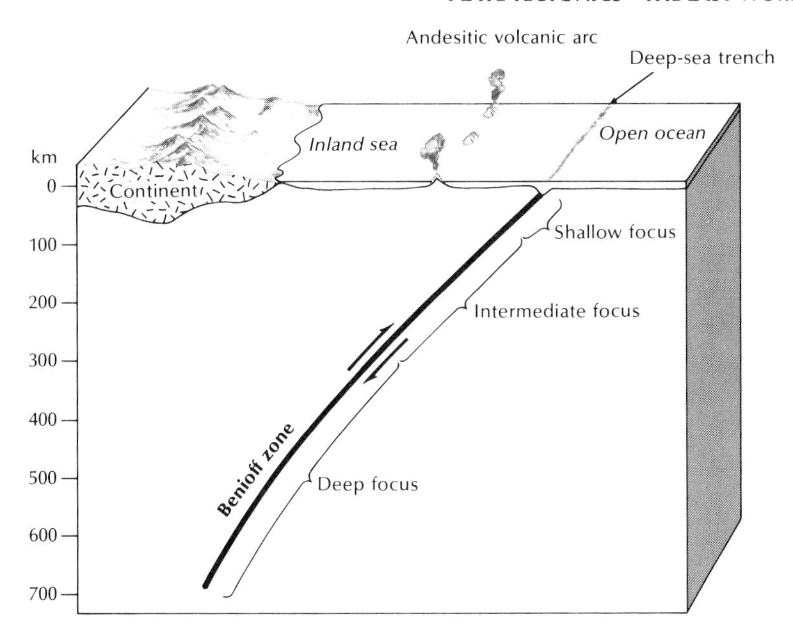

Fig. 20-14 The relation of a deep-sea trench, a volcanic island arc, and the Benioff zone.

along one zone and downwarping and consumption of the lithosphere along another zone.

PLATE TECTONICS—THE LAST WORD

Most earthquakes occur in the vicinities of the oceanic ridges and rises and along the Benioff zones of island arcs. In Figure 20-17 are plotted all of the foci for quakes that occurred in the world between 1961 and 1967. The pattern confirms that most of the world's seismic activity is confined to very narrow linear zones. Geologists have postulated from the pattern that the earth's crust is divided into a relatively small number of rigid blocks, or ***plates,*** that move and jostle one with respect to another (Fig. 20-18). Seismic activity takes place at the edges of the plates, but the plates themselves are nearly devoid of activity. The ***plate tectonic theory,*** as noted earlier is the dominant theory now used by earth scientists to explain the internally generated features of the earth. It, being the most recent theory, embodies all the concepts and features of continental drift and new global tectonics that went before.

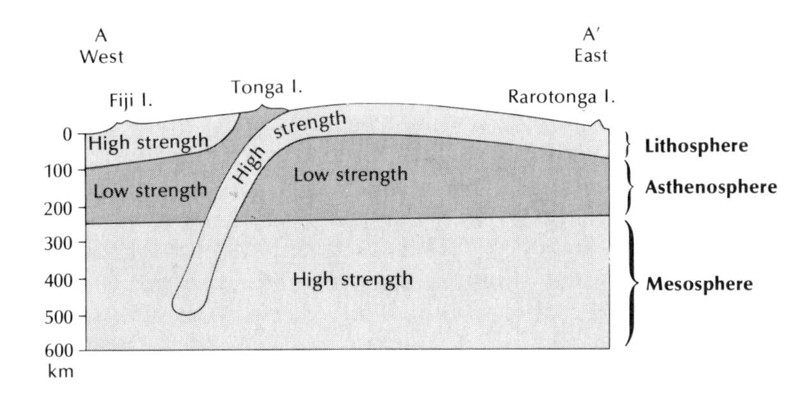

Fig. 20-15 A hypothetical cross section A-A′ extending from the Fiji Islands through the Tonga Islands to Rarotonga illustrates the disposition and strength of the rock layers beneath the islands, as determined from the interpretation of seismic waves.

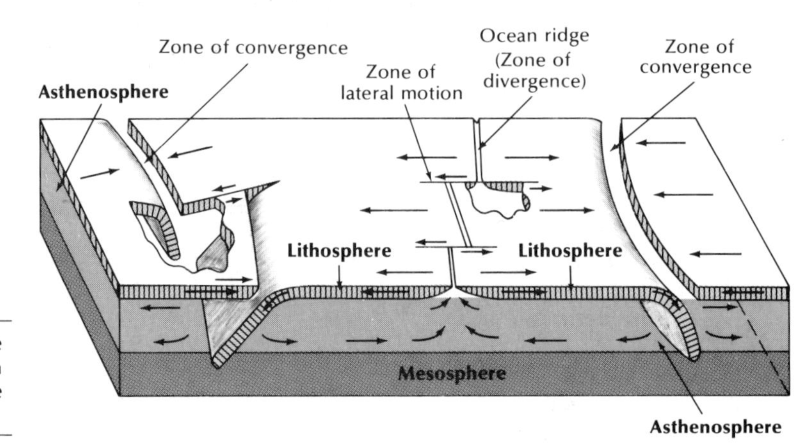

Fig. 20-16 The kinds of movement possible in the lithosphere and the relation of such movements to compensating flow in the asthenosphere.

Zones of divergence

Where plates move apart (***pull-apart zones***), as along the Mid-Atlantic Ridge, new lithospheric material is formed (see Fig. 20-18). They comprise zones of basaltic volcanism and shallow-focus earthquakes. Rocks, when heated, tend to expand. Therefore, the newly formed hotter rocks near the axis of the zones of divergence occupy more volume, which results in an elongate topographical high—an oceanic ridge or rise. As the lithosphere moves laterally outward from the ridge crest, it cools and contracts, and the ridge system is reduced in elevation. Earthquakes in the vicinity of the divergence zones occur at depths of less than 20 km (12 mi) as a rule and are restricted to the ridge axes or the fracture zones that cross them at nearly right angles.

Since 1970, divergence zones have been studied by submersibles in three different places—the Mid-Atlantic Ridge, the Galapagos spreading center off the coast of South America, and the East-Pacific Rise south of Baja California. The studies indicate that these zones are relatively narrow; in places the line of demarcation is almost knife-sharp. Newly formed pillow lavas (Fig. 20-19) are found in all three places, indicating that largely undetected submarine basaltic volcanism is common throughout such zones. Sensors showed that the temperature of the sea water near the divergence-zone axis was often higher than normal in response to interactions between the sea water and recently erupted ridge basalts. Unusual marine organisms (see Fig. 15-2) are concentrated in the heated waters at both the Galapagos center and the East-Pacific Rise. Sulfide minerals rich in copper, iron, zinc, cobalt, lead, silver, and cadmium also precipitate from hot-water vents along the axes of the divergence zones. Certain mineral deposits, such as those found on the Mediterranean island of Cyprus, are now thought to have formed in this way.

In a few places on earth, apparently demarcating old convergence zones, rocks are present that represent the ancient ocean floor. The explanation is that after formation at the zone of divergence, the sea floor moved progressively across the ocean basin and ultimately into an active ***subduction zone,*** where, during its consumption, thin, highly deformed slices were thrust onto the plate margin. These unusual rocks consist of deep-sea sediments, serpentinized pillow basalts, and mafic igneous intrusions. Such associations of rock have been called the ***ophiolite suite,*** from the Greek *ophis* (serpent) and *lithos* (stone), because the mafic rocks have largely been altered to serpentine.

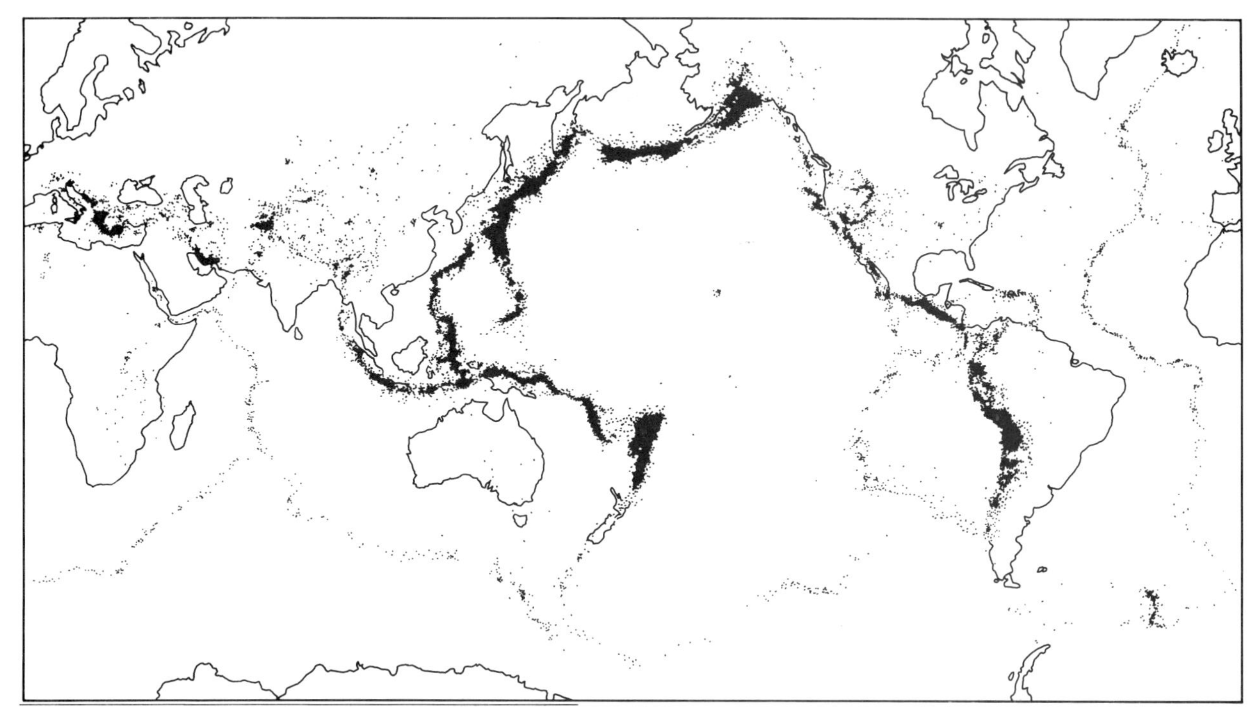

Fig. 20-17 Distribution of earthquake epicenters, 1961 to 1967. Most earthquakes occur in narrow belts that are now considered to mark the edges of moving lithospheric plates.

Fig. 20-18 The earth's lithospheric plates, zones of convergence and divergence. Arrows indicate the supposed movement of the plates.

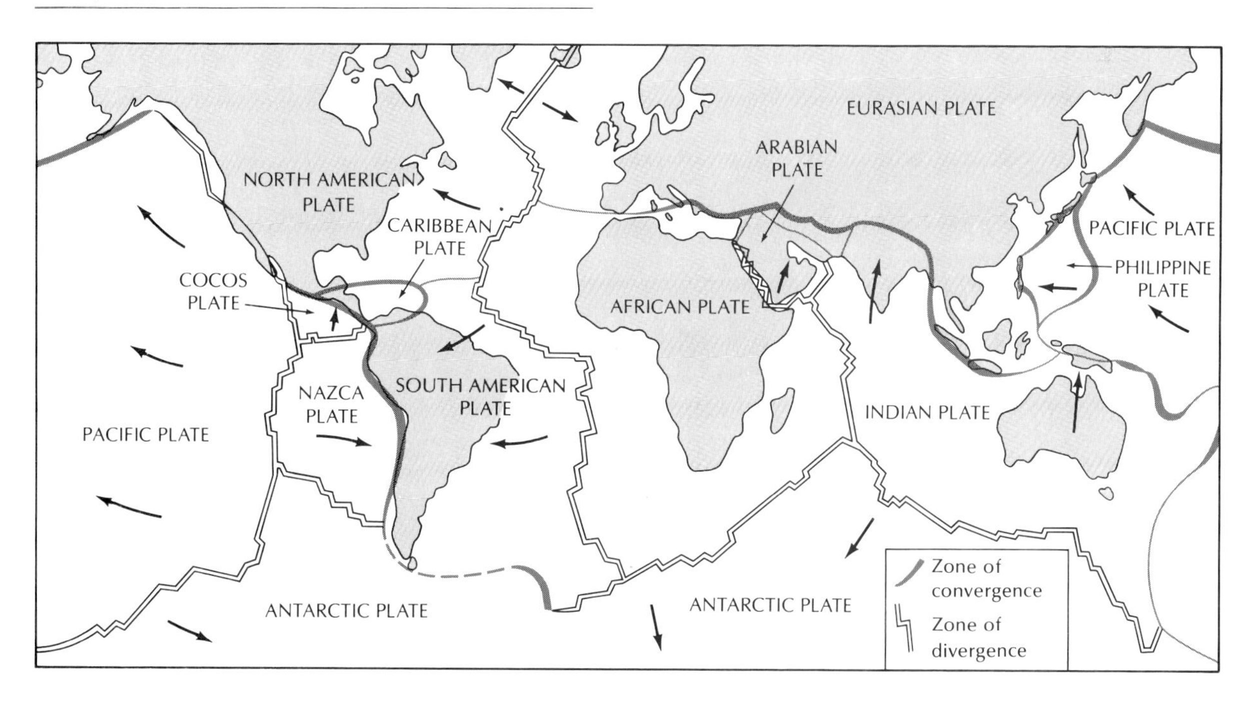

Fig. 20-19 Pillow basalt near the axis of a divergence zone (the Mid-Atlantic Ridge), photographed from the submersible *Alvin*. *Woods Hole Oceanographic Institution*

Zones of lateral movement—transform faults

Throughout the world oceanic ridges are transected and sharply offset by a great number of narrow fracture zones (Fig. 20-20). Because of the numerous offsets, the map traces of the rises and ridges (see Fig. 20-18) zig-zag across the oceans. In 1965, the Canadian geologist J. Tuzo Wilson postulated that offsets should lead to an unusual type of fault activity that he named ***transform faulting.*** He reasoned that along the transverse fracture zone and between the ends of the displaced ocean rise (Fig. 20-20) strike-slip faulting (and earthquake generation) should prevail, since the plates are moving laterally one past the other. Beyond the offset ridge ends, however, the plates on either side of the fracture zone move in the same direction, and earthquakes should not occur. Transform faulting, therefore, should be restricted only to that part of the fracture zone between the ends of the offset ridges.

Several years later, Lynn Sykes confirmed the existence of transform faulting by plotting the epicenters of many recent earthquakes that occurred in the Atlantic Ocean (Fig. 20-21). He found that they were concentrated either over the ridge crest or along the fracture zones between the ends of the dislocated ridge-crest ends, just as Wilson postulated. Sykes also studied the first motions of the quakes and verified that the relative displacement between blocks was indeed strike-slip (Fig. 20-21). The seismicity along transform faults is another proof that the oceanic ridges are active spreading centers.

Transform faulting does not cause separation of ridge axes, but occurs because they are already offset. The cause of the initial offset in the ridge lines is still unknown, although it is undoubtedly related to the generation of the spreading centers.

Zones of convergence

The curvilinear island arcs of the world mark the zones where lithospheric plates push together and where consumption of lithospheric

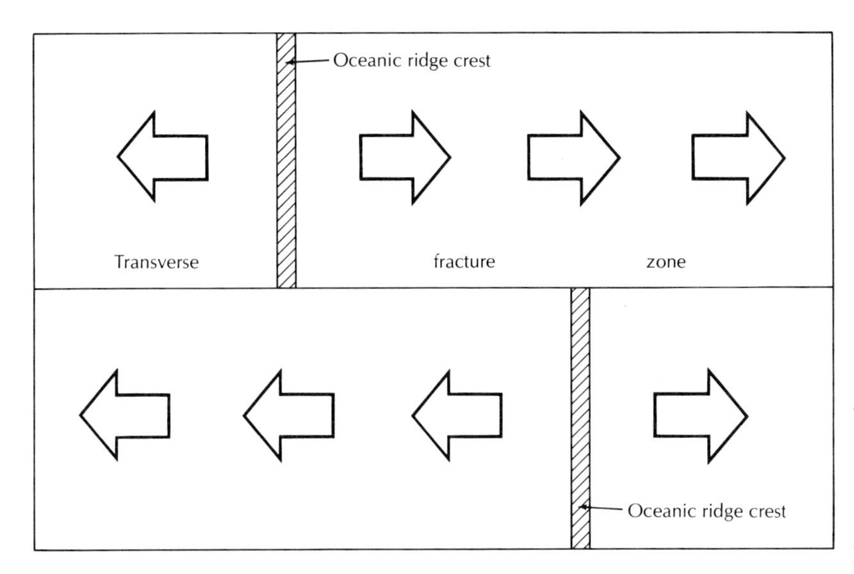

Fig. 20-20 A zone of divergence offset along a transverse fracture zone. Because of the offset, the diverging plate material moves in opposite directions along the fracture zone, but only between the offset ridge ends.

Fig. 20-21 Earthquake epicenters near the Mid-Atlantic Ridge and fault motions along fracture zones.

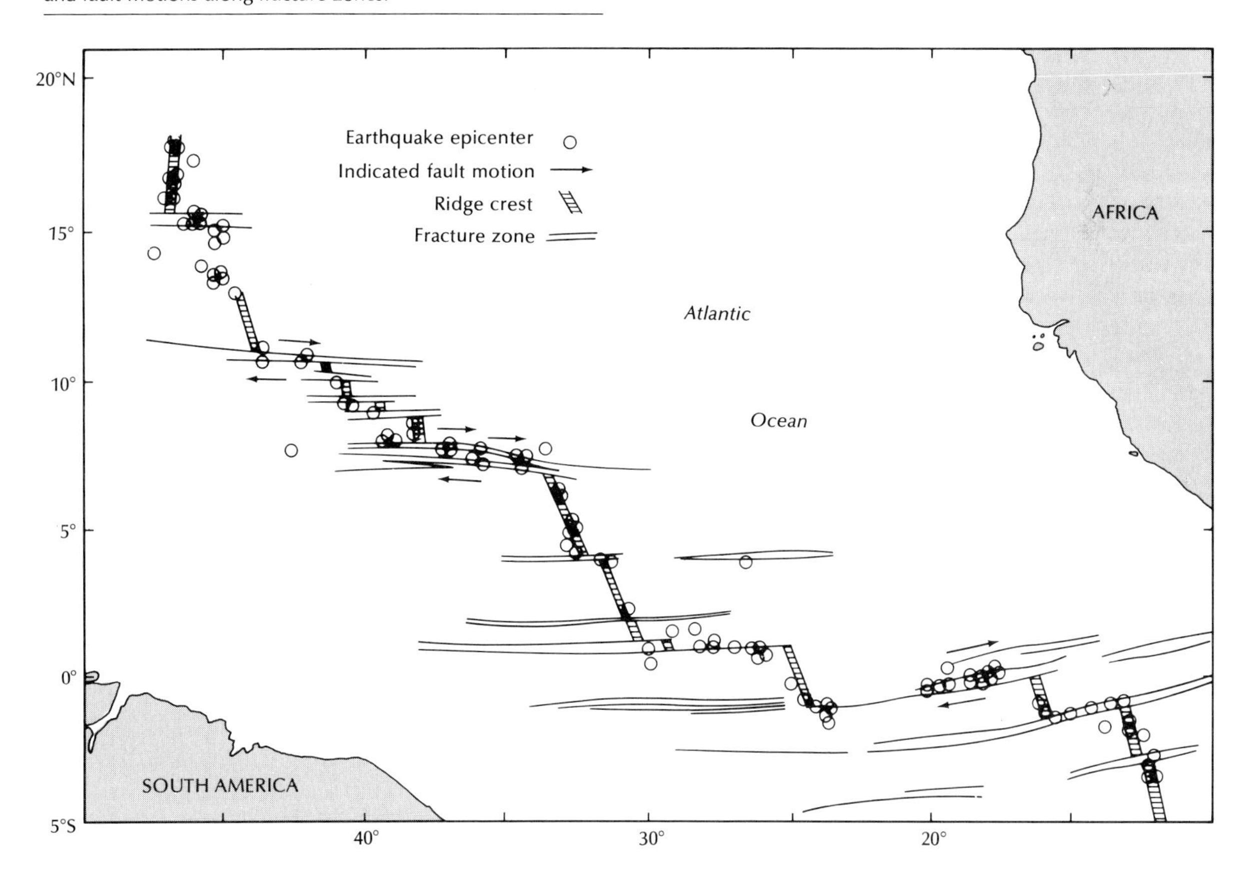

material occurs (see Fig. 20-18). As a slab of lithosphere bends over and slowly descends into the asthenosphere, it is heated. Eventually, temperatures will rise to the point at which the descending slab becomes indistinguishable from the mantle rocks that surround it (Fig. 20-22). The depth to which a plate edge can penetrate depends on the rate of movement, but no plate extends below 700 km (435 mi).

As the relatively cold lithospheric plate descends, frictional drag occurs. Sporadic release of the strain in the slab supposedly produces the shallow-, intermediate-, and deep-focus quakes characteristic of island arcs (Fig. 20-22).

Sediments derived from the nearby continental land masses and from erosion of the islands of the volcanic arc themselves can accumulate in the deep trench next to the arc. As the lithospheric plate moves downward, those stratified rocks are dragged down, or ***subducted.*** At first they will be folded and faulted; then, as they move to greater depths, they will be heated—and dynamothermal metamorphism and local melting occur. Eventually, a linear mountain chain will be formed, as described in Chapter 17. Magma derived from the melting of the sediments or parts of the descending slab can rise to the surface wherever cracks and fractures are present, to produce the linear chain of andesitic cones so common to island arcs. The exact nature of the material that is melted to produce the volcanic eruptions and the depth at which the melting takes place are still points of controversy among petrologists. Some even believe that some melting of the continental crustal rocks must be involved as well.

New wrinkles

The plate tectonic theory, as first stated, was relatively straightforward and elegant in its sim-

Fig. 20-22 Relation of a divergence zone to a convergence zone. New lithospheric material, formed along oceanic ridges, moves outward and bends sharply downward into a zone of convergence, where it is eventually consumed at depth. The axis of a deep-sea trench forms where the slab bends downward, and a line of andesitic volcanoes forms where heating adjacent to the descending slab is sufficient to melt rocks. Shallow-, intermediate-, and deep-focus earthquakes of the Benioff zone are thought to be generated along the upper surface of the slab.

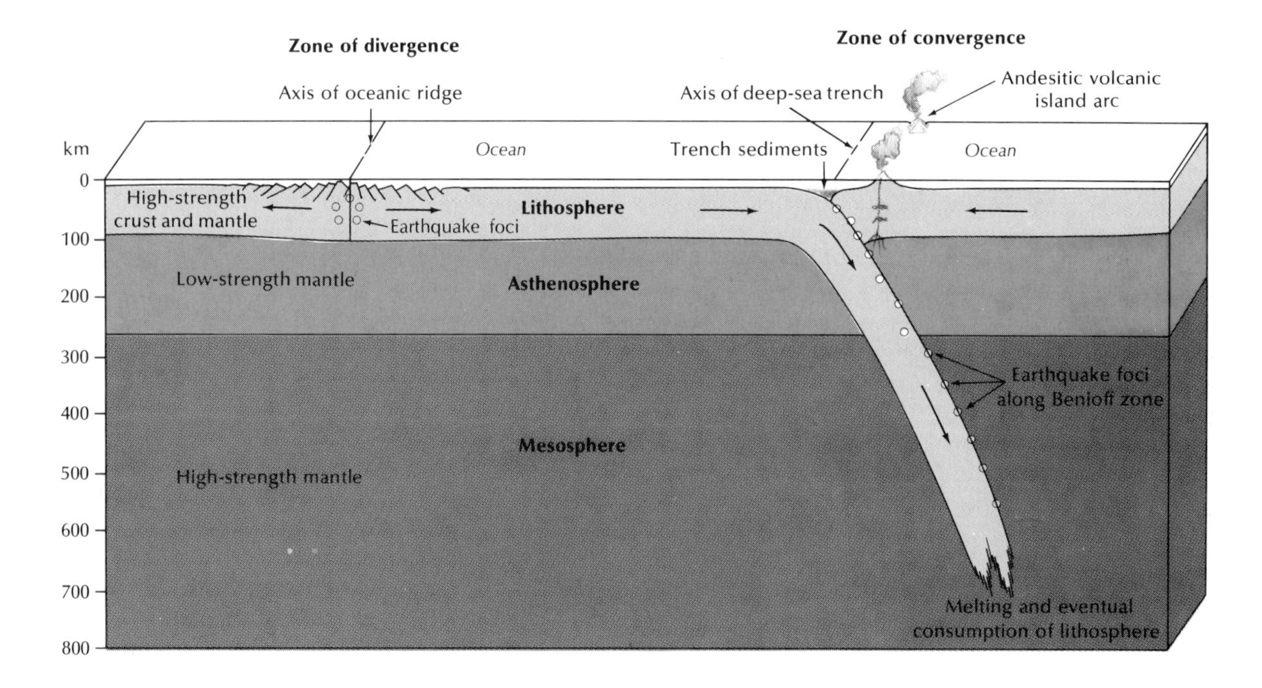

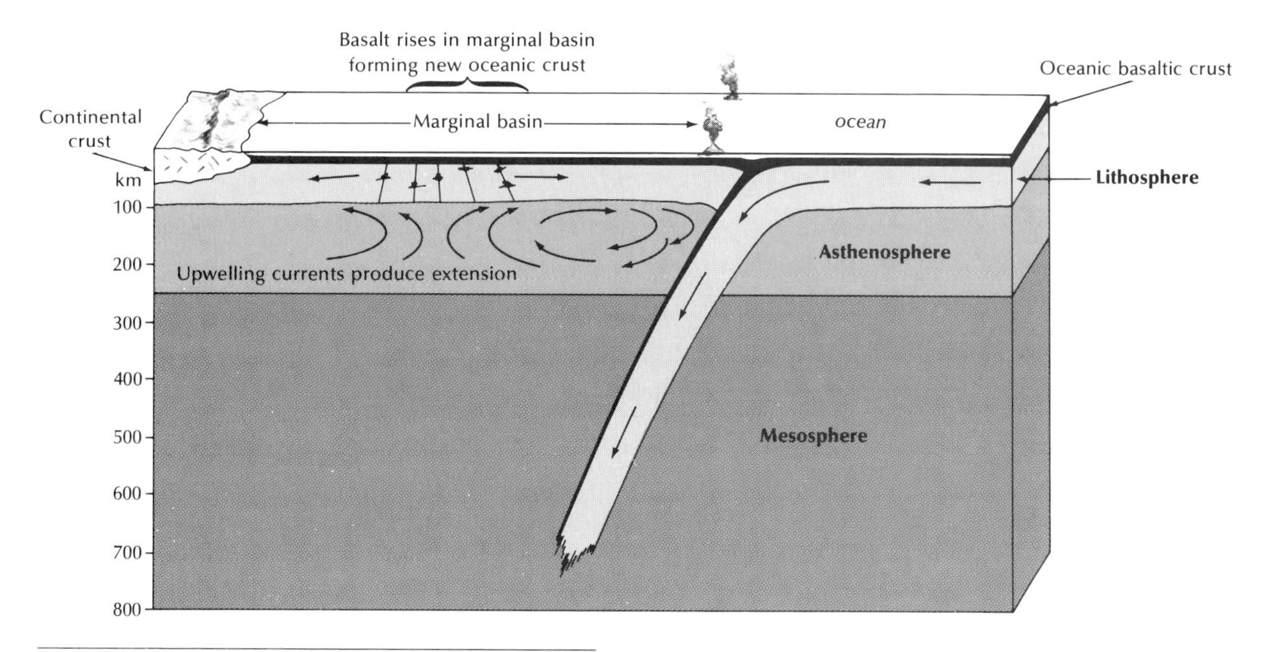

Fig. 20-23 Hypothetical model to account for the development of back-arc upwellings.

plicity. Much of its appeal stemmed from those attributes. During the years of its existence, however, it became apparent that the theory could not explain all the earth's geological features, and the theory has had to be modified. If exceptional phenomena become too numerous, however, there is a danger that the theory could be overturned. But at the same time, modifications must not be ignored. Dominant theories tend to steamroll facts and phenomena that may not be in accordance with them. A quote from the 1968 paper in which Isacks, Oliver, and Sykes first formulated the forerunner theory of new global tectonics, is pertinent: "At present there appears to be no evidence from seismology that cannot be reconciled with new global tectonics in some form."

Microplates The plate tectonic model was initially meant to apply to the movements of a small number of large lithospheric plates. It has become apparent, however, that in some parts of the world, the lithosphere is composed of numerous ***microplates,*** small plates which can move in very complicated ways. The region around the Mediterranean Sea, for example, appears to be composed of microplates, and its geological structure cannot be understood without some knowledge of the complex movement of those plates.

Back-arc upwelling Oceanographic studies, mostly by the American geologist Daniel Karig during the last several years, have shown that secondary zones of divergence commonly form between continents and island-arc systems, which leads to an oceanward migration, in time, of the arc and trench system. Such movement was not considered when the plate tectonic theory was initially formulated. It is now the opinion of geophysicists that such ***back-arc upwellings*** result from a convectional eddy current behind the island arc in response to heating of the mantle by the descending slab, as shown in Fig. 20-23.

Mantle hot spots and plumes Not all volcanism is restricted to the zones around plate margins; some, such as that of Hawaii, occurs out in the

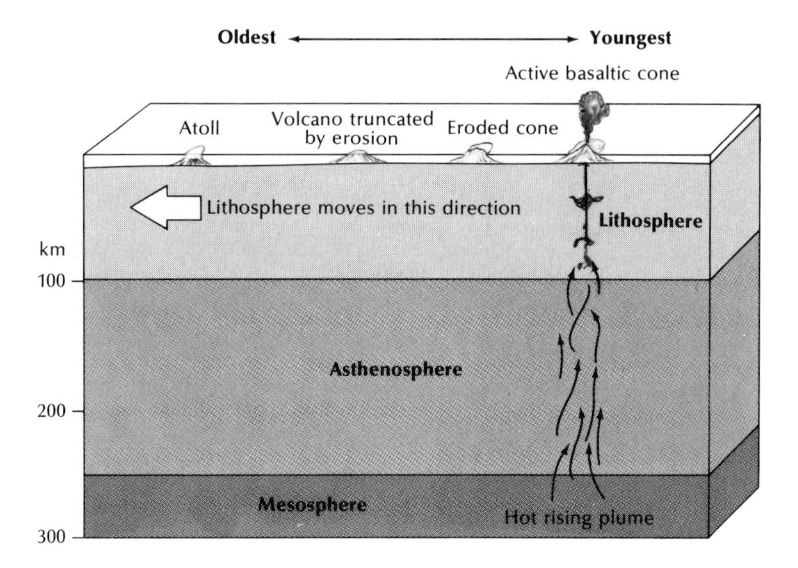

Fig. 20-24 Diagram, showing the postulated hot rising plume that melts rocks just below or within the lithosphere, with the formation of plate-center basaltic volcanoes. Once formed, the cones move laterally from the hot spot, carried piggyback on the moving lithospheric plate.

middle of the plates (see Chap. 6). It is difficult to understand the origin of such volcanoes within the framework of the plate tectonic model. J. Tuzo Wilson has speculated that somewhere at depth in the mantle, perhaps as far down as the core-mantle boundary, localized ***hot spots*** form. Heating reduces the density of the mantle material, which then begins to rise toward the surface in a finger-like ***plume.*** The rising plume of heated rock can partially melt parts of the asthenosphere, or the lower parts of the lithosphere, and it is the eruption of that magma that builds volcanoes in the center of a plate (Fig. 20-24).

Some geophysicists have taken the idea a step further, speculating that the rising plumes bend over as they reach the asthenosphere. Where they reach the underside of the lithospheric plates they push, like moving fingers, to cause lateral movement of the plates.

Since the plumes supposedly are generated in the relatively static mantle beneath the moving lithosphere above, the volcanoes, once formed, may be cut off from the magma source and carried away with the moving plate. The result is a formation of a line of volcanoes that increase in age away from the active hot spot (Fig. 20-24). The Hawaiian Islands form just such a chain of volcanoes, with the youngest ones over active vents on the island of Hawaii.

Not all geophysicists subscribe to Wilson's idea; some believe that the existence of

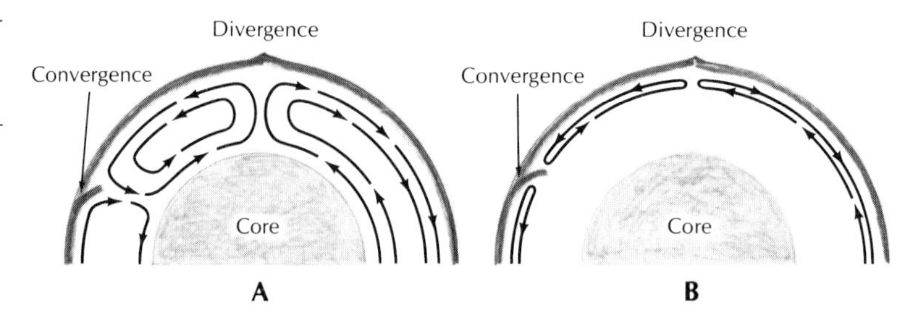

Fig. 20-25 Cross section of the earth with **(A)** a deep convection model and **(B)** a shallow convection model.

plumes in the mantle has not been demonstrated adequately.

Mechanism for plate tectonics

When Wegener first advanced his theory, many critics pointed out that he had failed to show how the continents might move from one place to another. In spite of recent progress in understanding the earth and its internal processes, earth scientists are still no closer to explaining the mechanism than was Wegener.

One idea, introduced first by Arthur Holmes, is that the earth's mantle is subject to inequalities in internal heating such that it can flow, or slowly convect, much like slowly boiling porridge in a pot (Fig. 20-25). Many advocates of plate tectonics have not only accepted the idea of convection, but consider it to be the force that drives plates over the earth's surface. Places where the currents rise and move apart correspond to zones of divergence, whereas places where the cooler currents converge and descend correspond to zones of convergence.

Yet we do not know whether the solid mantle is actually capable of convection, as postulated, and even if it were, we do not know how far down in the mantle convection can occur. We are also at a loss to explain how an ephemeral phenomenon like convection can maintain itself in narrowly restricted areas for hundreds of millions of years.

Before it was known that a low-velocity layer in the zone between 100 and 250 km (62 and 155 mi) existed, those advocating convection imagined the cells to originate far down in the mantle (Fig. 20-25A). Today, most advocates restrict the process to the relatively soft asthenosphere—which would make particularly difficult the generation of long, linear zones of divergence that stretch continuously for tens of thousands of kilometers (Fig. 20-25B). If con-

Fig. 20-26 The hypothetical sinking of the relatively cold and dense lithospheric slab into the asthenosphere and mesosphere. As the slab sinks, it pulls the oceanic lithosphere along behind it, causing an opening at the ocean ridge.

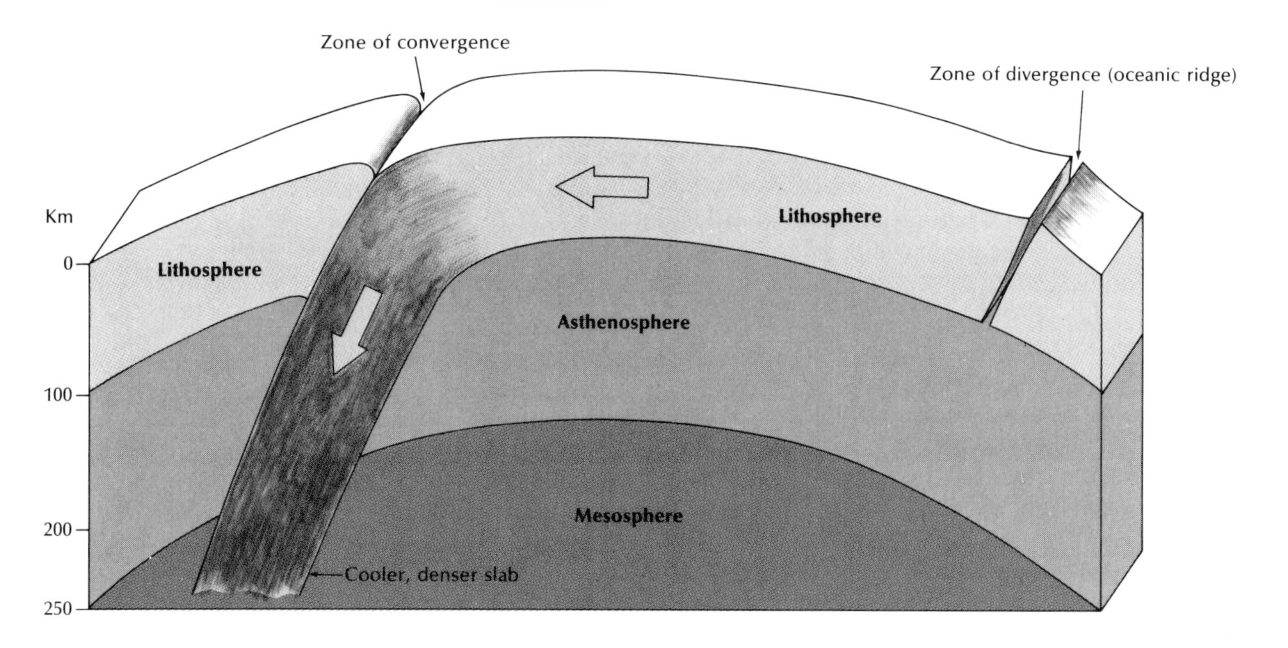

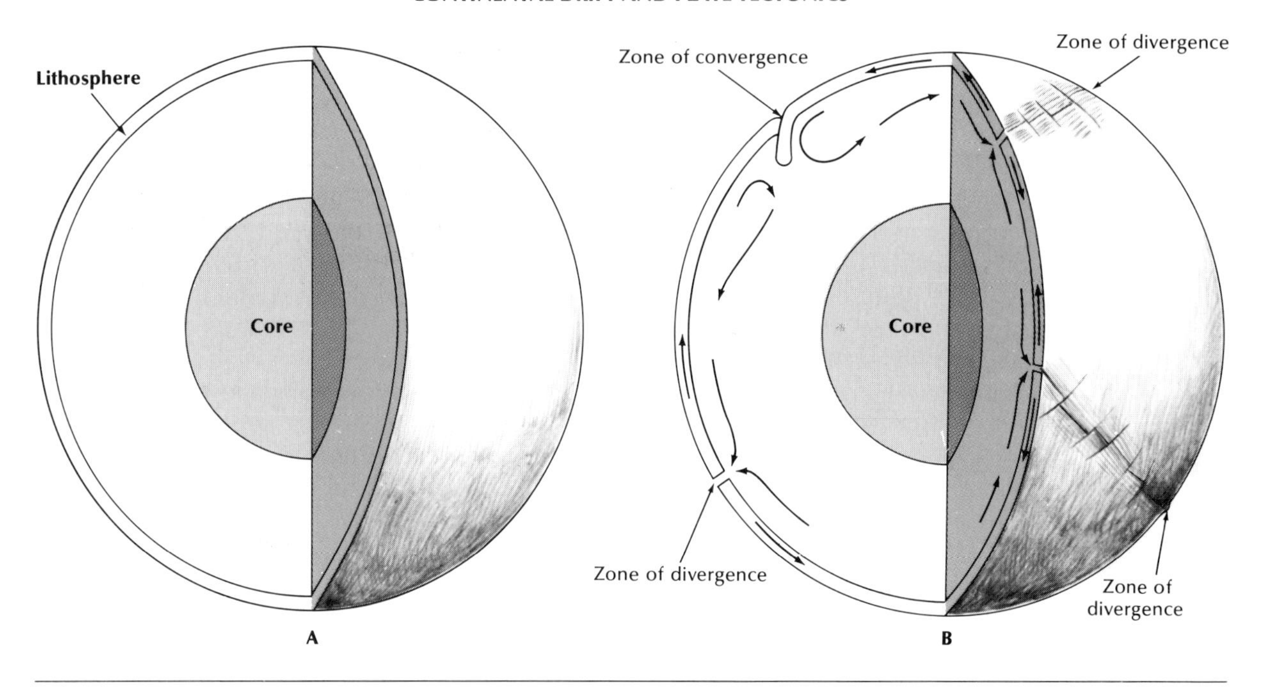

Fig. 20-27 A. Initial state of the earth. **B.** After expansion, the brittle lithosphere would crack apart and zones of divergence would form; as material flowed up to fill in the voids developing at the zones of divergence, mantle convection would be generated, creating the subduction zones.

vection really is the primary driving force, one might wonder also how the upwelling currents at the zones of divergence can be offset along transform faults for distances up to 1000 km (621 mi) or more.

Since the convection theory seems to present insurmountable difficulties, many geophysicists have turned toward alternative hypotheses. One is based on the relative differences in density between the descending lithospheric slab and the mantle into which it sinks (Fig. 20-26). It is thought that the cold, sinking slab is more dense than the mantle beneath and that the slab sinks under its own weight, thereby dragging the lithospheric plate along behind it, like an immense, flattened tail. Yet some geophysicists do not believe that the lithosphere is more dense than the mantle, and any differences that do exist would be small indeed. Moreover, in the plates diverging from the Atlantic and Indian oceans, there are no associated convergence boundaries and no sinking slabs, so it is hard to visualize how the mechanism could be responsible for plate motion in those two ocean basins.

Another theory now gaining support involves gravitational sliding of the lithospheric plates away from the topographically high oceanic ridge and rise crests. Opponents to that view, however, can point to several areas in the world where the downslope gradient apparently is inadequate for sliding to take place.

One other mechanism—expansion—deserves mention as a possible means of indirectly triggering plate movement, at least during the last 150 million years (Fig. 20-27). If the brittle lithosphere split apart as a result of expansion, then new material would have to flow from beneath and into the void to take up the space created by the splitting and would, in a way, effect some convection in the outer part of the mantle. But no one can really say whether or not the earth is expanding.

Almost all earth scientists subscribe to one or more of the above theories. Yet there are questions and problems concerning all of them. It well may be that the actual mechanism is not a simple one, but involves two or more processes working in unison or in tandem. Or perhaps the real mechanism is still not known. The forces work so slowly and over such large regions that it probably will be some time before a realistic model can be made.

Assessment of the plate tectonic theory

The theory of plate tectonics as it stands today is still relatively simple and elegant. By means of one philosophical approach it has been possible for earth scientists to explain reasonably many of the earth's internally generated geological features and processes. And that holds true both on a local and a regional scale. As a consequence, during the last several years, there has been an explosion of articles on plate tectonics.

In spite of the apparent success of the theory, most proponents readily admit that it still has deficiencies and inconsistencies and leaves some important questions unanswered.

One deficiency, as noted by the U.S. Geodynamics Committee, is that "the hypothesis has been more successful in providing explanations for phenomena in oceanic areas than in the continents." In a way this is understandable, since the theory was based largely on observations made in the ocean basins, which appear simpler in origin and relatively young (none older than Jurassic), whereas the continents reflect multiple rock-forming cycles that extend to ages of 3.5 billion years or more. Thus, continental processes have been spotlighted as one of the most important areas for investigation during the 1980s.

John Maxwell, an American structural geologist, points out that (1) there is a bilateral sym-

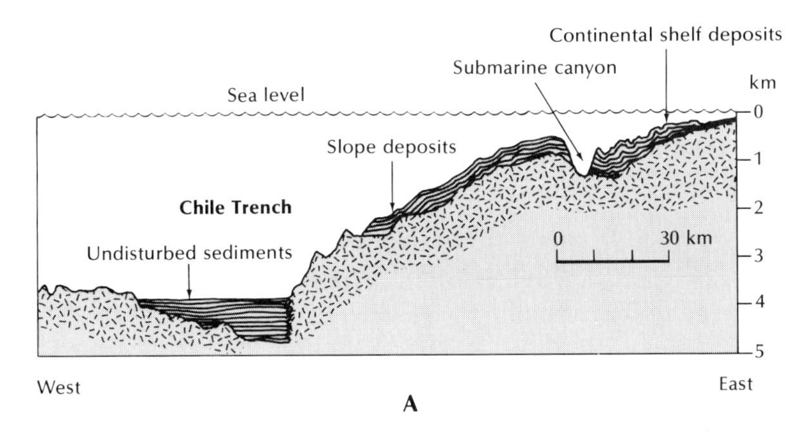

Fig. 20-28 Two deep-sea trenches showing undisturbed sediments in **(A)** the Chile Trench, and **(B)** the Aleutian Trench.

metry to the ridge systems and bordering ocean basins in the Atlantic and Indian oceans, but none in the Pacific, (2) the formation of new lithosphere occurs in all oceans, whereas its consumption takes place almost exclusively around the margin of the Pacific, and (3) many of the deep-sea trenches appear to have formed only recently and to be extensional, rather than compressional, in origin. Moreover, in some of the trenches the sediments are flat-lying (undisturbed), not folded and faulted as one might expect in an active zone of convergence (Fig. 20-28).

One structural feature that is inadequately explained by the plate tectonic theory is a set of active major strike-slip faults around the Pacific margin, such as the Alpine Fault in New Zealand, the Denali Fault in Alaska, the Atacama Fault in South America, and perhaps even the San Andreas Fault in California. Strike-slip motion on these faults, which parallel the linear zones of convergence, is at right angles to the motion of the converging plates (Fig. 20-29).

The origin of certain mountain ranges is also rather uncertain. For example, the prominent east-west mountain belt that includes the Alps and Himalayas appears to have been formed by north-south convergence of plates during the last 50 to 60 million years. Yet the evidence from the linear magnetic anomalies for that same time interval indicates that essentially all plate motion has been in an east-west direction.

According to the plate tectonic theory, the plates should act as if they were rigid and brittle. There are data, however (particularly from paleomagnetic studies), indicating that, in some instances, the plates are capable of large-scale bending. If this is true, some modification of the theory will be necessary.

Certainly, most geophysicists would admit that the tectonic situation in a wide band throughout the region of the Mediterranean Sea and eastward into southern Asia is difficult to categorize and describe in terms of plate tectonics. The deformation zone appears to be uncommonly wide and is composed of numerous microplates.

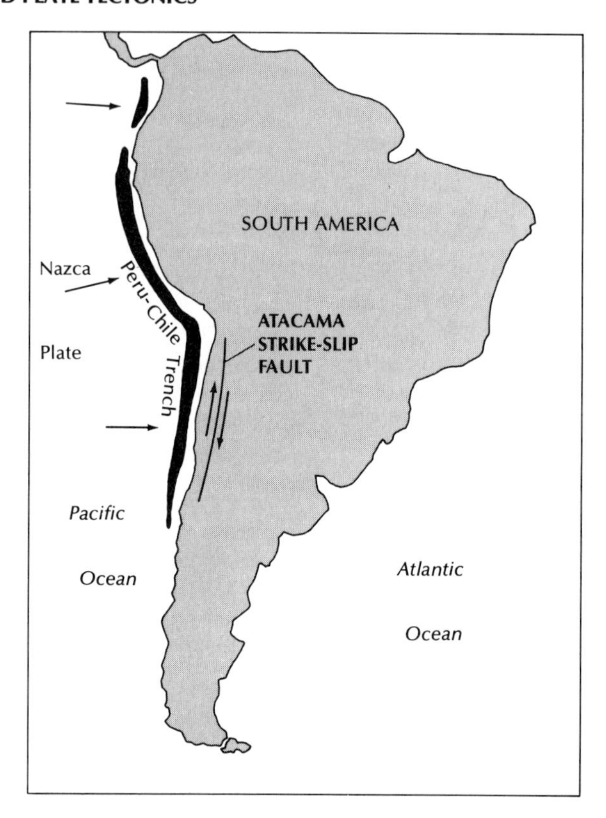

Fig. 20-29 South America, showing the Peru-Chile Trench and the Atacama strike-slip fault. The postulated movement of the Nazca Plate is shown by the three horizontal arrows. Notice that the direction of the plate motion is nearly at right angles to movement along the fault.

Questions also rise concerning the African Plate (Fig. 20-30). This plate is bounded on the west and on the east by active zones of divergence, and one might suppose that it is under compression and that a linear zone of convergence might form within it, parallel to the zones of divergence. In fact, the plate is being pulled apart, as evidenced by deformation, volcanism, and earthquakes in the East African Rift area. These conditions can be explained within the plate tectonic theory; however, to do so requires a delicate interplay between the relative lateral displacements of the Mid-Atlantic Ridge, East African Rift zone, and Mid-Indian rise, which has not yet been verified.

Some aspects of volcanism also raise questions. For example, andesitic volcanoes, according to the theory, are characteristically restricted to convergent zones. Yet active volcanoes in the Cascades of the Pacific Northwest, which resemble convergent-zone volcanoes in all other respects, are not associated with a well-defined active subduction zone. Similarly, the occurrence of numerous andesitic cones throughout the western United States during the last 30 million years or so, up to 1600 km (1000 mi) inland from the present coastal margin, is difficult to explain in terms of the plate tectonic theory.

Such deficiencies, inconsistencies, and questions require that further investigation of the nature of internal processes be made. And, as the theory is tested and reassessed, minor and perhaps even major revisions, a necessary part of the scientific process, will be required. Thus before a theory, particularly one as far-reaching as that of plate tectonics, can be widely accepted, it must be tested repeatedly in all of its parts through time.

One of the most significant tests will depend upon precise measurements between far-distant points to determine the inter-plate displacements. It now is possible to measure the distance between two points about 200 km (124 mi) apart with an error of only 2 cm (0.79 in.). Recently, such measurements have been made across the San Andreas Fault in an area near Los Angeles. The results, although only obtained over a two-year time span, indicate that the blocks on either side of the fault move jerkily, one with respect to the other, and that there has been twice as much displacement between points perpendicular to the fault trace than parallel to it. The last observation is somewhat unexpected and may merely reflect the short term over which the measurements were made.

By measuring distances between points in a network on different plates, we will be able to determine, unequivocally, the motions between and within plates. At that time, we will know whether the theory is basically sound or in need of major modifications.

Fig. 20-30 The African Plate and environs, showing relations of divergent, convergent, fracture, and deformation zones.

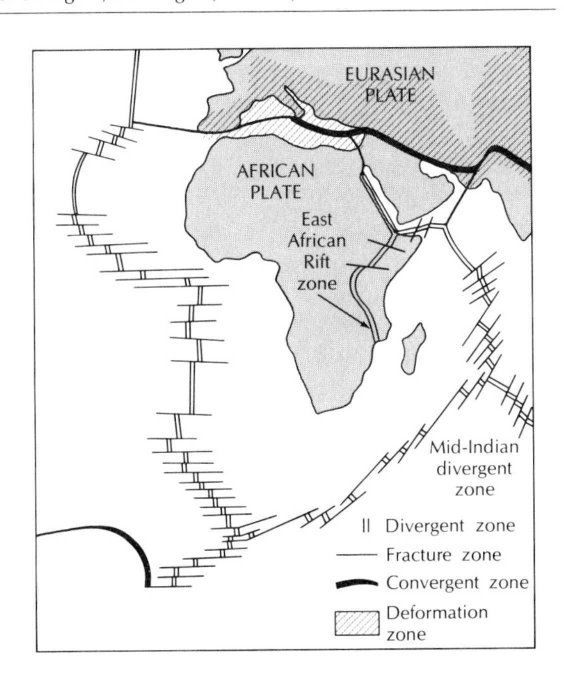

SUMMARY

1. Before the twentieth century, several theorists had suggested that some continents had split and drifted apart. In 1911, Alfred Wegener claimed that such drift did occur, and a great controversy over the theory of *continental drift* began.
2. The evidence that led to more acceptance of the hypothesis included fit of continental coastlines, stratigraphic and structural similarities between continents, paleoclimatic reconstructions and distribution of the late Paleozoic glacial deposits, distribution of similar fossil groups, paleomagnetic pole wander curves from different continents, linear magnetic anomalies over the oceanic ridges and rises, and age of cores from ocean sediments.
3. Most geologists believed in continental drift, but a unified theory was needed—one that would explain drift along with other in-

ternally generated phenomena, such as volcanism, mountain building, metamorphism, and seismicity.

4. *New global tectonics* was an attempt to provide a unified theory. The outer 100 km (62 mi) of the earth was defined as the *lithosphere,* the zone between 100 and 250 km (62 and 155 mi) as the *asthenosphere,* and the zone beneath 250 km as the *mesosphere.* The lithosphere is strong and brittle, the asthenosphere is relatively weak. New lithospheric material supposedly forms at the oceanic ridges and moves laterally away from the ridge systems toward the island arcs. At trench-island arc systems, lithospheric plates bend sharply over and plunge into the asthenosphere.
5. *Plate tectonics* is the most recent unifying concept. It is thought that the lithosphere is subdivided into a number of large plates that are moving. Some converge, some diverge, and some slide laterally past one another.

   Zones of divergence, where new lithospheric material is forming, are characterized by high heat flow, basaltic volcanism, and shallow-focus quakes.

   Zones of lateral movement (*transform faults*), which cross the ridges and rise nearly at right angles, are characterized by shallow-focus earthquakes.

   Zones of convergence are marked by deep oceanic tranches, andesitic volcanism, and a seismic zone, the *Benioff zone,* that extends down to 700 km (435 mi). Shallow-, intermediate-, and deep-focus quakes occur along that zone.
6. Modifications of the plate-tectonic theory include the recognition of *microplates,* back-arc upwelling, and mantle hot spots (*plumes*).
7. The mechanism that causes plates to move is unclear. *Mantle convection,* a density difference between the lithosphere and asthenosphere, gravity sliding, and expansion of the earth have all been suggested as possible candidates.
8. Some data do not fit into the plate-tectonic model convincingly.

QUESTIONS

1. What was Alfred Wegener's revolutionary idea? What led him to his conclusion?
2. What were the ideas used in opposition to Wegener's hypothesis? How well were they founded?
3. List and describe briefly all the evidence used to support continental drift up to 1960.
4. Describe new global tectonics. What was it based on? How did it differ from Wegener's theory?
5. What is the theory of plate tectonics, and what is its basis? How does it differ from new global tectonics?
6. Describe in detail, zones of convergence, divergence, and lateral movement.
7. What geological processes and features can be explained in terms of plate tectonics?
8. How has the plate tectonic theory been modified?
9. One of the problems with Wegener's initial ideas of continental drift was that it did not postulate a reasonable cause of the drift. Is this problem adequately addressed in the new plate tectonic theory?
10. What are some obvious flaws or inadequacies in the theory of plate tectonics?

SELECTED REFERENCES

Bird, J. M., and Isacks, B., eds., 1972, Plate tectonics, selected papers from the Journal of Geophysical Research, American Geophysical Union, Washington, D.C.

Beloussov, V. V., 1974, Sea-floor spreading and geologic reality *in* Plate tectonics: assessments and reassessments, C. F. Kahle, ed., American Association of Petroleum Geologists, Tulsa, pp. 155–66.

Bullard, E., 1969, The origin of the oceans, Scientific American, vol. 221, no. 3, pp. 66–75.

Cox, A., 1973, Plate tectonics and geomagnetic reversals, W. H. Freeman and Co., San Francisco.

Hallam, A., 1973, A revolution in earth science, Oxford University Press, Oxford.

Heirtzler, J. R., 1968, Sea-floor spreading, Scientific American, vol. 219, no. 6, pp. 60–70.

Hurley, P. M., 1968, The confirmation of continental drift, Scientific American, vol. 218, no. 10, p. 52.

James, D. E., 1973, The evolution of the Andes, Scientific American, vol. 229, pp. 60–69.

Marvin, W. B., 1973, Continental drift, The evolution of a concept, Smithsonian Institution Press, Washington, D.C.

Maxwell, J. C. 1974, The new global tectonics: an assessment, C. F. Kahle, ed., American Association of Petroleum Geologists, Tulsa, pp. 24–42.

Orowan, E., 1969, The origin of the oceanic ridges, Scientific American, vol. 221, no. 5, pp. 102–19.

Seyfert, C. K., and Sickin, L. A., 1973, Earth history and plate tectonics, Harper and Row, New York.

Sullivan, W., 1974, Continents in motion: the new earth debate, McGraw-Hill Book Vo., New York.

Takeuchi, H., Uyeda, S., Kanamori, H., 1970, Debate about the earth, revised edition, Freeman, Cooper, and Co., San Francisco.

Tarling, D., and Tarling, M., 1971, Continental drift, Doubleday and Co., Garden City, N. Y.

Toksöz, M. N., 1975, The subduction of the lithosphere, Scientific American, vol. 233, pp. 88–101.

Vine, F. J., and Matthews, D. H., 1963, Magnetic anomalies over oceanic ridges, Nature, vol. 199, p. 947.

Wegener, A., 1966, The origin of continents and oceans, Dover Publications, New York.

Wilson, J. T., ed., 1976, Continents adrift and continents aground, W. H. Freeman and Co., San Francisco.

Fig. 21-1 Oil derricks, Wheeler Ridge, southern California. *Exxon*

21

RESOURCES AND ENERGY

In recent years, resources have become a vital issue, as we have become increasingly aware that the energy sources and the raw materials upon which our society depends are not infinite. Western nations, in particular, and industrial nations, in general, must buy many resources from other nations—which are often developing countries—and this adds a political dimension to the problem of supply and demand. The role geologists must play, as energy sources and metal and mineral deposits are depleted, is an important one, for their task is to find and evaluate new supplies. Only during the past several decades have earth scientists and others quantitatively estimated the remaining resources of this planet. If their estimates are accurate, or even close to being accurate, it is evident that it may be necessary to change our way of life considerably and without too much delay.

In this chapter, we will discuss basic resources, their geological occurrences, and the problem of dwindling supplies relative to their geological setting as well as to the social issues of today.

BACKGROUND INFORMATION

By late 1980, Al Bartlett, a professor of physics at the University of Colorado, had given a speech on growth and resources 500 times and published the same information in two scientific journals. Bartlett has devoted much time and energy to explaining, in a simple and effective way, how economic growth affects resource usage, population, and life-spans. We will refer to the concepts he expounds and the examples he uses throughout this chapter, along with statements and reports from scientists and others in government and industry.

Rate of growth

When anything grows or increases at a fixed percentage rate, its growth is said to be *exponential,* and a curve plotted to show its growth rate will steepen with time. This idea should not be new to the reader. In Chapter 1, we discussed the decay of radioactive elements in relation to carbon-14 dating—a case in which change is exponential, although in the opposite direction (decrease) as that of growth. A good example of exponential growth is a savings account yielding 5 per cent interest yearly. If the interest is added to the account, the interest rate allows the principal to double every 14 years. If the interest rate is increased to 10 per cent, the principal will double in seven years, or one-half the time. Thus, a deposit of $1000 would grow to $2000 in seven years, to $4000 in 14 years, to $8000 in 21 years, and to more than $1,000,000 in 70 years. One can determine doubling time by using the equation

$$T = 70/P$$

where T is the doubling time in years and P the percentage growth per year.

Another example of exponential growth is population growth. In 1975, the annual world rate was estimated to be 1.9 per cent, apparently a low rate, but one that translates to a doubling of the world's population every 36 years. At this rate, population density would reach one person per square meter—excluding the Antarctic continent—in 550 years, if lifespan remains the same. Figure 21-2, a characteristically steep exponential growth curve, shows the increase in population in the United States since 1790, when the first census was taken, to 1980.

It is important to understand that while population is increasing, so too is the per capita demand for resources. Add to this the demand from countries hoping to develop industries, and one of the major problems of the future comes into focus: one exponential growth process (population) feeding another (resource use). Predicting when a population crisis will arise is difficult. Bartlett, as an illustration, hypothesizes a strain of bacteria that doubles its population every minute, and an empty bottle, which represents a non-renewable resource. A bacterium placed in the bottle multiplies.

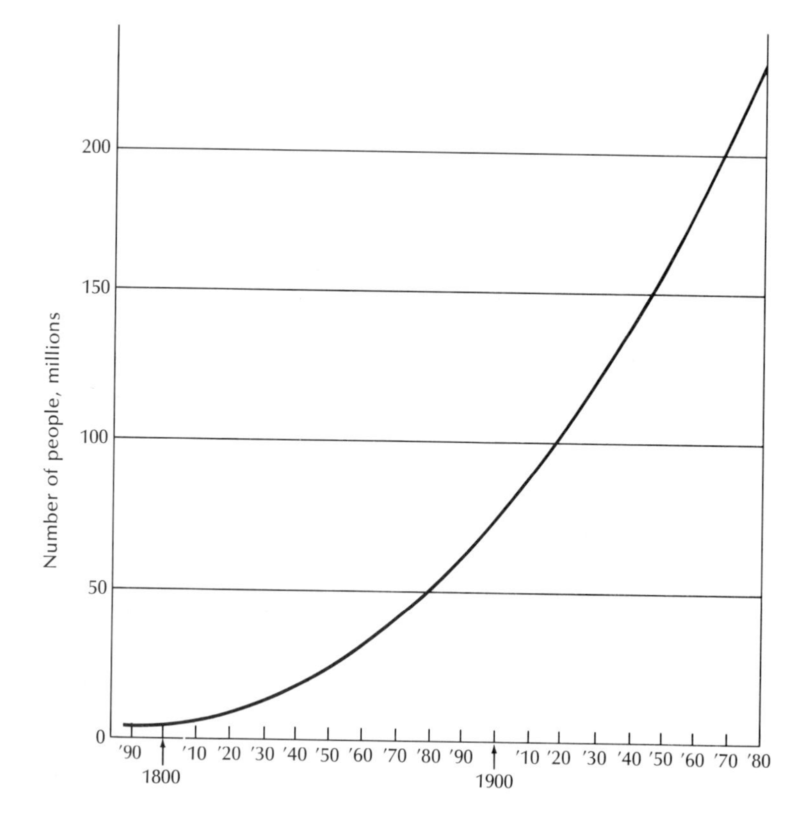

Fig. 21-2 Increase in population in the United States since 1790.

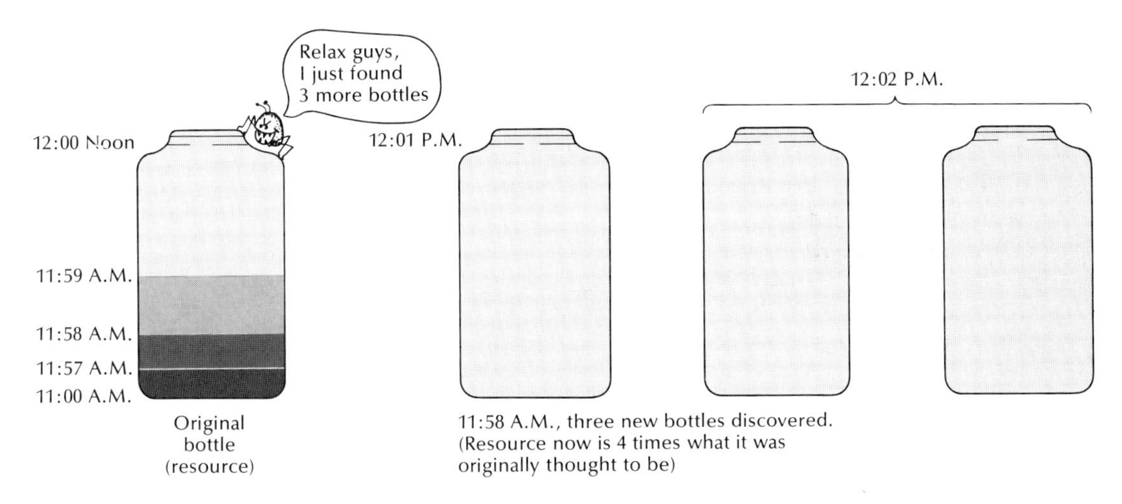

Fig. 21-3 Bartlett's example of steady growth: Bacteria as the population, or demand, and bottles as the resource.

When the bottle is full, the resource is gone (Fig. 21-3). If the bacterium is placed in the bottle at 11 A.M., the bottle will be filled to the brim by 12 noon. Only in the last few minutes before noon, however, does it become apparent that only a small amount of the resource remains. At 11:58 A.M., three more bottles are provided—a fourfold increase in resources in a little less than an hour. Yet, by 12:02 P.M., after only four more doubling times, all the bottles are full of bacteria, and again the resource is gone. The point to be made is that, given exponential consumption of a resource, that resource can be consumed in a very short time.

Bartlett's illustration makes several other points. First, it emphasizes our position in relation to resources, so that we may plan for the future. Second, it cautions us to view announcements of newly found or greatly increased resources in perspective, for in reality, they will not last long at an exponential consumption rate. For example, if the recently discovered Alaskan oil fields were our only source of oil, at the current rate of consumption, the proved reserves would last little more than a year, whereas the estimated amount of oil present might last only 6 or 7 years. Finally, Bartlett's illustration points up the fallacy of calculating resource use from current consumption, without taking growth into consideration.

A few more figures may help to underline the problem. Between 1972 and 1978, energy production in the United States decreased by 3 per cent, yet consumption increased by 9 per cent. To meet our increased demands, we imported more oil. Energy consumption worldwide is also rising, with an increase of three to four times the 1980 consumption forecast by the year 2010.

RESOURCES

Definitions of basic terms

By ***resource,*** we mean a commodity useful or commonly essential to people and thus to industry and to our life style. ***Reserve*** refers to the estimated amount of an identified resource recoverable with present-day technology (Fig. 21-4). However, reserves are usually a small fraction of the total resource. It is possible, though not always probable that undiscovered resources may be a sizable proportion of the total. A ***paramarginal*** resource is expensive to recover; the profit margin of its recovery is marginal, but recovery is feasible, technologically. Recovery of a ***submarginal*** resource is not profitable; it would involve financial loss un-

	Identified Resources			Undiscovered Resources	
	Proved	Probable	Possible	In known districts	In undiscovered districts
Recoverable	Reserves			Hypothetical Resources	Speculative Resources
Paramarginal	Identified Paramarginal and Submarginal Resources				
Submarginal					

← Degree of uncertainty

Feasibility of economic recovery →

Explanation

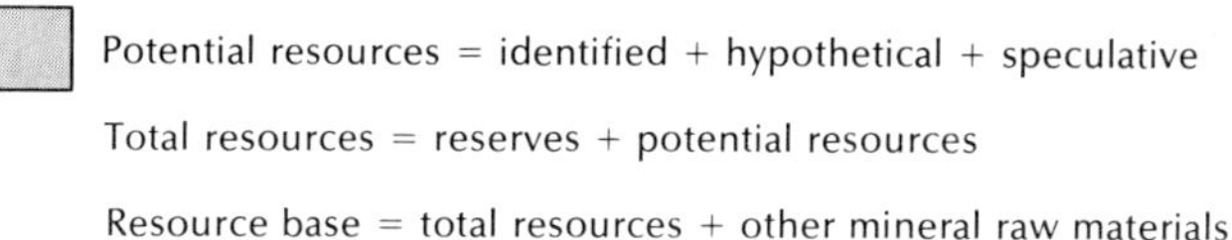

Fig. 21-4 Diagram, showing categories of a resource. The area of each box is proportional to the amount of the resource so classified.

less the technology could be improved or unless the value of the resource increased tremendously. The reworking of dumps from the placer-mining of gold illustrates the latter type of resource: an improved technology and a huge jump in the price of gold. The incentive to recover almost any resource is economic.

Renewable and non-renewable resources

Resources are divided into two main groups: renewable and non-renewable. ***Renewable*** resources may be replenished, and if carefully managed, continue to be available. Examples are farm produce and forest products. ***Non-renewable*** resources—the common ore minerals and fossil fuels, for example—once used, are gone forever. Within the latter group, some resources, such as metals, can sometimes be recycled, whereas others, such as oil, cannot be retrieved after use. Depending on how they are used, some resources can be placed in either category. In many areas, groundwater can be an important renewable resource, but only if it is replenished at the same or higher rate at which it is pumped from the earth. In many dry areas of the United States (parts of California and Arizona and the Great Plains), groundwater is being withdrawn faster than it is being renewed, causing the water table to drop at an alarming rate.

Similarly, farmland and forests can be exploited or they can be used judiciously so that they can renew themselves. If we allow our soils to erode, or our valuable croplands to be swallowed up in shopping centers or suburban development, or if we demolish our forests indiscriminately, the damage is irreversible. For example, the area paved over with asphalt since 1945 is about equal to the area of the state of Ohio—much of it good farmland that virtually cannot be retrieved.

What is a resource?

Most of the useful materials currently listed as resources can be found in ordinary rock. Gold and phosphorus, and even nuclear energy, are locked up in common basalt, for example, but not enough to make jewelry, fertilize a garden, or keep one warm on a cold winter's night. A resource represents a profitable concentration of a particular material. Some elements are found in concentrations only several times its average crustal abundance; the concentrations of other elements may be many thousands of times that (Table 21-1). The largest deposits, as well as the number of deposits of, say, a particular ore, correlate well with its crustal abundance (Fig. 21-5). It follows, then, that resources are unevenly distributed worldwide.

Each resource reflects one or more specific geological or biological processes, or both. The uneven occurrence of such processes explains why, for example, Saudi Arabia has an abundance of oil, yet nearby Egypt and Israel have little, or why some tiny Pacific islands supply vast quantities of phosphate for fertilizers (Fig. 21-6), from bird droppings, yet nearby islands with large bird populations supply none.

Table 21-1. Crustal abundance of some common ore-forming elements, and the concentration factor that indicates whether their mining would be economical

Element	*Crustal abundance, per cent weight*	*Concentration factor*
Aluminum	8.00	3–4
Iron	5.8	5–10
Copper	0.0058	80–100
Nickel	0.0072	150
Zinc	0.0082	300
Uranium	0.00016	1,200
Lead	0.00010	4,000
Gold	0.0000002	8,000
Mercury	0.000002	100,000

Data compiled from Skinner, 1969; Brobst and Pratt, 1973; and Howard and Remson, 1978

Locating economically profitable deposits of minerals and other resources is sometimes easy, sometimes not so easy. The holes that dot part of the Nevada landscape, for example, suggest that there are simple ways to search for an ore deposit. Other times, complex models must be used to establish an exploration strategy when location is difficult. And increasingly, economic factors affect the decision to

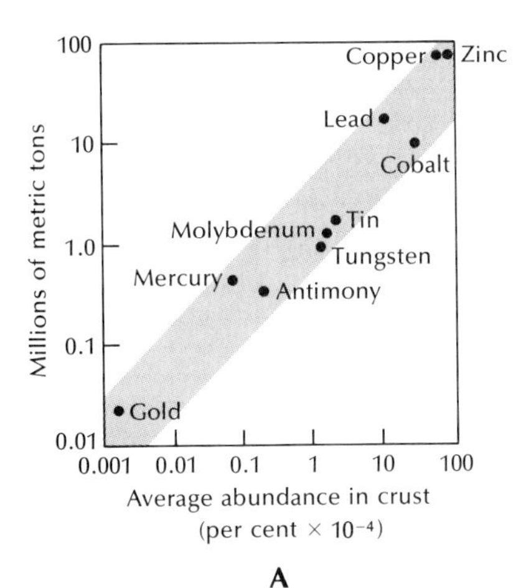

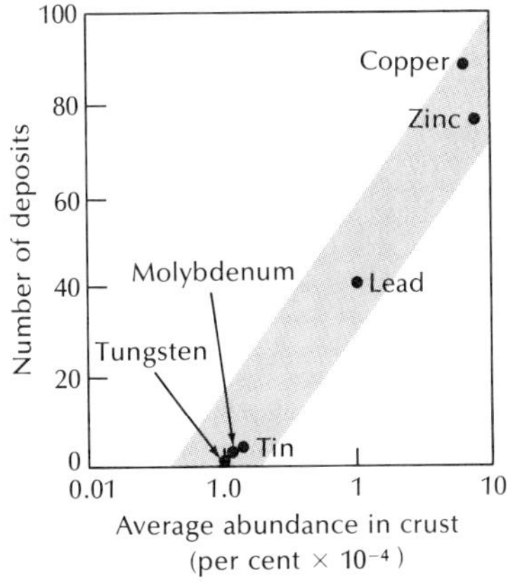

Fig. 21-5 Crustal abundance of scarce metals compared with **(A)** the largest known ore deposit of the metal; **(B)** the number of deposits.

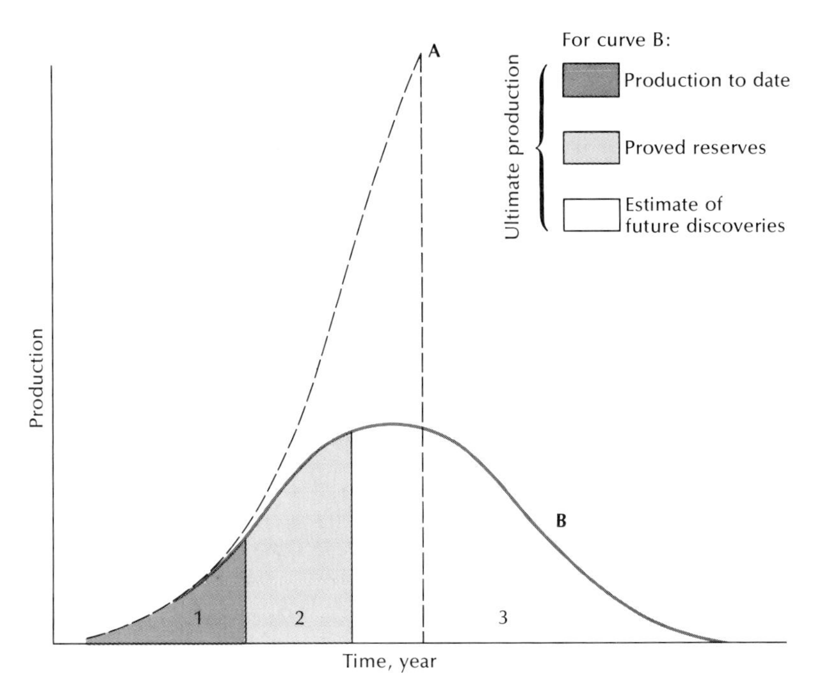

Fig. 21-7 Two curves, showing predicted production by year. The area under curve A is approximately equal to the area under curve B.

exploit a particular deposit. Extracting oil from oil shale, for example, is much more expensive and more energy and water intensive than offshore drilling of oil because the yield is so low. Another factor is the expense of transporting material to a processing plant or to the market.

So many variables are involved in determining resource reserves—rate of use, economic value, new technology—that estimates should be used only as guides to help us plan for the future, not as predictions of what the future might hold.

Rise and fall in non-renewable resource production

Assuming that many resources are non-renewable, we can predict how long a particular resource will last. Figure 21-7 shows two curves that predict resource production, in years. Curve A represents exponential growth, up to the day the resource is exhausted; so far as we know, such growth cannot be sustained. Taking fossil fuel oil as an example, a prominent geologist, M. King Hubbert, projects a more or less bell-shaped curve (curve B) with three parts: (1) total production to date, (2) proved reserves, and finally (3) hoped-for discoveries of further reserves. The area under curve B must equal the final total production of the resource under consideration. The curve falls to zero, even though material is still being extracted, because fewer deposits are being discovered or because the deposits are of low-grade ore or are more difficult to reach. However, major discoveries can alter the shape of the curve slightly. These curves have been used to predict peak production and the point at which the resource will be exhausted. Coal or oil curves show that even if the estimate of the amount of resource present is doubled, its life-span is changed but little, considering the rate of population growth and the increase in per capita demand (Fig. 21-8). Similar curves

Fig. 21-6 The per capita income of Nauru, an independent island-state in the South Pacific, is one of the world's highest. The source of the island's wealth, phosphate fertilizer, is shown here being mined; as the mining proceeds, large pillar-like carbonate structures are left behind. The fertilizer supply will last only a few more years, so Nauru is investing part of its income in real estate in Australia and in other ventures that will yield money for the future. *Peter W. Birkeland*

can also be constructed to predict the life-span of ore deposits.

Metallic resources—origins and availability

Metals have changed the course of history, so much so that the major prehistoric periods have been named for the metals that were developed and widely used during those periods. After the Stone Age came the Bronze Age, then the Iron Age. It has been observed, however, that we might still be in the Stone Age: based on tonnage of materials used in the United States, stone and gravel may well be our main resource. Of the many minerals that make up the earth, only about one in 20 is used by humans.

When we think of metal-containing ores, a prospector washing or panning gravel for gold comes to mind (Fig. 21-9), but although gold is a valuable resource, other resources are more important. Thus, a dusty gravel pit located near an urban area can be a far more valuable resource than a diamond mine.

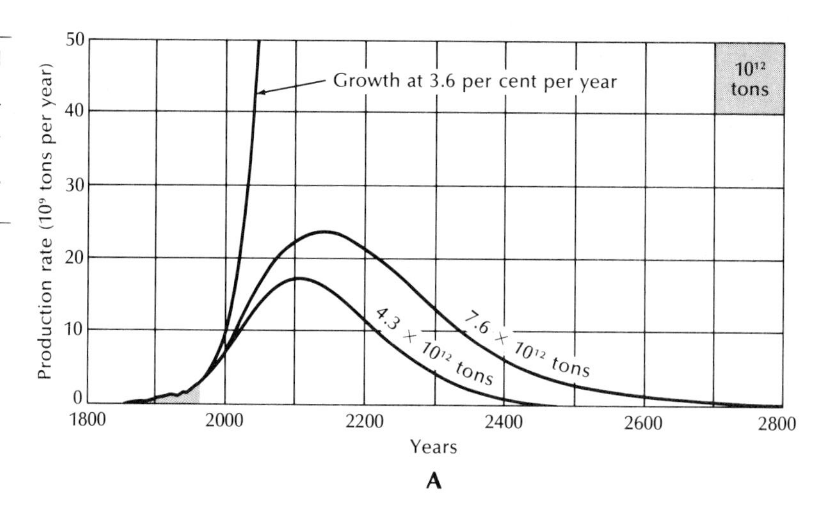

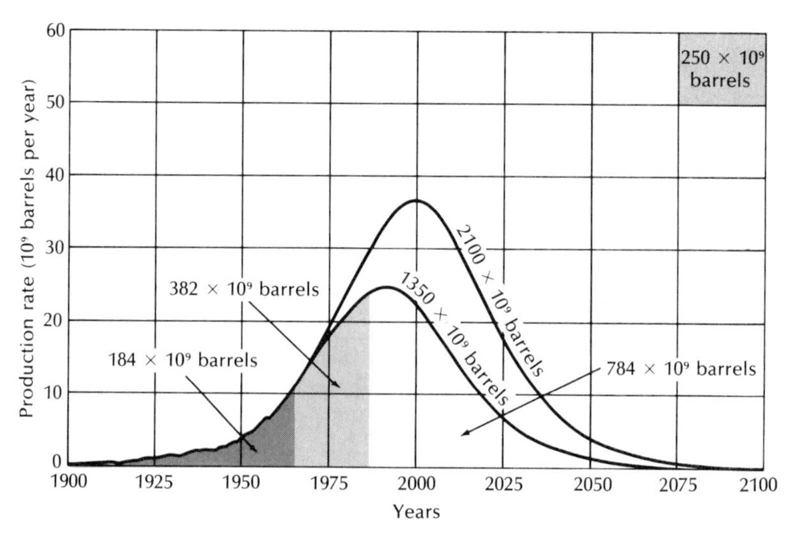

Fig. 21-8 Predicted cycles of world coal production (**A**) and crude-oil production (**B**), showing estimates of ultimate production for each resource. The steepest trend line on **A** represents the present rate of growth of coal production. Note how quickly the resource is depleted at that rate.

Fig. 21-9 Washing gravel for gold in 1898, on the banks of the Saskatchewan River, near Edmonton, Canada. *Geological Survey of Canada*

There are many complex ore-containing minerals, some of which are listed in Appendix I (Table 1). Most elements can combine with oxygen to form oxides (manganese, tin, uranium) or with sulfur to form sulfides (zinc, molybdenum, lead). Some elements, such as iron, can form both an oxide (e.g., magnetite) and sulfides (e.g., pyrite). Still others do not combine with other elements, but remain in what is called their ***native state.*** Familiar examples are gold, silver, and platinum.

Metallic ore deposits can be classified on the basis of abundance in the earth's crust. Two categories are recognized, ***abundant*** and ***scarce;*** if they are over 0.1 per cent of the total, they are abundant, if they are under, they are scarce. Iron, aluminum, manganese, magnesium, and titanium are the abundant metals; all the others are scarce. Deposits of scarce metals are both smaller and fewer (see Fig. 21-5).

THE GEOLOGICAL ORIGINS OF RESOURCES

Igneous and metamorphic processes

Igneous and metamorphic processes are important in the formation of many of the world's ore deposits (Fig. 21-10). Within a magma, several processes might occur to produce an ore deposit. Perhaps the simplest one is the separation of heavy minerals, some of which form early in the crystallization of the magma, settle to the bottom, and form a broad layer. This seems to have occurred in Canada, at Sudbury, Ontario, a major mining district, where a body of magma intruded between the basement rock and overlying sediments. Another important differentiating mechanism is the separation of liquids within the magma followed by crystallization. Deposits so formed contain minerals rich in iron, chromium, titanium, copper, and sulfur. Pegmatites can crystallize from the more volatile constituents of magma or from a very fluid magma, and mimic their granitic parent in their mineral content. They may contain such rare elements as lithium, boron, fluorine, or uranium. The rock is very coarse grained, and some single crystals are up to 15 m (49 ft) long.

Ores also may be formed by contact metamorphism, which occurs when a body of magma, commonly intermediate in composition, intrudes country rock. The alteration can range from the rearrangement of existing materials to complete replacement, by the addition of materials from the igneous source; the zone can be narrow or several kilometers wide. Limestone is a common host to ore deposits; the limestone is removed in the circulating hot solutions. Probably the most common material added to contact metamorphic zones is silica, which can either combine with other elements to form a variety of silicate minerals or merely replace the host materials to form quartz or chert. The ores commonly found at contact zones are simple oxides and sulfides of iron, zinc, copper, and molybdenum.

Contact features are well displayed in the Iron Springs District of southwestern Utah. Magma of intermediate composition intruded a body of limestone, cooled, and forced iron-enriched solutions into the limestone to form iron ore bodies up to 70 m (230 ft) thick. For every million tons of iron introduced to the limestone, 40,000 tons of silicon, 20,000 tons of magnesium, and 10,000 tons of aluminum were also introduced.

Ores may also be precipitated from fluids, in which case they are called ***hydrothermal*** deposits (see Fig. 21-10). Hot springs and geysers, the surface manifestations of this process, may be associated with nearby intrusive bodies. Hot solutions contain high concentrations of dissolved materials, and as chemical conditions and temperatures change, ore is deposited either in veins, as the fluids follow joints or faults, or as irregularly shaped bodies or very

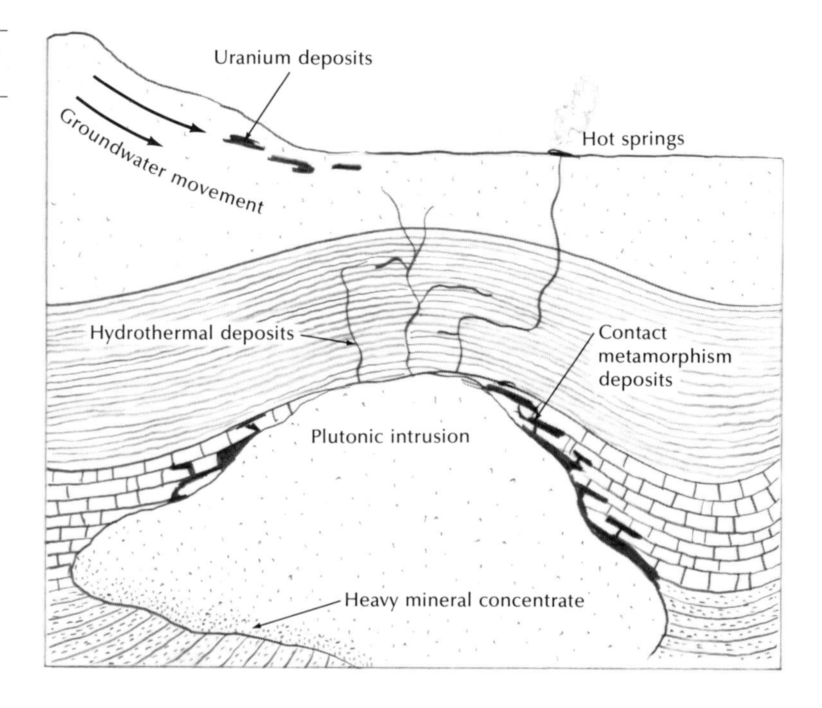

Fig. 21-10 Schematic diagram, showing the realms of various kinds of ore deposits.

Fig. 21-11 The open pit copper mine at Bingham Canyon, Utah.

small particles disseminated through the host rock. Discrete concentrations of ore are often easy to find, but disseminated ores, if they are fine grained, are more difficult to find.

Some ore deposits are precipitated from fairly cool solutions. Common are the uranium and vanadium deposits found in the sedimentary rocks of the Colorado Plateau, which are important future sources of fuel for atomic reactors. It is interesting to note that the organic matter in the sediments helps to create chemical conditions conducive to the precipitation of these minerals. Some prospectors have found buried logs almost completely replaced by uranium ore.

Porphyry copper, made famous by the huge mine at Bingham Canyon, Utah, is another example of a hydrothermal ore deposit (Fig. 21-11). In the Bingham deposit, ore is disseminated throughout a porphyritic plutonic rock. The magma moved to quite shallow depths, where, late in the intrusive stage, shattering occurred, creating extensive fracture systems in which ore minerals were later precipitated.

Although the amount of copper in these deposits is usually less than 1 per cent, relatively inexpensive mining techniques have made the extraction of such ores profitable.

In recent years, unusual hydrothermal deposits have been found in the seas and oceans. In the mid-1960s, newly discovered pools of hot brine (56°C; 133°F) at the bottom of the Red Sea, which has salinity second only to that of the Dead Sea, were found to be rich in many elements. The sediment on the sea floor is described as a black tar-like ooze, too hot to touch, containing the oxide and sulfide ore minerals of iron, manganese, zinc, and copper. Even more recently, scientists in the submersible *Alvin* have observed geysers at a depth of 2622 m (8600 ft) on the East Pacific Rise off the coast of South America (Fig. 21-12). The superheated water (over 300°C; 600°F) melted part of the *Alvin*'s temperature sensor. The ocean floor was multicolored, carpeted with sulfide minerals rich in copper, iron, zinc, cobalt, lead, silver, and cadmium.

The plate tectonics theory provides one explanation for such concentrations. Where plates are diverging, as at the mid-ocean ridges, magma ascends to fill the voids thus created. The cool oceanic water penetrates the cracked, newly formed crust some distance from the spreading center and migrates toward it; heating occurs, and the water rises into the overlying rocks (Fig. 21-13). As it flows through the crust, the water dissolves quantities of metals from the host rock. This dissolution process

Fig. 21-12 Black smoker geyser, East Pacific Rise, 1979. *Dudley Foster, WHO*

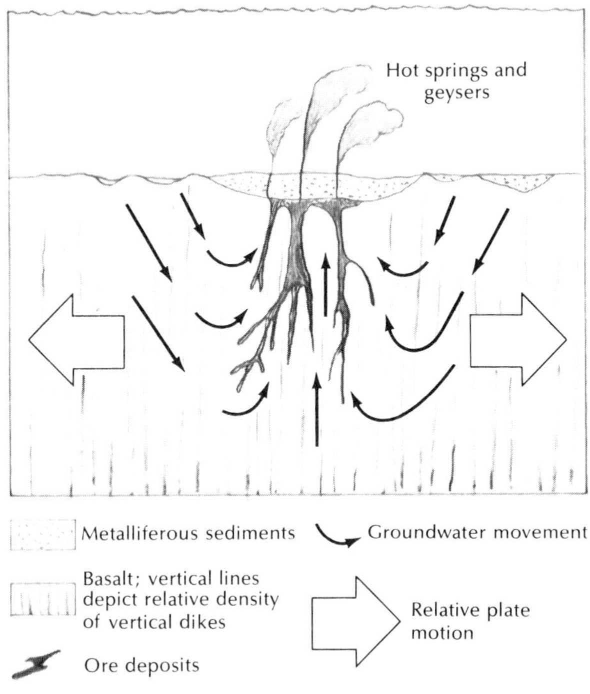

Fig. 21-13 Diagram, depicting deposits, groundwater motion, and hydrothermal activity at a mid-ocean ridge. Because the lavas were extruded beneath water, pillow structures are common.

is aided by the high chloride content of sea water. On rising, the ore is precipitated either as veins or as masses in the rock or directly on the ocean floor.

That some ores can originate at diverging plate boundaries is difficult to prove conclusively. If, as is thought, the plates are later consumed in a subduction trench, all evidence of such an origin would be lost. Yet, earth scientists think that they have found on the Mediterranean island of Cyprus a slab of ancient ocean floor laced with the original ore deposit. Cyprus, which has a long history of copper mining extending back to the Bronze Age, lies along a continental collision zone, but the rock containing the copper ore seems originally to have been a spreading center, and its cross section is similar to that shown in Figure 21-13. During subsequent continental convergence, a slab of ocean floor was upthrust to its present position on the island.

The plate tectonics theory can also explain the occurrence of the major ore deposits found in relatively young folded mountain belts (Fig. 21-14). For example, a generalized model has been used to explain the east-west zonation of ores in the Andes, which include the largest known porphyry copper deposit in the world. It is thought that the ores were first concentrated at a mid-ocean ridge (Fig. 21-15), then transported as part of an oceanic slab toward a subduction zone, carried downward, and melted. Eventually, the molten ores worked their way back toward the surface as hydrothermal deposits.

Island arcs are also areas with an ore-plate tectonic connection. In Japan, the sedimentary deposits enclosing ore bodies are of shallow-water marine origin, and the ores themselves are hydrothermal and associated with nearby arc volcanism.

Some major hydrothermal mineral deposits, such as the lead-zinc deposits of the Mississippi River Valley, however, are not associated with plate boundaries. Although the origin of these deposits is not clear, some theorists ascribe them to heat derived from a mantle-upwelling system accompanied by a spreading center that could show up as tension faulting of the continental mass.

The reader should bear in mind that the theories just described need further testing. Their application, however, has led to the discovery of ore bodies not otherwise suspected, and more discoveries may follow.

Sedimentary processes

Minerals and metals are concentrated by mechanical and chemical sedimentary processes to form ore deposits: the familiar ***placer*** deposits formed by a concentration of heavy mineral particles from weathered rocks, such as gold, which may be deposited in pockets along a river channel or in the sand of a beach. The Forty-niners who panned gold during the California Gold Rush washed sand and gravel in a pan to concentrate the metal—a process similar to that occurring in a stream. Diamonds found in beach placers were first brought to

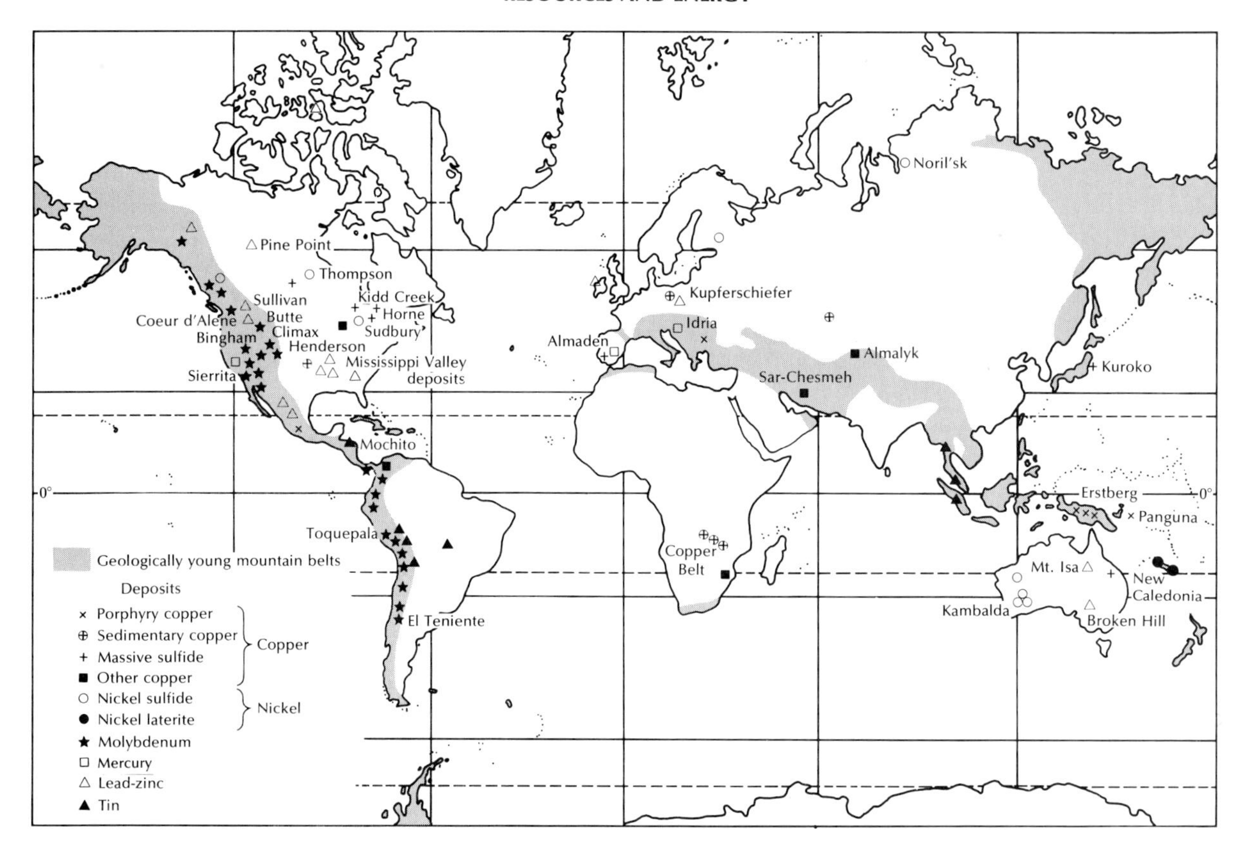

Fig. 21-14 The association between mountain belts formed during the last 200 million years and major deposits of the scarce metals.

the surface in narrow kimberlite pipes (see Chap. 5).

An unusual sedimentary deposit found in places on the deep ocean floor takes the form of large nodules of manganese dioxide (MnO_2)—some of them the size of cannon balls (Fig. 21-16)—first described a century ago by scientists on the HMS *Challenger* (see Chap. 15). The nodules, which superficially resemble cobbles or boulders, grew slowly, by accretion of ion upon ion on a simple substrate, such as a shark's tooth, over a period of millions of years. Estimates of their rate of growth vary from 1 to 100 mm (0.04 to 4 in.) per million years. Their origin is somewhat obscure, but precipitation was probably brought about by a combination of inorganic and organic processes. The nodules also contain appreciable amounts of iron, cobalt, nickel, and copper, and are so abundant in some places that work is now progressing on developing a technology for their recovery, although problems of legal ownership and the environmental impact of mining the ocean floor have not been resolved. It is estimated that billions of tons of nodules lie on the ocean floor—amounting to the largest mineral deposit on earth.

Other chemically precipitated sedimentary ores are the extensive banded iron deposits, some hundreds of kilometers across, that formed in Precambian seas. In these deposits, layers of iron commonly alternate with layers of silica (Fig. 21-17). Geologists are puzzled by their origin—no similar deposits are forming anywhere on earth today—and it is not clear how the iron reached the ancient seas, since

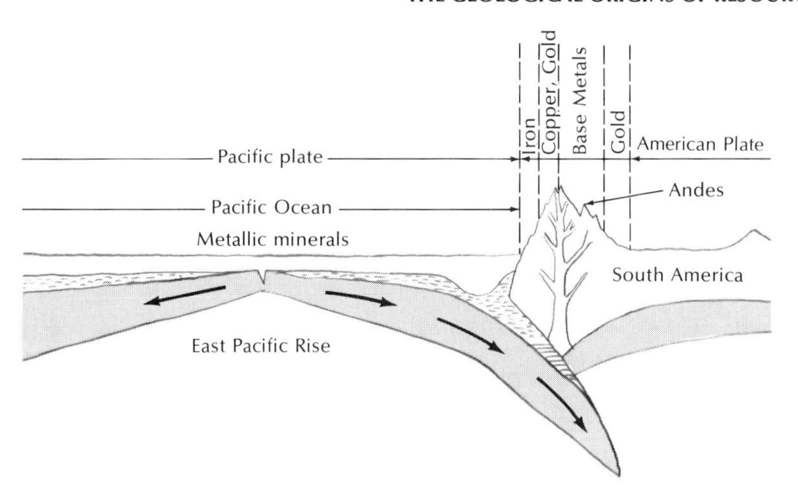

Fig. 21-15 Cross section from the East Pacific Rise to South America, showing how the ore minerals originally formed in the sediments and volcanic rocks at the rise were remobilized and then ascended into the overlying rocks of the Andes during a subduction phase.

iron released by the weathering of rocks is usually oxidized and trapped close to its site of release. However, it is hypothesized that in Precambrian times the atmosphere contained less oxygen, so that iron, instead of oxidizing, was carried in solution by rivers to the shallow seas, where it was precipitated as a nearly pure metal. The oxygen required for precipitation could have been derived from photosynthesis by the primitive organisms of that time.

Weathering processes

Weathering can help to produce soil, a major resource, but under special conditions it can also form an ore body. In humid tropical areas, soluble elements, such as potassium, calcium, and magnesium, are rapidly leached from the soil, as is silicon. What is left is a residue of insoluble materials containing sufficient quantities of nickel and iron and aluminum oxides to be designated ores.

In other places, weathering has solubilized metals such as silver and copper, which have been carried downward—sometimes to great depths—where the chemical conditions at the water table cause them to precipitate. This process is important because it increases ore proportions in the deposit, but it has also led investigators to look for disseminated ore bodies

Fig. 21-16 Manganese nodules, of cannonball size, at a depth of 5417 m (17,768 ft) in the southwest Pacific. *Smithsonian Oceanographic Sorting Center*

Table 21-2. Data on the principal metal ore resources

Element	Main Use	*Main process of ore formation*						*United States **		*World **	
		Igneous	*Contact metamorphic*	*Hydrothermal*	*Sedimentary*	*Weathering*	*Other*	*Reserves*	*Recoverable resource potential*	*Reserves*	*Recoverable resource potential*
ABUNDANT METALS											
Aluminum	Light-weight containers					X		8.1	203,000	1,160	3,519
Iron	Steel	X	X		X	X		1,800	118,000	87,000	2,035
Magnesium	Alloy, chemicals						Sea water	Unlimited			
Manganese	Steel				X	X		1	2,450	630	42
Titanium	White pigment in paints, alloy	X			X			25	13,000	117	225
SCARCE METALS											
Copper	Electrical wiring		X	X	X			77.8	122	200	2,120
Chromium	Alloy	X						1.8	189	696	3.26
Gold	Currency, jewelry			X	X			0.002	0.0086	0.011	0.15
Lead	Batteries, gasoline		X	X	X			31.8	31.8	0.54	550
Molybdenum	Alloy			X				2.8	2.7	2	46.6
Nickel	Alloy	X				X		0.18	149	68	2,590
Platinum	Metal, chemicals	X			X			0.00012	0.07	0.009	1.2
Silver	Currency, photographic process			X				0.05	0.16	0.16	2.75
Tin	Metal plating, alloy			X	X			—	3.9	5.8	68
Uranium	Weapons, energy			X	X			0.27	5.4	0.83	93
Zinc	Alloy, anti-corrosive			X				31.6	198	81	3,400

*Data are for millions of tons.
Source: USGS Prof. Paper 820, 1973.

in the seemingly less enriched rocks of the surrounding area.

MINERAL RESOURCES AND THE FUTURE

It is difficult to determine the extent of the world's mineral reserves and potential resources (Table 21-2) and even more difficult to predict how long they will last. The life-span of a resource depends upon geological factors, advances in mining and extraction, new technological applications, the economic and political climate of the times, and demand. The figures in Table 21-2, when compared with the production of some metals in the recent past (Table 21-3), provide a base for roughly estimating the life-span of a known resource. The predictions of Charles Park, Jr., noted economic geologist, should give us an idea of how long commonly used minerals will last. He forecasts plentiful supplies of iron, aluminum, and manganese far into the future; enough uranium for about a generation; and shortages of lead, zinc, and tungsten by the end of the century. He notes that silver and gold are already in short supply.

How does the United States fare, with respect to the minerals needed to run its industries? Table 21-4 shows that we are importing a large proportion of the minerals and the metals we need from a number of countries, but at the same time, we are depleting our national resources at an alarming rate. It has been evident for some time now that some of the countries we rely on for materials are politically unstable or unfriendly or that they no longer welcome foreign investment. Some have even nationalized their ore industries and have ousted U.S. and European industries.

Greater exploration and exploitation of our own resources have been urged by some people to remedy this situation. The minerals, it is said, are there—they simply have not been lo-

Table 21-3. Approximate annual world production of selected metals during the early 1970s

Metal	*Production, million of tons*
Aluminum	11.0
Copper	6.0
Gold	0.0014
Lead	3.4
Manganese	20.0
Molybdenum	0.078
Nickel	0.64
Zinc	5.5

cated. In some people's opinion, valuable resources are not available because they are protected within national parks or wilderness areas. Other people advocate increased production from lower-grade ore bodies and suggest that there is virtually no limit to the technology for extracting ore from such bodies, provided the financial incentive is there.

Before we start wholesale exploitation of our own resources, however, we should realize that our actions might severely damage the economies of developing countries that now supply us with minerals and other resources, undermining even further our relations with those countries. Or we might ask ourselves what the consequences would be if national parks, which are visited annually by millions of vacationers from the United States and abroad, and wilderness areas, which protect delicate ecosystems, were opened to commercial mining. It becomes apparent that the solution would in itself be extremely complex, involving sharing of resources among nations and in-

Fig. 21-17 Alternating bands of iron (dark) and silica (light) interlayered with gneiss, Kapiko Iron Range, Canada. *Geological Survey of Canada*

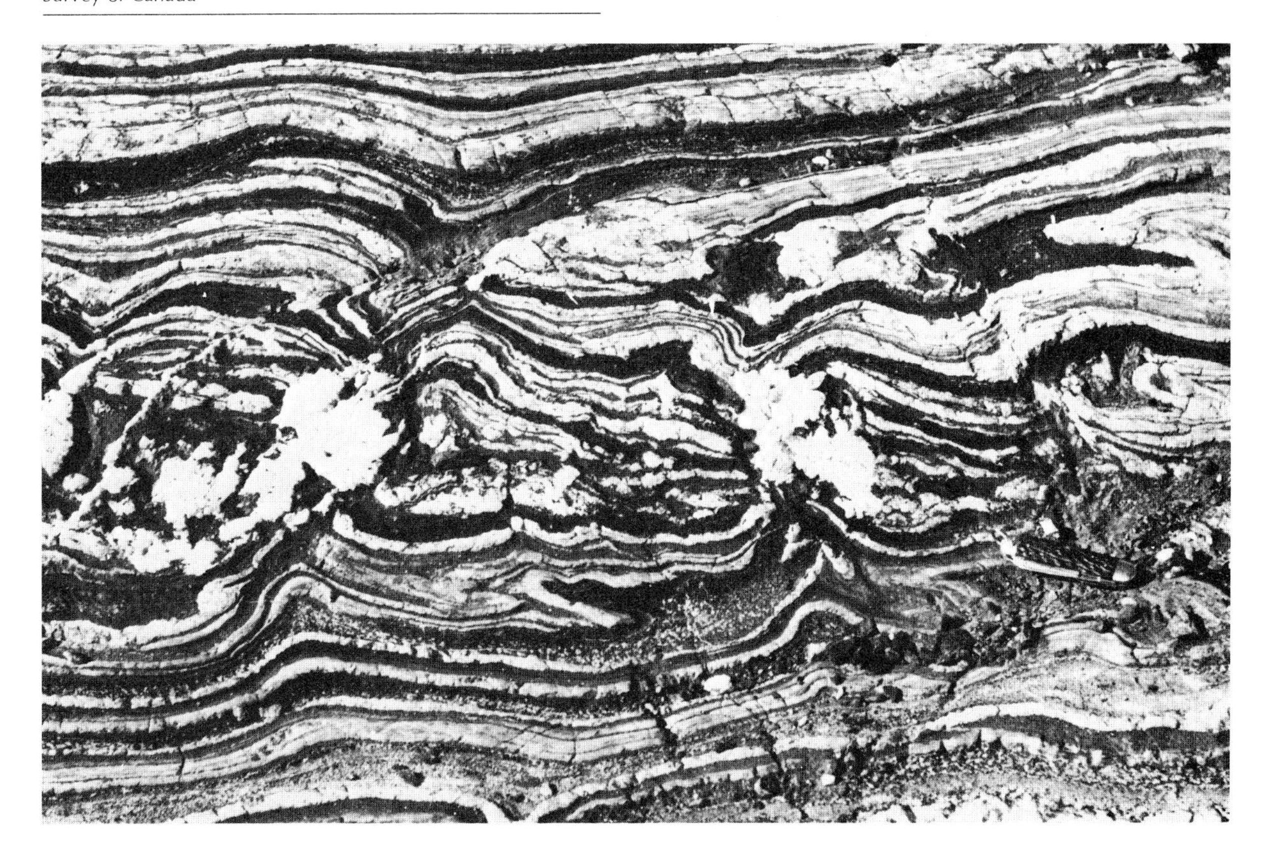

Table 21-4. United States net import reliance of metals and minerals as a per cent of apparent consumption *

Metals and minerals	*1978*	*1979*	*1980 (estimate)*	*Major foreign sources (1976–1979)*
Aluminum (metal)	11	4	E**	Canada, Ghana
Asbestos	85	83	76	Canada, Republic of South Africa
Barium	27	38	38	Peru, Ireland, Mexico, Morocco
Bauxite and alumina	93	93	94	Bauxite: Jamaica, Guinea Alumina: Australia, Jamaica
Chromium	91	90	91	Chromite: Rep. of South Africa, Philippines, USSR, Turkey Ferrochromium: Rep. of South Africa, Yugoslavia, Zimbabwe, Japan
Cobalt	95	94	93	Zaire, Bel.-Lux., Zambia, Finland
Copper	20	12	14	Canada, Chile, Zambia, Peru
Corundum	100	100	100	Rep. of South Africa, India
Diamond (industrial stones)	100	100	100	Ireland, Rep. of South Africa, Bel.-Lux., United Kingdom
Gem stones	99	99	99	Rep. of South Africa, Bel.-Lux., Israel, United Kingdom, India
Gold	53	50	28	Canada, USSR, Switzerland
Gypsum	32	35	38	Canada, Mexico, Jamaica
Iron ore	29	25	22	Canada, Venezuela, Brazil, Liberia
Lead	9	4	E**	Canada, Peru, Mexico, Honduras, Australia
Lime	3	3	2	Canada, Mexico
Manganese	97	98	97	Manganese ore: Gabon, Brazil, Australia, Rep. of South Africa Ferromanganese: Rep. of South Africa, France, Japan
Mercury	64	55	49	Spain, Algeria, Italy, Canada, Yugoslavia
Mica (natural) sheet	100	100	100	India, Brazil, Madagascar
Molybdenum	E**	E**	E**	Canada, Chile
Nickel	80	69	73	Canada, Norway, New Caledonia, Dominican Republic
Platinum–group metals	90	89	87	Rep. of South Africa, USSR, United Kingdom
Potash	64	66	62	Canada, Israel
Quartz crystal–industrial	NA†	NA†	NA†	Brazil
Selenium	43	28	40	Canada, Japan, Yugoslavia
Silicon	17	13	21	Norway, Canada, Rep. of South Africa, Yugoslavia
Silver	48	42	E**	Canada, Mexico, Peru, United Kingdom
Sodium sulfate	7	9	8	Canada, Belgium, Fed. Rep. of Germany
Strontium	100	100	100	Mexico
Sulfur	12	11	13	Canada, Mexico
Tin	79	80	84	Malaysia, Boliva, Thailand, Indonesia
Tungsten	56	58	54	Canada, Bolivia, Thailand, Rep. of Korea
Zinc	66	63	58	Ore & concentrates: Canada, Honduras, Peru Metal: Canada, Spain, Mexico, Fed. Rep. of Germany

Adapted from a U.S. Bureau of Mines publication.

* In per cent. Based on net imports of metal, minerals, ores, and concentrates. Net import reliance = imports − exports + adjustments for Government and industry stock changes; apparent consumption = U.S. primary + secondary production + net import reliance.
** Net imports.
† Not available.

creasing efforts to conserve and to recycle materials. Some warn that the energy necessary for extraction could be a major limiting factor for the future of some ores, some of which could be vital to civilization.

ENERGY: SOURCES AND PROBLEMS

Not too long ago, firewood was the principal source of energy in most places on earth, while animals and slaves did much of the hard, less desirable work. Our principal energy sources today are the fossil fuels, which came into their own only a little more than one hundred years ago, but few people realize how short their life-span will be—both on a geological and a human time scale (Fig. 21-18). It is predicted that we will exhaust our supply of fossil fuel in 1000 years or less. Hence, it seems clear that nuclear and solar energy do have a future and that we also will have to modify our standard of living and life-style. It is possible that wood might be used in countries where the yield can be sustained without harmful effects on the environment. In the section that follows, however, we will focus on geological energy sources.

Fossil fuels—coal and oil—are derived from organic remains. The ultimate source of energy for their formation is the sun, which is used by plants to convert water and carbon dioxide into various organic materials by ***photosynthesis.*** The sun's energy, therefore, is stored in living plants, but lost through oxidation when the plant decomposes. Fossil fuels form when some portion of the organic matter, however small, does not decompose and thus accumulates. Low-oxygen, or ***reducing,*** environments are necessary for this. You can well appreciate how slowly fossil fuels accumulate—in contrast to the extremely rapid rate at which we use them up.

Coal deposits

Coal deposits are very familiar to most of us. Coal can be thought of as a somewhat metamorphosed sediment, consisting of plant remains that accumulate in a low-oxygen swampy area. In such an environment, vegetation accumulates to form ***peat,*** a brown, soft organic deposit in which plant remains are easily recognized. When peat is buried under younger sediments, pressure and heat start the process of coal formation. In the first step, a brownish material called ***lignite*** is formed. Subsequently, the carbon content of lignite increases, and black ***bituminous*** coal, and then ***anthracite*** coal—the highest quality product—form. A good example of a present-day swamp that will eventually become a coal deposit is the Dismal Swamp of Virginia and North Carolina.

Most of the world's coal deposits were laid down during the Carboniferous Period (see Fig. 1-14), beginning about 350 million years ago, and lasting about 100 million years. In the eastern United States, individual coal seams can be traced for hundreds of kilometers, indicating the extent of the swamps of that period. Many of the eastern coal deposits are

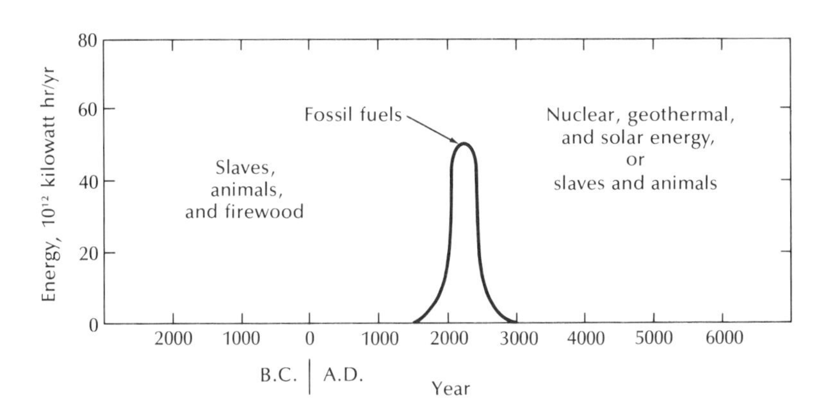

Fig. 21-18 Diagram, illustrating the energy sources used by people and showing the very brief time span over which fossil fuels will be important.

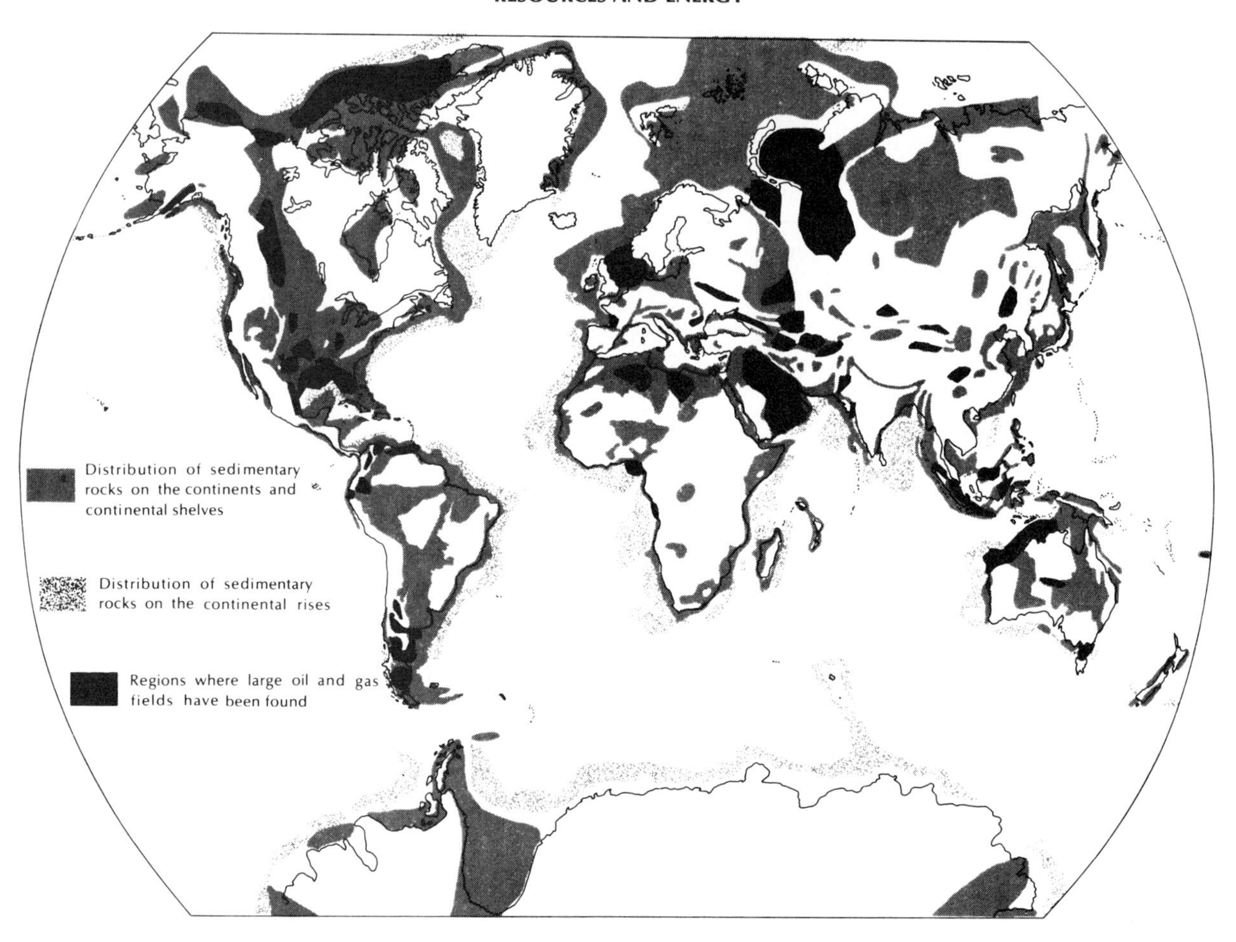

Fig. 21-19 World map showing areas of sedimentary rocks and regions with large oil and gas fields. Note the large potential areas offshore on the continental shelf and slope.

closely associated with sediments that reveal at least 50 cycles of marine and non-marine sedimentation, cycles that may record eustatic changes in sea level, since there is evidence for widespread glaciation at that time. It seems paradoxical that alternating cooling and warming cycles in the Carboniferous Period set the stage for the production of the material that keeps us warm today!

Oil and gas

Unlike coal, most oil and gas (petroleum) usually are not formed in the rock in which they are finally found. They form, instead, through complex chemical reactions and migrate to their final resting places in other rocks. It is the geologist's job to find these reservoir rocks.

Marine sedimentary rocks are the most common source of oil and gas (Fig. 21-19). During their deposition in offshore basins, a substantial amount of organic matter mostly in the form of microscopic plant remains was incorporated into the sediment. The organic matter was preserved and a series of complex reactions slowly converted it into oil and gas. Finally, the light, mobile oil worked its way toward the surface and eventually accumulated

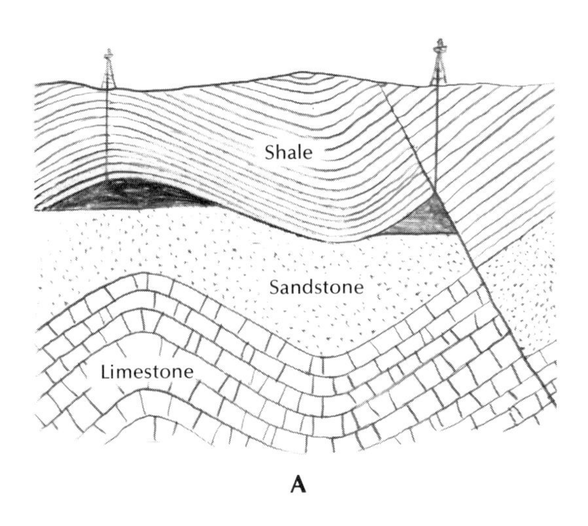

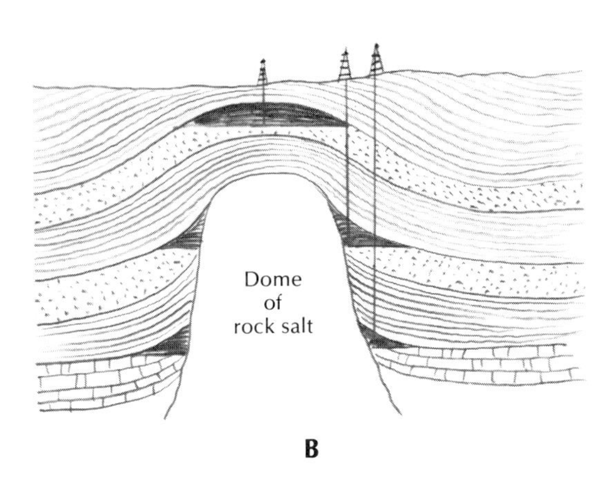

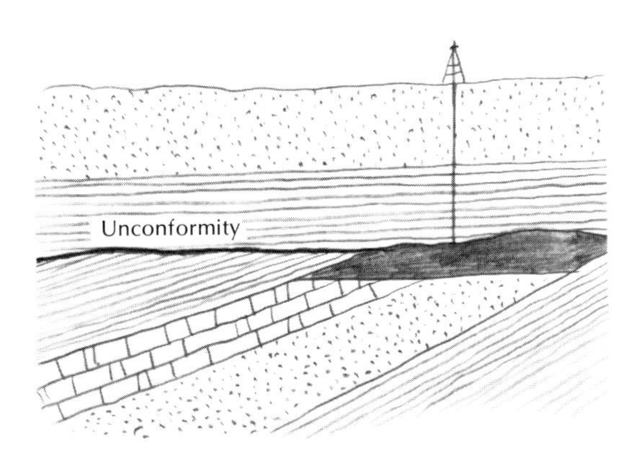

in permeable ***reservoir rocks.*** Such accumulations are called ***oil pools.*** An ***oil field*** is usually defined as a group of pools or an isolated pool. The latter term is a misnomer because oil does not collect in pool-like basins. Rather, it fills voids in the rock, such as the spaces between sandstone grains or cracks and hollow solution features in limestone and dolomite. Impermeable ***roof rocks*** or ***cap rocks,*** most commonly shale, prevent the hydrocarbons from migrating to the surface and being dissipated. Carbonates are important reservoir rocks; about 20 per cent of the hydrocarbons in North America, and 50 per cent worldwide, are found in such rocks.

Several kinds of geological structures can effectively trap hydrocarbons, provided that a cap rock is present (Fig. 21-20). The most simple trap is an anticline in which hydrocarbons migrate to the arch at the top of the structure (Fig. 21-21). More complicated are salt domes, which are formed by the intrusion of the salt through such permeable beds as sandstones. Faulting can also place impermeable beds against permeable beds and so provide favorable environments for the entrapment of oil. Still more complicated traps, largely because they are so difficult to locate, are ***stratigraphic traps.*** Usually these occur at unconformities where dipping reservoir beds beneath the unconformity are sealed by an overlying cap rock. Although oil and gas commonly are found together, gas can occur by itself.

Oil shale and tar sands

Oil shale is in the news these days because it may be a new source of oil in the United States

Fig. 21-20 Various geological settings for the occurrence of oil and gas (black). In each case, gas would rest on the oil, and any water present would underlie the oil, which is less dense than water. **A.** Oil trapped at the crest of an anticline capped by shale. To the right, down-faulted shale seals a sandstone unit. **B.** Rocks domed by a salt intrusion forming an anticlinal trap as in **A**; traps are also formed along the margins of the dome by an impermeable salt cap. **C.** Shales above a major unconformity form the cap for dipping reservoir rocks beneath the unconformity.

Fig. 21-21 An anticline with an oil well sited on its crest. East Los Angeles, California, 1931. *M.N. Bramlette, USGS*

(Fig. 21-22). In parts of the Rocky Mountains, shales laid down some 50 million years ago have attracted wide attention because they form the most extensive high-grade deposits. In that ancient environment, plant and animal life flourished. Subsequently it was incorporated into sediments, and in time, the organic matter in the sediments was converted to ***kerogen,*** which, on distillation and refining, yields oil. The deposits are well bedded (Fig. 21-23) and contain a wide variety of fossils, the most famous of which are the fossil fishes (Fig. 21-24).

Another oil-bearing deposit, the curious ***tar sands,*** contains oil so viscous that it cements the grains of sand together. Because the consistency of tar sands normally prevents their removal by pumping, to be recovered from depth they must first be heated to reduce the viscosity. Fortunately, the world's largest tar sand deposit, located in the Athabasca region of northern Alberta, lies close enough to the surface to be easily mined. Large deposits are also found in Trinidad, Venezuela, and the USSR.

FOSSIL FUEL RESOURCES AND THE FUTURE

Coal

Worldwide, coal is in fairly abundant supply, although the USSR and the United States have the bulk of the resource (Fig. 21-25).

In the United States, the distribution of coal is widespread (Fig. 21-26). Production grew be-

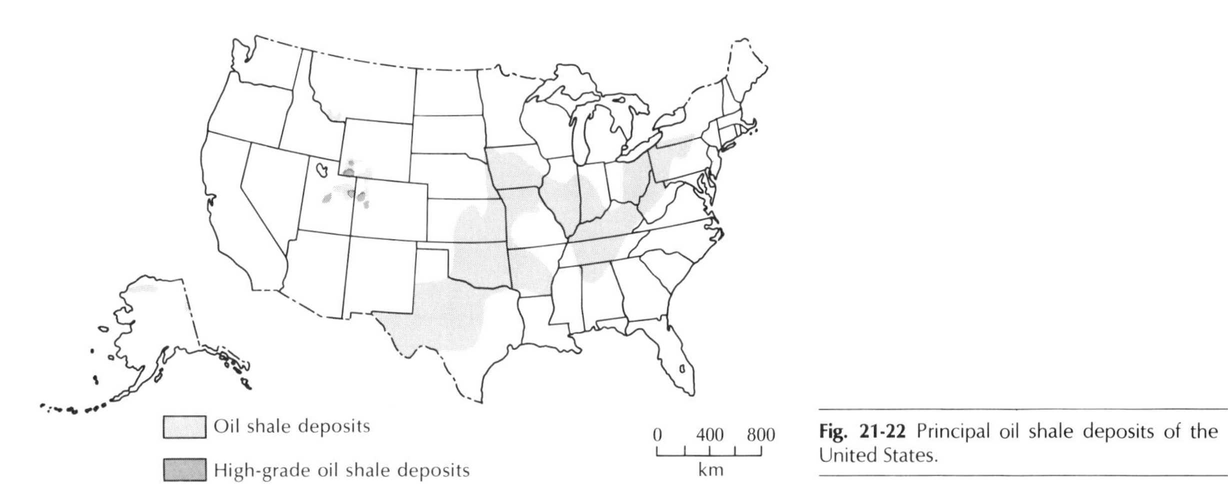

Fig. 21-22 Principal oil shale deposits of the United States.

Fig. 21-23 Oil shale exposed in valley walls near Grand Valley, Colorado. *W.C. Bradley*

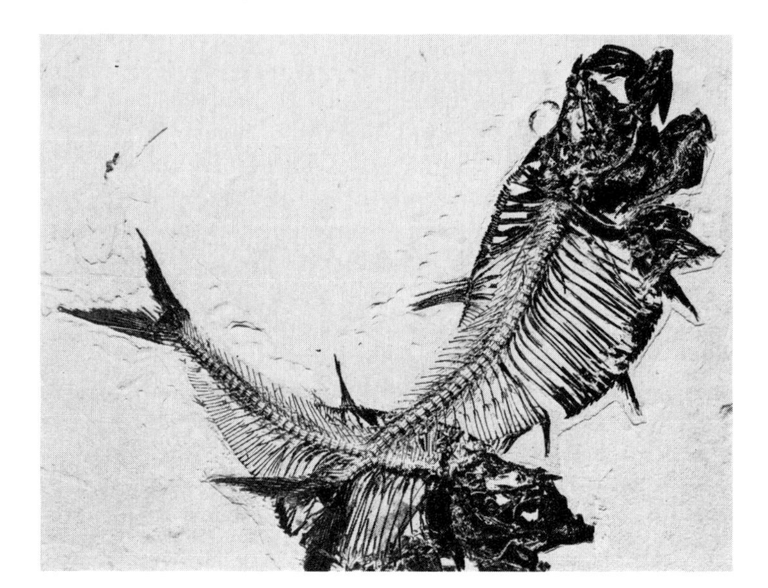

Fig. 21-24 Two armored herring (*Diplomystus dentatus*) from the Eocene Green River Formation in Wyoming. The fish were buried in this position by a shower of ash that fell on the lake. *Janet Robertson, Geological Society of America*

tween 6 and 7 per cent per year, until about 1910, when the switch to oil and gas began (Fig. 21-27). It then leveled off to a constant rate until 1972, when production again began to rise. How long our reserves—estimated at about 400 billion tons—will last will be determined by how much we increase the annual rate of production over the 1972 figure of 0.5 billion tons. Simple calculations show that with no increase in production rate, coal will last a long time. Recent Administrations have called for a 5 to 10 per cent increase in coal production (Fig. 21-27), but few people know what such growth will mean to the life expectancy of the coal industry. Al Bartlett, in his lectures, appropriately points out that supplies of coal are plentiful today because for 60 years no growth took place in the coal industry. Had we continued at the pre-1910 growth rate, we would be out of coal by 1990.

Increased use of coal in industry may have two adverse effects. When atmospheric carbon dioxide levels increase, the balance between incoming and outgoing solar radiation is disrupted, preventing the escape of long-wave radiation. Subsequently, the atmosphere is warmed—a phenomenon known as the ***greenhouse effect.*** Such warming could amount to a temperature increase of 3°C (5°F) by 2050, which could trigger the melting of glaciers resulting in a rise in sea level and flooding of coastal areas, where a large number of the world's peoples dwell.

The burning of high-sulfur coal releases sulfur dioxide, which combines with water vapor in the atmosphere to form sulfuric acid—the so-called ***acid rain*** now experienced in many parts of the United States and Canada. Acid rain has caused so much damage to crops, lakes and fish, and buildings and outdoor art, much of which is old and priceless (see Fig. 8-4), that millions of dollars are now being spent on acid-rain research.

Both the greenhouse effect and acid rain would occur on a wider scale if we increased our use of coal in an effort to cut down on our use of the other, scarcer fossil fuels. The Canadian Government has already registered complaints with Washington regarding the acid rain generated by U.S. industry.

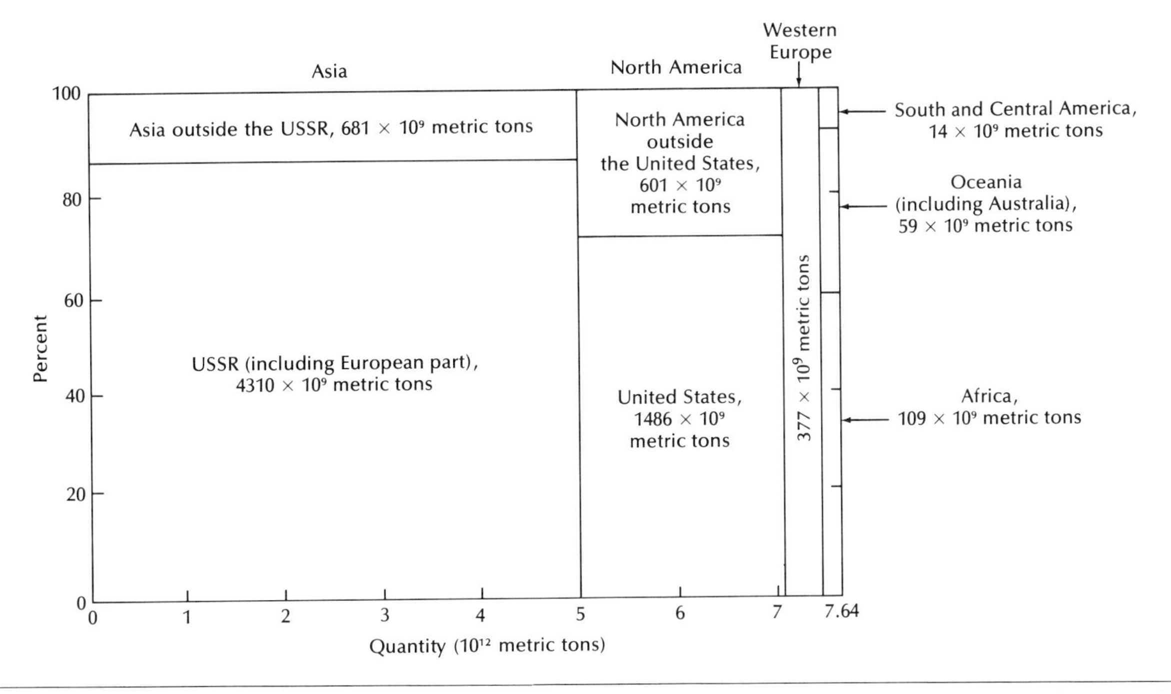

Fig. 21-25 Coal resources of the world. The horizontal scale gives the total quantity, worldwide, and by large regions. The vertical scale gives percentages in smaller political entities within the larger regions. The figures represent the recoverable or mineable coal, which is 50 percent of the total estimated amounts.

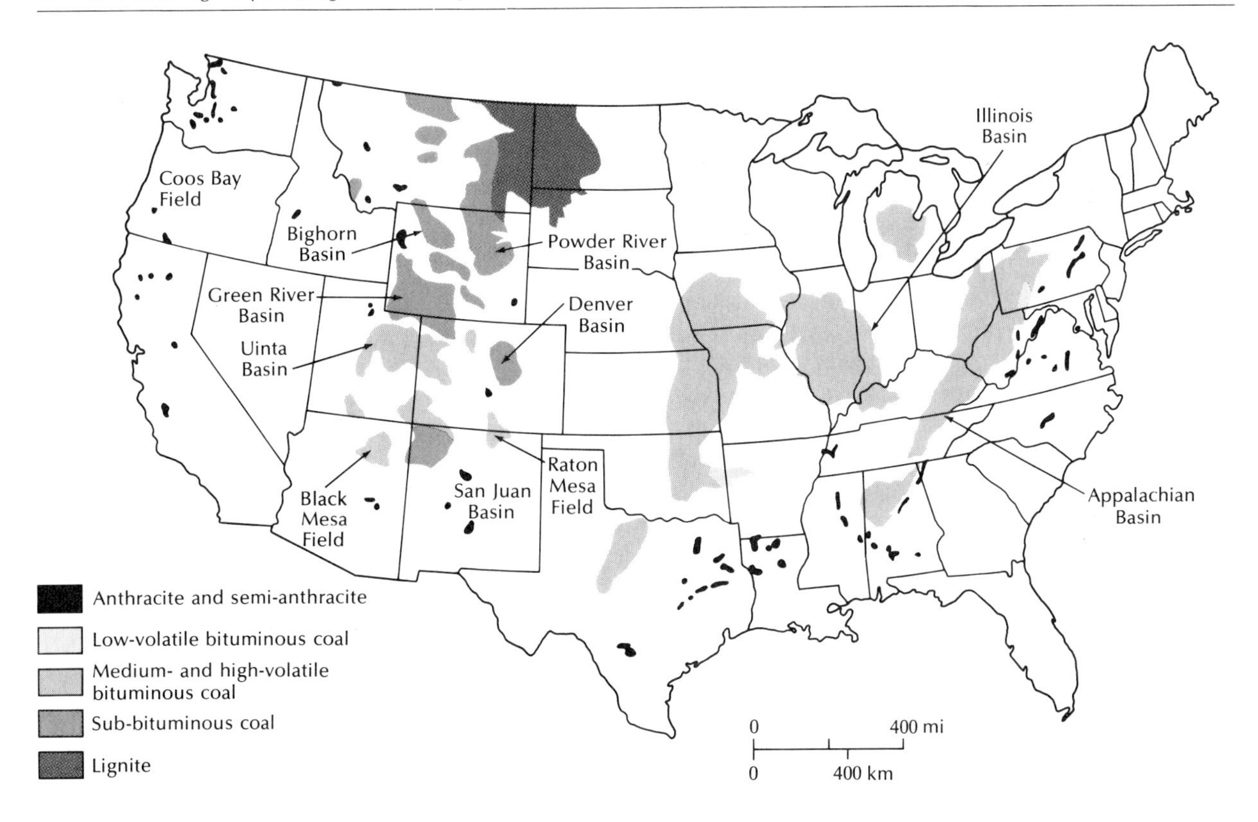

Oil

The outlook for oil is much worse than that for coal. Distribution is uneven, and the resources and proved reserves lie mainly outside the Western Hemisphere (Fig. 21-28). Even so, the demand for oil in the United States as well as in other countries continues to grow, and what we do not produce ourselves we import at skyrocketing prices. In the United States, the time of peak production has passed.

We can forecast what is in store by comparing oil production with the estimated total resource. From 1890 to 1970, world production grew at about 7 per cent per year, which means that the amount produced and used doubled every decade. Put even more dramatically, at this rate of growth, the amount used during any one decade is equal to the total consumption of all previous time (Fig. 21-29). Clearly, more successful exploration and greatly improved oil-field technology would be needed to ensure that every last drop of oil is squeezed from the reservoir rocks. M. King Hubbert, a geologist and an energy expert who has worked both with Shell Oil Company and the U.S. Geological Survey, has long been a leader in predicting the future of fossil fuel production. In 1969, he predicted that by the year 2000, if not sooner, maximum world oil production would be attained. Subsequently, production would decline. In 60 years, he estimates, 80 per cent of the available oil will have been produced, and beyond the year 2075 there will be little production (see Fig. 21-8).

A recent study, prepared for the U.S. Geological Survey and the Department of Energy and published by the Rand Corporation, is even more pessimistic. Based on geological study of individual oil fields, as well as likely oil-containing formations, and statistical analysis, the Rand study estimates that the United States can continue to supply oil for only 20 to 40 more years at the already inadequate 1979 rate of production.

What has happened since 1969 to extend our supply of oil? The discovery of large reserves in Alaska is important, but, put in perspective, if the Alaskan fields were our sole source of oil,

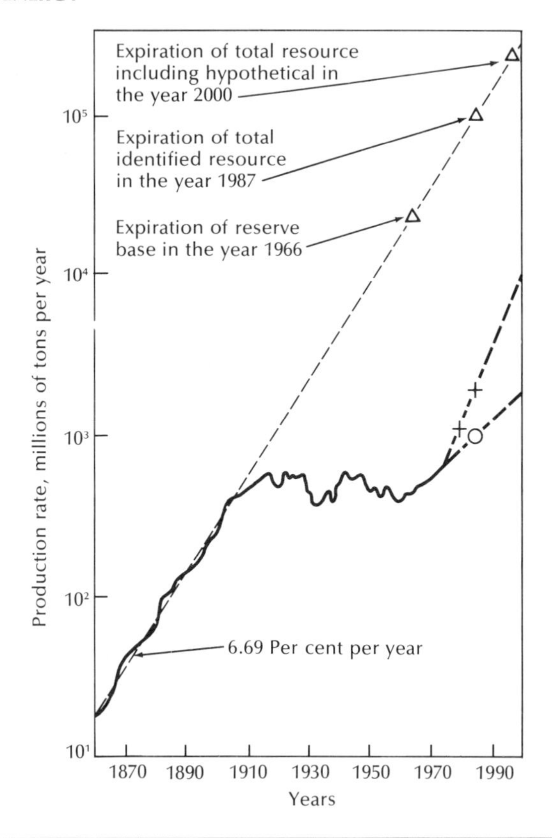

Fig. 21-27 History of coal production in the United States. For future production, the goals of the Ford administration (+) and the Carter administration (o) are indicated.

they would supply the needs of the United States for only about 6 years. Mexico is now a major oil producer following the discovery of an enormous field with estimated reserves of about 46 billion barrels, perhaps as much as 200 billion barrels. No one knows, however, whether the Mexican oil will be widely available to the world market. The impact of this or any other oil field can be estimated by a glance at Figure 21-29. The Mexican reserves could satisfy the world's needs for 2 to 3 years. Because of dwindling traditional supplies, there is increased interest in offshore drilling (Fig. 21-30), but such drilling is not without great risk, both financial and environmental (Fig. 21-31).

A continuing dependence on a finite resource such as oil is no longer feasible in a

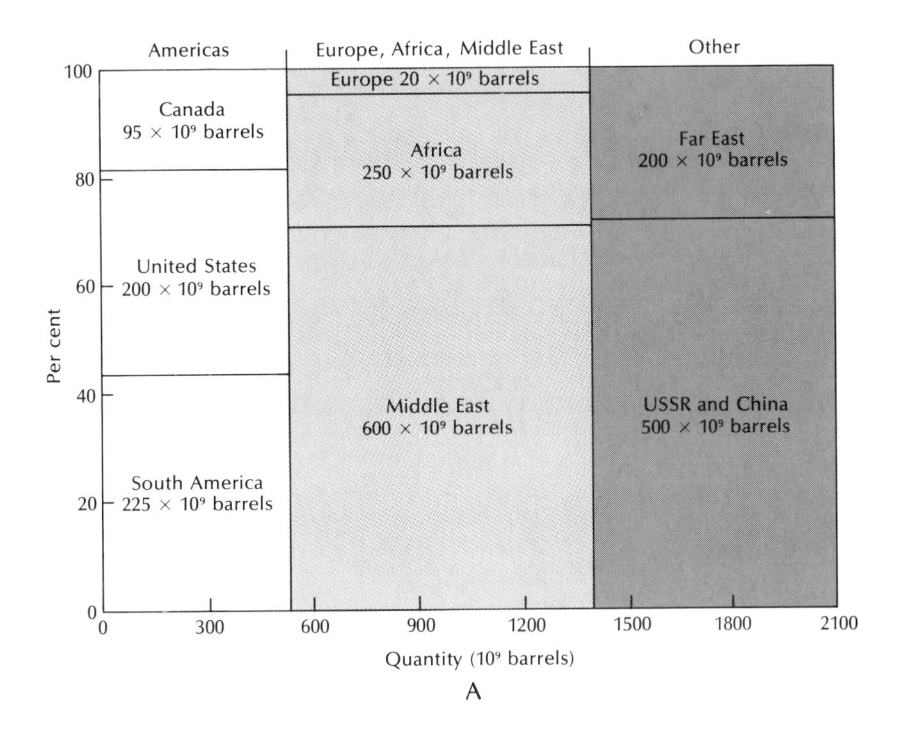

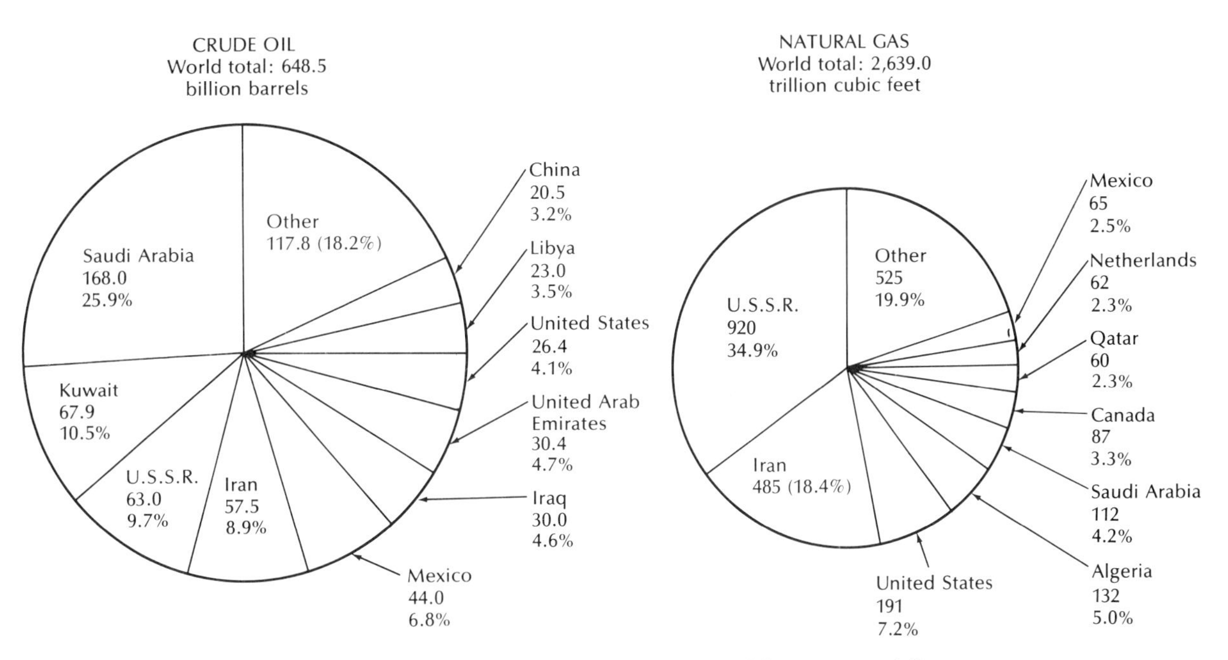

Fig. 21-28 Estimates of world oil resources (**A**) and proved reserves (**B**).

world with a growing population, increasing per capita demands for fuel, and a yearning for an adequate standard of living by people in developing countries. Yet the search for new fields continues in the face of an increasing number of dry holes and escalating drilling costs. For example, the cost of drilling a 3048-m (10,000-ft) well in Alaska has risen from $2.2 million in 1969 to $4 million today. One of the last exploratory wells drilled there cost more than $23 million, and the hole was dry.

Finally, a quote from Earl T. Hayes, former chief scientist of the U.S. Bureau of Mines, might be appropriate:

> It must be recognized that the United States never had and never will have the petroleum resources to sustain indefinitely the production levels of the last 25 years. In effect, we have been living off our capital all this time and cannot postpone the day of reckoning indefinitely. Talk of rising petroleum (and gas) production for long periods is both immoral and nonsensical. Whatever slight gain might be achieved for a very few years will be at the expense of the youth of today. Predictions of sustained increased production deny the records of 50 years of experience with the exploration, development, and extraction cycle of liquid hydrocarbons. There is a finite amount of easily recovered petroleum in this country, and no act of Congress or false optimism of government, industrial, or academic planners can add to our natural resource base.

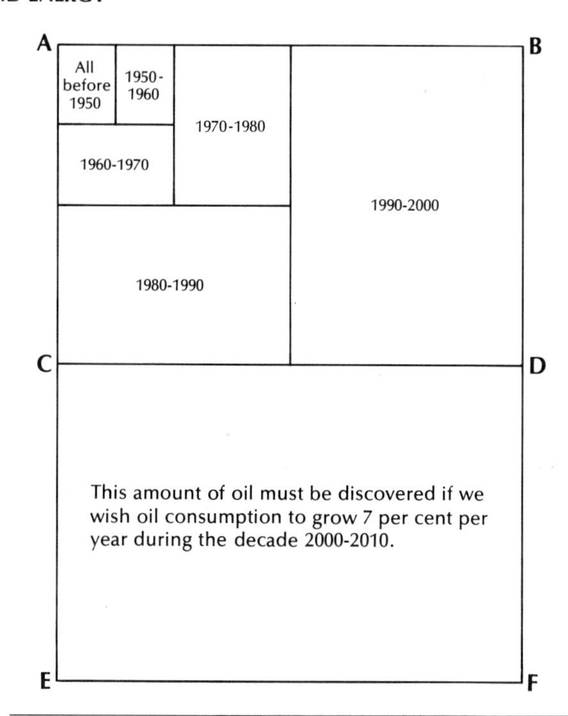

Fig. 21-29 Relative amounts of oil used each decade as a result of a 7 per cent annual increase in consumption. At this rate, the time required to double consumption is a decade; the diagram clearly shows that the quantity consumed in any one decade is equal to the total of all preceding years. For example, the amount represented by rectangle CDFE is equal to that represented by ABDC. Hence, production must double each decade to match consumption. The area of ABDC is an approximate representation of the known world oil reserve.

Tar sands and oil shales as oil sources are largely untapped. The estimated amount of oil in tar sands is large—some 1300 billion barrels, which is considerably more than the Middle East resource. Oil shales in the United States yield between 10 and 65 gallons per ton of rock, and in some thin layers the yield may be as high as 140 gallons per ton (42 U.S. gallons = 1 barrel). On average, the yield is close to 1 barrel of oil from 2 tons of rock. Present-day reserves in the United States are about 80 billion barrels; the potential resource could be 2 trillion barrels or more. For the world, the total reserve could be 2000 to 3000 trillion barrels, but in 1965 it was projected that only 190 billion barrels were recoverable.

What do these figures mean in terms of extensive excavating to reach the oil? The major oil company Exxon recently announced plans for an oil-shale plant with an output of 47,000 barrels per day, an amount, it is estimated, that would require the processing of roughly 94,000 tons of rock per day. President Carter had hoped for an industry producing 400,000 barrels per day (about 800,000 tons of rock per day) by 1990. Going one step further, we are currently using about 6.5 billion barrels of oil annually. To obtain all of our oil from shale means the processing of 13 billion tons of rock per year, or 130 million freight-train cars full. Bartlett makes another calculation of the volume of rock that must be handled: to extract 1 million barrels of oil per day requires the processing, each nine months, of an amount of rock equivalent to that removed in building the Panama Canal.

The problems involved in processing oil

Fig. 21-30 The central platform for the Ninian Field oil rig, being prepared for towing into position in the North Sea. *Courtesy Standard Oil Company of California*

shales are formidable. Considerable amounts of both water and energy are needed: the estimates are 3 to 5 barrels of water for each barrel of oil produced and the energy-equivalent of 0.25 barrel of oil to process 1 ton of shale. Avoiding stream pollution from the mining operations is important, just as are finding ways of re-establishing vegetation in mined areas and planning for the impact on local communities of increased population as workers and their families move in.

Natural gas

The recent Rand Study figures for natural gas production in the United States indicate only a 50 per cent chance of finding 170 trillion ft^3 of gas beyond the 570 trillion ft^3 already known. Earlier estimates of total world resource were 12,000 trillion ft^3 (340 trillion m^3); of U.S. resource, 1000 to 1300 trillion ft^3 (28–37 trillion m^3). It is forecast that gas production in the United States, exclusive of Alaska, will cease beyond the year 2000, and that we may already have passed the peak of production.

Fossil fuels in agriculture

Today, the agricultural industry consumes more petroleum than any other industry in this country—a fact that is probably not well known. Oil is used for running machinery on the farm, pumping water for irrigation, manufacturing and distributing fertilizers and pesticides, and transporting foodstuffs. The energy requirements of high-technology agriculture, or agribusiness, are so high that it is doubtful whether it can continue much longer. For example, the equivalent of about 112 gallons of gasoline was required to grow the food for the

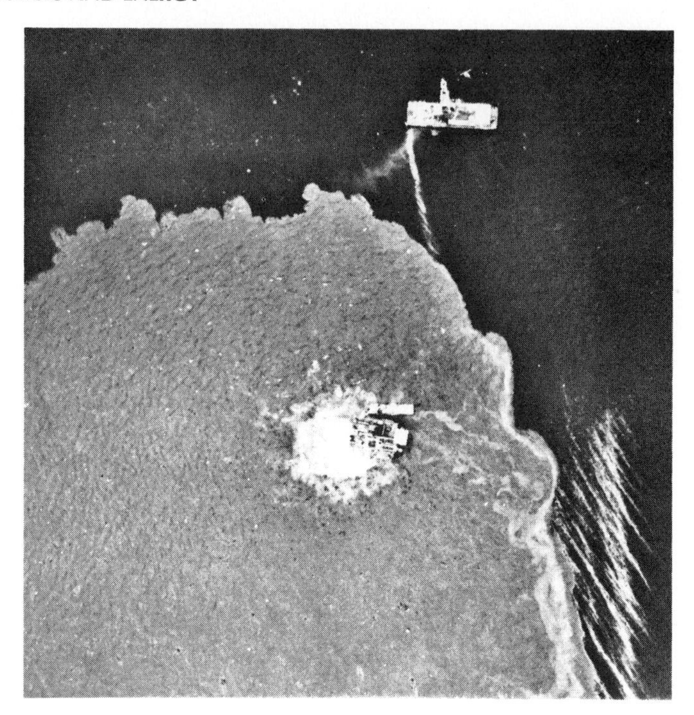

Fig. 21-31 Oil slick surrounding an offshore rig near Santa Barbara. An estimated 750,000 liters of oil escaped after a natural blowout of the drill hole. The barge at the upper right is about 50 m (164 ft) long. *Environmental Protection Agency*

average American in 1970. If the world were to eat the way Americans eat, and use our technology to grow their crops and produce their livestock, the world oil reserve would be depleted in less than 50 years. Or, put in a different way, by the end of the next population doubling (36 years), the percentage of people starving would remain the same, but the actual number would have doubled. The solution may be to return to the days of hand labor (which may become cheaper than the cost of buying, maintaining, and running machinery) and to fertilize with animal fertilizers that are still available in abundance.

GEOTHERMAL ENERGY

Areas in which hot water and steam occur naturally are known as ***geothermal regions,*** and people have long cast a speculative eye on them as a possible source of energy. The development of such areas is in its infancy, but geothermal energy has a high potential for driving turbines to produce electricity, as well as for other uses. Heat from geothermal resources can be used directly. Table 21-5 shows the temperature required for various domestic and industrial applications. In some instances, the same geothermal fluid can be cascaded through several different processes, as its temperature drops.

Producing geothermal areas, as well as those with potential for development, are all in areas of recent volcanism. Magma or hot masses of rock at depth are the primary sources of heat, which is transferred to the surface by rising hot water and steam. Of the water involved in such a system, fully 95 per cent originates at the surface and is circulating; the rest originates at the source of the heat in the form of magmatic steam.

The first commercial geothermal venture was in Italy, in 1904. Since that time, producing areas have been sited in Japan, the Kamchatka

Table 21-5. Ranges for Direct Uses of Various Temperatures

	Temperature °F	*Temperature* °C	*Use*	
Temperature of Saturated Steam (392 °F to 212 °F)	392	200		
	374	190		
	356	180	Evaporation of highly concentrated solutions Digestion in paper pulp	Temperature Range of Conventional Power Production (356 °F to 284 °F)
	338	170	Drying of diatomaceous earth	
	320	160	Drying of fish meal Drying of timber	
	302	150		
	284	140	Drying farm products at high rates Canning of food	
	266	130	Evaporation in sugar refining Extraction of salts by evaporation and crystallization	
Water Temperature (248 °F down to 68 °F)	248	120	Fresh water by distillation Concentration of saline solutions	
	230	110	Refrigeration by medium temperatures Drying and curing of light aggregate cement slabs	
	212	100	Drying of organic materials, seaweed, grass, vegetables, etc. Washing and drying of wool	
	194	90	Drying of stock fish Intensive de-icing operations	
	176	80	Space heating Greenhouses by space heating	
	158	70	Refrigeration by low temperature	
	140	60	Animal husbandry Greenhouses, combined space and hotbed heating	
	122	50	Mushroom growing Balneological baths	
	104	40	Soil warming	
	86	30	Swimming pools, biodegradation, fermentation Warm water for year around mining in cold climates De-icing	
	68	20	Hatching of fish, fish farming	

From: Lindal, B., 1974. Geothermal energy for process use: Proceedings of the International Conference on Geothermal Energy, Oregon Institute of Technology.

Fig. 21-32 (above) Steam wells at Yanbajain Thermal Field northwest of Lhasa, Tibet. The temperature in the wells is 172 °C at depths of 130 to 288 m. *Troy L. Péwé*

Peninsula (Soviet Union), China, New Zealand, Turkey, Mexico, El Salvador, and Iceland. There is also an experimental station in Tibet (Fig. 21-32). Geothermal energy is particularly appealing to such nations as Japan, where oil and coal are almost non-existent, hydroelectric power is limited, and about 70 per cent of the fuels consumed must be imported. In Iceland, volcanic steam is used to heat buildings, and steam pipes placed in the fields warm the soil so that crops that ordinarily would not survive in that severe climate can be grown.

In the United States, the first and the only successful large-scale exploitation of geothermal energy is at The Geysers in northern California, about 145 km (90 mi) north of San Francisco. In spite of its name, the area has no geysers, but steam rises from hot springs, wells, and fumaroles—vents from which gases and vapor rise. Steam from wells several thousand meters deep is piped to turbines that in 1980 produced 600 megawatts of electricity, enough to serve the electrical needs of about 600,000 homes. There are plans for an ultimate capacity of 2000 megawatts.

Like most resources, the production of geothermal energy presents problems. The hot waters carry materials in solution that can precipitate out as solids and clog pipes or pollute local water supplies. Thermal pollution or surface subsidence accompanying water withdrawal also can occur. Finally, if reserve hot water and steam are used before more are generated from the depths of the earth—an all too familiar story—then geothermal energy, too, will be in short supply.

Locally, geothermal energy is very important, but it has not yet had a significant impact on the total energy picture. In 1971, for example, geothermal energy accounted for only 0.08 per cent of the total world electrical capacity. It has been estimated that the figure might rise to as much as 10 per cent in the future.

Two new sources of geothermal energy are now being investigated. One involves the use of underground hot rocks, recently crystallized from magma, to heat water. Such rocks are not in contact with underground water, so they differ from the geothermal systems just described. What nature has not provided, however, geoscientists have. In Los Alamos, New Mexico, water under high pressure was forced down one of two holes drilled to a depth of 10 km (6.2 mi) where the temperatures were near 204°C (400°F) (Fig. 21–33). This created a system of cracks in the rock that previously was not porous enough to allow the transfer of water. Water was then pumped into the artificially cracked rocks, where it was heated, and then was returned to the surface by way of the other hole, ultimately to drive a steam turbine. Results thus far have been encouraging.

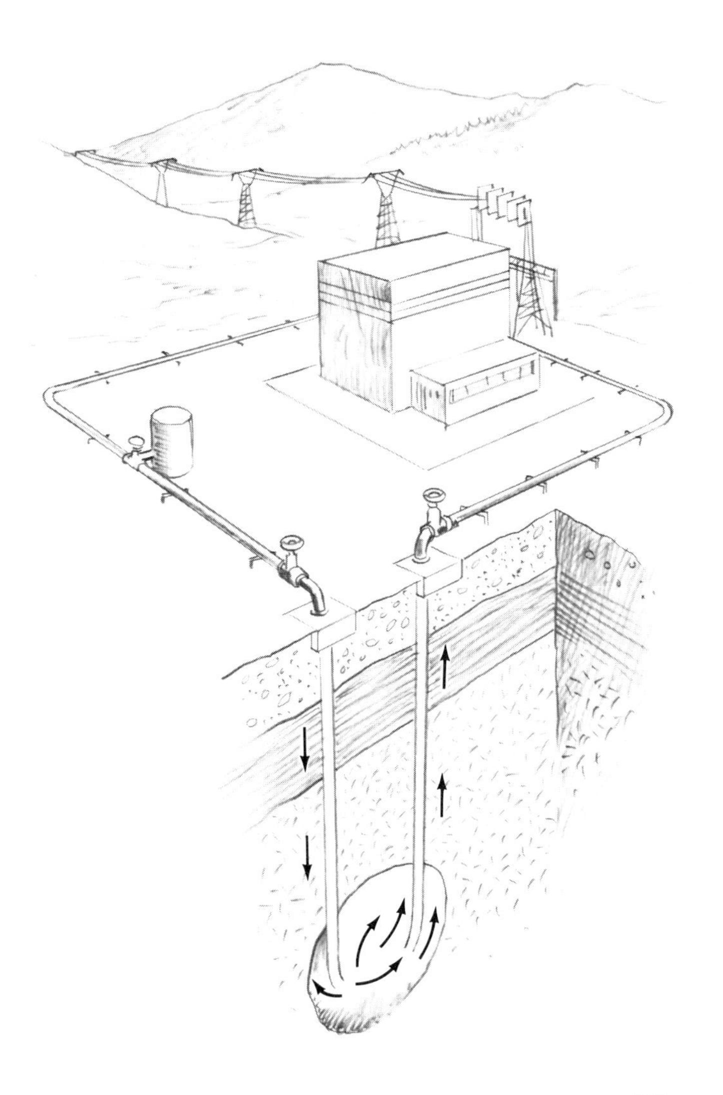

Fig. 21-33 Drawing, showing the concept of extracting geothermal energy from dry, hot rock. The arrows indicate directions of water flow.

Table 21-6. Estimated heat content of geothermal resources in the United States

	Heat content, 10^{18} calories	
	Identified sources	*Identified and estimated undiscovered sources*
Water and steam circulating systems	741	3,050
Hot dry rock	25,000	100,000

The units here are calories, but a generalized conversion is 10^{18} calories equals the heat of combustion of 690 million barrels of oil

Source: U.S. Geological Survey.

The other new source is the hot saline water contained in sedimentary rocks beneath the Texas and Louisiana Gulf coasts. When tapped with a drill hole, the water, which is under very high pressure, rises rapidly to the surface. The sources of energy are several: the force of the water itself, the heat contained in the water, and the natural gas dissolved in the water.

Because of the expanding technology of geothermal energy, it is difficult to assess this resource. Broad estimates for the United States are shown in Table 21-6. If cost is not considered, the recoverable electricity from known geothermal systems is equivalent to that produced by 140 Hoover Dams or about 140 modern nuclear power plants. As technology is improved and costs are reduced, geothermal energy will become an increasingly competitive energy source.

NUCLEAR ENERGY

In 1974, the U.S. Geological Survey was optimistic that nuclear power, which had provided 3.5 per cent of the nation's electrical energy in 1972, would by 1980 provide 21 per cent of the total, and by the year 2000, as much as 60 per cent of the total. These predictions, however, have not been met, for in 1979 only about 10 per cent of our electricity was generated by nuclear power. Here, we will investigate briefly how nuclear power is generated and the problems that the use of nuclear energy poses.

Nuclear energy derives from two reactions, fission and fusion. Fission in a nuclear reactor is achieved by bombarding uranium atoms with neutrons (Fig. 21-34), which produces lighter fission fragments, or different atoms, and at least two neutrons. The neutrons, in turn, encounter other uranium atoms, and the fission process is repeated. If continued, a chain reaction, such as that of an atomic bomb, occurs. Fusion, on the other hand, is just the opposite; that is, two light particles fuse, or join, to form a heavier element (Fig. 21-34). This same process fuels the sun, where hydrogen atoms fuse to form helium atoms, as in the hydrogen bomb. The fusion reaction is considered by some the best possible energy source, since hydrogen is so abundant, but the technology of controlling the reaction, so that it can be used safely, has still not been worked out. However, fusion has the distinct advantage over other nuclear reactions of leaving very little radioactive waste as a by-product.

In both fission and fusion reactions, energy is released. This energy can be used to heat water, produce steam, and generate electricity, but this can be done only if the reaction is controlled. Because fission of uranium can be controlled, it has become the basis for our present nuclear power plants. The advantage of uranium is that only small amounts of it are needed to produce the same amount of energy produced by large amounts of fossil fuels (Table 21-7).

Of the three different isotopes used in nuclear reactors, only uranium-235 occurs naturally. Uranium ores, however, contain only 0.7 per cent of uranium-235. If we continue to fuel U.S. power plants with uranium-235, our supply will probably run out before the end of this century—even though the mineral is plentiful in the United States.

One way to supplement our limited supplies of naturally occurring nuclear fuels is to create fissionable material artificially. This can be done by neutron bombardment of uranium-238 and thorium-232—both of which are more abundant in nature than uranium-235—to cre-

Table 21-7. Energy equivalence of various sources

Energy source	*Amount equivalent to one barrel of petroleum*
Petroleum	1 barrel
Natural gas	5620 cubic feet
Coal	0.223 tons
Uranium	0.076 grams
Uranium	0.000158 pounds

Source: U.S. Dept. of the Interior, Office of Economic Analysis.

ate the fissionable products plutonium-239 and uranium-233, respectively. Because the reaction produces more fissionable material than it starts with, it is said to "breed" new material, hence reactors based on this reaction are called ***breeder reactors.*** Many technological problems have to be overcome, however, before breeder reactors can be used to generate electricity on a large scale. And another problem, of international import, is that the plutonium produced can be used to produce nuclear weapons.

The number of new nuclear power plants built has not fulfilled earlier predictions (Fig. 21-35). By 1981 there were 74 operable nuclear power reactors and 87 test reactors in universities in the United States, about 90 plants under construction or with the necessary permits, and about 20 plants planned (Fig. 21-36). The operable plants produce energy equivalent to that produced by about a million barrels of oil per day. Worldwide there are about 250 oper-

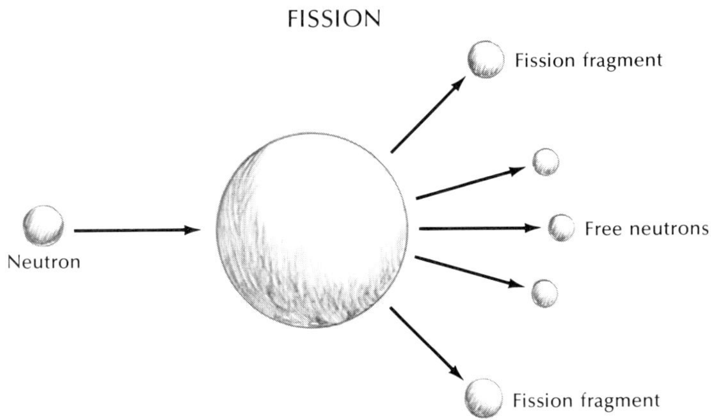

A

FUSION

Fusion reaction

Deuterium

Tritium

Helium

+

Neutron

Two isotopes of hydrogen

B

Fig. 21-34 Diagrams, illustrating fission (**A**) and fusion (**B**).

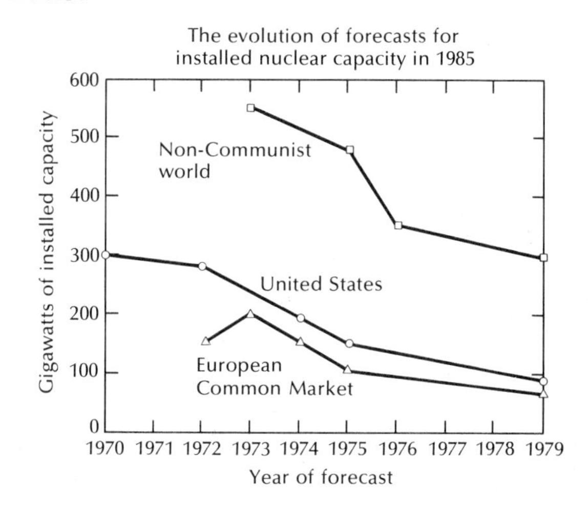

Fig. 21-35 Diagram, showing the changes in predicted installed nuclear capacity for 1985. The vertical scale is in gigawatts; one gigawatt is equal to 1 million kilowatts, the equivalent of 42,000 barrels of oil per day in an oil plant operating full time. All the curves show that, with time, the predicted capacity for the year 1985 has been drastically reduced.

Fig. 21-36 Nuclear power plants in the United States, 1981.

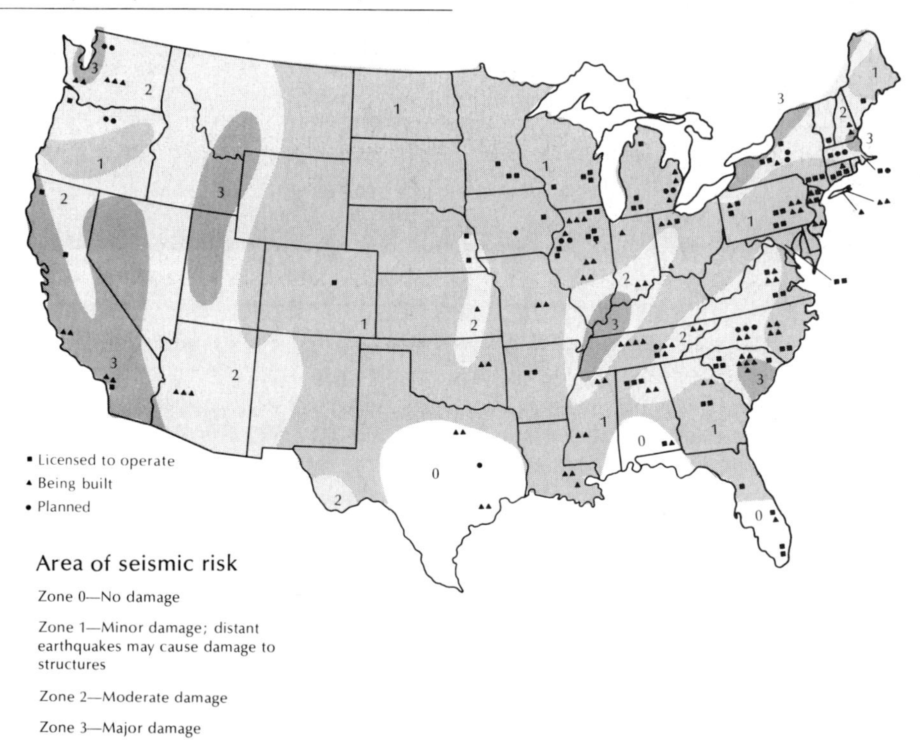

able plants, and another 300 are being built.

The major problems facing the nuclear power industry are safety and the safe disposal, for all time, of radioactive waste. All facilities must be located in tectonically stable, earthquake-free areas, although the stability of many sites in California has been questioned (Fig. 21-37). Safety also has to do with the operation of the plant itself, and there is virtually no room for error. The accident at the Three Mile Island plant near Harrisburg, Pennsylvania, in 1979, has led to questions on the effectiveness of safety systems in nuclear power plants and whether the risks involved in providing energy in this way are not altogether too great.

Fig. 21-37 Several thousand meters of trenches were dug adjacent to this nuclear reactor at Pleasanton, California, to determine if the reactor is near a potentially active fault. A zone of thrust faults that offset a buried late Pleistocene soil and the overlying Holocene soil was discovered near the reactor vessel. Although geologists with the U.S. Geological Survey and the Nuclear Regulatory Commission believe that the faults are tectonic in origin, geologists for General Electric, owners of the facility, contend that the faults are actually the slip surface of a great ancient landslide. *USGS*

The problem of how to dispose of radioactive wastes has not yet been solved. Various proposals, some of them rather far-fetched, have been made, including propelling waste material into space or burying it in the ice of Antarctica. At present, salt deposits, because of their impermeability, are being investigated as dumping sites. Even the clay deposits that blanket parts of the deep-ocean floor are currently being evaluated as repository sites.

CONCLUSION

The statistics for resources and reserves can be misleading, since they do not reveal that some resources cannot be profitably extracted and that the time gap between discovery of a reserve and the production of energy or materials from it could be years. Again, the environmental cost may be more than we care to pay now or to pass on to future generations. If we opt for rapid growth rates and all that they entail, how will other countries view our action, let alone history? And by how much more will we despoil our planet and destroy other, possibly more important natural resources? Earl Hayes's words on the subject are sobering:

> The facts point to the inescapable conclusion that exponential growth of energy supply is coming to an end in the United States. Energy and gross national

product have risen 3 to 3½ percent a year since 1940, and a decrease in the energy growth rate to less than 1 percent a year by 2000 will occasion some fundamental national problems for which we have no precedent. The involuntary conservation brought on by higher prices and decreased supplies will be exceedingly painful for an unprepared American public.

We have sufficient energy resources to supply our basic needs for many decades, but the costs will rise continually. The country still does not understand the problem. The layman wants to believe in inexhaustible, cheap gasoline and in this has been supported by many unsubstantiated claims. The time has come to realize that no miracle is imminent and we must make do with what we have. We will never again have as much oil or gas as we have today, nor will it be as cheap. Nuclear energy has been a major disappointment. Solar energy will be slow in developing and, contrary to popular opinion, quite expensive. Coal is the only salvation for the next few decades.

In the last analysis, we have entered into a massive experiment to determine what effect energy growth has on economic growth, or how much we can slow the machine down and still maintain a democratic, capitalistic form of government.

The United States plays a significant role in the future of world energy. With only six per cent of the world's population, we use one-third of the total current energy production. We could still enjoy life, as do many other people in many other countries, with a so-called "lower" standard of living. For example, there is now one car for about every two people in the United States—a ratio that drops to 1 to 6 in Japan, 1 to 25 in Mexico, and 1 to 23,378 in China. In contrast, our public transportation systems, in most big cities, compare unfavorably with those in most other industrial countries. One out of every nine barrels of oil used in the world is burned on America's highways. Surely, it would be possible to moderate our gluttonous use of world resources. By implementing some of the following ideas as outlined by Bartlett, this might be achieved.

1. Curtail and eventually stop population growth (achieve zero population growth) and curtail consumption. Even if we in the United States were able to heat one-half of the buildings in which we live and work with solar energy, the resulting 10 per cent savings in fossil fuels could be wiped out by two year's growth at 5 per cent.
2. Conserve and recycle resources and materials to the fullest extent possible. We might attempt to change some of our energy-consuming habits that do not contribute much to the quality of our way of life.
3. Intensify research in all forms of energy to determine how they best can be utilized.

More and more people are concluding that conservation is the key to the energy problem. A quote from *Energy Future,* a report from the energy project at the Harvard Business School seems an appropriate conclusion:

> The United States might use 30 or 40 percent less energy than it does, with virtually no penalty for the way Americans live—save that billions of dollars will be spared, save that the environment will be less strained, the air less polluted, the dollar under less pressure, save that the growing and alarming dependence on OPEC oil will be reduced, and Western society will be less likely to suffer internal and international tension. These are benefits Americans should be only too happy to accept.

SUMMARY

1. Resources are commodities that are useful or essential to us or our way of life. Resources are classed as *renewable,* those that can be replenished, and *non-renewable,* those that cannot be replenished. A *paramarginal* resource is technologically possible to recover; recovery of a submarginal resource is not profitable. A *reserve* is the estimated amount of a resource that can be recovered with present-day technology.
2. Population growth and consumption of resources are increasing exponentially, or at a fixed percentage rate.
3. Metallic ores deposits can be classified as *abundant* or *scarce.* The abundant metals are iron, aluminum, manganese, magnesium, and titanium. All other metals are scarce, and many are in short supply.
4. Igneous, metamorphic, sedimentary, and

weathering processes are responsible for the formation of the ore deposits.

5. Much of our energy is derived from coal and petroleum, which are both nonrenewable resources, and which could be depleted in a relatively short period of time.
6. Oil and gas collect in reservoir rocks, where they accumulate to form *oil pools*. An *oil field* is one or more such pools. Several different geological structures can trap oil and gas, if *roof rocks* (*cap rocks*) are present.
7. Geothermal energy is mainly available in areas of fairly recent igneous activity, but areas that can be exploited are limited in number.
8. Nuclear energy is considered a leading potential source of energy, but major problems, such as the possibility of catastrophic nuclear accidents, the safe disposal of nuclear wastes, and dwindling supplies of uranium fuel, must be solved before its potential can be realized.

QUESTIONS

1. What will be some of the environmental effects of attempts to increase production of fossil fuels and other minerals as the world's population increases exponentially?
2. How does the plate tectonics theory explain ore deposits?
3. What hypotheses can you advance to explain the lack of oil and oil shale in a folded mountain belt, such as the Alps, since other belts have an abundance of these resources?
4. Study the geological map of a particular region, state, or country (all geology departments have them on their walls), and outline prospective areas for mineral and oil exploration.

SELECTED REFERENCES

Abelson, P. H., and Hammond, A. L., eds., 1976, Materials: renewable and nonrenewable resources, American Association for the Advancement of Science.

Bartlett, A. A., 1980, Forgotten fundamentals of the energy crisis, Journal of Geological Education, vol. 28, pp. 4–35.

Committee on Nuclear and Alternative Energy Systems, National Research Council, 1980, Energy in transition, 1985–2010, W. H. Freeman and Co., San Francisco.

Hayes, E. T., 1979, Energy resources available to the United States, 1985 to 2000, Science, vol. 203, pp. 233–39.

Howard, A. D., and Remson, I., 1978, Geology in environmental planning, McGraw-Hill Book Co., New York.

Hubbert, M. K., 1971, The energy resources of the earth, Scientific American, vol. 224, pp. 60–70.

Kesler, S. E., 1976, Our finite mineral resources, McGraw-Hill Book Co., New York.

Levorsen, A. I., 1967, Geology of petroleum, W. H. Freeman and Co., San Francisco.

Lindholm, R. C., 1980, The oil shortage—a story geologists should tell, Journal of Geological Education, vol. 28, pp. 36–45.

Menard, H. W., 1974, Geology, resources, and society, W. H. Freeman and Co., San Francisco.

Miller, B. M., and others, 1975, Geological estimates of undiscovered and recoverable oil and gas resources in the United States, U.S. Geological Survey Circular 725.

Park, C. F., Jr., and MacDiarmid, R. A., 1975, Ore deposits, W. H. Freeman and Co., San Francisco.

Park, C. F., Jr., 1978, Critical mineral resources, Annual Review of Earth and Planetary Sciences, vol. 6, p. 305–24.

Pimental, D., and others, 1973, Food production and

the energy crisis, Science, vol. 182, pp. 443–49.

Ruedisili, L. C., and Firebaugh, M. W., eds., 1978, Perspectives on energy: issues, ideas, and environmental dilemmas, 2nd ed., Oxford University Press, New York.

Skinner, B. J., 1976, Earth resources, Prentice-Hall, Inc., Englewood Cliffs, New Jersey.

Stobaugh, R., and Yergin, D., eds., 1979, Energy future, Ballantine Books, New York.

White, D. E., and Williams, D. L., eds., 1975, Assessment of geothermal resources of the United States—1975, U.S. Geological Survey Circular 726.

APPENDICES

APPENDIX I
Minerals and Mineral Identification

Table 1. Principal occurrences of rock-forming minerals*

Igneous	*Sedimentary*	*Metamorphic*	*Alteration*
1. Olivine			
2. Pyroxene			
3. Amphibole		Amphibole	
4. Biotite } Micas		Biotite	
5. Muscovite } Micas	Muscovite	Muscovite	
6. Anorthite } Plagioclase		Anorthite } Plagioclase	
7. Albite } Plagioclase	Albite	Albite } Plagioclase	
8. Potassium feldspar	Potassium feldspar	Potassium feldspar	
9. Quartz	Quartz	Quartz	Fine-grained silica
	(a) Fine-grained silica		
10. Hematite	Hematite	Hematite	Hematite
	11. Limonite		Limonite
	12. Kaolinite } Clays		Kaolinite } Micas
	13. Montmorillonite } Clays		Montmorillonite } Micas
	14. Illite } Clays		Illite } Micas
	15. Calcite	Calcite	Calcite
	16. Dolomite	Dolomite	
	17. Gypsum		
	18. Halite		
		19. Serpentine	Serpentine
		20. Chlorite	Chlorite
		21. Epidote	Epidote
		22. Garnet	
		23. Talc	

*Listed in terms of the environments in which these minerals are stable under normal conditions of weathering, erosion, or metamorphism. They occasionally occur as metastable compounds in one or more of the other environments.

Luster	Color	Hardness	Cleavage	Description	Mineral
METALLIC LUSTER	DARK-COLORED MINERALS	SCRATCHES GLASS	NO CLEAVAGE	Conchoidal fracture. Luster vitreous. Color pale to olive green and brownish-green. Transparent to translucent. Moderately high specific gravity. H = 6.5 to 7	Olivine $(Mg,Fe)_2SiO_4$
				Conchoidal fracture. Luster vitreous to resinous. Color most commonly shades of red and brown. Transparent to translucent. Frequently occurs in well-formed 12-sided crystals. Moderately high specific gravity. H = 6.5 to 7.5	Garnet group $(Ca,Mg,Fe)_3(Al,Fe)_2(SiO_4)_3$
		DOES NOT SCRATCH GLASS	SHOWS CLEAVAGE	Perfect cleavage in six directions. Luster resinous to adamantine, also submetallic. Color most commonly yellow or brown to black. Transparent to translucent. Rubbed briskly on a streak plate, streak will give a faint H_2S odor. Moderately high specific gravity. H = 3.5 to 4	Sphalerite ZnS
				Perfect cleavage in one direction producing *elastic* folia. Luster vitreous to submetallic. Color dark green or brown to black. Translucent to opaque. Average to moderately high specific gravity. H = 2.5 to 3	Biotite $K(Mg,Fe)_3AlSi_3O_{10}(OH)_2$
				Perfect cleavage in one direction producing *flexible* folia. Luster vitreous to pearly. Color various shades of green to black. Translucent to opaque. Average to moderately high specific gravity. H = 2 to 2.5	Chlorite $(Mg,Fe)_5(Al,Fe)_2Si_3O_{10}(OH)_8$
				Perfect cleavage in one direction producing *flexible* folia. Luster bright to sometimes dull and earthy. Color and streak are black (readily marks paper and soils fingers). Greasy feel. Low specific gravity. H = 1 to 2	Graphite C
				Perfect cleavage in three directions at right angles. Luster is bright. Color and streak are lead-gray to silver-gray. Very high specific gravity. H = 2.5	Galena PbS
			NO CLEAVAGE	Conchoidal or irregular fracture. Luster dull. Color and streak are black. Strongly magnetic (lodestone variety acts as a natural magnet). High specific gravity. H =6	Magnetite Fe_3O_4
				Luster is bright in crystals, but dull and earthy in massive varieties. Color varies from red-brown to black. Light to dark red-brown streak. Subtranslucent to opaque. High specific gravity. H = 5.5 to 6.5 (as low as 1.5 to 2 in massive varieties)	Hematite Fe_2O_3
				Color varies from yellow to brown, occasionally black. Common luster is dull earthy, but may be shining metallic. Yellow-brown streak. Subtranslucent to opaque. Moderately high specific gravity. H = 5 to 5.5 (as low as 1.5 to 2 in dull earthy varieties)	Limonite FeO(OH)
				Conchoidal fracture. Splendent luster, but may be tarnished. Color pale brass-yellow. Greenish or brownish-black streak. High specific gravity. H = 6 to 6.5	Pyrite FeS_2
	LIGHT-COLORED MINERALS	DOES NOT SCRATCH GLASS		Uneven fracture. Brass-yellow color, but often tarnished to bronze or iridescent. Greenish-black streak. Moderately high specific gravity. H = 4	Chalcopyrite $CuFeS_2$

* Specific gravity ranges:
Low <2.5
Average = 2.5 to 3.0
Moderately high = 3.0 to 4.5
High = 4.5 to 6.0
Very high > 6.0

**H = hardness

Table 2. Physical Properties of Common Minerals

Luster	Color	Hardness	Cleavage	Properties	Variety properties	Mineral
NON-METALLIC LUSTER	LIGHT-COLORED MINERALS	SCRATCHES GLASS	SHOWS CLEAVAGE	Two directions of cleavage, one perfect and one good, at nearly right angles. Vitreous to pearly luster. Transparent to opaque. Average specific gravity.* H = 6**	White, gray or blue-gray; less commonly greenish. Striations on perfect cleavage	Feldspars: Plagioclase — Anorthite $CaAl_2Si_2O_8$; Albite $NaAlSi_3O_8$
					White, gray or flesh-pink; occasionally green	Feldspars: Potash — Orthoclase (Microcline) $KAlSi_3O_8$
			NO CLEAVAGE	Conchoidal to sub-conchoidal fracture. Color highly variable, commonly colorless to white. Average specific gravity. H = 7	Transparent to opaque. Vitreous to greasy luster	Silica: Quartz SiO_2
					Translucent to opaque. Waxy to dull luster	Silica: Fine-grained silica SiO_2
		DOES NOT SCRATCH GLASS	NO CLEAVAGE	Massive to fibrous habit resulting in a splintery to fibrous fracture. Waxy, greasy, or silky luster. Most commonly some shade of green, often mottled. Translucent to opaque. Average specific gravity. H = 3 to 5		Serpentine $Mg_3Si_2O_5(OH)_4$
			SHOWS CLEAVAGE	Perfect cleavage in four directions. Vitreous luster. Color variable, most commonly shades of yellow and lavender or purple. Transparent to translucent. Moderately high specific gravity. H = 4		Fluorite CaF_2
				Perfect cleavage in three directions at approximately 75°. Luster vitreous to earthy. Color highly variable, commonly colorless to white. Transparent to opaque. Effervesces rapidly in cold, dilute HCL. Average specific gravity. H = 3		Calcite $CaCO_3$
				Perfect cleavage in three directions at approximately 75°. Luster vitreous to pearly. Color available, but commonly white to pink or tan. Effervesces in cold, dilute HCL *only* when powdered. Crystal faces commonly curved forming "saddle-shaped" crystals. Average specific gravity. Translucent to opaque. H = 3.5 to 4.		Dolomite $CaMg(CO_3)_2$
				Perfect cleavage in three directions at right angles. Luster vitreous to dull. Colorless to white, but color somewhat variable. Transparent to translucent. Salty taste. Low specific gravity. H = 2.5		Halite NaCl
				Perfect cleavage in one direction producing *elastic* folia. Luster vitreous to pearly or silky. Colorless to light yellow or brown. Transparent to translucent. Average specific gravity. H = 2 to 2.5		Muscovite $KAl_3Si_3O_{10}(OH)_2$
				Perfect cleavage in one direction producing *flexible* folia. Luster usually vitreous, but also pearly and silky. Variable color, but commonly colorless to white. Transparent to translucent. Low specific gravity. H = 2		Gypsum $CaSO_4 \cdot 2H_2O$
			NO CLEAVAGE	Luster dull earthy. White, but commonly colored by various impurities. Opaque. Slight soapy or greasy feeling when pure. A clayey or earthy odor when moistened by breathing on it. Plastic when wet. Low to average specific gravity. H = 2		Clay Minerals: Illite $K(Al,Mg,Fe)_2(Al,Si)_4O_{10}$ $[(OH)_2,H_2O]$; Montmorillonite $(Na,Ca)(Al,Mg,Fe)_2Si_4O_{10} \cdot nH_2O$; Kaolinite $Al_2Si_2O_5(OH)_4$
	DARK-COLORED MINERALS	SCRATCHES GLASS	SHOWS CLEAVAGE	Imperfect cleavage in two directions at approximately right angles. Luster subvitreous to dull. Color dark green to black. Translucent to opaque. Commonly occurs in short or stubby four- or eight-sided crystals. Moderately high specific gravity. H = 5 to 6		Pyroxene group Augite is most common mineral $(Ca,Na)(Mg,Fe,Al)(Al,Si)_2O_6$
				Perfect cleavage in two directions at approximately 60° and 120°. Luster vitreous. Color dark green to black. Translucent to opaque. Commonly occurs in elongate six-sided crystals. Moderately high specific gravity. H = 5 to 6		Amphibole group Hornblende is most common mineral $(Ca,Na)_{2\text{-}3}(Mg,Fe,Al)_5(Al,Si)_8O_{22}(OH)_2$
				Cleavage in two directions. One perfect and one imperfect, at approximately 60° and 120°. Luster vitreous. Color yellow-green to dark green to black. Moderately high specific gravity. Translucent to opaque. H = 6 to 7		Epidote $Ca_2(Al,Fe)_3Si_3O_{12}(OH)$

APPENDIX II
Topographic and Geological Maps

A map is a small-scale representation of one or more features on the earth's surface. Road maps—representations of street and highway systems—are probably most familiar to us, although there are many other types of maps. Those maps most useful to geologists also represent landscape forms, or the type of rock outcrops in a particular area. ***Topographic maps*** are landscape maps that indicate elevations and shapes of landforms, or ***topography. Geological maps*** record lithology, orientation, and areal extent of bedrock and surficial deposits at or immediately below the earth's surface.

Scale, the ratio of a unit of distance on the earth's surface to the same distance represented on a map, allows us to measure distance between points on a map. There are three ways of representing scale on a map:

1. A fractional scale (Fig. 1) is a fixed ratio for the distance between any two points on a map and the real distance between these points. For example, a scale of 1 : 62,500 means that 1 linear unit on a map (1 meter, for example), represents 62,500 meters on the ground. Any unit of measure can be used—millimeters, inches, yards, meters, kilometers, or whatever is convenient—as long as the same unit is used on both the map and the ground.
2. A graphic scale, or bar scale, is a line on a map divided into units that represent a definite distance on the ground (Fig. 1).
3. On some simple maps, instead of a graphic scale, we might see the words "one inch represents one mile," for example. Of course, this means that one inch on the map represents one mile on the ground.

TOPOGRAPHIC MAPS

Topographic maps are essentially two-dimensional representations of three-dimensional features on the earth's surface. Symbols are used to represent topographic features. ***Contour lines,*** the most commonly used symbol, connect points of equal elevation above mean sea level and depict the third dimension otherwise missing on a map. Therefore, they are horizontal and define planes parallel to one another (Fig. 2). The vertical, even spacing of contour lines depends on the ***contour interval,*** which represents the vertical separation between lines. Such intervals are generally constant throughout a map.

A good example of a natural contour is the line defining the edge of a lake, because the surface of a body of water is horizontal, and all points around the shoreline are at the same elevation. If the water level drops in 1-meter increments over a period of years, it leaves behind a series of shorelines that represent multiple contour lines, with a contour interval of 1 meter.

Topographic maps also indicate the elevation (altitude) and height of a point, and the relief of a given area. ***Elevation*** is the vertical distance between mean sea level and a point. ***Height*** is the vertical distance between a point and some other local feature; for example, the distance from a valley floor to the top of a mountain. ***Relief*** is the difference in elevation between the highest and lowest points of a particular area or region. An area of high relief would be hilly or mountainous, with considerable distances between valley floors and hilltops or mountaintops, whereas an area of low relief would be smoother and more rolling, with hill tops not far above valley floors.

Certain rules apply to the use of contour lines, or contours, as they are commonly called:

1. All points on a contour have the same elevation.
2. The elevation of a contour will always be a multiple of the contour interval. For example, contours with an interval of 5 meters will have elevations of 5, 10, 15, 20 meters, etc. Generally, every fifth contour line is darker than the others and labeled with the elevation.
3. A contour separates all higher points of elevation from all lower points of elevation.
4. Contours never split or cross, although at vertical or undercut cliffs they will merge.
5. All contours will close on themselves, either within or beyond the limits of the map. In the latter case, the lines will simply end at the edge of the map.
6. Closely spaced contours indicate relatively steep slopes, and widely spaced contours indicate rel-

APPENDIX II

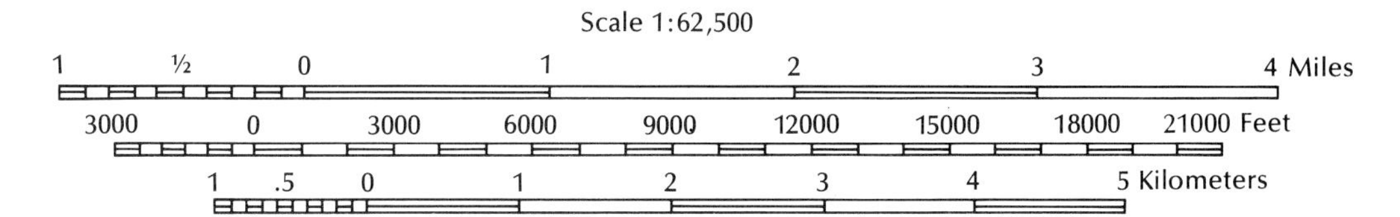

Fig. 1 An example of graphic and corresponding fractional scales used by the U.S. Geological Survey.

Fig. 2 (below) Hypothetical landform (above) as shown on the corresponding topographic map (below). All figures are in feet. Compare with the rules of contour lines.

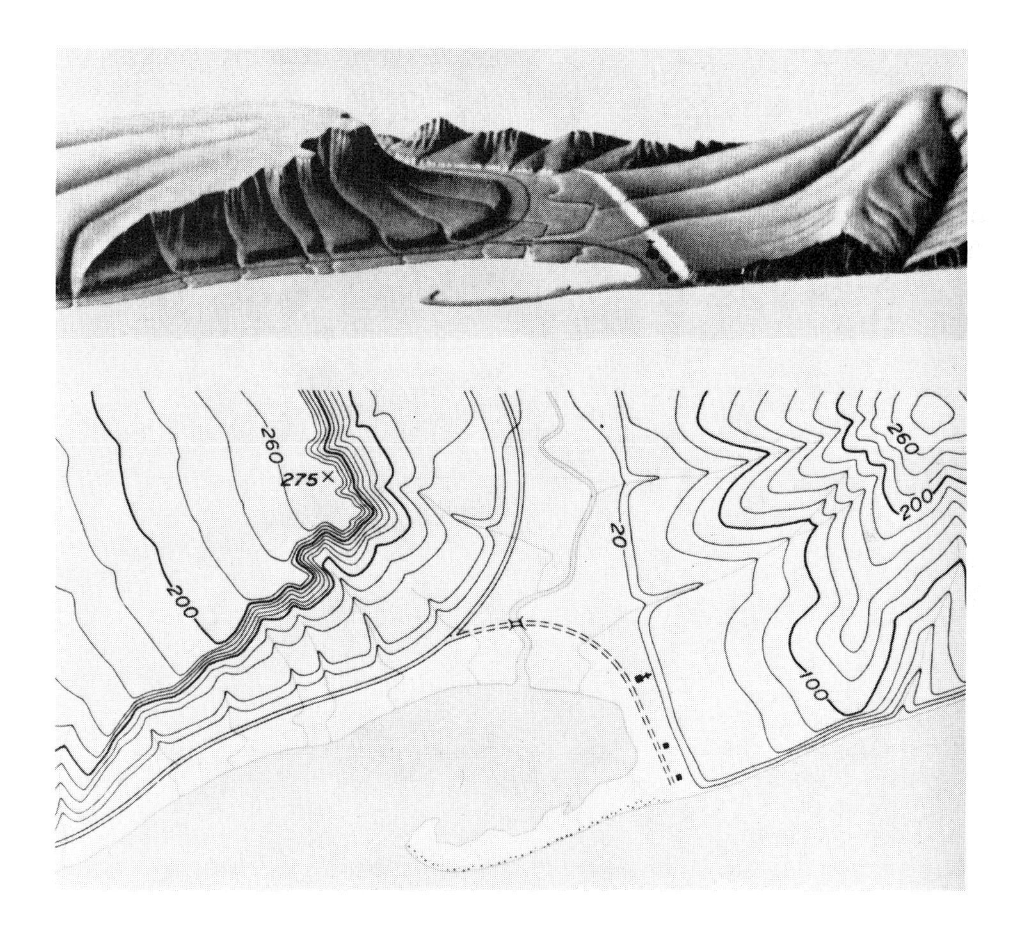

atively gentle slopes. Evenly spaced lines indicate a uniform slope, whereas uneven spacing depicts changes in slope angle.

7. Contour lines crossing a stream valley form a V pointing *up* the valley. In contrast, contour lines crossing a ridge form a V or a U pointing *down* the ridge.
8. Closed contours with hachures (short dashes at right angles to the contour line) indicate depressions. The area within a ***depression contour*** is lower than the area outside the contour.
9. The highest contours on ridges and the lowest ones in valleys always appear in pairs; that is, no single higher contour line can lie between two lower ones of the same elevation, and vice versa.

Spot elevations indicating hill summits, road intersections, lake levels, and other points may also be

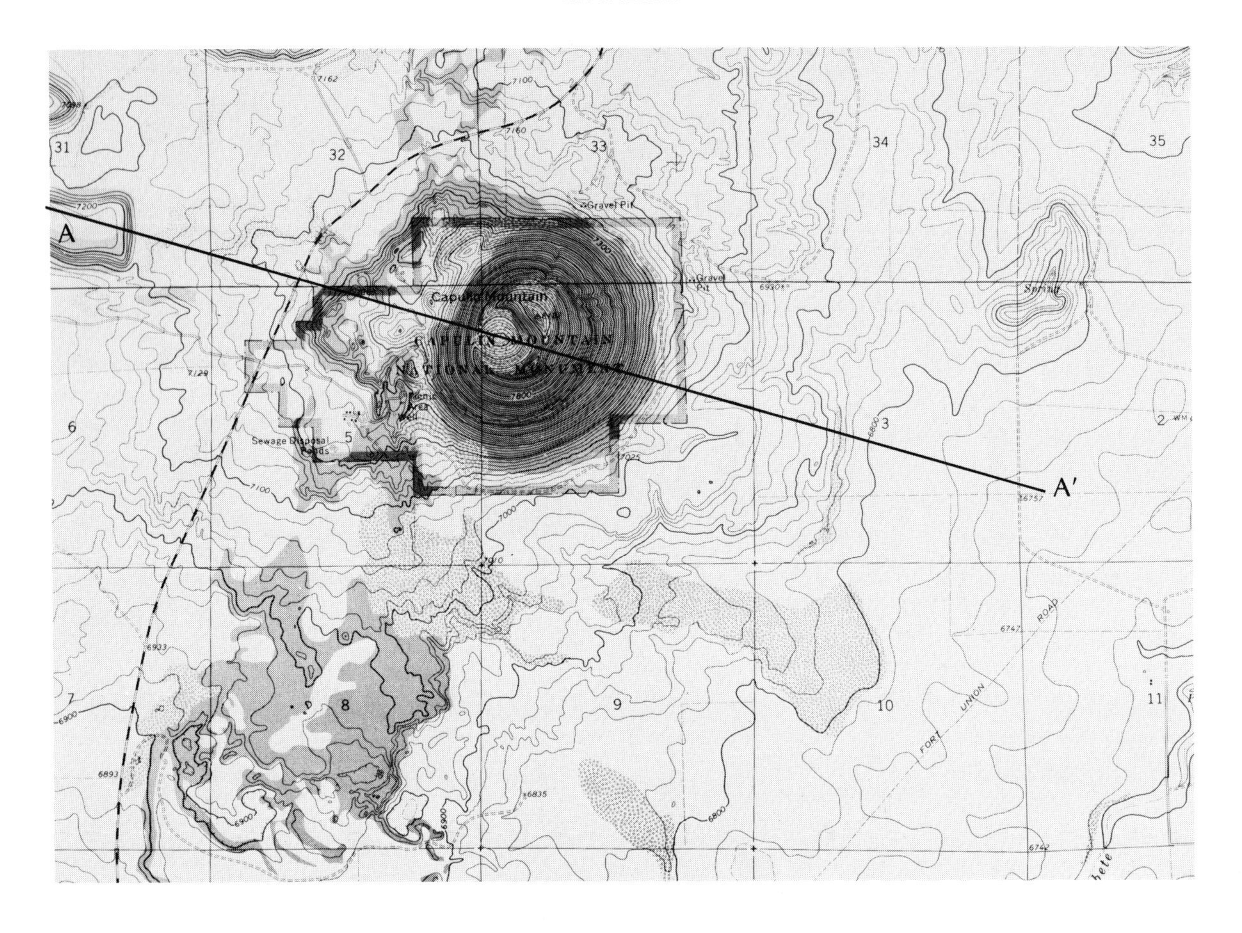

Fig. 3 Topographic map and profile of Capulin Mountain, a young (less than 10,000 years old) cinder cone in northeastern New Mexico. Compare the features on the profile, such as the crater, steep sides of the cone, and relatively level surrounding area, with the contour map. Note that the edge of the lava flow (which commonly extrudes from the base of cinder cones) can be followed for some distance on the map.

shown on topographic maps. Other points of elevation indicated are ***bench marks,*** the elevation of which is more precisely determined. Bench marks are brass plates permanently fixed in the ground, and on a map they are marked by an X followed by the elevation, which is sometimes prefaced by BM. On Figure 3, some bench marks are shown.

TOPOGRAPHIC PROFILES

A drawing of the land surface along a given line is called a ***topographic profile.*** Graphically, it shows the "skyline" from a distance. Features along a topographic profile are viewed along horizontal lines of sight, whereas map features are viewed along vertical lines. A profile can be drawn along any line on any topographic map.

Before a topographic profile is started, both horizontal and vertical scales are chosen. The most commonly used horizontal scale is the scale of the topographic map to be used. The vertical scale often is made somewhat larger than the horizontal scale, or

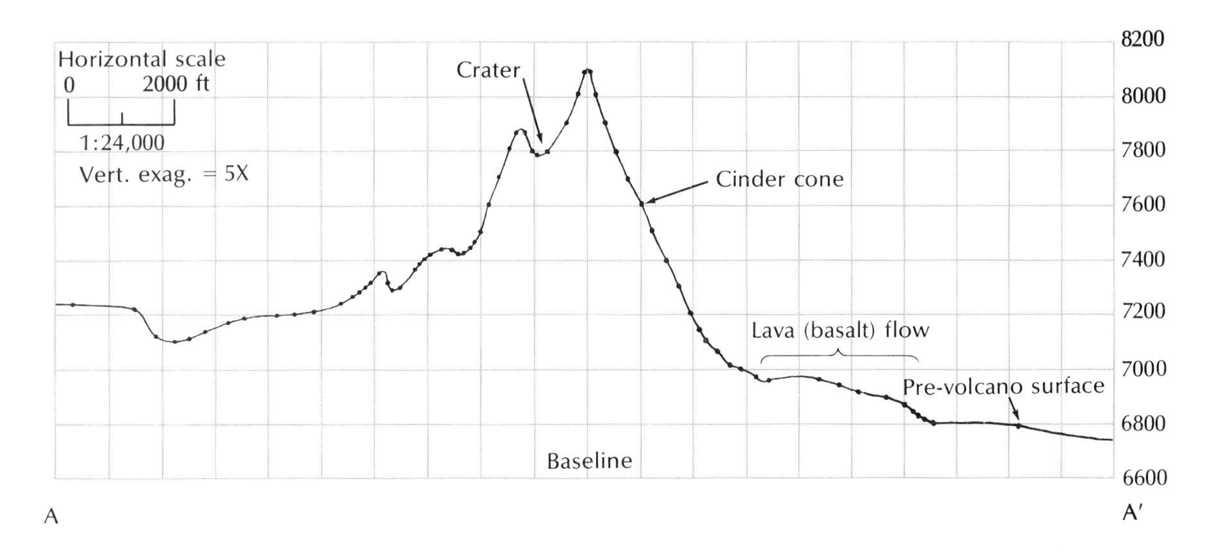

exaggerated, in order to accent topographic features. The difference between the horizontal and vertical scales is called the ***vertical exaggeration.*** For example, if the horizontal scale is 5000 feet to one inch and the vertical scale is 500 feet to one inch, the vertical exaggeration is 10 times the horizontal scale, written 10X. When vertical and horizontal scales are the same, the profile is said to be *true*.

Graph paper is most conveniently used to construct a profile. Once scales have been chosen, the procedure is as follows (Fig. 3):

1. Select one of the horizontal lines on the graph paper for a baseline. The base can be any established elevation, such as sea level, or the lowest point on the profile.
2. Number the vertical scale with the elevation of the baseline and other elevations appropriate to the purpose of the map.
3. Place the edge of the graph paper along the line of the profile on the map.
4. Plot the elevation of every contour line that crosses the profile line on the correct elevation line on the graph paper. The points should be plotted along lines that are perpendicular to the map profile line and the horizontal contour lines of the graph paper. The original vertical lines on the graph can be used as guides.
5. Complete the profile by connecting the points with a smooth line.
6. Indicate the vertical and horizontal scales and the vertical exaggeration on the profile.

GEOLOGICAL MAPS

The preparation of geological maps is usually the first stage in reconstructing the geological history of a given area, which may help to determine where to drill for oil, where to dig for ore, or where to locate a gravel deposit. Whereas topographic maps represent the landscape symbolically, geological maps illustrate how and where rock and sediments appear at the surface of the earth, as viewed from above. The data are gathered by extensive field work and interpretation of aerial photographs.

The units shown on a geological map are usually called ***formations.*** A formation is a rock unit or a deposit with an upper and lower boundary, and is large enough to be identified easily in the field. An example is an exposed white sandstone with drab-colored shales above and below it, as can be seen in a desert valley. Commonly, formations are given the name of the area in which they were first found. For example, the Lyons Sandstone is a formation composed primarily of a kind of sandstone first de-

APPENDIX II

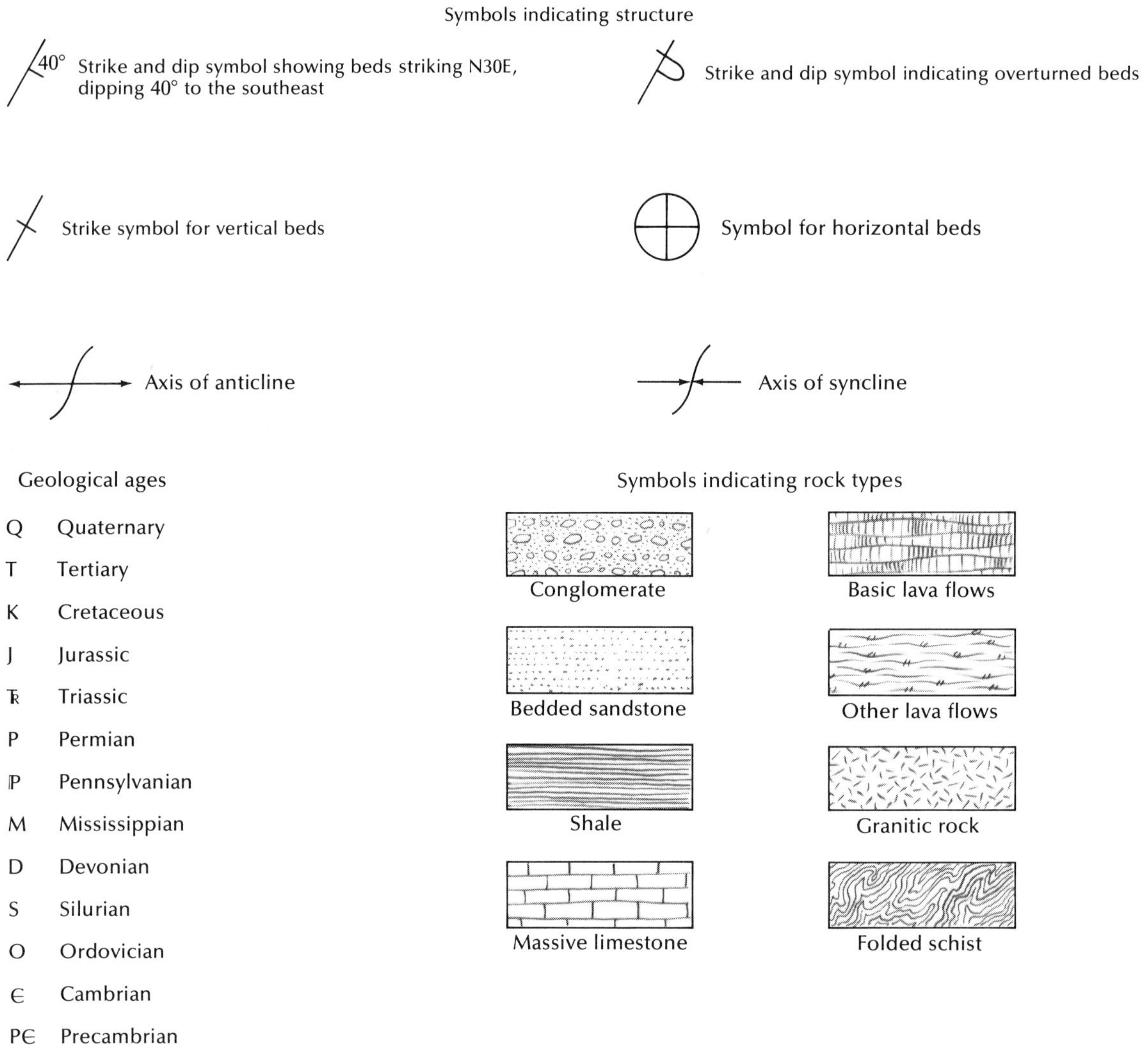

Symbols indicating structure

40° Strike and dip symbol showing beds striking N30E, dipping 40° to the southeast

Strike and dip symbol indicating overturned beds

Strike symbol for vertical beds

Symbol for horizontal beds

Axis of anticline

Axis of syncline

Geological ages

Q	Quaternary
T	Tertiary
K	Cretaceous
J	Jurassic
₸	Triassic
P	Permian
ℙ	Pennsylvanian
M	Mississippian
D	Devonian
S	Silurian
O	Ordovician
Є	Cambrian
PЄ	Precambrian

Symbols indicating rock types

Conglomerate

Basic lava flows

Bedded sandstone

Other lava flows

Shale

Granitic rock

Massive limestone

Folded schist

Fig. 4 Geological map symbols.

scribed near Lyons, Colorado. In contrast, if the composition of a unit is variable, the term "formation" is used instead of the rock type. The Ogallala Formation is composed of layers of clays, silts, sands, gravels, and caliche found in the Great Plains and first described near Ogallala, Nebraska.

On most geological maps, graphic symbols are often used to designate the more common rock types (Fig. 4). Letter symbols are commonly used to designate the geological age and formation name. The Lyons Sandstone, for example, is designated by the symbol ₽l; the ₽ indicating that the rock is of Permian age and the l standing for Lyons (note that the geological age is in capital letters and that the formation name is in lower case).

A ***contact,*** or plane separating two rock units, is shown on geological maps as a line between formations that represents the intersection of the plane and the ground surface.

An ***outcrop*** is exposed bedrock at the surface. Bedrock is commonly covered with soils, sediments, and vegetation, and whether a contact is easily found or inferred depends on the nature and scale of the map and the covering material. Indeed, in places, a certain amount of imagination is required to produce a map where outcrops are few and contacts are obscured. Often the overlying sediments and soils, if of sufficient thickness, extent, and geological significance, can be mapped as a separate unit.

The appearance of an outcrop on a map will vary according to the attitude of the rock (whether it has been folded, faulted, or otherwise deformed) and the nature of the landscape (whether streams have dissected it or not). If contours are shown on the map, considerable information can be gained by studying the outcrop pattern as it relates to the landscape. Some rules are:

1. If beds are horizontal, contacts will parallel the contours, for they are also horizontal. The Grand Canyon is an excellent example (Fig. 5).
2. If beds are tilted and crossed by a stream, the outcrop pattern will form a V pointing downstream if the beds tilt in the downstream direction, and form a V pointing upstream if the dip is upstream. If beds dip downstream at an angle more gentle than the stream gradient, however, the V will point upstream. (Do not confuse these rules with the rules pertaining to contours as they cross stream valleys or ridges.)
3. Beds dipping at the same angle as the stream gradient will crop out parallel to the slope of the stream.

If you wish to locate the older rocks in a particular sequence, in the field, you will need to consider both the structure and the topography of the area. If the beds are flat-lying and you begin at the top of the sequence, the older rocks could only be studied by viewing a canyon or a valley wall. In contrast, if the beds are tilted and the landscape is flat or undulating, all of the rock contacts will intersect the surface, making viewing fairly simple. Of course, you must be able to distinguish the tops of each bed from the bottoms or else you might think you are going down-section, when actually you are going up-section.

The nature of the rocks may also have an effect on the development and appearance of the landscape. This is particularly apparent in an area in which bedrock is composed of alternating layers of resistant beds (such as well-cemented sandstone) and more easily eroded beds (such as shale). If the beds are horizontal and extensive erosion has taken place, the more resistant beds will tend to form steep rims, and the easily eroded beds will form gentle slopes (see Chap. 9, Fig. 9-16). The Grand Canyon is a well-known example (see Chap. 11, Fig. 11-33). In contrast, if beds are tilted, the harder rocks may form ridges or ***hogbacks,*** and intervening softer sediments will form valleys (see Chap. 17, Fig. 17-1). These relationships often are apparent on combined topographic-geological maps.

GEOLOGICAL CROSS SECTIONS

When both topographic and geological information are provided on a map, it is a relatively simple matter to construct a geological cross section, which shows a side view of subsurface rocks and structure, as if the earth were split and separated along a given line. First we draw a topographic profile as outlined previously. Then, below the profile, the geological information found along the trace of the cross section is added. If the beds are horizontal, the cross section is rather simple (Fig. 5), but if the beds are tilted, folded, or faulted, the process of projecting the units at depth is more complicated. Using information on dip, we can accurately draw the projection of the beds, faults, and unconformities (Fig. 6).

A note of caution: vertical exaggeration of geological cross sections, because it will distort the attitude of such structural features as dipping beds and faults, should be avoided. For example, a low-angle cutting shallowly dipping beds would appear as a high-angle fault transecting steeply dipping beds if vertical exaggeration were applied.

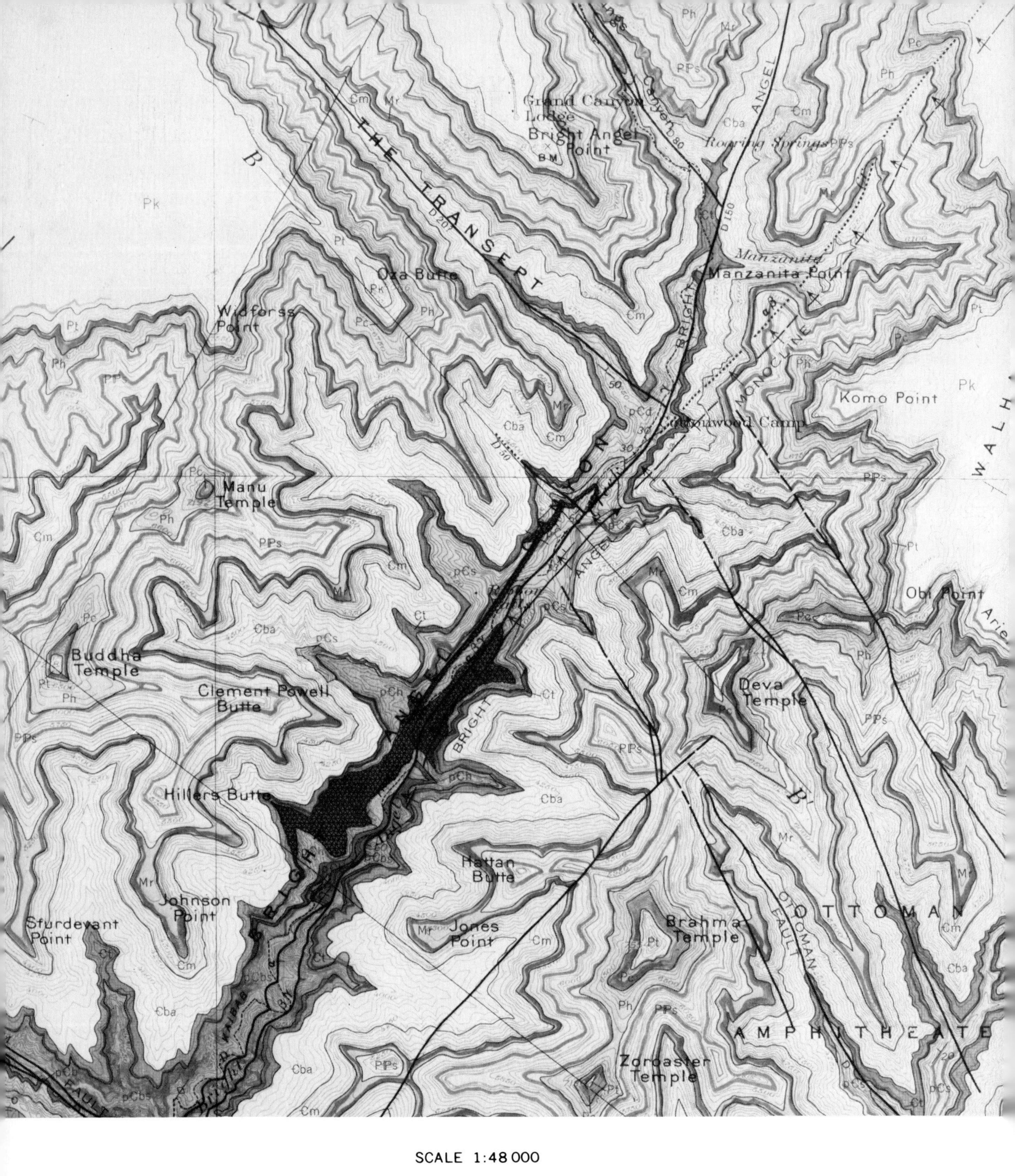

SCALE 1:48 000

1 ½ 0 1 2 3 MILES

CONTOUR INTERVAL 50 FEET
DATUM IS MEAN SEA LEVEL

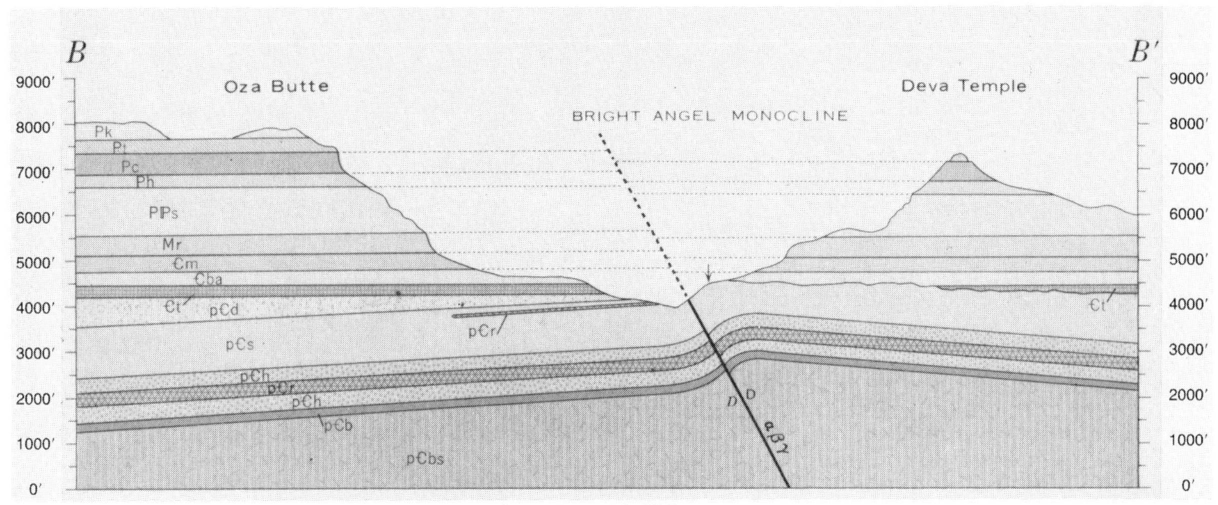

SCALE 1:48 000

Fig. 5 Geological and topographic map and cross section of a portion of the Grand Canyon, Arizona. Note that the contacts between formations approximately parallel the contours, indicating that the beds are horizontal. Also, some beds form steep rims, whereas others form wide benches, as a result of varying degrees of erosion.

LEGEND

Permian	Pk	Kaibab Formation
	Pt	Toroweap Formation
	Pc	Coconino Sandstone
	Ph	Hermit Shale
Permian-Pennsylvanian	P - IP	Supai Formation
Mississippian	Mr	Redwall Limestone
Cambrian	Єm	Muav Formation
	Єba	Bright Angel Formation
	Єt	Tapeats Sandstone
Precambrian	pЄs	Dox Formation
	pЄh	Hakatai Shale
	pЄb	Bass Limestone
	pЄbs	Brahma Schist

Fig. 6 Topographic map (**A**), geological map (**B**), and cross section (**C**) of a portion of the Appalachian Highlands in south-central Pennsylvania. A series of plunging anticlines and synclines is shown. Compare the outcrop pattern with the cross section. The topography shows a sinuous ridge that approximates the outline of the structure; this is due to varying degrees of erosion of the folded beds, with sandstone holding up or outlining the ridge. Note that there is no vertical exaggeration in the cross section.

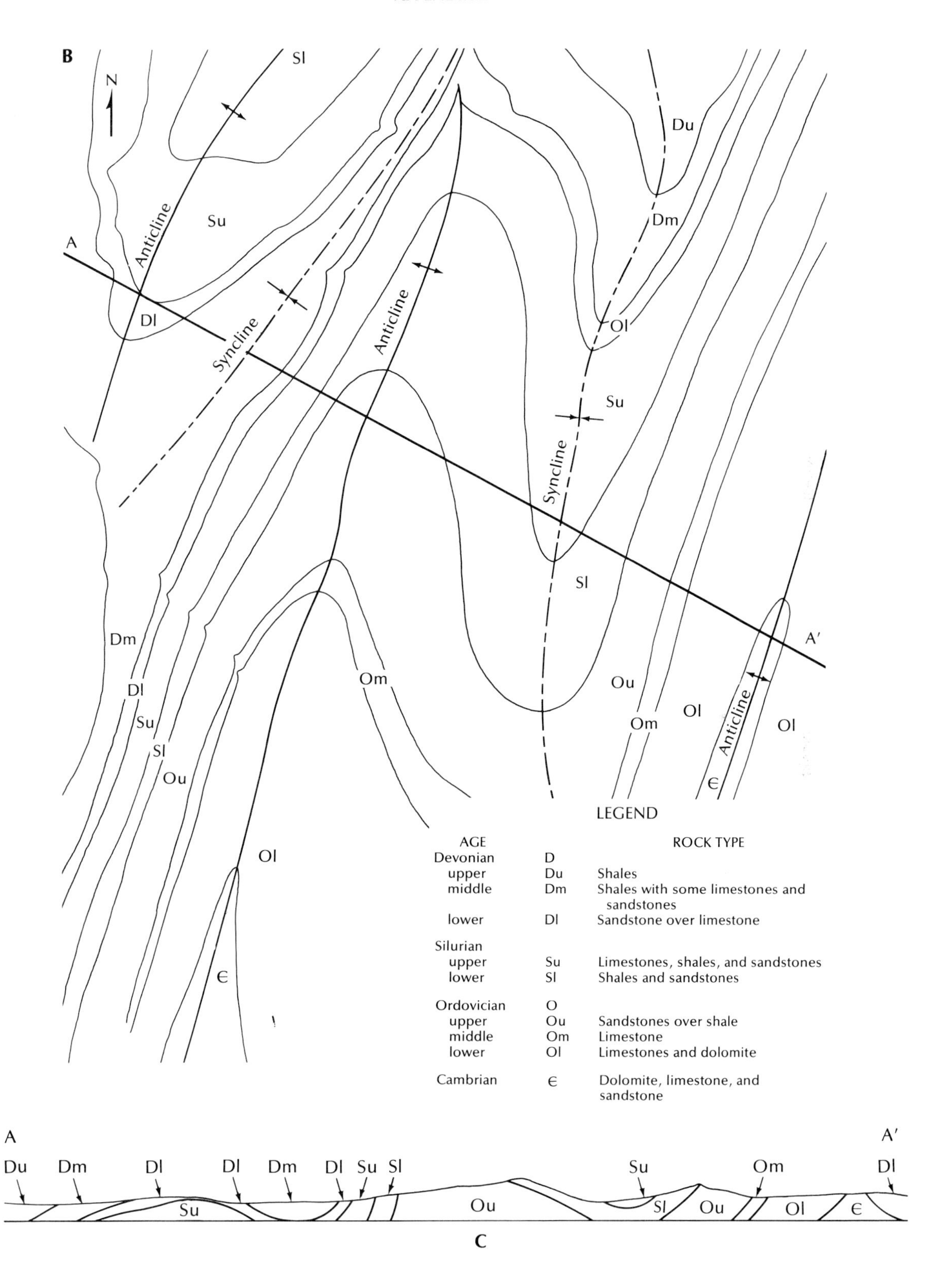
B
N
A
A′
Anticline
Syncline
Anticline
Syncline
Anticline
Sl
Su
Du
Dm
Dl
Ol
Su
Sl
Dm
Dl
Su
Sl
Ou
Om
Ou
Om
Ol
Ol
Ol
∈
∈
LEGEND
AGE
ROCK TYPE
Devonian D
upper Du Shales
middle Dm Shales with some limestones and sandstones
lower Dl Sandstone over limestone
Silurian
upper Su Limestones, shales, and sandstones
lower Sl Shales and sandstones
Ordovician O
upper Ou Sandstones over shale
middle Om Limestone
lower Ol Limestones and dolomite
Cambrian ∈ Dolomite, limestone, and sandstone
A
A′
Du Dm Dl Dl Dm Dl Su Sl
Su Om Dl
Su
Ou
Sl Ou Ol ∈
C

APPENDIX III
Conversion Tables

Length

Inches	×	2.54	=	Centimeters
	×	0.0254	=	Meters
	×	0.0833	=	Feet
	×	0.0277	=	Yards
Feet	×	0.3048	=	Meters
	×	12.	=	Inches
	×	0.3333	=	Yards
	×	0.00019	=	Miles
Miles	×	1609.34	=	Meters
	×	1.609	=	Kilometers
	×	63360.	=	Inches
	×	5280.	=	Feet
	×	1760.	=	Yards
Centimeters	×	0.3937	=	Inches
	×	0.0328	=	Feet
	×	0.0109	=	Yards
	×	0.0100	=	Meters
Meters	×	39.370	=	Inches
	×	3.2808	=	Feet
	×	1.1936	=	Yards
	×	0.0006	=	Miles
Kilometers	×	3280.83	=	Feet
	×	1093.61	=	Yards
	×	0.6214	=	Miles
	×	1000.	=	Meters

Mass

Ounces	×	28.3495	=	Grams
	×	0.0625	=	Pounds
Pounds	×	453.5923	=	Grams
	×	0.4536	=	Kilograms
Tons (long)	×	1016.05	=	Kilograms
	×	1.106	=	Tons (metric)
	×	2240.	=	Pounds
	×	1.12	=	Tons (short)
Tons (short)	×	907.185	=	Kilograms
	×	0.9072	=	Tons (metric)
	×	2000.	=	Pounds
	×	0.8928	=	Tons (long)
Grams	×	0.0352	=	Ounces
	×	0.0022	=	Pounds
	×	0.001	=	Kilograms
Kilograms	×	35.2739	=	Ounces
	×	2.2046	=	Pounds
Tons (metric)	×	2204.62	=	Pounds
	×	0.9842	=	Tons (long)
	×	1.1023	=	Tons (short)
	×	1000.	=	Kilograms

Area

Square inches	×	6.4516	=	Square centimeters
	×	0.000645	=	Square meters
	×	0.006944	=	Square feet
Square feet	×	929.03	=	Square centimeters
	×	0.0929	=	Square meters
	×	144.	=	Square inches
	×	0.1111	=	Square yards
	×	3.587×10^{-8}	=	Square miles
Square miles	×	2589988.	=	Square meters
	×	2.589988	=	Square kilometers
	×	2.7878×10^{7}	=	Square feet
	×	3.0976×10^{6}	=	Square yards
Square centimeters	×	0.1550	=	Square inches
	×	0.0011	=	Square feet
	×	0.00012	=	Square yards
	×	0.0001	=	Square meters
Square meters	×	1550.0031	=	Square inches
	×	10.7639	=	Square feet
	×	1.1960	=	Square yards
Square kilometers	×	1.0763×10^{7}	=	Square feet
	×	1.1960×10^{6}	=	Square yards
	×	$1. \times 10^{6}$	=	Square meters

Volume

Cubic inches	×	16.387	=	Cubic centimeters
	×	1.6387×10^{-5}	=	Cubic meters
	×	0.00058	=	Cubic feet
	×	2.1433×10^{-5}	=	Cubic yards
Cubic feet	×	28316.	=	Cubic centimeters
	×	0.0283	=	Cubic meters
Cubic centimeters	×	0.0610	=	Cubic inches
	×	3.5314×10^{-5}	=	Cubic feet
	×	1.308×10^{-6}	=	Cubic yards
	×	$1. \times 10^{-6}$	=	Cubic meters
Cubic meters	×	61023.74	=	Cubic inches
	×	35.3146	=	Cubic feet
	×	1.3079	=	Cubic yards
Cubic kilometers	×	0.240	=	Cubic miles

Velocity

Miles per hour	×	1.60934	=	Kilometers per hour
Kilometers per hour	×	0.62137	=	Miles per hour

Temperature

To convert from °Fahrenheit to °Celsius, subtract 32 and multiply by 5/9
To convert from °Celsius to °Fahrenheit, multiply by 9/5 and add 32

GLOSSARY

Aa A lava flow with a rough, blocky appearance.
ablation The processes by which snow or ice is lost from a glacier.
abrasion The mechanical wearing, grinding, or scraping of a rock surface by the friction and impact of moving rock particles.
absolute time The best estimate of the actual age, in years, of an object or event. Generally determined by the decay of radioactive elements.
abyss The deep ocean floor.
abyssal hills Hills that rise up to 1000 m above the deep ocean floor.
abyssal plains The relatively extensive low-relief plains of the deep ocean floor.
adobe A clayey deposit mixed with silt. Originally used to make sun-dried bricks in the Southwest.
aftershock A generally smaller earthquake that occurs after a larger earthquake and is closely related in origin.
aggradation (stream) Deposition of a stream's excess load on the channel bottom.
A horizon The dark-colored surface layer, or topsoil, of the soil profile.
A2 horizon A whitish layer in a podzol that lies between the A and B horizons and from which most of the iron oxides have been removed by downward-percolating water.
air-fall ash Volcanic ash that settles out of the air and can be deposited thousands of kilometers from the volcano's vent.
alluvial fan A fan-shaped deposit of sand and gravel deposited by a stream at the base of a mountain, usually in an area of arid or semi-arid climate.
alluvium Any clastic material deposited by a stream along its course.
alpine chain A complex linear mountain belt made up of batholithic igneous rocks, dynamothermal metamorphic rocks, and complexly folded and faulted sedimentary rocks.
alpine glacier A mountain glacier that moves downslope under gravity, often following a pre-existing stream valley. Also called a valley glacier.
amphibole group A large group of ferromagnesian silicates with similar physical properties. Their most distinctive property is the possession of two good cleavages intersecting at 124° and 56°.
amphiboles A group of dark, rock-forming ferromagnesian silicate minerals characterized by a double-chain arrangement of the SiO_4 tetrahedra. Hornblende, a black mineral, is one of the most common varieties.
andesite A gray to grayish-black volcanic rock composed mostly of intermediate plagioclase, augite, hornblende, and biotite. Volcanic equivalent of a diorite.
angle of repose The maximum slope at which relatively loose material will remain stationary without sliding downslope.
angular unconformity An *unconformity* in which the beds above and below the unconformity meet at an angle.
anion A negatively charged ion.
anorthosite A variety of gabbro consisting almost entirely of coarse crystals of calcium-plagioclase.
antecedent stream A stream that has maintained its original course despite the occurrence of local uplift or movement along its course.
anthracite The most highly metamorphosed form of coal, containing a higher percentage of carbon than bituminous coal.
anticline A fold in which the limbs dip away from the axis; older beds are found toward the axis.
aphanitic texture A crystalline rock texture in which most of

the crystals are too small to be seen by the unaided eye.

apparent polar wander curve The curve created by plotting on a map the succession of paleomagnetic poles for rocks of different ages from any one continent or lithospheric plate.

aquiclude A subsurface layer too impermeable or too tight to accept or transmit water.

aquifer A subsurface rock layer that readily yields water.

arête A jagged, thin rock wall or ridge separating two glaciated valleys, generally near their heads.

arkose A sandstone composed mainly of quartz and feldspar.

arroyo The steep-sided, flat-floored channel of an ephemeral or intermittent stream.

artesian well A well that taps a confined aquifer.

ash Volcanic fragments the size of dust blown into the air during an eruption.

ash flow See *glowing avalanche.*

asteroid belt A zone between Mars and Jupiter in which quantities of relatively small rock-like particles orbit the sun. Some meteorite showers originate in this zone.

asthenosphere The layer of the earth extending from 100 km below the surface to 250 km. It appears to be less strong than the zones above and below it and corresponds to the low-velocity earthquake zone, i.e., where *S* waves travel at reduced velocities.

atmosphere The gaseous envelope that surrounds the earth. It is composed largely of nitrogen (78 per cent by volume) and oxygen (21 per cent).

atoll A ring-shaped island with an interior lagoon and made up of the calcareous skeletons of marine animals.

atomic number A number characteristic of a chemical element and equal to the number of protons in an atom of that element.

atomic weight Of an element, a number that indicates, on the average, how heavy an atom of that element is compared to an atom of hydrogen.

augen Shear-resistant minerals in cataclastic metamorphic rocks; they appear as relatively large, rounded clots, or eyes.

augite A dark ferromagnesian mineral of the pyroxene group with two good cleavages that intersect at 87° and 93°.

aureole A halo of hydrothermally metamorphosed rocks that surrounds most batholiths.

axial plane A plane connecting the axial lines in successive beds in a fold.

axis (fold) A line drawn along the points of maximum curvature of a fold.

back-arc upwelling A secondary zone of divergence of lithospheric plates that commonly forms between a continent and an island arc system.

backswamp A section of low-lying ground between a natural levee and the river bluff.

backwasting A process by which valley slopes retreat essentially parallel to themselves.

badlands Rough, steeply gullied terrain usually found in dry areas.

bajada An apron of a desert mountain range formed from lateral coalescing of alluvial fans.

bar A spit that has grown almost long enough to close off a bay.

barchan A symmetrical, crescent-shaped dune that forms in deserts in which the wind direction is nearly constant, sand supply is limited, and the ground surface (bedrock) is hard.

barrier islands Long, narrow sand islands that are roughly parallel to the shore and are separated from it by a narrow body of water.

barrier reef A long, narrow coral reef roughly parallel to the shore and separated from it by a fairly deep and wide lagoon.

basal slip (glaciers) The sliding of glaciers over underlying material.

basalt A dark, volcanic rock, fine grained to aphanitic in texture and composed largely of pyroxene, calcium-plagioclase, and olivine. It is the most abundant volcanic rock.

base level (stream) The low point to which most streams flow, ultimately the ocean.

basin (structure) A saucer-shaped synclinal structure which lacks an axis and in which the beds dip toward the center.

batholith A pluton with an exposed surface area of at least 100 km; generally granitic in composition.

bauxite A chief ore of aluminum, formed as an end product of weathering in tropical climates.

bedding plane The surface that parallels the layers in sedimentary rock.

bed load The material moved along the bottom of a stream.

bedrock The rock that underlies soil or other unconsolidated surface material.

Benioff Zone The planar dipping zone of shallow, intermediate, and deep earthquakes that is common to the Pacific Ocean margins.

B horizon The subsurface horizon below the A horizon in the soil profile. Contains the most clay and is usually red to brown in color.

biotite A common ferromagnesian mineral of the mica group. Sometimes called black mica.

birdsfoot delta A delta formed by many distributaries of a river and resembling a bird's talons.

bituminous coal A soft black coal formed by the metamorphism of *lignite.*

block faulting A type of faulting in which the crust is broken into a number of sub-parallel blocks that are displaced with respect to each other, primarily by dip-slip motion.

body waves Seismic waves (*P* and *S* waves) that travel through the interior of the earth.

bottomset bed The horizontal strata at the bottom of a delta, deposited progressively in front of the advancing delta front.

boulder-clay A glacial deposit consisting of rocks the size of boulders set in a clayey matrix.

Bowen's reaction series The general sequence of crystallization in cooling magmas, going from minerals stable at higher temperatures to those stable at lower temperatures.

braided stream A stream that flows in a network of wide, anastomosing channels separated by low bars or islands.

breccia A clastic sedimentary rock composed of angular fragments cemented together.

breeder reactor A nuclear reactor which creates fissionable material artificially. It is called a breeder reactor because the reaction produces, or "breeds" more material than it begins with.

calcareous ooze Fine-grained sediment covering much of the ocean floor in tropical and warm-temperate seas and consisting of microscopic calcareous shells.

calc-silicate rocks Metamorphic rocks formed from sedimentary rocks rich in both calcite (or dolomite) and silicate clasts (predominantly quartz).
calcite A calcium-carbonate mineral with three directions of cleavage; effervesces in acid.
caldera A large, basin-shaped depression formed by the inward collapse of a volcano after an eruption.
caliche See *K horizon*.
calving The process by which a chunk of ice breaks away from an ice sheet into water.
cap rock Impermeable rock, commonly shale, that prevents oil and gas from escaping upward from the reservoir rock.
capacity The potential load that a stream can carry.
capillary fringe A relatively thin zone of water that migrates upward from the water table.
carbonates Sedimentary rocks made up of calcium or magnesium carbonate.
carbon-14 dating (radiocarbon dating) A method of radiometric dating in which decay of the isotope carbon-14 into nitrogen-14 is measured. The level of carbon-14 is continually readjusted in all organisms to an equilibrium level. Once they die, decay is not compensated for and steadily decreases, reaching an unmeasurable level in about 50,000 years.
cataclastic metamorphic rocks Rocks formed by shearing and granulation during fault movement. The process is called cataclastic or dynamic metamorphism.
catastrophism The belief that geological history occurs as a sequence of sudden, catastrophic events.
catenary A curve produced by a wire or chain suspended from two points. The cross section of a U-shaped valley approximates a catenary.
cation A positively charged ion.
Cca horizon In arid-region soils, a layer of calcium carbonate beneath the B horizon.
cementation The process by which sediments consolidate: a cement (usually $CaCO_3$ or SiO_2) is deposited in the pore spaces separating the sedimentary particles.
cenote In Yucatán, Mexico, a sinkhole formed by the collapse of the roof of an underlying cave.
Cenozoic Era The major division of geological time following the Mesozoic Era and beginning about 65 million years ago. Characterized by the rapid evolution of mammals.
channeled scabland The basalt plateau in eastern Washington crossed by deep, dry channels that formed during a huge prehistoric flood.
chemical remanent magnetization (CRM) Permanent magnetization that results from the growth of magnetic grains in a rock subsequent to its formation through chemical alteration.
chemical weathering The weathering of rocks and minerals by chemical reactions.
chert A dense, hard, sedimentary rock made up of cryptocrystalline silica.
chlorite A group of hydrous silicates containing magnesium, aluminum, and other metal ions. Occurs in thinly banded masses and is grass-green to blackish-green in color.
C horizon The slightly weathered layer in a soil profile beneath the B horizon.
cinder cone A relatively small volcanic vent commonly composed almost entirely of pyroclastic material.
cinders Pebble-sized, reddened pyroclastic fragments blown from a volcano. Abundant in cinder cones.
cirque A horseshoe-shaped, steep-walled glaciated valley head.
clast A fragment produced from the breakdown of a pre-existing rock.
clastic sediment Sediment made up of fragmental material derived from pre-existing rocks of any origin.
clastic texture A fragmental texture usually associated with sedimentary rocks in which the angular-to-rounded grains (clasts) are broken fragments of pre-existing minerals and rocks.
clay minerals A group of hydrous aluminosilicate minerals that result from the weathering of rocks.
clay-size A term used to designate the particle size of a sediment as very fine, and smaller than that of silt.
cleavage A planar surface or set of parallel surfaces along which a mineral will tend to split when broken.
coal A partially metamorphosed sedimentary rock formed from decomposed and altered plant remains.
columnar jointing The joint pattern that forms when lava flows or tabular, shallowly buried magma bodies cool.
compaction The squeezing together of particles in a sediment.
competence (stream) The ability of a stream to transport sedimentary particles, numerically measured as the diameter of the largest particle transported.
composite volcano See *stratovolcano*.
compositional zonation A partitioning of melt within the magma chamber such that the melt in the upper part of the chamber is relatively cooler and more felsic than that in the lower part.
conchoidal fracture A type of fracture that resembles the markings on a conch shell; the surface has a number of concentric irregularities.
concordant intrusion A pluton, typically a sill, that intrudes generally parallel to the layering of a metamorphic or sedimentary rock.
concretion The rounded bodies in sedimentary rock formed when cement preferentially collects in abundance around a small particle.
cone of depression A depression in the water table in the shape of an inverted cone; it develops around a heavily pumped well.
confined aquifer A layer of permeable material, such as sandstone, enclosed between layers of impermeable rock.
conglomerate A clastic sedimentary rock composed of rounded gravels cemented together.
contact The surface between two different types or ages of rocks.
contact metamorphic rocks Rocks recrystallized through heating marginal to a pluton.
continental drift The concept that the continents have moved apart relative to one another.
continental rise The area of coalescent sedimentary fans between the continental slope and the abyssal plain.
continental shelf The shallow, gradually sloping platform that surrounds almost all of the continents. Its width averages 80 km but ranges from 0 to 1500 km.
continental slope The slope that extends downward from the shelf break to the abyssal plain.
contour An imaginary line connecting points of the same value.
convection current The postulated circular movement of mantle material in a cell, as a result of local heating at the base of the cell.

coral Any of a large group of shallow-water, bottom-dwelling marine invertebrates having skeletons consisting of calcium carbonate. Common in warm seas.
core That position of the earth below the mantle, at a depth of about 2900 km. Apparently composed largely of iron, the core is divisible into two portions—an outer one, down to 5100 km, which is liquid, and an inner one, from 5100 to 6300 km, which apparently is solid.
core-mantle boundary The boundary, at a depth of about 2900 km, that marks a change from solid silicate rocks above to the fluid core below. The core is composed primarily of iron.
Coriolis force The apparent force that causes objects traveling over the earth's surface to deflect toward the right or left into a curved path.
corona A zone of rapidly moving ionized gases surrounding the photosphere of the sun and extending far out into space. The corona is visible only when the photosphere is blocked out, as during an eclipse of the sun by the moon.
country rock Pre-existing rocks into which plutonic rocks intrude.
covalent bonding A type of bonding in which the electron is shared by two atoms.
crater A funnel-shaped depression at the summit of a volcano marking the conduit through which volcanic products are erupted.
creep The slow movement of shallow soil material downslope.
crevasse A deep split, fissure, or crack near the surface of a glacier, caused by stresses during glacial movement.
cross-bedding The original layering of sedimentary rocks in which the layers are inclined, sometimes at steep angles to the horizontal.
crossing (river) The shallow part of a channel at the bend of a meander.
crust The rocks above the Moho (Mohorovicić discontinuity).
crystal A solid chemical compound or element having a regularly repeating atomic arrangement (crystal lattice) and commonly bounded by plane surfaces (crystal faces) that parallel prominent lattice planes.
crystal fractionation The physical separation of crystals from residual magma, producing one type of magmatic differentiation.
crystal lattice See *crystal.*
crystalline texture A texture of interlocking mineral grains formed as a result of crystallization from a magma, precipitation of minerals, or metamorphic recrystallization.
crystallinity See *long-range ordering*.
cuesta A ridge formed by a resistant bed that dips at a low angle; has a gentle slope on one side and a steep slope on the other.
Curie temperature The temperature above which a rock loses its permanent magnetism and becomes virtually non-magnetic. The Curie temperature of pure magnetite is 580°C.
current ripples The corrugated surface made by a water current flowing across a sandy base or by the wind blowing across desert sands. The ripples are asymmetric in cross section, with the more gentle slope facing the upcurrent or wind direction.
cut bank The outside bank in a meander, cut by lateral erosion of the stream.
cutoff A channel eroded through the neck of land between two bends, or meanders, of a river.
cut terrace A river terrace consisting of a thin layer of gravel resting on a fairly smooth bedrock surface.

daughter isotope The product of decay of a radioactive parent isotope.
debris flow A mass movement of high fluidity in which over half of the solid material is greater than sand size.
décollement A complexly folded overthrust sheet common to alpine chains.
deep-focus earthquake Originating in the depth zone between 300 and 700 km.
deep-sea trench See *trench*.
deflation Erosion of earth materials by the wind.
degradation (stream) The downcutting of a stream's channel.
delta A low, nearly flat body of sediment deposited near the mouth of a river.
dendritic drainage A drainage system in which tributary streams join the main stream in an irregular pattern resembling the branching of a tree.
dendrochronology The study of tree-ring patterns for dating of the recent past (back to a few thousand years ago).
density current A gravity-induced underflow of relatively more dense water. Density differences may be affected by temperature, salinity, or sediment content.
depositional remanent magnetization (DRM) Permanent magnetization resulting from the alignment of magnetic clasts with the earth's field during sedimentation processes.
depression contour A contour line with short dashes at right angles to the contour pointing toward a depression in the earth's surface.
desert pavement See *stone pavement*.
desert varnish A thin, shiny bluish-black coating composed largely of iron and manganese oxides formed in desert regions on stones and cliffs.
devitrification Conversion of volcanic glass to a fine, crystalline intergrowth, over a long period of time.
diabase Generally coarse-grained variety of basalt in which plagioclase occurs in an interlocking network of crystals and grains of pyroxene.
diapir A salt dome that has pushed through overlying sedimentary rock.
diastrophism Large-scale deformation of the earth's crust, such as folding, faulting, subsidence, or uplift.
diatom A marine plant that secrets silica. The remains form siliceous ooze.
differential weathering Variation in the rate of weathering in rocks. Ledges, recesses, and irregular forms are the result.
differentiation In cooling magma, the process that involves separation of earlier-formed minerals from the residual magma.
dike A discordant tabular intrusion of magma that cuts across the layering of the country rock.
dike swarm A group of dikes intruded at about the same time and possessing a geometrical relationship. Most commonly they occur in parallel alignment but also can occur concentrically or radially.
diorite A drab, gray coarse- to fine-grained plutonic rock with a composition midway between granite and gabbro.

Composed mostly of calcium-plagioclase, amphibole, pyroxene, and biotite.

dip The angle between a dipping bed or surface and a horizontal plane, measured perpendicular to the line of strike.

dipolar field A relatively simple magnetic field configuration consisting of two magnetic poles called the north and south poles. Analogous to a bar magnet.

dipping bed A layer of sedimentary rock tilted at an angle to horizontal.

dip-slip fault A fault in which movement (slip) is parallel to the dip of a fault.

discharge (stream) The quantity of water that passes a designated point in a given interval of time.

disconformity An unconformity in which the rock layers are parallel above and below the unconformity and a considerable gap in sedimentation is represented by the unconformity.

discordant intrusion A pluton, typically a dike, that cuts across the layering of a metamorphic or sedimentary rock.

disharmonic fold A complex fold in which the subsurface geometry has little resemblance to the surface geometry.

dissolved load (stream) The material a stream carries in solution.

distributaries The branches of a stream into which a river divides when it reaches a delta or alluvial fan.

dolomite A carbonate mineral with the composition $CaMg(CO_3)_2$.

dolostone A carbonate sedimentary rock in which the mineral dolomite predominates.

dome A shield-shaped anticlinal structure that lacks an axis and plunges nearly equally in all directions away from the dome crest.

doubly-plunging fold A fold in which the axis is either arched upward or bowed downward.

downwasting A process in which the slopes adjacent to a stream valley are gradually diminished by rock decomposition, soil creep, and other mass-movement processes.

drainage basin The entire area that gathers water and ultimately contributes it to a given river.

drumlin An elliptical, rounded low hill formed by glaciers and occurring in groups.

dynamic metamorphism See *cataclastic metamorphic rocks.*

dynamothermal metamorphic rocks Generally foliated rocks recrystallized by heat, pressure, and chemical solutions during alpine mountain-building processes. Also called regionally metamorphosed rocks.

earth flow A relatively slow earth movement usually with a spoon-shaped sliding surface and a crescent-shaped cliff at the upper end. Commonly breaks up internally.

earth reference ellipsoid The ellipsoid of rotation closely approximating the size and shape of the earth.

echo sounder An instrument that records ocean depths by measuring the time required for a sound impulse to travel to the sea floor and back.

ecliptic plane The plane of rotation of the moon about the earth.

effluent stream A stream that flows at the level of the water table and derives some of its water from the latter.

elastic rebound theory The theory that movement along a fault results from an abrupt release of elastic strain energy that has accumulated in the rock masses on either side of a fault as a result of their deformation.

elastic strain energy The energy stored in a rock body during deformation, which can be regained during faulting.

electron A negatively charged particle of low mass outside the nucleus of an atom.

ellipsoid of rotation The shape that is generated when an ellipse is revolved about one of its axes.

end moraine See *terminal moraine.*

eolian Pertaining to the wind.

ephemeral stream A stream that does not flow continuously. Also called an intermittent stream.

epicenter The point on the earth's surface above the focus of an earthquake.

epoch A geologic time-unit and subdivision of a period.

era A major division in geologic time; eras are divided into lesser units called periods.

erg A broad, dune-covered area in a desert.

erratic (glacial) A rock or boulder carried from its source by glacier ice or an iceberg, and deposited when the ice melts.

escarpment A steep cliff formed either directly as a result of erosion or as an erosional modification of a previously existing step-like topographic feature, such as a fault scarp.

esker A narrow, sinuous ridge of stratified sediment probably deposited by streams flowing in ice tunnels at the bottom of a glacier.

estuary A funnel-shaped mouth of a coastal river valley formed as a result of a rise in sea level or land subsidence. Also called a tidal river.

eustatic change (sea level) A change in sea level resulting from changes in volume of water in the ocean.

evaporite A sedimentary rock that results primarily from the evaporation of water containing dissolved solids.

evapotranspiration The transfer of water to the atmosphere through evaporation and through transpiration from plants and animals.

exfoliation (sheeting) The process by which concentric layers form at the surface of bare rocks.

fabric The way in which grains or crystals in a rock fit together. Involves consideration of relative sizes and shapes.

facies changes (sedimentary) The lateral variation in sedimentary rock bodies due to lateral variation in the depositional environment.

fault A fracture along which significant movement has occurred.

fault block A segment of crust bounded by a fault on one side (tilted fault block) or by faults on both sides.

fault-block mountains Mountain ranges bounded at one or both lateral margins by large, high-angle normal or reverse faults.

fault gouge A clayey, soft material formed when the rocks adjacent to a fault are pulverized during slippage.

fault scarp A low, linear cliff resulting from displacement of the earth's surface during fault movement.

fault-zone breccia Rocks in a fault zone that are broken and sheared as a result of fault movement.

feldspar A silicate mineral group that includes both potassium feldspar and plagioclase (rich in sodium and calcium). Has two good cleavages that intersect at 90°, striations on one of the cleavage faces, and a hardness of 6.

felsic rocks Igneous rocks containing a large proportion of feldspar and silica quartz.

fenster (window) An erosional window through a thrust sheet that displays the rocks beneath the sheet.

ferromagnesian mineral A silicate mineral containing relatively abundant iron and magnesium and that tends to be dark in color.

fetch The distance or area in an open body of water over which wind friction can affect waves.

fill terrace A river terrace made up of relatively thick deposits of gravel resting on a bedrock surface.

fiord A narrow glacial valley partly submerged by the sea.

firn The gritty, granular snow formed when snowflakes melt and refreeze several times.

fissility The tendency of sedimentary rocks to split along well-developed and closely spaced planes.

fissure eruption An eruption that takes place through a fissure, or large crack, rather than through a localized volcanic vent.

flint A variety of chert, usually dark colored.

flood lavas (flood basalts) Flows of basaltic lavas that erupt from innumerable cracks or fissures and commonly cover vast areas to thicknesses of 1000 m or more.

floodplain The flat surface adjacent to a stream, over which streams spread in time of flood.

flow banding In an igneous rock, alternating layers of different texture and minerals; it is the result of the flowing of viscous magma.

focus The initiation point of an earthquake within the earth.

foliated rocks Metamorphic rocks characterized by parallel orientation of tabular minerals and varying degrees of banding, or color layering. Common to regionally metamorphosed rocks.

foliation Layering, or banding, in some metamorphic rocks caused by the subparallel alignment of platy minerals.

footwall The face of the block below an inclined fault.

foraminifera Single-celled marine animals that secrete a calcite shell. Some limestones consist almost entirely of foraminifera.

forceful injection Name given to the process by which magma at depth pushes aside existing rocks and forces its way into cracks and fissures.

foreset bed The inclined layers that make up the front of a delta as it advances into a body of water.

formation A lithologically distinctive rock unit or deposit with an upper and lower boundary that is large enough to be mapped.

fossils The preserved remains or traces of prehistoric life. Most fossils consist of the bones or exterior shells of organisms.

fracture zone See *transform fault*

fringing reef A coral reef that is directly attached to the shore of an island or continent.

frost heave The uneven lifting of surface materials due to subsurface freezing of water.

frost wedging The mechanism by which jointed rocks are pried apart by ice acting as a wedge.

gabbro A dark, plutonic rock consisting typically of coarse-grained crystals of pyroxene, calcium-plagioclase, and olivine.

garnet A group of silicate minerals that lack cleavage and whose crystals are almost always well formed and equidimensional. They are common in some metamorphic rocks.

geoid A generalized earth shape in which high and low spots are represented but smoothed out and reduced. The attraction of gravity is everywhere perpendicular to the geoidal surface, which corresponds to the surface of mean sea level.

geologic time The vast period of time (about 4.5 billion years) covering the entire history of the earth.

geological map A map which records lithology, aerial extent, and orientation of bedrock and surficial deposits at or immediately below the earth's surface.

geosyncline A huge, elongated trough at the earth's surface in which thousands of meters of sedimentary rocks and volcanic rocks accumulate over many tens to hundreds of millions of years. Thought to represent the site of subsequent alpine mountain building.

geothermal region An area in which hot water and steam occur naturally underground and which is commonly characterized by hot springs and geysers.

geyser A hot spring that erupts jets of hot water and steam, resulting from the heating of ground water by hot rocks and steam.

glacial abrasion The scouring or wearing down of the rocks over which a glacier moves.

glacial marine (sediment) A marine sediment containing stones carried out to sea by icebergs. Found around glaciated areas and areas of pack ice.

glacial quarrying The process by which a glacier prys rocks loose from the surface over which it moves.

glacial striations Long grooves and scratches made by a glacier on the rocks it carries or on the bedrock over which it moves.

glacial surge A rapid movement in which a glacier becomes decoupled from its base and advances downvalley at high velocities, some approaching 6000 m/yr.

glacier A large mass of ice formed by the compaction and recrystallization of snow, which moves slowly under the stress of its own weight.

glassy texture A texture in which a large part of the rock is composed of volcanic glass, as in obsidian.

glide A movement of a large mass of intact rock downslope along a planar surface, such as a bedding plane.

globigerina A single-celled surface-dwelling marine organism with a shell made of calcium carbonate.

glowing avalanche (ash flow) A turbulent mass of pyroclastic fragments and some high-temperature gas erupted from a volcano.

gneiss A coarse-grained metamorphic rock in which bands of light-colored minerals (quartz, feldspar) alternate with dark-colored ones (amphibole, biotite).

Gondwana One of two hypothetical land masses consisting of all the continents in the Southern Hemisphere.

graben A relatively down-dropped fault block with linear margins and generally lying between two horsts (uplifted blocks).

graded bedding A type of stratification in which the particles in each layer are graded—the larger particles at the bottom grade into smaller ones at the top.

graded stream A stream in equilibrium: one that has the velocity and channel characteristics required to transport the load from its drainage basin.

gradient (stream) The downvalley slope of a stream channel.

granite A coarse- to fine-grained plutonic rock made up largely of potassium feldspar, sodium-rich plagioclase, quartz, and micas.
granitization The theory that most granitic rocks are formed in place by solid-state crystallization of preexisting rocks under the influence of chemical solutions.
gravity meter (gravimeter) An instrument used to determine the acceleration or force of gravity at a given spot on the earth's surface. Generally the gravimeter provides a measure of the acceleration which, of course, is proportional to the force of gravity.
graywacke A sandstone consisting mainly of detritus derived from mafic igneous rocks.
ground ice Ice in the thin soil layer overlying permafrost.
groundmass The fine material in a porphyritic rock that surrounds the phenocrysts.
ground moraine Till deposited beneath a glacier.
groundwater Subsurface water, generally occurring in the pore spaces of rock and soil.
gumbotil Highly weathered glacial till.
guyot A submerged volcano with a planed-off, nearly level summit.
gypsum A common evaporite mineral composed of hydrous calcium sulfate.
gyres Large, rotating current cells in the major ocean basins generated by the deflective effect of the Coriolis force on moving water currents.

half-life The time required for a mass of radioactive isotope to decay to one-half of its initial amount.
halite Sodium chloride, or common table salt; a common evaporite mineral.
hanging valley A tributary valley whose floor is higher than the main valley; produced when erosion deepens the main valley more rapidly than the tributary.
hanging wall The face of the block above an inclined fault.
harmonic tremor A near-continuous release of seismic wave energy associated with the movement or flow of magma in fissures and chambers at depth.
hematite A relatively non-magnetic oxide of iron having a red to red-brown streak.
high-grade metamorphic rocks Dynamothermal metamorphic rocks such as gneiss which are crystallized under high temperatures.
hogback A narrow, sharp-crested ridge formed by a resistant, steeply dipping bed.
horizon A soil layer with characteristic physical, chemical, and biological properties.
horn (glacial) A jagged peak formed in areas where mountain glaciers radiate away from a summit area and have removed most of the latter.
hornblende A dark ferromagnesian mineral of the amphibole group that resembles augite in color and luster, with two cleavages that intersect at 56° and 124°.
hornfels A dense, non-foliated contact metamorphic rock.
horst A fault block with linear margins. It rises above the blocks on either side of it.
hydration The volume expansion of salt minerals that occurs when water is added to their crystal structure. Pressures generated can break up a rock.
hydrologic cycle The complete cycling, or transfer process, of the earth's water.
hydrolysis A weathering process in which minerals are altered by chemical reaction with water and acids.
hydrosphere The waters of the earth.
hydrothermal metamorphic rocks Rocks formed by recrystallization through the action of hydrothermal solutions (hot fluids and gases) circulating marginal to large plutons.
hydrothermal ore An ore precipitated from a high-temperature fluid.

iceberg A large chunk of ice that breaks away from an ice sheet and floats away.
ice dome The summit of a continental glacier or ice sheet.
ice sheet (continental) Large, irregular sheets of ice that cover a large area and are not usually guided in their flow by the underlying topography.
icing The freezing of groundwater that reaches the surface in permafrost areas.
igneous rocks Rocks that have solidified from magma.
index fossil An easily identified fossil with a wide geographic range and limited span of time on earth. Useful in correlating and dating rocks.
influent stream A stream that flows above the water table and from which water flows in pores to the water table. Common in arid regions.
inselberg An isolated residual hill that rises abruptly above the surrounding plain in a dry region. Characteristic of the late stage of the erosion cycle.
interior drainage A pattern of streams that drain toward the center of a basin rather than toward the sea.
intermediate-focus earthquake An earthquake originating in the depth zone between 70 and 300 km.
intermediate rocks Igneous rocks intermediate in composition between felsic and mafic. Examples are andesite and diorite.
intermittent stream See *ephemeral stream.*
intrusive contact The surface of contact between a pluton and the surrounding rock (country rock).
ion An atom or group of atoms that carries a positive or negative charge.
ionic bond A bond that results from the electrostatic attraction between positively and negatively charged ions.
ionic-covalent bonding A type of bonding common in minerals; alternates between ionic and covalent.
island arc A volcanically active island archipelago.
isostasy The theory that blocks of the earth's crust are in a floating, gravitational equilibrium.
isotope A form of an element. Each isotope of a specific element has the same atomic number (contains the same number of protons and electrons) but contains a different number of neutrons; therefore each would have a different atomic weight.

joint A fracture in a rock along which no movement has taken place.
joint set A group of joints with a similar geometry. In most cases a set consists of parallel, concentric, or radial joints.
juvenile water Water derived directly from magma that comes to the surface for the first time.

karst The name given to hummocky landscapes characterized by the features caused by the solution of rocks by groundwater, such as sinkholes and caves.

kerogen The organic material usually found in shales which can be converted to petroleum products.

kettle Depression formed by the melting of a large block of ice trapped in glacial till or outwash.

K horizon (caliche) A hard, thick calcareous crust that forms beneath the B horizon in arid-region soils (i.e., a strongly developed Cca horizon).

kimberlite An ultramafic igneous rock which originated as deep as 250 km; commonly contains diamonds and forms pipe-like intrusive bodies.

klippe An erosional remnant of a thrust sheet, or nappe.

laccolith A concordant igneous body more or less circular in outline, with a flat base and a dome-shaped top.

lag deposit See *stone pavement.*

lagoon The body of water separating a barrier island from the mainland.

lahar A mud flow on the flanks of a volcano.

laminae Sedimentary layers whose thickness is less than 1 cm.

landslide A relatively rapid movement of soil and rock downslope.

lapilli Pyroclastic fragments about 2 cm in diameter.

lateral fault A fault in which the relative displacement is primarily strike slip. Also called a strike-slip fault.

lateral moraine A ridge of debris that continually accumulates along the side of a glacier.

laterite A highly weathered tropical soil rich in oxides of iron and aluminum. Hardens upon drying and can be used to make bricks.

lateritic soil A deep, highly weathered reddish soil, widespread in tropical climates. Most soluble elements (calcium, sodium, potassium, and silicon) have been leached out, leaving a residuum enriched in oxides of iron and aluminum.

Laurasia One of two hypothetical land masses consisting of the continents in the Northern Hemisphere.

lava The molten material that erupts from a volcanic vent or the rock that results from the solidification of the molten material.

law of gravitation The law formulated by Isaac Newton that describes the force of attraction between two bodies as a consequence of their masses and distance of separation:

$$F_g = \frac{Gm_1 m_2}{r_2}$$

law of original horizontality The geological law which states that all sediments are originally deposited horizontally or nearly so.

law of superposition One of the laws upon which relative geological chronology is based: in any sequence of sedimentary rocks that has not been disturbed any one layer will be older than the layer *above* it and younger than the layer *below* it.

leaching The dissolving out or removal of soluble materials from a rock or soil horizon by percolating water.

levee A low embankment adjacent to and confining a river channel.

lichen dating See *lichenometry.*

lichenometry The study of lichen sizes on rocks for dating of very recent surface deposits.

lignite A brownish coal-like material formed when peat is buried under younger sediments.

limb (flank) One of the two sides of a fold.

limestone A sedimentary rock in which calcium carbonate predominates.

lit-par-lit Term meaning "bed-by-bed." Applied to a mixed igneous-metamorphic rock, such as migmatite, composed of alternating layers of apparently metamorphic and igneous rock material.

lithosphere The outer 100 km of the earth. It appears to be relatively strong and brittle.

lithostatic pressure The all-sided pressure at depth exerted by the rock load above. Also called confining pressure.

lodestone See *magnetite.*

loess Wind-deposited silt originating in glacial outwash plains or in desert regions.

longitudinal dune A dune aligned parallel to the prevailing wind.

long-range ordering An ordering of atoms in a mineral that extends throughout the mineral grain. Also referred to as atomic ordering or crystallinity.

longshore current A coastal current, parallel to the shore, produced by wave refraction

low-grade metamorphic rocks Dynamothermal metamorphic rocks, such as slate, crystallized under moderate to low temperatures.

low-velocity zone (earthquake) See *asthenosphere.*

luster The subjectively evaluated character of the light reflected from the surface of a mineral.

***L* wave (long wave)** A seismic surface wave of relatively high amplitude and long wave length. It follows the *P* and *S* waves in arrival at a seismograph station.

mafic rocks Igneous rocks containing a large proportion of minerals rich in magnesium, iron, and calcium-plagioclase.

magma (melt) A liquid composed of molten material at high temperatures; generally rich in silicon and oxygen.

magma chamber A cavity beneath the earth's surface surrounded by solid rock and containing magma.

magmatic differentiation Term used to describe the separation of magma into fractions of differing chemical composition during cooling and crystallization.

magnetic anomaly Any departure from the normal magnetic field of the earth. In the ocean, anomalous highs and lows occur in alternating ridges and rises.

magnetic field A region in which magnetic forces are exerted.

magnetic polarity time-sequence The dated sequence of changes in magnetic field polarity as determined from paleomagnetism of rocks and ocean-floor anomaly stripes.

magnetic reversal The 180° reversal of the earth's magnetic field that has occurred intermittently throughout geological history.

magnetite A black, strongly magnetic iron oxide mineral, sometimes called lodestone.

magnetometer An instrument for measuring magnetism in rocks.

mantle The rocks below the Moho (Mohorovičić discontinuity) and above the core.

marble A fine- to coarse-grained non-foliated metamorphic rock recrystallized from limestone.

marine terrace (A) An accumulation of wave- and current-transported materials seaward of a wave-cut platform. (B)

The term also applies to old marine beach deposits and wave-cut platforms now found above sea level.

mass movement The movement of rock material downslope through the direct pull of gravity.

mass wasting See *mass movement*.

meander A large, curving bend in a river.

mechanical weathering Weathering by physical forces such as frost action or absorption of water.

medial moraine A moraine in the middle of a glacier formed by the merging of the lateral moraines of two coalescing valley glaciers.

median valley The keystone-like depression, or rift, in the crest of an oceanic ridge or rise.

melt See *magma*.

meltwater Melted ice and snow from a glacier.

mesosphere The mantle zone beneath the asthenosphere.

Mesozoic Era The era of geologic time between the Paleozoic and Cenozoic eras, lasting from about 225 to 65 million years ago. Characterized by the dominance of reptiles, especially dinosaurs.

metallic bonding A type of covalent bonding in which there are more metal atoms available than are necessary to satisfy bond requirements. Also called time-shared covalent bonding.

metamorphic rocks Rocks that form at depth through solid-state recrystallization of pre-existing rocks as a result of internal heat, pressure, and chemical activity of fluids.

metamorphism An isochemical process in which rocks are crystallized from pre-existing rock under the influence of higher temperatures, confining and directed pressure, and interstitial fluids.

metasomatism See *granitization*.

mica A mineral consisting of parallel sheets of silica tetrahedra strongly bonded at their bases but less strongly bonded across the sheets; the result is a well-developed cleavage in one direction.

microplate A relatively small lithospheric plate.

Mid-Atlantic ridge A submarine mountain range that stretches the length of the Atlantic Ocean.

mid-ocean ridge See *oceanic ridges and rises*.

migmatite A layered zone of igneous and metamorphic rocks between a batholith and the surrounding rocks. Generally found near the deeper parts of the batholith.

Milky Way Galaxy The large, rotating pinwheel-shaped celestial grouping of about 10 billion stars in which the solar system is located.

mineral A crystalline chemical compound (or element) that occurs naturally.

modified Mercalli scale (earthquake) A scale used to measure earthquake intensity based on damage caused.

Mohorovičić discontinuity A seismic-velocity boundary used to demarcate the earth's crust above from the mantle below. A relatively large increase in *P*- and *S*-wave velocities takes place at the boundary, which is also called the Moho or M-discontinuity.

Mohs hardness scale A scale used by mineralogists to judge the hardness of a mineral; ranges from 1 to 10.

monadnock An isolated hill of resistant rock that stands above the level of a peneplain.

monocline A one-limbed fold with horizontal strata on either side.

moraine See *lateral, medial,* or *terminal* moraine.

mud cracks The polygonal pattern produced by the drying and shrinking of wet, clayey mud.

mud flow A mass movement, usually of fine-grained earth materials containing a relatively high water content.

mudstone A shale without distinct bedding.

multiple working hypotheses Several equally plausible hypotheses which may be advanced to fit scientific observations. The formulation of such hypotheses is one of the most common approaches in geology.

muscovite (white mica) A mineral of the mica group; usually colorless, gray, or transparent. Common in metamorphic rocks and in many sedimentary rocks, especially sandstone.

mylonite A dark, hard, fine-grained cataclastic rock.

nappe A complex, large-scale recumbent anticlinal fold.

neutron An uncharged particle in the nucleus of an atom.

new global tectonics The hypothesis that new lithospheric material, formed at the ocean ridges, moves away from the divergent zones and toward convergent zones, where the lithosphere bends sharply downward into the asthenosphere and disappears below.

nonconformity An *unconformity* in which sedimentary rocks rest on plutonic or metamorphic rocks.

non-dipole center An area on the earth's surface where the magnetic field is greater or smaller than would be expected. These centers are non-dipolar inasmuch as any one center is unique and not paired with a center of the opposite polarity on the other side of the world.

non-plunging fold A fold in which the axis is horizontal.

normal fault A fault in which the footwall moves up relative to the hanging wall.

nuée ardente A turbulent gaseous cloud erupted from a volcano and containing primarily ash; accompanies eruption of a glowing avalanche.

oblique slip Fault movement (slip) oblique to the dip of the fault.

obsidian See *volcanic glass*.

oceanic ridges and rises Elongate suboceanic mountain ranges. See also *zones of divergence*.

oceanography The study of the ocean, including its chemical, physical, biological, and geological aspects.

oil field An underground accumulation of oil or a group of *oil pools*.

oil shale Shale rich in kerogen, which on distillation and refining yields oil.

olivine A ferromagnesian mineral that usually occurs as rounded and glassy green crystals.

oolite Small, round grains of calcium carbonate that form as layers of the mineral build up around a nucleus.

ooze The remains of microscopic free-floating organisms that drift down and cover the deep ocean floor.

ophiolite suite An association of mafic and ultramafic rocks largely altered to serpentine which appears to have formed initially at or close to an oceanic ridge.

original horizontality, law of The law stating that layers of sediments are deposited horizontally, or nearly so.

oscillation ripples The corrugated surface made in the bed of a shallow body of water. The ripples are symmetrical in cross section, in contrast to current ripples.

outwash plain Floodplains formed by streams draining from the front of a glacier.

overthrust fault A low-angle thrust fault in which the dip angle is less than 10°.

overturned bed A sedimentary bed that has been structurally rotated more than 90° from the horizontal.

overturned fold A fold in which one limb has been rotated past 90°.

oxbow lake A crescent-shaped lake formed when a stream bend, or meander, is cut off from the main stream.

oxidation A process of chemical weathering in which minerals or elements within the minerals combine with oxygen.

pahoehoe A lava flow with a ropy or corrugated surface and a glassy outer rind.

paleobotany The study of fossil plants.

paleoclimatology The study of past climates and the causes of their variations.

paleomagnetism The study of fossil magnetism in rocks.

paleontology The study of past geologic life based on plant and animal fossils.

Paleozoic Era The geologic time-period between the Precambrian and the Mesozoic eras, lasting from about 600 to 225 million years ago. Characterized by relatively simple invertebrates and backboned animals.

Pangaea The hypothetical single continent postulated by A. Wegener that split into fragments and began to drift apart during the Jurassic period.

parent isotope The initial, or starting, radioactive isotope that will decay spontaneously and progressively to a daughter isotope.

parent material The material from which a soil forms.

patterned ground Polygonal patterns formed in surface material subject to intensive frost action.

pedalfers Soils with a fairly high content of organic matter in the A horizon, and no calcium carbonate accumulation beneath the B horizon. Common in humid temperate regions.

pediment An erosional bedrock surface that slopes away from a desert mountain range.

pediplane An extensive erosion surface in deserts formed by the coalescence of two or more pediments.

pedocal An arid-region soil characterized by a thin A horizon and a layer of calcium carbonate beneath the B horizon known as the Cca horizon or the K horizon.

peneplain A broad, nearly featureless plain that has been eroded to nearly sea level by mass wasting and stream erosion.

period The fundamental unit of the geologic time scale and the subdivision of an era.

permafrost Ground that remains below 0°C and usually contains small to large quantities of ice.

permafrost table The upper surface of permafrost.

permeability The measure of the capability of a rock to transmit a liquid.

phenocryst The larger, usually well-formed crystals in a porphyritic igneous rock.

photosphere The luminous envelope of the sun visible from earth and composed largely of non-charged gas atoms at temperatures near 5400°C.

phreatic explosion A volcanic explosion caused by the heating and expansion of groundwater.

phyllite A common low-grade metamorphic rock, with a well-developed rock cleavage, between a slate and schist in stage of metamorphic development. Contains abundant aligned crystals of muscovite which produce a lustrous sheen.

pillow lava Lobes of lava resembling a bed of pillows. Results from extrusion into water or a water-rich environment.

placer deposit A mineral deposit, such as gold, formed by a concentration of heavy mineral particles in a beach or river environment.

plain A broad area of low relief occurring generally at low elevations.

plateau A broad, relatively flat or rolling region occurring at relatively high elevations.

plate-tectonic theory The theory that the earth is divided into rigid blocks, or plates, that move relative to one another.

plastic flow (glacial) The flow of the lower part of a glacier, caused by deformation of ice crystals under the pressure of the overlying ice.

playa A dried-up playa lake consisting of clay and silt or sand and deposits of soluble salts.

playa lake A seasonal lake in the center of a desert basin.

Pleistocene A recent epoch of the Cenozoic Era (and part of the Quarternary Period) in which 90–100 per cent of the fossil shells still exist today. Usually thought of as coincident with the ice age of the Cenozoic.

plume A column of heated rock rising from the mantle. It can bring about partial melting in the asthenosphere and the lithosphere.

plunging breaker A wave formed in shallow depths near the shore and whose crest takes the shape of a half-cylinder that curls over and breaks suddenly with a crash.

pluton A body of plutonic rock of any size or shape.

plutonic rocks Rocks formed from magma that cools and solidifies underground.

podzol A variety of pedalfer that forms in cooler climates toward the northern limit of trees. Podzols are characterized by a whitish layer, the A2 horizon.

point bar The bar on the inside of a meander.

porosity The percentage of the total volume of a rock that is occupied by open spaces.

porphyritic texture See *porphyry.*

porphyry An igneous rock containing two grain sizes: relatively large, well-formed phenocrysts imbedded in a finer-grained crystalline or glassy groundmass. Texture said to be porphyrytic.

porphyry copper A hydrothermal copper deposit in which copper-bearing minerals occur in veins throughout a large volume of rock.

Precambrian All geologic time before the beginning of the Paleozoic Era. Characterized in the record by scant primitive life or no life at all.

pressure surface The level to which water rises in a confined or unconfined aquifer.

proton A positively charged particle in the nucleus of an atom.

pull-apart zones. See *zones of divergence.*

pumice A frothy, porous volcanic glass that is generally rhyolitic in composition.

***P* wave (primary wave)** A relatively low-amplitude seismic body wave, the first to arrive at a seismograph station after an earthquake.

pyroclastic rocks Coarse- to fine-grained rocks formed from material hurled into the air during a volcanic eruption.

quartz A silicate mineral with a hardness of 7 that has a vitreous luster and commonly fractures conchoidally. Composed almost exclusively of silicon dioxide.

quartzite Unbanded metamorphosed sandstone or unmetamorphosed silica-cemented sandstone consisting of quartz grains. In the former, the rock breaks across the grains.

Quarternary The period of geologic time including the Pleistocene and Holocene epochs and covering the last 2–3 million years of earth history, up to the present.

quick clay Clay deposits with a high water content that become fluid when jarred, for example, by an earthquake.

radioactivity The spontaneous decay of an atom of one isotope into an entirely different isotope.

radiocarbon dating See *carbon-14 dating*.

radiolaria A microscopic marine organism that secretes a shell of silica. The remains form siliceous ooze.

radiometric date The age of a material as determined through measurement of radioactive decay.

recumbent fold An overturned fold; one in which one limb is overturned and roughly parallel to the normal limb and both limbs are nearly horizontal.

reducing environment An environment characterized by a low concentration of oxygen in the water.

reef A ridge of layered sedimentary rock built by the secretions and remains of marine organisms, usually coral.

regionally metamorphosed rocks See *dynamothermal metamorphic rocks*.

relative time sequence Sequence of geological events as established by their order of occurrence.

remanent magnetization The permanent magnetization acquired by rocks. In most cases it parallels the earth's magnetic field lines at the time of the rock's origin.

reserve The estimated amount of an identified resource recoverable with present-day technology.

reservoir rock A permeable rock in which oil accumulates.

resource A commodity useful or often essential to people.

reverse fault A fault in which the footwall moves down relative to the hanging wall. Also called a thrust fault.

rhyolite A light-colored fine-grained-to-glassy volcanic rock similar to granite in composition and commonly characterized by flow-banding.

ria coast A submerged coastline in which the sea extends inland, sometimes for long distances, in stream valleys.

Richter scale A scale of earthquake magnitudes developed by the seismologist C. F. Richter. The magnitudes can be determined from seismographs and are directly related to the amount of energy released during an earthquake.

rift valley A large, elongate trough, or graben.

Ring of Fire A zone containing about two-thirds of the world's active volcanoes and that girdles much of the Pacific Ocean.

river terrace Along a stream, an elevated flat area underlain by river deposits. Such a terrace is an abandoned floodplain.

rock avalanche A large mass of rock that slides very rapidly as a unit (100 km/hr) downhill, perhaps on an air cushion if the terrain traversed is flat.

rock cleavage The tendency for a rock to break along relatively smooth, closely spaced parallel surfaces. Slaty cleavage is an example.

rockfall The relatively free-falling movement of rock material from a cliff or other steep slope.

rock flour A fine silt that covers much of the surface of an outwash plain.

rock glacier A tongue-shaped slow-moving mass of rocks and ice in alpine areas.

salt A general chemical term that includes many compounds, the most familiar of which is table salt (sodium chloride).

saltation A mode of sediment transport in which the particles bounce along the floor of a stream or along a desert surface.

salt dome A structure dome produced by the upward movement of a body of salt through enclosing sediments.

salt playa A playa consisting of saline residues.

sandstone A clastic sedimentary rock consisting mainly of sand-size grains cemented together.

scarp A cliff or line of cliffs produced by faulting, erosion, or landsliding.

schist A foliated metamorphic rock of intermediate grain size. Individual folia are relatively thin; platy minerals commonly make up one-half of the rock; and color banding is not well developed.

schistosity The well-developed wavy or undulatory rock cleavage characteristic of schists.

scoriaceous Adjective applied to a mafic volcanic rock which is frothy and cellular; that is, filled with many vesicles.

sea cliff A cliff or slope produced by wave erosion.

sea floor spreading The hypothesis that continental drift occurs through cracking and spreading at the oceanic ridges and rises.

seamount A submerged volcano of basaltic composition.

sedimentary rock Rocks formed by the accumulation of layers of clastic and organic material or precipitated salts.

seiche A wave of oscillation set up in a lake, harbor, or bay and that is initiated by the motion of an earthquake or by local changes in atmospheric pressure.

seismic gap Segment of an active seismic zone that has not undergone recent strain release and is therefore considered a prime region of potential seismic activity.

seismic sea wave See *tsunami*.

seismic wave A wave or a vibration produced by an earthquake.

seismograph An instrument used to measure the vibrations, or waves, generated during an earthquake.

seismology The study of seismic waves, in its broadest sense. Includes studies of wave motion, the events that produce them (primarily earthquakes), and the nature of the materials of the earth through which the waves pass.

self-exciting dynamo A mechanism consisting of interrelated moving electronic currents and varying magnetic fields. Once primed, it is able to continually regenerate a magnetic field.

serpentine group A group of hydrous rock-forming minerals derived by the alteration of magnesium-rich silicate minerals (e.g., olivine) in water-rich environments at low temperatures.

serpentinization The process of forming serpentine from magnesium-rich silicates.

shadow zone A region from about 102° to 143° from the epicenter of an earthquake in which there is no reception of direct seismic waves.

shale A fine-grained laminated sedimentary rock composed of clay and silt.
shallow-focus earthquake An earthquake originating between the earth's surface and 70 km in depth.
sheet flood Muddy, turbulent water that fills and overflows the banks of an arroyo and eventually spreads out over the desert floor.
sheeting See *exfoliation*.
shelf break The outer edge of the continental shelf, characterized by a sudden steepening of the slope.
shield A large area of exposed igneous and metamorphic rocks, usually Precambrian, surrounded by sediment-platforms; e.g., the Canadian Shield.
shield volcano A shield-like volcanic cone built almost entirely of fluid lava flows.
silicate A mineral consisting of silicon, oxygen, and varying proportions of one or more metals. Most common minerals are silicates.
silieous rocks Rocks made up largely of silica.
siliceous sinter A hot-spring or geyser deposit composed of silica.
silicon tetrahedron The tent-shaped (tetrahedral) arrangement of four oxygen ions around one silicon ion; represents the basic building block of all silicate minerals.
sill A concordant tabular intrusion of magma that more or less parallels layers of the country rock.
sinkhole A pit in karst topography caused by the solution of surficial limestone or the collapse of a cave roof.
sinter A spongy, porous sedimentary rock formed from the chemical precipitation of silica from springs.
slate A common low-grade dynamothermal metamorphic rock, usually derived from fine-grained sedimentary rocks. Possesses slaty cleavage.
slickensides A polished and smoothly striated rock surface that results from movement along a fault plane.
slip The term used to denote actual relative displacement along a fault.
slip face The steep face of an asymmetrical dune.
slump The movement of rock material downslope as a unit along a concave-upward slip plane. Characterized by backward tilting of the mass.
snow Frozen water vapor, crystallized directly from the water vapor in the atmosphere.
snowline The line or altitude on a glacier separating the area where snow remains from year to year from the area where the snow of the previous season melts.
sodium chloride Common salt.
soil profile The three basic layers, or horizons, of most soils.
solar day The time between two consecutive passages of the sun at its zenith past a particular spot on earth. A mean solar day is an arbitrary unit of time, one that is always the same length, regardless of the season.
solar wind A continuous but spasmodic flow of ionized particles from the sun that moves out into space at speeds of 300 to 600 km/sec. The wind represents the outer portion of the sun's corona.
solid load The suspended load and bed load of a stream, taken together.
solid solution A mixed-crystal mineral composed of varying amounts of certain ions which can substitute for one another in the crystal lattice.
solifluction The slow downslope movement of water-saturated surface material occurring in permafrost areas.
solution A process of chemical weathering in which soluble minerals are dissolved.
sonar (Sound Navigation Ranging) A device that sends and receives sound signals underwater.
sorting The property that refers to the degree of similarity in particle size in a sediment or sedimentary rock.
specific gravity The weight of a specified volume of a mineral divided by the weight of an equal volume of water at 4°C.
spilling breaker A wave whose crest collapses gradually over a relatively long distance as water spills continuously down the wave front.
spit A curved embankment formed by a longshore current that trails down-current from the land.
stack A pillar-like rocky island or mass near a cliffy shore separated from the headland by wave erosion.
stalactites Icicle-like pendants of travertine that hang from the roof of a cave.
stalagmites Deposits of travertine built upward from a cave floor.
stellar day The time between two consecutive passages of a distant star past a particular spot on earth.
steppe An extensive dry region characterized by grass vegetation.
stock A large pluton, generally of granitic rock, less than 100 km^2 in exposed surface area.
stone pavement A thin veneer of stones several stones thick mantling a desert surface. Also called desert pavement or lag deposit.
stoping Enlarging of a magma chamber by the prying loose of small blocks or rocks from the roof and walls.
strata Layers of sedimentary rock.
stratovolcano A steep-sided volcano consisting of alternating layers of lava and pyroclastic materials. Also called a composite volcano.
streak The true color of a mineral as seen in its powdered form, generally as a result of rubbing the mineral on a streak plate, a small plate of unglazed porcelain.
strike The line of intersection made by a dipping bed or surface with an imaginary horizontal plane.
strike slip Fault movement (slip) parallel to the strike of the fault.
strike-slip fault See *lateral fault*.
subaerial erosion Erosion that takes place in the open air (compare subterranean and submarine erosion).
sub-bottom profile A profile of the ocean floor and a cross section of the stratigraphy and structure of the rocks below it obtained by a shipboard echo sounder.
subdelta A small delta forming a part of a complex of deltas.
subduction The pulling down, or sinking, of lithospheric plates into the asthenosphere at the convergent zone.
sublimation The process by which a solid substance vaporizes without passing through a liquid stage.
submarginal resource A resource that is not profitable to recover.
submarine canyon A steep-sided valley or canyon cut into the continental shelf or slope.
submarine fan A fan-shaped deposit of sediments located seaward of a submarine canyon.
submarine plateau A broad, relatively low relief plateau that rises, usually 200 m or more, above the ocean floor.
superposition, law of The law stating that in a layered sequence of rocks the age of any one layer will be greater than the age of that above it, and less than the layer below it.

surface waves Relatively slow seismic waves that travel close to or at the earth's surface. Also called *L* waves.
suspended load (stream) The material a stream carries in suspension, buoyed up by the moving water.
S wave (secondary wave) A seismic body wave, the second to arrive at a seismograph station. Characterized by sudden onset and is normally greater in amplitude than a *P* wave.
swell A regular, somewhat flat-crested wave that has traveled far from its generating area; made up of long-period waves.
syncline A fold in which the limbs dip toward the axis; younger beds are found toward the axis.

talus A loose pile of angular rocks at the base of a cliff.
tarn An alpine rock-basin lake commonly resulting from differential glacial scouring.
tectonic dam Refers to the origin of one type of continental shelf, in which shelf sediments are deposited behind a geologic uplift or lava; both act as dams.
terminal moraine A mass of debris that accumulates as a hummocky, rocky ridge around the snout of a glacier. Also called an end moraine.
terrestrial planet A planet composed chiefly of dense rocky material, like the earth.
Tertiary A period of the Cenozoic Era covering the time-span between 65 and about 2 million years ago.
Tethys sea The hypothetical oceanic waterway separating the super-continents, Gondwana and Laurasia.
texture The interrelations of the size, shape, and arrangement of the particles in a rock.
thermoremanent magnetization (TRM) Permanent magnetization that results from thermal cooling of a rock (generally igneous in type) from high temperatures through the Curie temperatures of contained magnetic crystals.
thrust fault See *reverse fault*.
till Material laid down directly by glacier ice.
tillite A consolidated glacial deposit consisting of sand, gravel, boulders, and clay.
time-shared covalent bonding See *metallic bonding*.
tombolo A strip of sand connecting a near-shore island to the mainland.
topographic profile A cross-sectional drawing of the surface of the land along a given line.
transform fault (zone of lateral movement; fracture zone) One of the numerous fracture zones in the ocean basin, along which the ridges and rises have been off-set and along which lateral movement occurs.
transverse dune A dune aligned at right angles to the wind.
travertine A limy cave or spring deposit formed by chemical precipitation of calcium carbonate from solution in surface and groundwater. Can also be precipitated by calcareous algae.
tree-ring dating See *dendrochronology*.
trench A narrow, elongate depression on the deep-sea floor paralleling the trend of an island arc or continental margin.
tsunami A destructive wave generated by disturbances on the ocean floor. Also called a seismic sea wave.
tufa A spongy or porous sedimentary rock formed by the precipitation of calcium carbonate around the mouth of a hot or cold spring or in a stream or lake.
tuff A fine-grained rock composed of pyroclastic fragments, primarily ash.
tuff breccia Rock consisting of relatively large pyroclastic fragments in an ashy matrix.
turbidite A sedimentary rock deposited by a turbidity current, characterized by graded bedding.
turbidity current Density currents of sediment-laden water triggered by the slumping of oversteepened and unconsolidated material.

ultramafic A term applied to magmas and igneous rocks that contain extremely large amounts of magnesium and iron.
unconfined aquifer A water-bearing surficial layer of permeable material, such as sand or gravel.
unconformity A surface where the sequence of rock units has been interrupted by either erosion or non-deposition. The time represented by the unconformity is variable.
uniformitarianism The doctrine that the geologic processes now modifying the earth's surface have acted in essentially the same way throughout geologic time, although possibly at different rates.
U-shaped valley A valley suggesting the shape of the letter "U," with steep sides and a flat floor carved by a glacier.

vadose zone Zone between the ground surface and the water table in which the pore spaces are not completely filled with water; also called the *aeration zone*.
valley glacier See *alpine glacier*.
varve A pair of thin, sedimentary layers made up of a coarse, silty lower layer and a fine-grained upper layer. Thought to represent seasonal variation in deposition.
velocity (stream) The direction and magnitude of displacement of a portion of a stream per unit of time.
ventifact A rock faceted by wind-driven particles.
vesicles Small, rounded cavities in lava formed by trapped gas bubbles.
viscosity Resistance of a liquid to flow.
vitreous luster A term applied to a mineral that reflects light to about the same degree as glass.
volcanic agglomerate An unsorted deposit of volcanic bombs, cinders, lapilli, and ash in crude layers. Sometimes called volcanic breccia.
volcanic bomb Irregular to spindle-shaped air-borne blocks of lava hurled from a volcanic vent during eruption.
volcanic breccia Any rock composed of angular fragments of volcanic rock. See also *volcanic agglomerate*.
volcanic cone (volcano) The accumulated eruptive products around a volcanic vent, generally in a steep- to flat-sided cone.
volcanic dome A rounded extrusion of glassy lava squeezed out from a volcano, forming a dome-shaped mound.
volcanic glass Magma that has cooled so quickly that the liquid quenched to glass without crystallization taking place. Most commonly associated with rhyolitic magma.
volcanic neck A pipe-like pluton of solidified lava that once connected a magma reservoir with a volcanic vent at the surface.
volcanic rocks Rocks formed from magma that erupts at the surface and cools and solidifies.
V-shaped valley A narrow valley with steep, sloping sides resulting from downcutting by a stream and mass movement of material down the side slopes.

water table The surface at which water stands in wells, the upper surface of groundwater.
wave base The lower effective limit of wave transportation and erosion.
wave-cut platform A planed-off rock bench cut by wave erosion at the base of a sea cliff.
wave length The horizontal distance separating two equivalent wave phases, such as two crests or two troughs.
wave of oscillation A water wave in which the individual particles move in orbits with little or no change in position, although the wave form itself advances.
wave period The length of time required for two crests or two troughs of a wave to pass a fixed point.
wave refraction The bending of a wave as it approaches the shore.
weathering The mechanical disintegration and chemical decomposition of rocks.
weight The force that gravity exerts on a body.
welded tuff A pyroclastic rock whose particles have been fused together by heat still contained in the deposit after it has come to rest; generally associated with large-scale caldera-forming events.

xerophyte A plant adapted to dry conditions.

yazoo A tributary river that runs parallel to the mainstream between a natural levee and the river bluff.

zone of aeration A subsurface zone above the water table where pore spaces in the ground may range from completely to partially full of water.
zone of convergence The zone where plates push together and lithospheric material is subducted, or pulled down into the asthenosphere.
zone of divergence The zone where plates move apart and new lithospheric material is formed; equivalent to active oceanic ridges and rises. Also called the pull-apart zone.

INDEX

Boldface page numbers refer to definitions; page numbers with asterisks to illustrations

SOURCES FOR LINE DRAWINGS

CHAPTER 1

Fig. 1-18. After W. M. Wendland and D. C. Donley, 1971, "Radiocarbon Calendar Age Relationships," *Earth and Planetary Science Letter*, v. 11, pp. 135–39. **Fig. 1-20.** From J. T. Andrews and P. J. Webber, 1969, "Lichenometry to Evaluate Changes in Glacial Mass Budgets, "*Arctic and Alpine Research*, v. 1, no. 3, Fig. 7. By permission of J. T. Andrews. **Fig. 1-21.** From J. T. Andrews and P. J. Webber, 1973, "Lichenometry: a Commentary," *Arctic and Alpine Research*, v. 5, no. 4., Fig. 1. By permission of J. T. Andrews.

CHAPTER 3

Fig. 3-15. After G. Dietrich, 1963, *General Oceanography*, John Wiley and Sons. **Fig. 3-22.** From B. C. Heezen and M. Tharpe, 1961, *Physiographic Diagram of the North Atlantic Ocean*, Geological Society of America. Reprinted by permission of B. C. Heezen and M. Tharpe. **Fig. 3-25.** From B. C. Heezen and M. Tharpe, 1976, *The Floor of the Oceans*, map painted by Tangy de Remur, American Geographical Society. Reprinted by permission of B. C. Heezen and M. Tharpe. **Fig. 3-28.** From B. C. Heezen and M. Tharpe, 1961, *Physiographic Diagram of the South Atlantic Ocean*, Geological Society of America. Reprinted by permission of B. C. Heezen and M. Tharpe.

CHAPTER 5

Fig. 5-32. After G. A. MacDonald, 1972, *Volcanoes*, Prentice-Hall, Inc. **Fig. 5-45.** After T. A. Steven, 1975, "Middle Tertiary Volcanic Field in the Southern Rocky Mountains," *Geological Society of America Mem. 144*, pp. 75–94. **Fig. 5-48.** After H. Cloos, 1931, "Neuer Jahrbuch fur Mineralogia, "Geologie und Paläontologie, Band 66, Abt. B.

CHAPTER 6

Fig. 6-3. After G. A. MacDonald, 1972, *Volcanoes*, Prentice-Hall, Inc. **Fig. 5-13.** After H. William, 1942, *Geology of Crater Lake National Park, Oregon*. By permission of the Carnegie Institution of Washington. **Fig. 6-20A.** After R. G. Luedke and W. S. Burbank, 1968, "Volcanism and Cauldron Development in the Western San Juan Mountains, Colorado," *Colorado School of Mines Quarterly*, v. 63, pp. 175–208. **B.** After P. W. Lipman and others, "Volcanic History of the San Juan Mountains, Colorado, as Indicated by Potassium-Argon Dating," *Geological Society of America Bulletin*, v. 81, pp. 2329–52. **Fig. 6-21.** From *Earthquake Information Bulletin*, 1980, v. 12, July–August, pp. 146–47. **Fig. 6-22B.** From *Earthquake Information Bulletin*, 1980, v. 12, July–August, pp. 146–147, Fig. 6. **Fig. 6-30.** After G. A. MacDonald and D. H. Hubbard, 1970, *Volcanoes of the National Parks in Hawaii*, 5th ed., Hawaii Natural History Association. **Fig. 6-31A.** After A. K. Baksi and N. D. Watkins, 1973. "Volcanic Production Rates: Comparison of Ocean Ridges, Islands, and Columbia Plateau Basalts," *Science*, v. 180, pp. 493–96. **B.** After A. C. Waters, 1955, *Volcanic rocks and the Tectonic Cycle*, Geological Society of America Special Paper 62, pp. 703–22.

CHAPTER 7

Fig. 7-19. From R. Trümpy, 1960, "Paleotectonic Evolution of the Central and Western Alps," *Geological Society of American Bulletin*, v. 71, pp. 843–908, p. 2. **Fig. 7-28A.** After C. O. Dunbar, 1960, *Historical Geology*, John Wiley and Sons, Fig. 215. **Fig. 7-31.** After A. J. Eardley, 1951, *Structural Geology of North America*, Harper and Bros., Fig. 21.

CHAPTER 8

Fig. 8-8. After H. L. James, 1955, "Zones of Regional Metamorphism in the Precambrian of Northern Michigan," *Geological Society of America Bulletin*, v. 66, pp. 1455–87. **Figs. 8-13 and 8-17.** After W. W. Moorhouse, 1959, *Study of Rocks in Thin Section*, Harper & Row.

CHAPTER 10

Figs. 10-2 and 10-28. After G. A. Kiersch, "The Vaiont Reservoir Disaster," *Mineral Information Service*, v. 18, no. 7. **Fig. 10-20.** From D. R. Crandell, and D. R. Mullineaux, 1967, "Volcanic Hazards at Mount Rainier, Washington," *U.S. Geological Survey Bulletin 1238*. **Fig. 10-23.** After R. L. Schuster and R. J. Krizek, 1978, *Landslides*, National Academy of Sciences. **Fig. 10-30A.** From O. J. Ferrians, Jr., and others, 1969, *U.S. Geological Survey Professional Paper 678*, Fig. 1. **B.** From R. J. E. Brown, 1970, *Permafrost in Canada*, University of Toronto Press, Fig. 4. By permission of the University of Toronto Press. **Fig. 10-31.** After C. F. S. Sharpe, 1938, *Landslides and Related Phenomena*, Columbia University Press, Fig. 5.

CHAPTER 11

Fig. 11-2. After A. L. Bloom, 1969, *The Surface of the Earth*, Prentice-Hall, Inc. **Fig. 11-5.** After L. B. Leopold and others, 1964, *Fluvial Processes in Geomorphology*, W. H. Freeman and Co., Figs. 6-1 and 6-9. **Fig. 11-6.** From U.S. Geological Survey Paper 44. **Fig. 11-10.** After Judson and Ritter, 1964, "Rates of Regional Denudation in the United States," *Journal of Geophysical Research*, v. 69, pp. 3395–3401. **Fig. 11-11.** After W. B. Langbeim and S. A. Schumm, 1958, "Yield of Sediment in Relation to Mean Annual Precipitation," American Geophysical Union *Transactions*, v. 39, pp. 1076-84. **Fig. 11-13.** After F. Hjulstrom, 1935, "Studies on the Morphological Activity of Rivers as illustrated by the River Fryis," *University of Upsala Geol. Institute Bulletin 25*, pp. 221–527. **Fig. 11-17.** From L. B. Leopold, 1968, *Hydrology for Urban Land Planning — a Guidebook on the Hydrologic Effects of Urban Land Use*, U.S. Geological Survey Circ. 554, Fig. 1. **Fig. 11-26.** After D. J. Easterbrook, 1969, *Principles of Geomorphology*, McGraw-Hill Book Co., Fig. 6-10. **Fig. 11-33.** From L. B. Leopold and J. P. Miller, 1954, "A Postglacial Chronology for Some Alluvial Valleys in Wyoming," *U.S. Geological Survey Water Supply Paper 1261*. **Fig. 11-40.** After *Water and Choice*, 1968, National Academy of Sciences, publication 1689.

CHAPTER 12

Fig. 12-3. After P. Meigs, 1956, *Future of Arid Lands*. By permission of the American Association for the Advancement of Science. **Fig. 12-17A.** After T. M. Oberlander, 1972, "Morphogenesis of Granitic Boulder Slopes in the Mojave Desert, California," *Journal of Geology*, v. 80, pp. 1-20, Fig. 14. **Figs. 12-21 and 12-25.** After R. A. Bagnold, 1941, *The Physics of Blown Sand and Desert Dunes*, Metheun and Co. **Fig. 12-35.** From R. B. Morrison, 1968, "Pluvial Lakes," in R. W. Fairbridge, ed., *Encyclopedia of Geomorphology*, Dowden, Hutchinson & Ross, Stroudsburg, Pa. By permission of Dowden, Hutchinson & Ross. **Fig. 12-39.** From M. J. Bovis and R. G. Barry, 1974, "A Climatological Analysis of North Polar Desert Areas," in T. L. Smiley and J. H. Zumberge, eds., *Polar Deserts and Modern Man*, University of Arizona Press, Fig. 2-4. By permission of the University of Arizona Press.

CHAPTER 13

Fig. 13-27. From R. F. Flint, 1971, *Glacial and Quaternary Geology*, John Wiley & Sons. By permission of John Wiley & Sons, Inc. **Fig. 13-31.** After V. K. Prest, 1970, *Quaternary Geology of Canada*, Canada Department of Energy, Mines, and Resources; and Bryson and others, 1969. "Radiocarbon Isochromes on the Disintegration of the Laurentide Ice Sheet," *Arctic and Alpine Research*, v. 1, pp. 1–13. **Fig. 13-38.** After D. J. Easterbrook, 1969, *Glacial Map of the United States East of the Rocky Mountains*, Geological Society of America. **Fig. 13-41.** After the *National Atlas of the United States of America*, U.S. Geological Survey, 1970.

CHAPTER 14

Fig. 14-7. After W. Bascom, 1964, *Waves and Beaches*, Doubleday & Co., Inc., Figs. 40 and 41. **Fig. 14-19.** After A. L. Bloom, and others, 1974, "Quaternary Sea-Level Fluctuations on a Tectonic Coast, New 230 Th 234 Dates from the Huon Peninsula, New Guinea," *Quaternary Research*, v. 4, pp. 185–205, Fig. 5. **Fig. 14-21.** From J. T. Andrews, 1975, *Glacial Systems*, Duxbury Press, Fig. 7-7. By permission of Duxbury Press. **Fig. 14-26.** After D. W. Johnson, 1919, *Shore Processes and Shoreline Development*, John Wiley and Sons, Fig. 88. **Figs. 14-30 and 14-31.** From J. H. Hoyt, 1967, "Barrier Island Formation," *Geological Society of America Bulletin*, v. 78, pp. 1125–36. By permission of the Geological Society of America and J. H. Hoyt.

CHAPTER 15

Fig. 15-3. From B. C. Heezen and M. Tharpe, 1976, *The Floor of the Oceans*, map painted by Tangy de Remur, American Geographical Society. Reprinted by permission of B. C. Heezen and M. Tharpe. **Fig. 15-4.** After K. O. Emery, 1969, *The Continental Shelves in the Ocean*, W. H. Freeman and Co., pp. 44–45. **Fig. 15-6.** From R. P. Shepard, 1963, *Submarine Geology*, 2nd ed., Harper & Row, Fig. 150. **Fig. 15-8.** From B. C. Heezen and M. Tharpe, 1976, *The Floor of the Oceans*, map painted by Tangy de Remur, American Geographical Society. Reprinted by permission of B. C. Heezen and M. Tharpe. **Fig. 15-11.** From H. W. Menard, 1964, *Marine Geology of the Pacific*, McGraw-Hill Book Co., Fig. 4-8. By permission of McGraw-Hill Book Co. **Fig. 15-14.** After W. M. Davis, 1928, *The Coral Reef Problem*, Special Pub. no. 9, American Geographical Society. By permission of the American Geographical Society. **Fig. 15-17.** From B. C. Heezen and C. D. Hollister, 1971, *The Face of the Deep*, Oxford University Press, Figs. 7-44, 45, and 46. By permission of B. C. Heezen. **Fig. 15-21.** After W. S. Broecker and J. van Donk, "Insolation Changes, Ice Volumes, and the 0^{18} Record in Deep-Sea Cores," *Reviews of Geophysics and Space Physics*, v. 8, no. 1, Fig. 3.

CHAPTER 16

Fig. 16-2. After Bybordi, 1974. Diagram by G. Olson, Dept. of Agronomy, Cornell University. **Fig. 16-7.** After A. N. Strahler, 1969, *Physical Geography*, John Wiley & Sons, Inc., Fig. 32-29. **Fig. 16-9.** From R. C. Heath and others, 1966, *The Changing Pattern of Ground-Water Development on Long Island, New York*, U.S. Geological Survey Circular 524. **Fig. 16-10.** After K. Cartwright and F. B. Sherman, 1969, *Evaluating Sanitary Landfill Sites in Illinois*, Illinois State Geological Survey Environmental Notes no. 27. **Fig. 16-14.** © William E. Davies.

SOURCES FOR LINE DRAWINGS

CHAPTER 18

Fig. 18-6. After O. Nuttli, 1973, "The New Madrid Earthquake of 1811 and 1812, Intensities, Ground Motion, and Magnitudes," *Bulletin of the Seismological Society of America*, v. 63, pp. 227–48, Fig. 1. **Fig. 18-7.** After R. M. Hamilton, 1980, Quakes along the Mississippi, *Natural History*, v. 89, pp. 70–74. **Fig. 18-9.** After G. Plafker and J. C. Savage, 1970, "Mechanism of the Chilean Earthquake of May 21 and 22, 1960," *Geological Society of America Bulletin*, v. 81, pp. 1001–30. **Fig. 18-11.** After R. Siever, 1973, "The Earth," *Scientific American*, v. 233, pp. 82–91. **Fig. 18-12.** After Gilluly and others, 1968, *Principles of Geology*, 3rd ed., W. H. Freeman and Co. **Fig. 18-15.** From S. T. Algermissen, 1969, "Seismic Risk Studies in the U.S.," *Proceedings of the Fourth World Conference on Earthquake Engineering*, v. 1, pp. 14–27. **Fig. 18-16.** After W. R. McCann and others, 1980, "Yakataga Gap, Alaska: Seismic History and Earthquake Potential," *Science*, v. 207, pp. 1309–15. **Fig. 18-17.** From C. Kissingler, 1974, *Earthquake Prediction*. By permission of C. Kissingler. **Fig. 18-18.** After D. M. Evans, 1966, *Geotimes*, v. 10, pp. 11–18. **Figs. 18-20 and 18-21.** From *The Earth Sciences*, 2nd ed. by Arthur N. Strahler, Fig. 23-14A and Fig. 23-14B (p. 394), Copyright © 1963, 1971 by Arthur N. Strahler. Reprinted by permission of Harper & Row, Publishers, Inc. **Figs. 18-25 and 18-26.** After O. M. Phillips, 1968, *The Heart of the Earth*, W. H. Freeman and Co. **Fig. 18-28.** After C. F. Richter, 1958, *Elementary Seismology*, W. H. Freeman and Co. **Fig. 18-36.** After F. Press and R. Siever, 1974, *Earth*, W. H. Freeman and Co. **Fig. 18-38.** After A. N. Strahler, 1971, *Principles of Physical Geology*, John Wiley & Sons, Inc. **Fig. 18-42.** After P. J. Wyllie, 1975, "The Earth's Mantle," *Scientific American*, v. 232, no. 3, pp. 50-63.

CHAPTER 19

Fig. 19-7. After Takeuchi, Uyeda, and Kanamori, 1970, *Debate About the Earth*, W. H. Freeman and Co. **Figs. 19-10 and 19-11.** After A. Cox, 1969, "Geomagnetic Reversals," *Science*, v. 163, pp. 237 and 202. **Fig. 19-12A.** After J. R. Heirtzler and others, 1965, "Magnetic Anomalies Over the Reykjanes Ridge," *Deep-Sea Research*, v. 13, p. 427. **B.** After F. J. Vine, 1960, "Magnetic Anomalies Associated with Mid-Ocean Ridges," in R. A. Phinney, ed., *History of the Earth's Crust, A Symposium*, Princeton University, p. 73, Fig. 6. By permission of Princeton University Press.

CHAPTER 20

Fig. 20-4. After S. W. Carey, 1958, "A Tectonic Approach to Continental Drift," in *Continental Drift, A Symposium*, Geology Dept., Univ. of Tasmania, pp. 177–355. **Fig. 20-5.** After E. C. Bullard and others, 1965, "The Fit of Continents Around the Atlantic," in P.M.S. Blackett and others, eds., *A Symposium on Continental Drift*, Phil. Tran. Royal Soc., London, v. 100. **Fig. 20-6.** After Tarling and Tarling, 1971, *Continental Drift*, Doubleday and Co. **Fig. 20-7.** After P. M. Hurley, 1968, "The Confirmation of Continental Drift," *Scientific American*, v. 229, pp. 60–69. **Fig. 20-8.** After Takeuchi, Uyeda, and Kanamori, 1970, *Debate About the Earth*, W. H. Freeman and Co. **Fig. 20-9.** After A. Cox and R. Doell, 1960, "Review of Paleomagnetism," *Geological Society of America Bulletin*, v. 71, p. 758. **Fig. 20-10.** After E. Orowan, 1969, "The Origin of the Ocean Ridges," *Scientific American*, v. 221, no. 5. **Fig. 20-11.** From W. C. Pitman and others, 1974, *Age of the Ocean Basins*, Geological Society of America map. **Fig. 20-12.** After A. E. Maxwell and others, 1970, "Deep-Sea Drilling in the South Atlantic," *Science*, v. 108, pp. 1047–59. © 1970 by the American Association for the Advancement of Science. By permission of A. E. Maxwell. **Fig. 20-13.** After B. Gutenberg and C. F. Richler, 1954, *Seismicity of the Earth*, 2nd ed., Princeton University Press. **Fig. 20-14.** After J. Oliver and B. Isacks, 1967, "Deep Earthquake Zones, Anomalous Structures in the Upper Mantle and the Lithosphere," *Journal of Geophysical Research*, v. 72, pp. 4259–75. **Fig. 20-15.** After B. Isacks and others, 1968, "Seismology and the New Global Tectonics," *Journal of Geophysical Research*, v. 73, pp. 5855–99. **Fig. 20-16.** After M. N. Toksoz, 1976, "The Subduction of the Lithosphere," in *Continents, Aground, Readings from Scientific American*, W. H. Freeman and Co., pp. 112–22. **Fig. 20-25.** After E. Orowan, 1969, "The Origin of the Oceanic Ridges," *Scientific American*, v. 221, no. 5. **Fig. 20-28A.** After D. W. Scholl and others, 1970, "Peru-Chile Trench Sediments and Sea-Floor Spreading," *Geological Society of America Bulletin*, v. 81, pp. 1339–60. **B.** After M. L. Holmes and others, 1972, "Seismic Reflection Evidence Supporting Underthrusting Beneath the Aleutian Arc Near Amchitka Island," *Journal of Geophysical Research*, v. 77, pp. 959–64.

CHAPTER 21

Fig. 21-2. U.S. Bureau of the Census. **Fig. 21-4.** U.S. Geological Survey. **Fig. 21-5.** After B. J. Skinner, 1976, *Earth Resources*, Prentice-Hall, Inc. **Fig. 21-7.** Modified from A. A. Bartlett, 1980, "Forgotten Fundamentals of the Energy Crisis," *Journal of Geological Education*, v. 28, pp. 4–35. **Fig. 21-8.** After M. K. Hubbert, 1968. **Fig. 21-13.** After J. B. Corliss, 1973. **Fig. 21-14.** After S. E. Kesler, 1976, *Our Finite Mineral Resources*, Fig. 8-2, McGraw-Hill Book Co. **Fig. 21-15.** After P. Rona, 1973. **Fig. 21-18.** After Hubbert, 1967. **Fig. 21-19.** After B. J. Skinner, 1976, *Earth Resources*, Prentice-Hall, Inc. **Fig. 21-22.** U.S. Geological Survey. **Fig. 21-25.** From Hubbert, 1969, based on data from Averitt, 1969. **Fig. 21-26.** U.S. Dept. of the Interior. **Fig. 21-27.** From A. A. Bartlett, 1980, "Forgotten Fundamentals of the Energy Crisis," *Journal of Geological Education*, v. 28, pp. 4–35, after Hubbert, 1968. **Fig. 21-28A.** After M. K. Hubbert, 1971, "The Energy Resources of the Earth," *Scientific American*, v. 224, p. 65. **B.** U.S. Dept of Energy. **Fig. 21-29.** After Bartlett, 1980, from M. Iona, 1977. **Fig. 21-33.** From the Univ. of California, Los Alamos Scientific Lab., 1979.